Human Bioarchaeology of the Transition to Agriculture

Human Bioarchaeology of the Transition to Agriculture

Edited by

Ron Pinhasi

Department of Archaeology, University College Cork, Cork, Ireland

Jay T. Stock

Department of Biological Anthropology, University of Cambridge, Cambridge, UK

WILEY-BLACKWELL

A John Wiley & Sons, Ltd., Publication

This edition first published 2011, © 2011 by John Wiley & Sons, Ltd.

Wiley-Blackwell is an imprint of John Wiley & Sons, formed by the merger of Wiley's global Scientific, Technical and Medical business with Blackwell Publishing.

Registered office
John Wiley & Sons Ltd, The Atrium, Southern Gate, Chichester, West Sussex, PO19 8SQ, UK

Other Editorial Offices
9600 Garsington Road, Oxford, OX4 2DQ, UK

111 River Street, Hoboken, NJ 07030-5774, USA

For details of our global editorial offices, for customer services and for information about how to apply for permission to reuse the copyright material in this book please see our website at www.wiley.com/wiley-blackwell

Library of Congress Cataloguing-in-Publication Data
Human bioarchaeology of the transition to agriculture / editors, Ron Pinhasi and Jay T. Stock.
p. cm.
Includes index.
ISBN: 978-0-470-74730-8 (cloth)
1. Human remains (Archaeology). 2. Agriculture—Origin. 3. Antiquities, Prehistoric.
I. Pinhasi, Ron. II. Stock Jay T,.
CC79.5.H85H857 2010
930.1—dc22

2010023376

A catalogue record for this book is available from the British Library.

This book is published in the following electronic formats: eBook [9780470670187]; Wiley Online Library [9780470670170]

Set in 10/12pt, Times Roman by Thomson Digital, Noida, India
Printed in Singapore by Ho Printing Singapore Pte Ltd

1 2011

Contents

Foreword

Clark Spencer Larsen
Department of Anthropology
The Ohio State University

INTERPRETING THE BIOARCHAEOLOGICAL RECORD OF THE FORAGING-TO-FARMING TRANSITION

Among the most important adaptive shifts in human evolution—along with hunting, meat eating, and cooking—was the transition from a lifeway based exclusively on hunting, gathering, and collecting of wild plants and animals to a lifeway involving dependence to varying degrees on domesticated plants or animals or both. This shift only occurred following the onset of the Holocene (after 10,000 B.P.). Today, virtually every member of our species depends to some degree on domesticated products of one form or another, especially plants. Many authorities regard the foraging-to-farming transition as a long, slow evolutionary process. When viewed in the long term of seven million years of human evolution, however, the process of domestication and the span of time involved in the global transition was a fast one, first originating in a dozen independent centers and spreading rapidly, during the last 10 000 years.

Much of the history of scientific study of this defining behavioral and economic shift has focused on causes, addressing the question: "*Why domestication in general and agriculture in particular emerged as the new and eventually the dominant economic system in most world regions?*" This question continues to fuel an active point of discussion in the social sciences (Cohen, 2009). An important subtext of this discussion is the *consequences* for human populations. Some three decades ago, a group of bioarchaeologists—those who study human remains from archaeological contexts—began comparing temporal sequences of skeletal samples from a range of settings around the world. Collectively, these sequences revealed evidence for a general decline in health (Cohen and Armelagos, 1984). My own work in the American Southeast, for example, showed an increase in the prevalence of dental caries, various infections, and a decline in body size. This documented reduction in health was linked to eating a more carbohydrate-rich diet and living in sedentary contexts among prehistoric agricultural populations (Larsen, 1982, 1984). Other studies documented a similar pattern of diachronic change. I was convinced that the bioarchaeological record showed a *universal* decline in health—wherever agriculture appeared and persisted, one should expect to see a pattern of increased morbidity.

These early research results focusing on regional variation spawned an entire generation of bioarchaeologists who directed their investigations to a range of circumstances and contexts. Since the publication of Cohen and Armelagos's (1984) seminal book, various additional settings have found that the transition provided a general but not universal picture of reduced health. In those settings where a decline in health has been documented, there is considerable variation in the mode, tempo, pattern, and degree of this process (e.g., Cohen and Crane-Kramer, 2007; Cohen, 2009; Steckel and Rose, 2002; Powell et al., 1991; Lambert, 2000; Steckel et al., 2001; Larsen, 1995, 2003; and others). While the results vary, all agree that understanding this variation lies at the root of better characterizing the consequences and costs of the foraging-to-farming transition.

In this volume, the contributors collectively make clear that there is no universal biological response to agriculture. This finding underscores our growing realization that we need to take a broad perspective to document, interpret, and understand the oftentimes local or at least regional variability relating to the kinds of plants and animals that were domesticated, climate and environment generally, the different strategies that human societies employed to manage and exploit these resources, and the social contexts for dietary change and how food is acquired. Moreover, the work makes it abundantly clear that the bioarchaeological record of adaptation and change must be viewed in archaeological, behavioral, and environmental contexts. The general picture that continues to emerge is one of health declining in some regions, but not in all. On the other hand, the picture also shows a period over the last 10 000 years of remarkable population growth. This strongly suggests that agriculture may have spread because it was a behavioral strategy that promoted reproduction, a fundamental element of adaptive success (Lambert 2009).

The chapters presented in this volume further our understanding of the foraging-to-farming transition, the outcomes of the transition, and the fund of data that human remains provide for developing a more informed perspective on the mode, tempo and processes that characterize later human biocultural evolution. The contributions reveal the remarkable advances in research made in bioarchaeological science, including stable isotope analysis and dietary reconstruction, nutrition, growth and development, behavioral reconstruction from biomechanical analysis, genetics and evolution, craniofacial adaptation, biodistance, and demographic transition. These contributions underscore the importance of not viewing the transition to farming as a simple pre- and post- comparison of bones and teeth. Rather, human remains provide us with a focal point of discussion; within the complex interplay of data derived from multiple fields, including archaeology and environmental science, all coming together in order to address areas of common interest.

This book is a crucial step towards building an understanding of Holocene human evolution as it is represented by the study of human remains. The advances presented in the following pages give us avenues and additional questions for promoting continued understanding of human variation in the last 10 000 years and the platform that biocultural adaptation provided for informing the human condition today. That is, to understand the causes, outcomes, and costs of the transition for foraging to farming is to understand the present human condition.

REFERENCES

Cohen, M.N. (2009). Introduction: rethinking the origins of agriculture. *Current Anthropology*, **50**, 591–595.

Cohen, M.N. & Armelagos, G.J. (eds.) (1984). *Paleopathology at the Origins of Agriculture*. Orlando, Florida: Academic Press.

Cohen, M.N. & Crane-Kramer, G.M.M. (eds.) (2007). *Ancient Health: Skeletal Indicators of Agricultural and Economic Intensification*. Gainesville, Florida: University Press of Florida.

Lambert, P.M. (ed.) (2000). *Bioarchaeologial Studies of Life in the Age of Agriculture: A View from the Southeast*. Tuscaloosa: University of Alabama Press.

Lambert, P.M. (2009). Health versus fitness: Competing themes in the origins and spread of agriculture? *Current Anthropology*, **50**, 603–608.

Larsen, C.S. (1982). *The Anthropology of St. Catherines Island: 3. Prehistoric Human Biological Adaptation.* Anthropological Papers of the American Museum of Natural History 57 (part 3).

Larsen, C.S. (1984). Health and disease in prehistoric Georgia: the transition to agriculture. In *Paleopathology at the Origins of Agriculture*, eds. M.N. Cohen & G.J. Armelagos pp. 367–392. Orlando, Florida: Academic Press.

Larsen, C.S. (1995). Biological changes in human populations with agriculture. *Annual Review of Anthropology*, **24**, 185–213.

Larsen, C.S. (2003). Animal source foods and human health during evolution. *Journal of Nutrition*, **133**, 1S–5S.

Powell, M.L., Bridges, P.S. & Mires, A.M.W. (eds.) (1991). *What Mean These Bones? Studies in Southeastern Bioarchaeology*. Tuscaloosa: University of Alabama Press.

Steckel, R.H. & Rose, J.C. (eds.) (2002). *The Backbone of History: Long-Term Trends in Health and Nutrition in the Americas*. New York: Cambridge University Press.

List of Contributors

Benjamin M. Auerbach Department of Anthropology, 250 South Stadium Hall, The University of Tennessee, Knoxville, Tennessee 37996, USA

Jeff Babb Department of Mathematics and Statistics, The University of Winnipeg, 515 Portage Avenue, Winnipeg, Manitoba, Canada R3B 2E9

Marien Béguelin División Antropología, Museo de La Plata, Universidad Nacional de La Plata., Paseo del Bosque s/n. 1900 La Plata, Argentina

Chelsea Budd Wetland Archaeology & Environments Research Centre, Department of Geography, University of Hull, Hull, HU6 7RX, UK

Joachim Burger Institut für Anthropologie, AG Palaeogenetik, Johannes Gutenberg-Universität, Colonel Kleinmann-Weg 2, D-55128 Mainz, Germany

Vered. Eshed Israel Antiquities Authority, P.O.B 1230, Tel Aviv, 61012, Israel

Ehud. Galili Israel Antiquities Authority, P.O.B 180, Atlit 30350, Israel

Jaime Ginter School of Community and Liberal Studies, Sheridan Institute of Technology & Advanced Learning, Oakville, ON, Canada and TUARC - Trent University, Archaeological Research Centre, Peterborough, ON, Canada

Gisela Grupe Staatssammlung für Anthropologie und Paläoanatomie, Karolinenplatz 2a, 80333 München, FRG

C. M. Haverkort The Department of Anthropology, 13-15 HM Tory Building, University of Alberta, Edmonton, Canada, T6G 2H4

M. A. Katzenberg Department of Archaeology, University of Calgary, 2500 University Drive NW, Calgary AB, T2N 1N4, Canada

Clark Spencer Larsen Department of Anthropology, 174 West 18th Avenue, 4034 Smith Laboratory, The Ohio State University, Columbus, OH 43210-1106, USA

Angela Lieverse Department of Archaeology and Anthropology, 55 Campus Drive, University of Saskatchewan, Saskatoon, S7N 5B1, Canada

Malcolm Lillie Department of Geography, University of Hull, Cottingham Road, Hull, HU6 7RX, UK

Anna Linderholm The Archaeological Research Laboratory, Stockholm University, Stockholm, Sweden

Damiano Marchi Department of Evolutionary Anthropology, Duke University, 04A Bio. Sci. Bldg., Science Drive, Box 90383, Durham, NC 27708-0383, USA

Christopher Meiklejohn Department of Anthropology, University of Winnipeg, Winnipeg R3B 2E9, Canada

Matthew C. O'Neill Center for Functional Anatomy and Evolution, 1830 E. Monument St., Johns Hopkins, University, Baltimore, MD 21205, USA

Anastasia Papathanasiou Ephorate of Paleoanthropology and Speleology, Greek Ministry of Culture, 11636 Athens, Greece

Joris Peters Staatssammlung für Anthropologie und Paläoanatomie, Karolinenplatz 2a, 80333 München, FRG

Ron Pinhasi Lecturer in Prehistoric Archaeology, Department of Archaeology, University College Cork, Cork, Ireland

Jerry Rose Department of Anthropology, Old Main 330, University of Arkansas, Fayetteville, AR 72701, USA

Christopher B. Ruff Center for Functional Anatomy and Evolution, 1830 E. Monument St., Johns Hopkins, University, Baltimore, MD 21205, USA

Marina L. Sardi División Antropología, Museo de La Plata, Universidad Nacional de La Plata., Paseo del Bosque s/n. 1900 La Plata, Argentina

Rick J. Schulting School of Archaeology, University of Oxford, UK

Laura Shackelford Department of Anthropology, University of Illinois at Urbana-Champaign, 109 Davenport Hall, MC-148, 607 S. Mathews Avenue, Urbana, IL 61801, USA

Colin N. Shaw Department of Anthropology & The Center for Quantitative Imaging, Pennsylvania State University, University Park, State College, PA 16802, USA

Vitale S. Sparacello Department of Anthropology, University of New Mexico, Albuquerque, NM 87131, USA

S. Stefanovic Department of Archaeology, Faculty of Philosophy, University of Belgrade, Cika Ljubina 18-20, 11000 Belgrade, Serbia and Montenegro

Jay T. Stock Lecturer in Human Evolution and Development, Leverhulme Centre for Human Evolutionary Studies, Department of Biological Anthropology, University of Cambridge, Fitzwilliam Street, Cambridge, United Kingdom, CB2 1QH

Daniel H. Temple Department of Anthropology, University of North Carolina at Wilmington, 601 South College RD, Wilmington, NC 28403-3201, USA

Mark G. Thomas Research Department of Genetics, Evolution and Environment, University College London, Gower Street, London, WC1E 6BT, UK

Melissa Zabecki Department of Behavioral Sciences, University of Arkansas, Fort Smith, 5210 Grand Ave., P.O. Box 3649, Fort Smith, AR 72913-3649, USA

1

Introduction

Changing Paradigms in Our Understanding of the Transition to Agriculture: Human Bioarchaeology, Behaviour and Adaptation

Jay T. Stock[1] and Ron Pinhasi[2]

[1] *Leverhulme Centre for Human Evolutionary Studies, Department of Biological Anthropology, University of Cambridge, Cambridge, UK*
[2] *Department of Archaeology, University College Cork, Cork, Ireland*

The evolution and history of our species is often considered as a series of major transitions and processes of evolution, which collectively 'make' us human (Klein, 2009). In this context, it is easy to view modern human origins and dispersals as the end of a long process of cultural and biological evolution, and the point of demarcation between the end of biological evolution and the period when socio-cultural evolution and diversity becomes the hallmark of our species (Dyson, 1997). The 'Neolithic Revolution', a term coined by Gordon Childe, is the central component in this perspective, referring to the transition from hunting and gathering to agricultural subsistence in the Holocene. It has been seen as perhaps the single most significant social, cultural and biological transition since the origin of our species, marking the development of human control over the reproduction and evolution of plants and animals (Childe, 1936). A natural conclusion of this perspective suggests that the Neolithic marks the period where humans shifted from being subject to changes in the natural environment, to become the agents of environmental change in which the natural world is modified to suit human needs.

The transition to agriculture is often viewed as the beginning of a series of significant changes in human social organization, on the basis of the rise of food production and the storage of food surpluses. These are interpreted as leading to property ownership, social hierarchy, task specialization and runaway technological evolution, which is fuelled by a surplus of food (Diamond, 1997). In this context, agriculture can also be viewed as a form of niche

Human Bioarchaeology of the Transition to Agriculture Edited by Ron Pinhasi and Jay T. Stock

colonization, which allows populations to enter a new adaptive niche within the same environment as hunter-gatherers. When this is combined with food surpluses, it results in reduced interbirth intervals; increased birth stacking associated with alloparenting, and increased fertility (Wells and Stock, 2007). It has long been speculated whether demographic shifts amongst hunter-gatherer societies stimulated this cultural change, because greater numbers could not be sustained on the basis of hunter-gatherer subsistence (Boserup, 1965; Cohen, 1977), and it has recently been argued that consensus falls on this 'push' model (Cohen, 2009). Regardless of whether demography was a causal factor in development of agriculture in different regions, it is apparent that major demographic change was a primary consequence of the transition to agriculture (Bocquet-Appel and Bar-Yosef, 2008). Whether population size was an important catalyst for, or a consequence of, the transition to agriculture, the positive feedback between demography and culture certainly underpinned subsequent urbanization and state formation. Agriculture remains the primary means of production underpinning the global population and economy today.

1.1 THE ORIGINS OF AGRICULTURE

The earliest evidence for the transition to agriculture occurs in the Levant, a region of the Eastern Mediterranean, including Syria, Lebanon, Israel, Palestine and Jordan. The late Epipalaeolithic 'Natufian' (about 14 500–11 600 calBP) period in this region is seen as reflecting a cultural precursor of the subsequent pre-pottery Neolithic, due to the extensive exploitation of wild grains and the use of groundstone, stone architecture and a variety of organized site structures, art and evidence for symbolic behaviour (Bar-Yosef, 1998; Belfer-Cohen and Bar-Yosef, 2000; Byrd, 2005; Goring-Morris, Hovers and Belfer-Cohen, 2009). These cultural characteristics are often interpreted as the earliest archaeological signature of the transition towards agriculture, with the final impetus for the Neolithic being the dramatic environmental cooling associated with the Younger Dryas climatic event (Bar-Yosef and Belfer-Cohen, 2002). The earliest evidence for plant cultivation comes from the site of Abu Hureyra at about 13 000 BP, and appears to be associated with a decline in wild plants associated with the Younger Dryas (Hillman *et al.*, 2001). The subsequent Pre-pottery Neolithic A period shows the first evidence for larger permanent human settlements with architecture, and demonstrates the first evidence for intensive use of grains, as evidenced by the 11 kya granaries at 'Dhra in Jordan (Kuijt and Finlayson, 2009). These PPNA villages represent the earliest expression of the Neolithic, but they also reflect an extension of trends in social complexity, longer-term site use, and extensive use of wild grains which occurred earlier in the Natufian (Byrd, 2005). While these late Pleistocene and early Holocene cultures in the Near East and Anatolia reflect the earliest transition to farming, it is now well established that agriculture originated independently in different regions of the world at different times throughout the first half of the Holocene (Smith, 1998; Diamond, 2002; Bellwood, 2005). Other regions of primary plant domestication include southern China, Ethiopia, New Guinea, and three different regions of the New World: Southeast North America, Meso-America and western South America (Bellwood, 2005), and there were also a number of independent centres of animal domestication (Diamond, 2002).

What explains the development of agriculture in different parts of the world remains an open question; however, it has been argued that there were a number of constraints on the domestication of plants and animals prior to the Holocene, including climate and social organization (Richerson, Boyd and Bettinger, 2001; Bettinger, Richerson and Boyd, 2009). Regardless of these issues, it is clear that the global dominance of agricultural subsistence occurred through a combination of regional innovation with locally domesticable plant and animal species, demographic expansion and cultural diffusion (Bellwood, 2005; Pinhasi, Fort and Ammerman, 2005). The result of this transition is that agriculture is the dominant mode of subsistence today, which supports the large global human population and the socio-economic and technological systems of our species in the modern world.

1.2 THE CONSEQUENCES OF AGRICULTURE

A considerable emphasis of research has been placed on understanding the impact of the adoption of agricultural subsistence on health. This is based on the premise that a shift from diverse diets based on hunting and gathering towards dependence on one or a few highly productive domesticated plants, with a diet based predominantly on complex carbohydrates, can lead to a number of negative health outcomes, including nutritional deficiencies and dental caries. In addition, increasing sedentism associated with permanent or semi-permanent villages, and living in close proximity to domestic animals, leads to poor sanitation and an increased prevalence of zoonotic disease. Palaeopathologial studies have provided a considerable body of evidence that the origins of agriculture often had a negative impact on human health (Cohen and Armelagos, 1984; Cohen, 1989). The palaeopathological paradigm has dominated most research on the impact of agriculture in recent decades; however, it presents a paradox: if agriculture clearly underpins the dramatic demographic expansion and success of our species in the Holocene, how do we explain patterns of pathology? Is there a trade-off between reproductive capacity and health? In this context, we need to ask how the impact of agriculture varies through time and space, and under what cultural conditions it varies.

Recent research is beginning to investigate these questions. A study of linear enamel hypoplasia (LEH), bands of poor quality dental enamel that form during periods of childhood illness or malnutrition, has demonstrated a dramatic increase in the frequency of LEH between the late Palaeolithic and Neolithic of Egypt (Starling and Stock, 2007). While this would be expected based upon models of nutrition and hygiene with the transition to agriculture, the study also showed a gradual recovery in the frequency of LEH with the formation of the Egyptian state, showing that the negative health consequences of agriculture were short-term, and mediated by cultural factors over several millennia. Recent studies investigating health and subsistence transitions across a range of populations have demonstrated a greater diversity of evidence than previously known (Cohen and Crane-Kramer, 2007). These studies demonstrate that there is no simple relationship between subsistence change and health and, while there is still evidence for a decline in health indicators amongst many populations, the emerging picture is more regionally specific and diverse than previously thought.

While research has predominantly focused on the impact of agriculture on human health (Cohen, 1989) and demography (Bocquet-Appel and Bar-Yosef, 2008), there has also been study of the impact of agriculture on other aspects of human biology. A portion of this work, primarily on human remains from North America, has focused on elucidating behavioural correlates of subsistence transitions (Larsen, 1995). This area of research was amongst the earliest to begin to show evidence for regional diversity of human biological change with the transition to agriculture (Bridges, 1989; Ruff, 1999, 2008). Another area of enquiry has investigated the idea of 'human domestication', that human populations underwent similar morphological (Leach, 2003) and behavioural changes (Wilson, 1991) as other species, following the transition to agriculture and the process of animal domestication. These approaches suggest a continuing feedback between cultural and human biological change (Durham, 1991). A frequently cited example of biological change associated with the origins of agriculture is dental and mandibular size reduction; however, it remains unknown whether this represents genetic evolution, a relaxation of directional selective pressures, or biological plasticity in response to changes in biomechanics associated with food preparation and dietary homogenization (Pinhasi, Shaw and Eshed, 2008).

1.3 AN ONGOING 'REVOLUTION' IN OUR UNDERSTANDING OF THE NEOLITHIC

The above discussion provides a brief, general and conservative picture of the origins of agriculture. A recent review of the issue of agricultural origins from a variety of perspectives, published as a special issue of *Current Anthropology*, generally supports these interpretations; namely that agriculture was a consequence of increasing population pressure, competition for resources and globally favourable climatic conditions, and it resulted in a general deterioration of health amongst agricultural populations (Cohen, 2009). However, a common theme in the commentary accompanying this issue is that these interpretations represent a broad-scale overview but do not explain regional and temporal variation that is apparent in the archaeological record (Denham, 2009; Belfer-Cohen and Goring-Morris, 2009; Zeder and Smith, 2009). On a surface level we could dismiss these disparities as inevitable conflict between the resolution of data found in specific archaeological contexts and the sort of generalizations that are necessary for understanding global trends. However, it begs the question, to what extent are broad-scale and global trends relevant to regional expressions of Holocene subsistence transitions? To what extent is regional variation important in understanding the 'big picture' of the causes and consequences of agriculture? If regional and temporal variation is so significant, can we even make such generalizations?

In recent years, simultaneous developments in our understanding of long-term trends in the archaeology of human populations, human genetic diversity, and animal and plant evolution, have begun to dramatically change our views of the transition to agriculture. The Late Pleistocene and Holocene archaeological record from the Levant presents amongst the most clear archaeological evidence for long-term cultural change associated with the transition to agriculture; however, recent research has demonstrated that the cultural and biological change associated with this transition is more complex than previously thought. In particular, there is evidence that the cultural characteristics of the Neolithic develop over a considerable span of time (Twiss, 2007) from precursors found in the Natufian (Belfer-Cohen and Bar-Yosef, 2000). However, recent excavations suggest that many of the characteristic features of the Natufian period developed gradually over a long period of time in the Late Pleistocene (Maher, 2007; Nadel and Hershkovitz, 1991; Belfer-Cohen and Goring-Morris, 2009). Collectively, the emerging evidence from the Levant suggests that the origins of agriculture did not occur as a rapid Neolithic revolution *per se*, but as a complex and long-term process of social change in the relationship between human behaviour and the natural environment. This suggests that investigation of subtle cultural, behavioural and dietary change amongst hunter-gatherers, pastoralists and early agricultural populations should be considered on a fine scale, with increased temporal and spatial resolution.

This 'revolution' in our understanding of the Neolithic is not restricted to issues of cultural change in the Levant, as there is clear evidence for social complexity and long-term behavioural change in other regions (Denham, 2009; Zeder and Smith, 2009). A major factor underpinning the longer temporal span of the process of domestication may be the evolution of plants. Recent archaeobotanical research has moved away from simple identification of plant remains at archaeological sites, to examine the evolution of the plants themselves (Fuller and Allaby, 2009). This research demonstrates that the process of plant domestication occurs over a longer temporal span, and occurs across plant taxa and in different centres of domestication (Fuller and Allaby, 2009; Fuller, Allaby and Stevens, 2010). This not only has implications for the timing of cultural change associated with the agricultural transition, but also the expression of culture and habitual behaviour associated with subsistence activity. The evolution of domestic plants from wild progenitors involves a narrowing of the period of ripening of seeds from several months to several weeks, presenting what has been called a 'labour bottleneck' (Fuller, Allaby and Stevens, 2010). Furthermore, wild grasses generally disperse seeds by the presence of an 'abscission scar', which is often lost in the process of domestication. As a result, domestic plants often require human activity, in the form of threshing and winnowing, to separate and disperse seeds. This has been called a 'labour trap' of domestication (Fuller, Allaby and Stevens, 2010) associated with the transition to agricultural food sources. While this does not necessarily mean that agricultural subsistence is more labour intensive than other subsistence strategies in all circumstances, it suggests that the transition to agriculture is behaviourally complex and likely fuelled technological innovation throughout much of the Holocene.

A further area where there has been major change in our understanding of the transition to agriculture has been in human genetics. Until very recently, many assumed that human evolution is at a standstill in the modern world, due largely to human control over the natural environment (Dyson, 1997). This perspective was based largely on the assumption the technological developments following the origins of agriculture led to rapid technological evolution and increasingly successful 'niche construction', where humans successfully modify the natural environment, and thus remove pressures of natural selection. This assumption was never justified by the niche construction model, and it is increasingly clear that modification of the environment not only buffers environmental stress but actually exerts new selective pressures on the genome (Laland and Brown, 2006; Stock, 2008). Selective pressure on the genome resulting from stresses associated with the transition to agriculture has been detected through evidence for selection in a number of genes, related to malarial resistance (Tishkoff *et al.*, 2001), lactase persistence (Burger *et al.*, 2007; Tishkoff *et al.*, 2007) and amylase gene copy variation (Perry *et al.*, 2007). The latter two cases appear to be the results of direct selection of particular genes in response to dietary stress associated with domestication of animals and plants. These cases relate to the use of milk as a fallback food amongst Neolithic populations, and the shift towards higher components of dietary starch, which may have driven selection for higher AMY1 copy numbers to aid starch hydrolysis, respectively. Recent research has dramatically extended our understanding of recent human evolution, on the basis of new methods for the detection of signatures of natural selection within the genome (Sabeti *et al.*, 2007). Further genetic analysis has identified greater genetic heterogeneity amongst modern humans than would be otherwise expected, leading to the speculation that the pace of human evolution has speeded up in recent prehistory (Hawks *et al.*, 2007). It will take a considerable amount of research to sort out what specifically this genetic diversity means in terms of evolution, drift and demographic factors; but it does seem clear that recent cultural changes are a major force driving evolution within our species (Laland, Odling-Smee and Myles, 2010).

The previous discussion has presented evidence that we are in the midst of a major change in our interpretation of the origin of agriculture. This includes several fundamental shifts in perspective:

1. that socio-cultural and dietary change likely occurred over a considerable range of time, involving change within socially complex hunter-gatherers, pastoralists and agriculturalists;
2. that the process of the transition to agricultural subsistence was regionally specific, and we cannot expect to find universal trends and characteristics of this transition;
3. that evolution of plants during the process of domestication posed both constraints upon the process of cultural change, and its own influence on behavioural adaptation through the 'labour trap' associated with winnowing and threshing; and
4. that cultural change associated with the transition to agriculture exerted its new selective pressures on human populations, driving continuing human evolution within the Holocene.

1.4 HUMAN BIOARCHAEOLOGY OF THE TRANSITION TO AGRICULTURE

Human remains comprise the primary evidence for human biology with the transition to agriculture. In this volume, we provide a synthesis of the bioarchaeological evidence for changes in mobility, behaviour, diet, growth, population dynamics and evolution associated with the transition to agriculture. We assemble the work of a number of researchers who have been independently tackling questions relating to human biology associated with major dietary transitions of the Late Pleistocene and early Holocene. Given recent and major shifts in our understanding of the complexity of the transition to agriculture in different parts of the world, it would be impossible for a volume of this sort to be exhaustive, or to provide a comprehensive review of all evidence. Instead we aim to provide a synthesis of current approaches to understanding the biological correlates and consequences of major subsistence transitions in the Late Pleistocene and Holocene, in the hope that these studies will stimulate further research.

The contributions presented here are innovative in several ways: they emphasize the complexity of social and cultural change, and often employ multidisciplinary approaches to understanding the context and consequences of the agricultural transition. Major themes in the book include:

- the direct evidence for dietary change through the use of stable isotope analyses; variation in growth associated with dietary and cultural change;
- skeletal biomechanics and evidence for variation in habitual behaviour; craniofacial morphology, population history and adaptation; and
- evidence for genetic adaptation relating to Holocene cultural change.

In addition, several chapters build upon the dismantling of the traditional hunter-gatherer/agriculturalist dichotomy, by investigating subtle variation in human biology amongst hunter-gatherers, pastoralists, and early cultivators and agriculturalists. Other studies take a very broad geographical or temporal approach to understanding change. Collectively the contributions emphasize the benefits of adopting multidisciplinary approaches to investigating change through time and space.

Given the multidisciplinary nature of many of the papers presented here, it is challenging to find ways to categorize approaches in order to organize the volume. In this context we have arranged the book into four sections that define general themes of the paper, but these are not mutually exclusive categories of the research and there is considerable overlap from one section to another. Section A focuses on evidence for subsistence transitions using stable isotopes and broad bioarchaeological perspectives. Section B focuses more closely on variation in growth, and variation in body size in relation to the transition to agriculture. Section C discusses biomechanical evidence for behaviour, through time and space across subsistence transitions. The final section combines a variety of approaches, which are interrelated: genetics and evolution; cranial morphology and adaptation; and demographic trends with the transition to agriculture.

The chapter by Schulting (Chapter 2) begins the volume with a broad and comprehensive overview of isotopic evidence for the Mesolithic to Neolithic transition throughout Europe. It provides compelling evidence for a shift from regional isotopic and dietary heterogeneity in the Mesolithic period, towards isotopic signatures that are relatively homogenous throughout Europe. This homogeneity is particularly striking in coastal regions where changes in isotopic signatures are most marked. The following chapter (Chapter 3) by Lillie and Budd investigates dietary change from hunting and gathering through the Neolithic in the Dniepr Rapids region of the Ukraine, using stable isotope data and radiocarbon dates. The study provides evidence of an increase in fish consumption in the later Mesolithic and into the Neolithic period, possibly stimulated by shifting environmental conditions in this area. Grupe and Peters (Chapter 4) assess stable carbon and nitrogen isotope indicators of the contribution of C_4-plants to the diet of fully domesticated animals and their human consumers during the Neolithic transition in Anatolia. Using oxygen isotopes as a proxy for climate, they demonstrate that early Neolithic farmers were able to take advantage of C_4-plants, which are more suitable for animal fodder than human consumption, to reduce food competition between domestic stock and the owners of the animals. Their work provides evidence of complex plant resource management in Neolithic human subsistence strategies. Human skeletal remains are used by Papathanasiou in Chapter 5, to investigate dietary health implications of the transition to agriculture in Greece. In this region, palaeodietary analyses provide evidence for a swift and complete shift from foraging to farming. Palaeopathological stress indicators indicate that Early Neolithic Greek populations had relatively low prevalence of stress and that their stature was close to the upper limits of the range found within the Late and Final Neolithic periods. The following chapter by Ginter (Chapter 6) shifts focus to the mid-Holocene Eastern Cape of Southern Africa. Her bioarchaeological analysis identifies a decrease in body size during a period of intensification of foraging between 3500 and 2000 BP, followed by a recovery of body size associated with the adoption of pastoralism by some groups. These changes are accompanied by general cranial homogeneity, which likely reflects population continuity, suggesting that cultural change, and in particular the intensification of foraging, was driving phenotypic variation amongst foragers prior to herding. These results are interpreted in the context of evidence for a gradual and incomplete adoption of herding as a delayed return subsistence technique amongst hunter gatherers over a considerable span of the late Holocene.

The second section of the book builds upon the theme of body size introduced by Ginter in Section A, but focuses more exclusively on variation in growth and body size associated with subsistence transitions in different regions. The chapter by Mieklejohn and Babb (Chapter 7) provides a systematic comparison of long bone lengths, and hence stature, throughout the late Pleistocene and Early Holocene of Europe. This paper demonstrates a clear decrease in long bone length between the Early and Late Upper Palaeolithic, but there is a general stasis in stature from the Late Upper Palaeolithic through the Neolithic. These results overturn the general impression that the most significant shifts in body size occurred between the European Mesolithic and Neolithic. Chapter 8, by Pinhasi and colleagues, investigates variability in growth trajectories of limb bones amongst Neolithic populations of the Danube Gorges and Greece, using variation in attainment of bone lengths relative to dental age. Bone lengths per age of subadults in these samples are compared to standards from the Denver growth study, and more recent skeletal growth variation within Europe. The results of these comparisons show variation throughout the skeleton in growth for age, between the Danube Gorges Mesolithic, Neolithic and Greek Neolithic, but also highlight general trends such as increased variation in distal limb segments. The general theme of body size variation and growth is continued by Auerbach (Chapter 9), with a specific focus on broad patterns of body size variation in North America throughout the Holocene. The analysis tests a number of hypotheses, but identifies a general trend towards different morphological patterns associated with the transition to agriculture in the Southeastern relative to Southwestern populations. Overall, the results suggest that the transition to agriculture in these regions did not lead to a general decline in health, measured by stature and body mass, as might be predicted by previous palaeopathological evidence. The theme of bone growth and size variation in prehistoric Japan is explored by Temple (Chapter 10), who provides evidence for shorter leg lengths amongst pre-agricultural Jomon foragers when compare to Yayoi farmers. This is linked to greater femoral growth rates in the latter population, which may indicate a reduction in chronic infectious disease and nutritional stress following the transition to wet rice economies. This provides further support that the biological impact of agriculture is regionally specific and dependent upon local conditions.

The third section includes four chapters focusing largely but not exclusively on biomechanical properties of bone and indicators of habitual activity in the context of subsistence transitions. The first of these (Chapter 11) by Lieverse and colleagues is linked closely to themes first discussed in Section A, by using stable isotope evidence from carbon, nitrogen and strontium isotopes, to investigate diet and mobility amongst mid-Holocene hunter-gatherers in the Lake Baikal region of Siberia. A multivariate approach is then applied to the interpretion of diet, mobility and behaviour, by combining the isotopic results with those obtained from the analyses of musculo-skeletal stress markers and long-bone cross-sectional geometry as indicators of skeletal biomechanics. The results are suggestive of a trend towards more frequent deep water fishing and use of watercraft to extend foraging ranges from the earlier to later mid-Holocene. While there were no domestic plants or animals in this region throughout this period, this study demonstrates the viability of multidisciplinary approaches for detecting subtle patterns of dietary and behavioural change amongst hunter-gatherers, prior to full-blown agriculture. This approach may be useful for detecting the behavioural precursors of agriculture amongst socially complex hunter-gatherers in other regions.

The following chapter by Larsen and Ruff (Chapter 12) also employs multiple approaches to the issue of understanding changes in habitual activity associated with the transition to agriculture. Here, osteoarthritis and cross-sectional geometric properties of long bones are used to investigate behavioural change in three regions of eastern North America: the Pickwick Basin, Georgia Bight and lower Illinois River Valley. The results demonstrate that osteoarthritic and cross-sectional geometric analyses do not always produce the same results. Despite this trend, they suggest that the transition from foraging to farming and subsequent farming intensification involved significant behavioural change that is regionally specific. An intriguing result is that preagricultural people from the Illinois River Valley show a shift in behaviour prior to the adoption of maize agriculture. Evidence that the adoption of agriculture was regionally specific is also presented by Marchi and colleagues (Chapter 13), who demonstrate that lower limb robusticity is very high in the Ligurian Neolithic of Italy, and similar to highly mobile European Late Upper Palaeolithic and Mesolithic populations. This, they argue, reflects a high level of terrestrial mobility in the Neolithic associated with pastoralism, but also considerable terrain relief in the region, demonstrating that morphological trends may be driven by regional variation in both habitual activity and topography. Section C concludes with a paper by Stock and colleagues (Chapter 14), which investigates trends in body size and biomechanics from the Late Palaeolithic through Neolithic and Dynastic periods of the Nile Valley, spanning over 10 000 years. The paper demonstrates a reduction and subsequent increase in body size through this period. Comparisons of biomechanical properties of long bones demonstrate a general reduction in humeral and femoral rigidity, but one that differs in timing between sexes. Male gracilization occurs between the Late Palaeolithic and Neolithic, but amongst females it occurs between the Neolithic and late Predynastic period.

The final section of the book includes contributions on current and ancient genetic signatures of the transition to agriculture, palaeodemography, and both skeletal and craniofacial change. Burger and Thomas provide the first of two genetic studies (Chapter 15), by investigating the relationship between modern genetic signatures and the history of Neolithic expansion in Europe. Their paper presents evidence that lactase persistence, as reflected by the –13 910-T allele, arose in Transdanubia with the LBK Culture (about 5700 BC), spreading to central Europe ca. 5500 BC. This ability to metabolize lactose into adulthood began to rise in frequency and by the Middle Neolithic after the emergence of specialized dairying cultures, lactase persistence spread throughout Northern Europe. The paper provides key evidence for gene-culture coevolution associated with subsistence change and demographic expansions in the Neolithic of Europe. An analysis of ancient DNA variation in the Neolithic of Sweden is provided by Linderholm (Chapter 16). Of primary interest in this region is the emergence of different Neolithic cultural traditions. The first agriculturalists were the Funnel Beaker culture, but their cultural expression lies in contrast to the marine resource based Pitted Ware Culture, which developed shortly after the initial Neolithic. This study provides evidence that these groups were genetically distinct, with the Funnel Beaker culture having its origin in continental Europe. In contrast, the people associated with the Pitted Ware culture seem to arrive from the east and then disappear without leaving a subsequent genetic signature.

Eshed and Galili (Chapter 17) present a paleodemographic study of Pre-Pottery Neolithic (PPN) B and C populations of the southern-central Levant, by constructing mortality curves for large samples of subadult and adult remains, and specific sites including Atlit-Yam, Kfar HaHoresh and Ain Ghazal. The results demonstrated a pattern of greater life expectancy at birth at the site of Atlit-Yam, while Kfar HaHoresh had higher mortality at younger ages (20–29), and Ain Ghazal had the highest rates of child mortality. These results demonstrate variation in mortality profiles for different sites within the same region. In Chapter 18, Sardi and Béguelin investigate variation in facial, humeral and femoral morphology between hunter-gatherers and farmers of the Diamante River in Argentina. The prehistoric people of this region adopted plant and animal domestication around 3000–2000 calBC, after which agricultural dependence intensified. The results illustrate a systemic reduction in size of the face and limb bones amongst farmers, but while facial variation affected both males and females, limb bone reduction primarily affected females.

The final chapter in Section D by Pinhasi and Mieklejohn (Chapter 19) investigates diachronic changes in the tooth size amongst Central European populations from the Upper Palaeolithic through Mesolithic to Neolithic, spanning the period from about 35 000 to 4500 calBC. The results indicate a significant reduction trend in Central Europe from the Late Pleistocene to the Middle Holocene (Upper Palaeolithic/Mesolithic/Early Neolithic). These differ from the pattern and pace of dental reduction identified in previous studies, and indicate that the magnitude and nature of the reduction trend in Europe differs from the change reported for early agricultural populations in the southern Levant. These results highlight that while the transition to farming resulted in significant changes in dental dimensions, the magnitude and nature of the changes need to be addressed on a case-to-case basis before it is possible to draw conclusions about universal evolutionary trends.

1.5 CURRENT AND FUTURE RESEARCH IN HUMAN BIOARCHAEOLOGY OF THE TRANSITION TO AGRICULTURE

As discussed earlier in this chapter, we are in the midst of a 'revolution' in our understanding of the transition to agriculture. This paradigm shift is coming from a number of perspectives, emphasizing:

- the long-term socio-cultural change associated with subsistence transitions;
- regional specificity of the process of the origin and adoption of agriculture;
- the relationship between plant and animal evolution with domestication, and their impact on our own species; and
- that cultural evolution stimulated biological evolution within the Holocene.

The chapters presented in this volume demonstrate that human bioarchaeology plays a central role in this paradigm shift. They demonstrate that the complexity of long-term cultural and dietary change associated with these subsistence transitions are manifest in variation of the human biological response to the agricultural transition. We cannot conclusively interpret the causes and consequences of the agricultural transition throughout the world in a single volume. However, the contributions assembled here provide new evidence that the subtleties and regional variation in dietary change, growth and morphological change, biomechanics and behaviour, and continuing human evolution were complex and region-specific. They also highlight the advantages of high resolution multidisciplinary approaches to the study of human biological and cultural variation associated with cultural and dietary transitions.

REFERENCES

Bar-Yosef, O. (1998) The Natufian culture in the Levant: Threshold to the origins of agriculture. *Evol. Anthropol.*, **6** (5), 159–177.

Bar-Yosef, O. and Belfer-Cohen, A. (2002) Facing environmental crisis: societal and cultural changes at the transition from the Younger Dryas to the Holocene in the Levant, in *The Dawn of Farming in the Near East, Studies in Early Near Eastern Production, Subsistence and Environment 6* (eds R.T.J. Cappers and S. Bottema), ex oriente, par Margareta Tengberg, Berlin, pp. 55–66.

Belfer-Cohen, A. and Bar-Yosef, O. (2000) Early sedentism in the near east: a bumpy ride to village life, in *Life in Neolithic Farming Communities: Social Organization, Identity, and Differentiation* (ed. I. Kuijt), Kluwer Academic/Plenum Publishers, New York.

Belfer-Cohen, A. and Goring-Morris, N. (2009) For the first time. *Curr. Anthropol.*, **50** (5), 669–672.

Bellwood, P. (2005) *First Farmers: The Origins of Agricultural Societies*, Blackwell Publishing, Oxford.

Bettinger, R., Richerson, P. and Boyd, R. (2009) Constraints on the development of agriculture. *Curr. Anthropol.*, **50** (5), 627–631.

Bocquet-Appel, J.P. and Bar-Yosef, O. (2008) Prehistoric demography in a time of globalization, in *The Neolithic Demographic Transition and its Consequences* (eds J.P. Bocquet-Appel and O. Bar-Yosef), Springer Science+Business Media.

Boserup, E. (1965) *The Conditions of Agricultural Growth*, George Allen and Urwin, London.

Bridges, P.S. (1989) Changes in activities with the shift to agriculture in the southeastern United States. *Curr. Anthropol.*, **30** (3), 385–394.

Burger, J., Kirchner, M., Bramanti, B. *et al.* (2007) Absence of the lactase-persistence associated allele in early Neolithic Europeans. *PNAS*, **104**, 3736–3741.

Byrd, B. (2005) Reassessing the emergence of village life in the Near East. *J. Archaeol. Res.*, **13**, 231–290.

Childe, V.G. (1936) *Man Makes Himself*, Watts and Co., London.

Cohen, M.N. (1977) *The Food Crisis in Prehistory: Overpopulation and the Origins of Agriculture*, Yale University Press, New Haven.

Cohen, M.N. (1989) *Health and the Rise of Civilization*, Yale University Press, London.

Cohen, M.N. and Armelagos, G.J. (eds) (1984) *Paleopathology at the Origins of Agriculture*, Academic Press, Orlando, FL.

Cohen, M.N. and Crane-Kramer, G.M.M. (2007) Ancient health: Skeletal indicators of agricultural and economic intensification, in *Bioarchaeologial Interpretations of the Human Past: Local, Regional, and Global Perspectives*, (Series editor C.S. Larsen), University Press of Florida, Gainesville.

Cohen, M.N. (2009) Introduction: Rethinking the origins of agriculture. *Curr. Anthropol.*, **50** (5), 591–595.

Denham, T. (2009) A practice-centered method for charting the emergence and transformation of agriculture. *Curr. Anthropol.*, **50** (5), 661–667.

Diamond, J. (1997) *Guns, Germs, and Steel: The Fates of Human Societies*, W.W. Norton and Company, New York.

Diamond, J. (2002) Evolution, consequences and future of plant and animal domestication. *Nature*, **418**, 597–603.

Durham, W.H. (1991) *Coevolution: Genes, Culture, and Human Diversity*, Stanford University Press, London.

Dyson, F. (1997) The era of Darwinian evolution is over. *New Perspect. Q*, **24**, 58–59.

Fuller, D.Q. and Allaby, R. (2009) Seed dispersal and crop domestication: shattering, germination and seasonality in evolution under cultivation, in *Fruit Development and Seed Dispersal, Annual Plant Reviews*, vol. **38** (ed. L. Ostergaard), Wiley-Blackwell, Oxford, pp. 238–295.

Fuller, D.Q., Allaby, R.G. and Stevens, C. (2010) Domestication as innovation: the entanglement of techniques, technology and chance in the domestication of cereal crops. *World Archaeology*, **42** (1), 13–28.

Goring-Morris, A.N., Hovers, E. and Belfer-Cohen, A. (2009) The dynamics of Pleistocene and Early Holocene settlement patterns and human adaptations in the Levant: an overview, in *Transitions in Prehistory: Essays in Honor of Ofer Bar-Yosef* (eds J. Shea and D. Lieberman), Oxbow Books, Oxford.

Hawks, J., Wang, E.T., Cochran, G.M. *et al.* (2007) Recent acceleration of human adaptive evolution. *PNAS*, **104** (52), 20753–20758.

Hillman, G., Hedges, R., Moore, A. *et al.* (2001) New evidence of Lateglacial cereal cultivation at Abu Hureyra on the Euphrates. *The Holocene*, **11** (4), 383–393.

Klein, R.G. (2009) *The Human Career: Human Biological and Cultural Origins*, 3rd edn, University of Chicago Press, Chicago.

Kuijt, I. and Finlayson, B. (2009) Evidence for food storage and predomestication granaries 11 000 years ago in the Jordan Valley. *PNAS*, **106** (27), 10966–10970.

Laland, K.N. and Brown, G.R. (2006) Niche construction, human behaviour, and the adaptive-lag hypothesis. *Evol. Anthropol.*, **15**, 95–104.

Laland, K.N., Odling-Smee, J. and Myles, S. (2010) How culture shaped the human genome: Bringing genetics and the human sciences together. *Nat. Rev. Genet.*, **11**, 137–148.

Larsen, C.S. (1995) Biological changes in human populations with agriculture. *Ann. Rev. Anthropol.*, **24**, 185–213.

Leach, H.M. (2003) Human domestication reconsidered. *Curr. Anthropol.*, **44** (3), 349–368.

Maher, L.A. (2007) Microliths and mortuary practices: New perspectives on the Epipalaeolithic in Northern and Eastern Jordan, in *Crossing Jordan: North American Contributions to the Archaeology of Jordan* (eds T.E. Levy, P.M.M. Daviau, R.W. Younker and M. Shaer), Equinox, London.

Nadel, D. and Hershkovitz, I. (1991) New subsistence data and human remains from the earliest Levantine Epipalaeolithic. *Curr. Anthropol.*, **32**, 631–635.

Perry, G.H., Dominy, N.J., Claw, K.G. *et al.* (2007) Diet and the evolution of human amylase gene copy number variation. *Nat. Genet.*, **39** (10), 1256–1260.

Pinhasi, R., Fort, J. and Ammerman, A.J. (2005) Tracing the origin and spread of agriculture in Europe. *PLoS Biology*, **3** (12), e410.

Pinhasi, R., Shaw, P. and Eshed, V. (2008) Changes in the masticatory apparatus following the transition to farming in the Levant. *Am. J. Phys. Anthropol.*, **135**, 136–148.

Richerson, P.J., Boyd, R. and Bettinger, R.L. (2001) Was agriculture impossible during the Pleistocene but mandatory during the Holocene? A climate change hypothesis. *Am. Antiq.*, **66**, 387–411.

Ruff, C.B. (1999) Skeletal structure and behavioral patterns of prehistoric Great Basin populations, in *Understanding Prehistoric Lifeways in the Great Basin Wetlands: Bioarchaeological Reconstruction and Interpretation*, (eds B.E. Hemphill and C.S. Larson), University of Utah Press, Salt Lake City, pp. 290–320.

Ruff, C.B. (2008) Biomechanical analyses of archaeological human skeletal samples, in *Biological Anthropology of the Human Skeleton*, 2nd edn (eds M.A. Katzenburg and S.R. Saunders), John Wiley and Sons, New York, pp. 183–206.

Sabeti, P.C., Varilly, P., Fry, B. *et al.*, The International HapMap Consortium (2007) Genome-wide detection and characterization of positive selection in human populations. *Nature*, **449**, 913–918.

Starling, A. and Stock, J.T. (2007) Dental indicators of heath and stress in early Egyptian and Nubian agriculturalists: Difficult transition and gradual recovery. *Am. J. Phys. Anthropol.*, **134** (4), 520–528.

Smith, B.D. (1998) *The Emergence of Agriculture*, Scientific American Library, New York.

Stock, J.T. (2008) Are humans still evolving? in *The Future of our Species*, EMBO Reports, **9**, Science and Society, Special Issue, pp. S51–S54.

Tishkoff, S.A., Varkonyi, R., Cahinhinan, N., *et al.* (2001) Haplotype diversity and linkage disequilibrium at human G6PD: Recent origin of alleles that confer malarial resistance. *Science*, **293**, 455–462.

Tishkoff, S.A., Reed, F.A., Ranciaro, A. *et al.* (2007) Convergent adaptation of human lactase persistence in Africans and Europeans. *Nat. Genet.*, **39** (1), 31–40.

Twiss, K. (2007) The Neolithic of the southern Levant. *Evol. Anthropol.*, **16** (1), 24–35.

Wells, J.C.K. and Stock, J.T. (2007) The biology of the colonizing ape. *Yearbook of Physical Anthropology*, **50**, 191–222.

Wilson, P.J. (1991) *The Domestication of the Human Species*, Yale University Press, New Haven, p. 201.

Zeder, M.A. and Smith, B.D. (2009) A conversation on agricultural origins: Talking past each other in a crowded room. *Curr. Anthropol.*, **50** (5), 681–691.

SECTION A

Subsistence Transitions

2

Mesolithic-Neolithic Transitions: An Isotopic Tour through Europe

Rick Schulting

School of Archaeology, University of Oxford, Oxford, UK

2.1 INTRODUCTION

The last decade has seen the increasing application of stable isotope analysis to Mesolithic and Neolithic human populations across many parts of Europe, in an effort to better understand aspects of the diets of both periods, and to investigate the transition between them. The question is an important one, as the timing and nature of changes in subsistence have wide-reaching implications for how the transition is interpreted. To what extent, for example, were communities living at the time – and this itself varies across Europe – presented with two different lifeways, one based on fishing, hunting and gathering, and the other based primarily on farming and herding? Where the process of Neolithization was gradual, with domesticated plants and animals initially adopted only as minor novel components within a largely 'traditional' economy, we might expect a more limited impact on the societies involved. If the proportion of domesticated plants and animals increased slowly, say over a period of centuries, people living through these changes may hardly have been aware of them. If, on the other hand, changes were very rapid (on the order of decades rather than centuries), then not only would people have been very conscious of these changes, but they would have consequences for most, if not all aspects of their lives. While it is possible to exaggerate the differences, they nevertheless are real enough: hunting and gathering is a very different way of making a living than farming and herding, with different demands on time and organization, often different settlement size and structure, different daily and seasonal rhythms, and differing potentials for the emergence and entrenchment of social inequality. Thus, the use of stable isotope analysis to investigate diet, while at one level only dealing with some aspects of subsistence, at another provides a means to explore a whole range of surrounding issues. The aims of the present chapter are to summarize and discuss what stable isotope analysis has contributed to our knowledge of Mesolithic and Neolithic diets and, more specifically, to the question of the transition across Europe (Figure 2.1).

Human Bioarchaeology of the Transition to Agriculture Edited by Ron Pinhasi and Jay T. Stock

Figure 2.1 Map of Mesolithic and Neolithic sites discussed in the text

2.2 STABLE ISOTOPES

The isotopes of most interest in dietary investigations are carbon ($\delta^{13}C$) and nitrogen ($\delta^{15}N$). Many summaries are available detailing the technical aspects of the method, and so these need not be repeated here (Ambrose and Krigbaum, 2003; Schoeninger and Moore, 1992), though a number of key points need to be emphasized. Many analyses focus on the collagen fraction of archaeological bone, which primarily reflects dietary protein, at least when protein consumption is moderate to high (Ambrose and Norr, 1993). While the mineral component, bioapatite, better reflects whole diet, there are issues with diagenesis that have limited its use (this applies only to carbon, as nitrogen only occurs in the collagen fraction). Dental enamel is less affected than bone, but is less available, and often subject to more stringent curatorial restrictions on sampling. In the context of northwestern Europe, $\delta^{13}C$ informs primarily on the consumption of marine protein, whether derived from seaweeds, shellfish, fish, sea mammals or sea birds. Its efficacy is therefore largely restricted to coastal and near-coastal areas, and this is indeed where it has seen most use. Some freshwater systems also have $\delta^{13}C$ values that distinguish them from purely terrestrial systems, but the effect is highly variable (Dufour, Bocherens and Mariotti, 1999). Stable nitrogen isotopes can address a range of other issues, primarily concerning an organism's position in the trophic web (Bocherens and Drucker, 2003; Hedges and Reynard, 2007; Minagawa and Wada, 1984). It has been particularly useful in identifying the consumption of non-marine aquatic resources (e.g. freshwater fish, shellfish, birds, etc.) (Bonsall *et al.*, 1997; Cook *et al.*, 2001; Lillie and Richards, 2000; Lillie *et al.*, 2009). Aside from this, $\delta^{15}N$ offers a means of investigating the relative contributions of plant vs. animal protein in human diet. This is of great interest in the light of ongoing debates over the relative importance of cereal growing compared with stock-keeping (Jones, 2000; Richards, 2000), though of course the same questions can be asked regarding the importance of plant foods in hunter-gatherer diets. However, because of the smaller difference in $\delta^{15}N$ values between cereals and meat – on the order of 3 to 4‰ compared to the about 9‰ difference in $\delta^{13}C$ between marine and terrestrial protein – and a series of other complicating factors, this is not straightforward (Hedges and Reynard, 2007).

The endpoints for $\delta^{13}C$ values in open water marine systems along the Atlantic façade seem to be reasonably consistent at about -12 ± 1‰, while terrestrial systems are typically -21 ± 1‰ (as are most freshwater aquatic systems, though with some significant exceptions). There is a slight north-south gradient for terrestrial systems, relating to environmental conditions, such that values in northern Europe tend to be 1 to 2‰ lower than those in southern Europe (Van Klinken, Richards and Hedges, 2000). The endpoints for $\delta^{15}N$ can be quite variable, though terrestrial herbivores usually range between about 4 and 7‰ (though note that this covers an entire trophic level as usually interpreted), with humans typically 3 to 5‰ higher (Bocherens and Drucker, 2003). Because of the potential variation in both $\delta^{13}C$ and $\delta^{15}N$, it is desirable to include a range of faunal remains from the same site and period in any isotopic study of humans.

2.3 AN ISOTOPIC TOUR THROUGH EUROPE

2.3.1 Northern Europe

The contribution of stable isotope analysis to the Mesolithic-Neolithic transition thus far has been greatest along the Atlantic façade and in the Baltic. The earliest studies were those undertaken by Tauber (1981, 1986) in Denmark, in which he drew attention to a marked disjunction in $\delta^{13}C$ values between later Mesolithic individuals, showing a very high reliance on marine protein, and Neolithic (and later prehistoric) individuals, showing strongly terrestrial signals. Further analyses have largely confirmed this pattern (Fischer *et al.*, 2007; Price *et al.*, 2007; Richards, Price and Koch, 2003), though, unusually in the context of western Europe, it is the Early Neolithic side of the equation that is hampered by small sample size. With this caveat, it remains the case that $\delta^{13}C$ values differ sharply between the later Mesolithic (as a result of changing relative sea levels, earlier Mesolithic measurements in Denmark are primarily on individuals from inland locations) and the Early/Middle Neolithic (Figure 2.2). However, Fischer *et al.* (2007) have recently indicated that the situation may be more complex, noting isotopic evidence for the continued consumption of small amounts of aquatic protein in

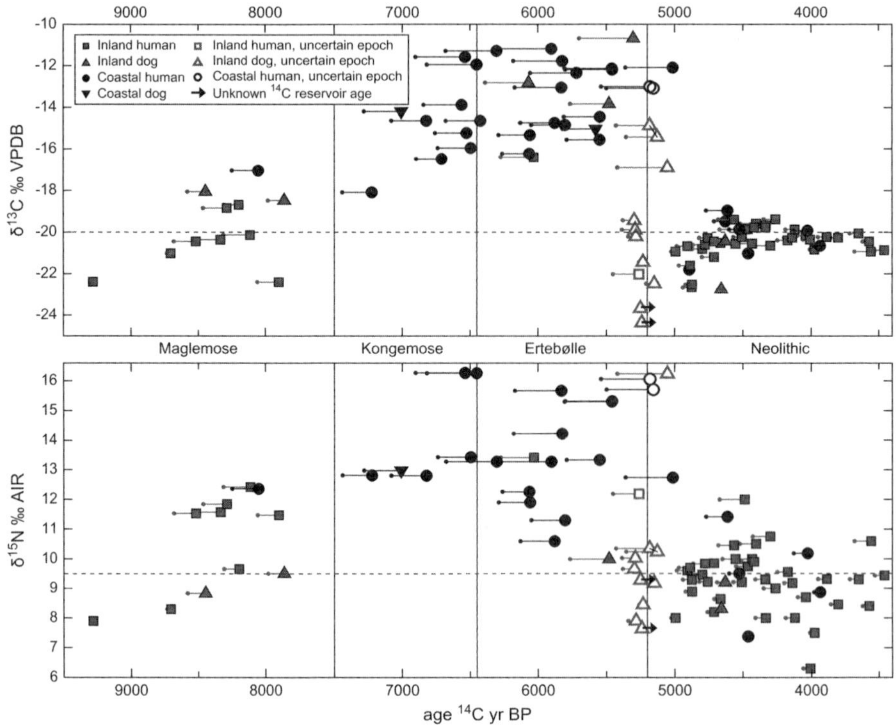

Figure 2.2 Stable N and C isotope values on Mesolithic and Neolithic human and dog remains from Denmark (reprinted from Fischer *et al.*, 2007, Figure 6, with permission from Elsevier). The lines refer to the estimated correction for reservoir effects. The vertical line at 5200 BP equates with about 4000 calBC, where the transition is usually placed (or in the range 4000–3800 BC). Individuals above the dashed horizontal lines are interpreted by Fischer *et al.* (2007) as regularly consuming at least small amounts of aquatic foods (marine in the case of $\delta^{13}C$, marine and/or freshwater for $\delta^{15}N$)

the Neolithic. Interestingly, this is seen more convincingly in some individuals with elevated $\delta^{15}N$ values rather than in $\delta^{13}C$ (Figure 2.2), suggesting that it is the consumption of freshwater species that is more important. The most obvious candidate would be the freshwater eel (*Anguilla anguilla*), a potentially very productive resource, rich in both protein and fats, that can be taken in large numbers during its migration from inland waters to the Sargasso Sea to breed. Because they spend most of their lives in fresh water, eels do not exhibit the elevated $\delta^{13}C$ values of marine organisms.[1] Their importance is confirmed archaeologically by the radiocarbon dating of large fish weirs – in some cases larger than their Mesolithic equivalents – to the earlier Neolithic (Pedersen, 1995). It is clear from how and where the traps are positioned that they are intended to take eels, rather than salmon (which would give a marine isotopic signal) (Pedersen, 1997). The richness and marked seasonality of this resource may have provided the incentive to continue to exploit eels at a time when marine resources were reduced to marginal importance.

Nevertheless, the greater part of the diet was probably still contributed by domesticated plants and animals in the Danish Neolithic. Eels alone could not possibly support these populations year-round; they did not do so even in the Mesolithic, where a large number of eel bones are found in the fish assemblages at sites in the Limfjord (Andersen, 1991; Enghoff, 1986), yet the overall protein was still predominantly marine, as seen in elevated human $\delta^{13}C$ values. A substantial proportion of Neolithic diets, then, must have come from terrestrial sources. While it is difficult to distinguish between wild and domesticated plants and animals isotopically, terrestrial game and wild plant foods (the latter rather limited in northwest Europe) would not be capable of sustaining sizeable human populations in the long term – consider that the elk and the aurochs had likely already been hunted to extinction on Zealand and the smaller Danish islands in the Mesolithic (Anderson *et al.* 1990) – nor can any convincing explanation be offered as to why coastal communities that had relied so heavily on marine resources for over a millennium, would suddenly abandon these in favour of less productive and less reliable alternatives, precisely at a time when domestic plants and animals are known to have been present.

An interesting exception is an adult male from the shellmidden of Rødhals on the small island of Sejerø, with $\delta^{13}C$ and $\delta^{15}N$ values of –11.7 and 12.7‰, respectively (Fischer *et al.*, 2007), indicating an essentially purely marine diet (though the $\delta^{15}N$ is surprisingly low). Omitting Rødhals as a clear outlier,[2] the overall $\delta^{15}N$ average of 10.1 ± 1.0‰ for coastal Danish Neolithic sites is well within the range seen in terrestrial locations across much of Europe, and where no particular emphasis on freshwater fish has been postulated (though this itself may need revisiting). Moreover, this value shows no significant difference with individuals from inland sites, averaging 9.6 ± 0.6‰, though the numbers remain small in both cases (Table 2.1).

The question of timing remains an important one, and there are complications in the direct dating of human remains relying on marine and/or freshwater sources of protein in Denmark, due to potentially highly spatially and temporally variable reservoir effects (Heier-Nielsen *et al.*, 1995). Thus, there is an issue over the assignment of some individuals falling near the transitional period about 4000 to 3800 calBC, when they lack diagnostic material culture associations. The above-mentioned individual from Rødhals is a case in point, with a radiocarbon age of 5360 ± 50 BP (AAR-8552); applying a standard marine reservoir correction of 400 years results in a date of 3913 to 3657 calBC (95%), placing this individual within the Early Neolithic, if defined purely on chronological grounds. This by no means obviates the observation of an abrupt shift in diet, with no indication of transitional isotopic values over time; rather, it blurs the temporal boundary somewhat and suggests that the transition was not instantaneous over all of Denmark. Nor, of course, would this be expected. Thus, while chronologically 'Neolithic', a more parsimonious interpretation of Rødhals would be that this individual was part of a community that persisted in a 'traditional' Late Ertebølle way of life at a time when many surrounding communities were rapidly becoming committed to farming and herding.

The Middle Neolithic (ca. 3300–2800 calBC) cemetery of Ostorf in northern Germany provides a rather different picture to that seen in Denmark. This population appears to have consumed significant amounts of freshwater fish, from the plentiful lakes and rivers of the region. Stable carbon isotope values are entirely 'terrestrial' (which can include freshwater fish and fowl) at -20.4 ± 0.8‰, while $\delta^{15}N$ values are high at 13.7 ± 1.0‰ (Lübke *et al.* 2007). Interestingly, values from single individuals from three other Neolithic sites in the wider area exhibit lower $\delta^{15}N$ values of ca. 11‰: why this should be so is not clear, though they may be some centuries earlier in date (there are issues with freshwater reservoir effects here that complicate ^{14}C results (Olsen and Heinemeier, 2007)). It is tempting to think of a degree of continuity in subsistence with the Mesolithic here, but the late dates for the site, and the absence of earlier comparative material make this problematic.

Table 2.1 Later Mesolithic and Early/Middle Neolithic human stable isotope values in selected areas of Europe

Site	Country	Location	Period	$\delta^{13}C$	±	$\delta^{15}N$	±	n	Comments	Source of data
various sites	Denmark	coastal	Meso	–13.5	2.2	14.4	1.5	25	Kongemose and Ertebølle sites	Fischer *et al.*, 2007; Price *et al.*, 2007
various sites	Denmark	coastal	EN/MN	–20.0	1.0	10.1	1.0	9	no significant difference from inland	Fischer *et al.*, 2007; Price *et al.*, 2007
various sites	Denmark	inland	EN/MN	–20.6	1.0	9.6	0.6	8	no significant difference from coastal	Fischer *et al.*, 2007
Västerbjers, Gotland	Sweden	coastal	MN	–15.1	0.5	15.6	0.5	18	seals average –16.0‰; Pitted Ware	Eriksson, 2004
Ajvide D, Gotland	Sweden	coastal	MN	–15.8	0.7	—	—	5	seals average –16.7‰; Pitted Ware	Lindqvist and Possnert, 1997
Köpingsvik, Öland	Sweden	coastal	MN	–14.5	0.5	16.8	0.6	19	Pitted Ware cemetery	Eriksson *et al.*, 2008
Resmo, Öland	Sweden	coastal	MN	–18.8	0.8	12.4	0.9	19	chambered tomb	Eriksson *et al.*, 2008
Torsborg, Kalleguta, Vickleby	Sweden	inland	LN	–20.5	0.2	9.3	0.4	8	note low variability in both C and N	Eriksson *et al.*, 2008
Zvejnieki	Latvia	inland	LM/EN	–23.1	1.1	11.9	1.1	14	large multi-period cemetery on inland lake	Eriksson, 2006
Zvejnieki	Latvia	inland	MN	–22.3	0.8	12.1	1.0	10	large multi-period cemetery on inland lake	Eriksson, 2006
Zvejnieki, Sarkani, Selgas	Latvia	inland	LN	–21.6	0.1	10.6	0.8	5	includes inland and coastal sites; no differences	Eriksson, 2006; Eriksson *et al.*, 2003
Ostorf	N Germany	inland	MN	–20.4	0.8	13.7	1.0	16		Lübke *et al.* 2007
various sites	N Germany	inland	EN/MN	–20.5	0.5	11.3	1.3	3	$\delta^{15}N$ significantly lower than Ostorf ($p = 0.002$)	Lübke *et al.* 2007
Hardinxveld	Netherlands	inland	LM	–21.9	1.0	14.5	1.6	15	excludes 4 results with C:N outside 2.6–3.6	Smits and van der Plicht, 2009
Swifterbant	Netherlands	inland	LM/EN							Smits *et al.*, 2010
Schipluiden	Netherlands	coastal	MN	–20.1	1.2	14.5	2.2	8	recalculated C:N values range 2.8–3.4	Smits and van der Plicht, 2009

Oronsay sites	Scotland	coastal	LM	–13.2	1.5	15.3	1.0	6	shellmiddens	Richards and Mellars, 1998
Holm of Papa Westray North	Scotland	coastal	MN	–19.9	0.6	10.5	0.6	5	chambered tomb	Schulting and Richards, 2009
Carding Mill Bay	Scotland	coastal	MN	–21.4	0.2	9.3	0.4	10	within Mesolithic shellmidden	Schulting and Richards, 2002b
Crarae	Scotland	coastal	MN	–21.6	0.3	9.2	0.3	3	chambered tomb	Schulting and Richards, 2002b
Creagnan Uamh	Scotland	coastal	MN	–20.8	0.4	—	—	4	cave site	Hedges *et al.*, 1998
Raschoille Cave	Scotland	coastal	EN/MN	–21.1	0.6	—	—	14	identified as Neolithic based on dates	Bonsall *et al.*, 2000
Preston Docks	England	coastal	EN/MN	–21.2	0.1	—	—	10	identified as Neolithic based on dates	Richards and Hedges, 1999
Broadsands	England	coastal	EN	–20.3	0.4	9.2	0.5	5	chambered tomb	Sheridan *et al.*, 2008
Parc le Breos	Wales	coastal	MN	–20.5	1.1	9.7	0.5	8	chambered tomb	Richards, in Whittle at Wysocki 1998
various sites, S Wales	Wales	coastal	EN/MN	–20.6	0.5	9.0	0.8	15	cave and rockshelter sites	Schulting, 2007; Schulting and Richards, 2002a
Ferriter's Cove	Ireland	coastal	LM	–14.0	0.1	16.8	—	3	only 1 $\delta^{15}N$ measurement available	Woodman, 2008
Poulnabrone	Ireland	coastal	EN/MN	–21.0				3	portal tomb	Woodman, 2004
various sites	Ireland	coastal	EN/MN	–21.2	0.3	10.6	0.8	6	chambered tombs	
Téviec, Brittany	France	coastal	LM	–15.1	0.7	13.1	1.6	9	excludes D1 and H1, contamination	Schulting and Richards, 2001; n.d.
Hoëdic, Brittany	France	coastal	LM	–14.0	0.5	13.2	1.1	10	excludes J1, contamination	Schulting and Richards, 2001; n.d.
Port Blanc, Brittany	France	coastal	MN	–20.0	0.4	11.0	0.5	9	chambered tomb	Schulting, 2005; this paper
Er Yoh, Brittany	France	coastal	LN	–19.3	—	12.0	—	1	Late Neolithic shellmidden	Schulting, 2005
Vierville, Normandy	France	coastal	MN	–20.3	0.1	—	—	4	chambered tomb	Schulting n.d. this paper
Le Déhus	Guernsey	coastal	MN	–20.3	0.1	14.1	0.9	3	chambered tomb	Schulting *et al.* 2010

(continued)

Table 2.1 (*Continued*)

Site	Country	Location	Period	$\delta^{13}C$	±	$\delta^{15}N$	±	n	Comments	Source of data
Cabeço da Amoreria	Portgual	estuarine	LM	–16.6	1.7	11.5	1.9	5		Roksandic, 2006; Umbelino, 2005
Cabeço da Amoreria	Portgual	estuarine	LM	–15.7	0.7	12.3	0.4	4	excluding terrestrial outlier (–19.7; 8.2‰)	Roksandic, 2006; Umbelino, 2005
Cabeço da Arruda	Portgual	estuarine	LM	–17.4	1.3	11.4	0.8	9		Lubell *et al.*, 1994; Roksandic, 2006; Umbelino, 2005
Moita do Sebastião	Portgual	estuarine	LM	–16.4	0.5	11.7	0.9	8		Lubell *et al.*, 1994; Umbelino, 2005
Muge sites combined	Portgual	estuarine	LM	–16.8	1.3	11.7	1.5	24		
Sado sites	Portgual	estuarine	LM	–18.3	0.6	10.3	1.2	3	markedly less marine than Muge sites	Umbelino, 2005
Muge sites	Portgual	estuarine	EN/MN	–19.1	1.5	9.2	1.5	10		Lubell *et al.*, 1994
Gruta do Lagar	Portgual	estuarine	MN	–14.9	—	13.1	—	1		Lubell *et al.*, 1994
Muge sites	Portgual	estuarine	EN/MN	–19.6	0.3	8.7	0.5	9	excluding outlier Gruta do Lagar	Lubell *et al.*, 1994
Cantabrian Mesolithic	Spain	coastal	LM	–16.4	0.5	12.1	0.6	3		Arias, 2005
Cantabrian Neolithic	Spain	coastal	EN/LN	–20.8	1.0	—	—	13		Arias, 2005
El Collado	Spain	coastal	Meso	–18.4	0.7	10.3	1.2	9	shellmidden, 3 km from modern coastline	Garcia Guixé, Richards and Subirá, 2006
Le Crès	France	inland	MN	–19.5	0.5	8.1	0.8	32		Le Bras-Goude, *et al.*, 2009
Pendimoun	France	coastal	EN	–19.9	0.2	7.9	0.7	5		Le Bras-Goude *et al.*, 2006
Arene Candide	Italy	coastal	EN/MN	–20.2	0.5	8.7	1.1	8		Le Bras-Goude *et al.*, 2006
Samari	Italy	coastal	EN/MN	–19.2	1.0	8.6	0.6	3		Giorgi *et al.*, 2005

various sites	Italy	inland	EN/MN	–19.9	0.3	8.4	0.9	5	sig. diff. in ^{13}C ($p = 0.02$); no difference in ^{15}N	Giorgi *et al.*, 2005
Brochtorff stone circle	Malta	coastal	LN	–19.2	0.3	9.7	0.9	7		Richards *et al.*, 2001
Franchthi	Greece	coastal	Neo	–18.7	0.8	9.2	1.8	11		Papathanasiou, 2003
Kephala	Greece	coastal	Neo	–19.1	1.2	9.2	1.0	5		Papathanasiou, 2003
Alepotrypa	Greece	coastal	Neo	–20.0	0.4	7.2	1.0	26	groups with inland sites in both ^{13}C and ^{15}N	Papathanasiou, 2003
Tharrounia	Greece	inland	Neo	–20.0	0.2	8.0	0.7	20		Papathanasiou, 2003
Theopetra	Greece	inland	Neo	–20.0	0.4	7.6	0.6	12		Papathanasiou, 2003
Kouveleiki	Greece	inland	Neo	–19.8	0.0	8.1	0.3	2		Papathanasiou, 2003
various sites, Dnieper	Ukraine	riverine	LM	–20.9	0.8	13.4	1.2	18	excludes Osipovka as extreme outlier in ^{15}N	Lillie and Richards, 2000; Lillie and Jacobs, 2006
Vasilyevka V, Dnieper	Ukraine	riverine	LM/EN	–22.8	0.6	11.5	1.2	2	transitional period	Lillie and Richards, 2000; Lillie and Jacobs, 2006
various sites, Dnieper	Ukraine	riverine	EN	–22.5	1.0	10.9	1.7	10		Lillie and Richards, 2000; Lillie and Jacobs, 2006
Lepenski Vir	Serbia	riverine	pre-6300	–19.1	0.7	14.7	1.1	7	adults and adolescents only	Borić *et al.*, 2004
Lepenski Vir	Serbia	riverine	post-6300	–19.7	0.8	13.4	2.1	19	adults and adolescents only	Borić *et al.*, 2004
Lepenski Vir	Serbia	riverine	I-II	–19.0	0.6	14.4	1.8	17	adults only; some overlap with Borić *et al.*, 2004	Bonsall *et al.*, 1997
Lepenski Vir	Serbia	riverine	III-Starçevo	–19.3	0.5	12.5	1.8	15	adults only; some overlap with Borić *et al.*, 2004	Bonsall *et al.*, 1997
Vlasac	Serbia	riverine	Meso	–19.4	0.5	14.2	0.8	31	adults and adolescents only	Borić *et al.*, 2004
Vlasac	Serbia	riverine	Meso	–19.0	0.3	14.8	0.6	29	adults only; some overlap with Borić *et al.*, 2004	Bonsall *et al.*, 1997
Schela Cladovei	Romania	riverine	Meso	–19.6	0.2	15.4	0.4	8	adults only	Bonsall *et al.*, 1997

Further east in the Baltic the situation is less clear, in that many populations demonstrate a continued reliance on marine resources, seen both through elevated $\delta^{13}C$ and $\delta^{15}N$ values and in the presence of large numbers of seal remains (Eriksson, 2004; Storå, 2001). Domestic animals, especially pig, are present but in smaller numbers. Human remains from the Middle Neolithic Pitted Ware sites of Västerbjers and Ajvide on Gotland and from Köpingsvik on Öland all exhibit strongly marine isotope signatures – particularly once the depressed, or at least variable, $\delta^{13}C$ baseline values for the Baltic are taken into account (e.g. seals from Västerbjers average –16‰). The extent to which this represents a continuation of Mesolithic lifeways is not entirely clear, however, due to the paucity of Mesolithic human remains and sites with faunal preservation in the same area but, at least based on the results from two individuals (Köpingsvik, Gotland and Alby, Öland) (Eriksson, 2004; Eriksson *et al.*, 2008), it is a plausible scenario.

Most intriguingly, at the same time (about 3500–2500 calBC), a very different culture was present on Öland's west coast, which did emphasize a farming and herding economy, made different pottery, and buried its dead in megalithic passage tombs (Papmehl-Dufay, 2006). The enclave of four surviving TRB passage tombs at Resmo is only some 45 km south of Köpingsvik, yet the stable isotope values are entirely distinct (Eriksson *et al.*, 2008). That these two cultures, with their very different lifeways, could exist in the same general region, demonstrates that the environment is not the sole determining factor, and that there was a strong element of cultural choice involved (Sjögren, 2003). Where the environment likely did play a role, however, was in providing a more level playing field. Farming became increasingly precarious further north and east in Europe, due to the shorter growing season and, in the latter case, increased distance from the warming influence of the Gulf Stream. At the same time, the marine environment remained relatively productive. The degree to which the groups following these different lifeways interacted is an interesting question, though not one that can be discussed here in any detail (see Papmehl-Dufay, 2006). There is no evidence that they formed part of an integrated, complementary system, and their spatial separation, while far from prohibitive, is such that contacts may have been intermittent. Indeed, recent aDNA evidence suggests that Pitted Ware and TRB populations were genetically distinct (Malmström *et al.*, 2009). By about 2000 calBC, towards the end of the Neolithic, diets for all groups on Öland had become predominantly terrestrial (Eriksson *et al.*, 2008).

Still further east, Zvejnieki is a large multi-period cemetery on Lake Burtnieks in northern Lativa, located some 50 km from the modern coastline. While this distance would have varied throughout the site's long history, it can be classed as essentially an inland site, though with access to the coast via the Salaca River. Burials range from the Middle Mesolithic (eighth millennium calBC) to the Late Neolithic (third millennium calBC) and beyond, and occur in distinct spatial clusters for each period (Larsson and Zagorska, 2006, and see Lillie and Budd, this volume). Given the site's distance from the coast, any changes in diet would be hard to detect isotopically, even if they did exist. But the Early Neolithic is defined here by the appearance of pottery, making it equivalent to the Late Mesolithic Ertebølle of southern Scandinavia, and the presence of domestic species is not attested archaeologically until many centuries later; indeed, with the exception of a single worked sheep/goat bone in a Late Neolithic Corded Ware burial, domestic fauna are entirely absent from all phases of Zvejnieki. No clear isotopic differences can be seen between the Mesolithic, Early and Middle Neolithic, with depressed $\delta^{13}C$ values and somewhat elevated $\delta^{15}N$ values of about 12‰ suggesting some contribution of freshwater fish from the lake (Eriksson, 2006; Eriksson, Lóugas and Zagorska, 2003). It is likely, as in the case of eastern Sweden, that there was a strong element of continuity in subsistence and lifeways more generally in the eastern Baltic (Zvelebil, 1996; Zvelebil and Lillie, 2000). As was seen on Öland, a limited number of available Late Neolithic humans show lower $\delta^{15}N$ values, which may indicate a dietary shift to a more committed farming and herding regime (Table 2.1).

2.3.2 Britain and Ireland

Britain and Ireland provide strong support for a relatively rapid and complete shift away from marine foods across the Mesolithic-Neolithic transition. This statement must be qualified by acknowledging that, due primarily to the relative scarcity of Late Mesolithic human remains, the sample sizes involved remain small. Most relevant are scattered human remains from the shellmiddens of Oronsay, on the west coast of Scotland (Richards and Mellars, 1998; Richards and Sheridan, 2000) and from Ferriter's Cove in southwest Ireland (Woodman, 2008; Woodman, Andersen and Finlay, 1999). In both cases, stable isotopic analysis indicates individuals whose dietary protein was comprised almost entirely of marine foods (Table 2.1). Crucially, for the present discussion, ^{14}C AMS dates place them at the very cusp of the transition as currently understood for Britain and Ireland, about 4000 to 3900 calBC. Most other Mesolithic human remains known from Britain and Ireland are substantially earlier, though the majority of those from coastal contexts support the importance of marine resources (Schulting and Richards, 2002a; Schulting, 2005).

Sample size is far less of an issue for the Neolithic side of the equation (even excluding individuals from inland locations, defined here as more than 5 km from the coast), and it is this that strengthens the argument for a rapid and complete transition. This position remains debated, with Thomas (2003) noting that the isotopic data are not capable of distinguishing wild and domestic terrestrial foods. This is true but, as already discussed above, it is difficult to conceive of a situation in which coastal resources would be abandoned in favour of wild terrestrial resources, just at the time when domestic plants and animals, together with substantial changes in material culture, make their first appearance. Such a position is even less tenable in Britain than in Denmark, since there are a number of large British Neolithic faunal assemblages, invariably and overwhelmingly dominated by domestic animals (Schulting, 2008); this is the case even in Orkney, from at least 3600 calBC (Tresset, 2003). Concerns raised over sample size and the possibility of regional variation remain (Milner *et al.*, 2004), but the accumulating data continue to support a rapid and complete dietary shift across Britain and Ireland. In addition to previously reported individuals from coastal Britain (Richards, Schulting and Hedges 2003; Schulting and Richards, 2002a, 2002b) and Ireland (Woodman, 2004), this now includes individuals from earlier Neolithic coastal sites in Orkney (Schulting and Richards, 2009), northern Ireland, and southwest England (Sheridan *et al.*, 2008). Nor are there significant isotopic differences between individuals from monumental and non-monumental mortuary contexts in terms of marine protein consumption (Schulting, 2007). There does remain the question of how rapid is rapid: most of the available directly dated humans fall after about 3800 calBC, leaving the crucial preceding one to two centuries less well-known. In archaeological terms, this still implies a rapid transition, while in terms of human experience, the possibility of a more gradual transition remains. If so, however, it is surprising that no more evidence for it has emerged.

2.3.3 The Netherlands

Recent isotopic research in the Netherlands has provided the first datasets from this part of northwest Europe (Smits *et al.* 2010; Smits and van der Plicht, 2009).[3] Three sites span the transition: Hardinxveld (Mesolithic, about 5450–4500 calBC), Swifterbant (transitional, about 4200–4000 calBC) and Schipluiden (Middle Neolithic, about 3600–3400 calBC). None show any significant use of marine protein, but all show elevated δ^{15}N values. The comparison is not straightforward, however, since Hardinxveld and Swifterbant are some 40 to 50 km inland, and so the habitual use of marine resources would not be expected. Even were coastal samples available, it might be questioned whether the nature of this coastline would be conducive to the exploitation of fully marine resources, as opposed to estuarine and tidal flats species. Schipluiden is the only site situated near the coast. Its combination of relatively depleted δ^{13}C values and high δ^{15}N values suggest the exploitation of freshwater aquatic species, such as the sturgeon evidenced in the faunal remains (with a single measurement of –21.6 and 12.3‰) (Smits and van der Plicht, 2009). But the mammalian fauna is still dominated by domestic species (Louwe Kooijmans, 2009), and so the overall importance of aquatic resources here remains uncertain, particularly since δ^{15}N values in plants and animals can be elevated in wetland habitats, as recently demonstrated by Britton, Müldner and Bell (2008). Thus the nature of the Mesolithic-Neolithic transition in the lowland zone of the Netherlands is not clear, but it does seem to be less marked than seen in many other parts of the Atlantic façade (Louwe Kooijmans, 2007, 2009). The loess zone of the Low Countries, by contrast, sees the westernmost extent of LBK settlement, with a fully formed Neolithic material culture and suite of domestic plants and animals (Modderman, 1988). The relationship between the two areas is of considerable interest, but remains poorly known (Vanmontfort, 2008).

2.3.4 Northwest France and the Channel Islands

Bone preservation is generally poor in northwest France and the Channel Islands, but there are important exceptions, mainly due to the buffering effects of shell. For the Late Mesolithic, the two outstanding sites are the small cemeteries of Téviec and Hoëdic in southern Brittany (Péquart and Péquart, 1954; Péquart *et al.*, 1937). The human remains here have been directly dated to about 5500 to 5000 calBC,[4] and yield isotope values showing a strong reliance on marine protein, ranging from about 60 to 80% (Schulting, 2005; Schulting and Richards, 2001). By contrast, the few Early Neolithic humans that have been measured show far less use of marine protein. The most relevant site is Port Blanc, a passage tomb located directly on the coast, near Téviec, with human remains directly dated to about 4000 calBC (Schulting, 2005). This is not the earliest Neolithic in southern Brittany, however, as the long mound at Erdeven has supplied earlier dates (Cassen, Boujot and Vaquero, 2000), and the central chamber of the great Carnac mound of Tumulus St Michel has yielded mid-fifth millennium dates on calcined bone and charcoal (Schulting, Lanting and Reimer, 2009); in neither case does unburnt bone survive. Thus a slower transition remains a possibility here, though the currently available evidence arguably appears to be most consistent with a rapid and sharp, though perhaps not complete, dietary shift. Continued use of marine resources is seen most clearly at Er Yoh, a Late Neolithic shellmidden on a small islet off the Morbihan coast, with substantial fish and seal remains, though its mammalian fauna is still dominated by terrestrial species (Schulting, Tresset and Dupont, 2004). Stable isotope data are available for two individuals from the site, the earlier dating to about 2910 to 2630 calBC; its δ^{13}C value of –19.3‰ combined with a $\delta^{15}N$ value of 12.0‰ suggests a minor contribution of marine protein, on the order of 10%, which could be seen as consistent with the faunal assemblage (ibid.).

The only known Mesolithic burial from Normandy is a complex multiple cremation from Les Varennes, with abundant faunal offerings (Billard, Arbogast and Valentin, 1999); in the absence of collagen, no stable isotope measurements are possible. Much more material is available from a number of Middle Neolithic mortuary monuments, but limited stable isotope studies have yet been undertaken. In any case, for the most part these are from inland locations, which tend to provide little information of relevance to an investigation of the Mesolithic-Neolithic transition (though see below). An exception is the near-coastal chambered tomb of Vierville (Verron, 2000), and measurements here are typical of purely terrestrial diets, with $\delta^{13}C$ for four individuals averaging -20.3 ± 0.1‰ (Schulting *et al.*, 2010). The associated dates on human remains fall at about 4200 calBC, placing the site early in the Middle Neolithic II (Chasséen). Again this is not the earliest Neolithic in Normandy, as sites with clear Villeneuve-St-Germain affinity are known (Marcigny, Ghesquiere and Desloges, 2007; Verron, 2000). These settlements appear fully formed with a suite of novel architectural and material culture elements and, presumably, the full complement of domestic plants and animals, though no faunal assemblages survive; they thus appear to be best interpreted as a direct movement of Neolithic farmers from the Paris Basin. What happened around the margins of these settlements, and their interaction with indigenous communities in the surrounding areas, particularly along the coasts, is of some interest, but again there is little in the way of relevant data that can be brought to bear at present. Further east, Limburg and La Hoguette pottery may provide some hints of interaction between LBK farmers and pottery-making foragers/herders (Gronenborn, 1999), though the details of this are far from clear, particularly as Limburg pottery has only been found on LBK sites (Vanmontfort, 2008).

The Channel Islands' position between northwest France and southern England make them of considerable interest. Unfortunately, bone survival is again poor due to predominantly acid soils. No Mesolithic sites with either human or faunal remains are known; any shellmiddens that were present have been lost to rising sea levels (Renouf and Urry, 1986). A number of passage tombs have yielded human remains, which survived due to the large numbers of limpet shells that were intentionally incorporated into the fills of their chambers (Kendrick, 1928). These monuments are almost invariably multi-period, however, and so direct AMS dates are required on all samples analysed in order to confirm their Middle Neolithic attribution (the tombs and the pottery they contain show clear affinity with this period in Brittany and Normandy). A recent study at Le Déhus on Guernsey has yielded dates of about 4100 to 3900 calBC on three individuals from a primary context. The stable isotope measurements show, as at Schipluiden, a combination of low $\delta^{13}C$ and high $\delta^{15}N$ values, averaging -20.3 ± 0.1 and 14.1 ± 0.9‰, respectively (Schulting, Sebire and Robb, 2010). Unlike the Netherlands site, however, it seems unlikely that freshwater aquatic resources could feature strongly on Guernsey. The explanation is not yet clear, but a combination of manuring (with dung and/or seaweed) and the use of wetland pastures – with their potentially elevated $\delta^{15}N$ values – has been suggested as one possibility (ibid.). Although not directly dated, isotope values on humans from other Neolithic chambered tombs on Guernsey, Jersey and Herm show comparable results, and confirm the absence of any significant use of marine protein (Bukach, 2005). Given the relative paucity of sustainable terrestrial animal resources on Guernsey (at least once it became an island in the mid-Holocene), it seems highly probable that marine resources featured strongly in Mesolithic diets, if indeed groups were permanently resident at all. Thus, the demonstration of terrestrial diets in the Neolithic suggests a sharp shift in subsistence practices. Given the suite of material culture and monumental mortuary architecture that appears at this time, and the presence of domestic fauna at a number of sites, the Neolithic here is best interpreted as featuring a strong element of direct colonization from northwest France (Schulting, Sebire and Robb, 2010). This leaves open the possibility of a period of interaction with indigenous Mesolithic communities (in fact, it arguably necessitates it), and of processes of incorporation and acculturation.

2.3.5 The Iberian Peninsula

The northern coast of Spain is well-known for Mesolithic sites of the Asturian culture, including a number of small shellmiddens. However, few have yielded human remains, an exception being three directly dated Late Mesolithic humans from the sites of La Poza l'Egua, J3, and Colomba, which provide some indication of coastal diets at this time, with an average $\delta^{13}C$ value of $-16.4 \pm 0.5‰$ suggesting that approximately 50% of the protein came from the sea (Arias, 2005). No such reliance on marine protein is seen in the limited Neolithic isotopic values that are available (ibid.). However, some incompletely analysed faunal assemblages (with uncertainties over dating), together with coastal site locations, do suggest the possibility of a degree of continuity in the use of marine resources into the Neolithic (Fano, 2007), though their place in the overall subsistence economy remains poorly understood (cf. Zapata Milner and Roselló, 2007).

By contrast with northern Spain, the shellmiddens of the Tagus estuary of central Portugal present one of the richest concentrations of Mesolithic burials in Europe (Arnaud, 1989; Ferembach, 1974; Roche, 1972; Zilhão, 2000). Early isotopic research here documented a clear shift away from estuarine resources (the sites were located some distance from the coast, but within the tidal reach of the Tagus) across the Mesolithic-Neolithic transition (Lubell *et al.*, 1994). This has been confirmed by more recent studies (Roksandic, 2006; Umbelino, 2005). Both $\delta^{13}C$ and $\delta^{15}N$ values are significantly lower for the Neolithic sites, though these are hampered by small sample sizes for the earliest Neolithic. In addition, there are two notable outliers: an individual directly dated to 6550 ± 70 BP (TO-10225) from the Late Mesolithic shellmidden of Cabeço da Amoreria with a typical terrestrial isotope signature (Roksandic, 2006) (Figure 2.3), and an individual from the coastal site of Gruta do Lagar, directly dated to the Middle Neolithic (TO2091: 5340 ± 70), but yielding a very strong marine signature (Lubell *et al.*, 1994). The exceptional status of the latter individual is apparent when compared to the remaining Neolithic sites (Table 2.1; Figure 2.3). While a single case like this is difficult to interpret, it may point to the presence of coastal communities specializing in the exploitation of marine resources well into the Neolithic. If so, it would be an unusual, if not unique, situation for the Atlantic façade.

2.3.6 The Mediterranean

Only limited isotopic data are as yet available for the Mediterranean, particularly for the Mesolithic side of the equation. A study on the small Mesolithic cemetery at El Collado on the east Spanish coast shows surprisingly minor use of marine protein, possibly as a result of the lower productivity of the Mediterranean compared to the Atlantic (Garcia Guixé, Richards and Subirá, 2006). An increasing number of Neolithic results are becoming available, and so far these follow the theme of little or no use of marine resources (Figure 2.1). The earliest of these is Khirokitia on Cyprus, dating to the seventh to sixth millennia BC, located some 6 km

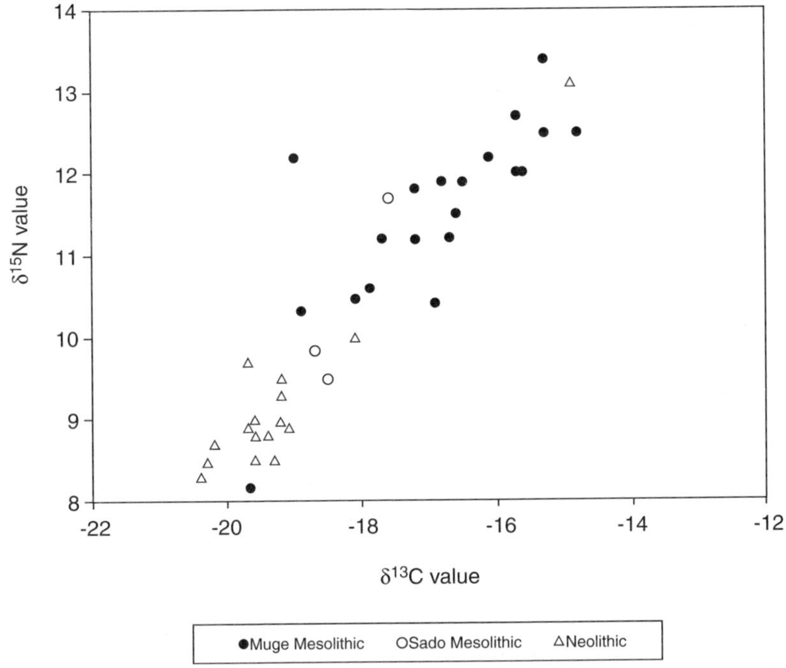

Figure 2.3 Stable N and C isotope values on Mesolithic and Neolithic human remains from Portugal (for sources, see Table 2.1). Note the single outliers at either end of the distribution.

from the coast, but with no isotopic evidence for the use of marine resources found in an analysis of 24 individuals (Lange-Badré and Le Mort, 1998). The results are important in demonstrating an early commitment to mixed farming. Further to the west is the Early Neolithic site of Pendimoun in southern France, and the Early/Middle Neolithic sites of Arene Candide and Samari in Italy (Giorgi *et al.*, 2005; Le Bras-Goude *et al.*, 2006). The Late Neolithic Brochtorff Circle on Malta shows a small contribution of marine protein, though given the small size of the island, it is surprising that it is not higher (Richards *et al.*, 2001). Franchthi Cave is an important Greek site, with both Mesolithic and Neolithic burials, and faunal evidence for the use of marine resources spanning the transition (Rose, 1995). Isotopic analysis of the Mesolithic material is currently underway (M. Mannino pers. comm.); coastal Neolithic humans from Franchthi and the site of Kephala show a minor contribution of marine protein while, interestingly, the coastal site of Alepotrypa clusters instead with a series of three inland Greek sites, with purely terrestrial diets (Papathanasiou, 2003; and Papathanasiou, this volume) (Table 2.1). This, incidentally, demonstrates that the minor use of marine resources, where present, is detectable, and contrasts with the situation in Britain, where no such differences are yet apparent between coastal and inland Neolithic sites (Richards, Schulting and Hedges, 2003; Richards and Schulting, 2006).

2.3.7 Eastern Europe

Two areas in eastern Europe have seen extensive isotopic studies relating to the Mesolithic-Neolithic transition: the Danubian Iron Gates and the Dnieper Rapids (Figure 2.1). Both share a proximity to constrained portions of large rivers, and so have the potential for highly productive fisheries. They thus provide greater possibilities for detecting differences between Mesolithic and Neolithic diets in inland settings, if the exploitation of fish varied significantly across the transition. The Dnieper Rapids are considered elsewhere (Lillie and Budd, this volume), and so are not discussed here, though it is worth noting that there is evidence for a decrease in the consumption of fish protein in the 'Neolithic' (Table 2.1). At issue here, however, is whether this can be attributed to the appearance of domesticated plants and animals, for which there is little archaeological evidence, though this needs to be tempered with the fact that few settlements have been excavated in the region (Zvelebil and Lillie, 2000). As with Denmark, direct dating of human bone is an issue for both the Danube and the Dnieper, as humans consuming fish from these particular rivers appear significantly older than terrestrial samples from the same contexts, demonstrating a freshwater reservoir effect (Cook *et al.*, 2001; Lillie *et al.*, 2009).

For the Iron Gates, the combination of low $\delta^{13}C$ values and high $\delta^{15}N$ values, on the order of 13 to 16‰ for adults, indicates the consumption of freshwater fish, the remains of which are also documented at a number of sites (Bonsall *et al.*, 1997). Bonsall *et al.* (1997, 2000, 2004) proposed that the appearance of the Neolithic about 6000 calBC saw a reduction in the proportion of fish consumed, though it remained significant (Proto Lepenski Vir (LV) and LV I-II average $\delta^{15}N = 14.4‰$, vs. 12.5‰ for LV-III/Starčevo; $t = 3.1$, $p = 0.004$) (Table 2.1). Borić *et al.* (2004) have disputed this, arguing for a more complex relationship between indigenous communities and incomers, and a minor role for dietary change. However, while the difference is not great, there are statistically significant shifts in adult $\delta^{15}N$ isotope values between the pre- and post-6300 calBC groups at Lepenski Vir in the data presented by Borić *et al.* (2004), in the direction of decreased consumption of fish in the later period (Table 2.1; $t = 2.3$, $p = 0.03$). Problems with directly dating human bone caused by a freshwater reservoir effect make fine chronological resolution problematic (Cook *et al.*, 2001). Moreover, understanding the Mesolithic-Neolithic transition in a location such as the Iron Gates is very difficult. The gorge itself would not be attractive for raising either crops or animals (Tringham, 2000), and so is likely to have seen very specialized use throughout prehistory, although bones of domestic species are found in Lepenski Vir III, making up some 20% of the mammalian faunal assemblage (Bonsall *et al.*, 2004). A new series of AMS determinations directly on domestic animal bones place these just after 6000 calBC (Borić and Dimitrijević, 2007), but the relative roles of fishing and farming/herding in the overall society are still poorly understood (cf. Boroneant and Dinu, 2006).

2.4 CONCLUSIONS

There is an increasing trend to look at regional variability in the Mesolithic-Neolithic transition across Europe, moving away from notions of a single process involving a farming juggernaut rapidly sweeping across the subcontinent, albeit with some minor stops and starts along the way. Nevertheless, it remains the case that the dominant impression provided by the stable isotope studies discussed here is of a shift from regional isotopic/dietary heterogeneity in the Mesolithic, to comparative homogeneity in the Neolithic. The strongest case can be made for coastal regions since, firstly, the technique is far more effective in distinguishing marine vs. terrestrial sources of protein and, secondly, many parts of inland Europe lack appropriate samples of human remains, particularly for the Mesolithic. Though undoubtedly a factor, this cannot be solely due to poor preservation, since some of these areas do have substantial quantities of Neolithic bone. Rather, it is probably the result of a combination of differential archaeological visibility, changing burial practices and, on average, substantially lower population densities in the Mesolithic away from the coasts.

However, it should be emphasized that isotopic homogeneity does not necessarily equate with dietary homogeneity, and this is a particularly important consideration for the Neolithic populations discussed here, many of which show limited isotopic variability. While stable isotope signatures falling within a terrestrial range may indicate – at least in coastal contexts – a clear shift away from preceding subsistence practices emphasizing marine resources, they do not identify the nature of the subsistence economy that replaced it. This applies equally in inland situations though, as discussed above, these are often more difficult to disentangle in terms of changes or continuities across the transition. From zooarchaeological evidence, for example, it is clear that ovicaprids generally played a much stronger role around the Mediterranean, while cattle tend to strongly dominate in northwest Europe (Tresset, 2003). The different sizes and demands of these animals can result in very different subsistence regimes, yet isotopically, humans will appear very similar, if not identical. Similarly, and with even more potential for divergent lifeways, the balance between crops and herds, and varying emphasis on milking, and so on will result in only subtly different isotope values for human consumers, that can be very difficult to interpret. For the Neolithic, it is these distinctions that undoubtedly formed the basis for many local and regional differences.

Furthermore, while a strong dietary shift across the transition may be the dominant impression, there are a number of important exceptions. The clearest of these comes from the islands of Gotland and Öland in eastern Sweden, where individuals belonging to the Middle Neolithic Pitted Ware Culture show a strong reliance on marine resources, perhaps reflecting continuity with the lesser known Late Mesolithic of the region. Alternatively, the clear presence of other Neolithic communities in the same region, following a farming way of life, may have provided the impetus for a greater degree of economic specialization (a kind of niche separation) for both groups. These scenarios need not be mutually exclusive. Potentially more complex and drawn-out transitions in subsistence practices can also be seen at Ostorf in northern Germany and at Schipluiden in the Netherlands, both suggesting the continued use of freshwater aquatic resources. By contrast, an isotopic shift suggesting a decrease in the use of freshwater fish appears to be seen along the Dnieper Rapids coterminous with the appearance of the 'Neolithic', though there may be an issue with terminology here, the period being defined largely by the presence of pottery (Zvelebil, 1996). The situation in the Iron Gates is complex, not only in terms of the interpretation of δ^{13}C and δ^{15}N results, but with regards to a reservoir effect, yet again there seems to be a decline in the use of freshwater fish, though this resource still appears to have made a substantial contribution to human diets after the initial appearance of domesticated animals. It is not possible to look at longer-term trends here, since the gorge seems to have been largely abandoned not long afterwards (Tringham, 2000). These exceptions to a strong isotopic/dietary shift in Europe may relate at least in part to the relative potential for productive mixed farming systems vs. the productivity of wild plant and animal resources in particular situations. When productivity and risk are not too dissimilar, other factors may come to the fore, such as cultural choice or historical contingency. This comparison needs to be explored in more detail, but it certainly provides one model to account for variation.

One of the key issues remaining to be better understood is the timing of the strong dietary shifts that have been observed in the coastal regions of Denmark, Britain and Ireland, Brittany, northern Spain, central Portugal (with the intriguing exceptions of single individuals from

Rødhals, Denmark and Gruta do Lagar, Portugal), and, given the absence or paucity of Mesolithic data, inferred for Normandy, the Channel Islands and the Mediterranean. In most of these cases, the impression is one of a 'rapid' dietary shift but, given the ranges involved in radiocarbon dating and calibration, this is a rather imprecise statement (cf. Whittle *et al.*, 2007). Across much of Britain, which has by far the most AMS determinations on earlier Neolithic human bone from coastal contexts, there is a crucial gap of one to two centuries around 4000 calBC that remains poorly represented. But, while this allows the possibility for a 'gradual' transition at the scale of the human lifespan, it nevertheless presents strong evidence for a rapid transition in archaeological terms, far more rapid than some revisionist positions would have it (Thomas, 1999, 2003). The gaps become rather larger than this in most other parts of Europe, mainly due to the paucity of available samples; there is also a need for more direct dating of the human remains that are present from known or suspected earlier Neolithic contexts. With the availability of more isotopic data and direct AMS dating, it will no doubt be possible to further refine the picture presented here, which for the most part remains at a rather coarse scale of analysis. That being said, the results that are available have made a considerable impact, and present a number of challenges for our understanding of the Mesolithic-Neolithic transition across Europe.

A wider comparison of the European situation to the agricultural transition in other parts of the world is a large but potentially very fruitful topic. Again, much will depend on the availability and characteristics of the domestic species involved, and how they compare to a region's 'natural' productivity. The most obvious points of departure are secondary centres for the spread of agriculture in the temperate northern hemisphere, for example the Jomon-Yayoi transition in Japan (Imamura, 1996), or the northwards spread of maize cultivation in eastern North America. But both these cases have an important difference in that only domestic cereals were involved, rather than the closely integrated package of cereals and animals that entered Europe (Bogaard, 2004), and this may help explain, for example, the apparently comparatively slow uptake of maize in parts of eastern North America (Smith, 1992), as well as the apparent absence across much, though not all, of Europe of the negative health consequences that have been widely documented elsewhere (Cohen and Armelagos, 1984). Subsistence change can be expected to have been more gradual in the primary centres of domestication (e.g. the Near East, Mesoamerica), and more difficult to detect isotopically, since by definition wild forms of the plants and animals are implicated in the process itself. Where isotopic analyses would have great potential is in the identification of practices such as herd management, penning (implying the provision of fodder) and manuring (Bogaard *et al.*, 2007; Grupe and Peters, 2007). The integration of the various lines of evidence relating to agricultural transitions – of which stable isotopic data are one – and comparisons at different temporal and spatial scales, is becoming increasingly feasible, and will no doubt both provide new insights and generate new questions and debates.

NOTES

1. A study of modern brackish/freshwater eels on the west coast of Ireland reports average flesh values of –23.6 and 12.1‰ for $\delta^{13}C$ and $\delta^{15}N$, respectively ($n = 37$) (Harrod *et al.*, 2005).
2. There are five additional samples that fall within the transitional period, but four are dogs and all are from inland contexts, and so are not relevant to the discussion. Their $\delta^{15}N$ values average 8.9 ± 0.8‰.

3. A small number of the reported measurements have C:N ratios falling outside of the generally accepted ranges of 2.9 to 3.6 (DeNiro,) or 2.6 to 3.4 (Schoeninger *et al.*,1989). When corrected for molecular weight (values are reported as a percentage ratio rather than following the usual practice taking molecular weights into account), most human samples do fall within the combined range of 2.6 to 3.6 (4 from Hardinxveld do not, and are omitted from the averages provided in Table 2.1).
4. A number of the AMS determinations originally reported were very late; these have proved to be in error, due to incomplete removal of contamination by a consolidant (Schulting, 2005). New samples subjected to ultrafiltration have made these older by some centuries, bringing them in line with the main cluster of dates from the site at about 5300 calBC. Two of the stable isotope values originally reported as outliers were also affected, and are now more in keeping with the majority of the samples. The issue is a complex one, and work is still ongoing.

REFERENCES

Ambrose, S.H. and Krigbaum, J. (2003) Bone chemistry and bioarchaeology. *J. Anthropol. Archaeol.*, **22**, 193–199.

Ambrose, S.H. and Norr, L. (1993) Experimental evidence for the relationship of the carbon isotope ratios of whole diet and dietary protein to those of bone collagen and carbonate, in *Prehistoric Human Bone: Archaeology at the Molecular Level* (eds J.B. Lambert and G. Grupe), Springer-Verlag, New York, pp. 1–37.

Andersen, S.H. (1991) Bjørnsholm. A stratified Køkkenmødding on the central Limfjord, north Jutland. *J. Danish Archaeol.*, **10**, 59–96.

Andersen, S.H., Bietti, A., Bonsall, C., *et al.* (1990) Making cultural ecology relevant to Mesolithic research: I. A database of 413 Mesolithic fauna assemblages, in *Contributions to the Mesolithic in Europe* (eds P.M. Vermeersch and P. Van Peer), Leuven University Press, Leuven, pp. 23–52.

Arias, P. (2005) Determinaciones de isótopes estables en restos humanos de la región Cantábrica. Aportación al estudio de la dieta de las poblaciones del Mesolítico y el Neolítico. *Munibe*, **57**, 359–374.

Arnaud, J.M. (1989) The Mesolithic communities of the Sado Valley, Portugal, in their ecological setting, in *The Mesolithic in Europe* (ed. C. Bonsall), John Donald, Edinburgh, pp. 614–631.

Billard, C., Arbogast, R.-M. and Valentin, F. (1999) Dépôts animaux et crémation dans une sépulture Mésolithique de Haute-Normandie. *L'Archéologue*, **40**, 50–51.

Bocherens, H. and Drucker, D. (2003) Trophic level isotopic enrichments for carbon and nitrogen in collagen: case studies from recent and ancient terrestrial ecosystems. *Int. J. Osteoarchaeol.*, **13**, 46–53.

Bogaard, A. (2004) *Neolithic Farming in Central Europe: An Archaeobotanical Study of Crop Husbandry Practices*, Routledge, London.

Bogaard, A., Heaton, T.H.E., Poulton, P. and Merbach, I. (2007) The impact of manuring on nitrogen isotope ratios in cereals: archaeological implications for reconstruction of diet and crop management practices. *J. Archaeol. Sci.*, **34**, 335–343.

Bonsall, C., Cook, G.T., Hedges, R.E.M. *et al.* (2004) Radiocarbon and stable isotope evidence of dietary change from the Mesolithic to the Middles Ages in the Iron Gates: new results from Lepenski Vir. *Radiocarbon*, **46**, 293–300.

Bonsall, C., Cook, G.T., Lennon, R. *et al.* (2000) Stable isotopes, radiocarbon and the Mesolithic–Neolithic transition in the Iron Gates. *Documenta Praehistorica*, **27**, 119–132.

Bonsall, C., Lennon, R., McSweeney, K. *et al.* (1997) Mesolithic and Early Neolithic in the Iron Gates: a palaeodietary perspective. *J. Eur. Archaeol.*, **5**, 50–92.

Borić, D. and Dimitrijević, V. (2007) When did the 'Neolithic package' reach Lepenski Vir? Radiometric and faunal evidence. *Documenta Praehistorica*, **34**, 53–72.

Borić, D., Grupe, G., Peters, J. and Mikic, Z. (2004) Is the Mesolithic–Neolithic subsistence dichotomy real? New stable isotope evidence from the Danube Gorges. *J. Eur. Archaeol.*, **7**, 221–248.

Boroneant, A. and Dinu, A. (2006) The Romanian Mesolithic and the transition to framing. A case study: the Iron Gates. *Studii de Preistoire*, **3**, 41–76.

Britton, K., Müldner, G. and Bell, M. (2008) Stable isotope evidence for salt-marsh grazing in the Bronze Age Severn Estuary, UK: implications for palaeodietary analysis at coastal sites. *J. Archaeol. Sci.*, **35**, 2111–2118.

Bukach, D. (2005) The Archaeology of Island Landscapes. Unpubl. DPhil thesis, Institute of Archaeology, University of Oxford.

Cassen, S., Boujot, C. and Vaquero, J. (eds) (2000) *Éléments d'Architecture. Exploration d'un Terre Funéraire à Lannec er Gadouer*, CNRS, Erdeven, Morbihan.

Cohen, M. and Armelagos, G.J. (eds) (1984) *Paleopathology at the Origins of Agriculture*, Academic Press, Orlando.

Cook, G.T., Bonsall, C., Hedges, R.E.M. *et al.* (2001) A freshwater diet-derived ^{14}C reservoir effect at the stone age sites in the Iron Gate Gorge. *Radiocarbon*, **43**, 453–460.

DeNiro. M.J.(1985) Post-mortem preservation and alteration of *in vivo* bone collagen isotope ratios in relation to palaeodietary reconstruction. *Nature*, **317**, 806–809.

Dufour, E., Bocherens, H. and Mariotti, A. (1999) Palaeodietary implications of isotopic variability in Eurasian lacustrine fish. *J. Archaeol. Sci.*, **26**, 617–627.

Enghoff, I.B. (1986) Freshwater fishing from a sea-coast settlement – The Ertebølle *locus classicus* revisited. *J. Danish Archaeol.*, **5**, 62–76.

Eriksson, G. (2004) Part-time farmers or hard-core sealers? Västerbjers studied by means of stable isotope analysis. *J. Anthropol. Archaeol.*, **23**, 135–162.

Eriksson, G. (2006) Stable isotope analysis of human and faunal remains from Zvejnieki, in *Back to the Origin. New Research in the Mesolithic–Neolithic Zvejnieki Cemetery and Environment, Northern Latvia* (eds L. Larsson and I. Zagorska), Acta Archaeologica Lundensia, Series 8, No. 52, Almqvist & Wiksell International, Lund, pp. 183–215.

Eriksson, G., Linderholm, A., Fornander, E. *et al.* (2008) Same island, different diet: cultural evolution of food practice on Öland, Sweden, from the Mesolithic to the Roman Period. *J. Anthropol. Archaeol.*, **27**, 520–543.

Eriksson, G., Lóugas, L. and Zagorska, I. (2003) Stone Age hunter-fisher-gatherers at Zvejnieki, northern Latvia: radiocarbon, stable isotope and archaeozoology data. *Before Farming* 1/2: 1–25, http://www.waspjournals.com.

Fano, M.A. (2007) The use of marine resources by the Mesolithic and Neolithic societies of Cantabrian Spain: the current evidence, in *Shell Middens and Coastal Resources Along the Atlantic Façade* (eds N. Milner and G. Bailey), Oxbow Books, Oxford, pp. 136–149.

Ferembach, D. (1974) *Le gisement Mésolithique de Moita do Sebastião, Muge, Portugal. II: Anthropologie*, Direccão Geral dos Assuntos Culturais, Lisbon.

Fischer, A., Olsen, J., Richards, M.P. *et al.* (2007) Coast–inland mobility and diet in the Danish Mesolithic and Neolithic: evidence from stable isotope values of humans and dogs. *J. Archaeol. Sci*, **34**, 2125–2150.

Garcia Guixé, E., Richards, M.P. and Subirá, M.E. (2006) Palaeodiets of humans and fauna at the Spanish Mesolithic site of El Collado. *Current Anthropology*, **47**(3), 549–556.

Giorgi, F., Bartoli, F., Iacumin, P. and Mallagni, F. (2005) Oligoelements and isotopic geochemistry: a multidisciplinary approach to the reconstruction of the palaeodiet. *Human Evolution*, **20**, 55–82.

Gronenborn, D. (1999) A variation on a basic theme: the transition to farming in southern central Europe. *J. World Prehist.*, **13**, 123–210.

Grupe, G. and Peters, J. (2007) Molecular biological methods applied to Neolithic bioarchaeological remains: Research potential, problems, and pitfalls, in *Non-Megalithic Mortuary Practices in the Baltic – New Methods and Research into the Development of Stone Age Society* (eds L. Larsson, F. Lüth, and T. Terberger), Schwerin: Bericht der Römisch-Germanischen Kommission, **88**, 275–306.

Harrod, C., Grey, J., McCarthy, T.K. and Morrissey, M. (2005) Stable isotope analyses provide new insights into ecological plasticity in a mixohaline population of European eel. *Oecologia*, **144**, 673–683.

Hedges, R.E.M. and Reynard, L.M. (2007) Nitrogen isotopes and the trophic level of humans in archaeology. *J. Archaeol. Sci.*, **34**, 1240–1251.

Hedges, R.E.M., Pettitt, P.B., Bronk Ramsey, C. and Klinken, C.J. (1998) Radiocarbon dates from the Oxford AMS SYSTEM: Archaeometry datelist 26. *Archaeometry*, **40**, 437–455.

Heier-Nielsen, S., Heinemeier, J., Neilsen, H.L. and Rud, N. (1995) Recent reservoir ages for Danish fjords and marine waters. *Radiocarbon*, **37**, 875–882.

Jones, G. (2000) Evaluating the importance of cultivation and collecting in Neolithic Britain, in *Plants in Neolithic Britain and Beyond* (ed. A.S. Fairbairn), Oxbow Books, Oxford, pp. 79–84.

Imamura, K. (1996) *Prehistoric Japan. New Perspectives on Insular East Asia*, UCL Press, London.

Kendrick, T.D. (1928) The Archaeology of the Channel Islands, *The Bailiwick of Guernsey*, Vol I, Methuen and Co., London.

Lange-Badré, B. and Le Mort, F. (1998) Isotopes stables du carbone et de l'azote et éléments traces indicateurs du régime alimentair de la population Néolithique de Khirokitia (Cyprus), in *L'Homme Préhistorique et al Mer* (ed. G Camps), Editions de Comité des Travaux Historiques et Scientifiques, pp. 417–426.

Larsson, L. and Zagorska, I. (eds) (2006) *Back to the Origin. New Research in the Mesolithic–Neolithic Zvejnieki Cemetery and Environment, Northern Latvia, Acta Archaeologica Lundensia, Series 8, No. 52*, Almqvist & Wiksell International, Lund.

Le Bras-Goude, G., Binder, D., Formicola, V. *et al.* (2006) Stratégies de subsistance et analyse culturelle de populations néolithiques de Ligurie: approche par l'étude isotopique (^{13}C et ^{15}N) des restes osseux. *Bull. Mém. Soc. Anthropol. Paris*, **18**, 45–55.

Le Bras-Goude, G., Schmitt, A. and Loiso, G. (2009) Comportements alimentaires, aspects biologiques et sociaux au Néolithique: le cas du Crès (Hérault, France). *Comptes Rendus Palevol.*, **8**, 79–91.

Lillie, M., Budd, C., Potekhina, I. and Hedges, R.E.M. (2009) The radiocarbon reservoir effect: new evidence from the cemeteries of the middle and lower Dnieper basin, Ukraine. *J. Archaeol. Sci.*, **36**, 256–264.

Lillie, M. and Jacobs, K. (2006) Stable isotope analysis of 14 individuals from the Mesolithic cemetery of Vasilyevka II, Dnieper Rapids region, Ukraine. *J. Archaeol. Sci.*, **33**, 880–886.

Lillie, M.C. and Richards, M. (2000) Stable isotope analysis and dental evidence of diet at the Mesolithic–Neolithic transition in Ukraine. *J. Archaeol. Sci.*, **27**, 965–972.

Lindqvist, C. and Possnert, G. (1997) The subsistence economy and diet at Jakobs/Ajvide and Stora Förvar, Eksta parish and other prehistoric dwelling and burial sites on Gotland in long-term perspective, in *Remote Sensing*, vol. **I** (ed. G. Burenhult), Dept. of Archaeology, Theses and Papers in North,-European Archaeology 13a, Stockholm, pp. 29–90.

Louwe Kooijmans, L.P. (2007) The gradual transition to farming in the Lower Rhine Basin, in *Going Over: the Mesolithic–Neolithic Transition in North-West Europe* (eds A. Whittle and V. Cummings), British Academy, London, pp. 287–309.

Louwe Kooijmans, L.P. (2009) The agency factor in the process of Neolithisation – a Dutch case study. *J. Arch. Low Countries*, **1**, 27–54. http://dpc.uba.uva.nl/jalc/01/nr01/a03.

Lubell, D., Jackes, M., Schwarcz, H. *et al.* (1994) The Mesolithic–Neolithic transition in Portugal: isotopic and dental evidence of diet. *J. Archaeol. Sci.*, **21**, 201–216.

Lübke, H., Lüth, F. and Terberger, T. (2007) Fishers or farmers? The archaeology of the Ostorf cemetery and related Neolithic finds in the light of new data, in *Non-Megalithic Mortuary Practices in the*

Baltic – New Methods and Research into the Development of Stone Age Society (eds L. Larrsen, F. Lüth and T. Terberger), Bericht der Römisch-Germanishen Kommission 88, Schwerin, pp. 307–338.

Malmström, H., Thomas, M., Gilbert, P. *et al.* (2009) Ancient DNA reveals lack of continuity between Neolithic hunter-gatherers and contemporary Scandinavians. *Curr. Biol.*, **19**, 1758–1762.

Marcigny, C., Ghesquiere, E. and Desloges, J. (2007) *La Hache et la Meule: les Premiers Paysans du Néolithique en Normandie*, Muséum d'Histoire Naturelle du Havre, Le Havre.

Milner, N., Craig, O.E., Bailey, G.N. *et al.* (2004) Something fishy in the Neolithic? A re-evaluation of stable isotope analysis of Mesolithic and Neolithic coastal populations. *Antiquity*, **78**, 9–22.

Minagawa, M. and Wada, E. (1984) Stepwise enrichment of ^{15}N along food chains: further evidence and the relation between $\delta^{15}N$ and animal age. *Geochimica et Cosmochimica Acta*, **48**, 1135–1140.

Modderman, P.J.R. (1988) The Linear Pottery Culture: diversity in uniformity. *Berichten van de Rijksdienst voor het Oudheidkundig Bodemonderzoek*, **38**, 63–140.

Olsen, J. and Heinemeier, J. (2007) AMS dating of human bone from the Ostoft cemetery in the light of new information on dietary habits and freshwater reservoir effect, in *Non-Megalithic Mortuary Practices in the Baltic – New Methods and Research into the Development of Stone Age Society* (eds L. Larrsen, F. Lüth and T. Terberger), Bericht der Römisch-Germanishen Kommission 88, Schwerin, pp. 339–352.

Papmehl-Dufay, L. (2006) Shaping an Identity: *Pitted Ware Pottery and Potters in Southeast Sweden*, Archaeological Research Laboratory, Stockholm University, Stockholm.

Papathanasiou, A. (2003) Stable isotope analysis in Neolithic Greece and possible implications on human health. *Int. J. Osteoarchaeol.*, **13**, 314–324.

Pedersen, L. (1995) 7000 years of fishing: stationary fishing structures in the Mesolithic and afterwards, in *Man and Sea in the Mesolithic* (ed. A. Fischer), Oxbow Books, Oxford, pp. 75–86.

Pedersen, L. (1997) They put fences in the sea, in *The Danish Storebælt since the Ice Age* (eds L. Pedersen, A., Fischer and B. Aaby), A/S Storebælt Fixed Link, Kalundorg Regional Museum, National Forest and Nature Agency, and the National Museum of Denmark, Copenhagen, pp. 124–143.

Péquart, M. and Péquart, S.-J. (1954) *Hoëdic, Deuxième Station-Nécropole du Mésolithique Côtier Armoricain*, De Sikkel, Anvers.

Péquart, M., Péquart, S.-J., Boule, M. and Vallois, H. (1937) *Téviec, Station-Nécropole du Mésolithique du Morbihan*, Archives de L'Institut de Paléontologie Humaine XVIII, Paris.

Price, T.D., Ambrose, S.H., Bennike, P. *et al.* (2007) New information on the Stone Age graves at Dragsholm, Denmark. *Acta Archaeologica*, **78**, 193–219.

Renouf, J.T. and Urry, J. (1986) The Channel Islands during the Neolithic: sea-level changes and patterns of exploitation. *Revue Archéologique de l'Ouest*, (Supplement 1), 13–23.

Richards, M.P. (2000) Human consumption of plant foods in the British Neolithic: direct evidence from bone stable isotopes, in *Plants in Neolithic Britain and beyond* (ed. A.S. Fairbairn), Oxbow Books, Oxford, pp. 123–135.

Richards, M.P. and Hedges, R.E.M. (1999) A Neolithic revolution? New evidence of diet in the British Neolithic. *Antiquity*, **73**, 891–897.

Richards, M.P., Hedges, R.E.M., Walton, I. *et al.* (2001) Neolithic diet at the Brochtorff Circle, Malta. *Eur. J. Archaeol.*, **4**, 253–262.

Richards, M.P. and Mellars, P. (1998) Stable isotopes and the seasonality of the Oronsay middens. *Antiquity*, **72**, 178–184.

Richards, M.P., Price, T.D. and Koch, E. (2003) Mesolithic and Neolithic subsistence in Denmark: new stable isotope data. *Curr. Anthropol.*, **44**, 288–295.

Richards, M.P., Schulting, R.J. and Hedges, R.E.M. (2003) Sharp shift in diet at onset of Neolithic. *Nature*, **425**, 366.

Richards, M.P. and Schulting, R.J. (2006) Against the grain? A response to Milner *et al.* (2004). *Antiquity*, **80**, 444–458.

Richards, M.P. and Sheridan, J.A. (2000) New AMS dates on human bone from Mesolithic Oronsay. *Antiquity*, **74**, 313–315.

Roche, J. (1972) *Le Gisement Mésolithique de Moita do Sebastião, Muge, Portugal. I: Archéologie*, Instituto de Alta Cultura, Lisbon.

Roksandic, M. (2006) Analysis of burials from the new excavations of the sites Cabeço da Amoreira and Arruda (Muge, Portugal), in *Do Epipapelolítico ao Calcolítico na Península Ibérica. Actas do IV Congresso de Arqueologia Peninsular* (eds N. Bicho and N.H. Veríssimo), Unviersity of Algarve Press, Faro, pp. 1–10.

Rose, M. (1995) Fishing at Franchthi Cave, Greece: changing environments and patterns of exploitation. *Old World Archaeology Newsletter*, **18**, 21–26.

Schoeninger, M. and Moore, K. (1992) Stable bone isotope studies in archaeology. *J. World Prehist.*, **6**, 247–296.

Schoeninger, M.J., Moore, K.M., Murray, M.L. and Kingston, J.D. (1989) Detection of bone preservation in archaeological and fossil samples. *Appl. Geochem.*, **4**, 281–292.

Schulting, R.J. (2005) Comme la mer qui se retire: les changements dans l'exploitation des ressources marines du Mésolithique au Néolithique en Bretagne, in *Unité et diversité des processus de néolithisation sur la façade atlantique de l'Europe (7-4ème millénaires avant J.-C.)* (eds G. Marchand and A. Tresset), Mémoire de la Société Préhistorique Française 36, Paris, pp. 163–171.

Schulting, R.J. (2007) Non-monumental burial in Neolithic Britain: a (largely) cavernous view, in *Non-Megalithic Mortuary Practices in the Baltic – New Methods and Research into the Development of Stone Age Society* (eds L. Larsson, F., Lüth and T. Terberger), Bericht der Römisch-Germanischen Kommission 88, Schwerin, pp. 581–603.

Schulting, R.J. (2008) Foodways and social ecologies from the Early Mesolithic to the Early Bronze Age, in *Prehistoric Britain* (ed. J. Pollard), Blackwell, London, pp. 90–120.

Schulting, R.J., Lanting, J.N. and Reimer, P.J. (2009) New dates from Tumulus Saint-Michel, Carnac, in *Explorations archéologiques et discours savants sur une architecture néolithique restaurée à Locmariaquer, Morbihan (Table des Marchands et Grand Menhir)* (ed. S. Cassen), CNRS and Université de Nantes, Nantes, pp. 769–773.

Schulting, R.J. and Richards, M.P. (2001) Dating women and becoming farmers: new palaeodietary and AMS data from the Breton Mesolithic cemeteries of Téviec and Hoëdic. *J. Anthropol. Archaeol.*, **20**, 314–344.

Schulting, R.J. and Richards, M.P. (2002a) Finding the coastal Mesolithic in southwest Britain: AMS dates and stable isotope results on human remains from Caldey Island, Pembrokeshire, South Wales. *Antiquity*, **76**, 1011–1025.

Schulting, R.J. and Richards, M.P. (2002b) The wet, the wild and the domesticated: the Mesolithic–Neolithic transition on the west coast of Scotland. *Eur. J. Archaeol.*, **5**, 147–189.

Schulting, R.J. and Richards, M.P. (2009) Radiocarbon dates and stable isotope values on human remains, in *On the Fringe of Atlantic Europe* (ed. A. Richtie), Society of Antiquaries of Scotland, Edinburgh, pp. 67–74.

Schulting, R.J., Sebire, H. and Robb, J. (2010) On the road to Paradise: new insights from AMS dates and stable isotopes at Le Déhus, Guernsey, and the Channel Islands, Middle Neolithic. *Oxford J. Arch.*, **29** (2), 149–173.

Schulting, R.J., Tresset, A. and Dupont, C. (2004) From harvesting the sea to stock rearing along the Atlantic façade of north-west Europe. *En. Arch.*, **9**, 143–154.

Sheridan, A., Schulting, R.J., Quinnell, H. and Taylor, R. (2008) Revisiting a small passage tomb at Broadsands, Devon. *Proc. Devon Archaeol. Soc.*, **66**, 1–26.

Sjögren, K.-G. (2003) Megaliths, settlement and subsistence in Bohuslän, Sweden, in *Stone and Bones. Formal Disposal of the Dead in Atlantic Europe during the Mesolithic–Neolithic Interface 6000–3000 BC* (eds G. Burenhult and S. Westergaard), BAR International Series 1201, Archaeopress, Oxford, pp. 167–176.

Smith, B.D. (1992) Prehistoric plant husbandry in Eastern North America, in *The Origins of Agriculture: An International Perspective* (eds C.W. Cowan and P.J. Watson), Smithsonian Institution Press, Washington, DC, pp. 101–120.

Smits, E., Millard, A.R., Nowell, G. and Pearson, D.G. (2010) Isotopic investigation of diet and residential mobility in the Neolithic of the Lower Rhine Basin. *Eur. J. Arch.*, **13**, 5–31.

Smits, L. and van der Plicht, H. (2009) Mesolithic and Neolithic human remains in the Netherlands: physical anthropological and stable isotope investigations. *J. Arch. Low Countries*, **1**, 55–85. http://dpc.uba.uva.nl/jalc/01/nr01/a04.

Storå, J. (2001) *Reading Bones. Stone Age Hunters and Seals in the Baltic*, Stockholm Studies in Archaeology, Stockholm, p. 21.

Tauber, H. (1981) ^{13}C evidence for dietary habits of prehistoric man in Denmark. *Nature*, **292**, 332–333.

Tauber, H. (1986) Analysis of stable isotopes in prehistoric populations. *Mitteilungen der Berliner Gesellschaft für Anthropologie, Ethnologie und Urgeschichte*, **7**, 31–38.

Thomas, J. (1999) *Understanding the Neolithic*, Routledge, London.

Thomas, J. (2003) Thoughts on the 'repacked' Neolithic revolution. *Antiquity*, **77**, 67–74.

Tresset, A. (2003) French connections II: of cows and men, in *Neolithic Settlement in Ireland and Western Britain* (eds I. Armit, E. Murphy, N. Nelis and D. Simpson), Oxbow Books, Oxford, pp. 18–30.

Tringham, R. (2000) South-eastern Europe in the transition to agriculture in Europe: bridge, buffer or mosaic, in *Europe's First Farmers* (ed. T.D. Price), Cambridge University Press, Cambridge, pp. 19–56.

Umbelino, C.I.S. (2005) Outros Sabores do Passado. As análises do oligoelementos e de isótopos estáveis na reconstituição da dieta das comunidades humanas do Mesolítico Final e fo Neolítico Final/Calcolítico do território Português. Unpubl. PhD thesis, Faculdade de Ciências e Technolgia da Universidade de Coimbra.

Van Klinken, G.J., Richards, M.P. and Hedges, R.E.M. (2000) An overview of causes for stable isotopic variations in past European human populations: environmental, ecophysiological and cultural effects, in *Biogeochemical Approaches to Palaeodietary Analysis* (eds S.H. Ambrose and M.A. Katzenberg), Kluwer Academic/Plenum Publishers, New York, pp. 39–63.

Vanmontfort, B. (2008) Forager–farmer connections in an 'unoccupied' land: First contact on the western edge of LBK territory. *J. Anthropol. Archaeol.*, **27**, 149–160.

Verron, G. (2000) *Préhistoire de la Normandie*, Éditions Ouest-France, Rennes.

Whittle, A., Barclay, A., Bayliss, A. *et al.* (2007) Building for the dead: events, processes and changing worldviews from the 38th to the 34th centuries calBC in southern Britain. *Camb. Archaeol. J.*, **17**, 123–147.

Woodman, P.C. (2004) The exploitation of Ireland's coastal resources – a marginal resource through time? in *The Mesolithic of the Atlantic Façade* (eds M.R. González Morales and G.A. Clarke), Arizona State University Anthropological Research Paper No. 55, Arizona State University Press, Tucson, pp. 37–55.

Woodman, P.C. (2008) Ireland's place in the European Mesolithic: why it's ok to be different, in *Mesolithic Horizons* (eds S.B. McCartan, R.J. Schulting, G. Warren, and P.C. Woodman), Oxbow Books, Oxford, pp. 36–46.

Woodman, P.C., Andersen, E. and Finlay, N. (1999) Excavations at Ferriter's Cove, 1983–95: Last Foragers, in *First Farmers in the Dingle Peninsula*, Wordwell, Bray.

Zapata, L., Milner, N. and Roselló, E. (2007) Picos Ramos cave shell midden: the Mesolithic–Neolithic transition in the Bay of Biscay, in *Shell Middens and Coastal Resources along the Atlantic Façade* (eds N. Milner and G. Bailey), Oxbow Books, Oxford, pp. 150–157.

Zilhão, J. (2000) From the Mesolithic to the Neolithic in the Iberian Peninsula, in *Europe's First Farmers* (ed. T.D. Price), Cambridge University Press, Cambridge, pp. 144–182.

Zvelebil, M. (1996) The agricultural frontier and the transition to farming in the circum-Baltic region, in *The Origins and Spread of Agriculture and Pastoralism in Eurasia* (ed. D.R. Harris), UCL Press, London, pp. 323–345.

Zvelebil, M. and Lillie, M. (2000) Transition to agriculture in eastern Europe, in *Europe's First Farmers* (ed. T.D. Price), Cambridge University Press, Cambridge, pp. 57–92.

3

The Mesolithic-Neolithic Transition in Eastern Europe: Integrating Stable Isotope Studies of Diet with Palaeopathology to Identify Subsistence Strategies and Economy

Malcolm Lillie[1] and Chelsea Budd[2]

[1] *Department of Geography, University of Hull, Hull, UK*
[2] *Wetland Archaeology & Environments Research Centre, Department of Geography, University of Hull, Hull, UK*

> . . .little is known about the importance of plant foods, shellfish, fish, or marine mammals in *Mesolithic* subsistence. . . (Price, 1989: 48)

3.1 INTRODUCTION

Since Price's 1989 observation above, in relation to Mesolithic diets, numerous improvements in sampling, recovery and analysis have considerably expanded our understanding of prehistoric subsistence strategies. Recent research in Eastern Europe has facilitated an holistic approach to the study of past human populations through the integration of multi-disciplinary analyses of the available archive. Amongst the approaches adopted, new radiocarbon dating, stable isotope studies and palaeopathological analyses have allowed a more nuanced understanding of individual life histories and general population based expressions of pathology and diet. This chapter presents a consideration of recent research, and evaluates the integration of stable isotope studies towards our understanding of prehistoric food procurement strategies.

Human Bioarchaeology of the Transition to Agriculture Edited by Ron Pinhasi and Jay T. Stock

In general terms there is a dichotomy between the exploitation of Mesolithic subsistence strategies, equated with variable expressions of hunter-fisher-gatherer diets, and those of Neolithic societies, usually assumed to equate to some degree of agro-pastoralist economy. However, as noted by Bartosiewicz, Bonsall and Şişu (2008), while the increasing importance of domesticated livestock occurs across the Neolithic in regions such as the Danube valley, between the Balkans and the Carpathian Basin, fishing retains some importance. The point is that from about 6000 BC onwards many factors influence the rate of spread, integration and ultimate adoption of the new/alternative subsistence strategies as 'farming' is disseminated, resulting in a process that is both piecemeal and protracted in certain areas (Zvelebil and Dolukhanov, 1991; Zvelebil and Lillie, 2000; Thomas, 2004). Unfortunately, there is no simple dichotomy when viewing the transition to farming in Europe, as the myriad factors influencing adoption inherent at the regional level all combine to produce subtle degrees of variation that are specific to the region and even the catchment being studied.

Zvelebil (2006:179) has noted that earliest evidence for the uptake of aspects of a food-production economy in the Baltic Sea region begins at about 4400 calBC, and continues over the following 5000 years with the gradual establishment of farming communities during this timeframe (Eriksson, Lõugas and Zagorska, 2003). Similarly, in Ukraine and Eastern Europe, the transition from Mesolithic to Neolithic does not necessarily indicate the adoption of farming practices, but is often simply attributed to the first appearance of pottery on archaeological sites (Lillie, 1998a, 1998b; Lillie and Richards, 2000). Based on radiocarbon dates from the cemetery of Vasilyevka V, it appears that the transition from the Mesolithic to Neolithic in the Dnieper Basin region occurs at about 5500 to 5000 calBC. However, this observation is made with the caveat that the first appearance of ceramics at a number of the cemetery sites in this region only occurs towards the end of this period, and that the start of the Neolithic does not necessarily imply the adoption of agro-pastoralism (Lillie, 1998a, b).

3.2 STABLE ISOTOPE STUDIES AND DIET

The regular application of carbon and nitrogen (C/N) analysis in palaeodietary reconstruction is a relatively recent, albeit well established, technique in bioarchaeology and prehistoric research (Keegan, 1989; Ambrose, 1990, 1993; Schwarcz and Schoeninger, 1991; Richards *et al.*, 2003). As noted by Müldner and Richards (2005), stable isotope analysis for palaeodietary reconstruction is based on the principle that the isotope values ($\delta^{13}C$ and $\delta^{15}N$) of the food consumed by animals and humans are stored in the individual's tissues. A caveat to this general observation is the fact that the isotopic composition of the mineralized tissues of vertebrates has been shown to exhibit variability at the intra-individual level of analysis (Balasse, Bocherens and Mariotti, 1999, and references therein).

As there are differences in the $\delta^{13}C$ values obtained when using the different tissues, for example bone collagen or apatite-carbonate, it is worth noting that the stable isotope analysis of bone collagen only reflects the protein component of the diet (Krueger and Sullivan, 1984; Ambrose and Norr, 1993; Tieszen and Fagre, 1993; Schulting and Richards, 2001). As a consequence, many plant foodstuffs are difficult to 'see' isotopically as (with the exception of certain nuts and seeds) the protein levels of unprocessed plants are usually quite low when compared to meat/fish. This results in a 'bias' towards meat/fish protein when using bone collagen in isotope studies, as opposed to reflecting the diet as a whole (Müldner and Richards, 2005 and references therein).

By contrast, the stable isotope analysis of bone apatite-carbonate (and tooth enamel carbonate) reflects a mixture of dietary proteins, carbohydrates and fats, thereby having the potential to produce a better overall approximation of the diet (Ambrose and Norr, 1993; Tieszen and Fagre, 1993). However, the relative percentages of protein/carbohydrates/fat identified using this technique are not currently known, and as apatite-carbonate is far more susceptible to diagenesis, the limitations of this technique are difficult to resolve (Schulting and Richards, 2001). Furthermore, Hedges (2003) has shown that it is not possible to provide a quantitative model for the process of bone formation, and also that factors such as methanogenesis (in ruminants), dietary determination of collagen isotopic composition, and non-equilibrated bone synthesis are all implicated as interlinked explanations for the observed differences in collagen vs. apatite-carbonate spacing with trophic level.

Despite the above observations, stable isotope analysis does provide a direct measure of the nature of past human diet, with the carbon isotope value ($\delta^{13}C$), indicating the amount of marine protein in the diet, as compared to terrestrial protein. Bone collagen analysis also distinguishes between different dietary components, such as C_3 and C_4 photosynthetic plants and the animals that consumed them (Schwarcz and Schoeninger, 1991; Lillie, Richards and Jacobs, 2003; Richards, 2002). Nitrogen stable isotope ratios ($\delta^{15}N$) are used to establish the trophic level of an organism in the food web as an increase of about 3‰ occurs as we move up the food chain (Schoeninger and DeNiro, 1984; Müldner and Richards, 2005). Isotope analysis of bone collagen is generally limited to the identification of dietary proteins during the last *ca.* 10 years of an individual's life, depending on the bone elements analysed (Richards *et al.*, 2003).

Humans with a diet where all of the protein is derived from marine sources have bone collagen $\delta^{13}C$ values of approximately -12 ± 1‰ (Chisholm, Nelson and Schwarcz, 1982; Richards and Hedges, 1999; Schoeninger, DeNiro and Tauber, 1983). In Europe, Holocene human bone collagen values of about –20‰ are indicative of terrestrial C_3 pathways plants, and the meat or milk of animals consuming these (Richards *et al.*, 2003).

As noted above, the nitrogen isotope value, $\delta^{15}N$, tells us about the trophic level of an organism in an ecosystem, as consumers have bone collagen $\delta^{15}N$ values that are 2 to 4‰ higher than the protein they consume (Schoeninger and DeNiro, 1984). Therefore, an herbivore that consumes low trophic level protein plant foods, will subsequently have lower $\delta^{15}N$ values than carnivores that consume higher trophic level herbivores. In addition, in marine ecosystems, $\delta^{15}N$ values can be much higher than in terrestrial systems, simply because there are more steps in the food chain (Lillie, Richards and Jacobs, 2003). For example, Bonsall *et al.* (1997) report Mesolithic human $\delta^{13}C$ values of about –20 to –19‰ and $\delta^{15}N$ values of about 14 to 15‰ from the sites of Vlasac, Lepenski Vir and Schela Cladovei in the Danubian Iron Gates region. The high $\delta^{15}N$ values indicate that almost all of the dietary protein was from fairly high trophic level freshwater fish.

In general, when studying the isotope composition of hunter-fisher-gatherer populations, we could anticipate carbon and nitrogen stable isotope ratios of –20 to –23‰ (up to –24‰ for individuals consuming a high proportion of freshwater resources) for $\delta^{13}C$ and ratios of 10 to 14‰ for $\delta^{15}N$ in a terrestrial/freshwater context. The lower $\delta^{15}N$ values would reflect a proportionately greater emphasis on the exploitation of terrestrial fauna, while higher values would reflect a greater reliance of freshwater resources (Lillie, Budd and Potekhina, in press).

Keegan (1989:224) summarizes the role of stable isotope analysis as providing 'a method for testing and refining dietary reconstructions that are generated from the interpretation of other sources of evidence.' This is important to note, as the synergy between various strands of analysis rests on the fact that 'subtle variations in the importance of particular foods can be almost impossible to establish on the basis of faunal or floral remains alone' (Cannon, Schwarcz and Kynf, 1999:399).

3.3 THE EASTERN EUROPEAN EVIDENCE

As recently as 1994 Lubell *et al.* noted that 'very little attention (had) been paid to the biological characteristics of the human populations involved' in the transition from Mesolithic forager-fisher to Neolithic, agricultural, subsistence strategies (1994:201). Since this date an increasing number of stable isotope studies of diet, often integrating radiocarbon dating and palaeopathological analyses, have been undertaken in central and eastern Europe (Antanaitis and Ogrinc, 2000; Antanaitis *et al.*, 2000; Bonsall *et al.*, 1997, 2000, 2002, 2004; Borić *et al.*, 2004; Eriksson, 2006; Eriksson, Lõugas and Zagorska, 2003; Larsson and Zagorska, 2006; Lillie and Richards, 2000; Lillie, Richards and Jacobs, 2003; Lillie and Jacobs, 2006; Lillie *et al.*, 2009; Lõugas, Lidén and Nelson, 1996; O'Connell, Levine and Hedges, 2000). The overviews presented below focuses on the Baltic region and Ukraine in order to highlight the considerable potential of these regions for developing integrated research designs aimed at gaining an holistic appreciation of prehistoric socio-economic trajectories.

3.3.1 The Baltic Region

This region (Figure 3.1) has been the subject of intensive study, due to the discovery of a number of significant cemetery sites such as, Skateholm, Vedbæk, Zvejnieki and Oleneostroviskii Mogilnik (Eriksson, Lõugas and Zagorska, 2003; Larsson, 1984; O'Shea and Zvelebil, 1984; Jacobs, 1995; Timofeev, 1998).

Radiocarbon dating, stable isotope and faunal studies of prehistoric diet have recently been carried out at Zvejnieki in Latvia (Eriksson, Lõugas and Zagorska, 2003), along with a detailed multi-disciplinary, multi-authored study of additional aspects such as geology, vegetation, palaeodemography, burial customs and general socio-political and ritual characteristics (Larsson and Zagorska, 2006). At this site more than 300 individuals were interred over a period that spans approximately four millennia, in a region where the diversity of the environment was ideally suited to the hunter-fisher-forager lifestyle (Eriksson, Lõugas and Zagorska, 2003). At the cemetery site, a range of hunting and fishing equipment was recovered, including harpoons, spears, arrowheads and fish-hooks. The *ca.* 40 radiocarbon determinations from this site indicate continuity in use across the period about 5500 to 3500 calBC.

Two thousand, four hundred and forty six tooth pendants, recovered primarily from grave contexts, indicate a preference for elk but some pendants were made of wild boar, red deer, dog, aurochs and seal teeth. Other species that were identified from the archaeofaunal assemblage of the Zvejnieki complex (settlements and cemetery) include beaver, marten, badger, wild horse, otter, brown bear, fox, wolf, wild cat, wildfowl, fish (including pike, perch, a range of cyprinids [bream, tench, asp, carp], wels, eel and some salmon) and sheep/goat (Eriksson, Lõugas and Zagorska, 2003). Fish remains occur only sporadically in the burials at Zvejnieki, but predominate at the settlements (Eriksson, Lõugas and Zagorska, 2003) and in general, the grave goods reflect both large and small-game hunting activities (Eriksson, 2006).

Palaeopathological studies of the Zvejnieki population have shown that cribra orbitalia, possibly indicative of iron deficiency related to infectious disease and parasitism, occurs across

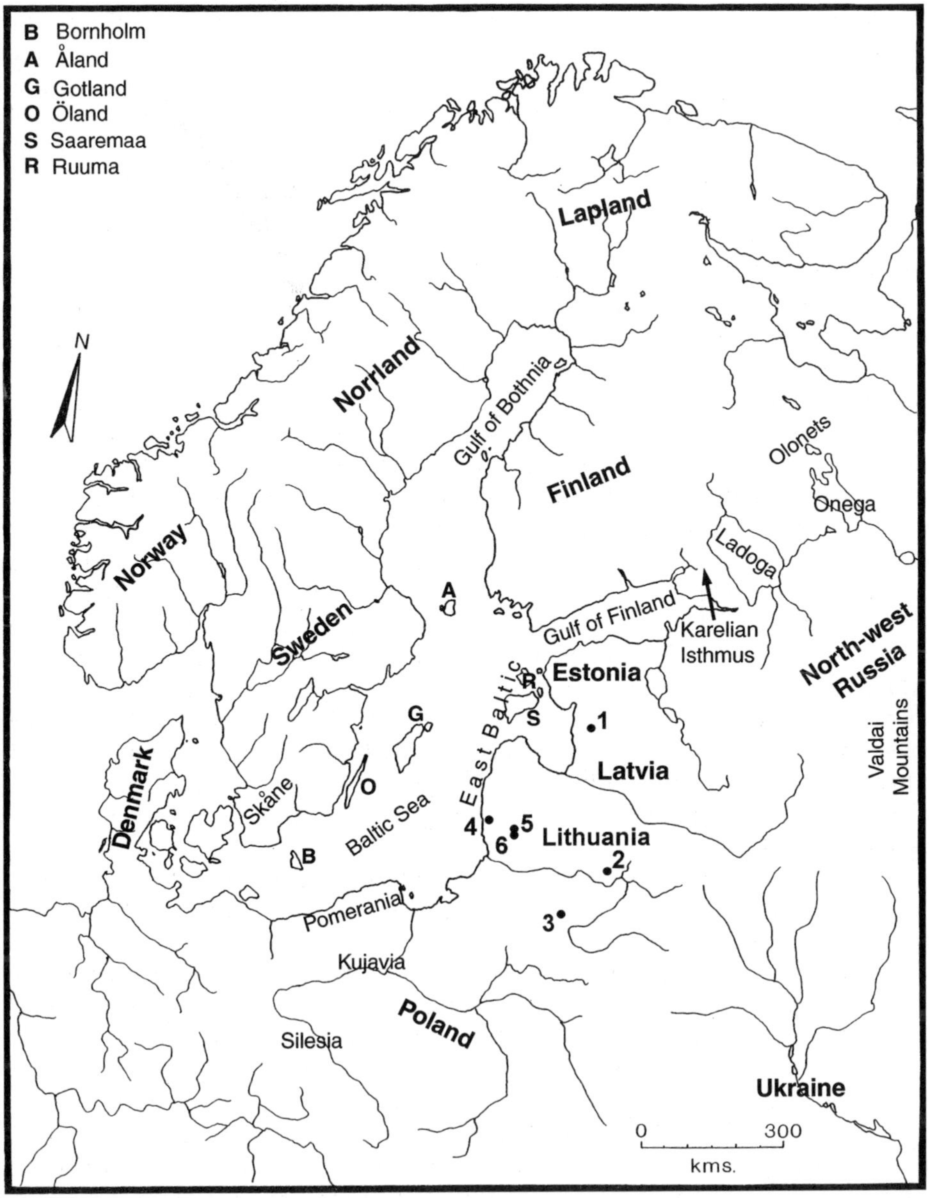

Figure 3.1 The Baltic Region: Showing locations of key sites discussed in text (*after* Larsson and Zagorska 2006 and Antanaitis-Jacobs *et al.* 2009). 1: Zvejnieki, 2: Kretuonas/Žemaitiškė, 3: Turlojiškė/Kisna, 4: Šventoji, 5: Donkalnis, 6: Spignas

the Mesolithic-Neolithic transition and during the Neolithic period (Jankauskas and Palubeckaitė, 2006). Cribra orbitalia prevalence was high amongst children in the cemetery population, while periosteal reactions were not recorded for children, possibly suggesting higher infant mortality in relation to these pathologies (Jankauskas and Palubeckaitė, 2006).

Other indicators of infection were recorded, including periosteal reaction of bone surfaces, which were reported as often being acute. Only 12 cases of suspected injury/violence were recorded on the 223 individuals analysed, and in general these appeared to reflect the general pattern of trauma occurring in forager populations, with males expressing more pathologies than females (Jankauskas and Palubeckaitė, 2006; and see Papathanasiou this volume). General degenerative joint disease was recorded for the population, with the most severe cases reported exclusively on males, perhaps suggesting that the forager lifestyle was more physically demanding for males (Jankauskas and Palubeckaitė, 2006).

Palubeckaitė and Jankauskas (2006) report that the overall pattern of dental pathology observed on 118 adult individuals from the Zvejnieki population is commensurate with forager populations, with high dental wear and low age-related antemortem losses (3% of all teeth). However, 49.2% of all individuals studied had at least one carious tooth. This is a very high prevalence rate for any forager population in northern regions. By contrast, Lillie (1996) found no evidence for caries on the dentitions of 36 individuals dated to the Mesolithic period and 105 individuals dated to the Neolithic period from the Dnieper Rapids region of Ukraine.

Interestingly, Palubeckaitė and Jankauskas (2006) found that Mesolithic individuals had the greatest number of carious teeth, followed by Bronze Age and then Neolithic individuals. Calculus was recorded for 97.5% of the individuals studied, with Neolithic and Bronze Age individuals exhibiting higher incidences than Mesolithic individuals (95.1 and 100%, respectively, compared to 55.4% in the Mesolithic) (Palubeckaitė and Jankauskas, 2006). The high incidence of caries at Zvejnieki is commensurate with levels recorded for southern Europe and the Mediterranean (Palubeckaitė and Jankauskas, 2006), where access to fruits rich in carbohydrates is suggested. As caries levels are broadly equivalent across the Mesolithic to Neolithic periods, the palaeopathological analysis appears to indicate continuity in diet across the transition. Rates of dental pathology at Zvejnieki suggest that males and females had similar access to subsistence resources during all periods studied (Palubeckaitė and Jankauskas, 2006).

The stable isotope analysis at Zvejnieki has shown that considerable variability occurs in diet across the late Mesolithic and up to the end of the middle Neolithic (Eriksson, 2006). The wide range of faunal remains analysed (mallard, bog tortoise, otter, fish, seal, beaver, badger, wolf, brown bear, wild boar, cervids, wild horse, sheep/goat and antler), produced a broad range of isotope ratios, with the faunal $\delta^{13}C$ ranging between –27.4 and –15.4‰, and the $\delta^{15}N$ ranging between 1.9 and 14.5‰. The herbivores studied in this analysis exhibited $\delta^{13}C$ ratios of –25.3 to –22.0‰, with $\delta^{15}N$ ratios ranging between 4.0 and 7.1‰ (Eriksson, 2006). Human stable isotope values range from –25.0 to –18.8‰ for $\delta^{13}C$ and 6.2 to 17.7‰ for $\delta^{15}N$. In terms of the $\delta^{13}C$, the average value is –22.7 ± 1.3, while the $\delta^{15}N$ ratio average value is 11.6 ± 1.8 per mg. Eriksson, Lõugas and Zagorska (2003) note that the outliers would indicate a marine diet of –18.8‰ ($\delta^{13}C$), while the $\delta^{15}N$ range is only 8.9 to 13.6‰ if the outliers are removed. The 6.2‰ nitrogen value indicates a diet that is almost exclusively composed of plant protein resources, while at the upper end of the range a $\delta^{15}N$ ratio of 17.7‰ suggests a heavy reliance on aquatic resources.

Eriksson (2006) interprets the average isotopic ratios as indicating an overall emphasis on a diet rich in terrestrial/freshwater proteins and of high trophic level species. The human isotope values clustered into two distinct groups, one with a diet similar to that of the otters analysed, the other showing a mixed freshwater fish and hunted animal diet (Eriksson, Lõugas and Zagorska, 2003) (Figure 3.2). This is not to suggest that humans and otters ate precisely the same diets, but simply that the exploitation of freshwater resources such as fish and other

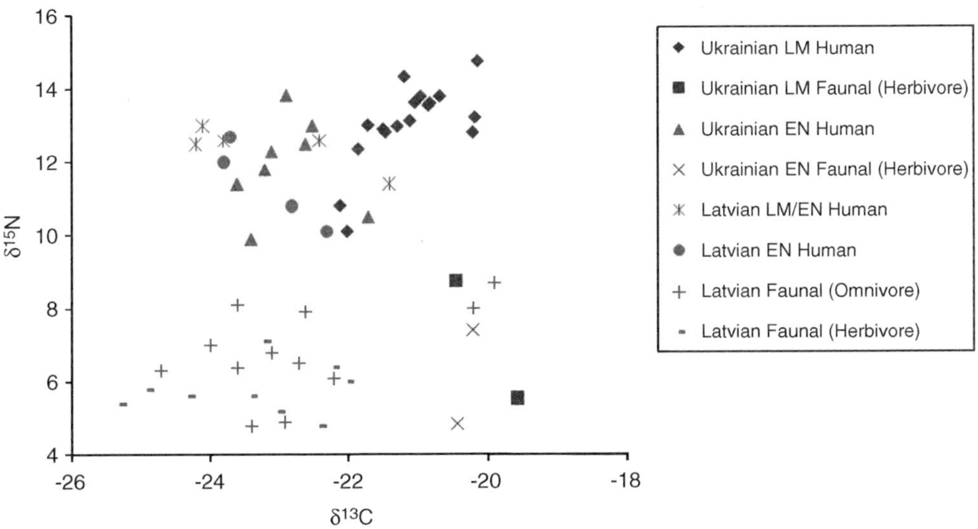

Figure 3.2 Stable isotope analysis of Ukrainian (Vasilyevka II, Marievka, Dereivka, and Yasinovatka) and Latvian (Zvejnieki) late Mesolithic (LM) and early Neolithic (EN) human and faunal remains (after Lillie and Jacobs, 2006; Budd, Lillie and Potekhina, in press); Eriksson, 2006; Eriksson, Lõugas and Zagorska, 2003)

freshwater species is resulting in humans exhibiting similar dietary isotope compositions to the otters that were analysed. Otters feed mostly on fish but occasionally also on crustaceans, amphibians, small mammals or birds (Eriksson, Lõugas and Zagorska, 2003; Taastrøm and Jacobsen, 1999). Hence, it is perhaps unsurprising that humans and otters can produce similar $\delta^{13}C$ and $\delta^{15}N$ isotope values when they are analysed.

Overall, it appears that the Mesolithic and early Neolithic populations consumed more freshwater fish than individuals of later periods, and that the later Neolithic and Bronze Age individuals had higher intakes of terrestrial animals and plants. The transition from specialized diets similar in composition to that of otters, towards more mixed terrestrial diets, occurs during the Middle Neolithic period (Eriksson, 2006).

Previous research, for example Lanting and van der Plicht (1998) and Bonsall *et al.* (2000, 2002, 2004), has shown that the consumption of freshwater resources can result in problems when radiocarbon dating human remains due to an 'over-ageing' effect. Cook *et al.* (2002: 78) have suggested that in cases where the human remains are consistently producing ages that are inconsistent with other dated materials, such as charcoal, the suggestion is that the diet of the human groups being studied 'may have included material from a reservoir that differed in ^{14}C specific activity from the contemporary atmosphere.' In the Danube region, the stable isotope data indicates that a significant aquatic component (i.e. freshwater fish) in the human diet was the source for the differing ^{14}C reservoir, and that an age correction may have to be applied to human bones used in ^{14}C dating at the sites.

Given the considerable inputs from freshwater resources in the diets of the Mesolithic to earlier Neolithic populations at Zvejnieki, the possibility exists that a freshwater reservoir effect may influence the absolute dating of at least some of the human remains from these periods (Lanting and van der Plicht, 1998; Cook *et al.*, 2001, 2002; Lillie *et al.*, 2009).

In Lithuania, recent research by Antanaitis-Jacobs and co-workers (Antanaitis-Jacobs *et al.*, 2009) has shown that the integration of stable isotope analysis alongside archaeological, zooarchaeological, chronological, palaeobotanical and bioarchaeological studies, is allowing a more nuanced approach to the characterization of Lithuanian prehistory in general, and subsistence practices in particular. Overall, hunter-fisher-forager subsistence strategies persist well into the Neolithic in this region (Zvelebil, 2006), and as is often the case in Ukraine, the only defining Neolithic 'signature' in the east Baltic is the appearance of pottery (at about 5600–5400 calBC in Lithuania and Latvia) (Antanaitis, 1999; Antanaitis-Jacobs and Girininkas, 2002).

In terms of the faunal remains being exploited by the populations of the region, domestic cattle, sheep/goat and pigs are present at low levels on middle Neolithic sites in Lithuania and Latvia (Antanaitis-Jacobs *et al.*, 2009). However, the Lithuanian evidence indicates that the hunting of elk, the predominant species exploited in the Mesolithic, occurs alongside red deer, aurochs, boar, marten, beaver and seal, with the exploitation of these species persisting into the Neolithic (when seal exploitation actually increases) (Antanaitis-Jacobs *et al.*, 2009). It is only during the Bronze Age that domesticates begin to dominate in faunal and floral assemblages (Antanaitis, 1999, 2001). The exploitation of pike and cyprinids by these populations may well have similar connotations to the Latvian evidence from Zvejnieki, wherein the potential for a reservoir effect in the radiocarbon dating of human remains from this region may require consideration in future research initiatives.

The most ubiquitous plant remains from Mesolithic and Neolithic sites in the Baltic region are hazelnut and water chestnut (Antanaitis-Jacobs *et al.*, 2009). The first domesticated plant recorded in western Lithuania at about 3300 to 2000 calBC, is in fact hemp (Antanaitis *et al.*, 2000). In general, the evidence for cereals is sparse in the east Baltic, with single finds of oat, barley, *Cerealia* and hemp/hops reported from middle Neolithic contexts (Rimantienė, 1992). In the later Neolithic period, emmer, barley and millet are recorded from archaeological contexts in the region.

In a recent attempt to further enhance the resolution of the palaeobotanical record for Lithuania, Antanaitis *et al.* (2000) and Antanaitis and Ogrinc (2000) undertook detailed analysis of the Neolithic and Bronze Age contexts at two habitation sites, Kretuonas in northeastern and Turlojiškė in southwestern Lithuania. At these sites, wild species such as raspberry, apple (?), and hazelnuts dominate in the assemblages studied, while of 166 samples investigated, only one instance of domesticated plants, millet from Turlojiškė (Antanaitis *et al.*, 2000), was recorded. The evidence suggests that despite the classification of many sites as 'Neolithic', the integration of domesticated species into existing subsistence strategies is very limited prior to the later Neolithic in the Baltic region. A similar protracted uptake of domesticates by the indigenous populations of Ukraine characterizes the earlier Neolithic period (see below). In essence, there may well be some value in 're-packaging the Mesolithic' in both the Baltic region and Eastern Europe, in order to account for the fact that in reality the shift from food extraction to food production often occurs at a chronologically later date than the currently accepted Mesolithic-Neolithic boundary (Antanaitis-Jacobs *et al.*, 2009).

New stable isotope studies undertaken by Antanaitis-Jacobs *et al.* (2009) comprise faunal material from the Neolithic-Early Bronze Age (EBA) Šventoji coastal site located in northwestern Lithuiania, and the inland Neolithic-EBA lacustrine sites of the Kretuonas/Žemaitiškė archaeological complex near the Kirsna river in southwestern Lithuania. Human bone samples for the stable isotopic studies of diet were obtained from six locations in

Lithuania, dating to the Mesolithic through to the Late Bronze Age (LBA) (Antanaitis-Jacobs *et al.*, 2009).

The fauna recovered from these Lithuanian sites, which include both herbivores and carnivores, represent a range of environments from marine, freshwater and terrestrial locations. Terrestrial herbivores (elk, red deer, aurochs) exhibit $\delta^{13}C$ ranges of –24.1 to –20.7‰ that are typical of C_3 browsers, with $\delta^{15}N$ values of 3.1 to 5.5‰. The carnivores studied exhibited $\delta^{13}C$ ranges of –20.7 to –18.5‰ with associated $\delta^{15}N$ values of 8.8 to 13.3‰. The marine $\delta^{13}C$ ratios were higher, as anticipated, at –18.7 to –15.5‰, with associated $\delta^{15}N$ values ranging between 10.6 to 13.9‰ (Antanaitis-Jacobs *et al.*, 2009).

The human isotope values at the possible Mesolithic/Neolithic site of Donkalnis, Lithuania, indicate the consumption of terrestrial animal proteins with inputs from freshwater resources (river fish). Similar diets are attested for Late Mesolithic individuals at Spiginas and Donkalnis, although the higher $\delta^{15}N$ average values of 12.6 ± 0.3‰ indicate elevated inputs from freshwater resources during the later Mesolithic (Antanaitis-Jacobs *et al.*, 2009). In the Baltic region, freshwater resources remain a significant component of the diet as late as the Middle Neolithic period. In contrast, the late Neolithic human bone samples from Lithuania exhibit $\delta^{13}C$ values of –21.9 to –21.4‰ and $\delta^{15}N$ values of 7.99 to 10.1‰, suggesting a shift away from the consumption of freshwater resources during this period, perhaps towards animal husbandry (Antanaitis-Jacobs *et al.*, 2009). The introduction of millet (*Panicum miliaceum*) is suggested for the Bronze Age period, on the basis of the evidence from the sites of Turlojiškė 1 and 4. The data obtained by Antanaitis-Jacobs *et al.* (2009) is limited due to preservation factors and the available archive, but the trends in the evidence do reinforce the observation that the integration of domesticates is a protracted process in the Baltic region (Zvelebil, 2006).

3.3.2 The Ukrainian Evidence

A considerable amount of new dating, dietary isotope and palaeopathological analyses, aimed at understanding subsistence across the period about 10 000 to 4000 calBC, has been undertaken in the past decade or so in the Ukrainian region (Lillie, 1996, 1998a, 1998b; Lillie and Richards, 2000; Lillie, Richards and Jacobs, 2003; Lillie and Jacobs, 2006; O'Connell, Levine and Hedges, 2000; Telegin *et al.*, 2002, 2003) (Figure 3.3). Until recently, the bulk of the analysis has focused on the dating, palaeopathology and isotope analysis of the Dnieper Rapids cemeteries, but more recent analysis of this dataset has been expanded to include sites from the Middle and Lower Dnieper basin (Budd, 2007; Lillie *et al.*, 2009).

When considering the evidence for resource availability in the Dnieper Rapids region, the excavations of the Eneolithic settlement site of Dereivka (Telegin, 1986), along with the data from Igren VIII, Sobachky and Buzky (Telegin and Potekhina, 1987), have provided some valuable insights into the potential range of animal and fish species that were available for exploitation by the indigenous populations. Amongst the species exploited, aurochs, red and roe deer, elk, wild boar, rabbit and beaver are reported (Telegin, 1986; Telegin and Potekhina, 1987). As noted by O'Connell, Levine and Hedges (2000), the inhabitants at the site of Dereivka consumed not only horse and dog species but also waterfowl, otter, beaver, European pond terrapin (*Emys orbicularis*), European catfish (*Siluris glanis*), asp (*Aspius aspius*), pike (*Esox lucius*), zander (*Lucioperca lucioperca*), rudd (*Scardinius erhythropthalamus*), mussel (*Unio*) and river snail (*Viviparus* sp.).

Finds from archaeological sites in the Dnieper Region provide evidence for fishing-related artefacts such as harpoons, net sinkers, fish-hooks and fish-tooth necklaces, which are, by their

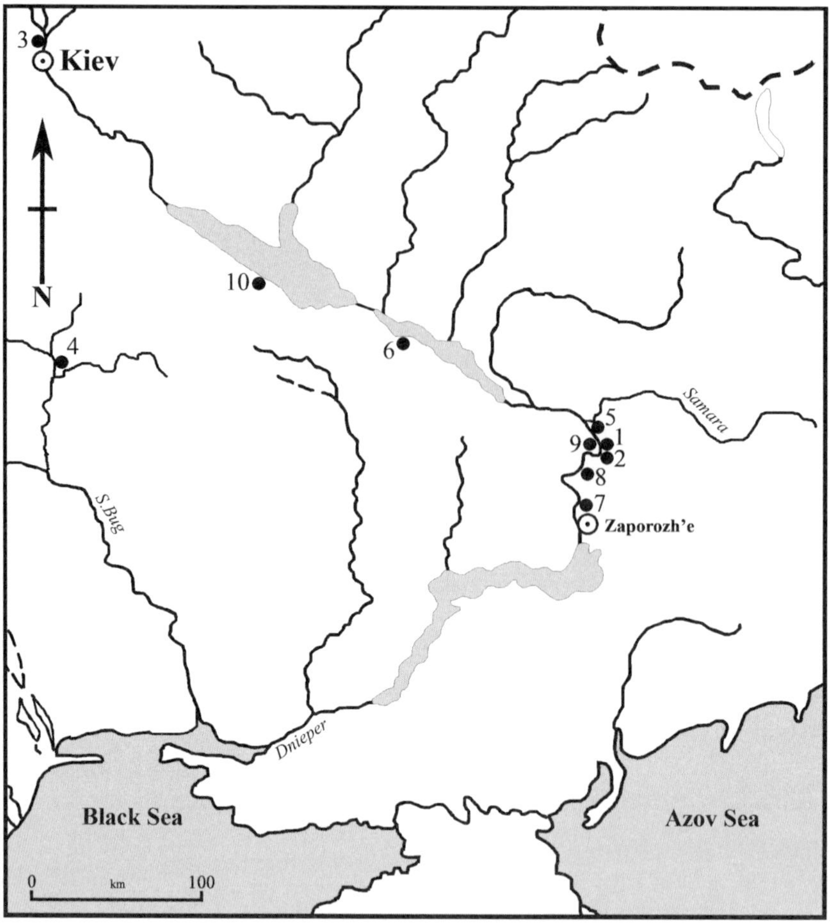

Figure 3.3 The Middle and Low Dnieper Basin region, showing key sites used in the recent research: 1 – Vasilyevka III and II, 2 – Vasilyevka V, 3 – Vyshgorod, 4 – Dobryanka, 5 – Igren VIII, 6 – Dereivka, 7 – Vil'nyanka, 8 – Yasinovatka, 9 – Nikolskoye, 10 – Molyukhov Bugor (after Lillie *et al.*, 2009)

very nature, more likely to be reported than fish remains themselves (Lillie and Richards, 2000; O'Connell, Levine and Hedges, 2000; Telegin, 1986; Telegin and Potekhina, 1987). As noted above, fish bones have also been recovered from a number of sites in the Dnieper region, despite the fact that no sampling strategies aimed specifically at the recovery of the smaller elements of food refuse have been implemented. However, it is the ubiquitous presence of fish-tooth pendants from freshwater species such as common carp (*Cyprinus carpio*) and pearl roach (*Rutlius frisii*) in burial contexts, alongside the artefacts from archaeological sites, that indicate that the exploitation of fish was an essential element of both diet and, as a corollary presumably, social constructs in this region (Lillie, 2003).

As the sites in this region span a considerable chronological timeframe, a brief summary of the dating, palaeopathological and recent stable isotope analyses for the Epipalaeolithic, later Mesolithic and earlier Neolithic cemeteries is presented below. As will be seen, the palaeopathological analysis of these populations confirms the presence of very low levels of dietary stress (Lillie, 1996, 1998a), which would equate to the European Mesolithic and Neolithic levels as outlined by Meiklejohn and Zvelebil (1991).

The most interesting sites, in terms of pathology, are the cemeteries of Vasilyevka III, Voloshkoe and Vasilyevka I, all of which are thought to date to the period about 10 400 to 9200 calBC (Lillie, Richards and Jacobs, 2003). The only cemetery in this group that has been radiocarbon dated, however, is Vasilyevka III, which is located to the south of the town of Dniepropetrovsk on the Dnieper (Figure 3.3). The three dates obtained by Ken Jacobs (1993) place the site between 10 080 and 9980 uncalBP, and when calibrated to 2σ using the OxCal program of Stuiver *et al.* (1998) these three radiocarbon determinations (10 080 $\pm$ 100 BP (OxA-3809), 10 060 $\pm$ 105 (OxA-3807) and 9980 $\pm$ 100 BP (OxA-3808)), indicate an age range of 10 400–9200 calBC.

The palaeopathological and dental studies undertaken at Vasilyevka III indicate a complete absence of caries and the frequent presence of dental calculus (Lillie, 1998a). In general, visual examination of the dentitions shows that attrition levels are low, and that where a high degree of crown wear occurs, this is proportional to the individual's age. In addition, the predominantly flat molar wear patterns in evidence across the Epipalaeolithic to Eneolithic periods are reflective of hunter-gatherer subsistence strategies (cf. Smith, 1984). Meiklejohn, Wyman and Schentag (1992) have suggested that caries and attrition can be independent variables; however, the absence of caries in the entire Ukrainian series studied to date must be considered to reflect an absence of cariogenic foodstuffs in the diets of the Ukrainian populations. This is especially relevant in light of the above observations in relation to attrition, and as Nikiforuk (1985) has suggested that with age, the loci for caries involvement can shift to the cemento-enamel junction of the interproximal tooth surface as attrition progresses, the absence of caries further supports an absence of cariogenic foodstuffs in the Epipalaeolithic to Eneolithic Ukrainian diets.

In total, 45.8% of the cemetery population at Vasilyevka III exhibit calculus deposits, with a statistically significant difference in occurrence between males and females. More than 63% of female teeth have no calculus and less than 17% of male teeth were recorded without calculus. In general, males had heavier calculus deposition by grade (after Hillson, 1979). Some age-dependent biases occur, in that the majority of those individuals for whom calculus could not be recorded were determined to be older than 45 to 55 years of age, based on the advanced wear of the dentition. However, despite this, all individuals above the age of 18 years exhibit some calculus. As calculus is assumed to have positive correlations with protein ingestion (Hillson, 1979, 1986), the observed gender bias in the expression of calculus could be associated with the unequal access to meat resources by males, as also noted by y'Edynak (1978, 1989) for the Mesolithic populations at Vlasac in the Danube Gorge region. The expression would also support Speth's (1990) suggestion that some degree of inequality in access to meat resources, as might be anticipated at kill sites, may occur in these hunter-fisher-forager populations.

Lillie (2004) outlined evidence for inter-personal violence at the Vasilyevka I and III and Voloshkoe cemeteries. At these cemeteries, a number of individuals have been shown to have injuries associated with projectile weapons, and there is some suggestion of the removal of skeletal elements. At the Vasilyevka III cemetery burial No. 33, a female aged 18 to 22 years at death, was found with an arrowhead in her rib cage. Similarly, burial No. 36, possibly a female aged 20 to 25, had fragments of a similar point in close proximity to the skeleton (Nuzhinyi, 1989), while burial No. 37, a male aged 25 to 35, had a microlith embedded in his lumbar vertebrae (Nuzhinyi, 1989). This evidence suggests that young individuals at

Vasilyevka III were targeted as victims of inter-personal violence. If this reflects violence due to competition for resources, then a discrete portion of the group was being singled out in this activity. At Vasilyevka I, only one individual exhibited a definite direct association between lithics and personal injury. A male individual, burial No. 17, had four weapon injuries, including a trapezoidal point, clearly shot into the body, fragments of two points with pronounced impact damage, and two further pieces that could be reconstructed, indicating impact damage (Nuzhinyi, 1989).

At Voloshkoe, Danilenko (1955) reported that one individual, No. 3, amongst the 19 burials, had been murdered. Interestingly, this individual was interred in the same crouched position as the majority of the other burials in the cemetery, yet this individual was buried away from the main concentration of burials, facing in the opposite direction to the remainder of the cemetery population. A backed bladelet was found embedded in the first cervical vertebra, indicating that the individual in question was most likely shot by someone using a bow and arrow. Nuzhinyi (1989) reports that in fact, three arrowheads were associated with this individual, and considers this to be the earliest evidence for the use of the bow in Ukraine. However, it should be remembered that Voloshkoe and Vasilyevka I remain undated in absolute terms. Burial No. 5 (male) at Voloshkoe also exhibits signs of violence in that, along with other examples, it appears that the hands were cut off prior to interment. Similarly, burial No. 16 at Voloshkoe was found in a complicated ritual positioning comprising a considerable degree of disarticulation, and with evidence to suggest that the right hand and adjoining long bones were cut off prior to burial. Finally, burial No. 15 had the hands and both legs below the knees missing (Danilenko, 1955).

Despite the fact that, to date, Voloshkoe and Vasilyevka I have not been available for stable isotope analysis, the stable isotope and palaeopathological analysis of Vasilyevka III suggests that this is an Epipalaeolithic hunter-fisher-gatherer population exploiting a non-cariogenic diet, with a strong indication of high protein intakes. The stable isotope data shows that all of the Vasilyevka III individuals have $\delta^{15}N$ values over 11.5‰, suggesting a relatively uniform diet, comprised of animal proteins with a significant input from freshwater resources such as fish, and plant resources (Lillie, Richards and Jacobs, 2003). Considering this fact, it is surprising that gender-specific variation is found in calculus levels. However, as differential calculus expression is not absolute evidence for differential consumption of meat proteins (as teeth can be cleaned), and as the isotope data is not precise enough to identify specific variation in the inferred variability in access to meat resources, this data is simply highlighting some potential disparity (or possible avenues for enhanced resolution) between the methods employed.

The later Mesolithic sites in Ukraine include Vasilyevka II (Lillie and Jacobs, 2006) and Marievka (Lillie and Richards, 2000), along with single internments at the cemeteries of Dereivka and Osipovka (discussed in Lillie and Jacobs, 2006).

At Vasilyevka II, recent Accelerator mass Spectrometry (AMS) radiocarbon dates have shown that this particular site was occupied during the late Mesolithic, about 7300 to 6220 calBC (OxA-3804 at 7920 ± 85 uncalBP; OxA-3805 at 7620 ± 80 uncalBP and OxA-3806 at 8020 ± 90 uncalBP). As in the case of the Epipalaeolithic cemetery of Vasilyevka III, the Vasilyevka II population does not display any evidence of dental caries. Conversely, of the 17 individuals studied, 13 exhibit calculus deposition, 3 individuals did not have dentitions available for study due to lack of preservation, and only one individual, a male aged 30 to 40, did not display any evidence of calculus deposits (Lillie, 1998a). It appears that a similar expression of calculus occurs at Vasilyevka II when contrasted against the earlier Epipalaeolithic cemetery of Vasilyevka III, with statistically significant differences in occurrence identified between males and females (Lillie, 1998a). Males exhibit more calculus of higher deposition grade (cf. Hillson, 1979) than females, perhaps indicating preferential access to dietary proteins, as has been suggested above.

By contrast, the stable isotope data suggests dietary equivalence between males and females, with the isotope ratios having a mean of –20.94 ± 0.49‰ $\delta^{13}C$ and 13.39 ± 0.62‰ $\delta^{15}N$, and with the individual with the highest $\delta^{15}N$ value being male. The Vasilyevka II individuals all have $\delta^{15}N$ values above 12.35‰, and the data are indicative of relatively uniform diets between males and females as a whole, with animal proteins and freshwater fish being important dietary elements (Lillie and Jacobs, 2006) (Figure 3.2).

The $\delta^{13}C$ and $\delta^{15}N$ human isotope values from Vasilyevka II contrast to those from the chronologically earlier cemetery of Vasilyevka III, with the $\delta^{13}C$ values being more positive (–20.94 ± 0.49‰) when compared to those from Vasilyevka III (–22.37 ± 0.31‰). Similarly, the $\delta^{15}N$ values from the Mesolithic site of Vasilyevka II differ from the values from the Epipalaeolithic Vasilyevka III cemetery, with the Mesolithic values at 13.3 ± 0.62‰ and Epipalaeolithic at 12.47 ± 0.61‰. The elevated nitrogen mean from Vasilyevka II may indicate a slightly higher input from freshwater fish in the later Mesolithic compared to the Epipalaeolithic period (Lillie and Jacobs, 2006).

The earlier Neolithic period in this region is represented by a number of cemetery sites, and the occurrence of significant numbers of individuals interred at habitation sites such as Dereivka, with about 173 individuals (Telegin and Zhilyaeva, 1964) and Yasinovatka, with 68 individuals (Telegin and Potekhina, 1987) (Figure 3.3). Ongoing assessment of the recent stable isotope analysis of about 100 samples of human, faunal and fish remains from the Dnieper region (Lillie, Budd and Potekhina, in press) will provide new insights into the diet of early Neolithic populations from this region. However, to date, limited analysis of material from the early Neolithic cemeteries of Dereivka, Yasinovaka and Nikoskoye (Lillie and Richards, 2000) has shown that a number of samples have $\delta^{13}C$ values that are between –22 and –24‰, which is more negative than we would expect for a purely terrestrial C_3 diet (*ca.* –20 to –21‰). These more negative values are indicative of the addition of aquatic resources to the diets, and based on the archaeological evidence, the resources responsible for these values likely include river fish, along with otter, beaver and tortoise. This interpretation is supported by the associated higher $\delta^{15}N$ values for these individuals (Lillie and Richards, 2000).

Palaeopathological and dental analysis of the Neolithic populations of the Dnieper region has shown that caries, indicative of the consumption of dietary carbohydrates, is completely absent from the entire dental sample studied (Lillie, 1998a). By contrast, dental calculus, associated with the consumption of dietary proteins, is consistently present on the dentitions of individuals from the Neolithic period. Analysis of about 32% of the population from the Neolithic cemetery of Vovnigi II, housed in St Petersburg (Lillie, 1998a), which has been dated by Jacobs to 5480 to 4750 calBC (OxA-5938, 6320 ± 80 uncalBP; OxA-5939, 6275 ± 70 uncalBP and OxA-5940, 6090 ± 100 uncalBP) (unpublished data), indicates that differential access to dietary proteins, in favour of males, may have been occurring in this population. This expression is supported by similar evidence from the Epipalaeolithic cemetery of Vasilyevka III, wherein heavy calculus deposition occurs on the anterior portion of the dentitions of three males from this cemetery, and there is evidence for the use of tooth picks (grooving) on the teeth of some of the male individuals from this cemetery.

Differential access to dietary protein in favour of males appears to be reduced at a number of the other early Neolithic cemeteries in the Dnieper region. The analysis of dentitions from Dereivka, Yasinovatka and Nikolskoye found similar male–female calculus expressions, suggestive of broadly equivalent levels of protein consumption between the sexes. This observation contradicts the general model of Mesolithic calculus expression whereby males tend to exhibit higher incidences of heavy calculus deposition than females (Lillie and Richards, 2000); although cleaning (oral hygiene) is likely to influence calculus expression.

To date there has been a paucity of faunal remains for use in the estimation of trophic levels in the Ukrainian skeletal series. However, an expanded isotope dataset from the cemeteries of Dereivka and Yasinovatka (Budd, Lillie and Potekhina, in press) indicates that, as might be anticipated, the faunal remains are significantly lower in terms of trophic level than both the human and fish samples (Lillie *et al.*, 2009). As a consequence of the exaggerated separation between the faunal and human samples studied, the $\delta^{15}N$ ratios clearly confirm the consumption of a diet by the human individuals (e.g. $\delta^{15}N$ values of 11.4 to 13.5‰ at Dereivka, and 12.5 to 14.62‰ at Yasinovatka) that is proportionately high in terms of freshwater protein inputs. By contrast, the deer sample from Dereivka has a $\delta^{13}C$ ratio of –20.43 and 4.86‰ for $\delta^{15}N$, while the deer sample from Yasinovatka is –20.2 for $\delta^{13}C$ and 7.4‰ for $\delta^{15}N$.

The available faunal sample size is clearly extremely limited. However, despite this, given the assumed *ca.* 2 to 4‰ trophic level shift between the animal protein consumed and the human consuming it, the ratios for the humans from Derievka and Yasinovatka indicate a significantly greater trophic level shift than would be anticipated if the diet was primarily focused on the consumption of animal proteins. This would be the case even where trophic level shifts of up to 5‰ occur (Jay and Richards, 2006; Hedges and Reynard, 2007).

The human isotope samples from the earlier Neolithic cemeteries contrast with the later Mesolithic values, as $\delta^{13}C$ values of -22.95 ± 0.61‰ and $\delta^{15}N$ values of 11.84 ± 1.21‰ are recorded for the earlier Neolithic. As mentioned above, the later Mesolithic $\delta^{13}C$ and $\delta^{15}N$ isotope values are -20.94 ± 0.49‰ and 13.3 ± 0.62‰, respectively. The later Mesolithic values exhibit a positive shift in $\delta^{13}C$ and an elevation of $\delta^{15}N$; a combination that suggests that the later Mesolithic populations at Vasilyevka II are exploiting different freshwater resources from those being exploited in the Epipalaeolithic and early Neolithic periods.

In addition, the greater standard deviation from the average values in the earlier Neolithic suggests that there is greater variability in individual dietary intakes in the earlier Neolithic populations of Ukraine. However, as noted by Lillie (2003), this is not unusual, as different subsistence regimes are to be expected where social situations may have been mediated through the procuring, allocating and controlling of resources. The evidence obtained to date suggests that some variation in either the exploitation or availability of certain freshwater resources is occurring across the Mesolithic to Neolithic periods in the Dnieper region.

3.4 DISCUSSION AND CONCLUSIONS

In this chapter we considered the nature of the available evidence for diet and subsistence in two contrasting regions, the Baltic Sea basin and the Dnieper River basin in Ukraine. The Baltic sites in Latvia and Lithuania are located in a region that Zvelebil (2006) considers to be ideally suited to the persistence of hunter-forager lifeways, due to the network of highly productive marine coastlines and freshwater shorelines, with extensive networks of estuaries, lakes and rivers, bays and archipelagos. Although perhaps more limited in terms of landscape diversity, the river basin of the Dnieper offers a broad range of freshwater resources, both aquatic and riparian, alongside terrestrial resources that appear to offer a similar potential in terms of the reliability of the available subsistence spectrum. In both regions, the shift from the exploitation of wild to domesticated plant and animal species is offset somewhat, as the availability of domesticates in the Neolithic does not necessarily prompt an immediate uptake of the newly available resources.

The evidence from Zvejnieki in Latvia (Eriksson, Lõugas and Zagorska, 2003; Larsson and Zagorska, 2006) has shown that in terms of the palaeopathology and dental analysis of the interred population, the general pattern of trauma is commensurate with that occurring in forager populations, with males having more pathologies than females (Jankauskas and Palubeckaitė, 2006). The human isotope analysis has shown that the stable isotope values cluster into two distinct groups, one with a diet similar to that of the otters analysed, and the other showing a mixed freshwater fish and hunted animal diet (Eriksson, Lõugas and Zagorska, 2003). The fact that specialized hunting strategies are followed depending on proximity to the coast, or inland habitats, is not surprising, given the availability of resources. In their studies Eriksson, Lõugas and Zagorska (2003) and Larsson and Zagorska (2006) used multiple strands of evidence to identify an increased focus on the consumption of freshwater fish by Mesolithic and early Neolithic populations than individuals in later periods. They also note that the later Neolithic and Bronze Age individuals had higher intakes of terrestrial animals and plants, with the shift towards more terrestrial diets occurring in the Middle Neolithic.

Building on the diversity in methodological approaches adopted by Eriksson, Lõugas and Zagorska (2003), Antanaitis-Jacobs and co-workers (Antanaitis-Jacobs *et al.*, 2009) have integrated stable isotope analysis alongside archaeological, zooarchaeological, chronological, palaeobotanical and bioarchaeological studies, in order to provide a more nuanced and holistic understanding of the Lithuanian Mesolithic to Bronze Age periods. Antanaitis (1999, 2001) has shown that it is only during the Bronze Age that frequencies of domesticates begin to dominate in faunal and floral assemblages. As might be anticipated, the human isotope ratios from Lithuania again suggest that in this region, dietary pathways consisting of terrestrial animal proteins with inputs from freshwater resources (Antanaitis-Jacobs *et al.*, 2009) occur across the Mesolithic and into the Middle Neolithic period. It is only during the Late Neolithic period that the human isotope data from Lithuania exhibits a shift away from the consumption of freshwater resources, perhaps towards animal husbandry (Antanaitis-Jacobs *et al.*, 2009).

The Ukrainian evidence has been studied in relation to radiocarbon dating, palaeopathology and stable isotope studies of diet since 1992 (Lillie, 1998a), and benefits to some degree from the extended temporal resolution afforded by the presence of large cemeteries in this region occurring from as early as the Epipalaeolithic period at about 10 400 to 9200 calBC (Jacobs, 1993; Lillie, 1998a).

Amongst the Ukrainian skeletal series, the rates of skeletal pathologies are low, and as with the Baltic evidence, the levels and nature of the observed pathologies are fully commensurate with those identified in hunter-fisher-forager populations. Caries, an indicator of the consumption of carbohydrates, and/or of sweet/sticky fruits, is universally absent from the Ukrainian skeletal series, while calculus deposition, suggestive of the consumption of dietary proteins, is present throughout. There is a marked contrast between the Ukrainian evidence and that from Zvejnieki, in Latvia, where high levels of caries occur from the Mesolithic period onwards. As cereals have not been recovered at Zvejnieki, the suggestion is that some form of sweet/sticky fruit/honey or similar foodstuffs may have been integral to the Latvian diet across the Mesolithic to Bronze Age periods.

With the exception of very limited (and equivocal) indirect evidence for cereal consumption, in the form of seed impressions on pottery fabrics (Kotova, 2003), there is very little archaeological evidence for the adoption and management of domesticated plants and animals in the region of Dnieper River system before about 4500 calBC. The palaeopathological evidence reinforces this observation (Lillie, 1998a). However, it is perhaps the stable isotope evidence that lends some additional support to the dental evidence regarding past diets, as the populations from the Epipalaeolithic through to Middle Neolithic period appear to have consumed diets commensurate with those of hunter-fisher-foragers. Overall, the Epipalaeolithic to late Mesolithic populations in the Dnieper Basin consumed terrestrial resources (e.g. red deer, roe deer, horse, wild boar and plants) with a significant input from freshwater fish. The contribution of the latter is less easy to assess for the early Neolithic, due to the apparent greater dietary breadth is evidence from the isotope studies (Lillie and Richards, 2000; Lillie, Budd and Potekhina, under review), but a similar range of wild animal species continued to be exploited into this period.

Finally, it is apparent that if we are to gain a holistic understanding of prehistoric diet and subsistence, a fully integrated research design comprising archaeological, chronological, zoological, palaeobotanical, bioarchaeological and stable isotope studies is necessary (see, for example, Stock *et al.* this volume); especially if we are to be able to disentangle the subtle variations in the importance of the particular foodstuffs being consumed (cf. Cannon, Schwarcz and Kynf, 1999; and many others).

ACKNOWLEDGEMENTS

Vladimir Timofeev and Ken Jacobs both helped MCL, in varying ways, during his early research years in Eastern Europe. As ever, this paper is dedicated to their memory. In addition, the first author would like to thank Dimitri Telegin and Inna Potekhina, Ukrainian Academy of Sciences, Kiev and Prof Gokhman and Alexander Kozintsev, Museum of Anthropology and Ethnography, St Petersburg, for invaluable assistance during my research visits to Eastern Europe. Inna Potekhina is currently working with the authors on the interpretation of the new isotope studies from the Dnieper Basin. As always, Malcolm Lillie would like to thank all friends and colleagues in Ukraine and Russia for the help and advice they have given since he began researching in Eastern Europe in 1992. Finally, a new discourse has begun with Indre Antanaitis-Jacobs, who has been kind enough to allow me access to her in press works and numerous papers on Lithuanian prehistory, I hope my summary of her work herein reflects the importance of her efforts to date in Lithuanian prehistory. On a more personal note, MCL would also like to dedicate this paper to the memory of two other individuals who have influenced his career in different ways since he entered academia in 1987, both of who passed away recently. As such, MCL would also like to dedicate this paper to Roger Jacobi and Pavel Dolukhanov.

REFERENCES

Ambrose, S.H. (1990) Preparation and characterisation of bone and tooth collagen for isotopic analysis. *J. Arch. Sci.*, **17**, 431–451.

Ambrose, S.H. (1993) Isotopic analysis of palaeodiets: methodological and interpretive considerations, in *Investigation of Ancient Human Tissue: Chemical Analysis in Anthropology* (ed. M.K. Stanford), Gordon & Breach Science Publishers, Langhorne, PA, pp. 59–130.

Ambrose, S.H. and Norr, L. (1993) Experimental evidence for the relationship of the carbon isotope ratios of whole diet and dietary protein to those of bone collagen and carbonate, in *Prehistoric Human Bone: Archaeology at the Molecular Level* (eds J.B. Lambert and G. Grupe), Springer-Verlag, New York, pp. 1–37.

Antanaitis, I. (1999) Concerning the transition to farming in the East Baltic. *Documenta Praehistorica*, **26**, 89–100.

Antanaitis, I. (2001) Eastern Baltic economic and social organization in the late Stone and early Bronze Ages. Unpublished PhD thesis summary, Vilnius University.

Antanaitis, I. and Ogrinc, N. (2000) Chemical analysis of bone: stable isotope evidence of the diet of Neolithic and Bronze Age people in Lithuania. *Istorija*, **45**, 3–12.

Antanaitis-Jacobs, I. and Girininkas, A. (2002) Periodization and chronology of the Neolithic in Lithuania. *Archaeologia Baltica*, **5**, 9–39.

Antanaitis, I., Riehl, S., Kisielienė, D. and Kelertas, K. (2000) The evolution of the subsistence economy and archaeobotanical research in Lithuania. *Lietuvos Archeologija*, **19**, 47–67.

Antanaitis-Jacobs, I., Richards, M., Daugnora, L. *et al.* (2009) Diet in early East Baltic prehistory and the new Lithuanian stable isotope evidence. *Archaeologia Baltica*, **12**, 12–29.

Balasse, M., Bocherens, H. and Mariotti, A. (1999) Intra-bone variability of collagen and apatite isotopic composition used as evidence of a change of diet. *J. Arch. Sci.*, **26**, 593–598.

Bartosiewicz, L., Bonsall, C. and Şişu, V. (2008) Sturgeon fishing along the Middle and Lower Danube, in *The Iron Gates in Prehistory. New Perspectives* (eds C. Bonsall, V. Boroneanţ and I. Radovanović), BAR, (Int. Series) 1893, Archaeopress, Oxford, pp. 39–54.

Bonsall, C., Cook, G., Lennon, R. *et al.* (1997) Stable Isotopes, Radiocarbon and the Mesolithic-Neolithic Transition in the Iron Gates: a palaeodietary perspective. *J. Eur. Arch.*, **5**, 50–92.

Bonsall, C., Cook, G., Lennon, R. *et al.* (2000) Stable isotopes, radiocarbon and the Mesolithic-Neolithic transition in the Iron Gates. *Documenta Praehistorica*, **27**, 119–132.

Bonsall, C., Macklin, M.G., Payton, R.W. and Boroneanţ, A. (2002) Climate, floods and river gods: environmental change and the Meso-Neolithic transition in southeast Europe. *Before Farming*, **3–4**, 1–15.

Bonsall, C., Cook, G.T., Hedges, R.E.M. *et al.* (2004) Radiocarbon and stable isotope evidence of dietary change from the Mesolithic to the Middle Ages in the Iron Gates: New results from Lepenski Vir. *Radiocarbon*, **46**, 293–300.

Borić, D., Grupe, G., Peters, J. and Mikić, Z. (2004) Is the Mesolithic-Neolithic subsistence dichotomy real? New stable isotope results from the Danube Gorges. *Eur. J. Arch.*, **7**, 221–248.

Budd, C.E. (2007) The Dnieper rapids region of the Ukraine: a re-assessment of prehistoric chronology and subsistence strategies. University of Oxford: Unpublished MSc Thesis.

Budd, C.E., Lillie, M.C. and Potekhina, I.D. (in press) Radiocarbon dating and isotope analysis of hunter-fisher-gatherers in Ukraine, in *From the Baltic to the Black Sea* (ed. T. Terberger), DAI, Frankfurt.

Cannon, A., Schwarcz, H.P. and Kynf, M. (1999) Marine-based subsistence trends and the stable isotope analysis of dog bones from Namu, British Columbia. *J. Arch. Sci.*, **26**, 399–407.

Chisholm, B.S., Nelson, D.E. and Schwarcz, H.P. (1982) Stable carbon ratios as a measure of marine versus terrestrial protein in ancient diets. *Science*, **216**, 1131–1132.

Cook, G.T., Bonsall, C., Hedges, R.E.M. *et al.* (2001) A freshwater diet-derived ^{14}C reservoir effect at the stone age sites in the iron gates Gorge. *Radiocarbon*, **43**, 453–460.

Cook, G.T., Bonsall, C., Hedges, R.E.M. *et al.* (2002) Problems of dating human bones from the iron gates. *Antiquity*, **76**, 77–85.

Danilenko, V.N. (1955) Voloshskiy Epipaleoliticheskiy Mogolnik. (The Voloshkoye Epipalaeolithic Cemetery). *Sovetskaya Etnografiya*, **3**, 56–61.

Eriksson, G. (2006) Stable isotope analysis of human and faunal remains from Zvejnieki, in *Back to the Origin: New Research in the Mesolithic-Neolithic Zvejnieki Cemetery and Environment, Northern Latvia* (eds L. Larsson and I. Zagorska), Almqvist, Wiksell, Stockholm, pp. 183–215.

Eriksson, G., Lõugas, L. and Zagorska, I. (2003) Stone age hunter-fisher-gatherers at Zvejnieki, northern Latvia: radiocarbon, stable isotope and archaeozoology data. *Before Farming*, **1**, 1–25.

Hedges, R.E.M. (2003) On bone collagen–apatite-carbonate isotopic Relationships. *Int. J. Osteoarch.*, **13**, 66–79.

Hedges, R.E.M. and Reynard, L.M. (2007) Nitrogen isotopes and the trophic level of humans in archaeology. *J. Arch. Sci.*, **34**, 1240–1251.

Hillson, S.W. (1979) Diet and dental disease. *World Arch.*, **11**, 147–162.

Hillson, S.W. (1986) *Teeth*, Cambridge University Press, Cambridge.

Jacobs, K. (1993) Human postcranial variation in the Ukrainian Mesolithic-Neolithic. *Current Anth.*, **34**, 417–430.

Jacobs, K. (1995) Returning to Oleni'ostrov: social, economic and skeletal dimensions of a boreal forest Mesolithic cemetery. *J. Anthropol. Archaeol.*, **14**, 58–61.

Jankauskas, R. and Palubeckaitė, Ž. (2006) Palaeopathological review of the Zvejnieki sample: analysis of cases and considerations of subsistence, in *Back to the origin. new research in the Mesolithic–Neolithic Zvejnieki cemetery and environment, northern Latvia* (eds L. Larsson and I. Zagorska), Acta Archaeologica Lundensia Series, 8, Almqvist and Wiksell, Stockholm, 149 Ž 63.

Jay, M. and Richards, M.P. (2006) Diet in the Iron Age cemetery population at Wetwang Slack, East Yorkshire, UK: carbon and nitrogen stable isotope evidence. *J. Arch. Sci.*, **33**, 653–662.

Keegan, W.F. (1989) Stable isotope analysis of prehistoric diet, in *Reconstruction of Life from the Skeleton* (eds M.Y. İşcan and K.A.R. Kennedy), Alan R. Liss Inc., New York, pp. 223–226.

Kotova, N.S. (2003) *Neolithisation in Ukraine, BAR (Int. Series)*, Archaeopress, Oxford, p. S1109.

Krueger, H.W. and Sullivan, C.H. (1984) Models for carbon isotope fractionation between diet and bone, in *Stable Isotopes in Nutrition* (eds J.R. Turnland and P.E. Johnson), Symposium Series 258, American Chemical Society, Washington DC, pp. 205–220.

Lanting, J.N. and van der Plicht, J. (1998) Reservoir effects and apparent ^{14}C ages. *J. Irish. Arch.*, **9**, 151–165.

Larsson, L. (ed.) (1984) *The Skateholm Project I: Man and Environment*, Almqvist and Wiksell, Stockholm.

Larsson, L. and Zagorska, I. (eds) (2006) *Back to the Origin. New Research in the Mesolithic–Neolithic Zvejnieki Cemetery and Environment, Northern Latvia*, Acta Archaeologica Lundensia 8, Almqvist and Wiksell, Stockholm.

Lillie, M.C. (1996) Mesolithic and Neolithic populations of Ukraine: indications of diet from dental pathology. *Curr. Anthropol.*, **37**, 135–142.

Lillie, M.C. (1998a) The Dnieper Rapids region of Ukraine: a consideration of chronology, dental pathology and diet at the Mesolithic-Neolithic transition. Sheffield University: Unpublished PhD thesis.

Lillie, M.C. (1998) The Mesolithic-Neolithic transition in Ukraine: new radiocarbon determinations for the cemeteries of the Dnieper rapids region. *Antiquity*, **72**, 184–188.

Lillie, M.C. (2003) Tasting the Forbidden Fruit: gender based dietary differences among prehistoric hunter-gatherers of Eastern Europe? *Before Farming*, **2**, 1–16.

Lillie, M.C. (2004) Fighting for your life? Violence at the Late-glacial to Holocene transition in Ukraine, in *Violent interactions in the Mesolithic: Evidence and Meaning* (ed. M. Roksandic), British Archaeological Reports International Series 1237, Archaeopress, Oxford, pp. 89–96.

Lillie, M.C. and Jacobs, K. (2006) Stable isotope analysis of fourteen individuals from the Mesolithic cemetery of Vasilyevka II, Dnieper Rapids region, Ukraine. *J. Arch. Sci.*, **33**, 880–886.

Lillie, M.C. and Richards, M.P. (2000) Stable isotope analysis and dental evidence of diet at the Mesolithic-Neolithic transition in Ukraine. *J. Arch. Sci.*, **27**, 965–972.

Lillie, M.C., Richards, M.P. and Jacobs, K. (2003) Stable isotope analysis of twenty-one individuals from the Epipalaeolithic cemetery of Vasilyevka III, Dnieper rapids region, Ukraine. *J. Arch. Sci.*, **30**, 743–752.

Lillie, M.C., Budd, C.E., Potekhina, I.D. and Hedges, R.E.M. (2009) The radiocarbon reservoir effect: new evidence from the cemeteries of the middle and lower Dnieper basin, Ukraine. *J. Arch. Sci.*, **36**, 256–264.

Lillie, M.C., Budd, C.E. and Potekhina, I.D. (in press) Stable isotope analysis of prehistoric populations from the cemeteries of the Middle and Lower Dnieper Basin, Ukraine. *J. Arch. Sci.*

Lõugas, L., Lidén, K. and Nelson, D.E. (1996) Resource utilisation along the Estonian coast during the Stone Age, in *Coastal Estonia: Recent Advances in Environmental and Cultural History* (eds T. Hackens, S. Hicks, V. Lang, U. Miller and L. Saarse), Conseil de l'Europe, Strasbourg, pp. 399–420.

Meiklejohn, C. and Zvelebil, M. (1991) Health status of European populations at the agricultural transition and the implications for the adoption of farming, in *Health in Past Societies: Biocultural Interpretations of Human Skeletal Remains in Archaeological Contexts* (eds H. Bush and M. Zvelebil), British Archaeological Reports International Series 567, Archaeopress, Oxford, pp. 129–144.

Meiklejohn, C., Wyman, J.M. and Schentag, C.T. (1992) Caries and attrition: dependent or independent variables? *Int. J. Anthropol.*, **7**, 17–22.

Müldner, G. and Richards, M.P. (2005) Fast or feast: reconstructing diet in later medieval England by stable isotope analysis. *J. Arch. Sci.*, **32**, 39–48.

Nikiforuk, G. (1985) *Understanding Dental Caries: Etiology and Mechanisms, Basic and Clinical Aspects*, S. Karger, New York.

Nuzhinyi, D. (1989) L'utilisation des microlithes géométrques comme armatures de projectiles. *Bull. Soc. Préhist. Fr.*, **86**, 88–96.

O'Connell, T.C., Levine, M.A. and Hedges, R.E.M. (2000) The importance of fish in the diet of central Eurasian peoples from the Mesolithic to the early Iron Age (conference organisers) (eds M.A. Levine, Y. Rassamakin, A. Kislenko and N. Tatarintseva), Late prehistoric exploitation of the Eurasian steppe: volume II. Proceedings of a symposium held at the McDonald Institute for Archaeological Research, Cambridge 12–16 January 2000: 303–327. Conference Papers: Cambridge University.

O'Shea, J. and Zvelebil, M. (1984) Oleneostrovskii Mogilnik: reconstructing social and economic organisation of prehistoric hunter-fishers in Northern Eurasia. *J. Anthropol. Archaeol.*, **3**, 1–40.

Palubeckaitė, Ž. and Jankauskas, R. (2006) Dental status of the Zvejnieki sample as a reflection of early ontogenesis and activities in adulthood, in *Back to the origin. new research in the Mesolithic–Neolithic Zvejnieki cemetery and environment, northern Latvia* (eds L. Larsson and I. Zagorska), Acta Archaeologica Lundensia Series 8, Almqvist and Wiksell, Stockholm, pp. 165–182.

Price, T.D. (1989) The reconstruction of mesolithic diets, in *The Mesolithic in Europe* (ed. C. Bonsall), John Donald, Edinburgh, pp. 48–59.

Richards, M.P. (2002) A brief review of the archaeological evidence for Palaeolithic and Neolithic subsistence. *Eur. J. Clin. Nutr.*, **56**, 1270–1278.

Richards, M.P. and Hedges, R.E.M. (1999) Stable isotope evidence for similarities in the types of marine foods used by Late Mesolithic humans at sites along the Atlantic coast of Europe. *J. Arch. Sci.*, **26**, 717–722.

Richards, M.P., Pearson, J.A., Molleson, T.I. *et al.* (2003) Stable isotope evidence of diet at Neolithic Çatalhöyv̌k, Turkey. *J. Arch.*, **30**, 67–76.

Rimantienė, R. (1992) The Neolithic of the East Baltic. *J. World Prehist.*, **6**, 97–143.

Schwarcz, H.P. and Schoeninger, M. (1991) Stable isotope analysis in human nutritional ecology. *Yearb. Phys. Anthropol.*, **34**, 283–321.

Schoeninger, M. and DeNiro, M. (1984) Nitrogen and carbon isotopic composition of bone collagen from marine and terrestrial animals. *Geochim. Cosmochim. Ac.*, **48**, 625–639.

Schoeninger, M., DeNiro, M. and Tauber, H. (1983) Stable nitrogen isotope ratios of bone collagen reflect marine and terrestrial components of prehistoric human diet. *Science*, **220**, 1381–1383.

Schulting, R.J. and Richards, M.P. (2001) Dating women and becoming farmers: new palaeodietary and AMS evidence from the Breton Mesolithic Cemeteries of Téviec and Hoëdic. *J. Anthropol. Archeol.*, **20**, 314–344.

Smith, B.H. (1984) Patterns of molar wear in Hunter-Gatherers and Agriculturalists. *Am. J. Phys. Anthropol.*, **63**, 39–56.

Speth, J.D. (1990) Seasonality, resource stress and food-sharing in so-called 'egalitarian' foraging societies. *J. Anthropol. Archaeol.*, **9**, 148–188.

Stuiver, M., Reimer, P.J., Bard, E. *et al.* (1998) INTCAL98 radiocarbon age calibration, 24 000–0 calBP. *Radiocarbon*, **40**, 1041–1083.

Taastrøm, H.-M. and Jacobsen, L. (1999) The diet of otters (*Lutra lutra* L.) in Danish freshwater habitats: comparisons of prey fish populations. *J. Zool.*, **248**, 1–13.

Telegin, D.Ya. (1986) *Dereivka: A Settlement and Cemetery of Copper Age Horse Keepers on the Middle Dnieper,* British Archaeological. Reports International Series 287, Archaeopress, Oxford.

Telegin, D.Ya. and Potekhina, I.D. (1987) *Neolithic Cemeteries and Populations in the Dnieper Basin,* British Archaeological. Reports International 383, Archaeopress Oxford.

Telegin, D.Ya. and Zhilyaeva, S.I. (1964) Dereivskiy neolitichniy mogilnik (Dereivka Neolithic cemetery). *Arkheologiya*, **14**, 144–171.

Telegin, D.Ya., Potekhina, I.D., Lillie, M.C. and Kovaliukh, M.M. (2002) The chronology of the Mariupol-type cemeteries of Ukraine re-visited. *Antiquity*, **76**, 356–363.

Telegin, D.Ya., Lillie, M.C., Potekhina, I.D. and Kovaliukh, M.M. (2003) Settlement and economy in Neolithic Ukraine: a new chronology. *Antiquity*, **77**, 456–470.

Thomas, J. (2004) Current debates on the Mesolithic-Neolithic transition in Britain and Ireland. *Documenta Praehistorica*, **31**, 113–130.

Tieszen, L.L. and Fagre, T. (1993) Effect of diet quality and composition on the isotopic composition of respiratory CO_2, bone collagen, bioapatite, and soft tissues, in *Prehistoric Human Bone: Archaeology at the Molecular Level* (eds J.B. Lambert and G. Grupe), Springer-Verlag, New York, pp. 121–155.

Timofeev, V.I. (1998) The east-west relations in the late Mesolithic and Neolithic in the Baltic region. *Baltic-Pontic Studies*, **5**, 44–58.

y'Edynak, G. (1978) Culture, diet, and dental reduction in Mesolithic forager-fishers of Yugoslavia. *Curr. Anthropol.*, **19**, 616.

y'Edynak, G. (1989) Yugoslav Mesolithic dental reduction. *Am. J. Phys. Anthropol.*, **78**, 17–36.

Zvelebil, M. (2006) Mobility, contact and exchange in the Baltic Sea basin 6000-2000 BC. *J. Anthropol. Archaeol.*, **25**, 178–192.

Zvelebil, M. and Dolukhanov, P. (1991) The transition to farming in Eastern and Northern Europe. *J. World Prehist.*, **5**, 233–278.

Zvelebil, M. and Lillie, M. (2000) Transition to agriculture in Eastern Europe, in *Europe's First Farmers* (ed. T.D. Price), Cambridge University Press, Cambridge, pp. 57–92.

4

Climatic Conditions, Hunting Activities and Husbandry Practices in the Course of the Neolithic Transition: The Story Told by Stable Isotope Analyses of Human and Animal Skeletal Remains

Gisela Grupe and Joris Peters

Staatssammlung für Anthropologie und Paläoanatomie, München, FRG

4.1 THE NORTHERN FERTILE CRESCENT – A CORE AREA OF THE NEOLITHIC TRANSITION

In large parts of the Fertile Crescent (Figure 4.1), the expansion of grasslands at the end of the Pleistocene had a profound impact on human subsistence. The effect of this invasion was to increase dramatically the gross yields of plant-foods per unit area, particularly potential starch-protein staples, and correspondingly to increase the carrying capacity of various ecotones. It has been suggested that these increases prompted significant extensions both in the storage of plant-foods and in sedentism of human groups, and that the ensuing increases in birth rate eventually affected the carrying capacity of regions. Prolonged site inhabitation necessitated prehistoric populations to adjust and optimize their subsistence strategies, illustrated for instance by the shift from harvesting and storing seeds of wild grasses to cereal cultivation and finally plant domestication (Hillman, 1996; Tanno and Willcox, 2006). The extension of wild cereals and other grasses also had a profound effect on the distribution of the different ungulate species in the Fertile Crescent, with a notable increase in population density of grazing or mixed grazing-browsing herbivores, that is of taxa associated with open landscapes (Uerpmann, 1987). It has therefore been hypothesized that ruminant domestication could

Human Bioarchaeology of the Transition to Agriculture Edited by Ron Pinhasi and Jay T. Stock

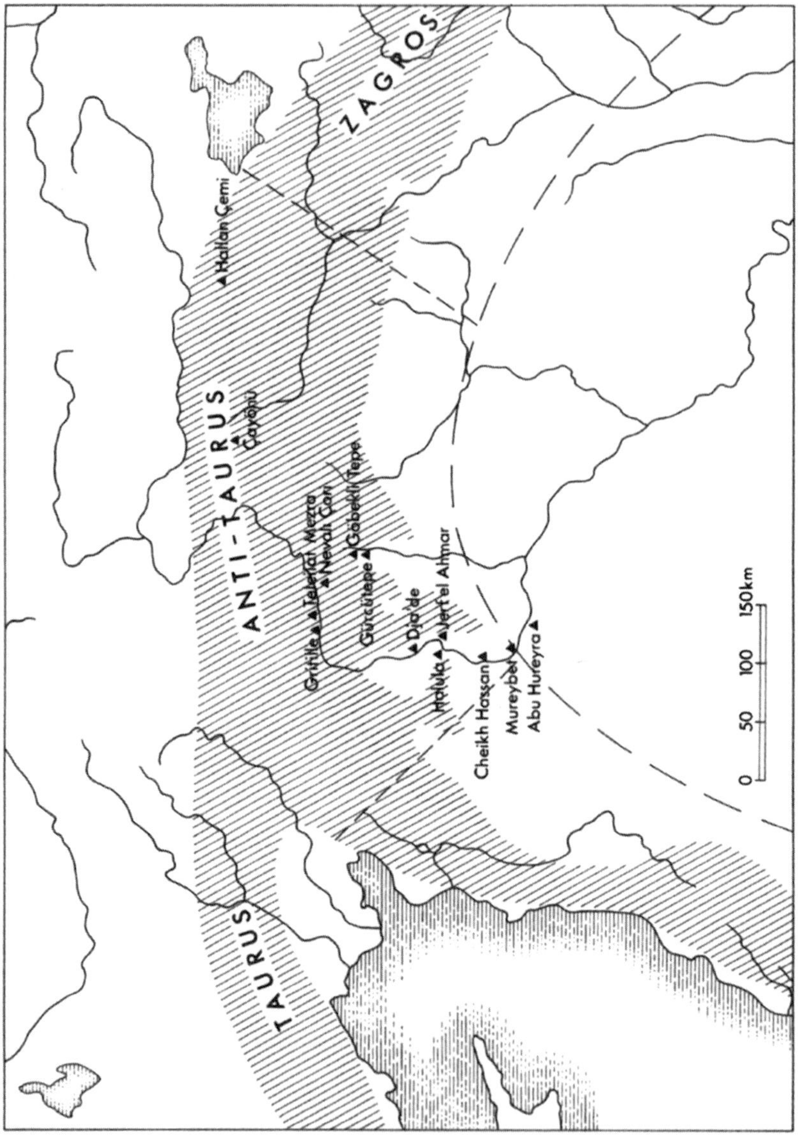

Figure 4.1 Location of the Anatolian and Syrian sites mentioned in the text

have been triggered by the growing dependence of human groups on cereals for their nutrition, thereby increasingly interfering with the lifecycles of the herbivore taxa *Ovis*, *Capra* and *Bos* populating simultaneously the same landscapes (Uerpmann, 1979; Harris, 2002; Peters, von den Driesch and Helmer, 2005a), but this assumption still needs to be verified.

The domestication of animals is but one stage in the transformation from incipient cultural control of a taxon by hunter-gatherers to livestock husbandry in farming societies, because from a functional point of view, the succession from (1) hunting to (2) cultural control of wild species, for example by keeping single or few (tamed) animals to serve as walking larders, to (3) domestication, where humans start controlling the reproduction of stock-on-the-hoof by keeping small herds of animals of the two sexes isolated from their wild progenitors and practising (un)conscious selection, and finally (4) livestock husbandry, can be considered a continuum involving increasing input of human energy per animal. It should be noted, however, that apart from a species' ability to be bred in captivity, other behavioural traits are necessary for a close, successful and long-term co-existence with humans. These include dietary flexibility, modifiable social hierarchy (human beings assuming the role as pack leaders), reasonable fast growth rate and early sexual maturity, pleasant disposition, and a temperament that makes it unlikely to panic (Gautier, 1990; Uerpmann, 1996; Diamond, 2002).

Our knowledge of the beginnings of animal domestication and development of livestock husbandry in the Upper Euphrates basin is based on archaeofaunal studies from a series of sites dating to the tenth-seventh millennia calibrated (cal) BC (Helmer *et al.*, 1998; Peters *et al.*, 1999; Peters, von den Driesch and Helmer, 2005a). The available faunal data suggest that in the ninth millennium calBC, the Upper Euphrates and Tigris regions were core areas for the domestication of food animals in the Near East. Ongoing research in Northern Syria and Southeast Anatolia suggests that the early Neolithic peoples inhabiting the southern Anti-Taurus played an active role in the domestication process relative to sheep, pig and probably also goat (Hongo and Meadow, 1998; Peters *et al.*, 1999; Peters, von den Driesch and Helmer, 2005a), whereas for *Bos*, though widely distributed throughout the Upper Euphrates and Tigris drainages, the North-Syrian Euphrates valley can be postulated as one of the core areas of domestication (Helmer *et al.*, 2005). Archaeozoological research in Central and West Anatolia (Martin, Russell and Carruthers, 2002; Russell, Martin and Buitenhuis, 2005; Russell and Martin, 2005) and Cyprus (Vigne, 2000) does not contradict the foregoing scenario, which is reinforced by mitochondrial DNA studies of present-day livestock breeds (Loftus *et al.*, 1994; Bradley *et al.*, 1998; Hiendleder *et al.*, 1998; Giuffra *et al.*, 2000; Luikart *et al.*, 2001; Troy *et al.*, 2001). These studies point to a Near Eastern origin of cattle, sheep, pig and goat, likely with polytopic domestication events (Larson *et al.*, 2005; Naderi *et al.*, 2008).

A first attempt to establish a more precise timeframe for early animal domestication at a regional scale was done by comparing herbivore bone size using a size-index scaling method (Uerpmann, 1979). Since then, this approach remained important when dealing with small samples of measurable animal remains (Meadow, 1999). At the time Uerpmann's study was published, however, the Upper Euphrates basin was still poorly investigated and generally considered of marginal importance relative to the core regions of the 'Neolithic Revolution'. Later it was realized that the region may have played a major role in the Neolithization process. Key archaeofaunal data for North Syria come from the sites of Mureybet, Abu Hureyra, Jerf el-Ahmar, Dja'de, Cheikh Hassan and Halula (Helmer *et al.*, 1998; Saña Seguí, 1999; Legge and Rowley-Conwy, 2000), and in Southeast Anatolia from the sites of Hallan Çemi, Çayönü,

Göbekli Tepe, Nevalı Çori, Mezraa Teleilat and Gürcütepe (Figure 4.1) (Rosenberg *et al.*, 1995; Hongo and Meadow, 1998; Peters *et al.*, 1999, 2005b; Peters, von den Driesch and Helmer, 2005a; Ilgezdi, 2008). The archaeological contexts considered in this study cover a timespan of more than 3000 years, from the end of the eleventh until the end of the eighth millennium calBC. Whereas most sites represent habitation areas with essentially domestic structures, excavations at Göbekli Tepe revealed a different architecture, charaterized by large curvilinear (earlier occupation) to rectangular (later occupation) enclosures lined by megaliths in the form of 2 to 5 m high, T-shaped stone pillars. More than 50% of these pillars show animal motifs in bas-relief (Schmidt, 2000; Peters and Schmidt, 2004; Peters *et al.*, 2005b). Snakes, red fox and wild boar are the species most commonly figured, but the repertoire also includes aurochs, goitered gazelle, mouflon, Asiatic wild ass, common crane and vulture as well as scorpion, beetle and other (in)vertebrates. Given the monumental aspect of its architecture and art, Pre-Pottery Neolithic (PPN) Göbekli Tepe likely served as a cult centre of supraregional importance (Schmidt, 2000). In this chapter, we will focus our discussion on three sites, Göbekli Tepe (Pre-Pottery Neolithic A (PPNA) and (Early Pre-Pottery Neolithic B(EPPNB), Nevalı Çori (Early and Middle PPNB (EPPNB/MPPNB)) and Gürcütepe (Late PPNB (LPPNB)), which are located in close proximity to each other (Figure 4.1), and their occupational phases provides a solid diachronic sequence documenting the Neolithic transition in this area.

Traditional hypotheses explaining the success of the neolithization process underline the enhanced security and predictability relative to food procurement brought about by the nutritional complementarity and productivity of the crop-livestock combination (Harris, 2002). Nevertheless, the initial stages of stock-keeping and breeding may have been characterized by drawbacks, for instance due to nutritional or health problems in animals that were raised in captivity. Major changes in human lifestyle and nutrition can be routinely assessed by the analysis of botanical and faunal remains recovered during excavation, but the picture of the human and animal palaeodiets becomes clearly more refined with the study of stable carbon and nitrogen isotope analyses of human bone collagen ($\delta^{13}C$ and $\delta^{15}N$, respectively). Collagen isotope composition relates primarily to the protein part of the diet, permitting the reconstruction of the average contribution of plants, animals, freshwater and marine fish to the daily diet on both individual and population levels (for a review, see Ambrose, 1993). Analysis of the animal bone finds associated with the human remains is a prerequisite for defining the trophic level of humans, which are generalist and opportunistic feeders, and for the quantification of the relative contribution of these food sources to the human diet by establishing mixing models (as introduced by Schwarcz, 1991). Unfortunately, the burial environments at the sites are not favourable for collagen preservation. Collagen could be retrieved only from bone samples collected at Nevalı Çori (but for less than 50% of the specimens studied). No more than a few single bones from the other two sites could be analysed in terms of collagen stable isotopic ratios, and hence the food web reconstruction for this period and region had to be predominantly based on the Nevalı Çori study. However, stable carbon and oxygen isotopic ratios from the bone structural carbonate ($\delta^{13}C$ and $\delta^{18}O$) could be measured from almost every bone and tooth find from all three sites and permitted the evaluation of the percentage of C_3- and C_4-plants in the diet of animals and humans which, as will be outlined below, provides essential information for the reconstruction of the local palaeoenvironment and the impact humans exerted on the landscape. In addition, carbonate $\delta^{18}O$ gave clues to the palaeoclimatic conditions prevailing at that time (see below). In total, 599 human and animal bones have been analysed for stable carbon and oxygen isotopic ratios in their structural carbonate, 258 from Göbekli Tepe, 204 from Nevalı Çori and 137 from Gürcütepe (Table 4.1). All carbonate

Table 4.1 Species and number of individuals per species analysed, and average stable isotopic ratios in the bone collagen and bone structural carbonate from the Anatolian sites of Göbekli Tepe, Nevalı Çori and Gürcütepe

	Göbekli Tepe			*Nevali Cori*						*Gürcütepe*		
species	no of individuals	$\delta^{13}C_{carb}$	$\delta^{18}O$	no of individuals	$\delta^{13}C_{carb}$	$\delta^{18}O$	no of individuals	$\delta^{13}C_{coll}$	$\delta^{15}N$	no of individuals	$\delta^{13}C_{carb}$	$\delta^{18}O$
asiatic wild ass	26	–10.9	–4.9				1	–20.4	7.2	1	–12.7	–8.9
aurochs	14	–10.9	–7.2	25	–11.4	–6.6	4	–20	8.2	1	–10.7	–8.7
badger				4	–12.8	–6.6	3	–18.6	9.5			
black kite	1	–12.6	–5.5									
black vulture	2	–10.6	–6.9									
catfish	1	–11.5	–13.3									
carrion crow	1	–13	–3.2									
cattle										18	–9.8	–8.7
common chukar	7	–11.5	–6.6									
crane	7	–11.1	–6.5									
demoiselle crane	4	–9.7	–5.4									
dog				2	–12.8	–8.5	2	–19.2	8.9	7	–10.8	–8.6
eagle owl	1	–12.4	–5.9									
fallow deer				8	–12.2	–6.9						
freshwater clam	6	–9.4	–6.2									
goitered gazelle	14	–11.6	–5.7	23	–11.6	–6.1	7	–20	6.6	8	–11.8	–4.7
golden eagle	1	–10.2	–5.2									
great bustard	4	–11.7	–5.9									
grey heron	2	–12.5	–8.4									
greylag goose	1	–11.4	–7.9									
grey partridge	2	–11.2	–4.9									
goshawk	7	–11.6	–6.8									
griffon vulture	5	–11.4	–6.6									
cape hare	6	–12.7	–6.4	22	–12.8	–6.9	7	–21.4	4.8			
hedgehog				1	–13.5	–5.1						
hooded crow	20	–12	–6.8									
human	21	–12.5	–8.4	17	–12.2	–7.3	42	–20.6	6.1	32	–10.4	–7.3
human infant				3	–13.3	–7.7	3	–20.7	8.5	1	–11.1	–8
jackal				1	–14.2	–6.2						
jackdaw	12	–12.9	–5.4									

(Continued)

jay	1	−11.8	−7.3									
kestrel	1	−13.4	−5.4									
leopard	2	−11.2	−8.4	3	−12.4	−7.4						
long legged buzzard	4	−11.6	−7.1									
pig				6	−12.3	−8.6	8	−20.3	5.5	20	−11.8	−9.1
pig?				2	−12.3	−7.2						
mallard	10	−12.7	−9									
Monk vulture	1	−11	−6									
Montagu's harrier	3	−12	−7.3									
raven	3	−12.1	−6.5									
red deer	4	−12	−7.1	19	−12.3	−7.6	5	−20.9	6.9	1	−13.6	−8.5
red fox	10	−12.8	−6.9	16	−12.6	−7	3	−21.1	9.6	1	−12.5	−4.1
red kite	1	−12.1	−5.6									
rook	13	−12.6	−6.2									
ruddy shelduck	2	−12	−9.1									
sand cat							1	−18.9	7.4			
sheep/goat										4	−10.9	−7.3
stock dove	1	−11.4	−5									
thrush	1	−11.8	−8.1									
whitefronted goose	1	−8.6	−6.9									
white stork	2	−12	−5.7									
tortoise	2	−11.8	−7									
white tailed eagle	2	−11.4	−5.5									
wild boar	11	−12.1	−8.4	22	−12.2	−7.5				1	−11.5	−7.5
wild boar?				3	−12.5	−8.5						
wild cat	4	−12.5	−7	3	−12.6	−6.6	1	−18.2	8.4			
wild goat				7	−11.8	−6.3						
wild goat?				4	−11.8	−7.1						
domesticated goat?				1	−11.4	−14.1	5	−20.4	6.3			
domesticated goat										18	−10.7	−6.7
wild sheep	8	−12	−5.4	3	−12.3	−7.8						
wild sheep?				3	−11.5	−5.3						
domesticated sheep?				3	−11.6	−7.7						
domesticated sheep							5	−20.3	6.1	24	−10.8	−8.4
wolf	3	−12.5	−6.9	3	−11.3	−7.3						
wood pigeon	2	−11.8	−7.9									
no of samples	**257**			**204**			**97**			**137**		
total	**696**											

stable isotope ratios are expressed against the *PDB* standard. Since the bone structural carbonate is rather susceptible to *post mortem* diagenesis, not only bone, but also tooth enamel was explicitly sampled in the course of the recent excavation campaigns. Due to its crystalline structure and the lack of considerable amounts of organic components, tooth enamel is regarded as much more resistant to diagenetic alteration (Lee-Thorp, 2002), an assumption which, however, should be tested for every skeletal series under study. Since rather far reaching implications for the Neolithization process were especially derived from the $\delta^{13}C$ values in the bone structural carbonate (see below), δ-values in bone were compared to those measured in dental enamel of vertebrate species to make sure that the isotopic variability was the same in both mineralized tissues.

4.2 WERE THE FIRST FARMERS VEGETARIANS? THE FOOD WEB AT NEVALI ÇORI

Nevalı Çori constitutes one of the earliest Pre-Pottery Neolithic settlements in the Upper Euphrates basin, with evidence for livestock husbandry. Site habitation probably started at the time of the transition from the PPNA to the PPNB (Lösch, Grupe and Peters, 2006). Given the fact that stock on the hoof still constituted a minor component in the diet compared to game species, it can be assumed that the 'Neolithic package' was still in its infancy during the Early PPNB (8700–8200 calBC) and at the beginning of the Middle PPNB (8200–7500 calBC). Based on stable carbon and nitrogen isotopic ratios of bone collagen, a rather detailed food web could be established. Surprisingly, the adult humans appeared to be nearly exclusively herbivorous, with a mean collagen $\delta^{15}N$ of 6.1‰ indistinguishable from those of, for example, goats and gazelles (Figure 4.2). Only three human infants who, based on their age, should still have been breastfed, are located on a higher trophic level (mean $\delta^{15}N = 8.50‰$). Three human bones from Göbekli Tepe (n = 2) and Gürcütepe (n = 1) had $\delta^{15}N$ values between 8 and 10‰, clearly indicative of a higher trophic level and hence a more regular meat supply. If this small sample is representative, then it suggests that the hunter-gatherers of Göbekli Tepe and the farmers of Gürcütepe consumed much more animal protein than the early farmers at Nevalı Çori, since $\delta^{15}N$ values between 8 and 10‰ match the isotopic ratios measured in omnivores and carnivores at the latter site (Figure 4.2).

Traditional hypotheses explaining the success of the Neolithization process underline the enhanced security and predictability of meat supply, brought about by the domestication of sheep and goat, but in hindsight this perspective may be flawed (Diamond, 2002). It is also contradicted to a certain extent by the archaeological and ethnographic record from other regions of the world, where the transition from a foraging way of life to agriculture was accompanied by nutritional shortages and a higher disease burden (Cohen and Armelagos, 1984; Cohen and Crane-Kramer, 2007; Diamond, 1992). However, at PPNB Nevalı Çori, a representative of an early farming community in the Fertile Crescent, Teegen and Schultz (1997) related a decreasing rate in human palaeopathological conditions from early to later phases to improved dietary conditions due to better mastering of cereal tillage and livestock husbandry techniques. According to our isotopic analyses, meat supply from early domesticates seem to have played only a minor role in the economic spectrum of these societies as these early domesticates were too few and probably too precious to provide a stable subsistence base. Our results therefore reinforce the idea that during the initial stages of food production, the 'Neolithic Revolution' may have been a time of experimentation with uncertain

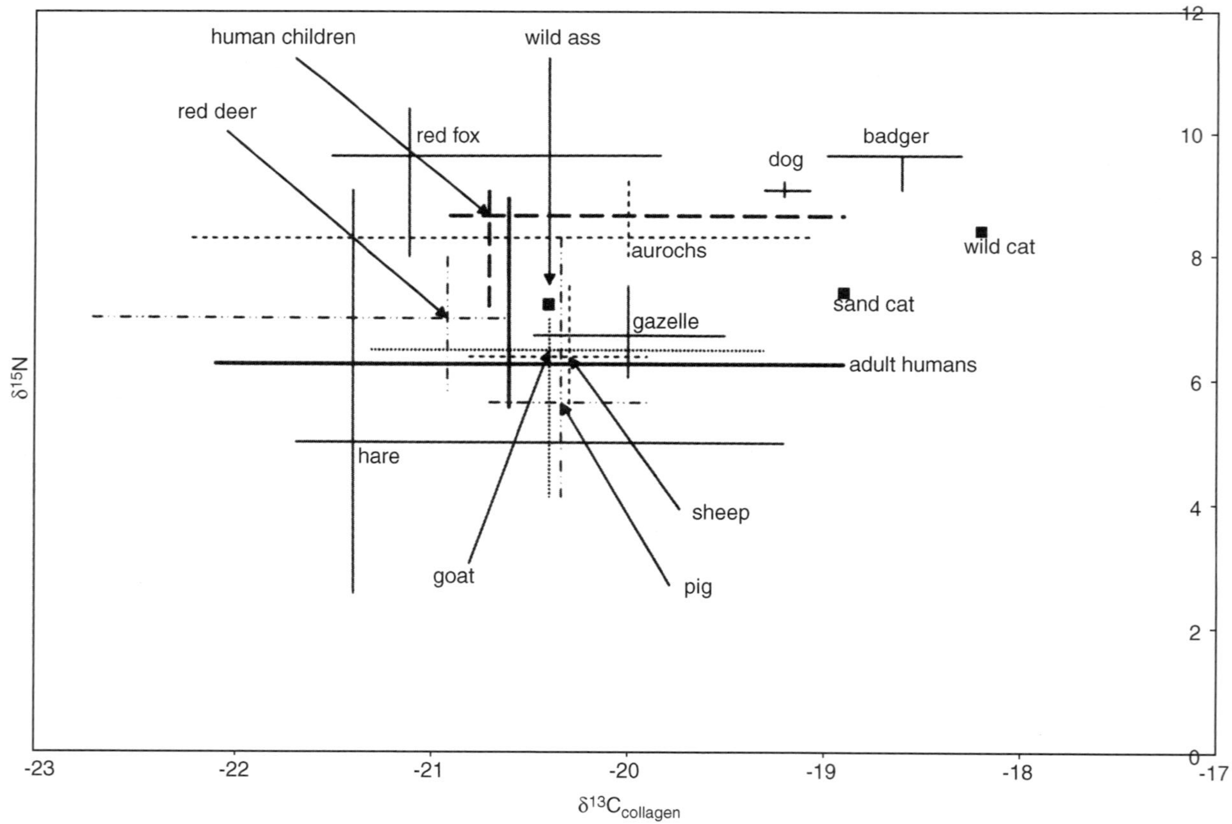

Figure 4.2 Food web of the Nevalı Çori site, based on stable isotope analysis of collagen ($\delta^{13}C$, $\delta^{15}N$) of animal and human bone finds (cf. Lösch, Grupe and Peters, 2006)

outcome and therefore a slower gradual transition, and the site of Nevalı Çori has preserved a rare snapshot of this process (Lösch, Grupe and Peters, 2006). At this site, the full reliance on livestock and plant domesticates was preceded by a transitional period, which probably involved experimentation, the partial dependency on domesticates, and a considerable variability in the local range of subsistence strategies (cf. Roberts, 2002).

In addition, we were able to show for the first time that animals considered to represent early domesticates (sheep, goat and pig) lived on a diet that had a different isotopic composition than that of their free ranging, wild relatives. It thus suggests that these early domesticates were already fed by the Neolithic agriculturalists (Figure 4.3). The variability of adult pig δ^{15}N-ratios was especially high, covering almost two full trophic levels, implying a much larger dietary spectrum compared to wild boar. Only a single, presumably domesticated goat was available from the Nevalı Çori faunal assemblage which, however, had an isotopic signature different from the ones recovered in bone remains from wild *Capra* specimens. Conversely, early domestic sheep had access to a less varied vegetation cover than their progenitors roaming the hills of the southern Anti-Taurus. This could offer an explanation for the observed decrease in body mass observed in this species (Peters, von den Driesch and Helmer, 2005a). Also, domesticated animals have, in general, lower collagen δ^{15}N-ratios, compared to their wild relatives (see Table 6 in Lösch, Grupe and Peters, 2006), which is conspicuous in terms of the considerable low δ^{15}N-ratios measured in the adult humans. This could indicate a proportionately high intake of pulses such as *Fabaceae* (plants with exceptional low δ^{15}N values due to their symbiosis with nitrogen-fixing bacteria) by humans, and the feeding of pulses and/or human food refuse to the animals. Pulses are very rich in protein, and they may have been cultivated even earlier than cereals (Pasternak, 1995, 1998; Miller, 2002). Remains of pulses are nearly as abundant as cereal grains in the archaeobotanical assemblage of Nevalı Çori. At least five species were identified, namely lentil *Lens* spp., field pea *Pisum* spp., grass-pea *Lathyrus* spp., bitter vetch *Vicia ervilia* and horse bean *Vicia faba* (Pasternak, 1995, 1998). Even today, pulses are still valuable as animal fodder. That humans at times had to compete over vegetable food with wild and domesticated animals could be confirmed by the carbonate analyses, which also would reveal how people might have coped with this problem.

4.3 δ^{13}C IN MAMMALIAN BONE STRUCTURAL CARBONATE: FOOD COMPETITION AND EARLY LANDSCAPE DEGRADATION

The carbon in the bone structural carbonate is derived from blood bicarbonate and reflects carbon from all dietary components (protein, carbohydrates, fat) in its isotopic composition (Ambrose, 1993). Relevant to our study is the fact that C_3-plants such as cereals exhibit much more negative δ^{13}C-values than C_4-plants. The latter do not thrive under close canopies, and prefer hot, arid habitats where they benefit from a selective advantage because of their more efficient water use, and which discriminate less efficiently against ^{13}C by the enzymes involved in photosynthesis. While C_3-plants typically have δ^{13}C-values around 27‰ – with considerable variation according to radiation, precipitation and woody cover-, variability is much less pronounced in C_4-plants, which exhibit δ^{13}C-values typically varying between –13 and –10‰ (Ambrose, 1993). This difference is maintained in the δ^{13}C of the consumer's bone structural carbonate. Therefore, one focus of interest is what the vegetation in the

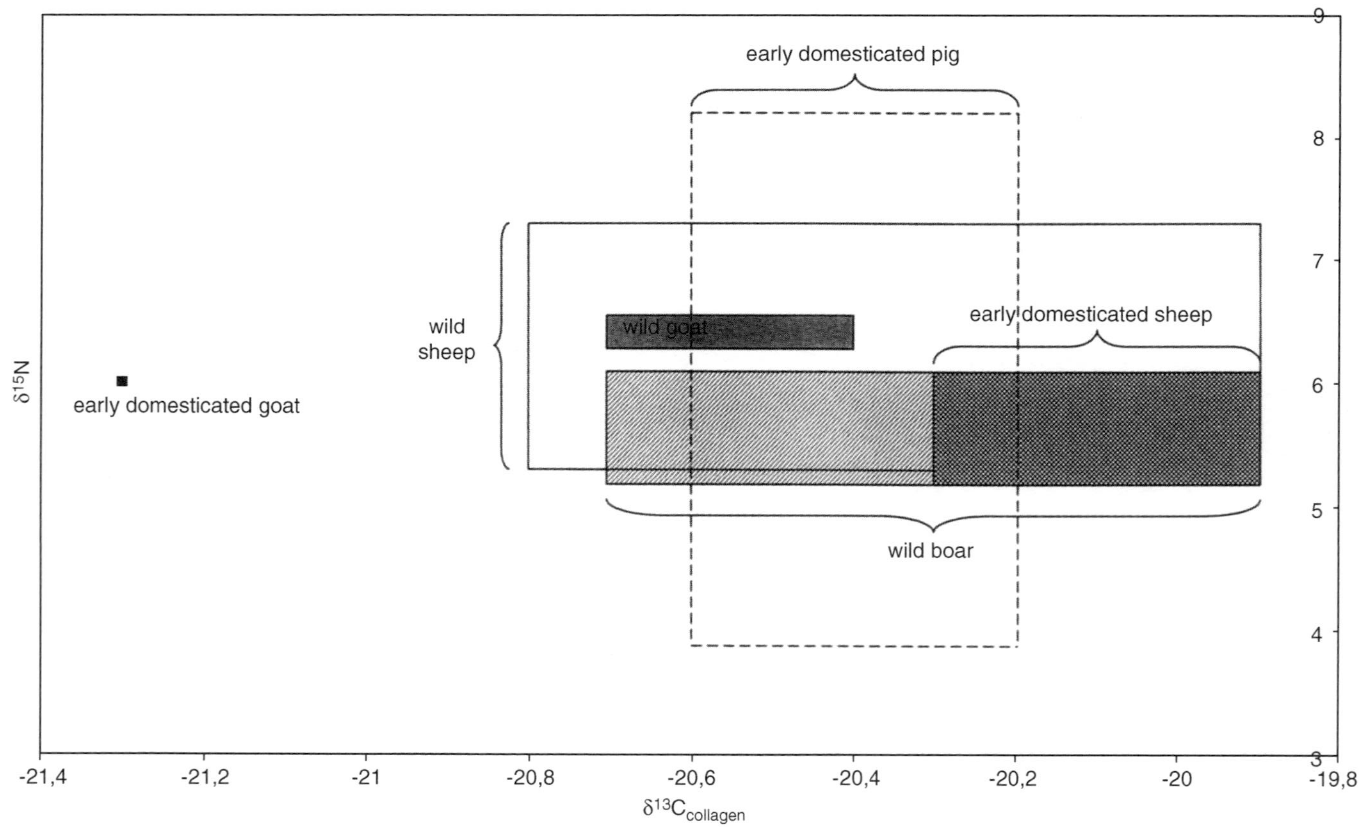

Figure 4.3 Variability of protein related stable isotopes in the bone collagen ($\delta^{13}C$, $\delta^{15}N$) of domesticated ungulates in comparison with their free ranging relatives from the Nevalı Çori site indicate a different dietary spectrum of the early domesticates

site environs looked like in Pre-Pottery Neolithic times and in particular whether C_4 plants were already present in the study area, and if so, which herbivore taxa took advantage of this.

In fact, the isotopic analysis indicates the statistical significance (an Analysis of Variance (ANOVA) analysis, confirmed by the non-parametric Kruskal-Wallis-test) of more positive $\delta^{13}C_{carbonate}$ values present in the bones of:

1. domestic cattle from Gürcütepe as opposed to wild aurochs from Göbekli Tepe and Nevalı Çori ($p = 0.001$);
2. humans from Gürcütepe compared to the humans from the two earlier sites ($p = 0.001$);
3. domestic sheep from Gürcütepe as opposed to wild sheep from the other two sites ($p = 0.01$);
4. domestic goats compared to wild goats and gazelles ($p = 0.0025$); and
5. of dogs from Gürcütepe as opposed to the canids and felids from the two other sites ($p = 0.01$).

Interestingly, no such changes to more positive $\delta^{13}C$-values in time and space were detectable in free ranging gazelles or deer (Figure 4.4).

With the onset of the Early Holocene, a phase marked by warmer and wetter climatic conditions than today (Robinson *et al.*, 2006), woodland vegetation expanded across the western and northern Fertile Crescent (Hillman, 1996). Accordingly, the $\delta^{13}C_{carbonate}$ values in the bones of the different herbivore species identified at Nevalı Çori and Göbekli Tepe are indicative of a C_3-biome, which is also supported by the archaeobotanical record (Pasternak, 1995; Neef, 2003). Plant and animal remains point to a mosaic of woodland and open areas dominated by annual grasses including wild cereals. At Late PPNB Gürcütepe, the situation is

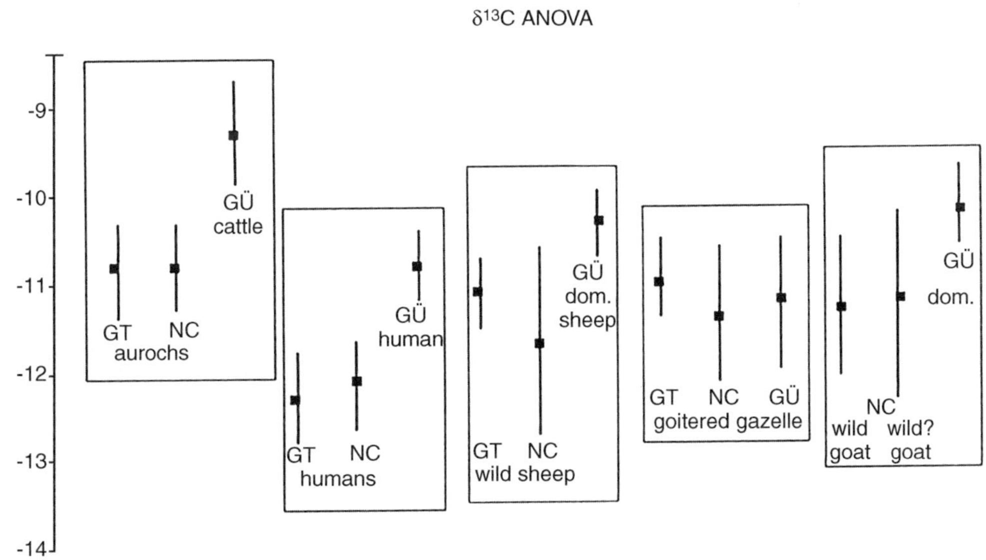

Figure 4.4 ANOVA results ($p \leq 0.05$) with regard to $\delta^{13}C$ in bone structural carbonate, group means with 95% confidence intervals. GT: Göbekli Tepe, NC: Nevalı Çori, GÜ: Gürcütepe. Data of dogs and wild ass are not shown.

very different. While the $\delta^{13}C_{carbonate}$ signatures of wild herbivores (gazelle, red deer, Mesopotamian fallow deer and Asiatic wild ass) still reflect an exclusive C_3-plant consumption, the bone specimens from domestic cattle, sheep, goat and dog exhibit more positive $\delta^{13}C$ values, reflecting the consumption of a certain quantity of C_4-plants (Figure 4.4).

Average vegetation $\delta^{13}C$ calculated from the signatures in wild herbivorous mammals is –23.3‰ at Göbekli Tepe (taxa: hare, Asiatic wild ass, red deer, gazelle, aurochs, wild goat), –24.2‰ at Nevalı Çori (taxa: hare, Mesopotamian fallow deer, red deer, gazelle, aurochs, wild sheep, wild goat) and –23.4‰ at Gürcütepe (taxa: Asiatic wild ass, red deer, gazelle, aurochs). It should be emphasized that this estimate can only be a gross approximation to the baseline plant $\delta^{13}C$ value, since animal species of different body size and metabolic peculiarities have been considered together. However, it is confirmed for Nevalı Çori, where the $\delta^{13}C_{carbonate}$ to $\delta^{13}C_{collagen}$ spacing could be controlled for, that such baseline values can be rather securely assessed from the carbonate analyses in herbivorous species (Lösch, Grupe and Peters, 2006). Since the average 'vegetation $\delta^{13}C$' values for the wild herbivores do not vary much between the sites, an average baseline value of about –23.5‰ for the C_3-plant cover between about 9000 to 7000 calBC is postulated for the study area, based on herbivore $\delta^{13}C_{carbonate}$ values (see above) and a fractionation factor of +12‰ (Lee-Thorp, Sealy and van der Merwe, 1989). With regard to the restricted variability of C_4-plant $\delta^{13}C$ (see above), we assume their average to be –11.5‰. Under the assumption of a linear mixing model, domestic cattle at Gürcütepe would have consumed on average 14% C_4-plants, with a maximum of 39% in the individual with the most positive $\delta^{13}C$-value (–6.8‰). Average C_4-plant contribution to the diet of small livestock must have been more moderate, approximately 9% in sheep and 7% in goats, occasionally reaching considerable percentages in single individuals, that is up to 18% in sheep (max. $\delta^{13}C_{carbonate}$ = –8.90‰) and 22% in goats (max. $\delta^{13}C_{carbonate}$ = –9.40‰). Even for humans and dogs, enriched $\delta^{13}C_{carbonate}$– values indicate an estimated C_4-plant contribution of about 9 and 6%, respectively. This is best explained by the consumption of the meat of domestic animals, and the dogs obviously had access to the leftovers of human meals. We conclude that even such a modest contribution of 10% or more of C_4-plants to the diet of livestock likely reduced competition over C_3-annuals, such as cereals, between humans and their flocks.

Although it is highly unlikely that *post mortem* diagenesis would have selectively affected only the bones of domesticated or free ranging animals respectively, the overall susceptibility of the bone structural carbonate towards diagenetic alteration – despite the application of appropriate sample processing protocols – deserves some consideration. In the course of the most recent excavation campaign at Göbekli Tepe, both enamel and bone specimens from individuals of the species *Bos*, *Equus* and *Gazella* were sampled and forwarded to analysis. $\delta^{13}C$-values in the bone structural carbonate fell entirely within the range of the respective values measured in dental enamel (*Bos taurus*: $\delta^{13}C_{enamel}$: –10.1 until –12.1‰, $\delta^{13}C_{bone}$: –10.9 until –11.5‰; *Equus hemionus*: $\delta^{13}C_{enamel}$: –9.6 until –12.2‰, $\delta^{13}C_{bone}$: –10.7 until –11.4‰; *Gazella subgutturosa*: $\delta^{13}C_{enamel}$: –11.2 until –13.5‰, $\delta^{13}C_{bone}$: –11.3 until –12.8‰, with the exception of a single, neonate bone exhibiting a more positive $\delta^{13}C$-value of –7.4‰). Therefore, we conclude that it is highly unlikely that the detected different feeding habits of domesticated and wild taxa with regard to the plant species consumed are an artefact of decomposition.

Interestingly, the investigation of the archaeobotanical remains from Gürcütepe has so far yielded no indication for the presence and/or human use of C_4-plants (R. Neef, DAI-Berlin, pers. comm.). However, the C_4-plant signature in the bones of domestic animals as opposed to wild herbivores implies a management of plant resources in the Upper Euphrates basin already at this early stage of mixed farming practises. Considering the probable geographic origin of the different livestock species in the study area, that is sheep and goat in the Anti-Taurus region (C_3-biome) and cattle along the Syrian Euphrates (C_3/C_4-biome) (Peters *et al.*, 1999; Peters, von den Driesch and Helmer, 2005a; Helmer *et al.*, 2005), it is plausible that sheep and goats were mainly pastured in the grassy, bushy foothills of the Anti-Taurus located north of the settlement, and that cattle keeping (and cereal tillage) took place closer to the settlement, that is in the lowlands of the present-day Harran Plain. Today, the Harran Plain represents the northernmost extension of the arid Syrian steppe and it is possible that in the Late PPNB, stands of C_4-plants were already thriving in the more sun-exposed locations within the site catchment. An additional explanation could be that the continuous presence of domestic ruminants near the settlement caused overgrazing and over-fertilization of the soils, resulting in a replacement of the C_3-vegetation by a C_4-plant cover. The repeated use of fire for clearing land may also have been advantageous for the spread of C_4-species (Bond, Woodward and Midgley, 2005). If the foregoing scenario applies, the C_4-plant component found in the diet of the early domestic ruminants would be indicative of a process of landscape degradation that became noticeable towards the end of the Pre-Pottery Neolithic B. The habitation at Gürcütepe had covered nearly 10 hectares, indicative of a rather large human community in need of adequate herds of livestock and pasture. Although still a little speculative at this point, landscape deterioration due to anthropogenic activities could be the key to our understanding why, after centuries of site occupation, this ecologically favourable setting had to be abandoned by its inhabitants.

4.4 PALAEOCLIMATE APPROXIMATION BY $\delta^{18}O$ IN THE BONE STRUCTURAL CARBONATE

Archaeological bone finds reflect palaeoclimates by stable oxygen isotope ratios of bone phosphate and bone structural carbonate (Aitken, 1991); however, the relationships are far from simple, let alone the taphonomic and diagenetic processes that may alter the original biological signal hidden in the isotopes (Grupe, 2007). In general, $\delta^{18}O_{carbonate}$ is related to the average temperature of surface water, which is meteorological water that went through the meteorological cycle of evaporation, condensation and precipitation. From a physiological viewpoint, it is plausible that bone remains of heterothermic animals are more suitable for climate reconstruction than those of thermoregulating vertebrates, although also the latter can provide valuable climate proxies (Kohn, 1996; Kohn and Cerling, 2002).

Oxygen isotope fractionation takes place during both evaporation and condensation, where ^{18}O is enriched in the liquid phase. Water vapour is therefore isotopically lighter than the ocean water it evaporated from, and rain water is isotopically heavier than the water vapour it has condensed from. With increasing distance from the coast, rain water becomes increasingly depleted in ^{18}O, a phenomenon known as the continental effect. While evaporation exceeds precipitation in the subtropical regions, the contrary holds for locations at higher distances from the equator. Continuous evaporation and rainfall lead to a steady depletion of $H_2{}^{18}O$ molecules in the vapour phase; at the same time, $\delta^{18}O$-values in precipitation also decrease (GNIP data bank, 2008). This latitude effect amounts to about 0.6‰ in temperate climates. Since precipitation is mainly caused by adiabatic cooling, that is without temperature exchange, continuing condensation processes are always due to lower air temperatures. This way, a relationship between air temperature and $\delta^{18}O$-values of precipitation becomes evident (Jouzel *et al.*, 1994). For the North Atlantic coast, a temperature gradient of 0.695‰ per centigrade has been observed (Dansgaard, 1964). Later investigations led to a temperature gradient $\Delta\delta^{18}O/\Delta T = 0.65 \pm 0.05‰$ per centigrade for the European continent (Rozanski, Araguas-Araguas and Gonfiantini, 1992), whereby the calculation of this temperature gradient was based on the Rayleigh equation (Gat, Nook and Meijer, 2000). Another effect that leads to oxygen isotope fractionation is the amount effect: especially heavy rainfalls, for instance the monsoon, are characterized by larger depletions of ^{18}O than expected from the temperature. Therefore, the amount effect is unrelated to temperature. In the course of palaeoclimatological approaches, these known spatial relationships between $\delta^{18}O$ and surface temperature are transferred into time in as much as differences in $\delta^{18}O$-values between different archaeological strata should be indicative of a change of overall climatic conditions.

The idea of reconstructing palaeotemperatures by the analysis of stable oxygen isotopes of fossil marine organisms dates back into the 1940s (Urey, 1947). In fact, the reversible precipitation of calcium carbonate from watery solutions ($Ca^{2+} + 2\,HCO_3^- \leftrightarrow CaCO_3 + CO_2 + H_2O$) is part of a system of equilibrium reactions, which finally lead to a fractionation of ^{16}O and ^{18}O from meteoric water ($CaCO_3 + H_2{}^{18}O \leftrightarrow Ca^{18}OCO_2 + H_2O$). The temperature dependency of the fractionation factor results from the temperature dependency of the equilibrium constant. With rising temperature, fractionation becomes less and approximates $\alpha = 1$ (Kim and O'Neil, 1997, Stosch, 2004). Likewise, oxygen isotope exchange with water happens in crystalline calcium phosphates such as hydroxyapatite, the main constituent of bone and teeth ($Ca_5(PO_4)_3OH + H_2\underline{O} \leftrightarrow Ca_5(PO_4)_3\underline{O}H + H_2O$; Longinelli, 1984, Kohn and Cerling, 2002). Again, a relationship exists between the precipitation temperature and the difference of the $\delta^{18}O$-values of apatite and water ($T(°C) = 111.4–4.3\ (\delta_{phosphate} - \delta_{water})$, (Longinelli and Nuti, 1973, Barrick, Fischer and Showers, 1999)). Stable oxygen isotope ratios in fossil carbonates and phosphates are therefore suitable for the reconstruction of the temperature at the spot of biomineralization. However, for a precise evaluation of palaeotemperatures, knowledge about the isotopic composition of the surface water would be a prerequisite. Nonetheless, temperature changes with time could be detectable if dealing with archaeologically stratified finds. It has to be kept in mind, though, that biomineralization is catalyzed by a variety of proteins and enzymes; hence isotope fractionation effects accompanying this complex metabolic pathway cannot be excluded. However, it has been shown (Grossman, 1984; and Bond *et al.*, 1993), that climate proxies established from marine fossil foraminifera agree very well with those obtained from ice cores, suggesting that biological effects might be of minor importance and not subjected to changes through time.

To assess the palaeotemperature in the Upper Euphrates basin, six shell fragments of the freshwater clam *Unio cf. tigridis*, and two carapace fragments of spur-thighed tortoises (*Testudo graeca*) from Göbekli Tepe were analysed. The $\delta^{18}O$ values in the carbonate of the *Unio* shell fragments varied from –5.9 to –6.9‰ with an average of –6.2‰. The basic isotopic variability in the region today can be estimated by the 'Online isotopes in precipitation calculator' (OIPC www.waterisotopes.org) which reveals a mean annual $\delta^{18}O$ value in meteoric water of $-6.8‰_{SMOW}$ with a potential variability of up to 6‰ in $\delta^{18}O_{\text{meteoric water}}$ ratios in samples from the higher mountain slopes and from the Syrian steppe. These freshwater clams can be assumed to be in isotopic equilibrium with the surrounding water body, and the mean $\delta^{18}O$ of the clams is only 0.6‰ more positive than today's average $\delta^{18}O_{\text{meteoric water}}$. The two tortoise carapaces had remarkably different $\delta^{18}O_{\text{carbonate}}$ values of –8.3 and –4.2‰, respectively. While the relationship between oxygen isotopic ratios of bone phosphate ($\delta^{18}O_{ph}$) in the carapace of turtles of the genus *Emys*, and the respective $\delta^{18}O$ in meteoric water ($\delta^{18}O_{mw}$), $\delta^{18}O_{mw} = 1.01 \times \delta^{18}O_{Ph} - 22.3$ [*SMOW*], has been experimentally established (Barrick, Fischer and Showers, 1999), no such equation exists for the terrestrial genus *Testudo*. Hence in our attempt to recover a palaeoclimate proxy by use of the carapace remains, we followed Barrick, Fischer and Showers (1999) in so far as bone growth should preferably take place at the upper margin of the generally broad variability of the body temperature of this heterothermic animal, and that bone precipitation would be restricted to a rather narrow temperature range. In analogy to the investigations of the genus *Emys*, we also assumed an average growth temperature for the *Testudo* carapace of around 32°C. According to Stosch (2004), a relationship exists between the precipitation temperature T of calcite and the δ^{18}O-values of the calcite and water, respectively: $T(°C) = 13.85–4.54 \times (\delta_c\text{-}\delta_w) + 0.04 \times (\delta_c\text{-}\delta_w)^2$. It is noteworthy that the numerical constants only slightly differ from the palaeotemperature equation originally established by Epstein *et al.* (1953) (Erez and Luz, 1983; Staudenbauer, 2008). Under the assumption of T (°C) = 32, and given the stable oxygen isotopic ratios of the two *Testudo* individuals, a $\delta^{18}O_{body\ water}$ of –4.33 and –0.33‰, respectively, would result. Given the relation of $\Delta\delta^{18}O/\Delta T \sim 0.7$‰ (Rozanski, Araguas-Araguas and Gonfiantini, 1992), the meteoric water of the tortoise's environments would have differed in temperature by 5.8°C. In contrast to the *Unio* shells, to which no systematic sampling had been applied and for which the δ^{18}O reflects the average temperature of the water body, carapace growth in tortoises occurs during the warmer months of the year. The two palaeoclimate proxies obtained from the *Testudo* specimens from Göbekli Tepe fall well within the range of variability of meteoric water δ^{18}O in the region today (see above). The reconstruction of definite temperature ranges for the past is contingent upon the acceptance of various presuppositions. Nevertheless, inter-individual comparison in time and place can reveal climatic changes, and the reconstruction of important palaeoenvironmental parameters.

Such a palaeoclimate approximation based on δ^{18}O in fossil and subfossil vertebrate remains could be of particular importance in tropical regions, since climate archives are very restricted in such environments. For instance, tropical trees do not preserve climate data in the form of tree rings, and 99% of the tropical glaciers are located in the Andean mountains (Thompson *et al.*, 2003), and so ice cores are not widely available. In a recent study (Staudenbauer, 2008), we tried to reconstruct changing palaeoclimates by bone structural carbonate analyses of carapace remains of 69 specimens of the leopard tortoise (*Geochelone pardalis* Gray, 1873), recovered from the Omungunda (99/1) site in the Cunene District of northwest Namibia. The archaeological strata at Omungunda (99/1) are exceptionally deep, covering the entire Holocene and beyond. It is the first and still only stratigraphy in the Cunene District that even reaches as far back as the final Pleistocene (Vogelsang, 2002). Based on radiocarbon dates obtained on charred wood, the exceptional stratigraphy covers the time spanning from 15 560 calBC until historical times. The bone finds were grouped into eight phases (phase 1 = historical times, phase 3 = 160–460 calAD, phase 4 = 20–80 cal AD, phase 5 = 380–230 calBC, phase 7 = 4590–4070 calBC, phase 8 = 7000–6580 calBC, phase 9 = 12 980–12 140 calBC, and phase 10 = 15 560–15 400 calBC) (Vogelsang, pers. comm.), and despite considerable variations, average δ^{18}O-values of the carapace samples differed significantly between phases (Kruskal-Wallis-test), decreasing from the deepest earlier to the topmost, latest archaeological level (Figure 4.5). A causal relationship between solar insulation and the periodical change between ice ages and interglacial periods has been suggested (Berger, 1988; Wang *et al.*, 2004; Thompson *et al.*, 2006), and the temperature change depicted from the tortoise bones in fact matches the changes in insulation (%) observed for the respective periods (Staudenbauer, 2008). $\delta^{13}C_{carbonate}$ analyses of the carapace remains revealed that parallel to a change in climate, the percentage of C_3-plants in the animals' diets increased with

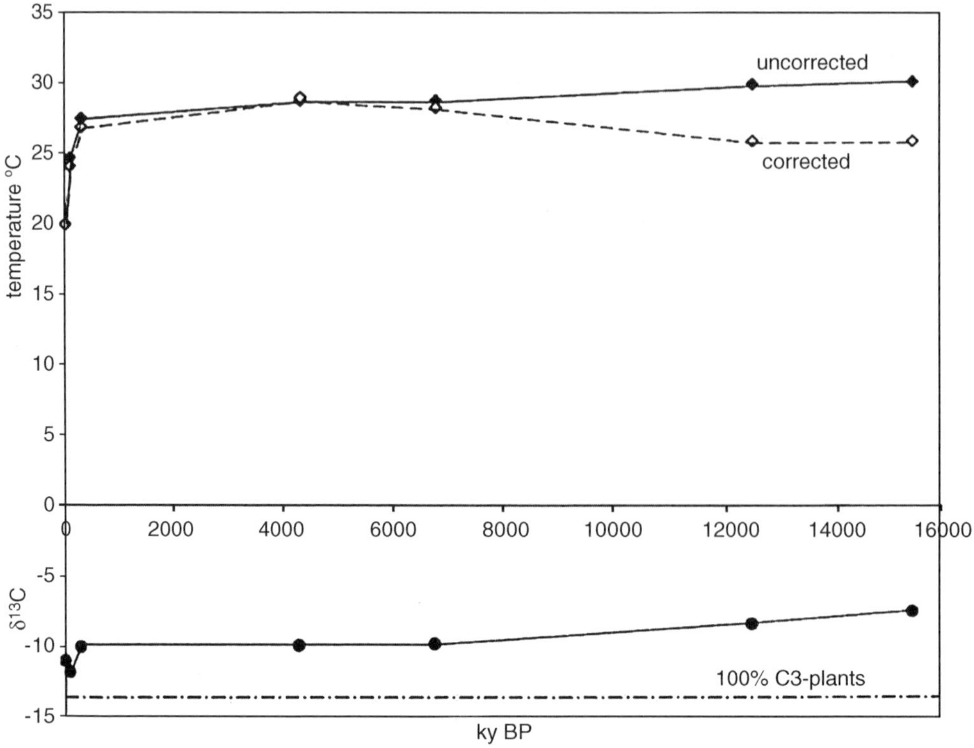

Figure 4.5 Diachronic temperature change at the Namibian site from the final Pleistocene until historical times, reconstructed by use of carapace remains of the leopard tortoise. Note that lower temperatures are accompanied by a larger proportion of C_3-plants in the animals' diet. Uncorrected data: based on the $\delta^{18}O$-values in the bone structural carbonate. Corrected data: the ice volume residual was taken into account, that is the depletion of ocean $\delta^{18}O$-values due to the melting continental ice shields at the turn from the Pleistocene to the Holocene (Holmgren *et al.*, 2003)

time. This is explained by the shift to cooler climates where C_3-plants have a selective advantage over C_4-plants (see above). Besides the lower temperatures, the slow but constant rise of atmospheric CO_2 in the course of the Holocene (Petit *et al.*, 1999; Thompson *et al.*, 2006) should also have been disadvantageous for C_4-plants, since at higher CO_2-concentrations, photorespiration of C_3-plants is depressed and the competitive advantage for C_4-plants is lost.

Finally, since skeletal remains of heterothermic animals are infrequent within the SE-Anatolian and Syrian archaeofaunal assemblages considered in this study, we tried to assess palaeoclimate proxies by measuring $\delta^{18}O$ in the structural carbonate of bones of a thermoregulating species – the goitered gazelle (*Gazella subgutturosa*) (Table 4.2). Gazelle finds from the three Anatolian sites mentioned above were studied together with gazelle bones from the Pre-Pottey Neolithic Syrian sites of Mureybet and Dja'de al Mughara (Figure 4.1). The 43 gazelle individuals (the same skeletal element had been sampled to avoid duplicate analysis and to exclude the possibility of measuring differences in temperature relating to the anatomical position in the body) from Mureybet were selected from the 8 cultural phases spanning the period between 12 000 to 7500 calBC. Gazelles are selective browsers and will feed on leaves that are enriched with ^{18}O because of preferential loss of ^{16}O through

Table 4.2 Number of individuals, archaeological stratification and stable isotopic data of the bone structural carbonate of gazelle bones (*Gazella subgutturosa*) from Syrian and Anatolian sites

site	date	no of samples	mean $\delta^{13}C$	1 sd	mean $\delta^{18}O$	1 sd
Mureybet	IA final Natufian 12 000–10 000 BC	7	−9.8	1.2	−0.6	1.4
	IB transition Natufian/Khiamian. 12 000–10 000 BC	5	−10.5	0.8	−2	0.7
	IIA Khiamian 10 000–9500 BC	5	−10.2	0.2	−1.5	0.5
	IIB Khiamian 10 000–9500 BC	5	−10.4	0.7	−1.1	0.8
	IIIA Mureybetian 9500–8700 BC	5	−10.8	0.5	−1.2	1.2
	IIIB Mureybetian 9500–8700 BC	5	−10.2	0.7	−2.2	0.5
	IVA early PPNB 8200–7500 BC	6	−10.7	0.4	−2	0.8
	IVB early PPNB 8200–7500 BC	5	−10.7	0.6	−2.3	0.6
Dja'de al Mughara	8700–8500 BC	6	−10.9	0.5	−4.1	0.7
Göbekli Tepe	9000–8500 BC	21	−11.4	0.8	−6.9	2.9
Nevalı Çori	8700–8000 BC	22	−11.8	0.9	−6.2	2.3
Gürcütepe	7500–7000 BC	8	−11.8	0.6	−4.7	2.1

the leaf stomata. Therefore, gazelle $\delta^{18}O_{carbonate}$ values are rather positive in general. Thermoregulation in gazelles, moreover, is brought about by panting, whereby preferentially $C^{16}O_2$ will be exhaled. In sum, the overall enrichment of gazelle bone $\delta^{18}O_{carbonate}$ should result from a combination of ecological, behavioural and metabolic features. A few more negative gazelle $\delta^{18}O$ values from Nevalı Çori do not necessarily contradict this preliminary interpretation, since these individuals need not have resided at higher elevations in the Anti-Taurus but may have utilized low temperature surface water from fast-running downhill creeks at the foot of the mountains.

As such, the assessment of palaeoclimatic conditions from the skeletons of thermoregulating animals is more complicated than the similar reconstruction based on reptiles, for example, since in order to monitor oxygen input and output, not only metabolic peculiarities and feeding preferences have to be taken into account, but also environmental parameters such as humidity. For East African gazelle species, Kohn (1996) established the following equation relating $\delta^{18}O_{phosphate}$ to the temperature sensitive $\delta^{18}O_{surface\ water}$, which was not specified in terms of species but should hold for the genus in general: $\delta^{18}O_{Ph} \sim 30.4\text{–}12.9h + 0.68{*}\delta^{18}O_{surface\ water}$, with h = humidity ranging from 0 to 1. He thereby emphasized that in drought-tolerant species such as gazelles, the dependency on humidity is much stronger than the dependency on meteoric water composition. Since it is impossible to reconstruct average humidity during the lifetime of the gazelle bone finds, an actualistic approach is necessary. At Aleppo, the modern hydrological station next to the Syrian sites, yearly relative humidity averages are 0.58 (i.e. 58%) with the lowest percentage in June (0.37), and the highest in December (0.82). Stable oxygen isotopes in meteoric water ($\delta^{18}O$ ‰$_{SMOW}$, www.waterisotopes.org) average –6.3‰, with highest values (–2.4‰) in August, and lowest values in January (–8.3‰). Converting the δ-notations ($SMOW = 1.03086 \times PDB + 30.86$), and $\delta^{18}O_{phosphate}$ into $\delta^{18}O_{carbonate}$ ($\delta^{18}O_{carbonate} = \delta^{18}O_{phosphate} + 8.7$‰, Bryant *et al.*, 1996), and applying the equation by Kohn (1996), expected $\delta^{18}O_{carbonate}$ in gazelle bones would be –3.2‰, assuming both average $\delta^{18}O$ values in meteoric water and average humidity. Such an assumption is readily justified since the structural carbonate in the bone apatite has a biological half-life of several years in larger mammals, and again, the isotopic data obtained from the archaeological bones are fully compatible with the overall climatic conditions in the region. Given the (unlikely) extremes of (1) maximum humidity and minimum $\delta^{18}O_{meteoric\ water}$ (winter season), and (2) minimum humidity and maximum $\delta^{18}O_{meteoric\ water}$ (summer season), expected $\delta^{18}O_{carbonate}$ of gazelles in this area may vary between –7.3 and +2.0‰. An ANOVA analysis applied to the $\delta^{18}O_{carbonate}$ values measured in bones from Syrian gazelles with those in the animals hunted by the inhabitants of the Anatolian sites (see above), revealed statistically highly significant differences ($p = 0.001$). It should be noted, however, that four individual measurements that resulted in $\delta^{18}O$-values of less than –10‰ (three individuals from Göbekli Tepe and one gazelle from Nevalı Çori) were excluded from the variance analysis, because these outliers are probably indicative of a very different source of water exploited by these individuals. One possible explanation is that these individuals acquired their drinking water from fast running creeks originating in the high mountains, which transported isotopically lighter meltwater. The climate sensitivity of $\delta^{18}O$ in the bone structural carbonate of the archaeological gazelle bones is illustrated in Figure 4.6. The isotope signatures reveal both a

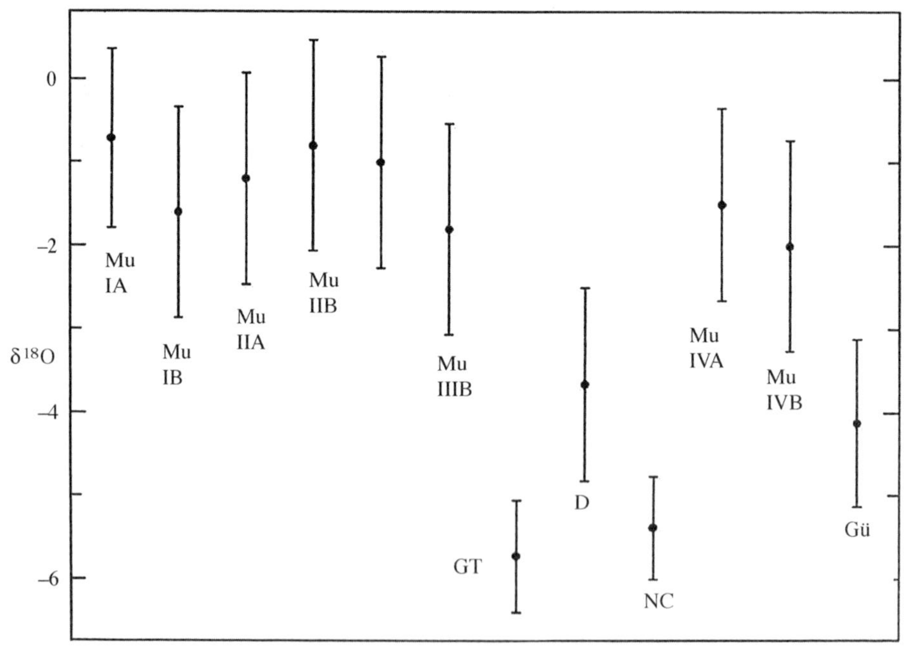

Figure 4.6 Results of a variance analysis of $\delta^{18}O$-values in the bone structural carbonate of archaeological finds of the goitered gazelles from Syrian and Anatolian sites. Mu: Mureybet, phases IA until IVB cf. Table 4.2; D: Dja'de al Mughara; GT: Göbekli Tepe; NC: Nevalı Çori; Gü: Gürcütepe

diachronic trend at one site (Mureybet) and clear allopatric trends in the other sites, indicative of the different environmental conditions to which the goitered gazelles were adapted. The actualistic comparison with the climatic indicators obtained from the Aleppo station further confirms the climate approximation based on the bones from gazelles retrieved from the archaeological occupational phases.

4.5 CONCLUSIONS

The reconstruction of climate proxies by means of stable oxygen isotope analysis in the bone structural carbonate of archaeological vertebrate finds necessitates a number of presuppositions and represents an actualistic approach. However, under the prerequisite that climate approximations are performed under controlled experimental conditions and by using the same vertebrate species and skeletal element, inter-individual and inter-group comparison will give clues to variability in palaeoclimatic conditions as well as changes in space and time. Since $\delta^{18}O$ and $\delta^{13}C$ in the bone structural carbonate are measured together, palaeoclimate proxies are supplemented with valuable palaeodietary information, which in turn are related to the floral spectrum that prevailed in the past. It could be shown for a restricted geographical area in the northern Fertile Crescent that in the course of the Neolithic transition, early farmers took advantage of the advancing C_4-plants, which are mostly unsuitable for human consumption but rather constitute valuable animal fodder, to reduce food competition between domestic stock and the owners of the animals. Since the results of isotopic analyses of a variety of vertebrate species are compatible with the dietary preferences and spatial distribution of extant animals, information on herding practices (e.g. in the vicinity or at a distance from the human settlements) can be deduced. On the basis of our isotopic data, it cannot be excluded that already at a very early stage of farming, overgrazing might have constituted a threat to the local palaeoenvironment. Plant resource management may have been one possibility to cope with the problem at Late PPNB Gürcütepe, but in the long run there might have been but one solution, namely site abandonment. Settlement at Gürcütepe likely ended at the turn of the eighth to the seventh millennium calBC.

REFERENCES

Aitken, M.J. (1991) *Science-Based Dating in Archaeology*, Longman, London, New York.

Ambrose, S.H. (1993) Isotopic analysis of paleodiets: methodological and interpretive considerations, in *Investigations of Ancient Human Tissue: Chemical Analyses in Anthropology* (ed. M.K. Sandford), Gordon & Breach, Langhorne, pp. 59–127.

Barrick, R.E., Fischer, A.G. and Showers, W.J. (1999) Oxygen isotopes from turtle bone: applications for terrestrial paleoclimates? *Palaios*, **14**, 186–191.

Berger, A. (1988) Milankovitch theory and climate. *Rev. Geophys.*, **26**, 624–657.

Bond, G., Broecker, W., Johnson, S. *et al.* (1993) Correlations between climate records from North Atlantic sediments and Greenland ice. *Nature*, **365**, 143–147.

Bond, W.J., Woodward, F.I. and Midgley, G.F. (2005) The global distribution of ecosystems in a world without fire. *New Phytol.*, **165**, 525–538.

Bradley, D.G., Loftus, R.T., Cunningham, P. and MacHugh, D.E. (1998) Genetics and domestic cattle origins. *Evol. Anthropol.*, **6**, 79–86.

Bryant, J.D., Koch, P.L., Froelich, P.N. *et al.* (1996) Oxygen isotope partitioning between phosphate and carbonate in mammalian apatite. *Geochim. Cosmochim. Acta*, **60**, 5154–5148.

Cohen, M.N. and Armelagos, G.J. (eds.) (1984) *Paleopathology at the Origins of Agriculture*, Academic Press, Orlando, FL.

Cohen, M.N. and Crane-Kramer, G.M.M. (eds) (2007) *Ancient Health: Skeletal Indicators of Agricultural and Economic Intensification*, Florida University Press, Gainsville, FL.

Dansgaard, W. (1964) Stable isotopes in precipitation. *Tellus*, **16**, 436–468.

Diamond, J. (1992) *The Third Chimpanzee: The Evolution and Future of the Human Animal*, HarperCollins, New York.

Diamond, J. (2002) Evolution, consequences and future of plant and animal domestication. *Nature*, **418**, 700–707.

Epstein, S., Buchsbaum, R., Lowenstam, H.A. and Urey, H.C. (1953) Revised carbonate-water isotopic temperature scale. *Geol. Soc. Am. Bull.*, **64**, 1315–1325.

Erez, J. and Luz, B. (1983) Experimental paleotemperature equation for planktonic foraminifera. *Geochim. Cosmochim. Ac.*, **47**, 1025–1031.

Gat, J.R., Nook, W.G. and Meijer, H.A.J. (2000) *Environmental Isotopes in the Hydrological Cycle: Principles and Applications. Vol. II: Atmospheric Water*, Atomic Energy Agency, Vienna.

Gautier, A. (1990) *La Domestication. Et l'homme créa l'animal*, Editions Errance, Paris.

Giuffra, E., Kijas, J.M.H., Amarger, V. *et al.* (2000) The origin of the domestic pig: Independent domestication and subsequent introgression. *Genetics*, **154**, 1785–1791.

GNIP Data Bank (2008) www-naweb.iaea.org/napc/ih/GNIP/IHS_GNIP.html.

Grossman, E.L. (1984) Carbonic isotopic fractionation in live benthic foraminifera – comparison with inorganic precipitate studies. *Geochim. Cosmochim. Ac.*, **48**, 1505–1512.

Grupe, G. (2007) Taphonomic and diagenetic processes, in *Handbook of Palaeoanthropology. Vol 1: Principles, Methods, and Approaches* (eds W. Henke and I. Tattersall), Springer, Berlin, pp. 241–259.

Harris, D.R. (2002) Development of the agro-pastoral economy in the Fertile Crescent during the Pre-Pottery Neolithic period, in *The Dawn of Farming in the Near East* (eds R.T.J. Cappers and S. Bottema), Ex Oriente, Berlin, pp. 67–83.

Helmer, D., Roitel, V., Saña Segui, M. and Willcox, G. (1998) Interprétations environnementales des données archéozoologiques et archéobotaniques en Syrie du nord de 16 000 BP à 7000 BP, et les débuts de la domestication des plantes et des animaux, in *Espace Naturel Espace habité en Syrie du Nord 10e-2e Millénaires av*, Travaux de la Maison de l'Orient 28 (eds M. Fortin and O. Aurenche), J.-C. Canadian Society for Mesopotamian Studies Bulletin, 33 and Lyon, pp. 9–33.

Helmer, D., Gourichon, L., Monchot, H. *et al.* (2005) Identifying early domestic cattle from Pre-Pottery Neolithic sites on the Middle Euphrates using sexual dimorphism, in *The First Steps of Animal Domestication* (eds J-.D. Vigne, J. Peters and D. Helmer), Oxbow Books, Oxford, pp. 86–95.

Hiendleder, S., Mainz, K., Plante, Y. and Lewalski, H. (1998) Analysis of mitochondrial DNA indicates that domestic sheep are derived from two different ancestral maternal sources: No evidence for contributions from Urial and Argali sheep. *J. Hered.*, **89**, 113–120.

Hillman, G. (1996) Late Pleistocene Changes in wild plant-foods available to hunter-gatherers of the northern Fertile Crescent: possible preludes to cereal cultivation, in *The Origins and Spread of Agriculture and Pastoralism in Eurasia* (ed. D.R. Harris), University College, London, London, pp. 159–203.

Holmgren, K., Lee-Thorp, J.A., Cooper, G.R.J. *et al.* (2003) Persistent millennial-scale climatic variability over the past 25 000 years in southern Africa. *Quat. Sci. Rev.*, **22**, 2311–2316.

Hongo, H. and Meadow, R.H. (1998) Pig exploitation at Neolithic Çayönü Tepesi (Southeastern Anatolia), in *Ancestors for the Pigs: Pigs in Prehistory*, MASCA Research Papers in Science and Archaeology 15 (ed. S.M. Nelson), University of Pennsylvania, Philadelphia, pp. 77–98.

Ilgezdi, G. (2008) The Domestication Process in Southeastern Turkey: The Evidence from Mezraa-Teleilat. Unpublished PhD dissertation, University of Tübingen.

Jouzel, J., Koster, R.D., Suozzo, R.J. and Russel, G.L. (1994) Stable water isotope behaviour during the last glacial maximum: a general circulation model analysis. *J. Geophys. Res.*, **99**, 25791–25801.

Kim, S.T. and O'Neil, J.R. (1997) Equilibrium and nonequilibrium oxygen isotope effects in synthetic carbonates. *Geochim. Cosmochim. Ac.*, **61**, 3461–3475.

Kohn, M.J. (1996) Predicting animal $\delta^{18}O$: accounting for diet and physiological adaptation. *Geochim. Cosmochim. Ac.*, **60**, 4811–4829.

Kohn, M.J. and Cerling, T.E. (2002) Stable isotope compositions of biological apatite, in Phosphates. Geochemical, geobiological, and materials importance, in *Reviews of Mineralogy and Geochemistry*, vol. **48** (eds M.J. Kohn, J. Rakovan and J.M. Hughes), Mineralogical Society of America, pp. 455–488.

Larson, G., Dobney, K., Albarella, U. *et al.* (2005) Worldwide phylogeography of wild boar reveals multiple centers of pig domestication. *Science*, **307**, 1618–1621.

Lee-Throp, J. (2002) Two decades of progress towards understanding: Fossilization processes and isotopic signals in calcified tissue minerals. *Archaeometry*, **44**, 435–446.

Lee-Thorp, J.A., Sealy, J.C. and van der Merwe, N.J. (1989) Stable carbon isotope ratio differences between bone collagen and bone apatite, and their relationship to diet. *J. Archaeol. Sci.*, **16**, 585–599.

Legge, A.J. and Rowley-Conwy, P.A. (2000) The exploitation of animals, in *Village on the Euphrates* (eds A.M.T. Moore, G.C. Hillman and A.J. Legge), Oxford University Press, Oxford, pp. 423–471.

Loftus, R.T., MacHugh, D.E., Bradley, D.G. *et al.* (1994) Evidence for two independent domestications of cattle. *Proc. Natl. Acad. Sci. USA*, **91**, 2757–2761.

Longinelli, A. (1984) Oxygen isotopes in mammalian bone phosphate: a new tool for paleohydrological and paleoclimatological research. *Geochim. Cosmochim. Ac.*, **48**, 385–390.

Longinelli, A. and Nuti, S. (1973) Revised phosphate-water isotopic temperature scale. *Earth Planet. Sci. Lett.*, **139**, 291–303.

Lösch, S., Grupe, G. and Peters, J. (2006) Stable isotopes and dietary adaptations in human and animals at Pre-Pottery Neolithic Nevalı Çori, SE-Anatolia. *Am. J. Phys. Anthropol.*, **131**, 181–193.

Luikart, G., Gielly, L., Excoffier, L. *et al.* (2001) Multiple maternal origins and weak phylogeographic structure in domestic goats. *Proc. Natl. Acad. Sci. USA*, **98**, 5927–5932.

Martin, L., Russell, N. and Carruthers, D. (2002) Animal remains from the central Anatolian Neolithic, in *The Neolithic of Central Anatolia. Internal Developments and External Relations During the 9th-6th Millennia calBC* (eds F. Gérard and L. Thissen), Yayınları, Istanbul, pp. 193–206.

Meadow, R.H. (1999) The use of size index scaling techniques for research on archaeozoological collections from the Middle East, in *Historia animalium ex ossibus. Festschrift für Angela von den Driesch zum 65. Geburtstag* (eds C. Becker, H. Manhart, J. Peters and J. Schibler), Leidorf Rahden, Westfalen, pp. 285–300.

Miller, N.F. (2002) Tracing the development of the agropastoral economy in Southeastern Anatolia and Northern Syria, in *The Dawn of Farming in the Near East* (eds R.T.J. Cappers and S. Bottema), Ex Oriente, Berlin, pp. 85–94.

Naderi, S., Hamid-Reza, R., Pompanon, F. *et al.* (2008) The goat domestication process inferred from large-scale mitochondrial analysis of wild and domestic individuals. *Proc. Natl. Acad. Sci. USA*, **105**, 17659–17664.

Neef, R. (2003) Overlooking the Steppe-Forest: a preliminary report on the botanical remains from early Neolithic Göbekli Tepe (Southeastern Turkey). *Neo-Lithics*, **2/03**, 13–16.

Pasternak, R. (1995) Die botanischen Funde aus Nevalı Çori, Türkei (akeramisches Neolithikum). Ein Vorbericht, in *Res Archaeobotanicae. Proceedings of the 9th Symposium* (eds H. Kroll and R. Pasternak), Oetker-Voges-Verlag, Kiel, pp. 345–358.

Pasternak, R. (1998) Investigations of botanical remains from Nevali Çori, PPNB, Turkey: A short interim report, in *The Origins of Agriculture and Crop Domestication* (eds A.B. Damania, J. Valkoun, G. Willcox and C.O. Qualset), ICARDA, Aleppo, pp. 170–177.

Peters, J., Helmer, D., von den Driesch, A. and Saña Segui, M. (1999) Early animal husbandry in the Northern Levant. *Paléorient*, **25/2**, 27–48.

Peters, J. and Schmidt, K. (2004) Animals in the symbolic world of Pre-Pottery Neolithic Göbekli Tepe, South-eastern Turkey: a preliminary assessment. *Anthropozoologica*, **391**, 179–218.

Peters, J., von den Driesch, A. and Helmer, D. (2005a) The Upper Euphrates-Tigris basin: Cradle of agropastoralism? in *The First Steps of Animal Domestication* (eds J-.D. Vigne, J. Peters and D. Helmer), Oxbow Books, Oxford, pp. 96–124.

Peters, J., von den Driesch, A., Pöllath, N. and Schmidt, K. (2005b) Birds in the megalithic art of Pre-Pottery Neolithic Göbekli tepe, Southeast Turkey, *Feathers, Grit and Symbolism. Documenta Archaeobiologiae 3* (eds G. Grupe and J. Peters), Leidorf, Rahden/Westfalen, pp. 221–232.

Petit, J.R., Jouzel, J., Raynaud, D. *et al.* (1999) Climate and atmospheric history of the past 420 000 years from the Vostok ice core, Antarctica. *Nature*, **399**, 429–435.

Roberts, N. (2002) *The Holocene. An Environmental History*, Blackwell, Oxford.

Robinson, S.A., Black, S., Sellwood, B.W. and Valdes, P.J. (2006) A review of palaeoclimates and palaeoenvironments in the Levant and Eastern Mediterranean from 25 000 to 5000 years BP: setting the environmental background for the evolution of human civilisation. *Quat. Sci. Rev.*, **25**, 1517–1541.

Rosenberg, M., Nesbitt, M.R., Redding, R.W. and Strasser, T.F. (1995) Hallan çemi tepesi: some preliminary observations concerning early neolithic subsistence behaviors in Eastern Anatolia. *Anatolica*, **21**, 1–12.

Rozanski, K., Araguas-Araguas, L. and Gonfiantini, R. (1992) Relation between long-term trends of oxygen-18 isotope composition of precipitate and climate. *Science*, **258**, 981–985.

Russell, N. and Martin, L. (2005) Çatalhöyük mammal remains, in *Inhabiting Çatalhöyük: reports from the 1995–99 seasons. Cambridge: McDonald Institute Monographs, Çatalhöyük Research Project Volume 4, British Institute of Archaeology at Ankara, Monograph 38* (ed. I. Hodder), McDonald Institute, Cambridge, pp. 33–98.

Russell, N., Martin, L. and Buitenhuis, H. (2005) Cattle domestication at Çatalhöyük revisited. *Curr. Anthropol.*, **46**, S101–S108.

Saña Seguí, M. (1999) Arqueología de la domesticación animal. La gestión de los recursos animales en Tell Halula (Valle del Éufrates-Siria) del 8. 800 al 7. 000 BP. Treballs d'Arqueologia del Pròxim Orient 1, Universitat Autònoma de Barcelona, Barcelona.

Schmidt, K. (2000) Zuerst kam der Tempel, dann die Stadt. Vorläufiger Bericht zu den Grabungen am Göbekli Tepe und am Gürcütepe 1995–1999. *Istanbuler Mitteilungen*, **50**, 5–40.

Schwarcz, H.P. (1991) Some theoretical aspects of isotope paleodiet studies. *J. Archaeol. Sci.*, **18**, 261–275.

Stosch, H.G. (2004) Einführung in die Isotopengeochemie, www.geologie.uni-freiburg.de/root/people/fschaft/Isotopengeochemie.pdf.

Staudenbauer, C. (2008) Klimaapproximation mittels $\delta^{18}O$ und $\delta^{13}C$ aus subfossilen Knochen der Pantherschildkröte (Geochelone pardalis). Diploma Thesis, LMU Munich.

Tanno, K. and Willcox, G. (2006) How fast was wild wheat domesticated? *Science*, **311**, 1886.

Teegen, W.R. and Schultz, M. (1997) Transverse linear enamel hypoplasia in the PPNB population of Nevali Cori (Turkey). *Am. J. Phys. Anthropol.*, **S24**, 226–227.

Thompson, L.G., Mosley-Thompson, E., Davis, M.E. *et al.* (2003) Tropical glacier and ice core evidence of climate change on annual to millennial time scale. *Climate Change*, **59**, 137–155.

Thompson, L.G., Mosley-Thompson, E., Brecher, H. *et al.* (2006) Abrupt tropical climate change: Past and present. *Proc. Natl. Acad. Sci. USA*, **103**, 10 536–10 543.

Troy, C.S., MacHugh, D.E., Bailey, J.F. *et al.* (2001) Genetic evidence for Near Eastern origins of European cattle. *Nature*, **410**, 1088–1091.

Uerpmann, H.-P. (1979) *Probleme der Neolithisierung des Mittelmeerraumes. Beihefte zum Tübinger Atlas des Vorderen Orients, Reihe B (Geisteswissenschaften), Nr. 28*, Dr. Ludwig Reichert, Wiesbaden.

Uerpmann, H.-P. (1987) *The Ancient Distribution of Ungulate Mammals in the Middle East. Beihefte zum Tübinger Atlas des Vorderen Orients, Reihe A (Naturwissenschaften), Nr. 27*, Dr. Ludwig Reichert, Wiesbaden.

Uerpmann, H.-P. (1996) Animal domestication – accident or intention? in *The Origins and Spread of Agriculture and Pastoralism in Eurasia* (ed. D.R. Harris), UCL Press, London, pp. 227–237.

Urey, H.C. (1947) The thermodynamic properties of isotopic substances. *J. Chem. Soc.*, 562–582.

Vigne, J.-D. (2000) Les débuts néolithiques de l'élevage des ongulés au Proche Orient et en Méditerranée: acquis récents et questions, in *Premiers Paysans Du Monde: Naissances Des Agricultures* (ed. J. Guilaine), Editions Errance, Paris, pp. 143–168.

Vogelsang, R. (2002) Migration oder Diffusion? – Frühe Viehhaltung im Koakoland, in *Interdisziplinäre Perspektiven zu Kultur- und Landschaftswechsel im ariden und semiariden nordwest-Namibia*, vol. **77** (eds M. Bolling, E. Brunotte and T. Becker), Kölner Geographische Arbeiten, pp. 137–144.

Wang, X., Auler, A.S., Edwards, R.L. *et al.* (2004) Wet periods in northeastern Brazil over the past 210 kyr linked to distant climate anomalies. *Nature*, **432**, 740–743.

5

Health, Diet and Social Implications in Neolithic Greece from the Study of Human Osteological Material

Anastasia Papathanasiou

Ephorate of Paleoanthropology and Speleology, Greek Ministry of Culture, Athens, Greece

5.1 INTRODUCTION

The Neolithic transition in prehistory has profound effects on human health, social organization and the economic infrastructure (Childe, 1936). During the Neolithic, economies changed to include domesticated plants and animals, which is closely linked to a range of technological developments, as well as an increase in sedentism. This led to corresponding changes in dietary habits, which in most cases initially resulted in lower nutritional quality and a deterioration of certain aspects of human health (Larsen, 1997). Although similar domesticated species and cultural developments were adopted in all areas across Europe, significant cultural and economic regional diversity has been documented. This diversity is associated to some extent with environmental conditions, the availability of local resources and ecological settings (Bogucki, 1996; Whittle, 1996; Renfrew, 1987; Pinhasi, Fort and Ammerman, 2005; Whittle and Cummings, 2007).

The regions of Thessaly, the Peloponnese, Cyprus and Crete, played a major role in the early part of the shift to agriculture in Europe. However, these regions vary in terms of both the timing and nature of their Neolithization process. The Greek Neolithic started around 6800 calBC and lasted until 3200 calBC (Andreou, Fotiadis and Kotsakis, 1996; Alram-Stern, 1996). Early Neolithic sites in Greece and the Balkans show some distinct similarities to their Near Eastern precursors, in their suite of domesticates, architectural features, pottery style and technology, and in other material culture attributes (Andreou, Fotiadis and Kotsakis, 1996; Alram-Stern, 1996; Bogucki, 1996; Whittle, 1996; Pèrles, 2001). However, from the very beginning, the Greek Early Neolithic phases also display some distinct characteristics that possibly reflect specific adaptation to Mediterranean environments and ecotones, outside the Near Eastern core regions. At present, the issue as to whether the Early Neolithic of Greece represents the adoption

Human Bioarchaeology of the Transition to Agriculture Edited by Ron Pinhasi and Jay T. Stock

of a farming economy by indigenous people or the intermixing of local foragers and groups of exogenous Near Eastern and/or Anatolian agriculturalists remains a topic of controversy (Ammerman and Cavalli-Sforza, 1971; Bogucki, 1996; Whittle, 1996; Renfrew, 1987).

This chapter focuses on changes in diet and health following the Neolithic transition in Greece. The first part of the chapter focuses on the reconstruction of subsistence practices based on the analysis of carbon and nitrogen stable isotope fractionations. The next part examines the effects of the Neolithic transition on human health and development, as well as on the interpersonal relations throughout the Neolithic by examining the human skeletal material from Greek Neolithic contexts.

5.2 SUBSISTENCE AND ECONOMY OF NEOLITHIC GREEK POPULATIONS

Stable isotope analysis provides direct, specific, and detailed dietary information based on the human remains themselves (cf. Grupe and Peters, Lille and Budd, this volume). Many foods, or groups of foods, are isotopically distinct, and when consumed and incorporated into body tissues, their isotopic signature is retained. Analysis of the stable carbon and nitrogen isotopes of collagen, which is the major protein in bone, provides information about the sources of protein in diets (DeNiro and Epstein, 1978, 1981; van der Merwe and Vogel, 1978; Norr, 1995; Ambrose and Norr, 1993; Ambrose *et al.*, 1997; Schoeninger, DeNiro and Tauber, 1983; Schoeninger, 1989). Broad categories can be discerned, such as terrestrial plant, herbivore, and carnivore, or marine protein. In general, C_3 resources average around –27‰ for $\delta^{13}C$, C_4 resources average around –13‰, while marine resources are enriched by about 10‰ when compared to terrestrial ones. Nitrogen accumulates along the food chain, with a 3‰ enrichment of the $\delta^{15}N$ value at each step (Hoefs, 1997). Terrestrial plants average about 1‰ and marine plants about 7‰. Therefore, marine resources have higher $\delta^{15}N$ values than terrestrial ones due to the elevated starting point and the longer marine food chains. However, it is not possible to distinguish isotopically different protein sources, such as meat and milk, coming from the same animal (Hedges and Reynard, 2007).

Previous work (Papathanasiou, 1999, 2001, 2003) has demonstrated that despite the apparent source diversity in the faunal and floral record, and despite the proximity of a number of sites to the sea, Neolithic populations lived on diets consisting predominantly of terrestrial foods, primarily grains and legumes. In this work, bone collagen $\delta^{13}C$ and $\delta^{15}N$ isotopic ratios were obtained from a sample of 101 human skeletal elements from three coastal (Alepotrypa, Franchthi, Kephala) and three inland (Theopetra, Tharrounia, Kouveleiki) Neolithic sites (Figure 5.1). Most of the specimens analysed produced valid values, as they were well preserved, gave good collagen yields, and C/N ratios between 2.9 and 3.6 (Table 5.1). Figure 5.2 is a scatter plot of $\delta^{15}N$ vs. $\delta^{13}C$ average stable isotope values for Greek Neolithic sites. All the $\delta^{13}C$ values of both coastal and inland populations cluster in the same area and indicates a primarily C_3 terrestrial diet. The $\delta^{15}N$ values are low and indicate that very little of the dietary protein is of marine origin. Furthermore, the very small range of values (4‰ in $\delta^{15}N$ and 2‰ in $\delta^{13}C$) implies a high degree of homogeneity in the diet for the populations under study.

Carbon and nitrogen isotope values for faunal samples were analysed for the Neolithic phases at the sites of Alepotrypa and Kephala, in order to provide a comparative range for the human isotopic values (Table 5.2). Results indicate similar $\delta^{13}C$ and $\delta^{15}N$ values for the mammalian

Figure 5.1 Map of Greek Neolithic sites with human osteological samples

species in both sites and the ratios for the mammalian species are overall similar in values to those reported for the human samples, pointing to a diet based on C_3 terrestrial resources.

A study of dietary variation amongst Neolithic human burials from the site of Makrygialos in Northern Greece (Triantaphyllou, 2001) provides similar results. Carbon and isotope fractionations were analysed for 22 human samples analysed. The results indicate that these humans had a very similar diet with a predominantly terrestrial C_3 pattern. This observation was further supported by the study of a number of animal samples from the same site. A dietary analysis of two human samples from Proskynas in central Greece, gave similar results with average values for $\delta^{13}C$ at $-19.72 \pm 0.70‰$ and for $\delta^{15}N$ at $7.51 \pm 1.57‰$, respectively, hence indicating a terrestrial diet with a focus on C_3 foodstuffs (Papathanasiou, Zachou and Richards, 2009).

Table 5.1 Carbon and nitrogen stable isotope ratios of human specimens from the Mesolithic Neolithic sites discussed in text

Site/year	Sample	δ^{12}N	δ^{13}Cco	C:N	Site/year	Sample	δ^{15}N	δ^{13}Cco	C:N
Alepotrypa 99′	AP 1100	8.13	–20.67	2.96		DA 2	7	–20	3.16
	AP 1101	6.74	–19.65	2.92		DA 3	8.12	–19.8	3.14
	AP 1102	5.8	–19.42	2.91		DA 4	7.65	–20.17	3.19
	AP 1103	8.09	–19.7	3.06		DA 5	8	–20	3.18
	AP 1104	7.92	–19.95	3.04		DA 6	6.9	–19.7	3.18
	AP 1105	6.7	–19.92	3.07		DA 7	5.8	–19.9	3.21
	AP 1106	5.64	–20.29	3.02		DA 8	7.4	–19.5	3.19
	AP 1107	7.47	–19.27	3.03		DA 9	8.1	–19.5	3.14
	AP 1108	8.48	–19.48	2.98		DA 10	6.9	–20	3.17
	AP 1109	7.21	–19.95	3.06		DA 11	7.2	–20	3.19
	AP 1110	6.62	–20.33	3.2	Kephala 99′	AP 1139			3.3
	AP 1112	7.82	–20.13	3.17		AP 1140			3.34
	AP 1113	4.46	–21.55	3.19		AP 1141			3.2
	AP 1114	7.47	–19.85	3.14		AP 1142			2.45
	AP 1115	8.73	–20.03	3.28		AP 1143			2.78
Franchthi 99′	AP 1122	8.16	–19.3	3.27		AP 1144			2.1
	AP 1123	9.74	–18.96	3.09		AP 1145			2.52
	AP 1124	8.16	–19.64	3.17		AP 1146			2.51
	AP 1125	9.08	–18.63	3.12		AP 1147			3.1
	AP 1127	10.44	–17.77	3.3		AP 1148			1.84
	AP 1128	9.52	–18.18	3.34		AP 1149	10.56	–19.26	3.32
	AP 1130	7.79	–19.23	3.28		AP 1150			2.6
	AP 1131			2.95		AP 1151			2
	AP 1132	8.38	–19.14	3.33		AP 1152			2.49
	AP 1133	7.84	–18.97	3.32		AP 1153			3.36
	AP 1134			2.18		AP 1154			2.67
	AP 1135	8.3	–18.43	3.37		AP 1155			3.34
	AP 1136			2.64		AP 1156			3.49
	AP 1137			2.1	Theopetra 99′	AP 1185	7.81	–20.39	3.46
	AP 1138	14.11	–16.96	3.32		AP 1186	7.69	–20.23	3.44
Tharrounia 99′	AP1165	8.75	–19.92	3.44		AP 1187	8.36	–19.32	3.41
	AP 1166	8.38	–20.16	3.47		AP 1188	7.55	–20.01	3.43
	AP 1167	7.55	–20.28	3.42		AP 1189	7.29	–20.15	3.39
	AP 1168	9.41	–19.55	3.45		AP 1190	7.68	–19.89	3.4
	AP 1169	6.93	–19.9	3.43		AP 1191	8.7	–19.8	3.42
	AP 1170	8.93	–19.99	3.47		AP 1192	7.24	–20.3	3.45
	AP 1171	8.6	–19.77	3.43		AP 1193	6.71	–20.51	3.42
	AP 1172	7.47	–20.31	3.43		AP 1194	7.14	–20.4	3.45
	AP 1173	8.24	–20.02	3.46		AP 1195			2.76
	AP 1174	8.2	–19.72	3.45		AP 1196	4.38	–17.22	3.39
	AP 1175	7.92	–19.97	3.43		AP 1197	7.48	–20.13	3.38
	AP 1176	7.57	–20.09	3.48		AP 1198	8.13	–19.46	3.4
	AP 1177	8.71	–20.27	3.46	Xirolimni 08′	6614	8.8	–19.7	3.2
	AP 1178	7.79	–20	3.43		6615	8.6	–19.7	3.2

(continued)

Table 5.1 (*Continued*)

Site/year	Sample	δ^{12}N	δ^{13}Cco	C:N	Site/year	Sample	δ^{15}N	δ^{13}Cco	C:N
	AP 1179	8.52	–19.83	3.43		6616	8.7	–19.6	3.2
	AP 1180	6.77	–20.09	3.43		6617	8.2	–19.8	3.3
	AP 1181	7.72	–19.55	3.44		6618	8.7	–20.3	3.5
	AP 1182	7.68	–20.17	3.46		6619	8.3	–20.3	3.1
	AP 1183	7.54	–20.24	3.44		6620	8.6	–19.9	3.1
	AP 1184	8.15	–19.98	3.45		6621	8.5	–19.7	3.1
Kouveleiki 99′	AP 1199		–21.78	3.33		6622	8.9	–19.4	3.2
	AP 1200			4.02		6623	8.1	–19.9	3.3
	AP 1201		–27.3	3.08		6624	9	–19.6	3.2
	AP 1202	8.32	–19.86	3.39		6625	8.8	–19.7	3.2
	AP 1203	7.85	–19.81	3.39		6626	9.7	–19.4	3.2
						6627	8.7	–19.7	3.2

These results support the hypothesis that Neolithic Greece, regardless of geographical location, was occupied by agricultural groups with a land-based economy and a diet involving only occasional or periodic exploitation of near-shore marine protein resources. Greek Neolithic populations focused their diets on C_3 plants (wheat, barley, legumes), dairy products, and meat, dominated by sheep and goat. The Greek farmers had only a minor amount of marine

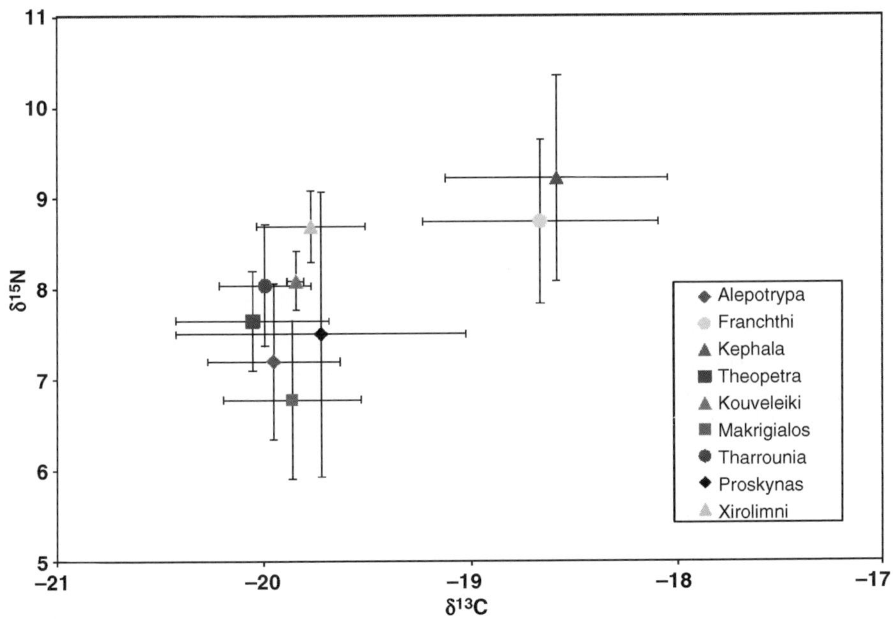

Figure 5.2 Scatterplot of δ^{15}N vs. δ^{13}C average stable isotope values for Greek Neolithic sites

Table 5.2 Average carbon and nitrogen stable isotope ratios for fauna from Neolithic phases at the sites of Alepotrypa and Kephala

Species	Alepotrypa	Kephala		
	$\delta^{13}C$	$\delta^{15}N$	$\delta^{13}C$	$\delta^{15}N$
Bovid	–19.78	4.41	–20.70	6.47
Deer	–21.45	4.32	–19.99	5.47
Sheep/Goat	–21.72	5.91	–20.18	7.02
Pig	–20.82	4.61		
Fox	–19.41	6.95		
Fish	–11.78	8.97		
Dog			–18.89	8.49

proteins, as is evident from low nitrogen values atypical of a diet that is based on a high proportion of marine foods (Papathanasiou, 2003).

Statistically significant differences exist between the average $\delta^{15}N$ values for Alepotrypa Cave (7.11‰), and the coastal site of Franchthi Cave (9.23‰) ($p < 0.0005$). In addition, significant differences exist between Franchthi and all the other inland sites ($p < 0.0012$). Franchthi's isotopic values do not indicate a typical C_3 terrestrial pattern but rather indicate a more varied diet. Results for these two sites suggest that the Neolithic populations in this location did not have diets that were only based on the consumption of C_3 foodstuffs, but also they included more meat, dairy products, and possibly marine foods, than in the case of other inland sites and Alepotrypa Cave. No other discernible differences in isotopic values – in terms of gender and social status as expressed in comparisons between males and females, adults and subadults, or amongst different burial types – were detected for the Late and Final Neolithic period.

It should be noted that all of the aforementioned sites refer to the later phases of the Greek Neolithic (Late and Final), as at the beginning of the palaeodietary reconstruction research in Greece there were no Early Neolithic human samples available. There were very few Mesolithic osteological series but all attempts at carbon and nitrogen stable isotope analysis failed due to poor collagen preservation. Because of this issue, in order to observe the Neolithic transformation in Greece, comparisons are only possible after excavations revealed the Early Neolithic settlements of Mavropigi, Xirolimni, and Pontokomi in the late 1990s.

The settlements of Mavropigi, Xirolimni and Pontokomi in Northern Greece (Figure 5.1) constitute some of the earliest agriculturalist communities in Greece and Europe in general (Karamitrou-Mendesidi, 1998, 2000, 2005; Ziota, 1995, 1998). They are located in Western Macedonia, on the Ptolemais plain at an elevation of 670 to 750 m above sea level. Mavropigi, is dated to ca. 6600 BC. The site extends over an area of about a hectare. It consists of 4 occupation horizons with irregular rectangular, semi-subterranean dwellings of 50 to 90 m^2 area, similar to the Nea Nikomedeia houses. The site yielded 18 *in situ* burials that are associated with mostly plain, coarse pottery, and a list of 2000 objects including stone and bone tools, loom weights, pendants, beads, 6 seals, and 132 female and animal figurines. Pontokomi is dated around 6200 BC and yielded a habitation area with 3 burials, flint stone tools, plain, coarse pottery and 120 fine clay figurines. Xirolimni is also an Early Neolithic (6500 BC) settlement with a high concentration of 90% plain, coarse, distinct ceramic ware and 14 pit burials. The final layer is a destruction level by fire (Karamitrou-Mendesidi, 1998, 2000, 2005; Ziota, 1995, 1998).

Palaeodietary reconstruction based on carbon and nitrogen stable isotope analysis has only been undertaken at Xirolimni. At this site, 14 samples from 14 different individuals were analysed. All of the samples are well preserved and yielded good quality collagen and valid results (DeNiro, 1985. The mean value for $\delta^{13}C$ at Xirolimni is $-19.37 \pm 0.26‰$, and for $\delta^{15}N$ is $8.41 \pm 0.39‰$. Both the $\delta^{13}C$ values and $\delta^{15}N$ values cluster in the range of values that are indicative of a primarily C_3 terrestrial diet with no marine input, as might be anticipated given the large distance of the sites from the sea. The Xirolimni population, as did the Later Neolithic populations, focused their diet on domesticated C_3 plants, such as wheat, barley, legumes and fruits. At this point, it cannot be determined how much terrestrial animal protein they incorporated into their diet, but the data suggest that the amount was significant. When the results for these sites are compared against other Greek Neolithic sites from later periods, there is a difference in the ^{15}N values, which suggests that this population incorporated more terrestrial animal protein in their diet, either from domesticated animals or wild game, than populations in later periods. The difference is small but it is statistically significant ($p < 0.00001$).

Bioarchaeological research on agricultural populations worldwide also agrees that while Mesolithic groups utilized a variety of wild species throughout the year, the tendency of many Neolithic societies was to reduce the diversity of their diets by focusing on a narrow range of crops and livestock species (Larsen, 1997). A number of isotopic studies for Mesolithic Europe (see contributions by Schulting,; Lillie and Budd, this volume) have shown a great reliance on marine foods by coastal Mesolithic populations in Denmark (Tauber, 1981, 1983, 1986), the British Isles (Schulting and Richards, 2001, 2002a, 2002b), Portugal (Lubell *et al.*, 1994) and the Baltic region (Liden, 1995). However, with the advent of agriculture, even in a number of the aforementioned regions, a shift in diet has been documented by isotopic studies, and a pattern of shifts towards the predominance of terrestrial diets is evident (Tauber, 1981; Liden, 1995; Lubell *et al.*, 1994). In the Mesolithic Mediterranean regions this shift was not as sharp, although the number of samples is admittedly very small. The available evidence suggests that Mediterranean Mesolithic populations never placed a heavy reliance on marine resources, perhaps because the Mediterranean regions were not as productive biotopes as the Atlantic regions (Garcia-Guixé, Richards and Subirà, 2006; Richards *et al.*, 2001; Schulting, this volume). Unfortunately, only one Mesolithic sample for Greece, from Theopetra Cave, yielded valid results, with carbon and nitrogen isotopic values that fall within the range of the Neolithic samples ($\delta^{15}N$ 7.81, $\delta^{13}C$ -20.39), therefore it is not possible to do draw any clear conclusions regarding the diets of Greek Mesolithic populations. During the Neolithic though, as documented both in Greece as well as other geographically, chronologically and culturally unrelated areas, farming populations exhibit a reliance on domesticated species with the adoption of agriculture (Larsen, 1997; Halstead, 2008). The overall dietary variety and quality of foodstuffs is decreasing and they are substituted by an investment on energy-rich plant foods with a significant contribution of domesticated animal products. Cereals became the main staple foods as they yielded food of high calorific value, the surplus of which could be stored and preserved for later consumption, a dependable resource for further population increase, which was the major issue of Neolithic societies (Halstead, 2008). The shift to intensive cereal consumption seems to have already been completed during the Early Neolithic, as evident from the study of the skeletal population of Xirolimni. In this population, the diet resembles the ones observed amongst the Later Neolithic populations and involves mixed plant/animal protein content but with a higher meat consumption, (Papathanasiou, 2008).

5.3 HEALTH

What was the interaction and effects of the above documented dietary changes on nutrition and health of the populations under study? In order to assess the impact of the Neolithic transition on Greek Neolithic populations, a number of health indicators were collected from the osteological assemblage. Demographic parameters such as minimum number of individuals (MNI), age, sex and stature were determined, as they provide the essential context for any further analysis. Data were collected following procedures described by Buikstra and Ubelaker (1994).

Specifically, sex determination was performed following the methods discussed in Buikstra and Ubelaker (1994), Ubelaker (1989), White (1991), Milner (1992), Krogman and Iscan (1986) and Phenice (1969), including patterns of robusticity, and cranial and pelvic morphology. Only adults with mature characteristics were sexed.

Age-at-death was estimated by applying a number of different methods, whichever were applicable for each individual due to the fragmentary nature of the sample. The method of Meindl and Lovejoy (1985) for ectocranial suture closures was used for crania belonging to adults. The method of Lovejoy (1985) for determining stages of dental wear was used for adult dentition. The methods of Todd (1920, 1921) and Brooks and Suchey (1990) for morphological changes of the pubic symphyseal face were used for adult os coxa. The method of Ubelaker (1989) for tooth formation and eruption was applied to subadult remains. The method of McKern and Stuart (1957) for epiphyseal fusion was used for ageing subadult and young adult postcranial bones. Subadult age was also estimated by long bone diaphyseal length and iliac breadth from the available complete bones, following Ubelaker (1989).

Stature was estimated according to formulas for white males and females (Trotter, 1970), which were applied on the preserved (complete) long bones of each individual. Pathological conditions were identified following Ortner and Putschar (1985), Resnick (1995), Buikstra and Ubelaker (1994) and Tyson and Alcauskas (1980). Observations were made by the naked eye under normal light conditions and without the aid of microscopy. Pathological conditions were recorded by presence-absence, and the percentages reflect the observed over the observable.

The MNI for Mesolithic Greece is 21 and includes 13 adults and 8 subadults, while the MNI for the Neolithic is 503, including 299 adults and 204 subadults (Table 5.2). The latter sample comprises of less subadults and especially infants and is hence below the anticipated 1:1 adult-subadult ratio that is common in a pre-industrial site with high child mortality. Regarding the demographic parameters, the mean adult age-at-death is around 30 years (Table 5.3) and it is deviating from life expectancy at birth, which is significantly lower. Thus, if a child survived until the age of 10, then there was a high probability that he/she would live until the age of 30 or longer. Average estimations of male and female stature for various Neolithic sites are shown in Table 5.4, and show limited inter-site variability for this time period, with Xirolimni men tending to be taller than average (Papathanasiou, 2008). However, this is not a statistically significant observation as the sample is too small. As these two demographic parameters, age-at-death and stature, are influenced by a large number of factors, such as genetic predisposition, hormonal deficiencies, physiological stress, disease and nutrition, they are at best indicators of a variety of environmental conditions such as overall welfare, growth, acclimatization and adaptation. The general pattern of growth in past populations is assumed to have been similar to that of modern populations. Thus, stress in past populations can be inferred from deviations in growth trajectories of long bones from the normal rates (Maat, 2003; Steckel, 1995). In order to

Table 5.3 Adult/subadult ratios based on the minimum number of individuals (MNI) analysis of Greek Neolithic sites

Period	Site	MNI	Adults/Subadults
Mesolithic	Franchthi	17	9/8
	Theopetra	4	4/0
Early and Middle Neolithic	Pontokomi	3	2/1
Neolithic	Xirolimni	14	11/3
	Mavropigi	20	9/11
	Nea Nikomedeia	34	13/21
Late and Final Neolithic	Makrygialos	72	56/16
Neolithic	Tharrounia	38	22/16
	Kouveleiki	14	11/3
	Theopetra	29	22/7
	Franchthi	46	16/30
	Kephala	65	51/14
	Proskynas	7	5/2
	Alepotrypa	161	81/80
Total		524	312/212

determine the health status of Neolithic people, a number of stress indicators were observed on osteological assemblages, including porotic hyperostosis and cribra orbitalia (sieve-like porous lesions on the cranial bones and the orbital roof, respectively) for the identification of anaemia, periosteal reaction (periosteal expansion and reactive surfaces) for inflammatory response, osteoarthritis (articular joint lipping and ebunation) and musculoskeletal stress markers (workload and activity), dental caries and antemortem tooth loss for dental health, and enamel defects (linear enamel hypoplasias) for physiological stress.

Palaeopathological analysis indicates that the lesions observed on the teeth (Table 5.4) include caries, linear enamel hypoplasias, antemortem tooth loss, calculus and severe tooth wear in a cupped shape characteristic of agricultural populations processing grains with grinding stones, a process in which grains are reduced to fine particles, and fine particles of the

Table 5.4 Mean age at death and stature estimation for Greek Neolithic sites

Site	Age	N	Stature (cm)			
			Males	N	Females	N
Xirolimni	33.3 ± 7.8	10	171.3 ± 2.7	3	154.4 ± 9.9	2
Mavropigi	29.2 ± 7.1	7	168.7 ± 0.7	2	153.8 ± 3.2	2
Nea Nikomedeia	30.4[a]		168.0[a]	9	155.5[a]	9
Kephala	31.7[a]		167.0[a]	3	153.0[a]	1
Franchthi	34.6 ± 4.6	6	164.0	1	156.5 ± 10.6	2
Alepotrypa (Femora only)	28.8 ± 8.8	45	169.7 ± 6.2	6	153.8 ± 4.1	37
			163.5 ± 3.5	2	153.5 ± 3.9	10
Other Neolithic Sites	N/A		165.5[b]	N/A	154.9[b]	N/A

[a]Angel 1977.
[b]Angel 1984.

Table 5.5 Prevalence rates of dental pathologies (by site)

Site	Caries	Linear Enamel Hypoplasia	Antemortem tooth loss
Xirolimni	9.6%	1.3%	13.5%
Mavropigi	1.6%	2.6%	2.2%
Nea Nikomedeia	2.0%	0.7%	
Franchthi	1.2%	7.5%	5.6%
Alepotrypa	3.2%	8.3%	18.4%
Proskynas	3.0%	6.4%	2.3%
Makrygialos	3.9%	5.6%	0.8%

grinding stones are introduced into food, promoting tooth wear (Smith, 1984). A significant observation is related to linear enamel hypoplasias. Their prevalence increases in the Late and Final Neolithic, as shown in Table 5.5 when, evidently, such disruptions were more frequent, and were possibly due to the deterioration of living conditions during those periods.

The prevalence of arthritic lesions and muscoloskeletal stress markers is also elevated following the transition (Table 5.6), and is indicative of physical stress, which may be due to changes in mobility and activity during the Neolithic. However, the percentage of bone infection is generally low throughout the Neolithic. It is important to note that inter-site differences in the prevalence of osteoarthritis and infection was not compared due to inter-observer error, as some researchers report the prevalence per individual while others report it per skeletal element.

Other observed pathological conditions include cranial trauma and long bone fractures (Table 5.7). However, the prevalence of long bone trauma is much lower than cranial trauma. Amongst the Neolithic populations, there is a high prevalence of anaemic conditions (Table 5.6), in the form of mild porotic hyperostosis and cribra orbitalia (Figures 5.3 and 5.4). Both conditions reflect reactions of the body in an effort to create more red blood cells by the expansion of the red blood cell producing areas of bone marrow (Hengen, 1971; Lallo, Armelagos and Mensforth, 1997; Angel, 1966; Larsen, 1997; Walker *et al.*, 2009). In a

Table 5.6 Prevalence rates of palaeopathological conditions discuss in text (by site)

Site	Period	Cribra orbitalia	Porotic hyperostosis	Osteoarthritis	Infection
Franchthi	Mesolithic	25.0%	16.7%		
Xirolimni	Early Middle Neolithic	11.1%	22.2%	42.9%	7.1%
Mavropigi	Early Middle Neolithic	0.0%	13.3%	20.0%	5.0%
Nea Nikomedeia	Early Middle Neolithic	50%		23.0%	0.0%
Franchthi	Early Middle Neolithic	50.0%	11.1%		
Franchthi	Late Final Neolithic	50.0%	26.3%		
Alepotrypa	Late Final Neolithic	60.0%	50.0%	12.2%	
Proskynas	Late Final Neolithic	30.0%		85.7%	
Makrygialos	Late Final Neolithic	100%		0.0%	0.53%
Kephala	Late Final Neolithic	8%			

Table 5.7 Prevalence rates of cranial trauma (by site)

Period	Site	Prevalence
Mesolithic	Franchthi	12.5% (16)
	Theopetra	0% (4)
Early and Middle Neolithic	Pontokomi	0% (3)
	Xirolimni	10% (10)
	Mavropigi	0% (20)
	Nea Nikomedeia	0% (34)
Late and Final Neolithic	Makrygialos	0% (72)
	Tharrounia	0% (38)
	Kouveleiki	0% (11)
	Theopetra	0% (29)
	Franchthi	0% (55)
	Kephala	6.5% (31)
	Proskynas	14% (7)
	Alepotrypa	13% (69)

healthy equilibrium state, the marrow production of red blood cells is equal to the number of the red blood cells being destroyed. Conditions causing anaemia could include haemorrhage, inadequate production of red blood cells or increased haemolysis. The necessary factors for maintaining the equilibrium are basic aminoacids, iron, vitamins A and B, and folic acid (Walker *et al.*, 2009). Recent studies have connected the clinical picture of megaloblastic and haemolytic anaemias with the lesions observed on archaeological populations (Walker *et al.*, 2009). In the cases of megaloblastic and haemolytic anaemias, there is compensatory marrow expansion to achieve overproduction of red blood cells so that the body can counter the effects of the anaemia, namely the low red blood cell count. Megaloblastic anemias result from chronic malnutrition and inadequate vitamin B_{12} absorption (Walker *et al.*, 2009). Both in the case of iron-deficiency anaemia and in the case of megaloblastic anaemia due to vitamin B_{12} deficiency, the primary underlying factor is very low or nonexistent animal protein consumption. This condition is exacerbated by high infection rates and diarrhoea caused by parasites and

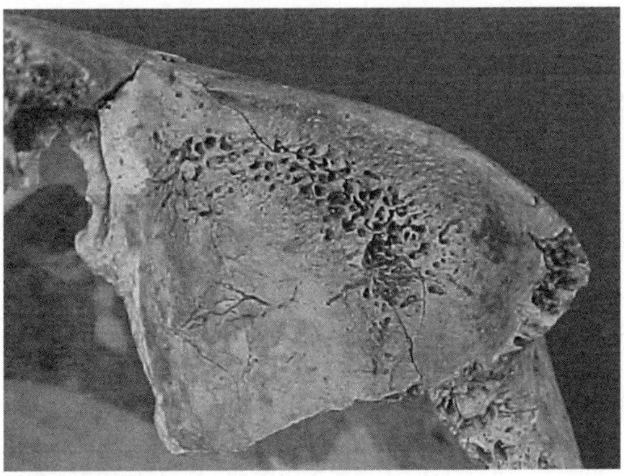

Figure 5.3 *Cribra orbitalia* from Alepotrypa Cave. (See Plate 5.3 for a colour version of this image)

Figure 5.4 *Porotic hyperostosis* from Alepotrypa Cave

high pathogen loads, which cause additional blood loss (Stuart-Macadam, 1985, 1989, 1992). Although a few uncommon conditions have a similar expression on the cranial bones, including hereditary spherocytosis, cyanotic congenital heart disease and polycythemia vera (Ortner and Putschar, 1985), their prevalence in all populations worldwide is negligible and certainly cannot account for the high prevalence observed in many archaeological populations. There are several reasons why genetic forms of anaemia, such as thalassaemia (Angel, 1966), are probably not affecting the Greek Neolithic populations, as was previously believed, including the absence of characteristic pathological lesions on postcranial bones, the slight to moderate expression of lesions, and the evidence of extensive healing and remodelling following the anaemic episode (Papathanasiou, 1999, 2001; Papathanasiou, Larsen and Norr, 2000; Lagia, Eliopoulos and Manolis, 2007). Therefore, the most probable cause for the lesions observed in the Greek Neolithic populations is nutritionally induced anaemia, in conjunction with chronic exposure to parasites and high pathogen loads.

These observations are consistent with the aforementioned stable isotope analysis, which documented a diet characterized by low animal protein intake and an agricultural, sedentary, land-based economy. It is noteworthy that in the Late and Final Neolithic there is a marked increase of the prevalence of anaemic conditions, which may be associated with an overall deterioration of living conditions (Table 5.6).

5.4 INTERPERSONAL VIOLENCE

There is little unambiguous archaeological evidence for trauma amongst Mesolithic and Neolithic Greek populations. There are no depictions of weapons or armaments but there are stone and bone spearheads and arrowheads, which could be used as tools or weapons. Also, a number of sites have been built on summits and promontories and some were encircled by protective walls, trenches or ditches. (Kotsakis, 1983, 1986; Andreou, Fotiadis and Kotsakis, 1996). These constructions, as well as others from materials that did not survive, could be regarded as evidence for fortifications, although they may have as well been symbolic or functional and protective.

Bioarchaeological data derived from lesions on skeletal material can provide direct and unambiguous evidence of violence. As it is difficult to distinguish between fractures caused by

Table 5.8 Classification of cranial trauma by site, sex and age

Site – Period	Sex	Age
Franchthi – Mesolithic	M	25–30
	M	30–35
Xirolimni – Early Neolithic	M	45–50
Proskynas – Late/Final Neolithic	M	45
Kephala – Late/Final Neolithic	M	35
	M	50+
Alepotrypa – Late/Final Neolithic	M	30–35
	M	18
	M	35
	M	35–40
	M	40
	M	30
	?	15

accidents or interpersonal violence, the focus will be on cranial traumatic lesions, either premortem or perimortem, which could be accurately interpreted as bearing evidence of violent interactions (Walker, 2001; Schulting and Wysocki, 2005; Lovell, 1997; Larsen, 1997). While deliberate aggression can affect any part of the body, the head is the most common target of interpersonal violence (Steckel and Wallis, 2007).

Results of the analysis of prevalence rates of trauma (Tables 5.7 and 5.8) (Papathanasiou, forthcoming) imply an overall low prevalence of violence in the Greek Neolithic (3.4% or 13 out of 379 cranial elements), although soft-tissue injuries cannot be ascertained. Also, there are no cases of embedded projectile points. Cranial trauma consists of depressed fractures. All fractures are small, circular, sometimes multiple, and well healed at the time of death (Figure 5.5). Out of the total of 13 individuals that exhibit cranial trauma, 10 are males, 2 are females, and 1 is an adolescent. The prevalence of cranial trauma in the Mesolithic is higher (10% or 2 out of 20), and is observed exclusively on males, although the sample size is very small (Papathanasiou, forthcoming). The incidence of postcranial trauma, either accidental or

Figure 5.5 Healed depressed cranial fracture from Alepotrypa Cave. (See Plate 5.5 for a colour version of this image)

deliberate, is very low and in several cases it affects individuals that exhibit cranial trauma as well. In most sites postcranial trauma affects primarily females, except for Makrygialos, where more males were affected, especially in the lower limbs (Triantaphyllou, 2001).

In summary, cranial depressed fractures are consistent with face-to-face fighting with blunt objects, and indicate that interpersonal conflict was not rare, especially amongst men during the Greek Neolithic (Papathanasiou, forthcoming). The elevated frequencies of cranial injuries in males indicate that violence was actually a 'male business' at these sites. The data also suggests sporadic violence and does not support a case for endemic warfare, as there is no age or sex-based differentiation, and there is a lack of frequent multiple trauma that is characteristic of many battlefield burials, which is present in other western European Neolithic sites (Beyneix, 2007; Teschler-Nicola *et al.*, 1999; Schulting and Wysocki, 2005; Jimenez-Brobeil, du Souich and Al Oumaoui, 2009). Most of the evidence for violence points to non-lethal, face-to-face combat amongst men, and could be unrelated episodes, either within the community or between groups. Again, the restricted sample size does not suggest any increase or decrease through time.

One of the sites provides evidence for a more elaborate biocultural interpretation is Alepotrypa Cave (Papathanassopoulos, 1971, 1996). The site yielded the largest Neolithic osteological series in Greece, which exhibits the highest prevalence of cranial trauma, 13.0% or 9 out of 69 individuals (Papathanasiou, 2001, 2005). Cranial depressed fractures are frequent in the series. They are small, circular, sometimes multiple and well-healed, which seems to indicate that they did not cause significant damage and are suggestive of interpersonal aggression and conflict but not of lethal encounters. Combining this data with other evidence from the archaeological context of the site, may lead to new interpretations and hypotheses.

In Alepotrypa Cave, the area of the living is separated from the area of the dead by the formation of the two ossuaries for secondary burials. The Alepotrypa population has a high prevalence of metopism. The population displays complete metopism in older subadults and adults at 14.7% or 5 out of 34 individuals. Interestingly, 4 out of 5 individuals with metopism come from Ossuary II, which has an incidence of 4 out of 13 or 30.8%. Metopism generally appears sporadically and infrequently in most populations. Some of the highest frequencies (11.3–13.7%) are found in eastern Asian populations and amongst European ones (8–12%) (Hauser and De Stefano, 1989). As inferred statistically, metopism is significantly higher in Ossuary II with respect to the rest of the population, and we may therefore speculate that the high frequency of metopism in this sample is associated with genetic affinities.

These assessments reveal:

1. a land-based agricultural economy and a terrestrial, mainly cereal, subsistence, as indicated by the stable isotope analysis (Papathanasiou, 2003);
2. a relatively high degree of interpersonal violence, as expressed by the elevated number of cranial fractures, especially in Ossuary II;
3. possible familial relationships between the individuals buried in Ossuary II, based on the high incidence of metopism; and
4. a population increase as indicated by the high (0.276) fertility rate (Papathanasiou, 2001).

High levels of cranial trauma (from 9% up to 27%) have been reported in various unrelated populations pursuing various adaptive strategies, ranging from agriculture in the late prehistoric American Midwest (Milner, 1995) to foraging in Australia (Webb, 1995) and the southern California coast (Walker, 1989). These investigations reveal a common theme for populations with elevated cranial trauma. That is, in these settings, populations were living in marginal circumstances involving environmental deterioration, populations stress and competition for increasingly limited resources, factors linked with violence (Milner, 1995; Larsen, 1997). The causes of interpersonal violence amongst the Alepotrypa Cave population cannot be determined but the entire cranial trauma observed shows evidence of antemortem healing, suggesting that in most cases the violence was not lethal (Webb, 1995). This pattern is similar to the presence of predominantly healed trauma in prehistoric native populations from the Channel Barbara Islands off the coast of southern California (Walker *et al.*, 2009).

Full-scale farming economies and year-round sedentism have emerged during the Neolithic. In this context, arable land, and possibly pasture and water, could have been critical but restricted in distribution, access and availability. It is within the context of subsistence-settlement systems, demographic and social processes and the invocation of the ancestors as a key element in social strategies, that a model was designed and justified. It states that groups are more likely to maintain formal disposal areas for the dead rather than dispersed grave sites, when control of restricted resources is crucial (Saxe, 1970; Goldstein, 1976; Charles and Buikstra, 1983; Chapman, 1990, 1995). Thus, this model links the appearance of formal disposal areas or cemeteries with increasing populations, changes in subsistence and reduced mobility (Cullen, 1995). Increasingly sedentary and stressed groups living in a small area, with limited resources, may have wished to mark social differences in a way that would preserve the community ties and at the same time sufficiently control access to these resources (Cullen, 1995). It seems tempting to suggest that creating specific places for successive placement of the dead, probably with familial relationships, may have served to emphasize one or more subgroups' claim or attachment to the site and the surrounding resources, while simultaneously creating a collective memory of the past. Within this context, where people had to defend their economic interests, the secondary burials are viewed as an attempt to create ancestral authority and inheritance, and establish control of critical resources (Papathanasiou, 2001, 2009).

5.5 CONCLUSIONS

This chapter provides a step towards understanding the interplay of the environment and human health and diet, the economy and land use, in the context of social interactions during the Neolithic in Greece. The archaeological record suggests that the first Neolithic people populated fertile plains with small dispersed settlements. Soon after, they appear to have been nucleated in core areas, and during the Late and Final Neolithic, when population size increased, colonized marginal areas in southern Greece, caves and islands had occurred (Kotsakis, 1999; Halstead, 2000, 2008; Whittle, 1996). The subsistence strategies of Greek Neolithic groups were based on intensive mixed (crop and stock) farming based mainly on plants, as livestock had limited potential as a stable food source, and reduced reliance on foraging (Halstead, 2000, 2008). These strategies are similar to the ones in Anatolia and the Near East, and are corroborated by the results of the bioanthropological analyses.

Inter-site comparison of period-specific variations in health indicates that the Early Neolithic populations exhibit a low prevalence of anaemic conditions, namely cribra orbitalia and porotic hyperostosis, linear enamel hypoplasias, as well as stature that is close to the upper limits of the range for later periods (Late and Final Neolithic). It is also evident that these early populations did not experience high levels of stress or deprivation during the years of growth and development. These results are compatible with the related observations from Çatalhöyük

(Koca *et al.*, 2006) and the Near East (Rathbun, 1984; Smith, Bar Yosef and Sillen, 1984; Smith and Kolska-Howritz, 2007; Larsen, 1997). The low levels of stress in the Neolithic state changes in later populations when these factors increase considerably, including high prevalence of dental caries and enamel hypoplasias, osteoarthritis, musculoskeletal stress markers and, mainly, anaemic conditions, which may affect the entire population. Both anaemic conditions and growth retardation are related to protein-poor diets (Walker *et al.*, 2009). Such diets have been documented by stable isotope analysis, which indicates that the earliest sample had adequate diet and protein intake as compared to the later populations.

The palaeodietary analysis contributes important information to the very poor southern European isotopic record for this time period, and provides evidence for a swift and complete shift from foraging to farming, a process for which there is an indication that may already have started during the Mesolithic (cf. Garcia-Guixé, Richards and Subirà, 2006). It also stresses the importance of the role of terrestrial domesticated foodstuffs for the Neolithic economy, even for the earliest ones. Such a diet, based primarily on terrestrial C_3 plants and animals, especially cereals and legumes and little incorporation of wild resources (game or marine foods), is in accordance with the archaeozoological observations, which also indicate that livestock was subsidiary to crop growing (Halstead, 2000, 2008).

This framework of a land-based economy, an increase in population size and density, narrowing viable subsistence options and decreasing communal cohesion (Halstead, 2008; Kotsakis, 1999), could have created a context for interpersonal violence which, however, was restricted to one amongst males, and expressed as non-lethal, face to face encounters, an observation that leans towards sporadic events and not generalized combat.

ACKNOWLEDGEMENTS

The author would like to thank the Wiener Laboratory of the American School of Classical Studies at Athens for funding the research on the Early Neolithic populations of Xirolimni, Mavropigi and Pontokomi, and Prof. Michael P. Richards for conducting the stable isotope analysis at the Max Planck Institute.

REFERENCES

Alram-Stern, E. (1996) *Die Agaische Fruhzeit.* Wien: Der Osterreichischen Akademie der Wissenschaften.

Ammerman, A.J. and Cavalli-Sforza, L.L. (1971) Measuring the rate of spread of early farming in Europe. *Man*, **6**, 674–678.

Andreou, S., Fotiadis, M. and Kotsakis, K. (1996) The Neolithic and Bronze Age of northern Greece. *Am. J. Archaeol.*, **100**, 537–597.

Angel, J.L. (1966) Porotic hyperostosis, anemias, malarias, and marshes in the prehistoric eastern Mediterranean. *Science*, **153**, 760–763.

Angel, J.L. (1977) Appendix 5: Human skeletons, in *Kephala: A Late Neolithic Settlement and Cemetery* (ed. J.E. Coleman), American School of Classical Studies, Princeton, pp. 133–156.

Angel, J.L. (1984) Health as a crucial factor in the changes from hunting to developed farming in the eastern Mediterranean, in *Paleopathology at the Origins of Agriculture* (eds M.N. Cohen and G.J. Armelagos), Academic Press, Orlando, pp. 51–73.

Ambrose, S.H. and Norr, L. (1993) Experimental evidence for the relationship of the carbon isotope ratios of whole diet and dietary protein to those of bone collagen and carbonate, in *Prehistoric Human Bone: Archaeology at the Molecular Level* (eds J.B. Lambert and G. Grupe), Springer, New York, pp. 1–37.

Ambrose, S.H., Butler, B.M., Hanson, D.B. *et al.* (1997) Stable isotopic analysis of human diet in the Marianas Archipelago, Western Pacific. *Am. J. Phys. Anthropol.*, **104**, 343–361.

Beyneix, A. (2007) Remarks about the beginnings of the war in the Neolithic of western Europe. *L' anthropologie*, **111**, 79–95.

Bogucki, P. (1996) The spread of early farming in Europe. *Am. Sci.*, **84**, 242–253.

Brooks, S.T. and Suchey, J.M. (1990) Skeletal age determination based on the os pubis: a comparison of the Ascadi-Nemeskeri and Suchey-Brooks methods. *Hum. Evol.*, **5**, 227–238.

Buikstra, J.E. and Ubelaker, D.H. (1994) Standards for data collection from human skeletal remains. Arkansas Archaeological Survey research series 44. Fayetteville: Arkansas Archaeological Survey.

Chapman, R. (1990) *Emerging Complexity: The Later Prehistory of South-East Spain, Iberia and the West Mediterranean,* Cambridge University Press, Cambridge.

Chapman, R. (1995) Ten years after – megaliths, mortuary practices, and the territorial model, in *Regional Approaches to Mortuary Analysis* (ed. L.A. Beck), Plenum, New York, pp. 29–51.

Charles, D.K. and Buikstra, J.E. (1983) Archaic mortuary sites in the central Mississippi drainage: distribution, structure, and behavioral implications, in *Archaic Hunters and Gatherers in the American Midwest* (eds J.L. Phillips and J.A. Brown), Academic Press, New York, pp. 117–145.

Childe, V.G. (1936) *Man Makes Himself,* Watts, London.

Cullen, T. (1995) Mesolithic mortuary ritual at Franchthi Cave, Greece. *Antiquity*, **69**, 270–289.

DeNiro, M.J. (1985) Postmortem preservation and alteration of *in vivo* bone collagen isotope ratios in relation to paleodietary reconstruction. *Nature*, **317**, 806–809.

DeNiro, M.J. and Epstein, S. (1978) Influence of diet on the distribution of carbon isotopes in animals. *Geochim. Cosmochim. Acta*, **42**, 495–506.

DeNiro, M.J. and Epstein, S. (1981) Influence of diet on the distribution of nitrogen isotopes in animals. *Geochim. Cosmochim. Acta*, **45**, 341–351.

Garcia-Guixé, E., Richards, M.P. and Subirà, E. (2006) Palaeodiets of humans and fauna at the Spanish Mesolithic site of El Collado. *Curr. Anthropol.*, **47**, 549–556.

Goldstein, L.G. (1976) Spatial structure and social organization: Regional manifestations of Mississippian society. Unpublished PhD dissertation, Northwestern University, Evanston, Illinois.

Halstead, P. (2000) Land use in postglacial Greece: Cultural causes and environmental effects, in *Landscape and Land Use in Postglacial Greece, Sheffield Studies in Aegean Archaeology 3* (eds P. Halstead and C. Frederick), Sheffield Academic Press, Sheffield, pp. 110–128.

Halstead, P. (2008) Between a rock and a hard place: coping with marginal colonisation in the later Neolithic and early Bronze Age of Crete and the Aegean, in *Escaping the Labyrinth: Cretan Neolithic in Context. Sheffield Studies in Aegean Archaeology 8* (eds V. Isaakidou and P. Tomkins), Oxbow, Oxford, pp. 229–257.

Hauser, G. and De Stefano, G.F. (1989) *Epigenetic Variants of the Human Skull,* E. Schweizerbart'sche Verlagbuchhandlung, Stuttgart.

Hedges, R.E.M. and Reynard, L.M. (2007) Nitrogen isotopes and the trophic level of humans in archaeology. *J. Archaeol. Sci.*, **34**, 1240–1251.

Hengen, O.P. (1971) Cribra orbitalia: pathogenesis and probable etiology. *Homo*, **22**, 57–75.

Hoefs, J. (1997) *Stable Isotope Geochemistry,* Springer-Verlag, Berlin.

Jimenez-Brobeil, S.A., du Souich, P. and Al Oumaoui, I. (2009) Possible relationship of cranial traumatic injuries with violence in the south-east Iberian Peninsula from the Neolithic to the Bronze Age. *Am. J. Phys. Anthropol.*, **140**, 465–475.

Karamitrou-Mendesidi, G. (1998) Xirolimni Kozanhs 1998. *To Archaiologiko Ergo sth Makedonia kai sti Thraki*, **12**, 472–474.

Karamitrou-Mendesidi, G. (2000) Nomos Kozanis 2000: Anaskafes en odois kai parodios. *To Archaiologiko Ergo sth Makedonia kai sti Thraki*, **14**, 625–640.

Karamitrou-Mendesidi, G. (2005) Mavropigi 2005: Lignitorycheia kai Archaiothtes. *To Archaiologiko Ergo sth Makedonia kai sti Thraki*, **19**, 511–539.

Koca, B., Guleç, E., Gultekin, T. *et al.* (2006) Implications of dental caries in anatolia: from hunting--gathering to the present. *Hum. Evol.*, **21**, 215–222.

Kotsakis, K. (1983) Kerameiki technologia kai kerameiki diaforopoiisi. Provlimata tis graptis kerameikis tis mesis epohis tou Sesklou. Unpublished Ph. D. thesis. Thessaloniki:Aristotelian University of Thessaloniki.

Kotsakis, K. (1986) Aspects of technology and distribution of MN Pottery of Sesklo, in *Science in Archaeology* (eds R.E. Jones and H.W. Catling), British School at Athens, Fitch Laboratory Occasional Paper 2, pp. 1–2.

Kotsakis, K. (1999) What tells can tell: social space and settlement in the Greek Neolithic, in *Neolithic Society in Greece* (ed. P. Halstead), Sheffield Academic Press, Sheffield, pp. 66–76.

Krogman, W.M. and Iscan, M.Y. (1986) *The Human Skeleton in Forensic Medicine*, C.C. Thomas, Springfield.

Lagia, A., Eliopoulos, C. and Manolis, S.K. (2007) Thalassemia: macroscopic and radiological study of a case. *Int. J. Osteoarchaeol.*, **17**, 269–285.

Lallo, J.W., Armelagos, G.J. and Mensforth, R.P. (1997) The role of diet, disease, and physiology in the origin of porotic hyperostosis. *Hum. Biol.*, **49**, 271.

Larsen, C.S. (1997) *Bioarchaeology: Interpreting Behavior from the Human Skeleton*, Cambridge University Press, Cambridge.

Liden, K. (1995) Megaliths, agriculture, and social complexity: a diet study of two Swedish megalith populations. *J. Anthropol. Archaeol.*, **14**, 404–417.

Lovejoy, C.O. (1985) Dental wear in the Libben population: its functional pattern and role in the determination of adult skeletal age at death. *Am. J. Phys. Anthropol.*, **68**, 47–56.

Lovell, L. (1997) Trauma analysis in paleopathology. *Yrbk. Phys. Anthrop.*, **49**, 139–170.

Lubell, D., Jackes, M., Schwarcz, H. *et al.* (1994) The Mesolithic-Neolithic transition in Portugal: isotopic and dental evidence of diet. *J. Archaeol. Sci.*, **21**, 201–216.

Maat, G.J.R. (2003) Male stature. A parameter of health and wealth in The Low Countries, 50–1997 AD, in *Wealth, Health and Human Remains in Archaeology. 25th Kroon-Lecture* (ed. W.H. Metz), Foundation for Anthropology and Prehistory in the, Netherlands, Amsterdam, pp. 57–88.

McKern, T.W. and Stuart, T.D. (1957) Skeletal age changes in young American males. US Natick, MA; Army Quartermaster Research and Development Command, Technical Report EP-45.

Meindl, R.S. and Lovejoy, C.O. (1985) Ectocranial suture closure: a revised method for the determination of skeletal age at death based on the lateral anterior sutures. *Am. J. Phys. Anthropol.*, **68**, 57–66.

Milner, G.R. (1992) *Determination of Skeletal Age and Sex*, Dickson Mound Museum, Lewiston.

Milner, G.R. (1995) An osteological perspective on prehistoric warfare, in *Regional Approaches to Mortuary Analysis* (ed. L.A. Beck), Plenum, New York, pp. 221–244.

Norr, L. (1995) Interpreting dietary maize from stable isotopes in the American tropics: the state of the art, in *Archaeology in the Lowland American tropics: Current Analytical Methods and Applications* (ed P. W. Stahl), Cambridge University Press, Cambridge, pp. 198–223.

Ortner, D.J. and Putschar, W.G.J. (1985) *Identification of Pathological Conditions in Human Skeletal Remains*, Smithsonian Institution Press, Washington.

Papathanasiou, A. (1999) A bioarchaeological analysis of health, subsistence, and funerary behavior in the Eastern Mediterranean basin: a case study from Alepotrypa Cave, Diros, Greece. Unpublished Ph.D. thesis. Iowa City: University of Iowa.

Papathanasiou, A. (2001) *Bioarchaeological Analysis of Neolithic Alepotrypa Cave, Greece, British Archaeological Reports – International Series 961*, Archaeopress, Oxford.

Papathanasiou, A. (2003) Stable isotope analysis in Neolithic Greece and possible implications on human health. *Int. J. Osteoarchaeol.*, **13**, 314–324.

Papathanasiou, A. (2005) Health status of the Neolithic population of Alepotrypa Cave. *Greece. Am. J. Phys. Anthropol.*, **126**, 377–390.

Papathanasiou, A. (2008) Bioarchaeological analysis of health and diet of Early Neolithic settlements in northern Greece. 17th European Meeting of the Paleopathology Association, Copenhagen, Denmark.

Papathanasiou, A. (2009) Mortuary behaviour in Alepotrypa Cave: assessments from the study of the human osteological material. (eds W.G. Cavanagh, C. Gallou and M. Georgiadis), Proceedings of the Conference 'Sparta and Lakonia from Prehistory to Premodern', Athens, British School at Athens Studies, pp. 21–28.

Papathanasiou, A. (Forthcoming) Evidence of trauma in Neolithic Greece (eds R. Schulting and L. Fibinger), Neolithic violence in a European perspective. Oxford.

Papathanasiou, A., Larsen, C.S. and Norr, L. (2000) Bioarchaeological inferences from a Neolithic ossuary from Alepotrypa Cave, Diros, Greece. *Int. J. Osteoarchaeol.*, **10**, 210–228.

Papathanasiou, A., Zachou, E. and Richards, M.P. (2009) Bioarchaeological analysis of the human osteological material from Proskynas, in *New Directions in the Skeletal Biology of Greece* (eds L.A. Schepartz, S.C. Fox and C. Bourbou), The American School of Classical Studies at Athens, Princeton, pp. 223–236.

Papathanassopoulos, G.A. (1971) Spelea Dirou, 1971. *Athens Annals of Archaeology*, **4**, 289–303.

Papathanassopoulos, G.A. (1996) *Neolithic Civilization in Greece*, Goulandris Foundation, Athens.

Pèrles, C. (2001) *The Early Neolithic in Greece*, Cambridge University Press, Cambridge.

Phenice, T.W. (1969) A newly developed visual method of sexing the os pubis. *Am. J. Phys. Anthropol.*, **30**, 297–302.

Pinhasi, R., Fort, J. and Ammerman, A.J. (2005) Tracing the origin and spread of agriculture in Europe. *PLoS Biol.*, **3**(12), e410.

Rathbun, T.A. (1984) Skeletal pathology from the Paleolithic through the Metal Ages in Iran and Iraq, in *Paleopathology at the Origins of Agriculture* (eds M.N. Cohen and G.J. Armelagos), Academic Press, Orlando, pp. 137–167.

Renfrew, C. (1987) *Archaeology and Language*, Cambridge University Press, Cambridge.

Resnick, D. (1995) *Diagnosis of Bone and Joint Disorders*, W.B. Saunders, Philadelphia.

Richards, M.P., Hedges, R.E.M., Walton, I. *et al.* (2001) Neolithic diet at the Brochtorff Circle. *Malta. Eur. J. Archaeol.*, **4**, 253–262.

Saxe, A.A. (1970) Social dimensions of mortuary practices. Unpublished Ph.D. thesis. Ann Arbor, Michigan: University of Michigan.

Schoeninger, M.J. (1989) Reconstructing prehistoric human diet, in *Chemistry of Prehistoric Human Bone* (ed. T.D. Price), Cambridge University Press, New York, pp. 38–67.

Schoeninger, M.J., DeNiro, M.J. and Tauber, H. (1983) Stable isotope nitrogen ratios of bone collagen reflect marine and terrestrial components of prehistoric human diet. *Science*, **220**, 1381–1383.

Schulting, R.J. and Richards, M.P. (2001) Dating women and becoming farmers: New palaeodietary and AMS dating evidence from the Breton Mesolithic cemeteries of Téviec and Hoëdic. *J. Anthropol. Archaeol.*, **20**, 314–344.

Schulting, R.J. and Richards, M.P. (2002) Finding the coastal Mesolithic in Southwest Britain: AMS dates and stable isotope results on human remains from Caldey Island, South Wales. *Antiquity*, **76**, 1011–1025.

Schulting, R.J. and Richards, M.P. (2002) The wet, wild, and the domesticated: The Mesolithic-Neolithic transition on the west coast of Scotland. *Eur. J. Archaeol.*, **5**, 147–189.

Schulting, R.J. and Wysocki, M. (2005) 'In this chambered tumulus were found cleft skulls…': an assessment of the evidence for cranial trauma in the British Neolithic. *Proc. Prehist. Soc.*, **71**, 107–138.

Smith, B.H. (1984) Patterns of molar wear in hunter-gatherers and agriculturalists. *Am. J. Phys. Anthrop.*, **63**, 39–55.

Smith, P., Bar–Yosef, O. and Sillen, A. (1984) Archaeological and skeletal evidence for dietary change during the late Pleistocene/early Holocene in the Levant, in *Paleopathology at the Origins of Agriculture* (eds M.N. Cohen and G.J. Armelagos), Academic Press, Orlando, pp. 101–136.

Smith, P. and Kolska-Howritz, L. (2007) Ancestors and inheritors: a bioarchaeological perspective on the transition to agropastoralism in the southern Levant, in *Ancient Health: Skeletal Indicators of Agricultural and Economic Intensification* (eds M.N. Cohen and M.M. Crane-Kramer), University of Florida Press, Gainesville, FL, pp. 207–222.

Steckel, R.H. (1995) Stature and the standard of living. *J. Econ. Lit.*, **33**, 1903–1940.

Steckel, R.H. and Wallis, J. (2007) Stones, bones, and states: A new approach to the Neolithic Revolution http://www.nber.org/confer/2007/daes07/steckel.pdf.

Stuart-Macadam, P. (1985) Porotic hyperostosis: representative of a childhood condition. *Am. J. Phys. Anthropol.*, **66**, 391–398.

Stuart-Macadam, P. (1989) Nutritional deficiency diseases: A survey of scurvy, rickets, and iron-deficiency anemia, in *Reconstruction of Life from the Skeleton* (eds M.Y. Iscan and K.A.R. Kennedy), Wiley-Liss, New York, pp. 201–222.

Stuart-Macadam, P. (1992) Anemia in past human populations, in *Diet, Demography, and Disease: Changing Perspectives on Anemia* (eds P. Stuart-Macadam and S. Kent), Aldine de Gruyter, New York, pp. 151–170.

Tauber, H. (1981) ^{13}C evidence for dietary habits of prehistoric man in Denmark. *Nature*, **292**, 332–333.

Tauber, H. (1983) ^{13}C dating of human beings in relation to dietary habits. *PACT*, **8**, 365–375.

Tauber, H. (1986) Analysis of stable isotopes in prehistoric populations, in *Innovative Trends in Prehistoric Archaeology. Mitteilungen der Berliner Gesellschaft für Anthropologie* (eds B. Hänsel and B. Herrmann), Ethnologie and Urgeschichte 7, Berlin, pp. 31–38.

Teschler-Nicola, M., Gerold, F., Bujatti-Narbeschuber, M. *et al.* (1999) Evidence of genocide 7000 BP – Neolithic paradigm and geo-climatic reality. *Collegium Antropol.*, **23**, 437–450.

Todd, T.W. (1920) Age changes in the pubic bone. I. The white male pubis. *Am. J. Phys. Anthropol.*, **3**, 285–334.

Todd, T.W. (1921) Age changes in the pubic bone. II, III, IV. *Am. J. Phys. Anthropol.*, **4**, 1–70.

Triantaphyllou, S. (2001) *A Bioarchaeological Approach to Prehistoric Cemetery Populations from Central and Western Greek Macedonia. British Archaeological Reports – International Series 976*, Archaeopress, Oxford.

Trotter, M. (1970) Estimation of stature from intact limb bones, in *Personal Identification in Mass Disasters* (ed. T.D. Stuart), National Museum of Natural History, Washington DC, pp. 71–83.

Tyson, R.A. and Alcauskas, E.S.D. (1980) *Catalogue of the Hrdlicka Paleopathology Collection*, San Diego Museum of Man, San Diego.

Ubelaker, D.H. (1989) *Human Skeletal Remains: Excavations, Analysis, Interpretation*, 2nd edn, Aldine, Washington DC.

van der Merwe, N.J. and Vogel, J.C. (1978) ^{13}C content of human collagen as a measure of prehistoric diet in Woodland North America. *Nature*, **276**, 815–816.

Walker, P.L. (1989) Cranial injuries as evidence of violence in prehistoric southern California. *Am. J. Phys. Anthropol.*, **80**, 313–323.

Walker, P.L. (2001) A bioarchaeological perspective on the history of violence. *Ann. Rev. Anthropol.*, **30**, 573–596.

Walker, P.L., Bathurst, R.R., Richman, R. *et al.* (2009) The causes of porotic hyperostosis and cribra orbitalia: a reappraisal of the iron-defiency-anemia hypothesis. *Am. J. Phys. Anthropol.*, **139**, 109–125.

Webb, S. (1995) *Palaeopathology of Aboriginal Australians: Health and Disease Across a Hunter-Gatherer Continent*, Cambridge University Press, Cambridge.

White, T.D. (1991) *Human Osteology*, Academic Press, Orlando.

Whittle, A. (1996) *Europe in the Neolithic*, Cambridge University Press, Cambridge.

Whittle, A. and Cummings, V. (2007) Going over the Mesolithic-Neolithic transition in North West Europe. Proceedings of the British Academy. Oxford.

Ziota, C. (1995) Kitrini Limni 1995: Nees erevnitikes drastiriotites. *To Archaiologiko Ergo sth Makedonia kai sti Thraki*, **9**, 47–58.

Ziota, C. (1998) Proistoriko nekrotafeio stin Koilada Kozanis. Mia proti analytikh parousiasi tis anaskafikis ereunas, in *Mneias Charin, Tomos sti mnimi Mairis Siganidou. Thessaloniki*, Archaeological Receipts Fund, pp. 81–102.

6

Using a Bioarchaeological Approach to Explore Subsistence Transitions in the Eastern Cape, South Africa During the Mid- to Late Holocene

Jaime K. Ginter[1,2]

[1] *School of Community and Liberal Studies, Sheridan Institute of Technology & Advanced Learning, Oakville, ON, Canada*
[2] *TUARC – Trent University Archaeological Research Centre, Peterborough, ON, Canada*

6.1 INTRODUCTION

Over the past century, much work has focused on the origins of food production in southernmost Africa. The first evidence of domestication occurs around 2000 BP, almost 4000 years later than in other parts of Africa. This process was not a local initiative as aspects of a food producing economy, first sheep and then agropastoralism, were introduced from outside the region. In South Africa, the new subsistence lifestyle of herding did not completely replace the existing way of life. Foraging continued to exist alongside herding, unlike many other parts of the world, where the hunting and gathering way of life terminated with the arrival of domesticates during the Neolithic. The incomplete nature of this subsistence transition has made determining the origins of the herders and their relationship to the foragers problematic.

The timing of the arrival of herding to southernmost South Africa at around 2000 BP is generally agreed upon (Sealy and Yates, 1994; Smith, 2005), yet the mechanisms responsible for this subsistence shift – a migration of peoples or a diffusion of ideas – as well as the impact of this novel form of food production remain open to debate. The most prominent theory, based on ethnographic, archaeological and linguistic information, suggests that herding groups migrated to South Africa from the north of the Zambezi River around 2000 BP

Human Bioarchaeology of the Transition to Agriculture Edited by Ron Pinhasi and Jay T. Stock

(cf. Boonzaier *et al.*, 1996; Deacon and Deacon, 1999; Smith *et al.*, 2000). However, recent excavations and re-analyses of archaeological data suggest that the idea of herding and the sheep, rather than herders themselves, spread throughout South Africa (Sadr, 2003). This led some foragers to make the transition to herding while others maintained their forager lifestyle. The speed at which herding travelled from the north to the southern part of the continent, as attested by a relative lack of variability in dates for sites up to 5000 km apart (Deacon, 1984; Kinahan, 1996), is puzzling. Although differences in artefact frequency and composition (ostrich egg shell bead diameters, pottery, sheep vs. wild faunal remains) (cf. Boonzaier *et al.*, 1996; Smith *et al.*, 1991, 2000) and burial style (Inskeep, 1986) have been proposed to distinguish forager and herder sites, social and cultural similarities, as well as the fluidity of the respective lifestyles (Kelly, 1995; Kusimba, 2002), have made this chapter of South Africa's prehistory difficult to resolve (Parsons, 2006). Genetic studies demonstrate that San foragers and Khoekhoe herders share the same ancient Southern African heritage (Chen *et al.*, 2000; Barkhan and Soodyall, 2006). However, genes cannot resolve the debate regarding the introduction of herding, as genetic diversity is the product of both evolutionary mechanisms, such as genetic drift, population bottlenecks and founder effects, and other factors, including climate and environment (Lahr and Foley, 1998; Cooper *et al.*, 2000). While evidence of sheep and the pastoral lifestyle has been uncovered at South African sites beginning at around 2000 BP (Binneman, 1998, 2000; Bousman, 1998; Sealy and Yates, 1994), the quantity of cultural residues supports an initial but minor pastoral influence around 2000 BP with an increased importance by 1000 BP (cf. Sadr, 2004). Because of their mobile lifestyle, the cultural and economic change of foragers and herders is often difficult to track, suggesting that archaeological evidence for sheep herding is possibly underestimated.

The complexity of this issue demands that research moves beyond the more traditional realms of enquiry of archaeology, ethnography and linguistics to explore additional lines of evidence in the hope of achieving clarity for this longstanding population interaction question. A bioarchaeological approach that incorporates information collected from the skeletal remains of the foragers and herders themselves, with multiple lines of evidence from other sources including the archaeological context, food residues, floral and faunal remains, linguistics and ethnography, holds much promise for teasing apart some of the complexities of subsistence change in southernmost Africa. This paper will address questions relating to the origin of food production in southernmost Africa, specifically the mechanisms responsible for the introduction and the biological relationship between herders and foragers, through the study of skeletal metric and cranial discrete data collected from a sample of Later Stone Age (LSA) adult skeletons dating between 8260 to 240 BP from the Eastern Cape region of South Africa.

6.2 BACKGROUND

Foraging was the first and most longstanding form of subsistence, existing exclusively in this region for 20 000 years (Deacon and Deacon, 1999; Phillipson, 2005). Later Stone Age foraging groups living across the Cape region of South Africa display some variation in their behaviour and lifestyle as a consequence of the different environments in which they lived, but these groups are thought to have been genetically and culturally homogeneous prior to the introduction of herding around 2000 BP (Deacon and Deacon, 1999; Mitchell, 2002a). Similarities in lithic technology, social organization and subsistence throughout southern Africa suggest that a shared foraging way of life and extensive trade network existed from the Zambezi River region (Zambia, Zimbabwe and Mozambique) to the South African Cape (Mitchell, 2002a). Foraging peoples made spiritual and symbolic connections with their environment and would tend to return to familiar areas year after year (Hall, 2000) and were buried in the home ranges that they occupied in life (Sealy, 1986, 2006; Sealy and van de Merwe, 1986). During the early Holocene (12 000–8000 BP) foraging groups relied upon a large flake and scraper lithic complex known as Oakhurst and hunted large migratory bovids (Deacon and Deacon, 1999). Foraging groups were distributed widely across the landscape, but group sizes were sufficient to support large game hunting. Around 8000 BP the focus on hunted resources shifted to smaller, territorial solitary browsers (Deacon, 1984; Mitchell, 2002b), but it is thought that gathered foods, primarily underground corms and bulbs, formed the main component of the diet (Deacon, 1993, 1976). This subsistence shift was accompanied by the development of a microlithic tool complex known as Wilton (Deacon and Deacon, 1999). A cooling and drying period that began around 8000 BP stimulated a widespread evacuation of interior regions of South Africa by foraging groups in favour of more habitable and productive localities to the south (Deacon, 1976, 1984).

Increasing population size and density across the southern region of South Africa coupled with climatic shifts seems to have put stress on existing resources, requiring inhabitants to diversify their subsistence behaviours. Resource scarcity and imbalance associated with increased population growth sets up conditions suitable for the intensification of productivity, (Price and Brown, 1985), increasing social complexity (Binford, 2001) and a move from an immediate to a delayed return economy (Woodburn, 1982). Around 4500 BP there is evidence that foraging groups throughout the Cape region began to intensify their subsistence approaches, initiating a shift in their dietary focus from terrestrial foods towards a greater emphasis on riverine and marine foodstuffs (Inskeep, 1986; Sealy and van de Merwe, 1988; Hall, 1990; Sealy *et al.*, 1992; Binneman, 1996a, b; Jerardino Wiesenborn, 1996; Jerardino and Yates, 1996; Mitchell, 2002a). The climate and vegetation in the Cape during the mid-Holocene was similar to present conditions (Rosen, Lewis and Illgner, 1999), yet around 4000 BP environmental conditions changed, initiating a period marked by lower temperatures and increased rainfall (Lee-Thorp *et al.*, 2001; Mayewski *et al.*, 2004; Lewis, 2005). These conditions favoured the growth of shrubs over grasslands in the southeastern region, resulting in a subsequent increase in smaller game in faunal assemblages (Mitchell, 2002). Foraging groups inhabiting the Cape region at around 3500 BP exhibited additional indicators of population pressure and resource stress. Evidence for increased territoriality and identity signalling is demonstrated by the individuation of tool production at coastal sites in the Eastern Cape (Binneman, 2004/2005). Other aspects of intensification, namely the establishment of food specialization and dietary niches have been noted at closely related sites in the southern Cape (Sealy, 2006), along with a general shift towards a reliance on smaller food packages, mainly smaller terrestrial mammals and a focus on marine resources, with plant foods constituting a greater proportion of the diet (Parkington, 1980; Buchanan *et al.*, 1984). Changes in the form and incidence of rock art, interpreted as indicators of population stress, are also observed in previously unoccupied (Manhire, Parkington and Robey, 1984) and existing rock shelters (Parkington *et al.*, 1986). Other signs of intensification, food production (Henshilwood, Nilssen and Parkington, 1994) and storage (Deacon, 1984; Hall, 1990; Binneman, 2004/2005) are also present.

The delayed onset of food production in this part of Africa has been attributed to the abundant, diverse and readily available wild plant and animal resources (Cunningham and

Davis, 1997). However, the current study and others suggest that during the period between approximately 4000 and 2000 BP the environment was no longer able to support the nutritional requirements of the existing population, requiring the exploitation of new and varied food sources (Hall, 1990; Binneman, 1996a; Sealy and Pfeiffer, 2000; Pfeiffer and Sealy, 2006; Stynder, Rogers-Ackerman and Sealy, 2007a). Environmental and social pressures created a situation that encouraged innovation, experimentation, and social and subsistence diversification as a means to cope with these stressors and ensure continued prosperity in this region. This widespread subsistence change has been suggested as the impetus for the introduction of animal domestication in this area (Sadr, 2004; Sealy, 2006).

The potential effects of subsistence, mobility, climate and environmental changes in other parts of sub-Saharan Africa during the Holocene must also be considered in any attempts to understand the sudden appearance of the herding lifestyle in southernmost Africa at around 2000 BP. On the African continent, root crop farming is believed to have originated in West Africa around the sixth millennium BP, while agricultural practices involving cereal cultivation and herding of domestic stock are thought to have developed in North-East Africa around the eighth millennium BP (Vansina, 1996; Wetterstrom, 1998). All forms of agriculture appear to have slowly spread towards the southeast, becoming firmly established during the succeeding millennia north of the Zambezi River (Marshall and Hildebrand, 2002; Vansina, 1996). The southeast spread of the agropastoral complex beyond the Zambezi River (also known as the Bantu Expansion) did not take place until the beginning of the second millennium BP (Vansina, 1996; Phillipson, 2005; Huffman, 2006). These genetically and culturally distinct Bantu-speaking mixed agriculturalists finally settled in the eastern fringes of the Eastern Cape area around 1300 BP, an area previously inhabited only by foragers (Silberbauer, 1979; Binneman, 1996a, b; Boonzaier *et al.*, 1996; Lewis, 2002). While the most likely source of sheep was Bantu-speaking agropastoralists, who settled in the Zambezi River area prior to their final southward migration into the eastern parts of South Africa around 1300 BP, the exact mechanism responsible for the transmission of sheep into South Africa remains to be determined.

6.3 MATERIALS

Adult skeletal remains of Holocene foragers recovered from various sites within the Eastern Cape Province of South Africa form the basis for this research. The Eastern Cape region of South Africa presents an excellent location to explore questions of subsistence transitions and subsistence variability throughout the Holocene, as an early sheep presence has been recorded at a number of archaeological sites in the region. The nature of the question being explored in this research demands a high resolution chronology within the Holocene. As such, only skeletal remains that had been radiocarbon dated, or were in direct archaeological association with radiocarbon dated individuals, were included in this study. The Eastern Cape sample includes the remains of 73 individuals dating between 8260 and 240 BP (uncalibrated) (Table 6.1). Of these, 40 individuals predate 2000 BP and 33 postdate the 2000 BP benchmark. The completeness of the skeletal remains included in the study sample is variable. Most were the product of controlled excavations and have been curated with accompanying information about the archaeological context, manner and style of burial and cultural affiliation. Yet, many museum specimens are the product of chance collections or donations by individuals unaware of archaeological protocols. Regardless of the variable

Table 6.1 Total Eastern Cape skeletal sample

Museum	Individual	Location	Date/ Association	Age[a]	Sex	Laboratory[#]	Reference
A	1139	Knkelbosch	4800 ± 50	YA	F	Pta-8816	[3]
A	1124	Port Elizabeth	4300 ± 32	MA	F	OxA-V-2056-42	[3]
A	1127	Jeffries Bay	1891 ± 29	MA	F	OxA-V-2066-36	[3]
A	1152	Amsterdam Hoek	1850 ± 35	OA	F	Pta-8757	[3]
A	1166	Humewood, Port Elizabeth	1818 ± 27	MA	F	OxA-V-2056-33 A	[3]
A	1117	Lime Bank/Loerie	1060 ± 50	MA	F	Pta-8727	[3]
A	2787A	Andrieskraal	3028 ± 32	OA	M	OxA-V-2161-55	[2]
ALB	119	Wilton Large Rock Shelter	8260 ± 720	YA	M	GaK-1541	[1]
ALB	121	Wilton Cave	4680 ± 60	MA	F	Pta-8566	[1]
ALB	124	Wilton Large Rock Shelter	Associated with ALB121	MA	M		[1]
ALB	128A	Spitzkop	Associated with ALB131	OA	F		[1]
ALB	128B	Spitzkop	Associated with ALB131	YA	F		[1]
ALB	128C	Spitzkop	Associated with ALB131	MA	F		[1]
ALB	129	Spitzkop	Associated with ALB131	MA	M		[1]
ALB	131	Spitzkop	4700 ± 60	MA	M	Pta-5979	[1]
ALB	136	Spitzkop	4930 ± 70	OA	F	Pta-8620	[1]
ALB	139	Spitzkop	5100 ± 70	YA	F	Pta-8626	[1]
ALB	150	Kabeljaaus, CaveA	1910 ± 60	MA	M	TO-10368	[2]
ALB	151A	Kabeljaaus, CaveB	2920 ± 45	YA	M	Pta-8570	[1]
ALB	160B	Kabeljaaus, CaveA	Associated with ALB161	YA	F		[1]
ALB	161	Kabeljaaus, CaveA	2730 ± 60	YA	M	Pta-8720	[1]
ALB	174	Kleinpoort farm	430 ± 50	YA	M	Pta-8574	[1]
ALB	177	Kleinpoort farm	390 ± 40	OA	F	Pta-8584	[1]
ALB	178	Kleinpoort farm	240 ± 45	OA	F	Pta-8599	[1]
ALB	180	near Uitenhage	330 ± 50	OA	F	Pta-8718	[1]
ALB	184	Dunbrody	320 ± 50	MA	M	TO-10374	[2]
ALB	195A	Melhoutboom	2870 ± 90	YA	F	Pta-706	[1]
ALB	198	Middlekop Kloof	5120 ± 70	AD	F	Pta-8618	[1]
ALB	200	Middlekop Kloof	5105 ± 20	YA	M	Pta-8638	[1]
ALB	204	Mooikrantz (Vygeboom)	3204 ± 60	MA	M	Pta-8690	[1]
ALB	206	Zuurberg	4610 ± 50	MA	M	Pta-8713	[1]
ALB	210	Corm Flats, Adelaide	1580 ± 50	MA	F	Pta-8734	[1]

(continued)

Table 6.1 (*Continued*)

Museum	Individual	Location	Date/ Association	Age[a]	Sex	Laboratory[#]	Reference
ALB	217	Kleinpoort farm	1670 ± 20	AD	F	Pta-8613	[1]
ALB	221	Kleinemonde	4180 ± 70	MA	F	Pta-8736	[1]
ALB	222	Seal Point, Cape St Francis	2640 ± 60	YA	M	Pta-8636	[1]
ALB	223	Seal Point, Cape St Francis	1650 ± 60	YA	F	Pta-8631	[1]
ALB	243	Farm near Conmadagga	290 ± 45	YA	F	Pta-9237	[1]
ALB	244(1)	Paardefontein near Jansenville	1180 ± 50	MA	F	Pta-8587	[1]
ALB	259	Addo	4400 ± 70	OA	F	Pta-9221	[1]
ALB	273	Teasdale 2	350 ± 50	MA	M	Pta-8671	[1]
ALB	277	Sea Vista/Cape St Francis	670 ± 50	MA	M	Pta-8685	[1]
ALB	293	Cannon Rocks	460 ± 50	MA	F	Pta-8679	[1]
ALB	296	Klasies Cave 5 burial 2	2180 ± 50	AD	M	Pta-8672	[1]
ALB	301	St Francis Bay	2570 ± 50	MA	M	Pta-8684	[1]
ALB	302	Sand River/ Geodgeloof	2620 ± 60	MA	F	Pta-8710	[1]
ALB	303	Sand River/ Geodgeloof	1550 ± 20	MA	M	Pta-8699	[1]
ALB	304	Sand River/ Geodgeloof	Associated with ALB302	MA	F		[1]
ALB	305	Hamburg	640 ± 40	MA	F	Pta-9224	[1]
ALB	307	Bokness River Mouth	400 ± 50	MA	M	Pta-8726	[1]
ALB	308	Welgeluk Shelter	5140 ± 70	MA	M	TO-10240	[2]
ALB	309	Welgeluk Shelter	Associated with ALB308	MA	M		[1]
ALB	314	Kleinemonde Island	2130 ± 50	MA	M	Pta-8693	[1]
ALB	316	Groot Kommandokloof	460 ± 60	OA	F	Pta-8600	[1]
ALB	319A	Joubertinia, Langkloof	2560 ± 60	YA	F	Pta-8682	[1]
ALB	323	Sand River/ Geodgeloof	1620 ± 35	MA	F	Pta-8578	[1]
ALB	328	Cape St Francis	1670 ± 60	MA	M	Pta-8655	[1]
ALB	338	Oyster Bay	2110 ± 45	MA	F	Pta-8721	[1]
ALB	339	Port Alfred	430 ± 45	MA	F	Pta-8674	[1]
ALB	341	Cape St. Francis	2950 ± 60	MA	F	Pta-8934	[1]
ALB	344	Gonube	1957 ± 26	YA	F	OxA-15077	[2]
ALB	354	Paradysstrand, Jeffreys Bay	3340 ± 60	YA	F	Pta-8680	[1]
NMB	86	Cape St Francis	2275 ± 40	MA	F	GrA-23657	[3]

Table 6.1 (*Continued*)

Museum	Individual	Location	Date/ Association	Age[a]	Sex	Laboratory[#]	Reference
NMB	82	Groot Kommandokloof	2355 ± 40	MA	M	GrA-23228	[3]
NMB	83	Cape St Francis	1590 ± 40	MA	M	GrA-23227	[3]
SAM	32	Humansdorp District	3754 ± 35	MA	F	OxA-V-2055-47	[3]
SAM	4179	Thys Bay, Humansdorp	1528 ± 27	MA	F	OxA-V-2065-45	[3]
SAM	4180	Thys Bay, Humansdorp	688 ± 27	MA	F	OxA-V-2056-23	[3]
SAM	4874	Cape St Francis	1426 ± 29	YA	M	OxA-V-2056-45	[3]
SAM	6032	Cape St Francis, near Seal Pt light house	5180 ± 65	MA	M	Pta-1089	[4]
UCT	78	Cape St Francis	2145 ± 40	OA	M	GrA-23241	[3]
UCT	109	Humansdorp	1590 ± 50	MA	F	GrA-23656	[3]
UCT	83	Cape St Francis	680 ± 40	YA	M	GrA-23072	[3]
UCT	114	Cape St Francis	650 ± 40	YA	M	GrA-23654	[3]

[a]Age: AD: Adolescent (16–20 years); YA: Young Adult (21–35 years); MA: Middle Adult (36–49 years); OA: Old Adult (50+ years).
References: [1] Sealy, pers. comm.; [2] Pfeiffer, pers. Comm.; [3] Stynder (2006); [4] Morris (1992a).

range of information available from each skeleton, the skeletal sample is sufficient to explore the questions posed in this study.

6.4 METHODS

6.4.1 Bioarchaeological Approach

The analytic approaches employed by bioarchaeologists enable information about biology, culture and the environment to be extracted from human skeletal remains. Because aspects of an individual's life are preserved within the skeleton, researchers are able to interpret various aspects of human life, such as diet and subsistence practices, health, body size, activity and population demography (cf. Blakely, 1977; Cohen and Armelagos, 1984; Cohen, 1989; Powell, Bridges and Wagner Mires, 1991; Larsen, 1997, 2002; Steckel and Rose, 2002; Williamson and Pfeiffer, 2003; Buikstra and Beck, 2006). Variation in skeletal morphology across time and space can be used to address questions that are at the centre of the southern African forager-herder debate: biological relatedness, population continuity and interactions, and cultural and subsistence change.

By examining both metric and discrete skeletal variables, behavioural and genetic information from the skeleton can be accessed, because differences in the size, shape and morphology of the human skeleton are influenced by both genetic and cultural factors. Both the nature and location of skeletal variation can inform about the factors responsible for any observed alterations to skeletal morphology, as the levels of plasticity and rates of development vary for the different regions of the skeleton. Although environmental factors are known to have some influence on cranial size and shape (cf. Guglielmino-Matessi, Gluckman and Cavalli-Sforza, 1979; Beals, Smith and Dodd, 1983; Franciscus and Long, 1991; Smith, Terhune and Lockwood, 2007), similarities in cranial form are generally assumed to reflect common biological affinity (cf. Howells, 1995; Lahr, 1996; Sparks and Jantz, 2002; Relethford, 2004; Roseman and Weaver, 2004). Cranio-dental development is believed to be more buffered against environmental stress than the postcranial skeleton, thus preserving genetic information more directly (cf. Kieser, 1990). However, environmental factors, such as nutritional stress, can produce changes in tooth size amongst genetically homogenous populations (Garn, Lewis and Walenga, 1968; Larsen, 1997; Stojanowski, 2005; Stojanowski *et al.*, 2007). While there is a genetic component to body size and physique (Trinkaus, Churchill and Ruff, 1994; Bogin, 1999), the postcranial skeleton is more subject to environmental influences and mechanical factors resulting from mobility and activity (Bogin and Loucky, 1997; Pearson, 2000; Bogin *et al.*, 2002). Skeletal discrete traits have been found to be relatively unaffected by external factors (Berry and Berry, 1967; Anderson, 1968; Berry, 1968; Molto, 1983; Hauser and De Stefano, 1989; Hanihara, Ishida and Dodo, 2003). Unlike osteometric measurements, discrete traits can be observed on incomplete and fragmentary skeletal remains, making this a favoured method of analysis for investigating topics of biological affinity and population relationships in archaeologically derived skeletal samples (cf. Ossenberg, 1969; Molto, 1983; DeLaurier and Spence, 2003; Rubini, Mogliazza and Corruccini, 2006). Because contributions from these different factors vary throughout the skeleton, it is important to integrate information from the entire skeleton and also to consider other sources to help interpret patterns in the skeletal data.

Stress, resulting from social and economic change associated with the shift to a new way of life, can manifest on the skeleton as growth disruption, disease or illness, or ultimately death. Studies that have explored the impact of the transition from foraging to agriculture on the human skeleton, have found that differences in diet, mobility, subsistence technology, and activity affect the growth and development of the skeleton in a variety of ways (cf. Blakely, 1977; Boyd and Boyd, 1989; Cohen and Armelagos, 1984; Cohen, 1989, 2008; Powell, Bridges and Wagner Mires, 1991; Larsen, 1995, 1997; Steckel and Rose, 2002). Yet, few investigations have examined the effects of the transition from foraging to herding. Regional and temporal differences in bone mass and upper arm asymmetry (Pfeiffer and Stock, 2002; Stock and Pfeiffer, 2004), body size and stature (Sealy and Pfeiffer, 2000; Pfeiffer and Sealy, 2006), positional behaviour (Dewar and Pfeiffer, 2004) and trauma (Pfeiffer, 2001, in press) have been identified amongst Later Stone Age (Holocene) forager populations from the southern and western Cape regions of South Africa. These differences, while not framed in terms of the introduction of herding, are all potentially indicative of subsistence variability. They add credence to the approach undertaken herein for examining the introduction of herding in southern Africa through the bioarchaeological study of Later Stone Age populations from the Eastern Cape region.

6.5 DATA COLLECTION AND ANALYSIS

6.5.1 Skeletal Identification

Sex was assessed by examining a number of morphological features of the cranium (Buikstra and Ubelaker, 1994) and pelvis (Phenice, 1969; Buikstra and Ubelaker, 1994). Later Stone Age populations from South Africa fall at the small end of the body size range, exhibit unique body proportions, relatively low cranial and pronounced pelvic sexual dimorphism, compared to other populations (Kurki *et al.*, 2010). Since most morphological methods of sex estimation have been developed on larger bodied and differently dimorphic populations, multiple methods were employed with primacy given to pelvic sex indicators.

Age was estimated using age-related morphological changes to the symphyseal face (Suchey and Katz, 1986; Brooks and Suchey, 1990) and auricular surface (Lovejoy *et al.*, 1985) of the pelvis, the closure of the cranial sutures (Meindl and Lovejoy, 1985), the obliteration of the palatal sutures (Mann *et al.*, 1991) and degenerative changes of the sternal rib ends (Iscan, Loth and Wright, 1984, 1985). The extent of dental wear (Smith, 1984) along with the overall condition of the skeleton, with respect to both age related degeneration and bone formation, at the joint surfaces of the elbow, shoulder, knee, hip and vertebral column, were considered in the final age estimate. For the purposes of analysis, the various age indicators were used to place each individual in a general age category (young adult, middle adult, and old adult), rather than assign an individually unique age range estimate.

6.5.2 Metric Skeletal Data

Quantitative variables specific to the cranium, dentition and postcranial skeleton were recorded to describe the morphology (size and shape) of the different functional units of the skeleton. The entire suite of metric and discrete variables could not be assessed for each skeleton in the sample due to differences in the degree of preservation of the skeletal remains. Variability exists in both the size of the sample and the number of variables that could be assessed per individual for each category of analysis.

Forty-seven craniometric variables were recorded following definitions employed by Howells (1973) and Martin (1957) in order to characterize the morphology of the vault, face and mandible. Measurements of paired cranial elements were recorded on the left side; if the left side was damaged, the right side was substituted. All cranial measurements were recorded to the nearest millimetre using sliding, spreading calipers and co-ordinate calipers.

Odontometric data was collected using the maximum dental method (Morrees and Reed, 1954) and the method developed by Hillson, FitzGerald and Flinn (2005) that measures the mesiodistal (MD) and buccolingual (BL) dimensions of the tooth at the cervico-enamel junction. However, cervical odontometric data was utilized for most analyses, because the significant degree of dental wear exhibited by most of the LSA would have precluded the application of the maximum dental method to a substantial portion of the skeletal sample. Teeth that exhibited any pathological or taphonomic alteration of the enamel surface at the measurement location were excluded. Measurements of all suitable teeth were recorded to the nearest 0.01 mm using Hillson-FitzGerald digital fine tipped digital calipers. Right and left antimeres of all available maxillary and mandibular incisors, canines, premolars and molars were measured.

Osteometric measurements were collected for all individuals with postcranial remains, using measurements listed in Buikstra and Ubelaker (1994) based on the definitions of Moore-Jansen *et al.* (1994). Fifty-eight measurements were selected to characterize the size and shape of the upper limb (humerus, ulna and radius), torso (clavicle, scapula, innominate and sacrum) and lower limb (femur, tibia, fibula and calcaneus, and first metatarsal). Size was assessed through measurement of bone lengths, breadths, heights and circumferences, while shape was evaluated through anterior-posterior (A-P) and medial-lateral (M-L) shaft diameters.

Principal Components Analysis (PCA) was employed to identify size and shape patterns in the metric data. PCA, as a multivariate statistical method, is very well suited to the composition of this skeletal sample, as well as the questions posed in this research, because it does not require the organization of the sample into *a priori* groups for analysis and is not adversely affected by small sample size, unlike other multivariate methods, such as discriminant analysis (Pimentel, 1992). Because the Eastern Cape sample does not include any skeletons that were complete enough to permit the collection of every metric variable, a reduced set of metric variables were selected to reflect the range of morphology of each skeletal region: skull (vault, face and mandible; 23 measurements); postcranial skeleton (upper limb, trunk and lower limb; 19 measurements); and 26 measurements of the dentition (maxillary and mandibular anterior (incisors and canine) and posterior (premolars and molars)) (Table 6.2). These resulted in different subsets of the total sample being used in different analyses: craniometric (total $N = 60$, PCA $n = 49$), postcranial metric (total $N = 56$, PCA $n = 45$), and odontometric (total $N = 57$, PCA $n = 34$). PCA was used to reduce these variables into factors that represent the maximum range of size and shape variation reflected in the original variable set (Pimentel, 1992). A component, typically the first principal component (PC1), is interpreted to reflect size variation if all of the scores are positive, while components with scores of mixed signs, usually the remaining principal components, are interpreted to reflect shape variation (Jolicoeur and Mosimann, 1960; Pimentel, 1992). Morphological differences in the sample were then explored by plotting principal component and examining potential patterns in the distribution of scores for the pre-2000 BP and post-2000 BP subgroups.

Curve estimation regression was used to assist in the interpretation of changes in skeletal morphology through time. Principal Component scores were regressed on the uncalibrated radiocarbon dates to explore relationships between these two variables. Regression analyses that yielded large coefficients of determination (r^2) values coupled with statistically significant ($p < 0.05$) values were interpreted to indicate the presence of a significant relationship between the dependent (PC score) and independent (^{14}C date) variables. Size and/or shape patterns in the metric data were further identified by grouping the skeletal sample by sex and ^{14}C date and plotting of the regression outputs graphically.

Metric data from other sub-Saharan African forager, herder and agricultural skeletal samples were compared with the Eastern Cape to further explore population continuity in this region. Comparative craniometric data (sample size, mean, standard deviation) is available for prehistoric sub-Saharan African agropastoralist populations from Malawi, East Africa, West-Central Africa, West Africa and Southern Africa (Ribot, 2002; Morris and Ribot, 2006) and for protohistoric herders and farmers from the Orange River region of South Africa (Morris, 1992b). Comparative odontometric data is available for Western and Southern Cape LSA foragers (Pfeiffer, 2007), early Griqua herders from South Africa (Kieser, 1985), South African Bantu-speaking agriculturalists (Kieser, Groeneveld and Cameron, 1987) and historic San foragers (Drennan, 1929). Because the cervical method of odontometric analysis has been developed recently, comparative data are not available for Sub-Saharan populations. Comparative data is also available for a selection of osteometric variables for East African, Central African (Pygmies), West African and southern African San skeletal samples dating to the nineteenth and twentieth centuries (Holliday, 1995), as well as three South African Bantu-speaking skeletal samples: Cape Nguni, Natal Nguni and Sotho (Lundy, 1986). The exploration of potential differences between the various sub-Saharan samples and the Eastern Cape subgroups (pre- and post-2000 BP) was facilitated through the computation of t-tests of

Table 6.2 Variables used in the craniometric, odontometric and postcranial metric Principal Component Analyses

Craniometric			Odontometric			Postcranial Metric		
Vault	GOL	Maximum Cranial Length	*Maxillary Anterior*	UI1CMD	First incisor cervical mesiodistal diameter	*Trunk*	CLAVXL	Maximum length clavicle
	BNL	Cranial Base Length		UI2CBL	Second incisor cervical buccolingual diameter		SACRANTB	Anterior breadth sacrum
	BBH	Basion Bregma Height		UCCBL	Canine cervical buccolingual diameter		ILIACBR	Iliac breadth
	XCB	Maximum Cranial Breadth		UCCMD	Canine cervical mesiodistal diameter		BIILIACBR	Bi-iliac breadth
	WFB	Minimum frontal breadth		UI2CMD	Second incisor cervical mesiodistal diameter	*Upper Limb*	HUMXL	Maximum length humerus
	ASB	Biasterionic breadth	*Maxillary Posterior*	UP1CBL	First premolar cervical bucolingual diameter		HUMMIDXD	Maximum midshaft diameter humerus
	FRC	Frontal Chord		UP1CMD	First premolar cervical mesiodistal diameter		HUMHEADD	Head diameter humerus
	PAC	Parietal Chord		UP2CBL	Second premolar cervical bucolingual diameter		HUMEPBR	Epicondylar breadth humerus
Face	NLH	Nasal height		UP2CMD	Second premolar cervical mesiodistal diameter		RADXL	Maximum length radius
	JUB	Bijugal breadth		UM1CBL	First molar cervical bucolingual diameter		RADHEADD	Maximum head diamter radius
	NLB	Nasal breadth		UMICMD	First molar cervical mesiodistal diameter		RADAPD	Anterior-posterior diameter midshaft radius
	MAB	Maxillo-Alveolar breadth		UM2CBL	Second molar cervical bucolingual diameter	*Lower Limb*	FEMXL	Maximum length femur
	OBH	Orbital Height		UM2CMD	Second molar cervical mesiodistal diameter		FEMMIDADL	Midshaft anterior-posterior diameter
	OBB	Orbital Breadth	*Mandibular Anterior*	LI1CMD	First incisor cervical mesiodistal diameter		FEMHHEDD	Horizontal head diameter femur

(continued)

Table 6.2 (*Continued*)

Craniometric			Odontometric			Postcranial Metric	
	DKB	Interorbital breadth		LI2CBL	Second incisor cervical buccolingual diameter	FEMEPICB	Epicondylar breadth femur
	FMB	Bifrontal Breadth		LI2CMD	Second incisor cervical mesiodistal diameter	TIBXL	Maximum length tibia
	EKB	Biorbital Breadth		LCCBL	Canine cervical buccolingual diameter	TIBPEPBR	Proximal epiphyseal breadth tibia
Mandible	GNI	Chin Height		LCCMD	Canine cervical mesiodistal diameter	TIBDEPBR	Distal epiphyseal breadth tibia
	GOG	Bigonial Width	*Mandibular Posterior*	LP1CBL	First premolar cervical bucolingual diameter	TIBAPNFD	Anterior-posteior diameter nutrient foramen tibia
	WRB	Minimum Ramus Breadth		LP1CMD	First premolar cervical mesiodistal diameter		
	XRH	Maximum Ramus Height		LP2CBL	Second premolar cervical bucolingual diameter		
	MXML	Mandibular Length		LP2CMD	Second premolar cervical mesiodistal diameter		
	MAN	Mandibular Angle		LM1CBL	First molar cervical bucolingual diameter		
				LM1CMD	First molar cervical mesiodistal diameter		
				LM2CBL	Second molar cervical bucolingual diameter		
				LM2CMD	Second molar cervical mesiodistal diameter		

summary data (i.e. sample size, mean and standard deviation) using the NCSS statistical software package.

6.5.3 Discrete Trait Data

Fifty-two cranial discrete traits, believed to reflect inheritance, were examined in the Eastern Cape sample (Table 6.3). Bilateral cranial discrete traits were recorded on both left and right elements when possible. Discrete variables were assessed for all available adult crania,

Table 6.3 Cranial discrete traits examined in this study

Trait	Code	Trait	Code
Facial		*Basilar*	
Infraorbital Suture	IOSUT	Condylar canal absent	CONDCANABS
Infraorbital Foramen	IOFORAM	Divided hypoglossal canal	DIHYPOC
Zygomaticofacial foramen	ZYFFOR	Tympanic dehiscence	TYMDIHS
Partial Metopic Suture	METOPPART	Marginal Foramen	MARGFOR
Complete Metopic Suture	METOPFULL	Foramen spinosum incomplete	FRSPINI
Supraorbital notch	SONOTCH	Foramen ovale incomplete	FOROVLI
Supraorbital foramen	SOFORAM	Pterygospinous bridge – trace	PTSBRTR
Supratrochlear notch	STNOTCH	Pterygospinous bridge – partial	PTSBRPT
Frontal Foramen	FRONTFOR	Pterygospinous bridge – complete	PTSBRCOM
Trochlear Spur	TROCSPR	Pterygospinous bridge – all	PTSBRALL
Superior		Pterygoalar bridge – trace	PTABRGTR
Frontal Lines	FRONTLN	Pterygoalar bridge – partial	PTABRGPT
Coronal Ossicle	CORONL	Pterygoalar bridge – complete	BTABRGCOM
Bregmatic bone	BREGMATB	Pterygoalar bridge – all	PTABRGALL
Sagittal ossicle	SATITOSSIC	Spinobasal bridge – trace	SPBASBRTR
Parietal Foramen	PFORAM	Spinobasal bridge – partial	SPBASBRPRT
Lateral		Spinobasal bridge – complete	SPBASBRCOM
Asterionic bone	ASTRINB	Spinobasal bridge – all	SPBASBRALL
Ossicle occipitomastoid suture	OMSUT	Ossified Apical Ligament	APICOSS
Parietal notch bone	PARNOTB	Palatine torus development	PALTOR
Parietal process of temporal	PARPROC	*Mandible*	
Auditory exostosis	AUDEXOS	Mylohyoid bridge development	MYLHBRD
Mastoid foramen temporal	MASTFRTEMP	Mental foramen absent	MENTFORABS
Mastoid foramen sutural	MASTFRSUT	Accessory mental foramen	ACCMENTFOR
Mastoid foramen occipital	MASTFROCC	Mandibular torus	MANDTOR
Epipteric bone	EPITER		
Os Japonicom	OSJAP		
Posterior			
Apical bone	APICALBN		
Inca Bone	INCABONE		
Lambdoid Ossicle	LAMBOSS		

regardless of the completeness or preservation. This approach allowed for the inclusion of 59 individuals in the cranial discrete trait analysis. The side method, whereby the right and left scores are treated as individual observations, was employed to maximize the number of individuals included in the analysis. Thus, the sample size (the number of right and left sides available) varies for each discrete trait. While most traits were assessed using a presence-absence approach, graded discrete data were converted to presence-absence following Molto (1983) and DeLaurier and Spence (2003). The Pearson chi square (χ^2) test is used to explore variation in the 48 cranial discrete traits between pre- and post-2000 BP individuals in the Eastern Cape sample.

Summary statistics (sample size and mean) for a selection of cranial discrete traits are available for South African Bantu-speaking samples (De Villiers, 1968; Rightmire, 1972, 1976), enabling comparisons to be made with the Eastern Cape sample. The discrete cranial sample is divided into subgroups by radiocarbon date (pre- and post-2000 BP) and relationships with the comparative samples are evaluated using an on-line chi-square calculator (Ball and Connor-Linton, 1996).

6.6 RESULTS

6.6.1 Principal Components Analysis

The results of the principal component analyses for all three skeletal anatomical regions examined (crania, dentition and postcrania) produced considerable overlap of the components representing both size (PC1) and shape (PC2) (Figures 6.1–6.3), suggesting that there are no significant changes in skeletal morphology through time in the Eastern Cape region. For the cranium, the greatest overlap of the PC1 and PC2 component scores is seen for the vault variables, suggesting significant homogeneity in vault size and shape throughout the Holocene (Figure 6.1a). Some separation of the four subgroups is evident in the face and mandible principal component scores, yet the scores for all subgroups still overlap considerably (Figures 6.1b and c). Plotting the maxillary and mandibular anterior and posterior teeth PC scores did not elucidate any patterns; there is considerable overlap amongst the sub-groups for both the size (PC1) and shape (PC2) components. The small size of the cervical odontometric sample may be a contributing factor to the absence of temporal patterns in the distribution of the PC scores. Since the principal component results are similar for the four dental regions, a plot of the PC1 and PC2 scores for the maxillary anterior cervical variables is provided to illustrate this trend (Figure 6.2).

The principal component results for the postcrania substantiate the cranial and dental findings. While the upper and lower limb PC scores for all four sub-groups overlap considerably, a slight increase in trunk size through time amongst both sexes is evident, as the pre-2000 BP PC1 scores cluster at the lower end of the axis, while the post-2000 BP PC1 scores overlap at the higher end (Figure 6.3a). The second principal component did not produce any meaningful trends pertaining to the shape of the postcranial skeleton (Figures 6.3b and c). The scores for all four sub-groups overlap and span the entire axis, signifying homogeneity in trunk, upper limb and lower limb morphology through time.

In sum, while there is minor separation of pre- and post-2000 BP principal component scores for a few of the principal component analyses denoting some temporal patterning in relation to

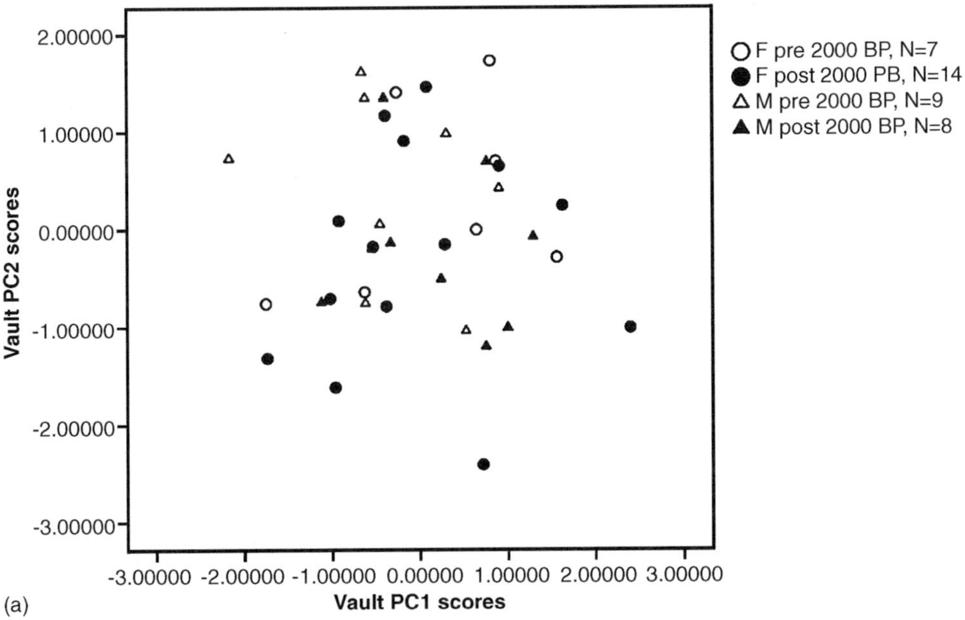

(a)

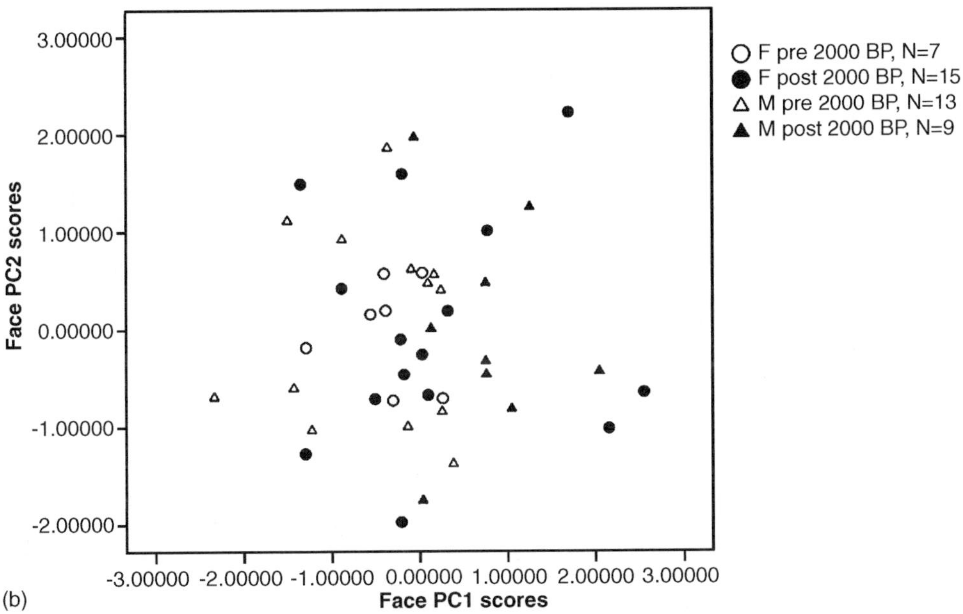

(b)

Figure 6.1 (a) Scatterplot of PC1 (40% total variance) and PC2 (17% total variance) scores for cranial variables – vault. (b) Scatterplot of PC1 (46% total variance) and PC2 (17% total variance) scores for cranial variables – face. (c) Scatterplot of PC1 (42% total variance) and PC2 (27% total variance) scores for cranial variables – mandible

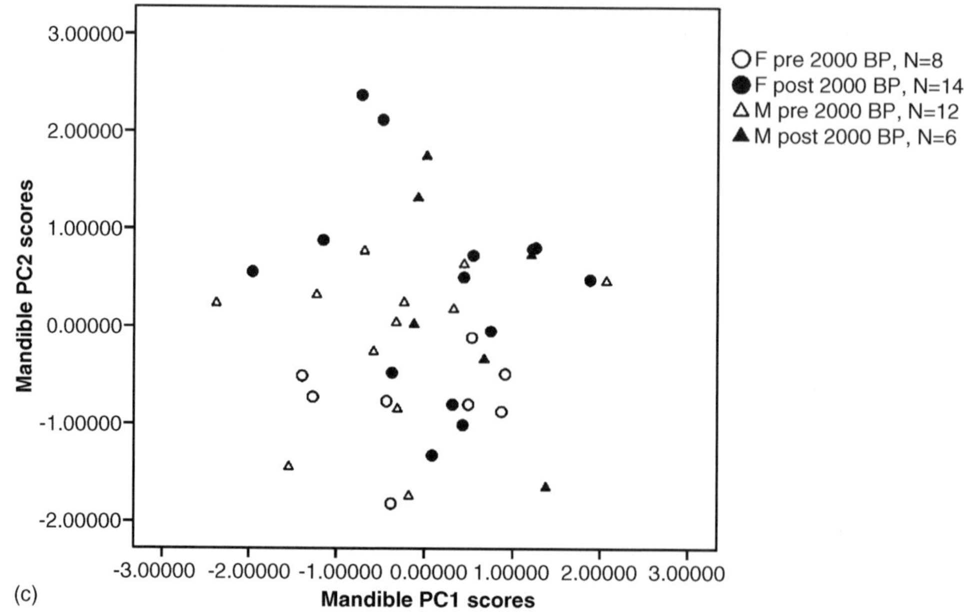

Figure 6.1 (*Continued*)

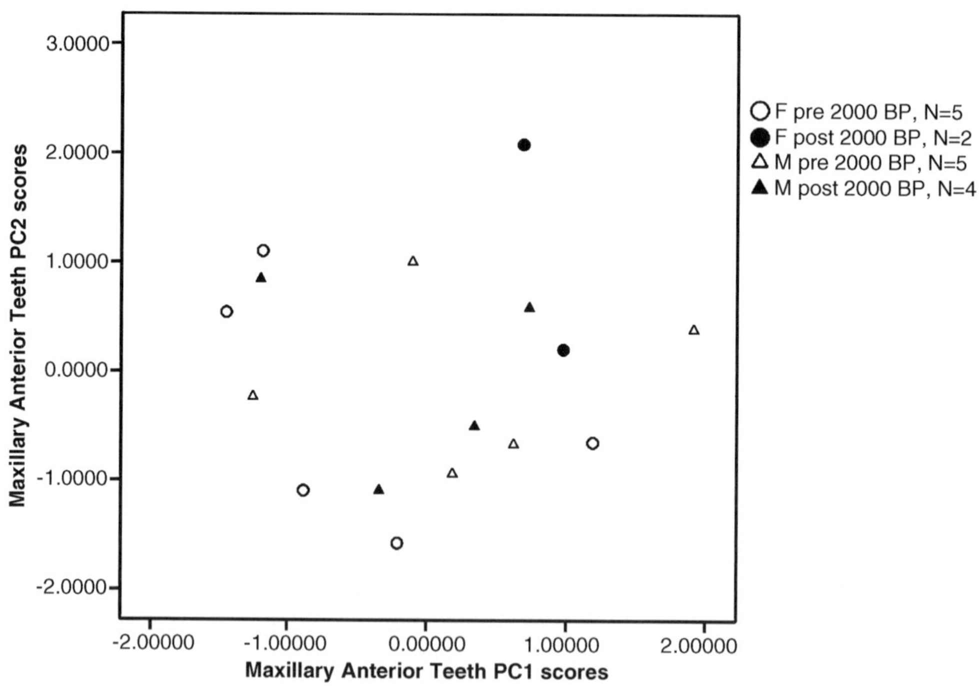

Figure 6.2 Scatterplot of PC1 (70% total variance) and PC2 (20% total variance) scores for maxillary anterior cervical variables

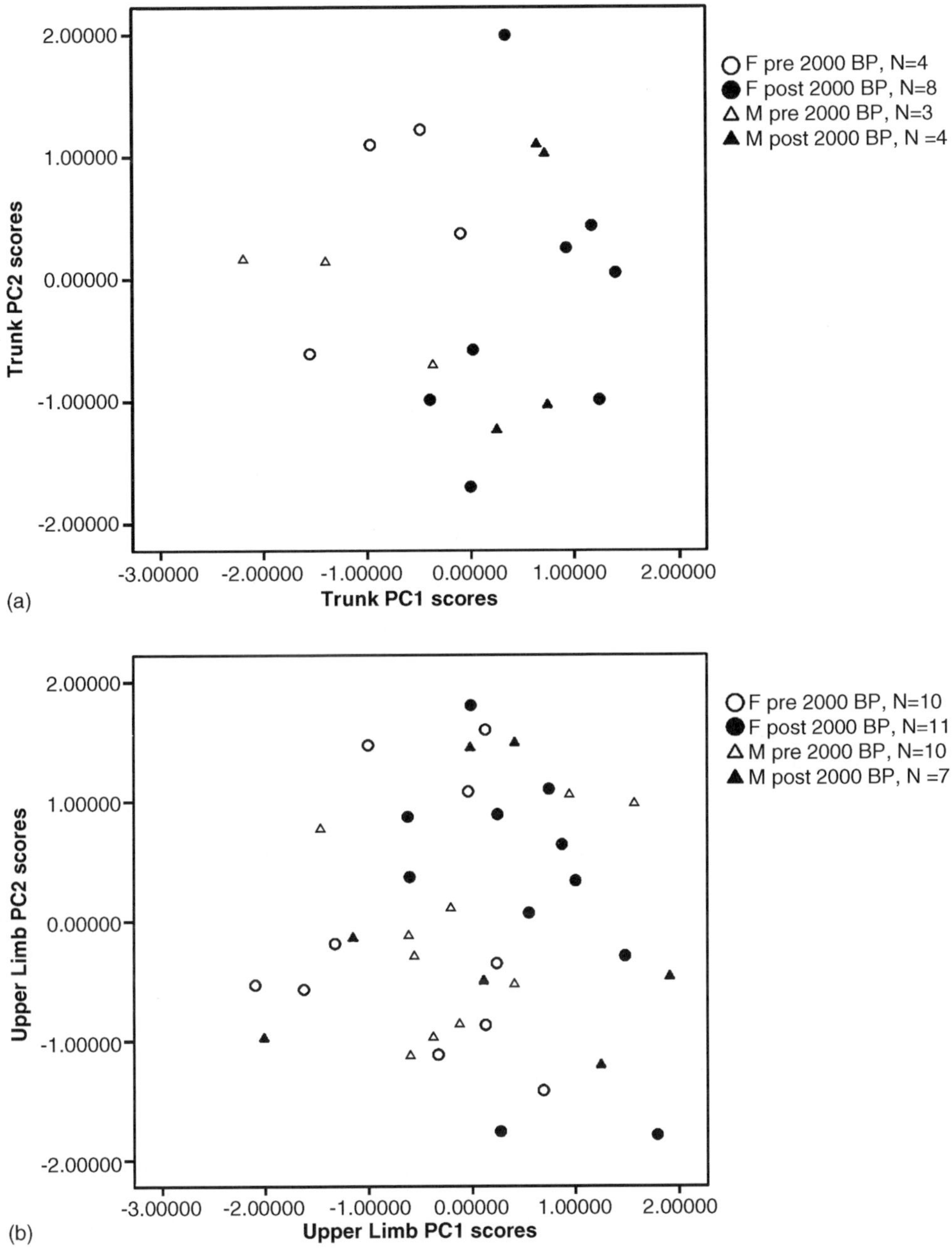

Figure 6.3 (a) Scatterplot of PC1 (77% total variance) and PC2 (13% total variance) scores for the trunk variables. (b) Scatterplot of PC1 (61% total variance) and PC2 (15% total variance) scores for the upper limb variables. (c) Scatterplot of PC1 (58% total variance) and PC2 (15% total variance) scores for the lower limb variables

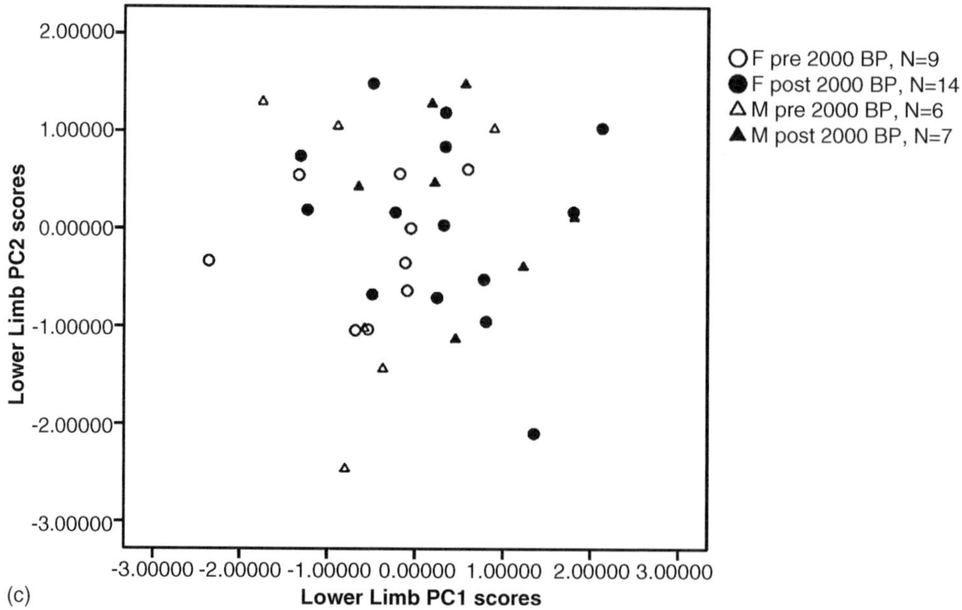

Figure 6.3 (*Continued*)

skeletal size, the general overlap of both PC1 and PC2 scores for all four subgroups suggests relative uniformity of size and shape across the skeleton throughout the Holocene.

6.6.2 Curve Estimation Regression

The absence of any clear patterns of cranial, dental and postcranial size and shape in the principal component results illustrates the difficulties of applying this statistical approach to a single, relatively small sample. However, the regression of principal component scores against radiocarbon date allows more subtle patterns to be identified and the quantitative results to be viewed along a continuum rather than as a dichotomy. A general pattern of size reduction beginning around 3500 BP with subsequent rebound in size around 2000 BP, to pre-3500 BP levels, is observed throughout the skeleton (Figures 6.4–6.6). Some skeletal regions exhibit stronger relationships between these two variables than others. Vault size appears to become more variable between 3500 and 2000 BP. However, a relationship between the vault PC1 scores and radiocarbon date is not statistically supported (Figure 6.4a). A weak trend of decreasing breadth and increasing length of the vault region was identified for both sexes, but is only statistically significant for males ($r^2 = 0.395$, $p = 0.007$). The region of the crania exhibiting the most significant change appears to be the face. A strong relationship between time and change in facial size is observed only for males ($r^2 = 0.507$, $p = 0.001$). Males with the smallest faces are observed between 3000 and 2000 BP, after which facial size returns to the maximum of the pre-3500 size range and continues to increase into the recent past. Female facial size appears to increase through time, becoming more variable just after 2000 BP, reflecting individuals with larger faces in the post-2000 BP sample (Figure 6.4b). Mandibular size remains relatively uniform through time, becoming less variable between 4000 and

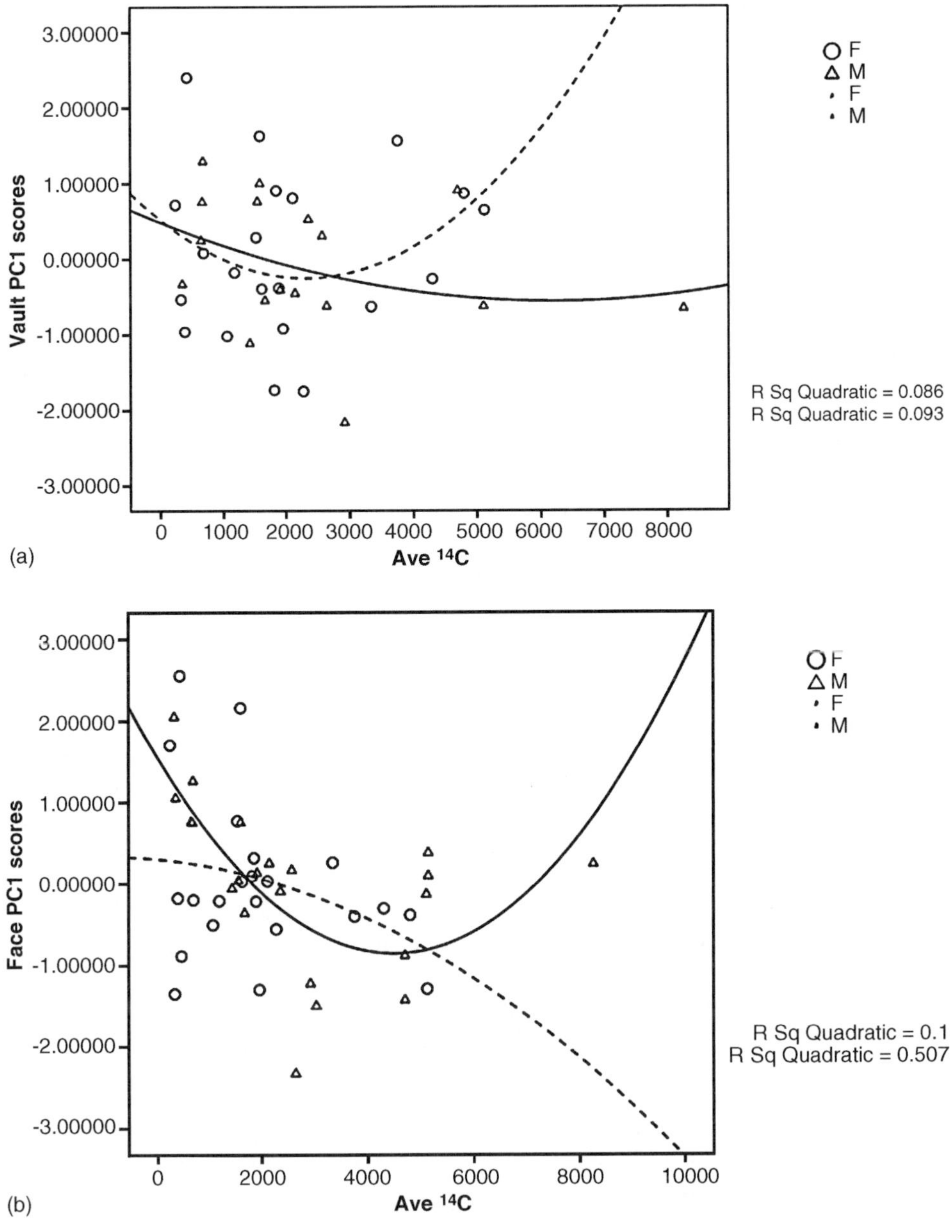

Figure 6.4 (a) Scatterplot of regression scores for vault variables: PC1 vs. ^{14}C. Females: dashed line, males: solid line. (b) Scatterplot of regression scores for face variables: PC1 vs. ^{14}C. Females: dashed line, males: solid line. (c) Scatterplot of regression scores for mandibular variables: PC1 vs. ^{14}C. Females: dashed line, males: solid line

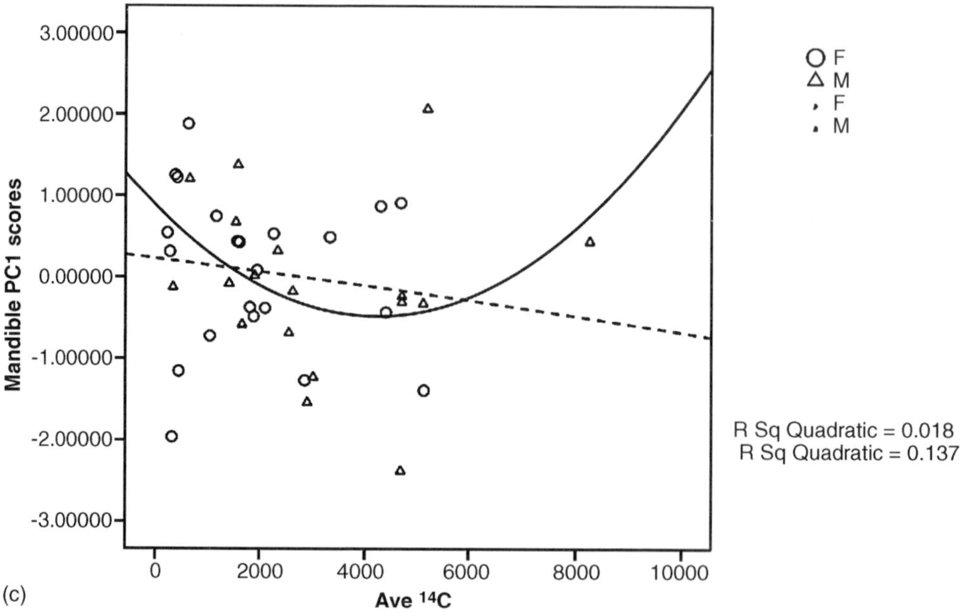

(c)

Figure 6.4 (*Continued*)

2000 BP (Figure 6.4c). Just prior to 2000 BP, it seems that male and female mandibular size increases, with the absence of mandibles falling at the small end of the range observed before 4000 BP. Statistically significant changes in mandibular shape are observed only for females ($r^2 = 0.222$, $p = 0.027$). Female mandibular shape appears to remain relatively stable, changing markedly just prior to 2000 BP.

Regression of the principal component scores against time reveals some changes in tooth size that were not detectable in the principal components results alone. A drop in PC1 scores between 2000 and 3500 BP, with a recovery in size to pre-3000 BP values, occurs in all of the odontometric regression analyses. Scatterplots depict a short-lived reduction in tooth size between approximately 3500 and 2000 BP for all mandibular and maxillary teeth (anterior and posterior), but this relationship is only statistically significant for male mandibular posterior teeth ($r^2 = 0.702$, $p = 0.048$) (Figure 6.5a) and female maxillary posterior teeth ($r^2 = 0.657$, $p = 0.014$) (Figure 6.5b). Although the regression analysis revealed only one significant relationship between female maxillary posterior PC1 scores and radiocarbon date, scatter plots depict a short-lived reduction in tooth size between approximately 3500 and 2000 BP. In comparison, no significant relationships resulted from the maxillary and mandibular PC2 regression analyses, indicating that male and female tooth shape remained stable through time.

The trunk, upper limb and lower limb follow the same decrease in size around 3500 BP, as was observed for the other skeletal regions, but the curve regressions produced a greater number of significant relationships (Figures 6.6a–c). Although the relationship between PC1 and ^{14}C date is significant for male ($r^2 = 0.971$, $p = 0.007$) and female ($r^2 = 0.631$, $p = 0.011$) trunks, female upper limbs ($r^2 = 0.337$, $p = 0.014$), and female ($r^2 = 0.321$, $p = 0.021$) and male ($r^2 = 0.521$, $p = 0.025$) lower limbs, scatter plots suggest that the reduction in postcranial

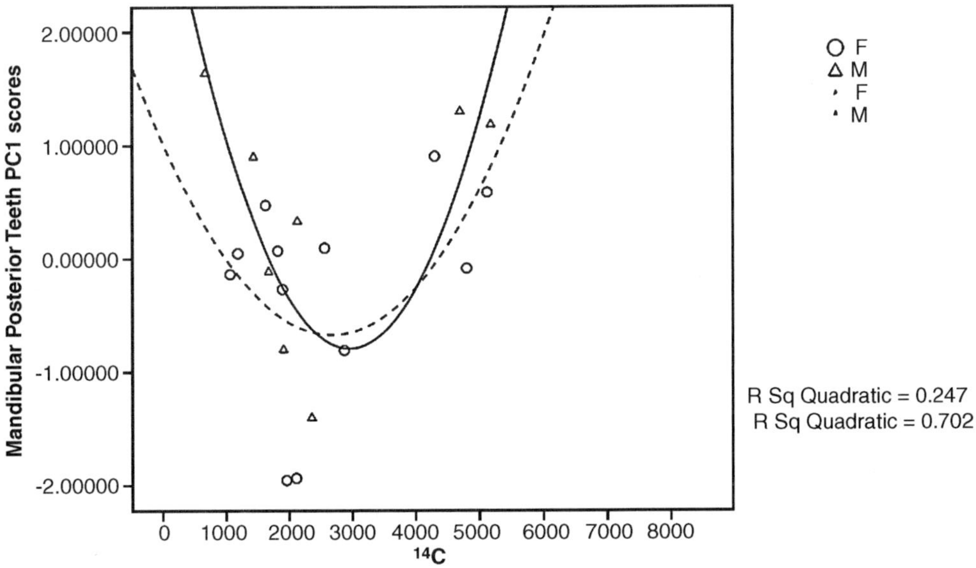

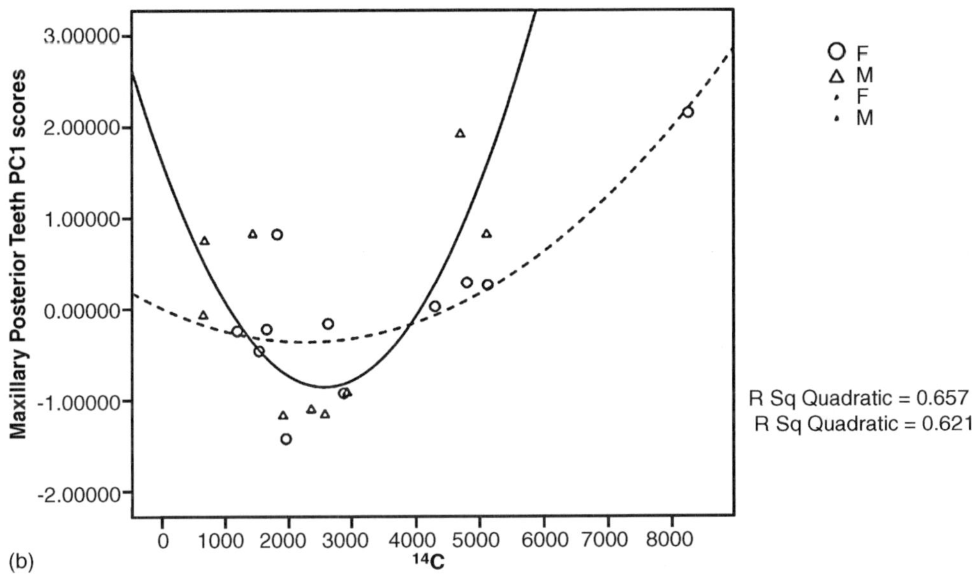

Figure 6.5 (a) Scatterplot of regression scores for mandibular posterior cervical variables: PC1 vs. ^{14}C. Females: dashed line, males: solid line. (b) Scatterplot of regression scores for maxillary posterior cervical variables: PC1 vs. ^{14}C. Females: dashed line, males: solid line

size extends to all regions, even those not statistically supported. All of the curve regression scatter plots depict a drop in body size between 3500 and 2000 BP, with size rebounding to and exceeding pre-3500 BP levels within the past 500 years. The regression of the PC2 scores against ^{14}C date did not produce any significant relationships indicative of changes in postcranial shape through time.

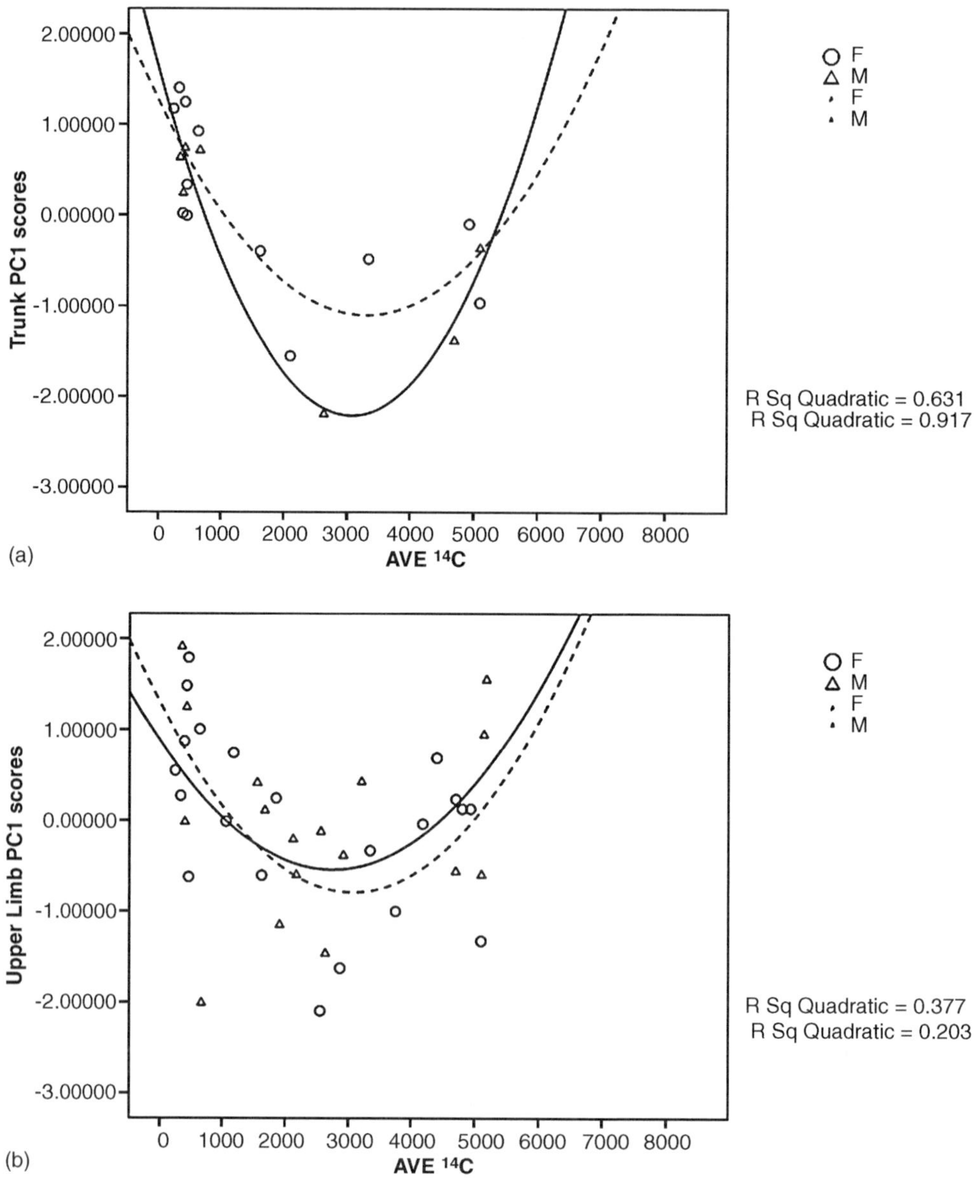

Figure 6.6 (a) Scatterplot of regression scores for trunk variables: PC1 vs. ^{14}C. Females: dashed line, males: solid line. (b) Scatterplot of regression scores for upper limb variables: PC1 vs. ^{14}C. Females: dashed line, males: solid line. (c) Scatterplot of regression scores for lower limb variables: PC1 vs. ^{14}C. Females: dashed line, males: solid line

Curve estimation regression helped to identify a general pattern of slight to moderate size changes, without significant accompanying changes in skeletal form throughout the skeleton. The congruity in the timing of the observed reduction in size between 3500 and 2000 BP, with the return to pre-3500 BP levels, suggests that environmental factors may be the cause.

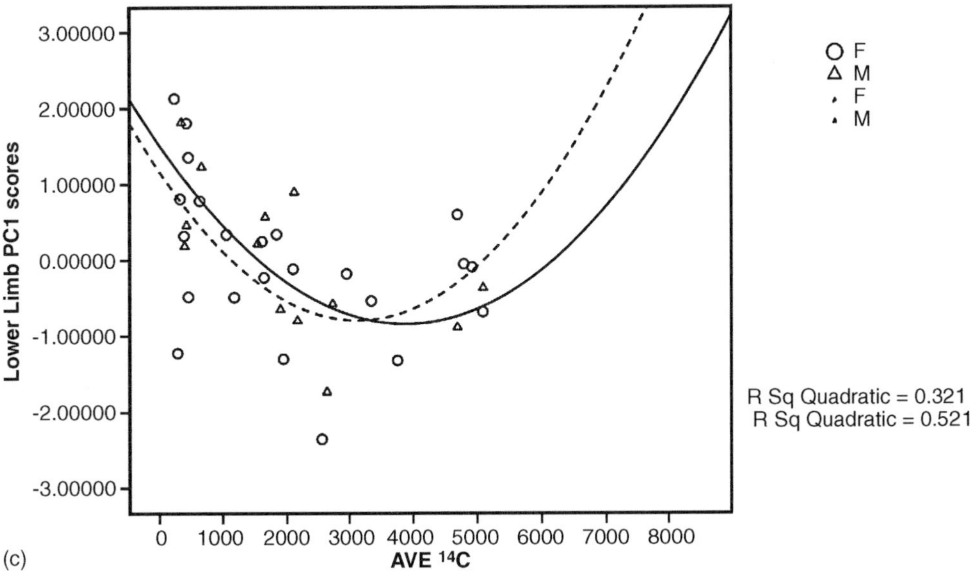

(c)

Figure 6.6 (*Continued*)

Although some shape changes are evident, these are weak and sex-specific. They are restricted to the cranial vault and mandible, and do not correlate strongly with time, signifying that genetic homogeneity was maintained in the Eastern Cape throughout the Holocene. Although the size and temporal distribution of some of the samples used in the various multivariate analyses may push the limitations of the respective statistical approaches and could be viewed as enhancing the magnitude of the observed size changes, the identification of a consistent temporal pattern of size reduction and rebound throughout the skeleton strengthens the results.

6.6.3 Results of Comparative Metric Analyses

Analysis of cranial, dental and postcranial metric data between the pre- and post-2000 BP Eastern Cape subsamples and other sub-Saharan comparative samples yielded significant differences (Tables 6.4-6.7). Significant differences in cranial size are observed between the Eastern Cape and the comparative Bantu-speaking samples (Tables 6.4 and 6.5). The greatest degree of cranial similarity is observed between the Eastern Cape and the South African Khoesan and forager and herder groups (Riet River, Kakamas, Abrahamsdam and Griqua) from the Orange River area of South Africa. While Eastern Cape male crania exhibit more similarities to the comparative samples than their female counterparts, the entire Eastern Cape sample exhibits little affinity with the Central African, East African and South African Bantu-speaking samples. Similarities in the dental dimensions between the Eastern Cape and LSA samples, as well as the prevalence of significant differences between the Eastern Cape and Bantu-speaking values, support the idea of genetic continuity in the Eastern Cape sample through time (Table 6.6). The results of the comparative osteometric analyses also support the

Table 6.4 Comparisons between Eastern Cape and comparative male craniometric data – t-tests using means and standard deviations

Variable	WCA-M[b] vs. M		CAB-M[b] vs. M		EA-M[b] vs. M		SAB-M[b] vs. M		SAK-M[b] vs. M		RR-M[c] vs. M		K-M[c] vs. M		A-M[c] vs. M		G-M[c] vs. M	
	t	p	t	p	T	p	t	p	t	p	t	p	t	p	t	p	t	p
PRE-2000 BP																		
GOL	–0.474	0.6431[a]	–0.094	0.3678[a]	0.858	0.4069[a]	1.753	0.1031[a]	–0.214	0.8317	0.148	0.8834	2.285	**0.0298**	0.515	0.6132	0.494	0.6289[a]
BBH	3.886	**0.0003**	3.744	**0.0030**	0.634	0.5285	4.386	**0.0000**	0.297	0.7676	0.558	0.5803	0.811	0.4263	0.215	0.8325	2.853	**0.0084**
XCB	–1.771	0.0966[a]	–3.413	0.0045[a]	–2.699	0.0173[a]	–2.530	**0.0138**	–2.290	0.0364[a]	–2.127	**0.0400**	–2.752	**0.0104**	–1.553	0.1436[a]	–2.166	**0.0390**
WFB	0.636	0.5281	0.188	0.8516	0.886	0.3786	1.446	0.1530	–1.623	0.1112								
NPH	4.598	**0.0000**	2.744	**0.0071**	6.209	**0.0000**	4.838	**0.0000**	1.363	0.1798	0.979	0.3343	3.265	**0.0028**	1.144	0.2669	–0.212	0.8334
NLH	5.456	**0.0000**	3.213	**0.0019**	5.365	**0.0000**	4.059	**0.0002**	0.918	0.3642	0.864	0.3933	2.131	**0.0417**	0.206	0.8388	1.801	0.0825
NLB	3.876	**0.0030**	3.713	**0.0003**	3.725	**0.0004**	3.041	**0.0034**	1.391	0.1709	1.096	0.2799	4.232	**0.0002**	1.096	0.2876	2.246	**0.0328**
OBB	3.910	**0.0003**	2.914	**0.0042**	3.575	**0.0007**	1.794	0.0773	0.206	0.8381	–0.170	0.8660	1.442	0.1600	0.499	0.6323[a]	–1.193	0.2427
OBH	9.058	**0.0000**	7.512	**0.0000**	8.951	**0.0000**	6.526	**0.0000**	3.217	**0.0023**	3.771	**0.0006**	2.663	**0.0125**	2.315	**0.0320**	2.994	**0.0056**
CDL	3.718	**0.0007**	3.682	**0.0005**	2.051	0.0649	4.787	**0.0000**	0.945	0.3513	1.443	0.1625	1.703	0.1058	0.835	0.4175	0.971	0.3413
GOG	2.207	**0.0334**	1.581	0.1194	0.306	0.7644	1.278	0.2056	–0.496	0.6224	–0.797	0.4314	–1.105	0.2799	–1.384	0.1843	–1.327	0.1954
MAN	–1.168	0.2501	1.702	0.0942	–0.510	0.6175	0.031	0.9753	1.563	0.1267	1.798	0.0802	0.070	0.9449	–0.635	0.5327	0.285	0.7773
POST-2000 BP																		
GOL	–1.657	0.1047	–2.392	**0.0092**	0.141	0.8887	1.33	0.1885	–0.944	0.3504	–0.727	0.4724	1.735	0.0946	0.021	0.9834	–0.356	0.7246
BBH	2.492	**0.0167**	2.173	**0.0318**	–0.532	0.5970	2.756	**0.0077**	–0.667	0.5080	–0.329	0.7446	–0.061	0.9517	–0.502	0.6231	1.357	0.1879
XCB	0.175	0.8616	–1.692	0.0934	–0.956	0.3424	–0.28	0.7805	–0.0481	0.6329	–0.014	0.9889	–1.062	0.2996	0.572	0.5817[a]	–0.315	0.4552
WFB	–1.56	0.1262	–1.992	**0.0488**	–1.254	0.2143	–0.627	0.5332	–3.229	**0.0023**								
NPH	1.183	0.2433	–0.456	0.6492	1.985	0.0520	0.993	0.3248	–1.183	0.2441	–1.43	0.1629	0.1	0.9215	–1.135	0.2755	–1.372	0.1837
NLH	2.515	**0.0157**	0.517	0.6171	2.375	**0.0209**	1.496	0.1430	–0.901	0.3740	–0.821	0.4180	–0.051	0.9598	–1.278	0.2219	–0.441	0.6635
NLB	1.351	0.1838	0.727	0.4686	0.571	0.5699	0.711	0.4797	–0.589	0.5587	–0.825	0.4151	1.346	0.1908	–0.936	0.3649	–0.0104	0.9177
OBB	3.12	**0.0032**	2.399	**0.0181**	2.884	**0.0056**	1.493	0.1404	0.24	0.8114	–0.052	0.9585	1.171	0.2533	0.51	0.6181	–0.843	0.4078
OBH	8.202	0.0000[a]	8.306	0.0000[a]	9.058	0.0000[a]	5.148	0.0000[a]	0.494	0.6239[a]	1.315	0.1975[a]	–0.223	0.8259[a]	0.093	0.9287[a]	0.446	0.6605
CDL	0.701	0.4887	0.454	0.6518	1.274	0.2433	1.226	0.2251	–1.316	0.1984	–0.734	0.4719	–0.758	0.4610	–0.971	0.3546	–1.167	0.2569
GOG	1.363	0.1826	0.702	0.4858	0.021	0.9835	0.481	0.6326	–0.813	0.4222	–1.161	0.2572	–1.435	0.1685	–1.801	0.1018	–1.633	0.1173
MAN	–0.83	0.4128	1.061	0.2939	–0.356	0.7323	–0.044	0.9649	0.966	0.3422	1.179	0.2478	–0.294	0.7721	–0.394	0.7004	0.133	0.8956

[a] Variance NOT Equal; **BOLD** denotes significance at $p < 0.05$.
[b] Ribot (2003): WCA-M: West-Central African, CAB-M: Central African Bantu, EA-M: East African, SAB-M: South African Bantu, SAK-M: South African Khoesan.
[c] Morris (1992b): RR-M: Riet River, K-M: Kakamas, A-M: Abrahamsdam, G-M: Griqua.

Table 6.5 Comparisons between Eastern Cape and comparative female craniometric data – t-tests using means and standard deviations

Variable	WCA-F[b] vs. F		CAB-F[b] vs. F		EA-F[b] vs. F		SAB-F[b] vs. F		SAK-F[b] vs. F		RR-F[c] vs. F		K-F[c] vs. F		A-F[c] vs. F		G-F[c] vs. F	
	T	p	t	p	T	p	t	p	t	p	t	p	t	p	t	p	t	p
PRE-2000 BP																		
GOL	–0.764	0.4504	–3.139	**0.0021**	–0.067	0.9467	0.225	0.8228	–1.01	0.3184	–0.782	0.4403	2.535	**0.0176**	0.328	0.7500	0.255	0.8021
BBH	5.184	**0.0000**	1.946	0.0537	0.935	0.3528	2.018	0.0507	0.541	0.5917	–0.138	0.8912	1.231	0.2374	0.908	0.3876	2.215	**0.0416**
XCB	–2.451	0.0194	–5.714	**0.0000**	–5.697	**0.0000**	–4.602	**0.0000**	–2.634	**0.0118**	–2.892	**0.0072**	–2.825	**0.0088**	–1.608	0.1388	–3.421	**0.0030**
WFB	2.4	0.0227[a]	–1.554	0.0735[a]	1.307	0.2072[a]	0.332	0.7413	0.394	0.6954								
NPH	7.81	0.0000[a]	3.822	0.009[a]	8.258	0.0000[a]	6.813	0.0000[a]	2.583	0.0144[a]	1.846	0.0769[a]	4.763	0.0001[a]	0.796	0.4765[a]	3.86	0.0023[a]
NLH	8.106	**0.0000**	5.28	0.0001[a]	7.221	0.0000[a]	6.833	0.0000[a]	2.365	**0.0247**	2.429	**0.0224**	3.98	0.0005[a]	1.965	0.1321[a]	2.942	0.0104[a]
NLB	4.249	**0.0001**	3.244	**0.0015**	2.357	**0.0211**	3.665	**0.0007**	2.508	**0.0160**	2.361	**0.0252**	2.694	**0.0118**	1.299	0.2166	1.323	0.2007
OBB	4.061	**0.0002**	0.967	0.3463[a]	1.448	0.152	–0.156	0.8770	0.743	0.462	1.689	0.1019	2.202	**0.0367**	1.467	0.1661	–0.829	0.4171
OBH	4.945	**0.0000**	2.491	0.0332[a]	3.597	0.0047[a]	3.19	**0.0027**	0.547	0.5877	0.527	0.6024	2.748	**0.0107**	1.261	0.2313	1.095	0.2872
CDL	6.678	**0.0000**	6.446	**0.0000**	3.732	**0.0097**	4.764	**0.0000**	3.982	**0.0004**	3.053	**0.0068**	3.841	**0.0009**	2.722	**0.0262**	4.384	**0.0005**
GOG	4.285	0.0010[a]	4.371	**0.0001**	3.154	0.0117[a]	2.119	**0.0402**	1.898	0.0651	14.147	0.262	5.371	0.0002[a]	0.861	0.4062	1.124	0.2751
MAN	–1.724	0.0957	1.032	0.3095[a]	–1.177	0.2642	2.805	**0.0076**	1.09	0.2834	0.823	0.4172[a]	–1.204	0.2391	0.609	0.5812[a]	0.216	0.8315
POST-2000 BP																		
GOL	–0.608	0.5464	–3.675	**0.0003**	0.319	0.7506	0.643	0.5234	–0.823	0.4144	–0.357	0.7246[a]	3.245	**0.0026**	0.594	0.5600	0.525	0.6043
BBH	8.052	**0.0000**	3.512	**0.0006**	2.151	**0.0346**	3.501	**0.0010**	1.422	0.1619	0.531	0.5991	2.248	**0.0344**	1.677	0.1119	3.662	**0.0012**
XCB	–0.76	0.4517[a]	–8.098	0.0000[a]	–5.8	0.0000[a]	–4.003	0.0002[a]	–1.094	0.2795	–1.542	0.1327[a]	–1.333	0.1325[a]	–0.653	0.5222	–2.213	**0.0362**
WFB	1.444	0.1560	–1.89	0.0608	0.251	0.8022	–0.391	0.6977	–0.415	0.6799								
NPH	2.906	**0.0058**	–0.329	0.7428	2.624	**0.0105**	2.152	**0.0365**	–0.496	0.6243	–0.506	0.6161	1.919	0.0634	–0.479	0.6374	1.671	0.1068
NLH	4.738	**0.0000**	1.895	0.0607	3.472	**0.0008**	3.606	**0.0007**	0.59	0.5584	0.669	0.5080	2.067	**0.0464**	1.145	0.2666	1.315	0.1999
NLB	3.171	**0.0028**	1.027	0.3062	0.428	0.6699	2.247	**0.0293**	0.626	0.5345	0.509	0.6143	1.22	0.2309	0.431	0.6712	–0.294	0.7708
OBB	5.5	**0.0000**	1.9	0.0596	2.986	**0.0038**	1.253	0.2166	2.1	**0.0414**	2.92	**0.0063**	3.298	**0.0025**	1.895	0.0752	0.242	0.818
OBH	4.502	**0.0001**	2.075	0.0544[a]	3.373	0.0034[a]	2.643	**0.0111**	–0.291	0.7737[a]	–0.245	0.8077	2.266	**0.0303**	0.803	0.4326	0.563	0.5785
CDL	2.202	**0.0353**	0.689	0.7496[a]	1.133	0.2762	0.281	0.7801	–0.395	0.695	–0.676	0.5047	0.235	0.8158	–0.281	0.7823	0.082	0.9356
GOG	1.713	0.0962	0.542	0.7048	–0.313	0.7590[a]	–1.736	0.0892	–2.057	**0.0456**	–2.492	**0.0182**	3.095	0.0057[a]	–1.119	0.2786	–1.83	0.0798
MAN	–5.132	**0.0000**	–1.577	0.121	–1.56	0.1397	–0.149	0.8823	–1.187	0.2429	–1.294	0.2046	–3.255	**0.0027**	–0.516	0.6128	–1.639	0.1143

[a]Variance NOT Equal; **BOLD** denotes significance at $p < 0.05$.
[b]Ribot (2003): WCA-F: West-Central African, CAB-F: Central African Bantu, EA-F: East African, SAB-F: South African Bantu, SAK-F: South African Khoesan.
[c]Morris (1992b): RR-F = Riet River, K-F = Kakamas, A-F = Abrahamsdam, G-F = Griqua.

Table 6.6 Comparisons between Eastern Cape and published maximum odontometric data – t-tests using means and standard deviations

		E Cape F vs. Griqua[b] F		E Cape F vs. South African Bantu (Skeletal)[c] F		E Cape M vs. Griqua[b] M		E Cape M vs. South African Bantu (Skeletal)[c] M		E Cape M vs. San (Skeletal)[d] M		E Cape vs. LSA[a]	
Variable		t	p	t	p	t	p	t	p	t	p	t	p[e]
UM1	BL	4.365	**0.0005**	–3.281	**0.0018**	4.32	**0.0002**	–2.265	**0.0274**	1.464	0.1528	–0.173	0.8631
	MD	0.395	0.6985	–3.301	**0.0018**	–0.543	0.5926	–2.568	**0.0131**	–0.369	0.7146	–0.288	0.7746
UM2	BL	5.428	**0.0000**	2.305	**0.0250**	4.072	**0.0003**	–1.632	0.1085	3.471	**0.0008**	0.55	0.5848
	MD	1.725	0.1016	–4.128	**0.0020**	0.557	0.5841	–6.162	**0.0000**	1.325	0.1888	0.792	0.6254
LM1	BL	0.762	0.4555	–5.05	**0.0000**	6.381	**0.0000**	–6.146	**0.0000**	2.071	0.0522	0.134	0.8869
	MD	0.576	0.5929	–2.617	**0.0117**	3.515	**0.0023**	–0.535	0.5949	12.21	**0.0000**	1.251	0.2173
LM2	BL	5.174	**0.0000**	–2.708	**0.0090**	6.682	**0.0000**	0	1.0000	4.497	**0.0000**	1.034	0.3046
	MD	4.281	**0.0008**	–1.05	0.2998	2.034	0.0548	–0.109	0.9139	2.524	**0.0143**	1.735	0.0885

[a]Pfeiffer (2007).
[b]Kieser (1985).
[c]Kieser, Groeneveld and Cameron (1987).
[d]Drennan (1929).
[e]Values in **bold** are significant at $p < 0.05$.

Table 6.7 Comparisons between Eastern Cape and published South African Bantu-speaking postcranial data – t-tests using means and standard deviations

	PRE-2000 BP											
	Cape Nguni[a] F vs. F		Natal Nguni[a] F vs. F		Sotho[a] F vs. F		Cape Nguni[a] M vs. M		Natal Nguni[a] M vs. M		Sotho[a] M vs. M	
Measurement	t	p	t	p	t	p	t	p	t	p	t	p[b]
Hum Mx Length	4.648	**0.0001**	6.709	**0.0000**	5.547	**0.0000**	7.281	**0.0000**	5.699	**0.0000**	4.815	**0.0000**
Hum Mx Vrt Diam Head	5.049	**0.0000**	6.953	**0.0000**	5.076	**0.0000**	5.099	**0.0000**	6.316	**0.0000**	3.959	**0.0002**
Hum – Epi Width	6.532	**0.0000**	6.992	**0.0000**	6.311	**0.0000**	8.66	**0.0000**	9.123	**0.0000**	8.031	**0.0000**
Ulnar Length	5.123	**0.0000**	8.105	**0.0000**	6.838	**0.0000**	5.884	**0.0000**	4.909	**0.0000**	3.383	**0.0013**
Radial Length	4.46	**0.0001**	6.384	**0.0000**	5.948	**0.0000**	6.925	**0.0000**	9.425	**0.0000**	6.48	**0.0000**
Fem Bicond Ln	3.811	**0.0006**	5.758	**0.0000**	4.655	**0.0000**	5.669	**0.0000**	3.702	**0.0004**	3.55	**0.0007**
Fem Mx Vrt Diam Head	4.541	**0.0001**	4.945	**0.0000**	4.605	**0.0000**	6.365	**0.0000**	7.124	**0.0000**	5.905	**0.0000**
Fem Epicond Wd	7.632	**0.0000**	8.032	**0.0000**	9.002	**0.0000**	5.263	**0.0000**	6.013	**0.0000**	4.885	**0.0000**
Tibial Length	2.965	**0.0058**	4.264	**0.0001**	3.605	**0.0006**	4.032	**0.0030**	2.929	**0.0044**	2.442	**0.0175**
POST-2000 BP												
Hum Mx Length	2.279	**0.0088**	2.863	**0.0001**	3.279	**0.0016**	6.088	**0.0000**	4.83	0.0000	**4.083**	**0.0001**
Hum Mx Vrt Diam Head	3.467	**0.0015**	4.538	**0.0000**	2.788	**0.0069**	2.073	0.0794	2.466	0.0047	1.733	0.1301
Hum – Epi Width	5.966	**0.0000**	6.053	**0.0000**	5.224	**0.0000**	5.883	**0.0000**	6.465	0.0000	**5.482**	**0.0000**
Ulnar Length	2.813	**0.0083**	3.019	**0.0098**	2.479	**0.0278**	5.307	**0.0000**	4.642	0.0000	**3.106**	**0.0030**
Radial Length	2.803	**0.0085**	2.903	**0.0119**	3.391	**0.0012**	5.119	**0.0000**	4.392	0.0000	**2.791**	**0.0069**
Fem Bicond Ln	3.174	**0.0032**	3.676	**0.0020**	3.909	**0.0002**	4.369	**0.0000**	2.652	0.0096	**2.516**	**0.0143**
Fem Mx Vrt Diam Head	4.419	**0.0001**	4.612	**0.0000**	4.155	**0.0001**	4.644	**0.0000**	4.287	0.0000	**3.31**	**0.0015**
Fem Epicond Wd	5.234	**0.0000**	6.401	**0.0000**	5.057	**0.0000**	2.828	**0.0074**	3.298	0.0014	**2.327**	**0.0234**
Tibial Length	2.023	0.0508	2.888	**0.0056**	2.247	**0.0278**	3.247	**0.0025**	2.328	0.0223	1.894	0.0629

[a]Data from Lundy (1986).
[b]Values in **bold** are significant at $p < 0.05$.

broader findings of this research. The pre-2000 BP males and females differ significantly from the South African Bantu-speaking samples (Cape Nguni, Natal Nguni and Sotho) for all osteometric variables (Table 6.7). A few variables show similarities between post- 2000 BP Eastern Cape and comparative samples, yet significant differences between the samples for the majority of osteometric dimensions suggest the body size of the South African Bantu-speaking samples is considerably larger.

The comparative metric results suggest that there is a weak relationship between the southern African Khoesan sample under study and other sub-Saharan populations. The fact that both the pre- and post-2000 BP sub-samples were significantly different from other sub-Saharan samples reinforces the idea that biological continuity existed amongst groups living in the southernmost region of South Africa throughout the Holocene.

6.6.4 Cranial Discrete Traits

Potential differences in cranial discrete traits in the Eastern Cape sample are explored by comparing trait frequencies by sex and time period. Out of the 44 bilateral and 8 midline traits, frequencies of only 1 trait, partial pterygoalar bridge ($\chi^2 = 5.612$, $p = 0.018$, phi $= 0.248$), differed significantly between males and females. The Pearson χ^2 value for this trait suggests that the differences are significant, yet the accompanying Phi values indicate that the magnitude of differences is low.

The analysis of cranial discrete trait frequencies across time also does not produce significant results. Pearson χ^2 values and accompanying Phi values suggest that while the change in frequency of tympanic dehiscence ($\chi^2 = 4.148$, $p = 0.042$, phi $= 0.21$), trace pterygoalar bridge ($\chi^2 = 4.284$, $p = 0.038$, phi $= 0.217$) and all manifestations of pterygoalar bridging ($\chi^2 = 5.123$, $p = 0.024$, phi-0.237) through time is significant, this correlation is not strong. Os japonicum ($\chi^2 = 7.93$, $p = 0.005$, phi $= 0.317$) is the only trait that supports a significant change through time. Only 1 out of the 34 pre-2000 BP individuals displays this trait, while it is present in 12 out of the 45 post-2000 BP individuals. All but three of the individuals displaying the os japonicum trait are buried in a small area along the coast (Cape St Francis). It is possible that the grouping of individuals displaying this trait represents related kin, as the single pre-2000 BP individual with the os japonicum trait is also located in the Cape St Francis grouping. This finding can be interpreted as a reflection of genetic homogeneity of South African LSA groups. However, the fact that this temporal and spatial patterning is observed for only one non-metric trait requires that this interpretation remain tentative.

Frequencies for seven discrete traits were compared between for the Eastern Cape pre- and post-2000 BP subsamples and four pooled-sex South African Bantu-speaking samples (Zulu, Xhosa, Sotho and Venda) (Rightmire, 1976). While similar frequencies were observed between both the pre- and post-2000 BP Eastern Cape sample and the four comparative samples for some of the selected discrete traits, significant differences in the frequencies of a greater number of discrete traits were observed between the two Eastern Cape and the comparative samples (Table 6.8). This suggests a high degree of genetic homogeneity in the Eastern Cape sample and corroborates the overall findings of this study.

In sum, sex and temporal differences in cranial discrete trait frequencies are negligible for the Eastern Cape sample. These results support the overall findings of the metric skeletal analyses, indicating that genetic homogeneity was maintained in the Eastern Cape region of South Africa.

Table 6.8 Results of Pearson chi-square analysis of Eastern Cape and comparative South African Bantu-speaking cranial discrete trait frequencies

	PRE-2000 BP (M and F combined)							
Trait	Zulu[a] vs. Pre- 2000 BP		Xhosa[a] vs. Pre- 2000 BP		Sotho[a] vs. Pre-2000 BP		Venda[a] vs. Pre-2000 BP	
	x^2	p	x^2	p	x^2	p	x^2	p^b
Supraorbital Foramen	**6.997**	**0.010**	**6.611**	**0.025**	2.4	0.200	**4.098**	**0.050**
Epiteric Bone	0.097	1.000	0.111	1.000	0.0153	1.000	0.0389	1.000
Parietal Notch Bone	1.493	1.000	1.224	1.000	1.342	1.000	0.14	1.000
Tympanic Dehiscence	**52.373**	**0.001**	**37.001**	**0.001**	**31.404**	**0.001**	**25.688**	**0.001**
Accessory Mental Foramen	0.0512	1.000	1.629	1.000	1.745	0.200	0.747	1.000
Mylohyoid Bridge	**33.119**	**0.001**	**27.222**	**0.001**	**29.932**	**0.001**	**43.716**	**0.001**
Os Japonicum	2.4336	0.200	0.054	1.000	2.669	0.200	2.079	0.200
POST-2000 BP (M and F combined)								
Supraorbital Foramen	1.773	0.200	1.655	0.200	0.0559	1.000	0.636	1.000
Epiteric Bone	1.398	1.000	0.1778	1.000	0.439	1.000	0.296	1.000
Parietal Notch Bone	0.0069	1.000	0.0283	1.000	0.0416	1.000	0.745	1.000
Tympanic Dehiscence	**27.575**	**0.001**	**16.479**	**0.001**	**12.462**	**0.001**	**9.918**	**0.010**
Accessory Mental Foramen	0.371	1.000	0.659	1.000	2.604	0.200	0.213	1.000
Mylohyoid Bridge	**28.953**	**0.001**	**23.572**	**0.001**	**26.143**	**0.001**	**40.168**	**0.001**
Os Japonicum	**24.148**	**0.001**	**19.286**	**0.001**	**26.341**	**0.001**	**20.841**	**0.001**

[a]Rightmire (1976).
[b]Values in bold are significant at $p < 0.05$.

6.7 DISCUSSION

The findings of this study provide insights about the origins of sheep and the herding lifestyle in southernmost Africa, as well as the factors and forces that may have contributed to the initiation of this subsistence shift. As such, this work sheds light on three key issues central to the debate on the origins of herding in South Africa:

1. the nature of this shift – as an *in situ* development amongst indigenous foragers or the introduction of a new way of life by foreign migrants;
2. the conditions which prompted existing populations to diversify their socioeconomic organization and subsistence; and
3. the impact of this transition.

6.7.1 What is the Nature of this Shift – Diffusion or Migration?

The balance of the findings of the skeletal analyses support the hypothesis that the introduction of sheep herding at around 2000 BP was the product of an indigenous development brought about by diffusion. The alternative hypothesis that a foreign migration of herders was responsible for introducing and establishing sheep herding in southernmost Africa can be confidently excluded. The introduction of new genetic material to a genetically and culturally homogeneous region would manifest physically as changes to skeletal size and form (shape). The results of both the principal components analysis and the curve estimation regression did not identify any novel changes in skeletal morphology (size and shape) at around 2000 BP, the generally agreed upon date for the introduction of sheep herding to southernmost Africa. Rather, body size returned to pre-3500 BP levels, suggesting that from 3500 to 2000 BP, foragers had difficulty securing the necessary resource requirements.

The results of the comparative metric skeletal analyses also support the interpretation of an indigenous development of sheep herding rather than a foreign introduction. The cranial and dental dimensions of the Eastern Cape sample are more similar to prehistoric and protohistoric South African foraging and herding groups and historic Khoesan, than to the African Bantu-speaking samples. Comparative postcranial metric analyses, although restricted to only South African Bantu-speaking skeletal samples, also support the broader findings of this research. However, the post-2000 BP Eastern Cape individuals exhibit some similarities in limb dimensions with the Bantu-speaking samples. This apparent body size affinity with the Bantu could be attributed to the large size of some of the Eastern Cape individuals who date to the last 500 years. Although differences in cranial morphology have not been identified for these more recent Eastern Cape individuals (last 500 years), it is possible that they could be migrants. However, in general, the post-2000 BP sample displays affinities with prehistoric, protohistoric or historic foraging groups from the region, yet differs significantly from South African Bantu-speaking and other sub-Saharan agricultural groups. Had the post-2000 BP sample included foreign herders, we would expect skeletal dimensions more in line with African Bantu-speaking samples. The fact that the post-2000 BP sample not only differs considerably from the non-local comparative samples, but also exhibits marked similarities with the pre-2000 BP sample, reinforces the acceptance of the *in situ* hypothesis.

The cranial discrete trait findings provide further grounds to reject the foreign migration hypothesis, and also argue against the possibility of a small-scale migration of genetically related foragers-turned-herders from other parts of southern Africa. Discrete traits are under tighter genetic control than human skeletal morphology, and can be used to detect subtle genetic differences between neighbouring populations (cf. De Villiers, 1967; Molto, 1983; Cybulski, 1992; Prowse and Lovell, 1996; DeLaurier and Spence, 2003). The general lack of differences in cranial discrete frequencies between the sexes suggests that Eastern Cape foragers sought mating partners from neighbouring foraging groups. If patrilocal or virilocal residence was practised by LSA groups, a higher level of variability in trait frequencies between the sexes could be expected (Spence, 1974). Had even small groups of foragers-turned-herders from other parts of southern Africa, such as Botswana, migrated south with sheep around 2000 BP, it should have been reflected in the frequencies of some discrete traits, if the study sample included crania representing the new immigrants or their offspring. There is no such evidence.

The lack of significant differences in cranial discrete trait frequencies over time argues against the introduction of sheep by foreign, yet genetically similar groups originating from other regions of southern Africa. While only 4 out of the 52 discrete traits examined yielded significant differences across time, a strong statistical difference between pre-2000 BP and post-2000 BP individuals was only produced for one trait, the os japonicum (3% vs. 27%, respectively). The distribution of the individuals exhibiting this trait is of particular interest as 10 out of the 13 individuals displaying this trait are buried in a relatively small area along the coast (Cape St Francis). The remaining three individuals are each interred a considerable distance from the coast. It is possible that, in the absence of radiocarbon dates, the patterning of this trait across the Eastern Cape landscape might have been interpreted to support the migration hypothesis, because the distribution of individuals with this trait loosely conforms to the north-south migration route of herders with sheep proposed by Elphick (1977). However, the fact that the individual exhibiting the os japonicum trait with the oldest ^{14}C date is part of the coastal concentration suggests that this trait emerged *in situ*. This interesting pattern, along with the lack of substantial temporal differences in trait frequencies between the pre- and post-2000 BP subgroups argues against the possibility that different, but closely related groups from the north could be responsible for the introduction of sheep herding. The results of the comparative discrete trait analysis corroborate this assertion and add further support to the argument that herding was not introduced by a migration.

6.7.2 What Factors are Responsible for the Short-Term Decrease in Skeletal Size?

The nature of the changes in skeletal size supports an environmental, in this case nutritional, rather than a genetic cause. This dramatic yet short-lived body size reduction, in the absence of changes in shape, corresponds with evidence of population stress and the intensification of foraging behaviour between 3500 and 2000 BP. The observed pattern of skeletal size change is consistent with a period of population stress and environmental deterioration followed by a return to more favourable conditions. The nature and magnitude of body size change observed in the Eastern Cape sample corresponds with reductions in stature (Wilson and Lundy, 1994; Sealy and Pfeiffer, 2000; Pfeiffer and Sealy, 2006) and cranial size (Stynder, 2006; Stynder, Rogers-Ackerman and Sealy, 2007a, 2007b) observed amongst contemporaneous Southern and Western Cape foraging populations. The fact that this decrease in size is observed not only in the regions of the postcranial skeleton known to be under strong environmental influence, but also in other areas the growth of which is believed to be more buffered against environmental stress, namely the cranium and dentition, speaks to the severity and magnitude of this period of stress.

The decline and subsequent rebound in size can be most reasonably linked to nutritional insufficiency experienced by a population whose energy demands exceeded that available from the local resources. Nutritional insufficiency can influence development and health, leading to individuals not achieving their genetic potential for overall body size. Selection might also be a factor in such a situation, as smaller individuals who require fewer calories for their normal development and maintenance would be more apt to survive, further resulting in a greater number of smaller individuals in a given population. Differences in the pattern and magnitude of the size changes observed throughout the skeleton conform to current understandings of skeletal growth, development and plasticity, as less marked change is observed in the areas of the skeleton, namely the vault and basilar regions that complete development earlier in life and are less influenced by external stress than the face, mandible and postcranial skeleton (Sinclair and Dangerfield, 1998; Wood and Lieberman, 2001).

The decline in tooth size observed throughout the dentition between approximately 3500 and 2000 BP is consistent with short-term population stress and food scarcity experienced during the period of forager intensification. Changes in tooth size have been identified amongst genetically homogeneous populations and have been attributed to modifications to the local environment, specifically changes in the quality of the diet (Garn, Lewis and Walenga, 1968). Research has demonstrated that negative environmental factors, such as nutritional stress, can produce a reduction in dental size (Larsen, 1997; Smith and Horwitz, 2007; Stojanowski, 2005; Stojanowski *et al.*, 2007). This pattern of tooth size reduction and rebound has also been observed in other prehistoric populations undergoing subsistence change. Bennike and Alexandersen (2007) observed a reduction in tooth size between Mesolithic and Early Neolthic Danish populations, which they attribute to poor adaptation to a new lifestyle (farming), with tooth size increasing again when populations became fully adapted to agriculture during the Middle/Late Neolithic. Similarly, Lukacs (2007) identified an opposite trend of an increase in dental size that he attributed to improved nutrition and health corresponding to the transition from farming back to foraging at a Chalcolithic site in western India.

The selective advantage that tooth size provides for different types of diets also requires consideration. Large teeth are believed to be advantageous in situations where tough and gritty diets and the use of the teeth as tools produce high rates of tooth wear, because their large size should protect against the adverse effects of extreme wear (pulp exposure, abscesses and tooth loss). Consequently, small teeth should be selected for in cases where tooth wear is reduced through subsistence innovation involving increased food processing (Sciulli, 1997). LSA South African foragers exhibit substantial dental wear (Sealy *et al.*, 1992; Pfeiffer, 2007). However, no changes in the rate of tooth wear, frequencies of carious lesions, antemortem tooth loss or dental abscesses associated with periods of subsistence change have been observed amongst the LSA inhabitants of the Eastern Cape (Ginter, 2005). This suggests that broader factors associated with changing subsistence behaviours, beyond the actual dietary components, may be responsible for the apparent reduction in tooth size.

Reduction in body size has been attributed to a number of complex environmental factors, such as climate change, dietary change, nutritional insufficiency, population density, disease load and population mortality, as well as genetic factors, including gene flow and admixture. Research has demonstrated an inverse relationship between population density and body size in animal species (cf. Damuth, 1981; Murdoch, 1994; Schmid, Tokeshi and Schmid-Araya, 2000), as well as humans (Walker and Hamilton, 2008). Similar patterns of skeletal change have been observed in other populations known to have experienced stress stemming from short-term environmental decline and deterioration of food resources. For example, stature rebounded rapidly with improvements in nutrition and health in the 1880s (Cohen, 1989). So, what possible explanations can be offered to account for the rebound in body size observed around 2000 BP?

6.7.3 Can Herding Explain the Rebound in Size at Around 2000 BP?

It has been argued, based on the various lines of evidence, that between 3500 and 2000 BP LSA foragers were finding it difficult to secure their caloric requirements, resulting in failure of many individuals to achieve their growth potential. During this period foragers altered the dietary focus towards marine resources and placed more emphasis on plant foods. Did South African foragers eventually adapt to this subsistence reorientation, like the Danish are believed to have done (Bennike and Alexandersen, 2007), or were additional dietary alternatives, such as sheep, sought to mediate against this stress? While food resources were available year-round in the Eastern Cape, sheep as a source of 'food on the hoof' could provide greater food security. Herders do not rely heavily on the meat from their herd for subsistence, because to do so would mean that a viable breeding population could not be maintained. Yet, significant, renewable nourishment can be gained from a living animal, in the form of milk and blood, without having to sacrifice a valuable member of a herd. It is unclear whether lactase persistence was common amongst these South African Holocene foragers. Addressing the linear increases in both upper and lower limbs, milk has been suggested to have a significant impact on the growth of undernourished children (Wiley, 2005). It can be argued that the addition of even a small amount of calories from sheep milk and blood to the diet could contribute positively to body size considering that it is high in both protein and fat, components that were likely lacking in the forager diet. It is therefore plausible that some forager groups might have incorporated sheep and aspects of the herding lifestyle into their existing forager subsistence approach as an innovation to relieve pressure from a stressed resource base. This minor refocusing of food resources might have freed up some wild resources for other foraging groups inhabiting the same area ultimately leading to increased dietary quality and diversity, and health, for all.

While the body size rebound at around 2000 BP could be attributed to the introduction of a more stable food source such as sheep, the strength of the association between skeletal change and the advent of sheep herding must be qualified, as it cannot be assumed that the initial introduction of sheep and herding was pervasive and immediate. Although there is evidence for sheep in the area around 2000 BP, it is not likely that this novel food source is exclusively responsible for the alleviation of the apparent nutritional stress. Faunal assemblages at sites with the earliest archaeological signatures of herding suggest that sheep constitute minor components of the diet (Sadr, 1998). However, basing such judgments on the presence and quantity of faunal remains can be problematic given that even at some Iron Age sites solidly associated with agriculture, wild remains predominate over domestic stock because foraging still factored heavily in subsistence (Voigt, 1986). A return to slightly warmer and drier conditions around 2000 BP, coupled with increased mobility and the redistribution of populations away from open coastal sites towards reoccupation of inland rock shelters (Binneman, 1998; Henshilwood, Nilssen and Parkington, 1994; Jerardino, Branch and Navarro, 2008), may have also helped to ease the effects of resource stress. Other studies (Sadr, 2004; Pfeiffer and Sealy, 2006; Stynder, Rogers-Ackerman and Sealy, 2007a, b) have suggested that foragers inhabiting other parts of the Cape region recovered from this stress event prior to the introduction of sheep herding.

A counter argument to the potential involvement of sheep as a subsistence innovation to relieve pressure on a stressed resource base has been raised. Smith (1998, 2006) asserts that foragers would not be able to make the transition to a way of life that was dependent on storage and conservation of resources, because of their egalitarian social organization, ideology and worldview. However, it could be argued that during the intensification period, changes to social and economic aspects of the foraging lifestyle indicative of a delayed-return economy (reduced mobility, increased connection to the land, territoriality and food storage) may have laid the groundwork for the adoption of herding by indigenous southern African foragers around 2000 BP (Sadr, 2004; Sealy, 2006). The subsistence transformations that accompanied the intensification of foraging behaviour already set foragers up for the integration of different more delayed food resources, as well as incorporating different ways of perceiving food into the existing foraging ethic. Thus, it is probable that foragers undergoing this period of intensification would have gained the skills and experiences necessary to adopt a more intensive delayed return economy, such as herding. Furthermore, the adoption or integration of a herding economy by foragers has been observed in other parts of Africa, especially the Sahara and Kenya (cf. Marean, 1992; Holl, 1998; Marshall, 1998; Wetterstrom, 1998).

The absence of a significant early pastoral archaeological signature across the Cape region (Sadr, 1998) suggests that sheep did not form a significant component of the diet, at least not immediately following 2000 BP. Archaeological deposits associated with groups known to practise a true herding lifestyle have been found to contain much greater quantities of domestic remains than has been observed at sites with sheep in the South African Cape region (Gifford-Gonzalez, 1998; Sadr, 1998). Smith (1992) suggests that sheep only became a significant presence in western and southern Capes around 1600 BP, calling into question the idea that sheep and herding had a significant positive impact on the lives of existing foragers that have been offered based on the Eastern Cape skeletal evidence. Sadr (1998) and others have suggested that the Khoekhoe herding groups that the first European explorers encountered at the Cape may represent a later migration of pastoralists at around 1000 BP. While this study did not identify a novel pattern of skeletal size and shape change (i.e. not different from the range of body sizes documented before the period of apparent resource stress beginning around 3500 BP), an increase in skeletal size identified 1000 years after the purported introduction of sheep supports this explanation, as physical differences pertaining mainly to stature were noted between historic Khoekhoe herders and San foragers inhabiting the Cape region (Schapera, 1930). Although Khoekhoe and San groups exhibited differences in stature and subsistence, cultural, linguistic and physical commonalities support an ancestral relationship. A scenario involving a more recent (sometime after 2000 BP) migration of foragers turned pastoralists from northern Botswana, who obtained sheep and possibly marriage partners from Bantu-speaking agro-pastoralists, is supported by the multivariate and cranial discrete findings of this study. Had a significant pastoral migration taken place at around 2000 BP, a sudden change in material culture and subsistence practices should be detected archaeologically, but this evidence is absent.

If sheep did not make a significant contribution to subsistence in the period immediately following 2000 BP, and evidence for a more complete pastoral lifestyle is not evident until almost 1000 years later, can the introduction of sheep around 2000 BP be classified as a 'subsistence transition?' Should this event be re-classified as subsistence variability within the broad foraging spectrum? In this case, it is possible that some foragers integrated sheep and aspects of a herding economy into their existing forager way of life, akin to the subsistence transformations that took place between 3500 and 2000 BP. Based on the culmination of biological and archaeological data, it is possible to propose that the intensification of foraging behaviour in the mid-Holocene had a more significant impact on the biology and quality of life of the foragers in this region than the advent of sheep herding. While we see an overall improvement in health and diet, as inferred from the rebound in overall skeletal size around 2000 BP, it is not possible, based on the available evidence, to attribute this trend solely to the introduction of sheep. In the Eastern Cape we see a rebound of skeletal size to pre-intensification levels, suggesting that the stresses that were present may have been alleviated with conditions returning to 'normal', rather than resulting from the introduction of a new domestic food resource and way of life.

6.8 SUMMARY AND CONCLUSIONS

The findings of the current study add new knowledge to our understanding of two important events in the Holocene prehistory of South Africa: the causes and impact of the intensification of foraging behaviour and the mechanisms responsible for the introduction of sheep herding. Skeletal metric and cranial discrete data indicate that cultural continuity was maintained in the Eastern Cape region throughout the Holocene. The multivariate regression results support interpretations of the continuity of the Eastern Cape inhabitants through the Holocene, but also identify a temporal pattern of body size change that correlates with the intensification of foraging behaviour. The nature and timing of the observed patterns of skeletal change in the Eastern Cape are consistent with a short-term period of environmental change and resource scarcity followed by a period of recovery. Although this recovery occurs around the time of the initial introduction of sheep, it is still unclear if the initial pastoral presence in this area could have been responsible for this recovery. When the results of this study are situated in a broader archaeological context, it seems that the intensification of foraging behaviour and subsistence between approximately 3500 and 2000 BP had a more significant impact on the lives of foragers than the introduction of food production. These findings suggest that the initial introduction of sheep herding should alternatively be viewed as subsistence variability within the foraging framework rather than a subsistence transition. While this study has shed some light on a complex period of South African prehistory, the assertions that have been made about the significance of the initial introduction of sheep on the lives of late Holocene groups must be viewed as tentative.

ACKNOWLEDGEMENTS

I am grateful to the following individuals for providing access to skeletal collections: Johan Binneman and Lita Webley (Albany Museum, Grahamstown, SA); James Brink (National Museum Bloemfontein, Florisbad Research Station, SA), Alan Morris and Caroline Powrie

(University of Cape Town, Cape Town, SA), Nalini Pather and Elijah Mofokeng (School of Anatomy, Witwatersrand University, Johannesburg, SA), and Sarah Wurz (Iziko Museums of Cape Town, Cape Town, SA). Thanks to Judith Sealy, Susan Pfeiffer and Deano Stynder for access to the radiocarbon dates for the Eastern Cape skeletal material, and to Isabelle Ribot and Alan Morris for access to the comparative African craniometric data. Thank you to the reviewers whose insightful comments helped to improve this paper. This research has been supported by the Social Sciences and Humanities Research Council of Canada (through grant funding to Susan Pfeiffer) and the University of Toronto.

REFERENCES

Anderson, J.E. (1968) Skeletal 'anomalies' as genetic indicators, in *The Skeletal Biology of Earlier Human Populations* (ed. D.R. Brothwell), Pergamon Press, Oxford, pp. 135–147.

Ball, C.N. and Connor-Linton, J. (1996) Web Chi Square Calculator http://www.georgetown.edu/cball/webtools/web_chi.html.

Barkhan, D. and Soodyall, H. (2006) Mitochondrial DNA (mtDNA) and Y chromosome DNA variation in African populations, in *The Prehistory of Africa: Tracing the Lineage of Modern Man* (ed. H. Soodyall), Jonathan Ball, Jeppestown, pp. 139–147.

Beals, K.L., Smith, C.L. and Dodd, S.M. (1983) Climate and the evolution of brachycephalization. *Am. J. Phys. Anthropol.*, **62**, 425–437.

Bennike, P. and Alexandersen, V. (2007) Population plasticity in southern Scandinavia: From oysters and fish to gruel and meat, in *Ancient Health: Skeletal Indicators of Agricultural and Economic Intensification* (eds M.N. Cohen and G.M.M. Crane-Kramer), University Press of Florida, Gainesville, p. 130.

Berry, A.C. and Berry, R.J. (1967) Epigenetic variation in the human cranium. *J. Anat.*, **101**, 361–379.

Berry, R.J. (1968) The biology of non-metrical variation in mice and men, in *The Skeletal Biology of Earlier Human Populations* (ed. D.R. Brothwell), Pergamon Press, Oxford, pp. 103–133.

Binford, L.R. (2001) *Constructing Frames of Reference: An Analytical Method For Archaeological Theory Building Using Hunter-Gatherer And Environmental Data Sets*, University of California Press, Berkeley.

Binneman, J. (1996a) The symbolic construction of communities during the Holocene Later Stone Age in the south-east Cape. PhD Dissertation. University of Witwatersrand.

Binneman, J. (1996b) Preliminary report on the investigations at Kulubele, an early Iron Age farming settlement in the Great Kei River valley, Eastern Cape. *S. Afr. Field Archaeol.*, **5**, 28–35.

Binneman, J. (1998) Results from a test excavation at Kleinpoort shelter in the Baviaanskloof, Eastern Cape Province. *S. Afr. Field Archaeol.*, **7**, 90–97.

Binneman, J. (2000) Results from two test excavations in the Baviaanskloof Mountains, Eastern Cape Province. *S. Afr. Field. Archaeol.*, **9**, 83–96.

Binneman, J. (2004/2005) Archaeological research along the South-Eastern Cape coast part 1: Open-air shell middens. *S. Afr. Field Archaeol.*, **13 & 14**, 49–77.

Blakely, R.L. (1977) *Biocultural Adaptation in Prehistoric America,* Proceedings No. 11, Southern Anthropological Society, University of Georgia Press, Athens.

Bogin, B. (1999) *Patterns of Human Growth*, University Press, Cambridge.

Bogin, B. and Loucky, J. (1997) Plasticity, political economy, and physical growth status of Guatemala Maya children living in the United States. *Am. J. Phys. Anthropol.*, **102**, 17–32.

Bogin, B., Smith, P., Orden, A.B. *et al.* (2002) Rapid change in height and body proportions of Maya American children. *Am. J. Hum. Biol.*, **14**, 753–761.

Bousman, C.B. (1998) The chronological evidence for the introduction of domestic stock into southern Africa. *Afr. Archaeolog. Rev.*, **15**(2), 133–150.

Boonzaier, E., Malherbe, C., Smith, A. and Berens, P. (1996) *The Cape Herders: A History of the Khoikhoi of Southern Africa*, David Philip Publishers, Cape Town, SA.

Boyd, D.C.M. and Boyd, C.C. Jr. (1989) A comparison of Tennessee Archaic and Mississippian maximum femoral lengths and midshaft diameters: Subsistence change and post cranial variability. *Southeast Archaeol.*, **8**, 107–116.

Brooks, S.T. and Suchey, J.M. (1990) Skeletal age determination based on the os pubis: A comparison of the Acsadi-Nemerskeri and Suchey-Brooks methods. *Hum. Evol.*, **5**, 227–238.

Buchanan, W.F., Parkington, J.E., Robey, T.S. and Vogel, J.C. (1984) Shellfish, subsistence and settlement: Some western Cape Holocene observations, in *Frontiers: Southern African Archaeology Today*, (eds M. Hall, G. Avery, D.M. Avery, M.L. Wilson and A.J.B. Humphreys), Cambridge Monographs in African Archaeology 10, BAR International Series 207, Oxford, pp. 121–130.

Buikstra, J.E. and Beck, L.A. (2006) *Bioarchaeology: The Contextual Analysis of Human Remains*, Academic Press, Burlington, MA.

Buikstra, J.E. and Ubelaker, D.H. (eds) (1994) Standards for Data Collection from Human Skeletal Remains: Proceedings of a seminar at the Field Museum of Natural History, organized by Jonathan Haas, Arkansas Archaeological Survey Research Series, No. 44. Fayetteville: Arkansas Archaeological Survey.

Chen, Y.S., Olckers, A., Schurr, T. *et al.* (2000) MtDNA variation in the South African !Kung and Khwe and their genetic relationship to other African populations. *Am. J. Hum. Genet.*, **66**, 1362–1383.

Cohen, M.N. (1989) *Health and the Rise of Civilization*, Yale University Press, New Haven, CT.

Cohen, M.N. (2008) Implications of the NDT for worldwide health and mortality in prehistory, in *The Neolithic Demographic Transition and its Consequences* (eds J.P. Bocquet-Appel and O. Bar-Yosef), Springer Science+Business Media B.V., pp. 481–500.

Cohen, M.N. and Armelagos, G.J. (1984) *Paleopathology at the Origins of Agriculture*, Academic Press, New York.

Cooper, R.S., Guo, X., Rotimi, C.N. *et al.* (2000) Heritability of angiotensin-converting enzyme and angiotensinogen: a comparison of US blacks and Nigerians. *Hypertension*, **35**, 1141–1147.

Cunningham, A.B. and Davis, G.W. (1997) Human use of plants, in *Vegetation of Southern Africa* (eds R.M. Cowling, D.M. Richardson and S.M. Pierce), University Press, Cambridge, pp. 474–501.

Cybulski, J.S. (1992) A Greenville Burial Ground: Human Remains and Mortuary Elements in British Columbia Coast Prehistory. Archaeological Survey of Canada, Mercury Series Paper No. 146. Ottawa: Canadian Museum of Civilization.

Damuth, J. (1981) Population density and body size in mammals. *Nature*, **290**, 699–700.

Deacon, H.J. (1976) *Where Hunters Gathered: A study of Holocene Stone Age people in the Eastern Cape,* South African Archaeological Society Monograph Series, No. 1, Claremont, SA.

Deacon, H.J. (1993) Planting an idea: An archaeology of Stone Age gatherers in South Africa. *S. Afr. Archaeol. Bull.*, **48**, 86–93.

Deacon, J. (1984) Later Stone Age people and their descendants in southern Africa, in *Southern African Prehistory and Paleoenvironments* (ed. R.G. Klein), A.A. Balkema Publishers, Rotterdam, ND, pp. 221–328.

Deacon, H.J. and Deacon, J. (1999) *Human Beginnings in South Africa: Uncovering the Secrets of the Stone Age*, AltaMira Press, Walnut Creek, CA.

DeLaurier, A. and Spence, M.W. (2003) Cranial genetic markers: Implications for postmartial residence patterns, in *Bones of the Ancestors: The Archaeology and Osteobiography of the Moatfield Site*, Archaeological Survey of Canada Mercury Series Paper 163 (eds R.F. Williamson and S. Pfeiffer), Canadian Museum of Civilization, Ottawa, 263–294.

De Villiers, H. (1967) *The Skull of the African Negro*, Witwatersrand University Press, Johannesburg.

De Villiers, H. (1968) *The Skull of the South African Negro: A Biometrical and Morphological Study*, Witwatersrand University Press, Johannesburg.

Dewar, G. and Pfeiffer, S. (2004) Postural behaviour of Later Stone Age people in South Africa. *S. Afr. Archaeol. Bull.*, **59**, 52–58.

Drennan, M.R. (1929) The dentition of a Bushman tribe. *Ann. S. Af. Mus.*, **24**, 61–87.

Elphick, R. (1977) *Kraal and Castle: Khoikhoi and the Founding of White South Africa*, Yale University Press, New Haven.

Franciscus, R.G. and Long, J.C. (1991) Variation in human nasal height and breadth. *Am. J. Phys. Anthropol.*, **85**, 419–427.

Gifford-Gonzalez, D. (1998) Early pastoralists in East Africa: Ecological and social dimensions. *J. Anthropol. Archaeol.*, **17**, 166–200.

Garn, S.M., Lewis, A.B. and Walenga, A. (1968) Evidence for a secular trend in tooth size over two generations. *J. Dent. Res.*, **47**(3), 503.

Guglielmino-Matessi, C.R., Gluckman, P. and Cavalli-Sforza, L.L. (1979) Climate and the evolution of skull metrics in man. *Am. J. Phys. Anthropol.*, **50**, 549–564.

Hall, S. (1990) Hunter-fisher-gatherers of the Fish River Basin: A contribution to the Holocene prehistory of the Eastern Cape. PhD Dissertation. University of Stellenbosch.

Hall, S. (2000) Burial and sequence in the Later Stone Age of the Eastern Cape Province. *South Africa. S. Afr. Archaeol. Bull.*, **55**, 137–146.

Hanihara, T., Ishida, H. and Dodo, Y. (2003) Characterization of biological diversity through analysis of discrete cranial traits. *Am. J. Phys. Anthropol.*, **121**, 241–251.

Hauser, G. and De Stefano, G.F. (1989) *Epigenetic Variants of the Human Skull*, Schweizerbart, Stuttgart.

Henshilwood, C.S., Nilssen, P. and Parkington, J. (1994) Mussel drying and food storage in the Late Holocene, SW Cape. *South Africa. J. Field Archaeol.*, **21**(1), 103–109.

Hillson, S., FitzGerald, C. and Flinn, H. (2005) Alternative dental measurements: Proposals and relationships with other measurements. *Am. J. Phys. Anthropol.*, **126**, 413–426.

Holl, A.F.C. (1998) Livestock husbandry, pastoralisms, and territoriality: The West African record. *J. Anthropol. Archaeol.*, **17**, 143–165.

Holliday, T.W. (1995) Body size and proportions in the Late Pleistocene Western Old World and the origins of modern humans. PhD Dissertation. Department of Anthropology, University of New Mexico.

Howells, W.W. (1973) *Cranial Variation in Man: A Study by Multivariate Analysis of Patterns of Difference Among Recent Human Populations*, Paper of the Peabody Museum of Archaeology and Ethnology, Vol. 67, Peabody Museum Harvard University, Cambridge.

Howells, W.W. (1995) *Who's Who in Skulls: Ethnic Identification of Crania from Measurements*, Harvard University Press, Cambridge, MA.

Huffman, T.N. (2006) Bantu migrations in southern Africa, *The Prehistory of Africa: Tracing the Lineage of Modern Man* (ed. H. Soodyall), Jonathan Ball, Jeppestown, pp. 97–108.

Inskeep, R. (1986) A preliminary survey of burial practices in the Later Stone Age, from the Orange River to the Cape coast, in *Variation, Culture, and Evolution in African Populations: Papers in Honour of Dr. Hertha de Villiers* (eds R. Singer and J.K. Lundy), Witwatersrand University Press, Johannesburg, pp. 212–219.

Iscan, M.Y., Loth, S.R. and Wright, R.K. (1984) Age estimation from the rib by phase analysis: White males. *J. Forensic Sci.*, **29**(4), 1094–1104.

Iscan, M.Y., Loth, S.R. and Wright, R.K. (1985) Age estimation from the rib by phase analysis: White females. *J. Forensic Sci.*, **30**(3), 853–863.

Jerardino Wiesenborn, A.M.S. (1996) Changing social landscapes of the Western Cape coast of South Africa over the last 4500 years. PhD Dissertation. University of Cape Town.

Jerardino, A. and Yates, R. (1996) Preliminary results from excavation at Seenbokfontein Cave: Implications for past and future research. *S. Afr. Archaeol. Bull.*, **51**, 7–16.

Jerardino, A., Branch, G.M. and Navarro, R. (2008) Human impact on precolonial West Coast marine environments of South Africa, in *Human Impacts of Marine Environments: A Global Perspective* (eds J.M. Erlandson and T.C. Rick), University of California Press, Berkeley, pp. 279–296.

Jolicoeur, P. and Mosimann, J.E. (1960) Size and shape variation in the painted turtle. A principal component analysis. *Growth*, **24**, 339–354.

Kelly, R.L. (1995) *The Foraging Spectrum: Diversity in Hunter-Gatherer Lifeways*, Smithsonian Institution Press, Washington, D.C.

Kieser, J.A. (1985) An odontometric analysis of the early Griqua dentition. *Anthropolgischer Anzeiger*, **43**, 51–58.

Kieser, J.A. (1990) *Human Adult Odontometrics: The Study of Variation in Adult Tooth Size*, Cambridge University Press, Cambridge.

Kieser, J.A., Groeneveld, H.T. and Cameron, N. (1987) Evidence for a secular trend in the Negro dentition. *Ann. Human. Biol.*, **14**, 517–532.

Kinahan, J. (1996) Alternative views on the acquisition of livestock by hunter-gatherers in southern Africa: a rejoinder to Smith, Yates, and Jacobson. *S. Afr. Archaeol. Bull.*, **51**, 106–108.

Kurki, H.K., Ginter, J.K., Stock, J.T. and Pfeiffer, S. (2010) Body size estimation of small-bodied humans: applicability of current methods. *Am. J. Phys. Anthropol.*, **114**, 169–180.

Kusimba, S.B. (2002) *African Foragers: Environment, Technology, Interactions*, AltaMira Press, Walnut Creek, CA.

Lahr, M.M. (1996) *The Evolution of Modern Human Diversity: A Study of Cranial Variation*, University Press, Cambridge.

Lahr, M.M. and Foley, R.A. (1998) Towards a theory of modern human origins: geography, demography and diversity on recent human evolution. *Yrbk. Phys. Anthropol.*, **41**, 137–176.

Larsen, C.S. (1995) Biological changes in human populations with agriculture. *Ann. Rev. Anthropol.*, **24**, 185–213.

Larsen, C.S. (1997) *Bioarchaeology: Interpreting Behaviour from the Human Skeleton*, University Press, Cambridge.

Larsen, C.S. (2002) Bioarchaeology: The lives and lifestyles of past people. *J. Archaeol. Res.*, **10**(2), 19–166.

Lee-Thorp, J.A., Holmgren, K., Lauritzen, S.E. *et al.* (2001) Rapid climate shifts in the southern African interior throughout the mid to late Holocene. *Geophys. Res. Lett.*, **28**(23), 4507–4510.

Lewis, C.A. (2002) Radiocarbon dates and the Late Quaternary palaeogeography of the Province of the Eastern Cape, South Africa. *Quatern. Int.*, **89**, 59–69.

Lewis, C.A. (2005) Late Glacial and Holocene palaeoclimatology of the Drakensberg of the Eastern Cape, South Africa. *Quatern. Int.*, **129**, 33–48.

Lovejoy, C.O., Meindl, R.S., Pryzbeck, T.R. and Mensforth, R.P. (1985) Chronological metamorphosis of the auricular surface of the ilium: A new method for the determination of adult skeletal age at death. *Am. J. Phys. Anthropol.*, **68**, 15–28.

Lukacs, J.R. (2007) Climate, subsistence and health in prehistoric India: The biological impact of a short-term subsistence shift, in *Ancient Health: Skeletal Indicators of Agricultural and Economic Intensification* (eds M.N. Cohen and G.M.M. Crane-Kramer), University Press of Florida, Gainesville, pp. 207–219.

Lundy, J.K. (1986) Intertribal metrical and morphological variation in the postcranial skeleton of the South African Negro, in *Variation, Culture, and Evolution in African Populations* (eds R. Singer and J.K. Lundy), Witwatersrand University Press, Johannesburg, pp. 85–99.

Manhire, A.H., Parkington, J.E. and Robey, T.S. (1984) Stone tools and Sandveld settlement, in *Frontiers: Southern African Archaeology Today*, (eds M. Hall, G. Avery, D.M. Avery, M.L. Wilson and A.J.B. Humphreys) Cambridge Monographs in African Archaeology 10, BAR International Series 207, Oxford, pp. 111–120.

Mann, R.W., Jantz, R.L., Bass, W.M. and Willey, P.S. (1991) Maxillary suture obliteration: A visual method for estimating skeletal age. *J. Forensic Sci.*, **36**, 781–791.

Marean, C.W. (1992) Hunter to herder: Large mammal remains from the hunter-gatherer occupation at Enkapune Ya Moto rock-shelter, Central Rift. *Kenya. Afr. Archaeol. Rev.*, **10**, 65–127.

Marshall, F. and Hildebrand, E. (2002) Cattle before crops: the beginnings of food production in Africa. *J. World Prehist.*, **16**(2), 99–143.

Marshall, F. (1998) Early food production in Africa. *The Review of Archaeology*, **19**(2), 47–58.

Martin, R. (ed. K. Saller) (1957) *Lehrbuch der Anthropologies in Systematischer Darstellung, Revised 3rd edn*, Gustav Fischer Verlag, Stuttgart.

Mayewski, P.A., Rohling, E.E., Stager, J.C. *et al.* (2004) Holocene climate variability. *Quatern. Res.*, **62**, 243–255.

Meindl, R.S. and Lovejoy, C.O. (1985) Ectocranial suture closure: A revised method for the determination of skeletal age at death based on the lateral-anterior sutures. *Am. J. Phys. Anthropol.*, **68**, 57–66.

Mitchell, P. (2002) Hunter-gatherer archaeology in southern Africa: Recent research, future trends. *Before Farming*, **1**(3), 1–36.

Mitchell, P. (2002) *The Archaeology of Southern Africa*, University Press, Cambridge.

Molto, J.E. (1983) Biological Relationships of Southern Ontario Woodland Peoples: The Evidence of Discontinuous Cranial Traits. National Museum of Man Mercury Series, Archaeological Survey of Canada, Paper 117. Ottawa: National Museum of Man.

Moore-Jansen, P.M., Ousley, S.D. and Jantz, R.L. (1994) Data Collection Procedures for Forensic Skeletal Material. Report of Investigations No. 48. Department of Anthropology, University of Tennessee, Knoxville.

Morrees, C.F.A. and Reed, R.B. (1954) Correlations among crown diameters of human teeth. *Arch. Oran Biol.*, **9**, 685–697.

Morris, A.G. (1992a) *A Master Catalogue: Holocene Human Skeletons from South Africa*, Witwatersrand University Press, Johannesburg.

Morris, A.G. (1992b) *The Skeletons of Contact: A Study of Protohistoric Burials from the Lower Orange River Valley, South Africa*, Witwatersrand University Press, Johannesburg.

Morris, A.G. and Ribot, I. (2006) Morphometric cranial identity of prehistoric Malawians in the light of Sub-Saharan African diversity. *Am. J. Phys. Anthropol.*, **130**, 10–25.

Murdoch, W.W. (1994) Population regulation in theory and practice. *Ecology*, **75**, 271–287.

Ossenberg, N.S. (1969) Discontinuous Morphological Variation in Human Skeletal Remains. Unpublished PhD dissertation. University of Toronto.

Parkington, J.E. (1980) Time and place: Some observations on spatial and temporal patterning in the Later Stone Age sequence in southern Africa. *S. Afr. Archaeol. Bull.*, **35**, 73–83.

Parkington, J.E., Yates, R., Manhire, A.H. and Halkett, D. (1986) The social impact of pastoralism in the south-western Cape. *J. Anthropol. Archaeol.*, **5**, 313–329.

Parsons, I. (2006) Later Stone Age socio-economic variability in the Northern Cape, South Africa. Calgary, AB: Unpublished paper presented at the 18th biennial conference of the Society of Africanist Archaeologists, June 23–26.

Pearson, O.M. (2000) Activity, climate and postcranial robusticity: Implications for modern human origins and scenarios of adaptive change. *Curr. Anthropol.*, **41**(4), 569–589.

Pfeiffer, S. (2001) Patterns of skeletal trauma during the Holocene Later Stone Age, Southern Africa. Paper presented at the Pan-African Congress of Prehistory, Bamako, Mali, February 2001.

Pfeiffer, S. (2007) The health of foragers: People of the Later Stone Age, southern Africa, in *Ancient Health: Skeletal Indicators of Agricultural and Economic Intensification* (eds M.N. Cohen and G. Crane-Kramer), University of Florida Press, Gainesville, pp. 223–236.

Pfeiffer, S., Cranial trauma as evidence of a stressful period among southern African foragers, in *Volume in Honour of Michael Spence* (eds P. Timmins and N. Ferris), Ontario Archaeological Society, London, ON, in press.

Pfeiffer, S. and Sealy, J. (2006) Body size among Holocene foragers of the Cape Ecozone, southern Africa. *Am. J. Phys. Anthropol.*, **129**, 1–11.

Pfeiffer, S. and Stock, J. (2002) Upper limb morphology and the division of labour among southern African Holocene foragers. *Am. J. Phys. Anthropol.*, **S34**, 124.

Phenice, T.W. (1969) A newly developed visual method of sexing the os pubis. *Am. J. Phys. Anthropol.*, **30**, 297–302.

Phillipson, D.W. (2005) *African Archaeology*, Cambridge University Press, Cambridge.

Pimentel, R.A. (1992) An introduction to ordination, principal components analysis and discriminant analysis, in *Ordination in the Study of Morphology, Evolution and Systematics of Insects: Applications and Quantitative Genetic Rationals* (eds J.T. Sorensen and R. Foottit), Elsevier Press, Amsterdam, pp. 11–28.

Powell, M.L., Bridges, P.S. and Wagner Mires, A.M. (1991) *What Mean these Bones? Studies in Southeastern Bioarchaeology*, University of Alabama Press, Tascaloosa.

Price, T.D. and Brown, J.A. (1985) Aspects of hunter-gatherer complexity, in *Prehistoric Hunter-Gatherers: The Emergence of Cultural Complexity* (eds T.D. Price and J.A. Brown), Academic Press, Orlando, pp. 3–20.

Prowse, T.L. and Lovell, N.C. (1996) Concordance of cranial and dental morphological traits and evidence for endogamy in ancient Egypt. *Am. J. Phys. Anthropol.*, **101**, 237–246.

Relethford, J.H. (2004) Boas and beyond: Migration and craniometric variation. *Am. J. Hum. Biol.*, **16**, 379–386.

Ribot, I. (2002) Craniomandibular variation in sub-Saharan Africa: Sexual dimorphism, geography, ecology and history. PhD Dissertation. Cambridge, University of Cambridge.

Rightmire, G.P. (1972) Cranial measurements and discrete traits compared in distance studies of African Negro skulls. *Hum. Biol.*, **44**, 263–276.

Rightmire, G.P. (1976) Metric versus discrete traits in African skulls, in *The Measures of Man: Methodologies in Biological Anthropology* (eds E. Giles and J.S. Freidlaender), Peabody Museum Press, Cambridge, pp. 383–407.

Roseman, C. and Weaver, T.D. (2004) Multivariate apportionment of global human craniometric diversity. *Am. J. Phys. Anthrol.*, **125**, 257–263.

Rosen, D.Z., Lewis, C.A. and Illgner, P.M. (1999) Paleoclimatic and archaeological implications of organic rich sediments at Tiffindel Ski Resort, near Rhodes, Eastern Cape, South Africa. *T. Roy. Soc. S. Afr.*, **54**(2), 311–321.

Rubini, M., Mogliazza, S. and Corruccini, R.S.T. (2006) Biological divergence and equality during the first millennium BC in human populations of central Italy. *Am. J. Hum. Biol.*, **19**, 119–131.

Sadr, K. (1998) The first herders at the Cape of Good Hope. *Afr. Archaeol. Rev.*, **15**(2), 101–132.

Sadr, K. (2003) The Neolithic of southern Africa. *J. Afr. Hist.*, **44**, 195–209.

Sadr, K. (2004) Feasting on Kasteelberg? Early herders in the west coast of South Africa. *Before Farming: The Archaeology of Old World Hunter-Gatherers*, Online] 2004/3 (2), Available at: http://www.waspress.co.uk/journals/beforefarming/ (Accessed 12 November 2007).

Schapera, I. (1930) *The Khoisan Peoples of South Africa: Bushmen and Hottentots*, Routledge & Kegan Paul Ltd, London.

Schmid, P.E., Tokeshi, M. and Schmid-Araya, J.M. (2000) Relation between population density and body size in stream communities. *Science*, **289**, 1557–1560.

Sciulli, P.W. (1997) Dental evolution in prehistoric Native Americans of the Ohio Valley Area. I. Wear and pathology. *Int. J. Osteoarchaeol.*, **7**, 507–524.

Sealy, J. (1986) *Stable Carbon Isotopes and Prehistoric Diets I the South-Western Cape Province, South Africa*. Cambridge Monographs in African Archaeology 15, BAR International Series 293, Oxford.

Sealy, J.C. (2006) Diet, mobility, and settlement patterns among Holocene hunter-gatherers in southernmost Africa. *Curr. Anthropol.*, **47**(4), 569–595.

Sealy, J.C., Patrick, M.K., Morris, A.G. and Alder, D. (1992) Diet and dental caries among Later Stone Age inhabitants of the Cape Province, South Africa. *Am. J. Phys. Anthropol.*, **88**, 123–134.

Sealy, J.C. and Pfeiffer, S. (2000) Diet, body size and landscape use among Holocene people in the Southern Cape, South Africa. *Curr. Anthropol.*, **41**(4), 642–655.

Sealy, J.C. and van de Merwe, N.J. (1986) Isotope assessment and the seasonal mobility hypothesis in the Southwestern Cape of South Africa. *Curr. Anthropol.*, **27**(2), 135–150.

Sealy, J.C. and van de Merwe, N.J. (1988) Social, spatial and chronological patterning in marine food use as determined by $\delta^{13}C$ measurements of Holocene human skeletons from the south-western Cape, South Africa. *World Archaeol.*, **20**(1), 87–102.

Sealy, J.C. and Yates, R. (1994) The chronology of the introduction of pastoralism to the Cape, South Africa. *Antiquity*, **68**, 58–67.

Silberbauer, F.B. (1979) Stable Carbon isotopes and prehistoric diets in the Eastern Cape Province, South Africa. MA thesis. Department of Archaeology, University of Cape Town.

Sinclair, D. and Dangerfield, P. (1998) *Human Growth After Birth*, 6th edn, Oxford University Press, Oxford.

Smith, A.B. (1998) Keeping people on the periphery: The ideology of social hierarchies between hunters and herders. *J. Anthropol. Archaeol.*, **17**(2), 201–215.

Smith, A.B. (2005) *African Herders: Emergence of Pastoral Traditions*, AltaMira Press, Walnut Creek.

Smith, A.B. (1992) *Pastoralism in Africa: Origins and Development Ecology*, Witwatersrand University Press, Johannesburg.

Smith, A.B. (2006) Ideological inhibitors to hunters becoming food producers in Africa. Calgary, AB: Unpublished paper presented at the 18th biennial conference of the Society of Africanist Archaeologists, June 23–26.

Smith, A.B., Sadr, K., Gribble, J. and Yates, R. (1991) Excavations in the south-western Cape, South Africa, and the archaeological identity of prehistoric hunter gatherers within the last 2000 years. *S. Afr. Archaeol. Bull.*, **47**, 62–64.

Smith, A., Malherbe, C., Guenther, M. and Berens, P. (2000) *The Bushmen of Southern Africa: A Foraging Society in Transition*, David Philip Publishers, Cape Town, SA.

Smith, B.H. (1984) Patterns of molar wear in hunter-gatherers and agriculturalists. *Am. J. Phys. Anthropol.*, **63**, 39–56.

Smith, H.F., Terhune, C.E. and Lockwood, C.A. (2007) Genetic, geographic, and environmental correlates of human temporal bone variation. *Am. J. Phys. Anthropol.*, **134**, 312–322.

Smith, P. and Horwitz, L.K. (2007) Ancestors and inheritors: A bioanthropological perspective on the transition to agropastoralism in the southern Levant, in *Ancient Health: Skeletal Indicators of Agricultural and Economic Intensification* (eds M.N. Cohen and G.M.M. Crane-Kramer), University Press of Florida, Gainesville.

Sparks, C.S. and Jantz, R.L. (2002) A reassessment of human cranial plasticity: Boas revisited. *PNAS*, **99**, 14636–14639.

Spence, M.W. (1974) Residential practices and the distribution of skeletal traits in Teotihuacan, Mexico. *Man*, **9**, 262–273.

Steckel, R.H. and Rose, J.C. (2002) *The Backbone of History: Health and Nutrition in the Western Hemisphere*, Cambridge University Press, Cambridge.

Stock, J.T. and Pfeiffer, S. (2004) Long bone robusticity and subsistence behaviour among Later Stone Age foragers of the forest and fynbos biomes of South Africa. *J. Archaeol. Sci.*, **31**, 999–1013.

Stojanowski, C.M. (2005) *Biocultural Histories in La Florida: A Bioarchaeological Perspective*, University of Alabama Press, Tuscaloosa.

Stojanowski, C.M., Larsen, C.S., Tung, T.A. and McEwan, B.G. (2007) Biological structure and health implications from tooth size at Mission San Luis de Apalachee. *Ame. J. Phys. Anthropol.*, **132**, 207–222.

Stynder, D.D. (2006) A quantitative assessment of variation in Holocene Khoesan crania from South Africa's western, south-western, southern, and south-eastern coasts and coastal forelands. PhD Dissertation. University of Cape Town, Cape Town, South Africa.

Stynder, D.D., Rogers-Ackerman, R. and Sealy, J.C. (2007) Craniofacial variation and population continuity during the South African Holocene. *Am. J. Phys. Anthropol.*, **134**, 489–500.

Stynder, D.D., Rogers-Ackerman, R. and Sealy, J.C. (2007) Early to mid-Holocene South African Later Stone Age human crania exhibit a distinctly Khoesan morphological pattern. *S. Afr. J. Sci.*, **103**, 349–352.

Suchey, J. and Katz, D. (1986) Skeletal age standards derived from an extensive multiracial sample of modern Americans. *Am. J. Phys. Anthropol.*, **69**, 269.

Trinkaus, E., Churchill, S.E. and Ruff, C.B. (1994) Postcranial robusticity in Homo: Humeral bilateral asymmetry and bone plasticity. *Am. J. Phys. Anthropol.*, **93**, 1–34.

Vansina, J. (1996) A slow revolution: Farming in subequatorial Africa, in *The Growth of Farming Communities in Africa from the Equator Southwards, Azania special volume XXIX-XXX* (ed. J.E.G. Sutton), The British Institute in Eastern Africa, pp. 15–26.

Voigt, E.A. (1986) Iron Age herding: Archaeological and ethnoarchaeological approaches to pastoral problems. *S. Afr. Archaeol. Society Goodwin Series*, **5**, 13–21.

Walker, R.S. and Hamilton, M. (2008) Life-history consequences of density dependence and the evolution of human body sizes. *Curr. Anthropol.*, **49**(1), 115–122.

Wetterstrom, W. (1998) The origins of agriculture in Africa: With particular reference to sorghum and pearl millet. *Rev. Archaeol.*, **19**(2), 30–46.

Wiley, A.S. (2005) Does milk make children grow? Relationships between milk consumption and height in NHANES 1999–2002. *Am. J. Hum. Biol.*, **17**, 425–441.

Williamson, R.F. and Pfeiffer, S. (eds) (2003) *Bones of the Ancestors: The Archaeology and Osteobiography of the Moatfield Ossuary, Mercury Series Archaeology Paper 163*, Canadian Museum of Civilization, Gatineau.

Wilson, M.L. and Lundy, J.K. (1994) Estimated living statures of dated Khoisan skeletons from the south-western coastal region of South Africa. *S. Afr. Archaeol. Bull.*, **49**, 2–8.

Wood, B. and Lieberman, D.E. (2001) Craniodental variation in *Paranthropus boisei*: A developmental and functional perspective. *Am. J. Phys. Anthropol.*, **116**, 13–25.

Woodburn, J. (1982) Egalitarian societies. *Man*, **17**, 431–451.

SECTION B

Growth and Body Size Variation

7

Long Bone Length, Stature and Time in the European Late Pleistocene and Early Holocene

Christopher Meiklejohn and Jeff Babb

Departments of Anthropology & Mathematics and Statistics, University of Winnipeg

7.1 INTRODUCTION

In 1984, C. Meiklejohn *et al.* used trends in stature as a proxy measure of population stress in the transition from Upper Palaeolithic through Mesolithic to Neolithic (Meiklejohn *et al.*, 1984). The 1984 study confirmed work of the time that showed a decrease in stature in Europeans from the Upper Palaeolithic through the Mesolithic, suggesting that the end of this trend might relate to the agricultural transition.

This chapter revisits the 1984 conclusions in the light of currently available data and a further 25 years of work on approaches to stature calculation. A key result is that the use of stature as the primary variable, as opposed to individual long bone length, adds unnecessary noise to the analysis. Within this light, we begin with a review of stature studies, and of the variables involved in its calculation, both generally and in relation to the issue in European Neolithic and earlier groups. Focus is placed on trends in individual long bone length. We review studies since publication of Cohen and Armelagos' (1984a) volume and show that though there has been a dramatic increase in the quality and quantity of both Upper Palaeolithic and Mesolithic datasets, problems still exist in the Neolithic set. After discussion of the current dataset and our approach to quality control, our analysis of long bone length follows. We conclude that the current dataset confirms the overall decline in European stature from Upper Palaeolithic through Neolithic, but find that the decline largely occurs between earlier and later portions of the Upper Palaeolithic. Within-period analysis of Late Upper Palaeolithic, Mesolithic and Neolithic datasets shows overall stasis in long bone length, and by extension stature, over the three periods. We conclude by looking at how our European results compare with those of other regions and how this relates to studies of the agricultural transition.

Human Bioarchaeology of the Transition to Agriculture Edited by Ron Pinhasi and Jay T. Stock

7.2 THE THEORETICAL STUDY OF STATURE

Stature is a primary variable in human osteology, with methods for calculation being a central topic. Our initial review centres on theory and method, especially in a European context. We show that stature calculation is largely focused on forensic reconstruction *sensu lato*, with study of the individual central. Comparative study of populations in space and time is tangential. As a result, use of stature as the primary variable in time related studies, rather than individual long bone lengths, creates noise obscuring rather than clarifying trends. For further detail on the use of stature in forensic applications, see Ubelaker (2008).

Methods relating stature and long bone length have evolved as a set of steps, each of which solved an existing problem but, in turn, raised further issues. The recognition that body proportionality permitted calculation of stature from long bone length was initially developed in the late nineteenth century on a series of 100 French cadavers, by Etienne Rollet (1888), and Leonce Manouvrier (1892, 1893). Tables for each long bone compared bone length and stature. Both used the same data but presented their results differently; 'Manouvrier determined the average stature of those individuals who presented the same lengths for a given long bone, whereas Rollet determined the average length of a given long bone from those who presented the same stature' (Trotter and Gleser, 1952: 464). This approach was widely applied for the next half century (see e.g. Vallois, 1943 and below) and some use continues today (see discussion in Iscan and Quatrehomme, 1999; Formicola, 1983).

The second methodological step, by Karl Pearson (1899), introduced the use of regression equations for calculation of stature, though still using Rollet's limited dataset. Pearson recognized and discussed a number of issues limiting stature calculation, including sex differences, the use of equations on bones of different origin, and the limited validity of equations to the range of the original dataset. A half century later Dupertuis and Hadden (1951) demonstrated that Rollet's dataset was both limited and from a short population. The basic use of regression equations still dominates stature calculation today and more systematic approaches were not published for over half a century, though studies published between 1900 and 1950 introduced population-based formulae (Stevenson, 1929; Telkka, 1950) and raised issues of sample size (Breitinger, 1937).

It was only in 1948 that the use of the Terry (St Louis) and Todd (Cleveland) cadaver-based collections of known age and origin was suggested, resulting in two studies (Dupertuis and Hadden, 1951; Trotter and Gleser, 1952). For several technical reasons, only the latter gained acceptance and today the regression equations of Trotter and Gleser (Trotter, 1970; Trotter and Gleser, 1952, 1958) provide the most widely used approach to stature calculation. However, the original framework and need was forensic, neither designed to be used beyond the original populations studied nor the chronological present. As pointed out by Feldesman (Feldesman, Kleckner and Lundy, 1990: 360), most applications of Trotter and Gleser equations fail to recognize that they were 'devised for entirely different purposes'.

Trotter and Gleser's original study (1952), with a sample of 2055 individuals, used material from the Terry Collection and remains of deceased American soldiers from World War II. Compared to earlier work, the new equations produced better fit to original recorded stature, and equations for all six limb bones were calculated for the first time. The second paper (Trotter and Gleser, 1958) had a larger sample of 5517 individuals (4572 identified as white and including data from Korean War dead). For both, the primary equations are for white and black series (the latter identified as Negro as *per* practice at the time). There were minor technical differences but stature means and bone length correlations did not significantly differ between the two.

Much of the discussion stemming from the Trotter and Gleser papers is technical and includes matters such as precision of the estimates. Though steps were made in generating different equations for different populations, issues involving secular trends also arose, seen in differences between the World War II and Korean War data. The approach used in 1984 (Meiklejohn *et al.*, 1984) followed Trotter's (1970) conclusion that better results were obtained from the 1952 equations.

By 1960, an empirical methodology existed for stature calculation, though with a number of inherent problems. The current general acceptance of the Trotter and Gleser equations (Bass, 1995; Buikstra and Ubelaker, 1994; Byers, 2007) indicates that acceptable results are produced for skeletal evaluation of recent remains, although only the original white male sample is of substantial size. Though other equations exist, all are based on limited samples of narrow utility. Other large datasets, such as those used by Trotter and Gleser, have not materialized.

Work since 1960 has largely focused on issues that include differences between populations and inherent issues in the use of regression equations. Recent publications highlight issues in applying equations to samples from other than the original population (Ruff, 2002; Kurki *et al.*, 2010; Auerbach and Ruff, 2010).

Recent work clearly recognizes the limitations of stature reconstruction but largely focuses on specific issues rather than development of new methods. A brief review of work since 1980 demonstrates the difficulties involved. Feldesman, Kleckner and Lundy (1990) were one group of researchers to focus on the femur rather than derived stature as the main variable of interest, mainly since the femur has a remarkably constant ratio to adult stature, independent of sex. Our position that follows from this and related findings is that the direct study of long bone length avoids the added error involved in moving from a direct to an indirect measure of overall height.

Work in the last two decades largely fortifies older issues. Stature-focused papers in the *American Journal of Physical Anthropology*, a principal source in work of this kind, generally fall into one area, provision of more accurate stature regression equations for specific populations or parts of populations. Examples include regional studies on Europeans (De Mendonça, 2000; Giannecchini and Moggi-Cecchi, 2008; Vercellotti *et al.*, 2009) and Native Americans (Sciulli, Schneider and Mahaney, 1990; Sciulli and Giesen, 1993), the specific issue of small-bodied populations (Kurki *et al.*, 2008, 2010) and the accuracy of historical records (Hernández, García-Moro and Lalueza-Fox, 1998). The issue of whether the variable to be reported is stature or long bone length is largely ignored.

7.3 STUDIES OF EUROPEAN STATURE FROM THE UPPER PALAEOLITHIC TO THE NEOLITHIC

The assessment of stature in European Upper Palaeolithic through Neolithic samples began with the use of Rollet and Manouvrier's tables to archaeological skeletal material. However, though early studies by the French (and others) included Verneau's (1906) study of the *Grotte des Enfants* (Grimaldi), there was little methodological progress in later studies, with the same sources used by Vallois (1943) in studying French late Neolithic samples. Studies of stature prior to 1950 are largely based on methods that we would now fault. Published measurements from before 1950 that meet clearly defined standards can be used; however, early stature estimates should be used with caution, if at all. In addition, early studies are largely, if not totally, descriptive, with minimal explanation. Vallois (1943) thus provides data, but beyond noting that postcranial studies linger behind those of crania, goes no further than showing series differences, which he assumed to be 'racial'.

Only in the 1980s was there a shift in approach in European stature studies, coincident with the realization that pre-Neolithic populations are important to the study of human evolution. The works of Patrick Key and David Frayer act as prologue; each shows the shift from description to explanation. Key (1980) examined trends in European sexual dimorphism from the Mesolithic to the Middle Ages, using linear regression on femoral dimensions. Dimorphism, already raised by Frayer (1978) in studying dental dimensions, increased over time as a function of different rates of increase in femoral length (and, by extension, stature). Frayer (1980, 1981) looked at postcranial size and dimorphism through long bone length and stature and (with Key) was amongst the first to suggest stature trends in the European pre-Neolithic. He found no apparent stature difference between earlier and later Upper Palaeolithic male samples, but a decrease in females, resulting in increased dimorphism, a trend continuing through the Mesolithic. Change was seen within an evolutionary context and suggested mode of selection, related to increasingly sophisticated hunting technology and demands on human biomechanics.

The work of Key and Frayer was a core background for the 1984 study of stature (Meiklejohn *et al.*, 1984). The Upper Palaeolithic sample was the same size as Frayer's and largely similar, the Mesolithic sample twice the size (see below). To these was added a Neolithic/post-Neolithic sample. Each individual or sample was provided with a date, either based on radiocarbon, if available, or an archaeological estimate. Stature (based on Trotter and Gleser equations) was regressed on date in a linear model: on date and date^2 in a quadratic (curvilinear) model. For the total series, the linear model was highly significant, sexes combined or separate. Trends were more marked in the female sample. The conclusion was for 'a significant stature decline from the Upper Palaeolithic through the Neolithic, with possible reversal from that point onwards, in agreement with Frayer (1980) and Key (1980), (Meiklejohn *et al.*, 1984: 90). The pattern within phases of the sample was less clear. The female Upper Palaeolithic sample showed highly significant stature decrease ($p = 0.0189$), but no other sub-sample showed within-period change. The linear regression results were clear, the quadratic analysis less so. We interpreted the trend as a proxy for increasing stress, alleviated by the introduction of food production technologies (Meiklejohn *et al.*, 1984). We could not refute Frayer's activity pattern model and were aware that the pattern varied by sex.

The Cohen and Armelagos (1984a) volume presaged a spate of papers, regional and theoretical, looking at stature and the European Mesolithic-Neolithic transition. Jacobs' (1985) examination of postcranial evolution in late glacial and early postglacial Europe (independent of Meiklejohn *et al.*, 1984) stressed the lack of earlier postcranial studies and limited availability of postcranial data (other than the work of Trinkaus (1976), on femoral diaphysis shape, and the survey of von Bonin (1935)). Frayer (1980, 1981) was seen as alone in using postcranial variation to address European late Upper Palaeolithic and Mesolithic issues. Jacobs' sample was divided into three groups, pre-glacial maximum (Aurignacian and Gravettian), late glacial (Magdalenian) and postglacial (Epipalaeolithic and Mesolithic). The Upper Palaeolithic sample was larger than that of Meiklejohn *et al.* (1984) but the Mesolithic sample was only half the size. Pooled sample from these three time periods, rather than the individual specimens, were treated as entities. By considering Epipalaeolithic (primarily Epigravettian) material as postglacial, his breakdown was not based on absolute date. His logic was that Epipalaeolithic populations were experiencing postglacial environmental change, assuming that selective forces would not be balanced by gene flow. With the overlap of the datasets, it is not surprising that Jacobs' (1985) conclusions mirrored those of Meiklejohn *et al.* (1984). Reduction in long bone length occurred between the combined Upper Palaeolithic and Mesolithic samples, with reduction also found between earlier and later Upper Palaeolithic groups. Greatest change occurred in the female sample.

Jacobs' later work (1993) reflected his interest in Eastern Europe. Mesolithic and Neolithic samples from the Dnieper Rapids region (southern Ukraine, see Lillie and Budd, this volume) showed no significant difference in Mesolithic and Neolithic long bone length, and hence results were in agreement with the more broadly based conclusion of Meiklejohn *et al.* (1984). Nevertheless, stature estimations showed Ukrainian samples to be significantly taller than Western Europeans in both periods.

Formicola (1983) was the first to raise the issue of the applicability of published stature tables to European prehistoric samples, stressing the need for population-specific methods. Formicola's intention was to find an approach resulting in the 'least divergence amongst stature values calculated from different limb segments', measured as the 'average of the differences in the stature values obtained from the individual bones' (1983: 35). He found that the best fit for Italian samples came from Trotter and Gleser's (1952, 1958) formulae for blacks (despite their European origin). However, the Italian data showed a similar decrease from Upper Palaeolithic to Neolithic as seen in other work. Two further points were also made. The first, predictable from earlier work, was that Trotter and Gleser's standards for blacks result in lower stature estimates than do those for whites, a function of divergent body proportions. The other, more subtle, drew attention to very high stature values obtained from some early Upper Palaeolithic samples, especially from Grimaldi. Recognition of issues with the Trotter and Gleser (1952) standards for whites was also made by Constandse-Westermann, Blok and Newell (1985). However, rather than calculating new equations, they suggested correcting distal segment stature estimates by a constant derived from their European Mesolithic sample, thus further enhancing distortions stemming from the use of stature equations instead of directly comparing long bone measurements.

By 1985 issues had been raised about calculation of European stature. Of the workers discussed only Formicola further probed European trends and their explanation. Formicola (1993) pointed out issues in predicting stature, given that the real value cannot be determined. He pointed out strengths of the Fully method, requiring the full skeleton (Fully, 1956; Fully and Pineau, 1960), but it can seldom be applied to prehistoric material, given issues of completeness (Vercellotti *et al.*, 2009). Of more obvious utility was the identification of generic problems associated with least-squares regression, the approach of both Pearson and Trotter and Gleser (and taken further by Formicola and Franceschi, 1996). Least-squares (Model I) and major-axis (Model II) regression approaches were compared. Sex-specific samples were maintained despite Sjovold's (1990) argument, questioned by Formicola, that sex differences in limb proportions were a simple function of height rather than sex. With major-axis regression providing better results for short and tall individuals, and working well with those of intermediate stature, they concluded that, 'in our opinion, the latter (Model II) equations could prove particularly suitable in analyses focused on temporal or spatial stature variations and more in general when large variation in body size is foreseen amongst the samples studied' (Formicola and Franceschi, 1996: 87). From this base Formicola and Giannecchini (1999) reviewed stature in European Upper Palaeolithic and Mesolithic samples, which were divided into pre-glacial maximum, late glacial and postglacial groups, using major-axis regression (Model II). Comparison with results from the least-squares approach shows similar overall trends. Differences are internally consistent, and most obvious in individuals at the stature extremes. They also (Formicola and Giannecchini, 1999: 322) refer to Holliday's (1997) finding that European Upper Palaeolithic, Mesolithic and Neolithic populations share 'relative elongation of distal parts of the extremities'. Comparison of the three periods produced generally similar results to earlier work. Early Upper Palaeolithic individuals were taller than those of the late glacial, with no east/west partitioning of the results, though few samples were from the east. In comparison, the Mesolithic sample was separated into shorter western and taller eastern groups, with homogeneity within the subsets. Western Mesolithic individuals were not significantly shorter than those of the late Upper Palaeolithic (the small eastern Upper Palaeolithic sample voided attempts at regional comparison).

Formicola (2003) responded to Holliday (1997), who found no stature change between early and late Upper Palaeolithic samples. However, Holliday used different regression equations for early and late samples, the black and white least-squares methods of Trotter and Gleser, arguing that the former have elongated (tropical) distal limb segments, the latter shorter (temperate) segments. To us this approach has two faults. As Formicola notes, it fails to recognize the poor behaviour of white standards on early European samples (see above). We also see a failure to recognize that use of different formulae creates a 'fault line', with individuals on either side of the early/late boundary appearing different in stature whether the shift is gradual or rapid.

The above review highlights methodological and practical problems associated with the use of stature as the primary variable, which interposes noise between data and interpretation. We therefore focus below on raw bone length. We also have improved chronological quality control of the skeletal samples though we thereby sacrifice sample size in order to obtain what we refer to as a cleaner signal.

7.4 CURRENT STUDY

7.4.1 Database

As discussed above, this paper revisits conclusions made in Meiklejohn *et al.* (1984), with a database of both increased size and better cultural and chronological quality control (Table 7.1). To show the growth in sample size, we compare five samples, that of Frayer (1980), the 1984 study, the most recent review of Formicola (Formicola and Giannecchini, 1999), and two versions of the sample used here, the full dataset and the abbreviated set dated directly by radiocarbon. We exclude 75 further individuals of unknown sex. The sample size increase is especially clear for the Upper Palaeolithic and Mesolithic, showing samples with significantly improved chronological control. Issues with the Neolithic sample are discussed below.

Table 7.1 Comparative sample sizes

Sample	Frayer, 1980	Meiklejohn *et al.*, 1984	Formicola and Giannecchini, 1999	This study total (known sex)[a]	This study ^{14}C dated (known sex)[a]
Upper Palaeolithic	29	29	66	68	56
Mesolithic	41	82	289	173	171
Neolithic	—	190	—	467	222
Total	70	301	355	708	449

[a] 75 individuals of unknown sex have been removed from the current sample.

With regards to quality control, we applied a similar process to that described by Pinhasi and Meiklejohn (this volume), in order to determine what samples should be included in the analysis. For this paper, sites were sorted into four levels of chronological security as follows:

1. *Level A*: human skeletal material from a site is directly dated by radiocarbon.
2. *Level B*: though human skeletal material is not directly dated, there are radiocarbon dates from material directly associated with the burial (e.g. material from the burial environment).
3. *Level C*: there are no dates from the burial itself or its direct association, but dates exist from the general cultural level of the site.
4. *Level D*: there are no dates from the site, but the cultural association is good and can be dated by reference to other sites of similar cultural affinities.

Material that failed to meet any of these levels was excluded from the analysis for this paper. The discussion that follows compares previous samples with each other and with the current sample.

For the Upper Palaeolithic, samples from the 1980s showed slight differences in detail, but few individuals had direct radiocarbon dates (level A), the majority identifiable only by archaeological association (e.g. Gravettian, Magdalenian) (level D). Associations often depended on early assumptions, some from the nineteenth century. Assigned temporal placements, essential to regression analysis, were often little better than educated guesses. Especially problematic were assumed Aurignacian specimens, often reflecting the understanding of Oakley, Campbell and Molleson (1971). Our current sample of 68 individuals is approximately 2.4 times larger than in 1984, with the majority (56) directly or indirectly dated by radiocarbon (levels A or B). Of the 12 cases with age independent of radiocarbon (level D), some were excavated in the nineteenth or early twentieth century (e.g. the Magdalenian material from Cap Blanc and Chancelade in France). Some material used earlier is now removed (e.g. Combe Capelle). Others have altered archaeological association (e.g. Cro-Magnon) (for recent reviews of Upper Palaeolithic material, see Holt and Formicola, 2008; Trinkaus, 2007). Only one Early Upper Palaeolithic (EUP) site (Mladec) remains, all other pre-glacial maximum finds having Gravettian or equivalent association.

The 1984 Mesolithic sample size was twice that of Frayer (1980), with more clearly controlled archaeological context. In comparison with the Upper Palaeolithic sample, a high percentage of individuals had direct or indirect radiocarbon dates (levels A or B), with quality control assisted by the work of Newell, Constandse-Westermann and Meiklejohn (1979). In 1984 this was the best-controlled sample. The current Mesolithic sample of 173 individuals is twice the size of the 1984 sample, with all but 2 individuals associated with radiocarbon dates (levels A or B). This remains the best-controlled group, though the difference between it and the Upper Palaeolithic sample is reduced.

In 1984 the Neolithic sample size of 190 was the largest of the 3 but with only a small minority of sites being directly dated (levels A, B or C). Most archaeological assessments predated the introduction of radiocarbon. More recent reports were less likely to include raw data, reflecting publication practice. However, as important was that little to no benefit was seen in direct dating of either burials or associated artefacts. Archaeological typologies were viewed as sufficient in themselves. The current sample size of 467 is approximately 2.5 times that of 1984. However, only 222 individuals have associated radiocarbon dates (levels A, B or C). Much of our analysis below uses this smaller sub-sample in order to remove the uncertainty involved in date estimation, a sample size with only 32 more individuals than in 1984. Chronological control of the Neolithic sample has therefore not improved greatly. New samples tend to be geographically restricted, and many cases with good chronological control only provide skeletal data in summary format.

As discussed above, the following analysis focuses on long bone length rather than stature, thereby reducing analytic noise created by the process of converting bone length to stature. Though this avoids issues raised above that surround regression formulae, it introduces further questions arising from use of individual bone lengths. Do different long bones provide similar results? If similar, then a case may be made for use of a single bone or bone-set in the analysis. Also at play are issues of the appropriateness of using upper or lower limb bones for analysis itself or for simple trends in length of individual bones. The results below must be viewed with these caveats in mind.

7.5 RESULTS

The measurements used are the maximum long bone lengths as defined by Martin (1957), M1 of all bones except tibia where the defined maximum length is M1a (M1 of the tibia is to the lateral condyle rather than the proximal spine). Though our dataset includes physiological lengths for femur (M2) and tibia (M1), they are less available and not used here.

The available sample sizes per bone are shown in Table 7.2, divided into two subsamples, one containing all available bones, the other those with associated radiocarbon dates. As discussed above, the primary difference relates to the low number of dated Neolithic finds. Distribution of data is based on four bones, the humerus and radius of the arm and femur and tibia of the leg, present in from 45 to 63% of individuals in the database. Ulna and fibula samples are less than half the size, and our discussion below is primarily restricted to the more common bones.

Summary statistics of the two datasets by skeletal part and sex are provided in Table 7.3. The figures for the two less common bones are kept for the sake of completeness and possible later comparison by others.

Table 7.2 Distribution of long bone maximum length data

Bone	All long bones (n = 708)		All long bones with direct or indirect radiometric dates (n = 449)	
	Available measurements	Percent available	Available measurements	Percent available
Humerus	408	57.6	272	60.6
Radius	320	45.2	216	48.1
Ulna	145	20.5	114	25.4
Femur	443	62.6	269	59.9
Tibia	364	51.4	219	48.8
Fibula	89	12.6	74	16.5

Table 7.3 Summary statistics for long bone maximum length data – all periods

Bone	All long bones (n = 708)		All long bones with direct or indirect radiometric dates (n = 449)	
	Male Mean and SD	Female Mean and SD	Male Mean and SD	Female Mean and SD
Humerus	316.41 ± 22.49 (n = 240)	289.03 ± 17.78 (n = 168)	315.32 ± 22.32 (n = 154)	288.51 ± 17.37 (n = 118)
Radius	242.66 ± 17.52 (n = 187)	220.26 ± 15.07 (n = 133)	241.59 ± 16.93 (n = 123)	219.70 ± 15.22 (n = 93)
Ulna	261.99 ± 19.36 (n = 88)	238.24 ± 13.55 (n = 57)	261.48 ± 18.84 (n = 67)	238.04 ± 14.44 (n = 47)
Femur	442.44 ± 31.16 (n = 256)	408.84 ± 26.41 (n = 187)	441.02 ± 29.68 (n = 153)	407.39 ± 26.03 (n = 116)
Tibia	368.28 ± 26.31 (n = 211)	333.86 ± 21.38 (n = 153)	367.30 ± 25.44 (n = 124)	331.99 ± 20.78 (n = 95)
Fibula	355.47 ± 27.41 (n = 63)	326.85 ± 19.08 (n = 26)	354.67 ± 24.94 (n = 54)	324.35 ± 16.01 (n = 20)

7.6 ANALYSIS OF THE TOTAL SAMPLE

We now consider the total bone sample, excluding only those without sex identification (see above). When divided by period, a clear pattern emerges for the four periods: Early/Middle Upper Palaeolithic (MUP), Late Upper Palaeolithic (LUP), Mesolithic (MESO), and Neolithic (NEO) (Table 7.4). The Early and Middle Upper Palaeolithic samples are merged since there is only a single site in the former. In each case and for each sex the bone lengths are obviously greater in the MUP sample than in any of the other three.

A somewhat different way to depict these patterns is seen in Figure 7.1a–d, giving error bar plots for bone lengths for each sex by archaeological period. The figures show graphically a number of points already mentioned. Perhaps most obvious is the dramatic difference between the MUP and all later samples for all bones. For the humerus and radius plots, the general flatness of the trend for values from LUP through NEO samples is clearly shown. This is less obvious in the femur and tibia, where the female plots show apparent continuing decline in bone lengths over these three periods in contrast with the male samples (see further below).

For each of the long bones, a two-factor analysis of variance (ANOVA) was conducted on the entire dataset to assess the influence of sex and archaeological period on bone length. A complete factorial model with fixed factor effects was applied, assuming a completely randomized design (Montgomery, 2001). At $\alpha = 0.05$, there were no significant interactions between sex and archaeological period, although the p-value for the interaction term was 0.061 for the humerus. The results are summarized in Table 7.5. For each of the six bones, both sex and archaeological period were highly statistically significant factors with p-values ≤ 0.001. Mean long bone lengths were significantly greater for males, consistent with the error bars depicted in Figure 7.1a–d.

For each sex, the Tukey multiple comparison procedure was used to compare mean bone lengths for the archaeological periods; to avoid being overly conservative, we adopted

Table 7.4 Summary Statistics (mean ± sd) by bone and period for (a) males and (b) females

Bone	Period			
	MUP	LUP	MESO	NEO
(a)				
Humerus	347.42 ± 216.51 (n = 19)	304.70 ± 16.08 (n = 15)	310.25 ± 24.55 (n = 61)	316.15 ± 18.66 (n = 145)
Radius	269.10 ± 12.58 (n = 15)	240.08 ± 14.99 (n = 13)	237.95 ± 15.48 (n = 50)	241.50 ± 16.28 (n = 109)
Ulna	293.00 ± 15.33 (n = 8)	259.28 ± 14.46 (n = 18)	257.08 ± 17.52 (n = 39)	261.63 ± 17.81 (n = 23)
Femur	475.39 ± 36.41 (n = 23)	437.72 ± 17.99 (n = 15)	437.23 ± 30.80 (n = 58)	440.03 ± 28.81 (n = 160)
Tibia	398.53 ± 24.28 (n = 18)	370.77 ± 19.48 (n = 11)	367.87 ± 30.18 (n = 47)	364.19 ± 23.00 (n = 135)
Fibula	385.33 ± 28.35 (n = 12)	361.73 ± 16.93 (n = 12)	340.52 ± 20.09 (n = 26)	352.04 ± 24.78 (n = 13)
(b)				
Humerus	311.94 ± 13.50 (n = 9)	284.90 ± 14.57 (n = 10)	291.02 ± 16.81 (n = 45)	286.58 ± 17.44 (n = 104)
Radius	241.50 ± 11.75 (n = 7)	217.00 ± 14.59 (n = 7)	221.15 ± 15.01 (n = 37)	218.33 ± 14.13 (n = 82)
Ulna	263.40 ± 13.13 (n = 5)	238.25 ± 16.34 (n = 8)	236.67 ± 9.51 (n = 26)	233.50 ± 10.39 (n = 18)
Femur	436.38 ± 31.06 (n = 8)	415.12 ± 13.56 (n = 8)	409.71 ± 26.76 (n = 42)	406.46 ± 25.79 (n = 129)
Tibia	366.38 ± 21.67 (n = 8)	348.25 ± 10.74 (n = 6)	329.75 ± 20.86 (n = 28)	331.77 ± 19.82 (n = 111)
Fibula	355.00 ± 18.88 (n = 4)	350.00 ± 0.0 (n = 1)	322.80 ± 12.82 (n = 10)	318.18 ± 13.62 (n = 11)

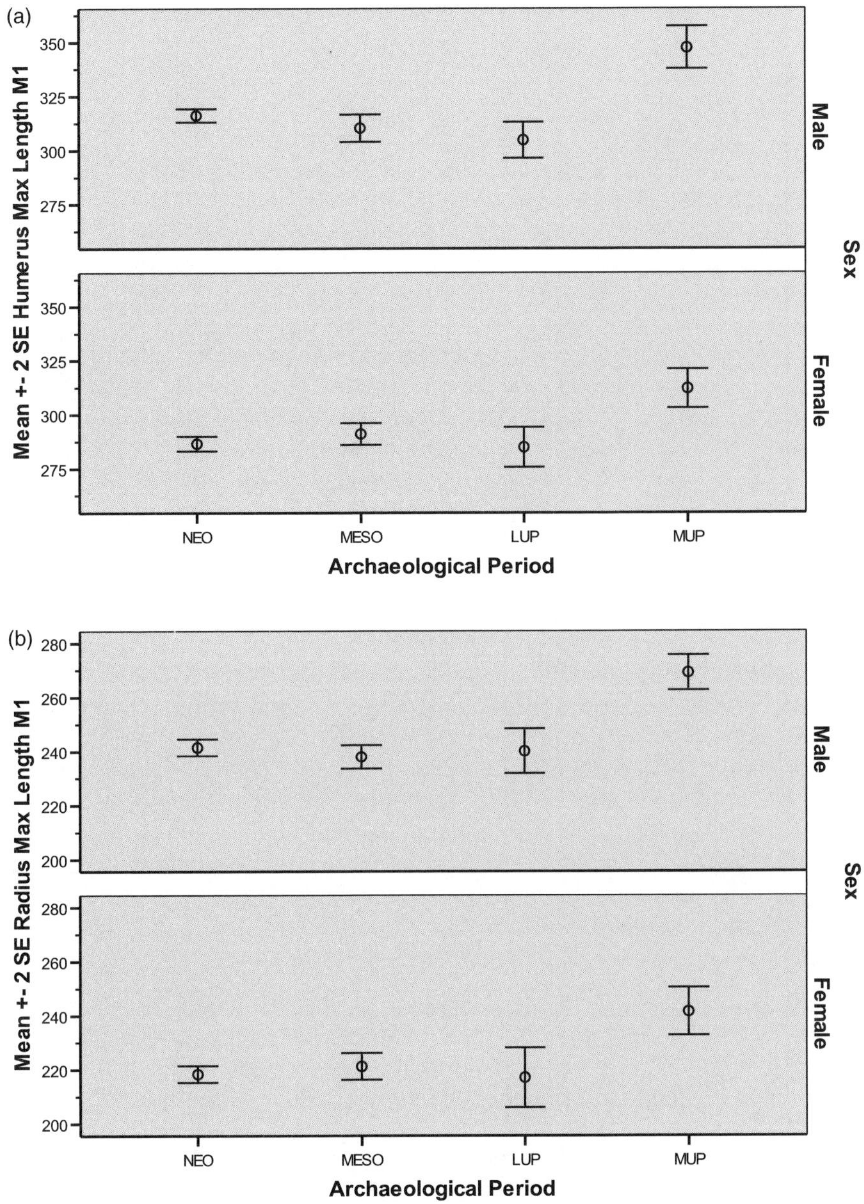

Figure 7.1 (a) Error bar plots for Humerus Maximum Length M1 by archaeological period. Circles denote means. (b) Error bar plots for radius maximum length M1 by archaeological period. (c) Error bar plots for femur maximum length M1 by archaeological period. (d) Error bar plots for tibia maximum length M1 by archaeological period

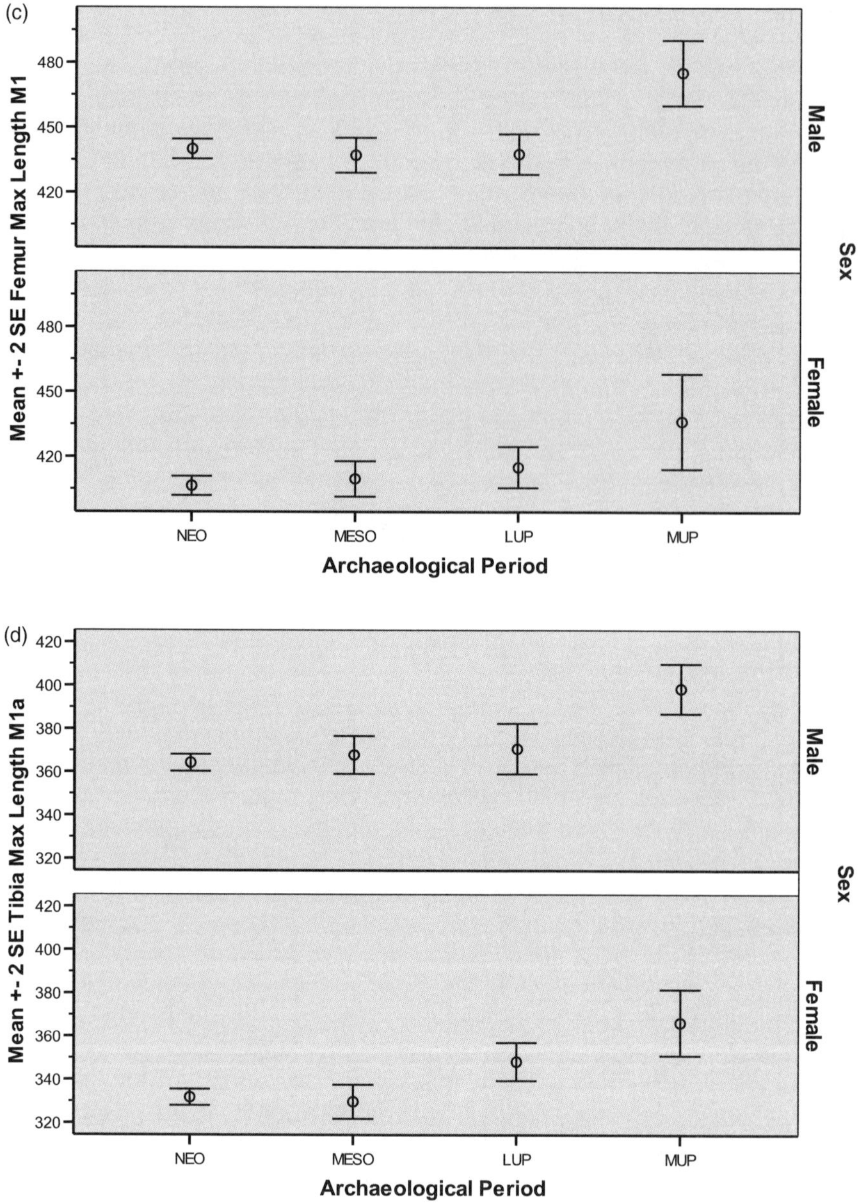

Figure 7.1 *(Continued)*

a family-wise significance level of $\alpha_{FW} = 0.10$, which Keppel and Wickens (2004) identify as a popular choice. As there were unequal sample sizes for the periods, the Tukey procedure used the harmonic mean (Keppel and Wickens, 2004). A summary of the analyses is presented in Table 7.6.

Male and female samples show different results, consistent with earlier findings. For males the results are consistent; all bones except the fibula show similar results. In females the

Table 7.5 Two-Factor ANOVA summaries for all long bone maximum length data

		Humerus (n = 408)		Radius (n = 320)		Ulna (n = 145)	
Source of Variation	df	F	p-value	F	p-value	F	p-value
Sex	1	77.320	<0.001	71.852	<0.001	66.066	<0.001
Arch Period	3	18.476	<0.001	17.056	<0.001	15.968	<0.001
Sex * Arch Period	3	2.474	0.061	1.140	0.333	0.769	0.513
		Femur (n = 443)		Tibia (n = 364)		Fibula (n = 89)	
Source of Variation	df	F	p-value	F	p-value	F	p-value
Sex	1	46.072	<0.001	58.843	<0.001	12.173	0.001
Arch Period	3	10.066	<0.001	16.372	<0.001	11.459	<0.001
Sex * Arch Period	3	0.601	0.614	0.582	0.627	0.875	0.458

consistency of results is slightly less clear. The earliest (MUP) sample has significantly longer bone length than the three later samples with three exceptions, the MUP vs. LUP comparison for male fibula and the same comparison for female femur and tibia. Examination of the summary statistics in Table 7.4 suggests that all three exceptions may stem from the small sample sizes involved.

However, in the other three comparisons, LUP vs. both MESO and NEO, and MESO vs. NEO, all but one comparison are non-significant. The exception involves the fibula in the LUP

Table 7.6 Multiple comparisons of archaeological period means of long bone maximum lengths

Bone	MUP vs. LUP	MUP vs. MESO	MUP vs. NEO	LUP vs. MESO	LUP vs. NEO	MESO vs. NEO
	Males					
Humerus (n = 240)	<0.001	<0.001	<0.001	0.781	0.166	0.233
Radius (n = 187)	<0.001	<0.001	<0.001	0.972	0.990	0.552
Ulna (n = 88)	<0.001	<0.001	<0.001	0.968	0.971	0.734
Femur (n = 256)	0.001	<0.001	<0.001	1.000	0.991	0.926
Tibia (n = 211)	0.019	<0.001	<0.001	0.985	0.831	0.816
Fibula (n = 63)	0.057	<0.001	0.002	0.041	0.701	0.434
	Females					
Humerus (n = 168)	0.004	0.005	<0.001	0.731	0.991	0.460
Radius (n = 133)	0.009	0.004	<0.001	0.895	0.995	0.752
Ulna (n = 57)	0.001	<0.001	<0.001	0.985	0.751	0.792
Femur (n = 187)	0.358	0.041	0.010	0.949	0.795	0.894
Tibia (n = 153)	0.333	<0.001	<0.001	0.168	0.201	0.963
Fibula (n = 26)	na	na	na	na	na	na

Notes: Table entries are p-values for two-sided comparison by Tukey procedure. We suggest using a family-wise significance level of $\alpha_{FW} = 0.10$ to assess the p-values.
Multiple comparisons of means for female fibulae were not performed as LUP had only one case.

vs. MESO comparison. In this case the very short length of the Mesolithic sample is implicated; the result may be an aberrancy deriving from small sample size. The clear interpretation is that the samples from the Later Upper Palaeolithic through Neolithic are strikingly similar.

7.7 ANALYSIS OF THE CORE SAMPLE

The core sample consists of those individuals dated directly or indirectly by radiocarbon (levels A, B and C). In this analysis we correlate long bone length directly with radiometric age. A significant p-value refers to the correlation coefficient, or equivalently, to the slope and indicates either a significant increase or decrease in bone length over time. Our first step was to carry out statistical analysis of the samples by sex for two different samples. The first sample included the earliest period, the MUP. The second excluded MUP. The logic here is that Figure 7.1 through 7.3 and related analyses suggest that MUP bone lengths are consistently longer than those of later periods. The results are seen in Table 7.7.

As expected, in the full dataset, the correlations are all highly significant. There is clear reduction in bone lengths for all bones and both sexes. A possible conclusion from the total set is that the decrease in bone length continues throughout all periods. However, in the smaller dataset, excluding MUP, the situation is more complex. For males the correlation of length on age is non-significant for all bones. However, for females this is only true for the proximal limb bones, the humerus and femur. For distal segments there is a further significant change. We take this to confirm earlier findings that male and female samples follow different evolutionary trajectories.

Table 7.7 Correlations of long bone maximum lengths with radiometric date bp by Sex: (a) including the MUP and (b) excluding the MUP

	Males			Females		
	r	p-value	n	r	p-value	n
(a)						
Humerus	0.401	<0.001	154	0.395	<0.001	118
Radius	0.448	<0.001	123	0.452	<0.001	93
Ulna	0.273	0.025	67	0.656	<0.001	47
Femur	0.305	<0.001	153	0.213	0.022	116
Tibia	0.401	<0.001	124	0.437	<0.001	95
Fibula	0.493	<0.001	54	0.526	0.017	20
(b)						
Humerus	−0.101	0.239	138	0.060	0.534	110
Radius	−0.109	0.256	111	0.243	0.023	87
Ulna	−0.199	0.121	62	0.335	0.028	43
Femur	−0.023	0.792	134	0.117	0.225	110
Tibia	0.146	0.130	109	0.229	0.031	89
Fibula	0.181	0.240	44	0.536	0.022	18

The pattern produced by least squares regression of bone length on date bp for the humerus and femur is seen in Figures 7.2 and 7.3, respectively.

For each of the six long bones, regression lines of bone length vs. date are approximately parallel for the two sexes; Figures 7.2 and 7.3 are fairly typical in this regard. Therefore, for each bone, a one-factor analysis of covariance (ANCOVA) was conducted using SPSS (Green and Salkind, 2005) according to a completely randomized design (Montgomery, 2001), in order to assess the influence of sex and date on maximum length. Sex was regarded as a fixed factor, date was the covariate and the analyses were restricted to the core dataset.

In each case sex and the common slope of the lines were highly significant, with p-values $<$ 0.0001. As well, the proportion of total variation attributable to the ANCOVA model was relatively high, and slightly higher for distal than proximal bones:

- $R^2 = 0.468$ for the tibia vs. $R^2 = 0.316$ for the femur
- $R^2 = 0.450$ for the radius vs. $R^2 = 0.412$ for the humerus

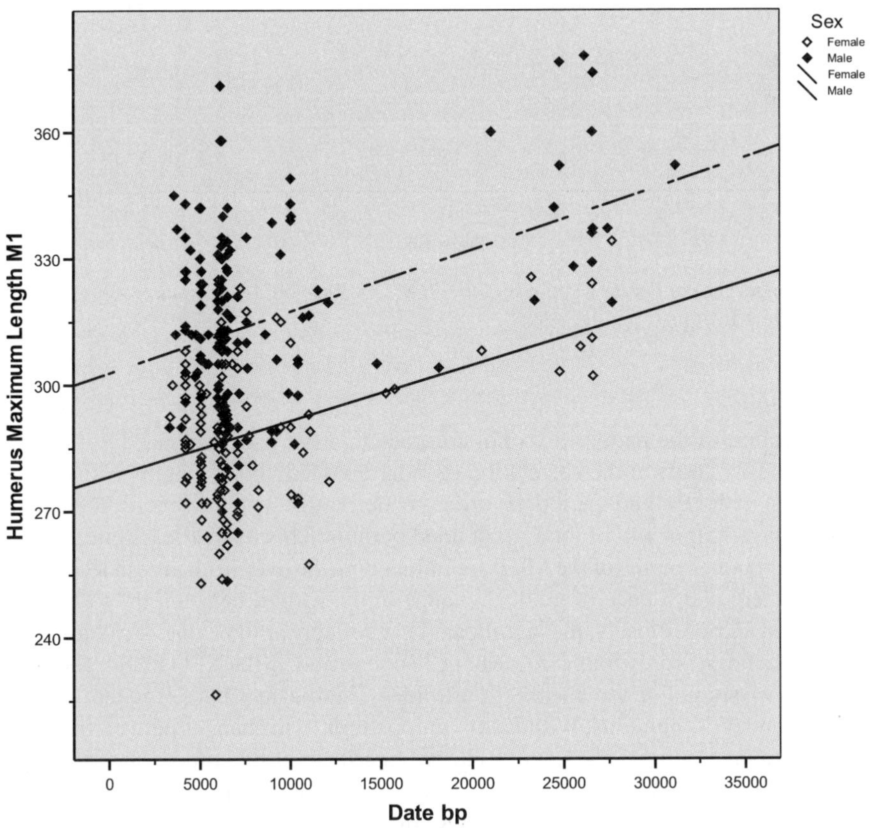

Figure 7.2 Scatter plot of Humerus Maximum Length M1 vs. Date bp. Least squares regression lines are depicted separately for males and females

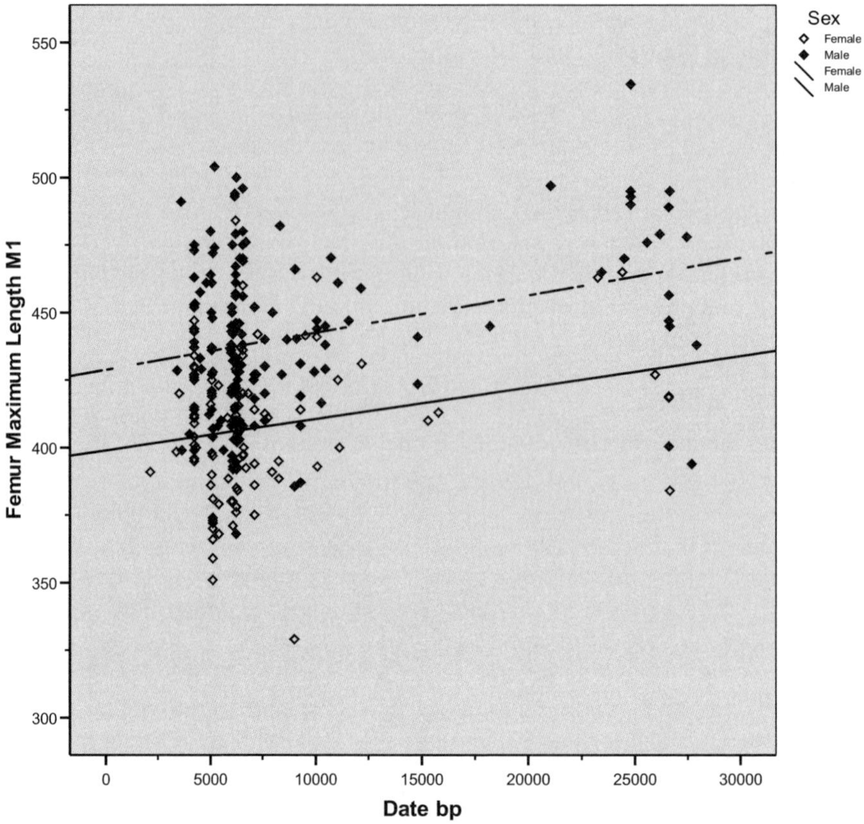

Figure 7.3 Scatter plot of Femur Maximum Length M1 vs. Date bp. Least squares regression lines are depicted separately for males and females

The general nature of the pattern does not contradict earlier conclusions.

Finally, Table 7.8 looks at the correlation of individual long bone lengths with chronology, controlling for both sex and period in order to determine whether there is evidence for significant change within any of the four defined periods. This further sorts out the degree to which the longer bone lengths of the MUP are influencing the overall change in length, and also asks whether there is evidence of gradual change in the pattern between the MUP and LUP.

When looked at more closely, the significant r-values appear to be due to random sampling fluctuation. In males it is only found in one bone in one period, in the NEO fibula, perhaps due to sample size aberrancy. For females there are three significant values, for the MUP femur, MESO radius and NEO humerus. Without obvious pattern, as in changes between proximal and distal limb segments, we are wary about interpreting them at this time.

The results indicate a lack of clear trends in bone length within individual periods, perhaps not surprising for the later three periods, as it confirms earlier results. However, the lack of a meaningful trend within the MUP and the LUP separately means that no apparent common pattern of change links the two, suggesting a break between the two as noted by Formicola and Holt (2007), and discussed further below. Data resolution may play a role. At this time there is

Table 7.8 Correlations of long bone maximum lengths with radiometric date bp, by sex and period

	MUP			LUP			MESO			NEO		
Bone	r	p-value	n	r	p-value	n	r	p-value	n	r	p-value	n
Males												
Humerus	0.275	0.302	16	0.328	0.324	11	0.114	0.387	60	−0.142	0.250	67
Radius	0.150	0.641	12	0.121	0.740	10	−0.165	0.257	49	−0.432	0.213	10
Ulna	0.486	0.406	5	0.201	0.491	14	−0.154	0.356	38	−0.199	0.157	52
Femur	−0.235	0.333	19	−0.051	0.881	11	−0.168	0.211	57	0.008	0.947	66
Tibia	0.003	0.993	15	0.143	0.714	9	−0.068	0.659	45	0.139	0.311	55
Fibula	−0.070	0.848	10	0.230	0.522	10	−0.036	0.861	26	−0.802	0.017	8
Females												
Humerus	0.250	0.550	8	0.653	0.057	9	0.219	0.149	45	−0.323	0.015	56
Radius	0.059	0.912	6	0.189	0.761	5	0.551	<0.001	37	0.176	0.249	45
Ulna	0.388	0.612	4	0.544	0.265	6	0.265	0.192	26	0.325	0.330	11
Femur	−0.875	0.022	6	0.206	0.695	6	0.109	0.494	42	−0.024	0.853	62
Tibia	−0.656	0.157	6	−0.445	0.555	4	0.206	0.294	28	0.143	0.287	57
Fibula	na	na	2	na	na	1	0.455	0.186	10	−0.059	0.901	7

no evidence to suggest population discontinuity between the MUP (largely Gravettian) and LUP (largely Magdalenian), though the break is roughly coincident with the Late Glacial Maximum (LGM).

7.8 DISCUSSION

How do our results relate to the 1984 analysis of Meiklejohn *et al.* (1984), and the related themes of the present volume? The general trend in stature shown in 1984 is confirmed through the analysis of individual long bones, namely that a general decline in long bone length occurs from the Upper Palaeolithic through Neolithic. However, the current analysis clarifies the direction of change for the period after the last glacial maximum, the LUP as defined here, through the Neolithic. In 1984 (Meiklejohn *et al.*, 1984: 90) the conclusion was that '. . . we suspect a manifest curvilinear pattern, a trend toward decreasing stature from the Upper Palaeolithic to the Neolithic being replaced by increasing stature within the Neolithic.' This is refuted here with our analysis suggesting general stasis from the LUP through the Neolithic. Our sample does not allow us to address the issue of post-Neolithic increase, suggested in the 1984 paper. Whatever is happening also refutes the further conclusion (Meiklejohn *et al.*, 1984: 92) that '. . . view(ed) the decline as related to increasing stress, alleviated by the introduction of food production.' We have no evidence for a change in trajectory of stature or long bone length associated with the agricultural transition.

How do our results fit with other studies of general trends in stature and long bone length? The general conclusion of Ruff (2002, 216) was that 'declines in . . . Late Pleistocene and early

Holocene . . . body dimensions, including estimated stature . . ., have been demonstrated in many areas of the world, including Europe, the Mediterranean region, sub-Saharan Africa, South Asia, and Australia.' Our results show that, at least for Europe, the alteration in trajectory is earlier. However, few if any samples other than the one studied here have similar sample size, temporal continuity and chronological control. Ruff's decline in stature (and body size) covers a longer period than is considered here, including the general transition to anatomically modern populations. The issue of decline in pre-LGM Europe is especially difficult to compare since no other region has a large and well-dated sample for this period. Most reported sequences are Holocene in age, post-date 10 kya and are compared to region specific cultural and environmental issues.

The general issue of stature trends and their causation has been examined in several studies. As noted above, Ruff (2002) sees the trend as widespread and lists several possible causative mechanisms. One is technological improvement, leading to decreased size due to lack of need for large body size and associated metabolic costs (Frayer, 1980, 1984). Another is nutritional stress, suggested as a possible mechanism by Meiklejohn *et al.* (1984). Others include general response to postglacial climatic amelioration, reduced gene flow and inbreeding related to postglacial population increase. In reference to the focus of this volume, the transition to agriculture, Ruff notes general body size reduction associated with the transition and especially the introduction of intensive agriculture, referring especially to Cohen and Armelagos (1984b). However, caution is suggested, since the same trend occurs in areas such as Australia without a Holocene agricultural transition.

A general conclusion from comparative work is that Holocene stature variation correlates strongly with nutritional intake. An example of such a conclusion, drawn from Latin American data, is that of Bogin and Keep (1999). With data for approximately 8000 years, they note general long-term decline in stature for the majority of the Holocene. More recent periods show more complicated change, due to the combination of more complete records and associated historical data. However, their conclusion is clear, that 'secular trends in Latin American . . . stature . . . reaffirm the use of growth data as a "mirror for society" and a metric for the biological standard of living.' The patterns are noted as compatible with results from South African and Australian populations.

A similar conclusion comes from Richard Steckel and colleagues, albeit in a framework directed at the general health status of Western Hemisphere populations over the last six millennia and a sample exceeding 6,000 skeletons (Steckel *et al.*, 2002). They relate stature to general health and their matrix of health measures indicates that the variate with highest correlation to stature is anaemia. Their findings validate the general use of stature as a proxy for health status.

However, no other population studies are directly comparable to the one presented here in terms of chronological control and duration. Specific comparisons are therefore, by necessity, limited. It can be argued that the pattern seen in our results, with populations prior to the glacial maximum clearly taller than those after the event, fits Ruff's comment about loss of robusticity in early modern humans compared to previous populations. However, as shown in Table 7.8, we have no evidence of this trend within the Early to Middle Upper Palaeolithic (MUP) sample itself, though this may stem from relatively small sample size. Rather the trend is manifest in the fact that the MUP sample is taller than all later samples. No trend occurs through the other three chronological samples as a group or within any of the three analysed periods alone. In this regard our sample does not corroborate findings from other regions as noted above, either from the late glacial into the Holocene (LUP and MESO samples) or through the agricultural transition (MESO and NEO samples). We stress that this finding refutes the conclusion of Meiklejohn *et al.* (1984), and the combination of larger sample size and better chronological control presented here means that it is the current study that has the greater validity. We have no evidence for change in long bone length and, by extension, stature from the Late Upper Palaeolithic (Magdalenian and Epigravettian) through the Neolithic.

The above finding is as much in need of explanation as would a clear decline and rebound around the agricultural transition, as possibly predicted from reading the sources discussed above. Our results do suggest that the transition was not accompanied by general decline in health status as measured by long bone length, in general agreement with earlier surveys of the transition by Meiklejohn (Meiklejohn and Zvelebil, 1991; Jackes, Lubell and Meiklejohn, 1997). It is also consistent with the conclusion that the European transition was not a response to rapid population growth (and by extension declining health) during the later Mesolithic (Jackes and Meiklejohn, 2008; Jackes, Roksandic and Meiklejohn, 2009). Whether this means that the nature of the transition differed from those where decline in stature was recorded is beyond the mandate of this paper or the quality of the current dataset.

Finally, there is further evidence in our dataset for differing male and female trajectories in long bone length decline, with females continuing the decline in distal long bone segments after it had ended in males. While treated as a function of technological change by Frayer (see above) we wonder whether it simply fits the model suggested by Ruff (2002, 219) that dimorphism in later populations reflects 'more subtle differences in subsistence strategy, diet, and possibly sex-related buffering against the environment.' Clearly this needs to be looked at further with much more tightly controlled samples, both geographically and chronologically.

7.9 CONCLUSIONS

Our analysis above confirms the general stature trend put forward in 1984, a decline in long bone length from the Upper Palaeolithic through Neolithic in Europe. However, our further conclusion (Meiklejohn *et al.*, 1984, 90) is refuted, that there is 'a manifest curvilinear pattern, a trend toward decreasing stature from the Upper Palaeolithic to the Neolithic being replaced by increasing stature within the Neolithic.' We show clear decrease in long bone length between the Early and Late Upper Palaeolithic samples though, somewhat paradoxically, the results within each of the periods show no change. The results from the Late Upper Palaeolithic through the Neolithic show general stasis, both for the three periods (LUP, MESO, NEO) as a unit, and within each of the periods themselves. The issue of possible post-Neolithic increase, suggested in 1984, cannot be addressed with the current sample. Whatever is happening also fails to confirm the further overall conclusion of 1984 (p. 92), that '. . .view(ed) the decline as related to increasing stress, alleviated by the introduction of food production.' We have no evidence that long bone length undergoes an alteration in trajectory related to the agricultural transition, in apparent contradiction to results from other world regions. Whether this reflects a different dynamic at the transition from that of other regions is unclear at present. Our results do confirm that male and female patterns differ over the period, and a further study of limb proportions, beyond the purview of this paper, might assist. However, the small samples currently available for individual combinations of humerus/radius and femur/tibia might make conclusions problematic. One thing that the above results do show is clearly needed, however, is a Neolithic sample with better chronological control. Also needed is a more rigorous examination of variation in space, again beyond the purview of this paper.

ACKNOWLEDGEMENTS

We thank Ron Pinhasi and Jay Stock for suggesting that this paper would fit the volume. The roots for our database were put together over many years. Early versions were created by Catherine T. Schentag and Alexandra Venema, co-authors of the 1984 paper. Jeffrey M. Wyman did later work. Support was made possible by grants from the Social Sciences and Humanities Research Council of Canada to Meiklejohn. The current database was created by Meiklejohn, in part in conjunction with Ron Pinhasi and Winfried Henke, using the earlier work as a base. Many colleagues have supplied papers and data over the years, and without them the database would not exist. Similarly the radiocarbon database has been a work in progress and we thank the numerous colleagues who have supplied papers, data and dates, in some cases unpublished. Without you this paper could not have been written.

REFERENCES

Auerbach, B.M. and Ruff, C.B. (2010) Stature estimation formulae for indigenous North American populations. *Am. J. Phys. Anthropol.*, **141**, 190–207.

Bass, W. (1995) *Human Osteology: A Laboratory and Field Manual*, 4th edn, Missouri Archeological Society, Springfield.

Bogin, B. and Keep, R. (1999) Eight thousand years of economic and political history in Latin America revealed by anthropometry. *Ann. Hum. Biol.*, **26**, 333–351.

Breitinger, E. (1937) Zur Berechnung der Korperhohe aus den langen Gliedmassenknochen. *Anthropol. Anz.*, **14**, 249–274.

Buikstra, J.E. and Ubelaker, D. (eds) (1994) *Standards for Data Collection from Human Skeletal Remains: Proceedings of a Seminar at the Field Museum of Natural History*, Arkansas Archaeological Survey Press, Fayetteville.

Byers, S.N. (2007) *Introduction to Forensic Anthropology: A Textbook*, 3rd edn, Pearson, Toronto.

Cohen, M.N. and Armelagos, G.J. (eds) (1984a) *Paleopathology at the Origins of Agriculture*, Academic Press, Orlando.

Cohen, M.N. and Armelagos, G.J. (1984b) Paleopathology at the origins of agriculture: editors' summation, in *Paleopathology at the Origins of Agriculture* (eds M.N. Cohen and G.J. Armelagos), Academic Press, Orlando, pp. 585–601.

Constandse-Westermann, T.S., Blok, M.L. and Newell, R.R. (1985) Long bone length and stature in the Western European Mesolithic. I. Methodological problems and solutions. *J. Hum. Evol.*, **14**, 399–410.

De Mendonça, M.C. (2000) Estimation of height from the length of long bones in a Portuguese adult population. *Am. J. Phys. Anthropol.*, **112**, 39–48.

Dupertuis, C.W. and Hadden, J.A. Jr. (1951) On the reconstruction of stature from long bones. *Am. J. Phys. Anthropol.*, **9**, 15–54.

Feldesman, M.R., Kleckner, J.G. and Lundy, J.K. (1990) Femur/stature ratio and estimates of stature in mid-and late-Pleistocene fossil hominids. *Am. J. Phys. Anthropol.*, **83**, 359–372.

Formicola, V. (1983) Stature in Italian prehistoric samples, with particular reference to methodological problems. *Homo*, **34**, 33–47.

Formicola, V. (1993) Stature reconstruction from long bones in ancient population samples: an approach to the problem of its reliability. *Am. J. Phys. Anthropol.*, **90**, 351–358.

Formicola, V. (2003) More is not always better: Trotter and Gleser's equations and stature estimates of Upper Paleolithic European samples. *J. Hum. Evol.*, **45**, 239–243.

Formicola, V. and Franceschi, M. (1996) Regression equations for estimating stature from long bones of early Holocene European samples. *Am. J. Phys. Anthropol.*, **100**, 83–88.

Formicola, V. and Giannecchini, M. (1999) Evolutionary trends of stature in Upper Paleolithic and Mesolithic Europe. *J. Hum. Evol.*, **36**, 319–333.

Formicola, V. and Holt, B.M. (2007) Resource availability and stature decrease in Upper Palaeolithic Europe. *J. Anthropol. Sci.*, **85**, 147–155.

Frayer, D.W. (1978) *The evolution of the dentition in Upper Palaeolithic and Mesolithic Europe,* Publications in Anthropology 10, University of Kansas.

Frayer, D.W. (1980) Sexual dimorphism and cultural evolution in the Late Pleistocene and Holocene of Europe. *J. Hum. Evol.*, **9**, 399–415.

Frayer, D.W. (1981) Body size, weapon use and natural selection in the European Upper Paleolithic and Mesolithic. *Am. Anthropol.*, **83**, 57–73.

Frayer, D.W. (1984) Biological and cultural change in the European Late Pleistocene and Early Holocene, in *The Origins of Modern Humans: A World Survey of the Fossil Evidence* (eds F.H. Smith and F. Spencer), Wiley-Liss, New York, pp. 211–250.

Fully, G. (1956) Une nouvelle methode de determination de la taille. *Ann. Med. Leg.*, **6**, 266–273.

Fully, G. and Pineau, H. (1960) Determination de la stature au moyen du squelette. *Ann. Med. Leg.*, **40**, 145–154.

Giannecchini, M. and Moggi-Cecchi, J. (2008) Stature in archeological samples from central Italy: methodological issues and diachronic changes. *Am. J. Phys. Anthropol.*, **135**, 284–292.

Green, S.B. and Salkind, S.J. (2005) *Using SPSS for Windows and Macintosh: Analyzing and Understanding Data*, 4th edn, Pearson Prentice Hall, Upper Saddle River (NJ).

Hernández, M., García-Moro, C. and Lalueza-Fox, C. (1998) Stature estimation in extinct Aónikenk and the myth of Patagonian gigantism. *Am. J. Phys. Anthropol.*, **105**, 545–551.

Holliday, T.W. (1997) Body proportions in late Pleistocene Europe and modern human origins. *J. Hum. Evol.*, **32**, 423–447.

Holt, B.M. and Formicola, V. (2008) Hunters of the Ice Age: The biology of Upper Paleolithic people. *Yearb. Phys. Anthropol.*, **51**, 70–99.

Iscan, M.Y. and Quatrehomme, G. (1999) Medicolegal Anthropology in France. *Forensic Sci. Int.*, **100**, 17–35.

Jackes, M.K. and Meiklejohn, C. (2008) The palaeodemography of central Portugal and the Mesolithic-Neolithic transition, in *Recent Advances in Paleodemography: Data, Techniques, Patterns* (ed. J.P. Bouquet-Appel), Springer Verlag, New York, pp. 179–229.

Jackes, M.K., Lubell, D. and Meiklejohn, C. (1997) Healthy but mortal: human biology and the first farmers of Western Europe. *Antiquity*, **71**, 639–658.

Jackes, M.K., Roksandic, M. and Meiklejohn, C. (2009) Demography of the Đerdap Mesolithic-Neolithic Transition, in *The Iron Gates in Prehistory: New Perspectives,* Brit Archaeol Rep S1893 (eds C. Bonsall, V. Boroneant and I. Radovanovic), Archaeopress, Oxford, pp. 77–88.

Jacobs, K.H. (1985) Evolution on the postcranial skeleton of Late Glacial and early Postglacial European hominids. *Z. Morphol. Anthropol.*, **75**, 307–326.

Jacobs, K.H. (1993) Human postcranial variation in the Ukrainian Mesolithic-Neolithic. *Curr. Anthropol.*, **34**, 311–324.

Keppel, G. and Wickens, T.D. (2004) *Design and Analysis: A Researcher's Handbook*, 4th edn, Prentice-Hall, Englewood Cliffs.

Key, P. (1980) Evolutionary trends in femoral sexual dimorphism from the Mesolithic to the late Middle Ages in Europe. *Am. J. Phys. Anthropol.*, **52**, 244.

Kurki, H.K., Ginter, J.K., Stock, J.T. and Pfeiffer, S. (2008) Adult proportionality in small-bodied foragers: a test of ecogeographic expectations. *Am. J. Phys. Anthropol.*, **136**, 28–38.

Kurki, H.K., Ginter, J.K., Stock, J.T. and Pfeiffer, S. (2010) Body size estimation of small-bodied humans: applicability of current methods. *Am. J. Phys. Anthropol.*, **141**, 169–180.

Manouvrier, L. (1892) Determination de la taille d'apres les grands os des membres. *Revue Mensuel de l'École d'Anthropologie*, **2**, 227–233.

Manouvrier, L. (1893) La determination de la taille d'apres les grands os des membres. *Mem. Soc. Anthropol. Par Série 2*, **4**, 347–402.

Martin, R. (1957) *Lehrbuch der Anthropologie*, Revised 3rd edn (ed. K. Saller), Gustav Fischer Verlag, Stuttgart.

Meiklejohn, C. and Zvelebil, M. (1991) Health status of European populations at the agricultural transition and the implications for the adoption of farming, in *Health in Past Populations,* Brit Archaeol Rep S567 (eds H. Bush and M. Zvelebil), Archaeopress, Oxford, pp. 129–145.

Meiklejohn, C., Schentag, C.T., Venema, A. and Key, P. (1984) Socioeconomic change and patterns of pathology and variation in the Mesolithic and Neolithic of Western Europe: some suggestions, in *Paleopathology at the Origins of Agriculture* (eds M.N. Cohen and G.J. Armelagos), Academic Press, Orlando, pp. 75–100.

Montgomery, D.C. (2001) *Design and Analysis of Experiments*, 5th edn, John Wiley & Sons, Toronto.

Newell, R.R., Constandse-Westermann, T. and Meiklejohn, C. (1979) The skeletal remains of Mesolithic Man in Western Europe: an evaluative catalogue. *J. Hum. Evol.*, **8**, 1–228.

Oakley, K.P., Campbell, B.G. and Molleson, T.I. (eds) (1971) *Catalogue of Fossil Hominids, Part II: Europe*, British Museum (Natural History), London.

Pearson, K. (1899) IV. Mathematical contributions to the theory of evolution. V. On the reconstruction of stature of prehistoric races. *Philos. Trans. R. Soc. Lond. B*, **192**, 169–244.

Rollet, E. (1888) De la mensuration des os longs des membres. *Thèse pour le doctorat en Medecine, Série 1*, **43**, 1–128.

Ruff, C.B. (2002) Variation in human body size and shape. *Annu. Rev. Anthropol.*, **31**, 211–232.

Sciulli, P.W. and Giesen, M.J. (1993) An update on stature estimation in prehistoric native Americans of Ohio. *Am. J. Phys. Anthropol.*, **92**, 395–399.

Sciulli, P.W., Schneider, K.N. and Mahaney, M.C. (1990) Stature estimation in prehistoric native Americans of Ohio. *Am. J. Phys. Anthropol.*, **83**, 275–280.

Sjovold, T. (1990) Estimation of stature from long bones utilizing the line of organic correlation. *Hum. Evol.*, **5**, 431–447.

Steckel, R.H., Rose, J.C., Larsen, C.S. and Walker, P.L. (2002) Skeletal health in the western hemisphere from 4000 BC to the present. *Evol. Anthropol.*, **11**, 142–155.

Stevenson, P.H. (1929) On racial differences in stature long bone regression formulae, with special reference to stature reconstruction formulae for the Chinese. *Biometrika*, **21**, 303–321.

Telkka, A. (1950) On the prediction of human stature from the long bones. *Acta Anat.*, **9**, 103–117.

Trinkaus, E. (1976) The evolution of the hominid femoral diaphysis during the Upper Pleistocene in Europe and the Near East. *Z. Morphol. Anthropol.*, **67**, 291–319.

Trinkaus, E. (2007) Early modern humans. *Annu. Rev. Anthropol.*, **34**, 207–230.

Trotter, M. (1970) Estimation of stature from intact long limb bones, in *Personal Identification in Mass Disasters* (ed. T.D. Stewart), Smithsonian Institution, Washington, pp. 71–83.

Trotter, M. and Gleser, G.C. (1952) Estimation of stature from long bones of American Whites and Negroes. *Am. J. Phys. Anthropol.*, **10**, 463–514.

Trotter, M. and Gleser, G.C. (1958) A re-evaluation of estimation of stature based on measurements of stature taken during life and of long bones after death. *Am. J. Phys. Anthropol.*, **18**, 79–123.

Ubelaker, D. (2008) Forensic Anthropology: methodology and diversity of applications, in *Biological Anthropology of the Human Skeleton*, 2nd edn (eds M.A. Katzenberg and S.R. Saunders), Academic Press, Orlando, pp. 41–70.

Vallois, H.V. (1943) Les caractères différentiels des os longs chez certaines populations préhistoriques de la France. *Bull. Mem. Soc. Anthropol., Par, Série 9*, **4**(4), 1–24.
Vercellotti, G., Agnew, A.M., Justus, H.M. and Sciulli, P.W. (2009) Stature estimation in an early medieval (XI-XII c.) Polish population: Testing the accuracy of regression equations in a bioarcheological sample. *Am. J. Phys. Anthropol.*, **140**, 135–142.
Verneau, R. (1906) Anthropologie, in *Les Grottes de Grimaldi, 2* (eds L. de Villeneuve, M. Boule, R. Verneau and É. Cartailhac), Imprimerie de Monaco, Monaco, pp. 1–212.
von Bonin, G.V. (1935) European races of the Upper Paleolithic. *Hum. Biol.*, **7**, 196–221.

8

Variability in Long Bone Growth Patterns and Limb Proportions Within and Amongst Mesolithic and Neolithic Populations From Southeast Europe

Ron Pinhasi[1], S. Stefanović[2], Anastasia Papathanasiou[3] and Jay T. Stock[4]

[1] *Department of Archaeology, University College Cork, Cork, Ireland*
[2] *Faculty of Philosophy, Department of Archaeology, Čika Ljubina 18-20. 11000 Belgrade, Serbia*
[3] *Ephorate of Paleoanthropology and Speleology, Greek Ministry of Culture, Athens, Greece*
[4] *Leverhulme Centre for Human Evolutionary Studies, Department of Biological Anthropology, University of Cambridge, Cambridge, UK*

8.1 INTRODUCTION

The scientific study of human growth trajectories began more than 200 years ago with the longitudinal growth study by Montbeillard (1759–1777), who periodically recorded the stature of his son. In the early part of the nineteenth century new growth studies appeared, with a focus on the development of epidemiological growth standards that can be used in the assessment of healthy vs. unhealthy growth patterns in individuals and populations, known as *auxological epidemiology* (Tanner, 1998). During the last 50 years, worldwide variations in growth have been reported by various international programmes and data are assessed against published growth standards, such as those used by the World Health Organization (WHO) (http://www.who.int/childgrowth/en/).

Many physical anthropologists are interested in different aspects of growth, such as:

Human Bioarchaeology of the Transition to Agriculture Edited by Ron Pinhasi and Jay T. Stock

1. the evolution of human growth and development, and its contextualisation in the broader fields of human and mammalian life history (Bogin, 1999; Bogin and Smith, 1996); and
2. changes in growth velocity at different ages, differences between sexes in growth trajectories, and variations within and between populations in growth curves.

The latter approaches are mainly propelled by the interest into whether variations in human growth can be attributed to genetic differences or environmental factors.

The anthropological interest in growth patterns in archaeological samples began with Johnston's study of the Indian Knoll Native American population (Johnston, 1962) and continued with a series of studies during the late 1960s to 1980s on Native American, European and Nubian samples (Hummert and Van Gerven, 1983; Jantz and Owsley, 1984; Mensforth, 1985; Merchant and Ubelaker, 1977; Stloukal and Hánaková, 1978; Sundick, 1978; and see review of additional studies in Saunders, 2000). Most of these studies focused on inter-population comparisons of growth trajectories of long bones amongst archaeological 'populations', and in some cases, in comparison to growth curves from modern studies (mainly the Denver Study, see below). Less attention has been placed on the study of temporal and within-site variations in growth (Saunders, 2000).

Various growth studies were carried out during the 1990s to 2000s, many of which focused on European medieval and post-medieval archaeological samples. The availability of relatively large samples, from proximate geographical regions and from short time periods (i.e. measured in 100s rather than 1000s of years), allowed researchers to investigate the social, economic, cultural and nutritional factors that may have played a role in the observed variations in the growth trajectories. One major factor that was only addressed in a few publications is the effect of variability in health (as indicated from the study of palaeopathological indicators), sanitation, weaning and socioeconomic status on the growth patterns of archaeological populations (Lewis, 2002a, 2002b, 2007; Mays, Brickley and Ives, 2008; Mays, Ives and Brickley, 2009b; Mays, 1999; Ribot and Roberts, 1996). More recently, a few studies examined the specific role of vitamin D deficiency and rickets on growth (Mays, Brickley and Ives, 2009a; Pinhasi *et al.*, 2006). These studies highlight the potential of bioarchaeological investigations that combine palaeopathological, osteological, physiological and isotopic methods with the study of anthropometric variations during growth.

Despite these trends, there are several major limitations that affect the majority of archaeological growth studies:

1. While the actual mortality ratio of subadults-to-adults is often as high as 50:50, the actual number of subadults in a sample is often considerably smaller due to taphonomic factors (Humphrey, 2000, 2003; Saunders, 1992, 2000). Subadult bones are often more poorly preserved due to their relative fragility, but different mortuary treatment of subadults may also be a factor affecting preservation (as individuals may have been cremated, incorporated into refuse pits, buried in a separate location, etc.) (Pinhasi and Bourbou, 2008).
2. In most archaeological subadult samples from a given cemetery, the majority of individuals are newborns and infants and there is a severe under-representation of children and adolescents. Consequently, most growth studies cannot provide equal representation of all subadult age groups.
3. All archaeological samples are cross-sectional and therefore their analysis cannot reveal variations in the individual's growth rates as in the case of longitudinal studies.

Any variation in the slope of growth curves of such samples may not necessarily reflect actual variation in the rate of growth of the analysed bone dimension, due to the smoothing out of growth events that are imperfectly synchronized between individuals (Humphrey, 2003).

4. It is often desirable to assess the possible synergy between growth patterns, disease and nutritional status, since it has been well-documented in both archaeological and living populations that poor health, poor nutrition and high disease load have major impact on the observed growth patterns (Berti, Leonard and Berti, 1998; Cook, 1984; Larsen, 1997; Leonard *et al.*, 2000; Lewis, 2002a, 2002b; Ribot and Roberts, 1996). However, the interpretation of the health profile of skeletal populations is not straightforward due to the effects of selective mortality and hidden intra-population variability in the susceptibility to illness (Wood *et al.*, 1992; Wright and Yoder, 2003). As a result, the mortality pattern of the sample may not reflect the parameters of the living population. These complicating factors have been described as the 'osteological paradox', a further complication of which is that skeletons with palaeopathological lesions may have been in overall good health, after surviving periods of episodic stress. However, several studies on morbidity and demography (Bennike *et al.*, 2005; Saunders and Hoppa, 1993) suggest that the majority of specimens with such lesions are those that are more likely the non-survivors. Hence, high prevalence of palaeopathological lesions of stress is an indicator of the overall poor health of the studied population.
5. A final major concern is the accuracy, reliability and replicability of skeletal and dental age estimation methods (as further discussed below).

Relatively few anthropological studies have investigated intra-sample variation in growth amongst past populations, and male-female differences in growth trajectories provide insights regarding complex patterns of ontogenetic allometric and isometric variability in growth proportions and dimensions (c.f. Ruff, 2003). As pointed out by Maresh (1955, p. 732):

> When assessing the growth of the major long bones, perhaps we should be willing to allow greater variability in the pattern by which these bones increase in length than we have heretofore considered desirable of "healthy". If one accepts these hypotheses, the growth curves of the long bones in infancy become more reasonable than if one continues to expect the growth of all segments to be affected equally by growth stimuli-inheritance of body built, nutritional building materials, hormone-enzyme systems, or whatever the factors may be.

Indeed, the majority of anthropological studies have focused on inter-population variation in growth, and examined differences in average bone dimension-per-age. Less attention has been placed on intra-sample variability which, when studied in conjunction with archaeological evidence, can provide important clues about within-sample and sex-specific differences in nutrition, disease load, living conditions and so on.

Bioarchaeological growth studies have focused on the analysis of skeletal samples from the relatively recent past and the great majority of studies have focused on Native American populations. This temporal and geographical bias is mainly due to the paucity of subadult samples from European and Asian Palaeolithic, Mesolithic and Neolithic periods. Yet, the assessment of growth in prehistoric populations from these periods is of particular importance as it can provide insight into whether:

- pre-agricultural hunter-gatherers had growth trajectories similar to those of later prehistoric and historical (agricultural) samples;
- there are notable differences in growth patterns of pre-agriculturalists and agriculturalists, and if so, whether this can be attributed to genetic differences, environmental factors, or both.

In this context it is interesting to chart the differences in subadult size and then attempt to correlate these with a growing body of literature on physiological differences in skeletal robusticity, activity patterns, behaviour, mobility, sexual dimorphism and so on (see various contributions in this volume).

In her review of the literature on archaeological growth studies, Saunders (2000) questions whether the trend amongst most past populations to have shorter limb dimensions-per-age when compared to most modern populations, reflects genetic differences, or harsher environmental conditions that were more common in the past. This question is particularly relevant when we consider the Mesolithic-Neolithic transition in various world regions. In regions where we can assume population continuity, such as in the Danube Gorges, and for which we have both Mesolithic and Neolithic subadult skeletal samples, it is possible to examine the effect of the agricultural transition on growth patterns. The underlying expectation is that in regions where the archaeological record suggests *in situ* transition, any differences between Mesolithic (hunter-gatherers) and Neolithic (farmers) should be attributed to changes in environmental and culturally mediated factors. These include variations in disease load, mortality, morbidity, population density and temperature, as well as changes in age of weaning, type of weaning foods, quality and quantity of diet, housing, activity and access to resources.

In this chapter we investigate the following:

1. Whether there are any differences in long bone dimensions between Mesolithic and Neolithic newborns from the Danube Gorges region;
2. The growth profiles of Mesolithic and Neolithic Danube Gorges populations from the sites of Vlasac and Lepenski Vir, Serbia, in comparison to those of Neolithic subadults from Greece, medieval and post-medieval populations from Austria and Britain and the modern population of Denver, Colorado;
3. Changes of lower and upper limb proportions during growth and development of subadults from birth to 14 years of age; and
4. The degree of inter-bone variability in growth, by examining differences in percentiles of dimensions-per-age based on the corrected Denver reference sample (see below).

Trends observed in these comparisons are interpreted in the context of studies on inter-population variation in growth.

8.2 MATERIALS

8.2.1 Danube Gorges Samples

The site of Lepenski Vir is located in the middle of the Upper Gorge of the right bank of the Danube, Serbia on a semicircular terrace. The site was first discovered during a survey in 1960, and excavation began in 1965 under the direction of Dragosalv Srejović. In subsequent years, an area of 2500 m^2 was excavated to reveal architecture, monumental sculpture and graves of Lepenski Vir culture (8200–5500 calBC, cf. Borić, 2002).

Human remains for the site as a whole amount to 190 individuals. Skeletal representation ranges from small fragments to complete skeletons. Of them, 101 individuals were buried in single graves, 58 in double or triple graves, and 4, 5, 6 and 7 individuals were buried in one instance each. The total number of adults and subadults are almost identical (83 compared to 84); however, age could not be assigned to 23 individuals (Roksandić, 1999). In this study, we analyse the dimensions of 45 subadult skeletal remains: 39 of these individuals are newborn babies (35–50 gestational weeks), 3 are children 1 to 5 years of age and 3 are young adolescents 12 to 14 years of age (Table 8.1).

The site of Vlasac is located 3 km downstream from Lepenski Vir and it was submerged by the Danube owing to the creation of the accumulation lake of the Djerdap Hydro-plant during the 1960s. It was first discovered in 1970 and was partially excavated in 1970/1971 (Srejović and Letica, 1978). During these two seasons of excavations, 43 dwelling structures, 87 graves and more than 35 000 mobile objects were unearthed and dated to the period between 9800 and 6900 calBC; (cf. Borić and Stefanović, 2004). The monograph of the site was published in 1978 and it is the most comprehensive publication on archaeological, environmental and anthropological data on any individual site of the Lepenski Vir culture (Srejović, 1981; Tringham, 2000). New excavations at the site started in 2006, and have confirmed that certain portions of the site are still preserved and accessible for research (Borić, 2007).

Human skeletal remains from Vlasac comprise 164 individuals from the 87 reported graves. Adults represent the majority of the sample: 108 individuals or 66%. Forty-seven

Table 8.1 Frequencies of individuals per dental age cohort for the four studied sites

Age[a]	DG Mesolithic	DG Neolithic	Greek Neolithic
Totals	28	49	14
Sites	Vlasac	Vlasac, Lepenski Vir	Xirolimni, Alepotrypa, Mavropigi, Franchthi
Location	Danube Gorges, Serbia	Danube Gorges, Serbia	cf. Figure 6.1 in Chapter 6, this volume
birth –1	22	41	8
1–2	1	1	0
2–3	0	1	2
3–4	0	1	1
4–5	0	2	0
5–6	0	0	0
6–7	0	0	1
7–8	2	0	0
8–9	1	0	0
9–10	1	0	2
10–11	0	0	0
11–12	1	0	0
12–13	0	2	0
13–14	0	1	0

[a]Each number marks the latest age for a given cohort.

individuals are determined as subadults and nine individuals are of undetermined age. In this study we analyse the dimensions of 31 subadult skeletal remains: 28 from the Mesolithic phases and 3 from the Neolithic phases. Of these individuals, the majority – 22 Mesolithic and 3 Neolithic individuals – are newborn babies, 5 are children (1–10 years of age) and 1 is an adolescent (11–14 years of age) (Table 8.1).

8.2.2 Greek Neolithic Samples

Mavropigi and Xirolimni are two of the earliest agricultural communities in Greece and Europe. Mavropigi (Karamitrou-Mendesidi, 2005) is dated around 6600 calBC and covers about one hectare. It consists of 4 occupation levels with irregular rectangular dwellings of 50 to 90 m^2 similar to the Nea Nikomedeia houses. The site yielded 18 *in situ* burials, mostly plain, coarse pottery, and over 2000 objects including stone and bone tools, loom weights, pendants, beads, six seals and 132 female and animal figurines. The specimens analysed in this study include one newborn and three children.

Xirolimni (Karamitrou-Mendesidi, 1998) is an Early Neolithic (6500–5700 calBC) settlement with a high concentration of 90% plain, coarse, distinct ceramic ware and 14 pit burials. Data from stable isotope analyses point to a swift and complete transition to agricultural practices, even in the very early Neolithic, and a diet based mainly on cereals and significantly higher consumption of meat compared to later Neolithic sites. One child (2.5 years of age) from this site is included in this study.

The site of Franchthi consists of a large, 150 m long cave and an open settlement at the surrounding area, located at the coast of the southern tip of the Argolid peninsula in eastern Peloponnese (Jacobsen, 1969; 1973a,b). The well-documented stratigraphic sequence has revealed evidence of human occupation, starting from 22 000 to 3000 calBC. Apart from the habitation debris, the site has yielded mortuary evidence and fairly well preserved human osteological material, dating from 9000 to 3000 calBC, and consisting of formal burials and extensive human bone scatter. The specimens analysed in this study include four newborns, two infants and one child.

Alepotrypa Cave is one of a group of three caves located around the rocky, limestone Diros Bay on the western coast of the Tainaron Peninsula of southern Greece. The cave is situated about 50 m above present-day sea level, in a rocky limestone environment. It is about 300 m long, extending along an east-west axis, and contains a large freshwater lake, and several smaller lakes that attracted the Neolithic inhabitants (Papathanasiou, 2001). Based on ceramic typological data, the cave was occupied from approximately 5000 to 3200 calBC, corresponding to the Late and Final Neolithic Periods. More than 50 activity areas have been identified, with cultural deposits ranging between 50 cm and 5.5 m in thickness, including both habitation areas and mortuary loci (Papathanasiou, 1996, 2001). The specimens analysed in this study include one newborn and one child.

8.2.3 Comparative Samples

Archaeological samples from Lower Austria and Great Britain were used for comparison. The Austrian samples include those from the tenth century AD Slavic sites of Gars-Thunau and Zwentendorf and the Avar Period site (7th–9th centuries AD) of Zwölfaxing in Lower Austria (Pinhasi *et al.*, 2005) (Table 8.2). The British samples consist of subadults from four medieval English sites: the Anglo-Saxon site at Raunds Furnells (Boddington and Cadman, 1981); the

Table 8.2 Description of the comparative samples

Sample	N	Location	Period	Reference
Libben	12[a]	Ohio, USA	Late Woodland 800–1100 AD	(Lovejoy, Russell and Harrison, 1990)
Slavonic Mikulcice	17[a]	Bohemia, Czech Republic	7–9 centuries AD	(Stloukal and Hánaková, 1978)
Zwentendorf	19	Lower Austria	tenth century AD	(Pinhasi *et al.*, 2005)
Gars-Thunau	11	Lower Austria	tenth century AD	(Pinhasi *et al.*, 2005)
C.C. Spitalfields	42	East London, England	1729–1859 AD	(Lewis, 2002a, 2002b; Pinhasi *et al.*, 2006)
Broadgate	42	East London, England	1569–1720 AD	(Lewis, 2002a, 2002b; Pinhasi *et al.*, 2006)
Raunds	25	Northamptonshire, England	850–1100 AD	(Lewis, 2002a, 2002b; Pinhasi *et al.*, 2006)
St Helen-on-the-Walls	9		950–1550 AD	(Lewis, 2002a, 2002b)
twentieth century Caucasians	24	Denver, Colorado, USA	Modern	(Maresh, 1943, 1955)

[a]Mean values by age cohort.

late medieval cemetery of St Helen-on-the-Walls from York (Dawes and Magilton, 1980); the post-medieval cemetery of Broadgate, Central London (Schofield and Maloney, 1998) (Table 8.2); and the post-medieval cemetery of Christ Church Spitalfields, Central London (Molleson and Cox, 1993) (Table 8.2). The Broadgate and Christ Church Spitalfields cemeteries are only about 500 m apart, but they differ greatly in the socioeconomic status of their respective populations. The Broadgate cemetery, located at the inner eastern flank of the London Wall, was founded in 1569 by the City to relieve the congestion occurring in parish burial grounds and contained many primary uncoffined burials at a high density of 8 per m^3 (Schofield and Maloney, 1998). The Christ Church Spitalfields cemetery is located at the outer eastern flank of the London Wall and contained coffined crypt burials of individuals of medium-high socioeconomic status (Molleson and Cox, 1993).

Data on the diaphyseal length of limb bones for age amongst modern (twentieth century) subadults was obtained from publications on the Denver Growth Study, which was carried out between 1927 and 1967 (Maresh, 1943, 1955, 1970). All subjects were of European ancestry, in good health and of middle-class socioeconomic status (Ruff, 2003). They were examined, measured and radiographed from two months of age at two-month intervals to six months, semi-annually from six months to early adolescence, and annually thereafter until late adolescence (Ruff, 2003).

8.3 METHODS

8.3.1 Bone Measurements

The metric data utilized in this study include diaphyseal length dimensions (excluding epiphyses) of the femur, tibia, humerus, ulna and radius. In additional, maximum and minimum midshaft diameters of these long bones and distal metaphyseal breadth dimensions of the femur and humerus were recorded for the Danube Gorges samples, in order to carry out the comparison between the growth dimensions of these two samples during the period of birth-infancy (see below). Measurements were taken according to the procedures described by Buikstra and Ubelaker (1994).

8.3.2 Dental age

Chronological age was estimated on the basis of the examination of postnatal dental formation. The most commonly applied method is based on comparing each cheek tooth to a series of 14 formation stages, using published plots (Moorrees, Fanning and Hunt, 1963a, 1963b). This system was reworked by Smith (1991) to account for the cross-sectional rather than longitudinal nature of archaeological samples. Ages for a given growth stage were based on the average age of subjects at a given stage rather than the actual age when that stage was first attained. Consequently, in longitudinal studies in which subjects are examined at specific interval lengths, observation of a growth stage usually postdates the actual onset of attainment by several months or even years. Smith devised new charts for males and females, in which a given age for a specific formation stage can be interpreted as the actual average age of attainment of a growth stage. The male and female values can then be averaged to obtain estimates for the age of subadult individuals from archaeological samples.

A prevailing assumption amongst anthropologists who study growth patterns in past skeletal populations is that the ontogenetic process of dental root and crown formation is genetically controlled and is thus much less affected by environmental factors (Saunders, 2000; Smith, 1991). However, this assumption has been questioned by researchers who examined clinical samples of modern American children (Garn, Lewis and Polacheck, 1959) and in more recent comparative studies on past skeletal populations (Saunders, 2000; Tompkins, 1996) that yielded varied results (cf. Pinhasi *et al.*, 2005).

8.3.3 Analytical Methods

Analysis of Variance (ANOVA) was applied to study inter-population variations in limb dimensions. Gompertz curves were calculated following the methods described by Pinhasi (Pinhasi *et al.*, 2005, 2006). Nonlinear interpolations were carried out in the SPSS statistical package. In the case of each long bone dimension, the initial value of the point of inflection (m) was set to zero, the value of the slope (b) was set to 0.2, and the initial value of each long bone diaphyseal length dimension was calculated from the average value amongst Lepenski Vir adults.

The radiographic data of the subadults from Denver Colorado (Maresh, 1943, 1955, 1970) were corrected for parallax (magnification) following Ruff (2007; p. 700):

> ...for femoral, humeral, and tibial lengths 217 mm or greater, adjusted length $= 0.949 \times$ original length $+ 5.63$. For smaller bones (except the radius), adjusted length $= 0.975 \times$ original length. For the radius, adjusted length $= 0.98 \times$ original length. Femoral distal metaphyseal and head breadth magnification factors were the same (2.5%) for bones with lengths under 217 mm, and the same slope (0.949) was used for larger specimens, with intercepts determined by average size differences between dimensions: for distal metaphyseal breadth, adjusted $= 0.949 \times$ original $+ 0.965$; for head breadth: adjusted $= 0.949 \times$ original $+ 0.555$'.

A comparison between the corrected and uncorrected curves indicates that the latter's interpolated dimension-per-age values are on average between 2 and 4% larger than the corrected values (Maresh, 1943, 1955, 1970; Ruff, 2007).

Bone lengths were used to investigate limb proportions by calculating the following indices, which reflect distal-proximal lower and upper limb proportions, respectively:

1. The Tibio-femoral index = 100 × (femoral bicondylar length [or shaft length if epiphyses are not fused]/total length of tibia); and,
2. The Radio-humeral index = 100 × (greatest length of radius/greatest length of humerus).

These intralimb indices were analysed for the Danube Gorges Mesolithic, Neolithic and Greek Neolithic samples. Newborns and infants (0–1 years of age) of these skeletal samples were compared to newborns and infants from the post-medieval sites of Broadgate and Christ Church Spitalfields, East London, the medieval sites of Raunds in Northamptonshire and St Helen-on-the-Walls, in York (Lewis, 2002a, 2002b), the Slavic sites of Zwentendorf and Gars-Thunau (Pinhasi *et al.*, 2006), and mean proportions per age cohort calculated for subadults from the Slavic site of Mikulcice (Stloukal and Hánaková, 1978), and the Native American site of Libben (Lovejoy, Russell and Harrison, 1990; Mensforth, 1985) (Table 8.2).

8.4 RESULTS

Statistical comparisons of Mesolithic vs. Neolithic neonates and infants (aged between 0 and 1) for each variable were carried out using the Mann-Whitney nonparametric test. Results indicate that the two groups differ only for the following dimensions for which the Danube Gorges Mesolithic had greater dimensions per age: Tibial maximum (anterior-posterior (a–p)) width ($p < 0.001$) and minimum (medio-lateral (m-l)) width ($p < 0.003$) at midshaft and tibio-femoral ratio ($P < 0.01$).

8.4.1 Femur Length

Gompertz curves were calculated for the Danube Gorges Mesolithic, Greek Neolithic and Broadgate archaeological samples. In addition, corrected–2 standard deviations (SD), average and +2 SD curves were plotted for the modern Denver sample. Due to the small number of individuals in the Danube Gorges Neolithic sample, these were plotted as single individual cases without interpolation. The attained femoral length of Danube Gorges and the Greek Neolithic newborns and infants fall below the –2SD value for the Denver reference sample (Figure 8.1). In the case of the Danube Gorges Mesolithic group, the curve intersects the lowest Denver (–2SD) curve only around the age of 8 to 9 years, which coincides with the same trend found amongst the Broadgate sample. Each of the Danube Gorges Neolithic subadults falls below the –2SD Denver reference curve. This pattern contrasts with the femoral growth curve of the Greek Neolithic sample that intersects the lower Denver range at around the age of 2 years and attains dimensions close to those of the Denver average by the age of 5. If we accept any values below the lowest –2SD line of Denver to represent low growth attainment per age, then it is clear that in the case of the femur, the two Danube Gorges samples resemble the pattern observed at Broadgate, which is known to be a stressed population

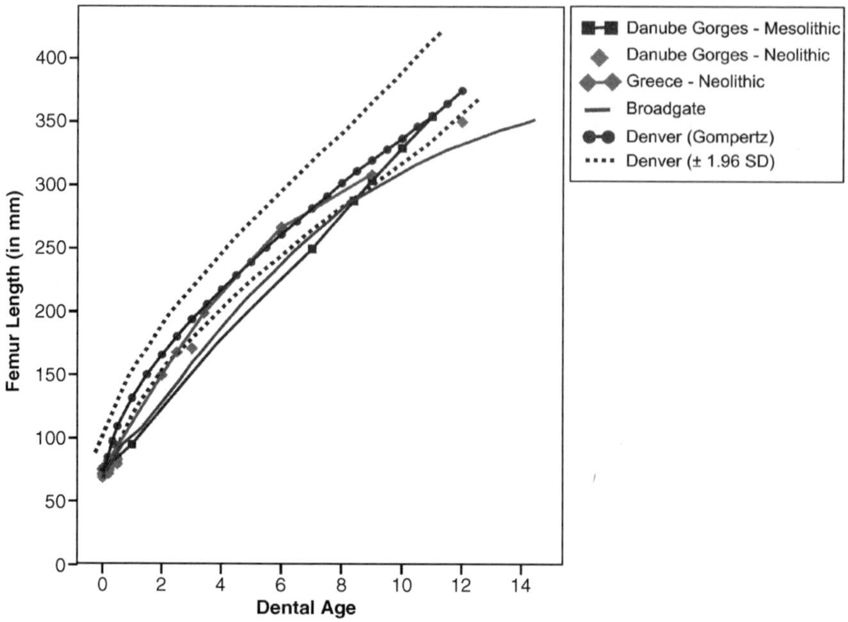

Figure 8.1 Growth patterns of the femoral diaphysis of the Danube Gorges (Mesolithic and Neolithic samples), Greek Neolithic and Broadgate in relation to the curve of the modern (Denver) reference sample

(cf. Pinhasi *et al.*, 2006). This pattern contrasts with the Greek Neolithic curve which, after the age of two, falls within the Denver reference –2SD-average range.

8.4.2 Humerus Length

Inter-population growth patterns of the humerus diaphyseal dimensions are different from the pattern observed for the femur. In general, all populations show less marked deviations from each other and from the Denver reference curve (Figure 8.2). However, as in the case of the femur, the archaeological samples have low humerus length-per-age values for newborns and infants. Of the four Danube Gorges Neolithic cases aged between 1 and 12 years, two fall within the Denver 2SD range and two below it. In contrast, the Danube Gorges Mesolithic curve intersects the lower Denver curve at around the age of 7 and remains within the lower –2SD –average range of the Denver sample. The Greek Neolithic curve intersects the lower Denver curve around the age of 10. Both the Greek Neolithic and the Danube Gorges Mesolithic curves are relatively close to the dimensions-per-age values of the Broadgate curve.

8.4.3 Tibia Length

Less data points were available for the study of inter-population variation in tibial diaphyseal length dimensions. The Greek Neolithic curve falls within the lower range of the Denver dimensions-per-age (Figure 8.3). The Broadgate curve falls below this range till the age of 7. The one Danube Mesolithic child, aged 7, falls on the–2SD Denver curve, as does a 2-year-old Neolithic child from the Danube. The one Danube Neolithic adolescent falls below the–2SD curve and closest to the Broadgate sample.

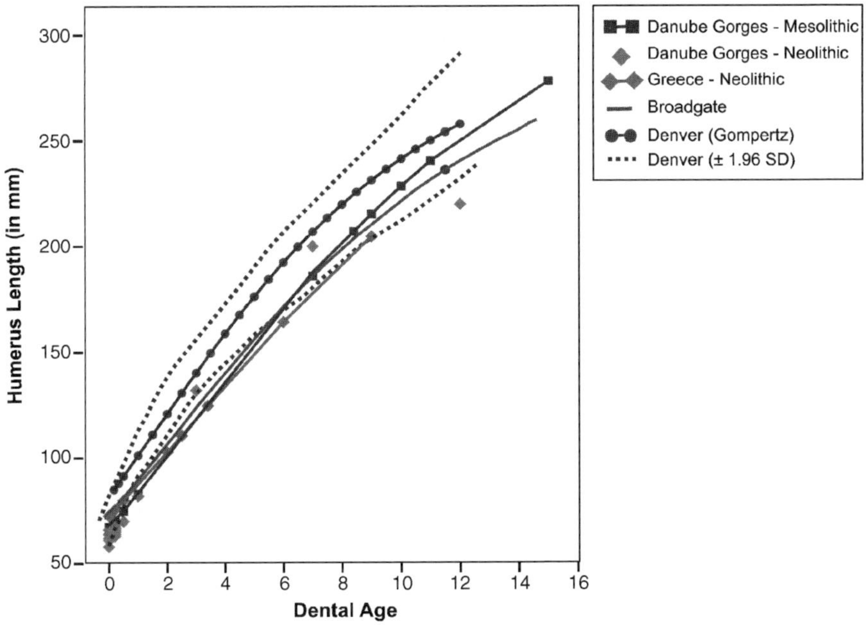

Figure 8.2 Growth patterns of the humeral diaphysis of the Danube Gorges (Mesolithic and Neolithic samples), Greek Neolithic and Broadgate in relation to the curve of the modern (Denver) reference sample

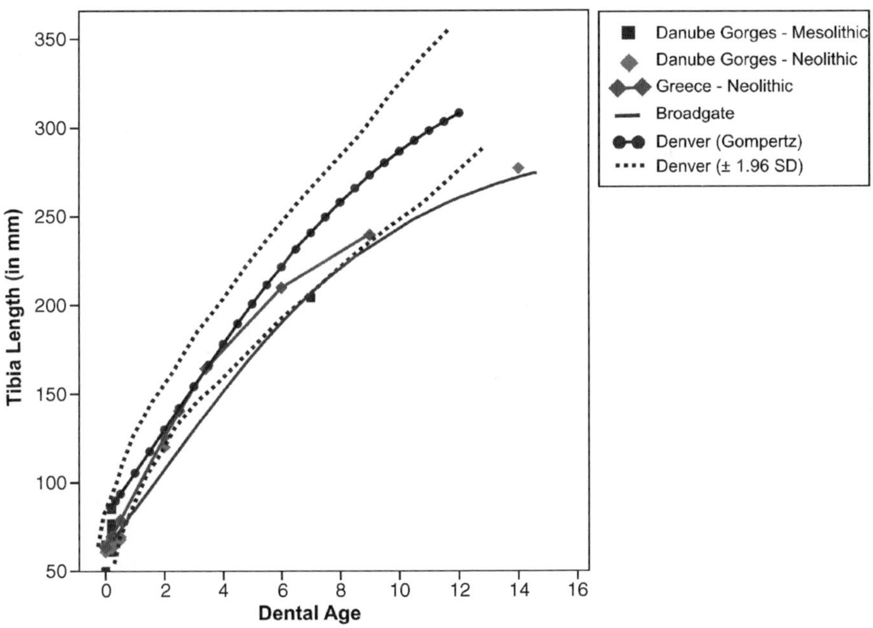

Figure 8.3 Growth patterns of the tibial diaphysis of the Danube Gorges (Mesolithic and Neolithic samples), Greek Neolithic and Broadgate in relation to the curve of the modern (Denver) reference sample

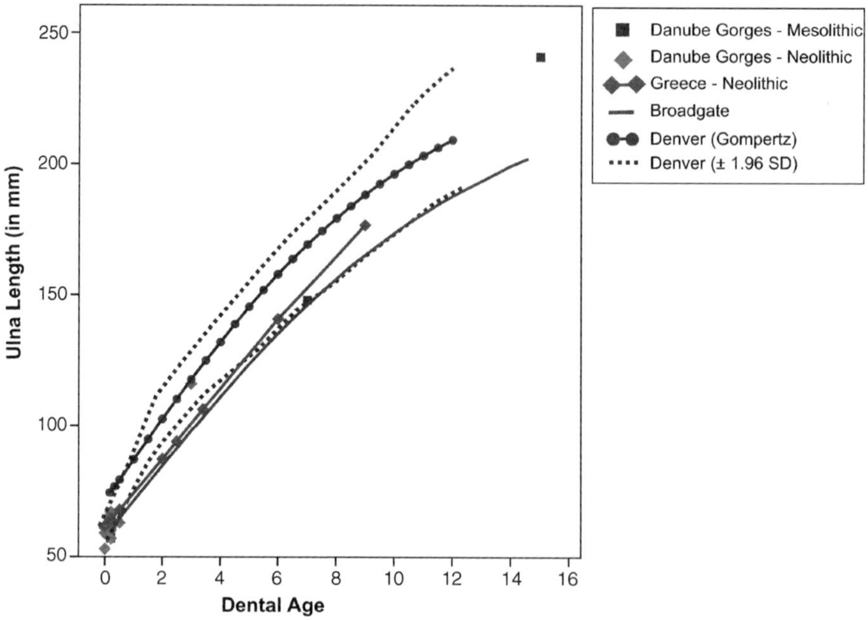

Figure 8.4 Growth patterns of the ulnar diaphysis of the Danube Gorges (Mesolithic and Neolithic samples), Greek Neolithic and Broadgate in relation to the curve of the modern (Denver) reference sample

8.4.4 Ulna Length

The ulnar diaphyseal dimensions of the Greek Neolithic sample between the age of 1 and 5 fall below the Denver –2SD-average range (Figure 8.4). However, Greek Neolithic children and adolescents older than 5 attain values within this range. The few Danube Gorges Mesolithic and Neolithic cases fall within the Denver range and some attain dimensions that are very close or even larger than the Denver average. As in the case of the tibia, the Broadgate curve is very close to the lower Denver range.

8.4.5 Radius Length

The Greek Neolithic curve is positioned below the Denver–2SD curve until the age of 7; after which it attains values within the–2SD-average Denver range (Figure 8.5). The Danube Gorges Mesolithic and Neolithic cases have lower radius length-per-age than the Denver range at birth and infancy. However, the two Danube Gorges Neolithic children attain lengths-per-age that are close to the Denver average values. In contrast, the two Danube Gorges Mesolithic children attain dimensions-per-age that are close to the lower (–2SD) Denver curve, while the adolescent case attains dimensions that are close to the Denver average dimensions-per-age curve. The Broadgate curve is positioned below the lower Denver curve but intersects it at the dental age of 12.

In sum, there are some clear inter-population and intra-skeletal variations between diaphyseal growth curves. If we take the Denver–2SD-average- + 2SD range as the modern reference standard and the Broadgate curve as the comparative example of values attained by a stressed

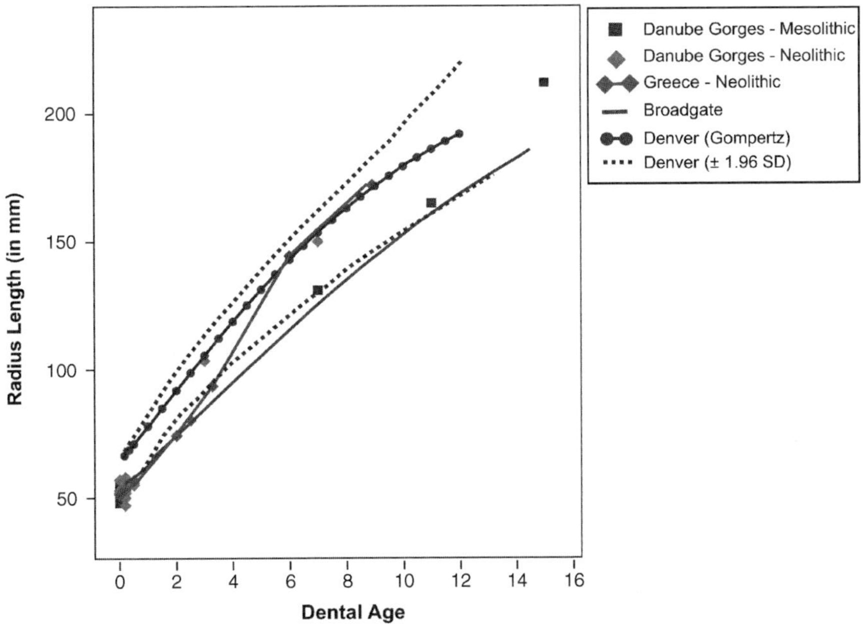

Figure 8.5 Growth patterns of the radial diaphysis of the Danube Gorges (Mesolithic and Neolithic samples), Greek Neolithic and Broadgate in relation to the curve of the modern (Denver) reference sample

population, it is apparent that the two Danube Gorges and the Greek Neolithic samples attain dimensions-per-age that are intermediate between the two. In general, the Mesolithic and Neolithic samples intersect the lower Denver curve during childhood. This pattern may indicate that individuals from stressed populations manage to compensate for lower dimensions-per-age during childhood or early adolescence, and attain dimensions close to those of the Denver reference sample following amelioration in nutrition and/or improved health. While catch-up growth usually occurs before the age of 2 (Tanner, 1986) a study of more than 2000 children from the Philippines demonstrated that while 63% of children were stunted at 2 years of age, about 30% were no longer stunted at 8.5 years, and 32.5% were no longer stunted at 12 years (Adair, 1999). However, it is important to note that all interpolated curves of archaeological populations are of cross-sectional data and hence cannot capture actual 'catch-up' growth or changes in growth velocity. While catch-up growth may explain the observed trends, it is also possible that apparent 'catch up' amongst archaeological samples during older childhood is due to different causes of death that may affect older children or adolescents. In contrast to small newborns and infants, where malnourishment and infectious disease may be primary causes of mortality, the human immune system functions much more efficiently during childhood (Sinclair and Dangerfield, 1998), suggesting that death during childhood or adolescence is rare, and more often occurs in response to severe and aggressive illnesses rather than chronic conditions or malnutrition that would affect growth.

It is clear that greater variations between samples are noted in femoral than humeral dimensions. In the case of the Greek Neolithic sample, tibial and femoral long bone dimensions-per-age are close to those of the Denver average values, while those of the ulna and radius fall closer to the Denver lower (–2SD) curve. This pattern contrasts with the Danube Gorges where the majority of subadults have dimensions-per-age, which are closer to the Denver average for the distal upper limb bones (radius, ulna) but not for the tibia. The Broadgate curve does not differ in its general position in relation to the Denver curves for any of the three distal limb bones, as it is always close to or just below the lower Denver curve.

8.4.6 Limb Proportions

8.4.6.1 The Tibio-Femoral Index

The index values of all Greek Neolithic and Danube Gorges individuals with a complete tibia and femur (a right pair with the substitution of left pair for four individuals missing one or both of the left bones) are plotted against age, together with individual cases from Broadgate, Christ Church Spitalfields, Raunds and St Helen-on-the-Walls. Average indices per age were also calculated for the Slavic sample from Mikulcice, the Native American Late Woodland population from Libben, Ohio and for the Denver reference sample. The inter-population comparison (Figure 8.6) indicates that individuals from the Mesolithic and Neolithic Danube Gorges populations have tibio-femoral indices-per-age that are always larger than the Denver average (between 80 and 83 for any age category between 0 and 14). Tibio-femoral indices were only calculated for six Greek Neolithic specimens. All infants and one (2.5 years of age) of the two children have indices higher than the Denver average. In the case of the medieval populations, most individuals greater than six years of age tend to have tibio-femoral indices

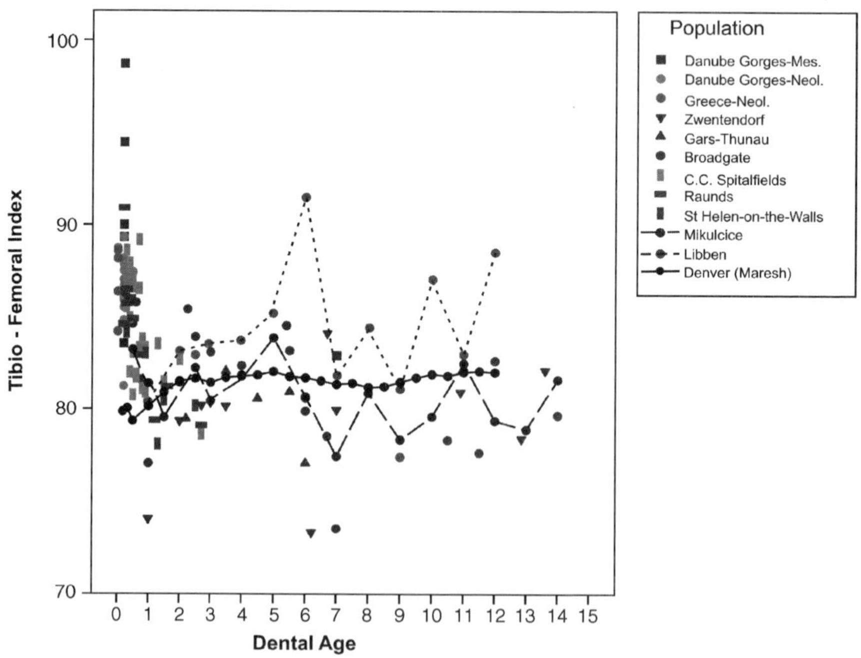

Figure 8.6 Age and population specific differences in tibio-femoral indices (age 0–15)

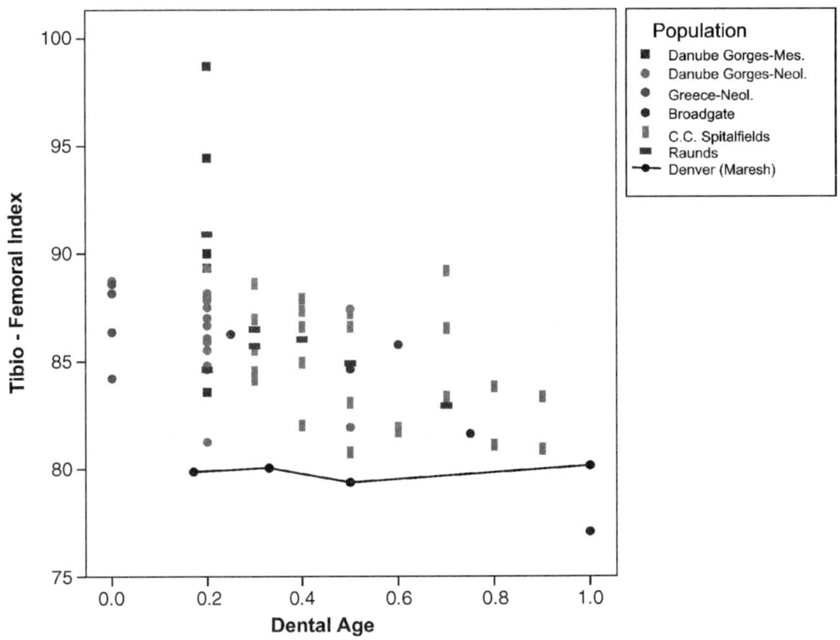

Figure 8.7 Age and population specific differences in tibio-femoral indices (age 0–1)

that are smaller than the Denver average. Another interesting observation is the comparison between the indices-per-age of the Native American population from Libben and the one for the Slavonic Mikulcice population. The Libben population have on average indices that are greater than the values recorded for the Denver population, while the Mikulcice population tend to have on average indices that are smaller than the Denver average. However, both lines share a similar peak and troughs pattern, which contrasts with the limited fluctuations observed in the case of the Denver line. An examination of the same graph with a focus on the birth-infancy age interval (Figure 8.7) indicates that all individuals, with the exception of a single case from Broadgate, London, have tibio-femoral indices that are greater than the Denver average for this age interval. Amongst the Mesolithic and Neolithic populations, the Greek sample display less variation than the Danube Gorges Neolithic and Mesolithic populations. Moreover, the Danube Gorges Mesolithic population has the greatest variability in tibio-femoral dimensions at birth and during infancy, with two individuals having indices that are greater than 90. The Danube Gorges Neolithic population displays a similar range of variation.

The pattern of inter-population variation for the radio-humeral index (Figure 8.8) is different from the one observed for the lower limbs. The Denver reference sample shows a decrease in index values from around 82 at birth to around 75 after the age of 2. At the age interval between 2 and 12, the ratio of the Denver population is relatively stable at a value of approximately 75. For the most part, the graphs of both the Libben and Mikulcice indicate higher index-per-age values than in the case of Denver. Broadgate and Christ Church Spitalfields, on the other hand, have lower than average indices in general (in comparison to Denver). An examination of the same graph with a focus on the age interval of birth to 1 (Figure 8.9) indicates that the

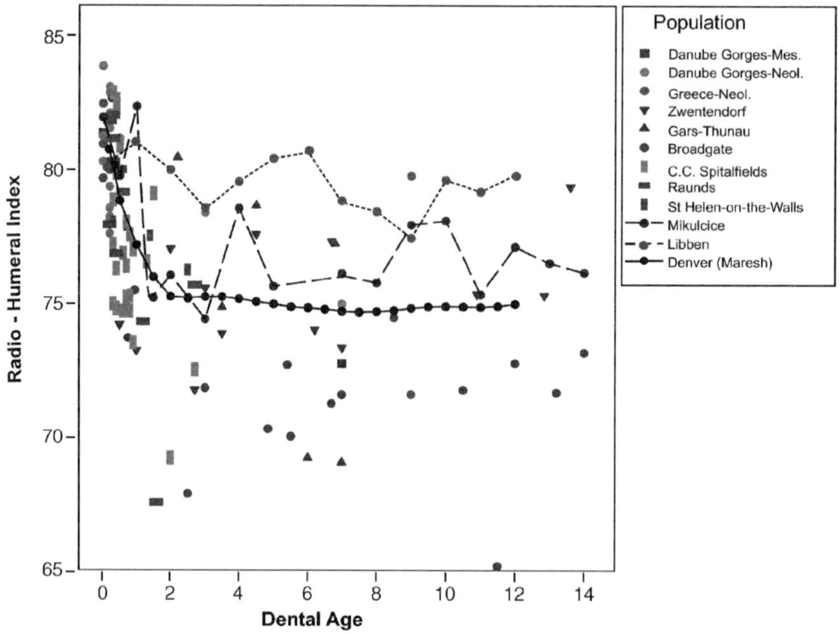

Figure 8.8 Age and population specific differences in radio-humeral indices (age 0–15)

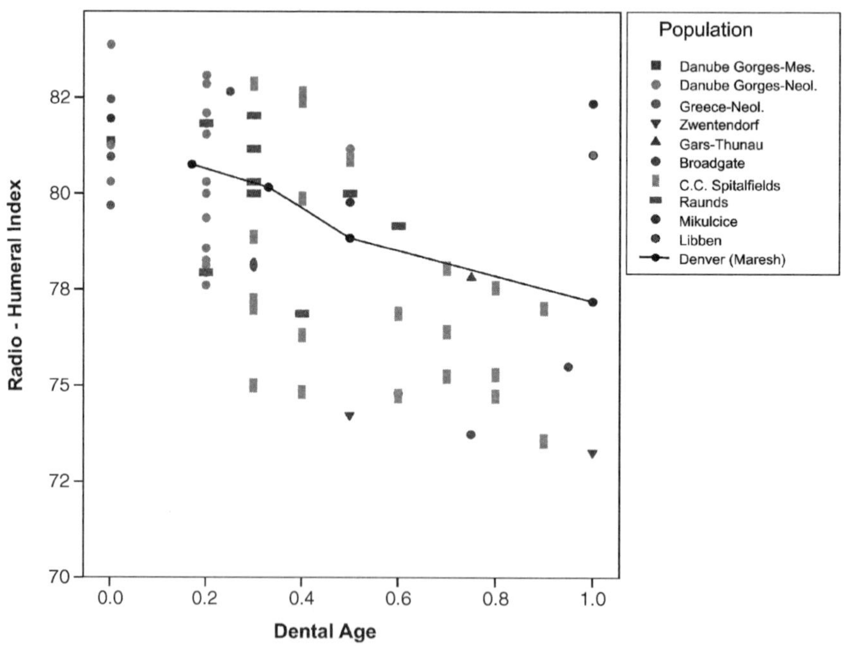

Figure 8.9 Age and population specific differences in radio-humeral indices (age 0–1)

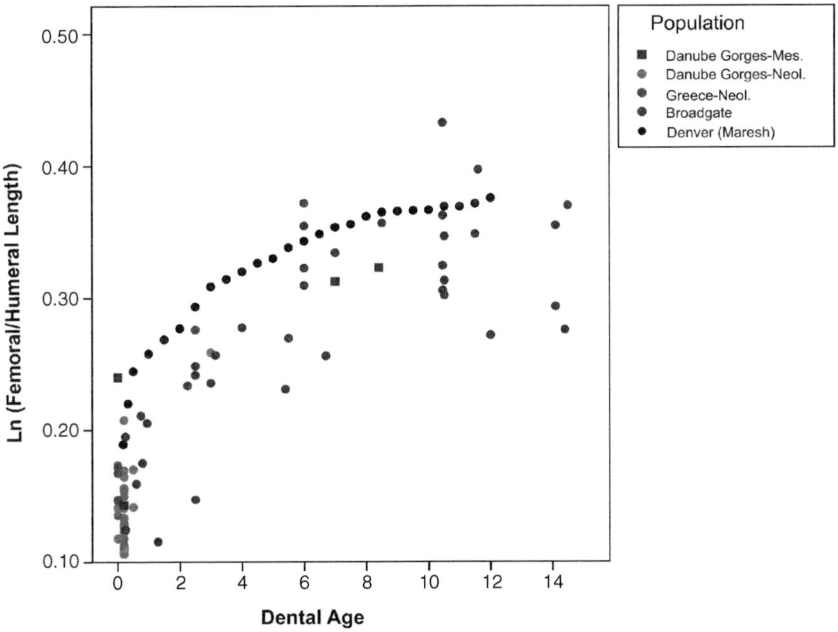

Figure 8.10 Femoral/humeral length proportions for all Mesolithic and Neolithic individuals aged between 0 and 14 with complete (paired) bones in comparison to proportions of the Broadgate and average proportions of the Denver reference sample

individuals from the Danube Gorges Mesolithic and Neolithic and the Greek Neolithic population have indices that are within the range of the Denver reference curve and those of other archaeological populations.

Femoral/humeral length proportions for all individuals in this study with complete (paired) bones are plotted together with the Denver average proportions and individual proportions for the Broadgate subadults following the methods of Ruff (Ruff, 2003) (Figure 8.10). The main observation is the difference in the proportions of the two individuals from the Danube Gorges Mesolithic sample and those of the Neolithic, which contrast with the less pronounced variability amongst the Greek Neolithic infants (although only three cases). In addition, proportions of Mesolithic and Neolithic specimens aged between 2 and 12 years are all lower than the Denver averages. Similarity, while the Broadgate sample displays a considerable variability in proportions for a given age cohort, the great majority of individuals have proportions that are lower than the Denver average.

Intra-individual variations in attainment of diaphyseal long bone dimensions-per-age in comparison of the Denver percentiles are provided in Table 8.3. The results indicate a significant degree of intra-limb variation and hence reaffirm Maresh's (1955) observation from his study of the Denver sample. These variations may reflect intra-population differences in health, nutrition and other environmental factors, but they may also reflect biological differences between individuals in terms of growth velocities and response to stress.

Table 8.3 Individual patterns of variability in attainment of the various limb disphyseal length dimensions-per-age

Site (Period)	Grave	Age	Femur	Tibia	Humerus	ulna	Radius	Variability
Franchthi	104	birth	10–25%	50%		<−2SD		high
Franchthi	48	birth	0–2.5%	10–25%	0–2.5%	<−2SD	<−2SD	high
Mavropigi	15	birth	0–2.5%	25–50%	0–2.5%	<−2SD	<−2SD	high
Alepotrypa	A1921	birth	0–10%			<−2SD	<−2SD	low
Franchthi	407	birth	<−2SD	<−2SD	<−2SD	<−2SD	<−2SD	low
Lepenski Vir (Neol.)	117	0.2	25–50%	25–50%	10%			high
Lepenski Vir (Neol.)	133	0.2	10–25%	25–50%		<−2SD		high
Lepenski Vir (Neol.)	95	0.2	10–25%	50–75%	25–50%			high
Lepenski Vir (Neol.)	120	0.2	10–25%	10%		10%	<−2SD	high
Lepenski Vir (Neol.)	131	0.2	0–10%	25–50%	10–25%		<−2SD	high
Lepenski Vir (Neol.)	63	0.2	50–75%	50–75%	50–75%	10–25%	10–25%	high
Vlasac (Mes)	42a2	0.2	0–10%	>90%	50%			high
Vlasac (Mes)	61	0.2	25–50%	75–90%	50–75%			high
Vlasac (Mes)	35a2	0.2	0–2.5%	10%		<−2SD	25%	high
Lepenski Vir (Neol.)	111	0.2	0%			10%	10%	low
Lepenski Vir (Neol.)	103	0.2	10–25%	25–50%	10–25%	10–25%	10–25%	low
Lepenski Vir (Neol.)	113	0.2	10–25%	25–50%	10%	10–25%	10%	low
Vlasac (Neol.)	U62	0.2	10–25%	25–50%	10–25%	25–50%	10–25%	low
Lepenski Vir (Neol.)	115	0.2	10–25%		10–25%		<−2SD	medium
Lepenski Vir (Neol.)	130	0.2	0–2.5%	10–25%	10–25%			medium
Lepenski Vir (Neol.)	123	0.2	0–10%	25–50%	10–25%		10%	medium
Lepenski Vir (Neol.)	129	0.2	10–25%	25–50%	10–25%		<−2SD	medium
Lepenski Vir (Neol.)	94	0.2	10–25%	50–75%	10–25%	10–25%	10–25%	medium
Lepenski Vir (Neol.)	96	0.2	10–25%	25–50%	25–50%	10–25%	10%	medium
Lepenski Vir (Neol.)	125	0.2	0–2.5%	25–50%	10–25%	10%	10%	medium
Vlasac (Mes)	35a1	0.2	0–10%	25–50%	0–10%			medium
Xirolimni	12	2.5	0–2.5%	10–25%	0–10%			medium
Lepenski Vir (Neol.)	92	3	<−2SD	0–10%	0–10%	10–25%	0–10%	medium
Franchthi	12	6	50–75%	25–50%			50–75%	medium
Vlasac (Mes)	53	7	<−2SD	0–10%		0–10%	0–10%	low
Alepotrypa	2198	9	25%	0–10%	10–25%	25%	10–25%	medium

Results are provided as percentiles based on those of the corrected Denver reference sample. Variability pattern is reported in relation to the overall variation in percentiles for a given specimen.

8.5 DISCUSSION

The Mann-Whitney nonparametric analysis of Danube Gorges Mesolithic and Neolithic neonates and infants indicated that there are no significant differences in limb dimensions between these populations, with the exception of differences in tibial width dimensions and the tibio-femoral ratio. When comparing the growth profiles of diaphyseal long bone dimensions of Mesolithic and Neolithic Danube Gorges populations and Greek Neolithic subadults to the Broadgate and Denver samples, it became apparent that amongst the Danube Gorges samples, growth attainment is below the –2SD Denver curve until the age of 7 to 9, when attained dimensions fall within the 5 to 50 (lower) range of the Denver percentiles. The Greek Neolithic growth curves intersect the –2SD Denver curve at an earlier stage, between the age of 2 and 4. The Greek sample also attains dimensions-per-age that are close to the Denver average during childhood in the case of the femur, tibia and radius, but attain lower dimensions-per-age in the case of the humerus and ulna. Hence it appears that the growth pattern of the Greek sample differs from the Danube Gorges samples in two distinct ways:

1. the attainment of dimensions-per-age which fall within the Denver –2SD-average range during early childhood; and
2. the attainment of dimensions-per-age which correspond to higher percentiles of the Denver reference sample (10–50%) for the lower limb bones and the radius.

While these curves should be interpreted with caution, due to the small sample sizes of the Mesolithic and Neolithic groups and the unequal presentation of some age cohorts, it does appear that the Greek Neolithic sample shows more inter-limb differences in growth attainment than the Danube Gorges samples. This may reflect inter-site variations in diet, disease load or other environmental factors that were not tested in this study. This pattern also contrasts with the Broadgate curves, which overall show higher attainment of dimensions-per-age (although still just above the Denver–2SD curve) for the proximal limbs than for the distal limbs. Since the same measuring methods and dental age values have been applied to all archaeological samples, and these are plotted against the Denver data, it appears that these inter-limb differences are not likely to be the outcome of methodological biases, and may therefore reflect inter-population variations in growth patterns.

Previous cross-sectional studies of diaphyseal limb bone growth amongst archaeological populations (see reviews by Saunders, 2000; Humphrey, 2003) confirm a general trend, in which attainment of dimensions-per-age of all long bones are well below those of the Denver modern reference sample (Humphrey, 2003; Johnston, 1962; Miles and Bulman, 1994; Pinhasi *et al.*, 2005). Humphrey (2003) examined fluctuations in femoral growth trajectories amongst archaeological samples from various periods and regions of the world, dated between 3000 BC to nineteenth century AD, relative to the Denver sample (c.f. Humphrey, 2003, Table 6.1). Her results indicate marked fluctuations in growth trajectories of the femur, which are most pronounced during the first three years of life. She reports a notable similarity between the growth trajectories of the three Native American populations, which was not found amongst other samples: an increase in percentage of adult size attained relative to the Denver sample in the first three years of life. This trend reflects a more rapid rate of femoral growth in the Native American samples during infancy and early childhood than amongst the other samples and therefore there may be a genetic component to this pattern (Humphrey, 2003). However, this pattern may be also attributed to other factors, such as a systematic underestimation of age of infants due to delays in dental development formation stages, and variations in weaning and dietary patterns.

Pinhasi and colleagues (2005) examined inter-population variability in long bone and pelvic bone growth (ilium, ischium and pubis) during the Early Medieval period by comparing the growth trajectories of four archaeological populations: two Slavonic (Gars–Thunau, Zwentendorf, Austria, tenth-century AD), one Avar (Zwölfaxing, Austria, eighth-century AD), and one Anglo-Saxon (Raunds, England, tenth-century AD). In this comparison, it was expected that:

1. the greatest differences in growth patterns would be found between the Anglo-Saxon and the Austrian samples due to their distinct genetic and biocultural background; and
2. minimal differences would be detected between the two Slavonic populations, as these were approximately contemporaneous, recovered from geographically close locations, and shared relatively similar archaeological contexts.

The results showed significant differences in long bone growth between populations in the growth patterns of distal diaphyseal dimensions of the femur and humerus and the dimensions of the ilium. Inter-population variability in growth curves for femoral and humeral dimensions were most pronounced during infancy (0–2 years). The most consistent differences in bone growth and related dimensions were those between Zwölfaxing and the other samples and these may be attributed to inter-population differences in weaning practices, overall health and other environmental factors (Pinhasi *et al.*, 2005). No significant differences in growth were detected between the Anglo-Saxon and the Austrian populations, despite the fact that these were less likely to share genetic attributes than the Central European samples. The results therefore suggest that varying growth patterns are associated with inter-population differences in absolute dimensions relative to age, but that these differences are not uniform for all bones. This may reflect differences in growth velocities at a given age interval for different bones, and perhaps also differences in the extent to which the growth of a given bone dimension is affected by fluctuations in environmental conditions such as diet and disease, and variation in the timing of susceptibility of growth to environmental perturbations during development.

In another study, Pinhasi and colleagues (2006) assessed the effects of vitamin D deficiency related rickets on long bone growth in post-medieval skeletal populations from East London (Broadgate and Christ Church Spitalfields). The study revealed that rickets had no effect on the growth curves for any of the long bones studied but a pronounced variation in growth between the populations was noted to occur during infancy and early childhood (from birth to 4 years of age), which was followed by 'catch-up' growth in which growth trajectories paralleled those attained by the children from the Denver reference sample. These similar trends were observed in the East London samples, despite differences in socioeconomic status.

The analysis of changes of lower and upper limb proportions during growth and development from birth to 14 years of age illustrates a contrast between the tibio-femoral index (Figures 8.6 and 8.7) and the radio-humeral index (Figures 8.8 and 8.9). The Mesolithic and Neolithic samples have higher tibio-femoral indices than the Denver sample, while their radio-humeral indices fluctuate both above and below the Denver reference line. However, the Denver index for the age of 2 months is 0.799, while the forensic assessment of 138 modern Hungarian foetuses (Fazekas and Kósa, 1978) gave average indices of 0.877, 0.870 and 0.876 for foetal age of 9, 9.5 and 10 lunar months, respectively. The Hungarian values are therefore closer in value to those of the Neolithic newborns (both the Danube Gorges and Greek samples) but are still lower than those of some of the Mesolithic newborns. It is possible that the high tibio-femoral indices for the Mesolithic newborns reflect different relative growth velocities of the femur and tibia than the average values reported for modern reference samples. The relatively low tibio-femoral values of the Denver sample, which was comprised of children from above-average socioeconomic status and good health (Maresh, 1943, 1955), may be due to the effects of reduced oxygen intake, thermal variations and other high-elevation conditions on the children from Denver county (~1600 m above sea level), as has been reported in the case of Highland Ecuadorian infants (Leonard *et al.*, 2000).

The assessment of intra-individual variation in the attainment of dimension-per-age (Table 8.3) does not indicate any clear pattern of inter-sample differences but suggests that there is a high degree of variability between individuals within a population that perhaps exceeds inter-population variability (although not tested here). Smith and Buschang (2004) examined the Denver mixed longitudinal data for boys and girls between 3 and 10 years of age and found that inter-bone differences in growth velocity parallel inter-bone differences in size. Hence, greater velocities and sizes are recorded for proximal limb bones (femur, humerus) in comparison to distal limb bones (tibia, radius, ulna). They also report an overall uniform pattern of childhood growth for this age interval. On average, the size of one bone explains 71% of the variation in the size of the other bones as well as a close coordination of inter-bone (diaphyseal) growth velocities. They state that:

> despite the aforementioned growth gradients and the allometric difference between limbs, the growth changes of all long bones are closely associated. For example, growth of the girls' tibia explains 79% of variation in the growth of the girls' radius. This suggests that the same control mechanisms, hormonal or otherwise, regulate childhood long bone growth in a parsimonious fashion (Smith and Buschang, 2004: p. 654).

The results of the comparison of archaeological samples in this study suggest somewhat different results. Indeed, the overall homogeneity of growth patterns of the various limb bones of the Denver (Maresh) samples is evident in Figures 8.6 to 8.9, whereas the Denver sample has nearly a constant tibio-femur index for all ages, and a decelerating (age 0–2) and then relatively constant (age 2–12) radio-humeral index. But the assessment of individual inter-limb dimensions-per-age from the Greek Neolithic and Danube Gorges Mesolithic and Neolithic sites indicates a medium-to-high degree of variability. In the case of the tibio-femoral index, all the studied individuals have index values that exceed those of the Denver sample. To some extent, variation in this trend may be driven by the fact that tibial lengths appear to be more variable than femoral lengths in response to environmental stress and individual variation (Holliday and Ruff, 2001), but they may also be the result of sampling bias due to small sample sizes.

In his assessment of variability in femoral/humeral limb proportions by age in the Denver Study sample, Ruff (2003) reports a relatively constant log-linear increase in proportions from the age of 6 months to the age of 5, followed by a more moderate increase from the age of 5 to 12.5 years. His study also indicates that growth trajectories of length, strength and proportions of proximal limb bones are largely independent. Femoral strength patterns are related to biomechanical demands that do not seem to affect length proportions. In fact, Ruff states that:

> ...the characteristically longer human femur begins to develop even prior to birth and femoral/humeral proportions are already not far from adult proportions (within 10%) in infancy. Thus, it seems likely that long bone length proportions are highly heritable, although partially environmentally modifiable (Ruff, 2003: 338).

Hence, the observed variability in femoral/humeral proportions amongst the Danube Gorges individuals may reflect intra-population variations in heritability of the genotypic complex traits that control and modify diaphyseal limb length growth, although we cannot rule out the possibility of environmental influences.

8.6 CONCLUSIONS

This study focused on the analysis of inter-population variability in the growth trajectories of limb bones, and intra-population and intra-individual variations in growth proportions and the attainment of dimensions-per-age amongst Danube Gorges and Greek Neolithic populations. The aim was not only to compare and contrast pre-agricultural vs. agricultural growth trajectories (in the case of the Danube Gorges region) but also Neolithic subadults from two regions in southeast Europe–Greece and Serbia. The main limitation of this study is the small sample sizes utilized and the paucity of children and early adolescents. Consequently, much emphasis has been placed on the visual interpretation of interpolated growth curves and dimensions-per-age of individual cases. In the case of the Danube Gorges region, the relatively large samples of newborns and infants from the Mesolithic and Neolithic contexts facilitated a comparison of pre-agricultural vs. agricultural variations in long bone dimensions that revealed no significant differences in most dimensions with the exception of tibial midshaft minimal and maximal breadth dimensions. However, it is important to note that in the Danube Gorges, the Mesolithic-early Neolithic period (i.e. samples included in this study) reflects a continuum of Mesolithic economic and cultural patterns, while the Neolithic pottery of the Starčevo-Körös-Cris only appears in the later stages of the Neolithic, reflecting the gradual nature of cultural change in this region (Borić *et al.*, 2004). In contrast, the Greek Neolithic samples are all from fully agricultural communities and hence the subadult burials are associated with fully Neolithic subsistence (see Papathanasiou, this volume). It is therefore interesting to note that the Gompertz growth curves of diaphyseal dimension-per-age revealed differences between the Danube Gorges Mesolithic, Greek Neolithic, and those of the Broadgate, the Denver samples and the individual Danube Gorges Neolithic cases. The Greek Neolithic subadults show a contrast between tibial and femoral dimensions-per-age, which are altogether close to those of the average Denver curve, while those of the ulna and radius fall closer to the Denver's –2SD curve. In the case of the femur, attained dimensions close to those of the Denver average dimensions-per-age occur in the Greek Neolithic sample at the age of 2 to 4, while they occur in the Danube Gorges Mesolithic sample at around the age of 8 to 10.

8.6.1 The Analysis of the Tibio-Femoral and Radio-Humeral Indices

Individuals from the Mesolithic and Neolithic Danube Gorges populations have indices-per-age that are always larger than the Denver average. The Greek samples display less variation than the Danube Gorges Neolithic and Mesolithic populations, while the Danube Gorges

Mesolithic population has the greatest variability in tibio-femoral dimensions at birth or during infancy. In contrast, the individuals from the Danube Gorges Mesolithic and Neolithic and the Greek Neolithic population have radio-humeral indices that are within the range of the Denver reference curve and those of other archaeological populations. The analysis of femoral/humeral length proportions for all individuals in this study highlights the difference in the proportions of the two individuals from the Danube Gorges Mesolithic sample and those of the Neolithic, which in turn contrasts with the less pronounced variability amongst the Greek Neolithic infants (although only three cases).

Finally, intra-individual variation in attainment of diaphyseal long bone dimensions-per-age show a significant degree of intra-limb variation and hence reaffirms Maresh's (1955) observation that even amongst healthy individuals, there is a significant degree of inter-bone differences in attainment of dimensions-per-age. Since the Denver Study, only included healthy infants and children from middle-class Caucasian American families, it remains unclear whether these variations can be attributed to genetic variability that affects ontogenetic and growth processes, differences in other biological aspects such as metabolism, immunity, or perhaps some environmental differences in diet and other conditions between the subadults from the different households.

The observations made in this study must be considered as preliminary due to the small sample size. Nonetheless, the comparison of inter-bone variation in growth of archaeological cases to 'standard' eferences (Denver Study) point out the potential of archaeological growth studies that go beyond the interpretation of average trends and seek the more complex patterns of variation in growth between individuals and population variation inherent to the growth process. Further research, which combines growth analysis with health status (paleopathological markers) and diet (stable isotope analysis), of these Mesolithic and Neolithic populations, may reveal important information about whether the transition to agriculture in southeast Europe has caused some alternations in the growth trajectories and, if so, chart the specific nature of these changes.

REFERENCES

Adair, S. (1999) Filipino children exhibit catch-up growth from age 2 to 12 years. *J. Nutr.*, **129**, 1140–1148.

Bennike, P., Lewis, M.E., Schutkowski, H. and Valentin, F. (2005) Comparison of child morbidity in two contrasting medieval cemeteries from Denmark. *Am. J. Phys. Anthropol.*, **128** (4), 734–746.

Berti, P.R., Leonard, W.R. and Berti, W.J. (1998) Stunting in an Andean community: prevalence and etiology. *Am. J. Hum. Biol.*, **10**, 229–240.

Boddington, A. and Cadman, G. (1981) Raunds Furnells: An interim report on excavations 1977–1980, in *Anglo-Saxon Studies in Archaeology and History Cambridge* (eds J. Campbell and S.C. Hawkes), British Archaeological Reports, British Series, Oxford, pp. 103–122.

Bogin, B. (1999) Evolutionary perspective on human growth. *Annu. Rev. Anthropol.*, **28**(1), 109–153.

Bogin, B. and Smith, B.H. (1996) Evolution of the human lifecycle. *Am. J. Hum. Biol.*, **8**, 703–716.

Borić, D. (2002) The Lepenski Vir conundrum: reinterpretation of the Mesolithic and Neolithic sequences in the Danube Gorges. *Antiquity*, **76**, 1026–1039.

Borić, D. and Stefanović, S. (2004) Birth and death: infant burials from Vlasac and Lepenski Vir. *Antiquity*, **78**, 526–546.

Borić, D., Grupe, G., Peters, J. and Mikić, Ž. (2004) Is the Mesolithic-Neolithic subsistence dichotomy real? New stable isotope evidence from the Danube Gorges. *Eur. J. Archaeol.*, **7**(3), 221–248.

Borić, D. (2007) Mesolithic-Neolithic Interactions in the Danube Gorges, in *Mesolithic-Neolithic Interactions in the Danube Basin* (eds J.K. Kozlowski and M. Nowak), British Archaeological Reports, International Series, Oxford, pp. 31–45.

Buikstra, J.E. and Ubelaker, D.H. (1994) Standards for Data Collection from Human Skeletal Remains. Fayetteville: Arkansas Archeological Survey.

Cook, D.C. (1984) Subsistence and health in the Lower Illinois Valley: osteological evidence, in *Paleopathology at the Origins of Agriculture* (eds M.N. Cohen and G.J. Armelagos), Academic Press, Orlando, FL, pp. 237–270.

Dawes, J.D. and Magilton, J.R. (1980) *The Cemetery of St Helen-on-the-Walls*, The archaeology of York. The Medieval cemeteries 12/1 Tork: Council for British Archaeology for York Archaeological Trust, Aldwark.

Fazekas, I.G.y. and Kósa, F. (1978) *Forensic Fetal Osteology*, Akadémiai Kiadó, Budapest.

Garn, S.M., Lewis, A.B. and Polacheck, D.L. (1959) Variability of tooth formation. *J. Dent. Res.*, **28**, 135–148.

Holliday, T.W. and Ruff, C.B. (2001) Relative variation in human proximal and distal limb segment lengths. *Am. J. Phys. Anthropol.*, **116**, 26–33.

Hummert, J.R. and Van Gerven, D.P. (1983) Skeletal growth in a medieval population from Sudanese Nubia. *Am. J. Phys. Anthropol.*, **60**, 471–478.

Humphrey, L.T. (2000) Growth studies of past populations: an overview and an example, in *Human Osteology in Archaeology and Forensic Science* (eds S. Mays and M. Cox), Greenwich Medical Media, London, pp. 25–38.

Humphrey, L.T. (2003) Linear growth variation in the archaeological record, in *Patterns of Growth and Development in the Genus Homo* (eds J.L. Thmpson, G.E. Krovitz and A.J. Nelson), Cambridge University Press, Cambridge, pp. 144–169.

Jacobsen, T.W. (1969) Excavations at Porto Cheli and vicinity, preliminary report, II: the Franchthi Cave, 1967–1968. *Hesperia*, **38**, 343–381.

Jacobsen, T.W. (1973a) Excavation in the Franchthi Cave, 1969–1971, part I. *Hesperia*, **42**, 45–88.

Jacobsen, T.W. (1973b) Excavation in the Franchthi Cave, 1969–1971, part II. *Hesperia*, **42**, 253–283.

Jantz, R.L. and Owsley, D.W. (1984) Long bone growth among Arikara skeletal populations. *Am. J. Phys. Anthropol.*, **63**(1), 13–20.

Johnston, F.E. (1962) Growth of long bones of infants and young children at Indian Knoll. *Am. J. Phys. Anthropol.*, **20**, 249–254.

Karamitrou-Mendesidi, G. (1998) Xirolimni Kozanhs 1998. *To Archaiologiko Ergo sth Makedonia kai sti Thraki*, **12**, 472–474.

Karamitrou-Mendesidi, G. (2005) Mavropigi 2005: Lignitorycheia kai Archaiothtes. *To Archaiologiko Ergo sth Makedonia kai sti Thraki*, **19**, 511–539.

Larsen, C.S. (1997) *Bioarchaeology*, Cambridge University Press, Cambridge.

Leonard, W.R., Dewalt, K.M., Stansbury, J.P. and McCaston, M.K. (2000) Influence of dietary quality on the growth of highland and coastal Ecuadorian children. *Am. J. Hum. Biol.*, **12**, 825–837.

Lewis, M.E. (2002a) The impact of industrialisation: Comparative study of child health in four sites from medieval and post-medieval England (850–1859). *Am. J. Phys. Anthropol.*, **119**, 211–223.

Lewis, M.E. (2002b) *Urbanisation and Child Health in Medieval and Post-Medieval England*, BAR, Oxford.

Lewis, M.E.P.D. (2007) *The Bioarchaeology of Children: Perspectives from Biological and Forensic Anthropology* (ed. Mary E. Lewis), Cambridge University Press, Cambridge, p. 255.

Lovejoy, C.O., Russell, K.F. and Harrison, M.L. (1990) Long bone growth velocity in the Libben population. *Am. J. Hum. Biol.*, **2**, 533–542.

Maresh, M.M. (1943) Growth of the major long bones in healthy children. *Am. J. Dis. Child.*, **66**, 227–257.

Maresh, M.M. (1955) Linear growth of long bones of extremities from infancy through adolescence. *Am. J. Dis. Child.*, **89**, 725–742.

Maresh, M.M. (1970) Measurements from roentgenograms, in *Human Growth and Development* (ed. R. W. McCammon), Charles C. Thyomas, Springfield, pp. 157–200.

Mays, S., Brickley, M. and Ives, R. (2008) Growth in an English population from the Industrial Revolution. *Am. J. Phys. Anthropol.*, **136**(1), 85–92.

Mays, S., Brickley, M. and Ives, R. (2009a) Growth and vitamin D deficiency in a population from 19th century Birmingham. *England. Int. J. Osteoarchaeol.*, **19**(3), 406–415.

Mays, S., Ives, R. and Brickley, M. (2009b) The effects of socioeconomic status on endochondral and appositional bone growth, and acquisition of cortical bone in children from 19th century Birmingham, England. *Am. J. Phys. Anthropol.*, **140**(3), 410–416.

Mays, S.A. (1999) Linear and appositional long bone growth in earlier human popultaions: a case study from medieval England, in *Human Growth in the Past: Studies from Bones and Teeth* (eds R.D. Hoppa and C. Fitzgerald), Cambridge University Press, Cambridge, pp. 290–312.

Mensforth, R.P. (1985) Relative tibia long bone growth in the Libben and Bt-5 prehistoric skeletal populations. *Am. J. Phys. Anthropol.*, **68**(2), 247–262.

Merchant, V.L. and Ubelaker, D.H. (1977) Skeletal growth in the protohistoric Arikara. *Am. J. Phys. Anthropol.*, **46**, 61–72.

Miles, A.E.W. and Bulman, J.S. (1994) Growth curves of immature bones from a Scottish island population of 16th to mid-19th century: limb-bone diaphyses and some bones of the hand and foot. *Int. J. Osteoarchaeol.*, **4**, 121–136.

Molleson, T. and Cox, M. (1993) *The Spitalfields Project, Volume 2: The Anthropology*, Council for British Archaeology, York, CBA Research Report.

Moorrees, C.F.A., Fanning, E.A. and Hunt, E.E. (1963a) Age variation of formation stages for ten permanent teeth. *J. Dent. Res.*, **42**, 1490–1502.

Moorrees, C.F.A., Fanning, E.A. and Hunt, E.E. (1963b) Formation and resorption of three deciduous teeth in children. *Am. J. Phys. Anthropol.*, **21**, 205–213.

Papathanasiou, A. (ed.) (1996) *Neolithic Civilization in Greece*, Goulandris Foundation, Athens, Greece.

Papathanasiou, A. (2001) *A Bioarchaeological Analysis of Neolithic Alepotrypa Cave*, Greece, International Series 961, BAR, Oxford.

Pinhasi, R. and Bourbou, C. (2008) How representative are human skeletal assemblages for population analysis? in *Advances in Human Palaeopathology* (eds R. Pinhasi and S. Mays), Wiley-Liss, Chichester, UK, pp. 31–44.

Pinhasi, R., Teschler-Nicola, M., Knaus, A. and Shaw, P. (2005) Cross-population analysis of the growth of long bones and the os coxae of three early medieval Austrian populations. *Am. J. Hum. Biol.*, **17**, 470–488.

Pinhasi, R., Shaw, R., White, B. and Ogden, D. (2006) Morbidity, rickets, and long bone growth in post-medieval Britain – a cross-population analysis. *Ann. Hum. Biol.*, **33**, 372–389.

Ribot, I. and Roberts, C. (1996) A study of non-specific stress indicators and skeletal growth in two medieval subadult populations. *J. Anthropol. Sci.*, **23**, 67–79.

Roksandić, M. (1999) *Transition from Mesolithic to Neolithic in the Iron Gates Gorge: Physical Anthropology Perspective*, Simon Fraser Universit, Burnaby.

Ruff, C. (2003) Ontogenetic adaptation to bipedalism: age changes in femoral to humeral length and strength proportions in humans, with a comparison to baboons. *J. Hum. Evol.*, **45**(4), 317–349.

Ruff, C. (2007) Body size prediction from juvenile skeletal remains. *Am. J. Phys. Anthropol.*, **133**(1), 698–716.

Saunders, S.R. (1992) Subadult skeletons and growth related studies, in *Skeletal Biology of Past Peoples: Rtesearch Methods* (eds S.R. Saunders and M.A. Katzenberg), Wiley-Liss, New York, pp. 1–19.

Saunders, S.R. (2000) Subadult skeletons and growth-related studies, in *Biological Anthropology of the Human Skeleton* (eds M.A. Katzenberg and S.R. Saunders), Wiley-Liss, New York, pp. 135–161.

Saunders, S.R. and Hoppa, R.D. (1993) Growth deficit in survivors and non-survivors: biological correlates of mortality bias in subadult skeletal samples. *Yearb. Phys. Anthropol.*, **36**, 127–151.

Schofield, J. and Maloney, C. (eds) (1998) *Archaeology in the City of London, 1907–1991: A Guide to Records of Excavations by the Museum of London and its Predecessors*, Museum of London, London.

Sinclair, D. and Dangerfield, P. (1998) *Human Growth After Birth*, 6th edn, Oxford University Press, Oxford.

Smith, B.H. (1991) Standards of human tooth formation and dental age assessment, in *Advances in Dental Anthropology* (eds M.A. Kelley and C.S. Larsen), Wiley-Liss, New York, pp. 143–168.

Smith, S.L. and Buschang, P.H. (2004) Variation in longitudinal diaphyseal long bone growth in children three to ten years of age. *Am. J. Hum. Biol.*, **16**, 648–657.

Srejović, D. (1981) *Lepenski Vir: Menschenbilder Einerfrühen Europäischen Kultur*, Verlag Philipp von Zabern, Mainz am Rhein.

Srejović, D. and Letica, Z. (1978) *Vlasac: A Mesolithic settlement in the Iron Gates*, Serbian Academy of Sciences and Arts, Belgrade.

Stloukal, M. and Hánaková, H. (1978) The Length of long bones in ancient Slavonic populations with particular consideration to the questions of growth. *Homo*, **29**, 53–69.

Sundick, R.I. (1978) Human skeletal growth and age determination. *Homo*, **29**, 228–248.

Tanner, J.M. (1986) Growth as a target-seeking function; Catch-up and catch-down growth in man, in *Human Growth: A Comprehensive Treatise*, Vol. **3**, 2nd edn (eds F. Falkner and J.M. Tanner), Plenum, New York, pp. 167–179.

Tanner, J.M. (1998) A brief history of the study of human growth, in *The Cambridge Encyclopedia of Human Growth and Development* (eds S. Ulijaszek, F.E. Johnston and M.A. Preece), Cambridge University Press, Cambridge, pp. 2–7.

Tompkins, R.L. (1996) Human population variability relative to dental development. *Am. J. Phys. Anthropol.*, **99**, 79–102.

Tringham, R. (2000) Southeastern Europe in the transition to agriculture in Europe: bridge, buffer or mosaic, in *Europe's first Farmers* (ed. T.D. Price), Cambridge University Press, Cambridge, pp. 19–56.

Wood, J.W., Milner, G.R., Harpending, H.C. and Weiss, K.M. (1992) The osteological paradox. Problems of inferring prehistoric health from skeletal samples. *Curr. Anthropol.*, **33**, 343–370.

World Health Organization growth standards http://www.who.int/childgrowth/en/.

Wright, L.E. and Yoder, C.J. (2003) Recent progress in bioarchaeology: approaches to the osteological paradox. *J. Archaeol. Res.*, **11**, 43–70.

9

Reaching Great Heights: Changes in Indigenous Stature, Body Size and Body Shape with Agricultural Intensification in North America

Benjamin M. Auerbach

Department of Anthropology, The University of Tennessee, Knoxville, TN, USA

9.1 INTRODUCTION

Variation in stature, body mass and body shape associated with shifts in subsistence economies has not been examined extensively amongst indigenous groups living in North America prior to European contact. Since the publication of Blakely's 1977 edited volume, *Biocultural Adaptation in Prehistoric America*, a regular succession of papers and books have presented biological examinations of the effects of lifeway changes associated with the advent of agriculture (Cohen and Armelagos, 1984a; Cohen and Crane-Kramer, 2007; Lambert, 2000; Powell, Bridges and Mires, 1991). Studies of diachronic change in skeletons amongst these publications have cited several environmental influences, including alterations in nutrition, infectious disease, activity levels and differential limb use. These effects and potential causal factors have been explored in few samples as extensively as amongst the indigenous human remains of North America from both before and after European colonization, especially in the Eastern[1] and Southwestern regions (Bridges, 1992; Danforth, Cook and Knick, 1994; Larsen, 1981, 1995; Rose, Marks and Tieszen, 1991; Ruff, Larsen and Hayes, 1984). Yet, despite the breadth of studies on these geographical areas, the only morphological (body shape and size) aspects of the skeletons regularly studied were those associated with stature or with bone robusticity and mechanics. Indeed, a majority of analyses have focused on evidence of

Human Bioarchaeology of the Transition to Agriculture Edited by Ron Pinhasi and Jay T. Stock

stress or disease processes on the skeleton. In addition, only within the last 20 years have biological anthropologists recognized that the utilization of various forms of plant cultivation depended on local subsistence economies (Hutchinson *et al.*, 1998; Larsen, 1995), and so the changes observed in any archaeological skeletons were likely regionally or population specific. This study examines these within- and between-region variations in a large number of samples. in order to elucidate the relationships of morphological variation to changes in diet over time.

9.1.1 North American Agriculture and Morphology in Context

The adoption of agriculture in the Americas, as elsewhere on the globe, was a complex and gradual process (Fritz, 2007). Recent publications have indicated that the purposeful growth of plants as predictable food sources has a potentially great temporal depth, even as early as 8000 BP in the Andean regions of South America (Dillehay *et al.*, 2007; Piperno and Dillehay, 2008; Scheinsohn, 2003). The precise mechanisms of plant management, cultivation and eventual domestication have long been a topic of debate (Carter, 1946; Casas *et al.*, 2007). However, there is general agreement that a least three major, independent centres of domestication occurred in the Americas (Diamond and Bellwood, 2003), and likely in far more locations (Iriarte, 2007).

One of these major centres was in the Eastern region of North America, where archaeobotanical evidence indicates preferential harvesting (by presence of seed remains in middens), planting and artificial selection (determined by seed size) of specific plants, including species of squash, sunflower, sumpweed and goosefoot (Fritz, 2007; McLauchlan, 2003). However, the timing and locations for the earliest purposeful cultivation of these plants are difficult to ascertain. This is in part because:

1. distinguishing wild from cultivated or domesticated plant remains is often difficult;
2. the organic remains of plants have not consistently or systematically been collected at archaeological sites;
3. there is growing evidence that the domestication of the plants started during the late Archaic (about 4000 BP or before), where poor preservation and low site visibility are complications;
4. some plant genera, and even species, may have been domesticated more than once across separate locations or times; and
5. the presence of cultivars varied within regions, as their adoption was a local decision based on politics as much as ecology or economics (Anderson, Russo and Sassaman, 2007; Hart and Sidell, 1997; Hutchinson *et al.*, 1998; Pickersgill, 2007; Scheinsohn, 2003; Smith, 1992).

An additional problem of defining group subsistence – 'agriculture' vs. 'foraging' or 'horticulture' – is concomitant with those issues enumerated above (Bridges, Blitz and Solano, 2000; Smith, 2001), especially because subsistence economies are variable within categories and collectively form a continuum; there is no clear delineation between hunter-gatherers and agriculturalists.

Despite these sources of uncertainty, a picture of the development of cultivars in eastern North America is emerging from archaeological studies. By 4000 years ago, populations in the middle Mississippi, Tennessee and Ohio River Valleys developed an increasing dependence on wild plants either purposefully managed or locally planted (Boyd and Boyd, 1989; Casas *et al.*, 2007; Fritz, 1997). Whether this was associated with the florescence of mound building in the lower Mississippi River Valley (Anderson, Russo and Sassaman, 2007; Russo, 1996) that led to the Poverty Point culture[2] is not known, though there is little evidence for the use of domesticated plants at sites associated with this late Archaic tradition (Gibson, 2000; Sassaman, 2010). However, evidence for high dependence on a number of plant species amongst the Woodland period Hopewell is established (McLauchlan, 2003), including local squashes and the first, low-concentration appearance of maize from tropical Mesoamerica by 2000 BP (Hart, 1999). With this greater reliance on cultivars, populations became more sedentary year-round and developed new technologies for the processing of food, while also maintaining wild game hunting and the gathering of non-domesticates.

Against this backdrop, the penetration of maize into eastern North America as a focal domesticated crop was complex, affected by its perceived utility in comparison with locally managed and cultivated plants (Hutchinson *et al.*, 2000), the efficacy of its cultivation in the temperate environments of the Eastern region (Diehl, 2005) and its social importance within local and regional cultures (Blake *et al.*, 1992; Danforth, 1999a). With the Mississippian 'explosion' just over a millennium ago (Pauketat, 2004), maize emerged as a ritually important crop that increasingly supplanted local domesticates as a dietary staple in some areas of the Eastern region. The management of maize cultivation, surplus storage and dissemination during shortages occurring with climatic anomalies may have helped lead to the development of the Mississippian polity (Anderson, Stahle and Cleaveland, 1995; Pohl *et al.*, 1996), and its eventual spread. It is essential to underscore that there was variability in the adoption of maize across the region (Hutchinson *et al.*, 1998), as well as differential use of maize amongst social strata (Danforth, 1999a; Fritz, 2007; Rose, Marks and Tieszen, 1991). As Buikstra (1991: 175–176) stated concerning the spread of culture, crops and technologies, 'we must now consider Mississippian adaptations in the plural and that the impact of these upon the human condition varied considerably, given other social and environmental variables.'

Although evidence argues for important local and regional variation in subsistence shifts to agriculture, some of the first broadly synthetic observations on its effects on human skeletons yielded a nearly universal picture of coinciding degradations in health, mortality and bone strength amongst populations (Cohen and Armelagos, 1984b). In the Eastern region, researchers argued that a heavy reliance on maize consumption, as well as demographic and activity changes associated with increased sedentism and food production, appeared to have driven this decline (Buikstra, Konigsberg and Bullington, 1986; Cassidy, 1984; Larsen, Shavit and Griffin, 1991). More than two decades of research, however, have supplanted this view by a pluralistic paradigm, in which past populations fared differently from each other, even neighbours, as biological anthropologists have appreciated the complexities of subsistence economy change in eastern North America and other regions (Cohen, 2007). For example, diachronic changes in bone mechanical properties significantly differed amongst not only agricultural populations, but also between sexes within those populations (Bridges, 1991, 1996; Larsen, 1995; Ruff *et al.*, 1994). Likewise, although there are multiple examples of stature decreasing as agriculture intensified within regions (Cook, 1984; Goodman *et al.*, 1984; Larsen *et al.*, 2007; Perzigian, Tench and Braun, 1984), there are as many indicating that stature remained stable, if not increased, with the adoption of agriculture (Boyd and Boyd, 1989; Cook, 2007; Danforth *et al.*, 2007; Rose *et al.*, 1984).

9.1.2 Morphologies Considered and Research Goals

As noted above, stature has been the only body size or shape variable examined extensively in comparisons between past 'forager' and 'agricultural' populations, or amongst agriculturalists, mostly as a proxy for health. Adult stature is considered a good indicator of health because of its association with nutrition and stress encountered during primary growth (Bogin, 1999; Malina *et al.*, 2004; Newman, 1962; Tanner, 1986), especially when there is chronic disease or nutritional deprivation. That is, individuals who are malnourished or stressed during primary growth fail to achieve their maximum potential statures. Although genes certainly contribute to the determination of ultimate stature attained at the end of growth, some research has suggested that these environmental perturbations have a more significant effect on development (Frishancho, 1993; Frishancho and Housh, 1988), and that human populations (with notable exceptions, such as pygmoid groups) ideally have similar patterns of growth and, ultimately, similar statures (Danforth, 1999b; Martorell and Habicht, 1986). The timing of these perturbations during primary growth is crucial though, especially because stunted growth encountered in early childhood may be 'corrected' by catch-up growth during adolescence (Martorell, Kahn and Schroeder, 1994; cf. Golden, 1994). In fact, as Larsen (1995) noted, some of the variation in adult statures observed amongst agriculturalists may have resulted in part from group differences in adolescent provisioning.

It is possible that body mass predicted from skeletal remains provides another indicator of overall population health. Numerous studies of living populations have used lean body mass as a measure of nutrition and overall health (Bogin, 1999; Eveleth and Tanner, 1990). Prior to the current volume, there are no studies, to the author's knowledge, that have attempted to compare estimated body masses amongst past populations engaged in different subsistence economies (see chapters by Temple and Stock *et al.*, in this volume for other investigations of body size and mass variaton). There is some evidence that femoral head size – which may be used to reliably predict body mass in adults (Ruff, Scott and Liu, 1991) – follows a similar growth trajectory as femoral length, which is either used as a proxy for stature or is used to predict it (see Figure 9.3 in Ruff, 2007). Although body mass amongst juveniles does not follow the same trajectory, living body mass at the end of growth closely matches mass estimated from femoral head diameter (Ruff, 2007). Thus, it is conceivable that insults encountered during primary growth that stunt the longitudinal growth of long bones may likewise reduce the size of the femoral head at the end of primary growth.

Some caveats to this potential relationship should be noted. Under the cylindrical model of human body shape (Ruff, 1991), stature and body mass (volume) are linked, though the correlation is low (Auerbach, 2007) because variation in volume under the model is also determined by body breadth; two individuals of equal stature but different body breadths will not have the same masses. Thus, variation in body breadth amongst populations must be taken into account when making any comparisons amongst groups. Also, as Ruff (2007) states, the growth of femoral head diameter and femoral length outpace actual body mass, which in turn indicates that morphological integration between these two metrics confounds the potential use of femoral head size (and therefore masses estimated from it) as an independent indicator of health. However, this is not regarded as a problem for two reasons:

1. even if femoral head size is genetically 'programmed' to reach a specific dimension, it still follows a similar trajectory as femoral length and therefore likely responds similarly to environmental perturbations that alter the programmed growth; and

2. other analyses on morphological integration between these two femoral dimensions indicates independence between them (DeLeon and Auerbach, 2007).

Finally, body mass has been long recognized as a morphological trait amongst humans that relates to ecogeographic patterns in association with climate (Holliday, 1997; Ruff, 1994). However, Auerbach (2007) found that the relationship between climatic factors and body mass amongst a broad sample of New World groups was inconsistent and may have been influenced by subsistence, especially within climatically similar regions. The examination of body mass in comparison with stature, then, is a worthwhile avenue of research as a potential evaluation of changes amongst groups practising different subsistence economies within circumscribed regions.

The main hypothesis of this study is that body mass should demonstrate similar patterns of variation as stature amongst groups from eastern North America engaged in distinctive subsistence economies. Furthermore, changes in these two morphologies between and within subsistence groups are not expected to follow identical patterns throughout the Eastern region, as multiple prior studies indicate both stature increases and decreases with agricultural intensification amongst eastern groups (Boyd and Boyd, 1989; Cassidy, 1984; Danforth *et al.*, 2007; Larsen, 1995; Rose *et al.*, 1984). Also, a basic understanding of variation in stature and body mass will be compared between groups from the Eastern region and samples from the Southwest, the other major centre of agricultural development north of Mesoamerica.

Limb proportions and bi-iliac breadth will also be examined to ascertain general population continuity in the Eastern region, under the assumption that recent population replacements from other regions would be indicated by significant differences in these morphologies between samples temporally but not geographically distinct (Auerbach, 2010; Holliday, 1999; Temple *et al.*, 2008). As argued by Holliday (1999), limb proportions demonstrate stability over time, as these dimensions likely are genetically determined, and so significant changes in them amongst groups occupying the same geographic area are likely to be the result of gene flow. Auerbach (2007) corroborated this pattern in indigenous past groups from the Americas, and the crural index was implicated as more stable over time than the brachial index in this same research. Furthermore, bi-iliac breadth appears to change slowly over time, likely due to multiple factors (thermoregulation, obstetrics, locomotion) influencing its shape (Ruff, 1994; Auerbach, 2007). In fact, Auerbach (2010) has demonstrated that body breadth, as represented by bi-iliac breadth, has remained wide in the Americas compared with groups from similar environments in Europe and Africa, perhaps as a result of a retained ancestral characteristic; significant differences in body breadth within the Eastern region would therefore be of special interest.

A brief note should be made about the use of bi-iliac breadth in this study. Auerbach and Ruff (2004) argued that body mass estimated from bi-iliac breadth and stature is potentially more precise and accurate than estimations utilizing the femoral head. However, as also shown in that study, femoral head body mass estimations have good comparability to bi-iliac breadth/stature estimations. As pelvic breadth shows stability over time in the Americas, it stands to reason that any variation observed in body mass – assuming little change in body breadth – would be driven more by changes in stature. Any use of the bi-iliac breadth/stature body mass estimations would inherently reflect changes in stature, and would therefore tautologically correlate with changes in stature. Therefore, femoral head diameters are used for body mass estimation to avoid this circular reasoning.

9.2 MATERIALS AND METHODS

9.2.1 Sample and Measurements

Two large samples were measured for use in this study to yield a total of 1161 skeletons (609 males, 552 females). Data were collected to form a primary sample for examining temporal and subsistence economy changes within a circumscribed geographical area in the Southeastern subregion of the Eastern region. These constitute 15 sites (including 2 sets of geographically and temporally proximate sites from modern Arkansas and Louisiana) that broadly represent the Middle to Late Archaic periods and the Middle to Late pre-contact Mississippian period. The sampled sites are listed in Table 9.1 and their locations are depicted

Table 9.1 Skeletal sample. Shaded cells designate comparative samples from the Prairie and the Southwest

Site name (Reference #)[a]	N (males/ females)	General subsistence[b]	Mean temporal depth[c]	Source[d]
Windover (1)	74 (44/30)	Forager	7000 BP	FSU
Palmer (2)*	39 (19/20)	Forager	2000 BP	FLMNH
Bayshore (3)*	23 (10/13)	Forager	2000 BP	FLMNH
Indian Knoll (4)	61 (31/30)	Forager	4800 BP	WOAC
Eva (5)	32 (19/13)	Forager	6000 BP	MM
Cherry (6)	20 (15/5)	Forager	3500 BP	MM
Ledbetter Landing (7)	17 (13/4)	Forager	3500 BP	MM
St Francis and Black River Sites (8)[e]	22 (11/11)	Agricultural	500 BP	NMNH
Ouachita River Sites (9)[e]	30 (15/15)	Agricultural	400 BP	NMNH
Irene Mound (10)	32 (13/19)	Agricultural	500 BP	NMNH
Averbuch (11)	55 (27/28)	Agricultural	650 BP	UTK
Hiwassee Island (12)	40 (20/20)	Agricultural	400 BP	MM
Ledford Island (13)	47 (24/23)	Agricultural	400 BP	MM
Thompson Village (14)	26 (13/13)	Agricultural	700 BP	MM
Toqua (15)	37 (18/19)	Agricultural	450 BP	MM
EASTERN REGION SUBTOTAL	555 (292/263)			
Canyon del Muerto	30 (18/12)	Agricultural	750 BP (?)	AMNH
Carter Ranch	16 (9/7)	Agricultural	750 BP	FMNH
Grasshopper	48 (27/21)	Agricultural	600 BP	ASM
Ma'ip'ovi	25 (13/12)	Agricultural	600 BP	ASM
Point of Pines	18 (9/9)	Agricultural	600 BP	ASM
Pueblo Bonito	26 (9/17)	Agricultural	1000 BP	NMNH
Paa-Ko	42 (21/21)	Agricultural	600 BP	SDMM
Hawikuh	75 (30/45)	Agricultural	400 BP	NMNH
Pottery Mound	43 (25/18)	Agricultural	550 BP	UNM
Puye	40 (17/23)	Agricultural	550 BP	NMNH
Glen Canyon Sites	48 (29/19)	Agricultural	700 BP	NMNH/UMNH
Pecos Pueblo	60 (30/30)	Agricultural	400 BP	Christopher Ruff
Dickson Mound	53 (26/27)	Agricultural	650 BP	ISM

Table 9.1 (*Continued*)

Site name (Reference #)[a]	N (males/ females)	General subsistence[b]	Mean temporal depth[c]	Source[d]
Kuhlman Mound	14 (8/6)	Horticultural	1100 BP	ISM
Middle Woodland Hopewell Sites[e]	68 (46/22)	Horticultural	1500 BP	NMNH/ISM
TOTAL	1161 (609/552)			

[a] Reference numbers refer to site locations indicated in Figure 9.1. Most sites are temporally and geographically constrained, with the notable exceptions of Windover, Indian Knoll and Eva, which demonstrate occupations that range more than a millennium. All Eva burials were taken from the Benton phase.

[b] As noted in the text, no subsistence category adequately represents the variation amongst locations within each economy. These are broad categories provided for general reference and used in some analyses.

[c] The average site antiquity is based on direct dating of sites (in the majority of cases) as reported in site reports or peer-reviewed literature. In cases where no direct dates were reported, mean dates for archaeological traditions and variants for sites are provided.

[d] **AMNH**, American Museum of Natural History, New York City, New York; **ASM**, Arizona State Museum, Tucson, Arizona; **FLMNH**, Florida Museum of Natural History, Gainesville, Florida; **FMNH**, Field Museum of Natural History, Chicago, Illinois; **FSU**, Florida State University Department of Anthropology, Tallahassee, Florida; **ISM**, Illinois State Museum, Springfield, Illinois; **MM**, Frank H. McClung Museum, Knoxville, Tennessee; **NMNH**, National Museum of Natural History (Smithsonian Institution), Washington, D.C.; **SDMM**, San Diego Museum of Man, San Diego, California; **UMNH**, Utah Museum of Natural History, Salt Lake City, Utah; **UNM**, University of New Mexico, Albuquerque, New Mexico; **UTK**, University of Tennessee – Knoxville Department of Anthropology, Knoxville, Tennessee; **WOAC**, Webb Osteology and Archaeology Collection, Lexington, Kentucky.

[e] Three sites listed are amalgamations of temporally and geographically similar, smaller sites: Caddo Mississippian sites from Arkansas (along the St Francis and Black rivers) and Louisiana (along the Ouachita River) and the Middle Woodland Hopewell sites from along the Illinois River.

*Sites combined in analyses, as they are temporally and geographically in the same location, and share the same archaeological tradition.

in Figure 9.1. Sites dating to the Woodland period – during which time agricultural practices were developing and local food production increased (see above) – were purposefully excluded from this primary sample in order to emphasize the long-term changes to physique that were incurred by agriculturalists. In addition to these Southeastern samples, two comparative samples were measured from the Southwest and from the Illinois River Valley in the western Prairie; these samples are listed in Table 9.1 and designated in the shaded rows. All samples were measured by the author with the exception of the sample from Pecos Pueblo, which were measured by Christopher Ruff, who has generously shared his data with the author.

Table 9.1 also provides the general subsistence category for each site. Subsistence categories were assigned based on archaeological site reports or published data, which may be referenced in Appendix I of Auerbach (2007). Sites that show evidence for incipient horticulture or low-level food production are included in the 'forager' category. All sites in which intensified agriculture and year-round occupation are evident are categorized as 'agricultural'. In the case of the Hopewell sites from the western Prairie, these sites are listed as 'horticultural', given their intermediate food production compared with the foragers and agriculturalists. As explained in the introduction, broad subsistence categories do not adequately demonstrate the variations in food choice, procurement methods or preparation practised amongst populations. For example, Mississippian period sites in Florida do not demonstrate the high percentage of maize consumption associated with contemporary sites

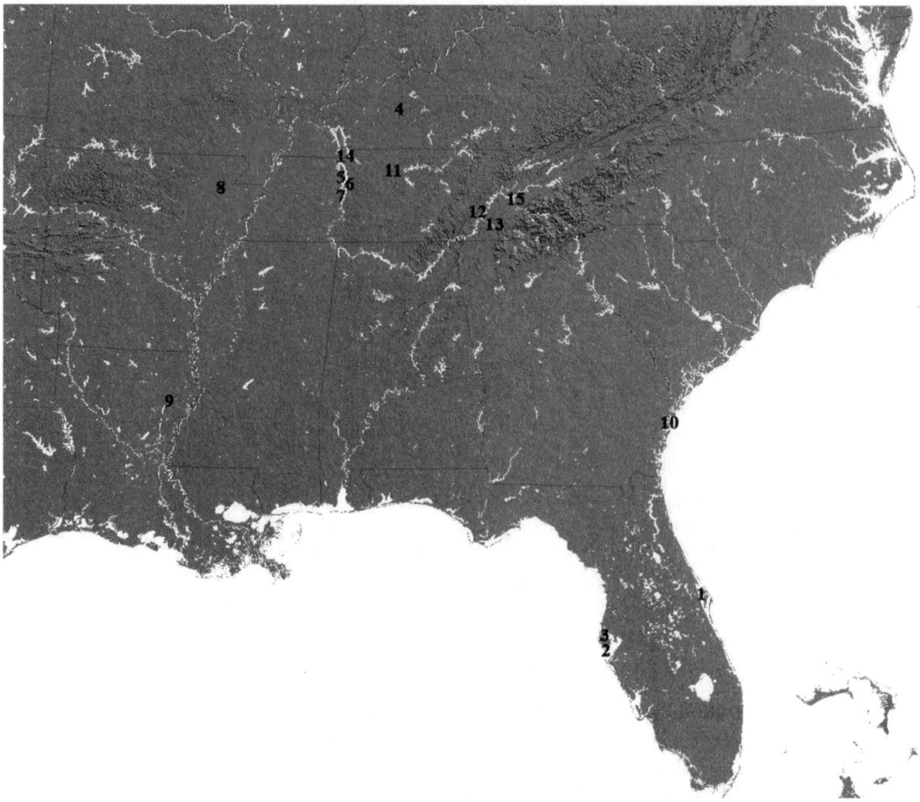

Figure 9.1 Southeastern sites sampled in this study. Numbers correspond with site names listed in Table 9.1. Note that the St Francis and Black River sites (8) and Ouachita River sites (9) are both combinations of multiple cemeteries, and so the locations designated on the map are geographical centres for these locations

found in the Georgia Bight (Hutchinson *et al.*, 1998). Comparisons therefore were made amongst sites listed within each subsistence category for the primary and comparative data, in addition to analyses between subsistence categories within the Southeastern subregion (see below).

All skeletons included in the study were ascertained to be adults (determined by epiphyseal fusion on all vertebral elements). The sexing and ageing of skeletons were determined using methods previously described for this dataset (Auerbach, 2007; Auerbach and Ruff, 2010). Individuals of indeterminate sex were excluded from analyses. Measurements were not taken from bones that were pathological or exhibited trauma.

The minimum criteria for the inclusion of skeletons were the presence of one humerus, radius, femur and tibia, and these bones were measured bilaterally when possible. Maximum lengths were taken on the four limb bones, in addition to bicondylar length and anteroposterior head diameter from femora. All measurements were averaged bilaterally to minimize the effects of bilateral asymmetry (Auerbach and Ruff, 2006). Intact os coxae and sacra were present as well for the majority (68.1%) of the observed skeletons, allowing for the measure of bi-iliac breadth. It should be additionally noted that revised Fully stature estimation osteometrics (Raxter, Auerbach and Ruff, 2006) were likewise taken on many of these skeletons, which have previously been used to devise new stature estimation regression formulae (Auerbach and Ruff, 2010).

The raw measurements were used to determine a series of derived morphological dimensions. Body mass was estimated for all skeletons using the femoral head estimation equation derived by Grine *et al.* (1995), as this regression formula provides body mass estimations with the most similarity and least systematic bias when compared with body masses obtained using the stature and bi-iliac breadth morphometric method (Auerbach and Ruff, 2004; Auerbach, 2007). This result is attributable to the high mean body masses for males (63.45 kg) and females (53.01 kg) in this sample. Statures were estimated using the 'Temperate' sex-specific equations developed by Auerbach and Ruff (2010) for all samples, except for the western Prairie sites, where their 'Great Plains' equations were instead used based on recommendations by the authors. The performance of these equations for many sites in this study's sample may be observed in that publication. Bi-iliac breadth was directly measured from reconstructed os coxae and sacra, and the raw measurement was used for body breadth. Finally, brachial (radius $\div$ humerus maximum lengths $\times$ 100) and crural (tibia maximum length $\div$ femur bicondylar length $\times$ 100) indices were calculated.

9.2.2 Statistics

Raw dimensions were compared with each other before examining variation in the derived morphological dimensions. Pearson's correlation coefficients for geometric mean-scaled measurements regressed against the geometric mean of long bone lengths and femoral head size were used to evaluate allometry in skeletal dimensions (Holliday, 1995), both in the total sample and within subsistence groups, by sex. In addition, the relationship of femoral head size and femoral bicondylar length was assessed by ordinary least squares (OLS) regression of natural log-transformed values of these dimensions. These measurements – especially the latter – have often been used in previous studies as proxies for body mass and stature, respectively, without the estimation of either dimension from the raw measurements.

Comparisons of body mass and stature were first conducted amongst samples from Southeastern sites within the 'forager' and 'agricultural' broad subsistence categories. Bi-iliac breadth and intralimb indices were also compared amongst sites within these categories to determine whether significant patterns in these dimensions occurred within the geographically circumscribed area of the Southeast subregion. As intralimb indices are ratio data, and therefore violate the assumptions of parametric statistics (Sokal and Rohlf, 1995), comparisons were made by regressing the two component bone lengths using reduced major axis (RMA) regression and conducting a Quick-Test (Tsutakawa and Hewett, 1977). Brachial index has previously been shown to be sexually dimorphic (Auerbach, 2007; Holliday and Ruff, 2001), so these intralimb RMA regressions were conducted by sex. For the three other dimensions, an ANCOVA assessed whether sex had a significant effect in these comparisons; where sex was a significant covariate, analyses were conducted amongst males and females separately. Further comparisons used ANOVAs with *post-hoc* Tukey's T' tests, which take unequal sample sizes into account. Subsequent analyses between sites by subsistence categories within the Southeastern subregion and amongst agriculturalists were conducted using similar methods.

All analyses were performed using Stata 10.1 SE and Excel 2008 for Macintosh.

9.3 RESULTS

9.3.1 General Results and Raw Measurement Comparisons

Means, standard deviations and coefficients of variation for stature, mass and bi-iliac breadth for each sample are presented in Table 9.2, by sex. Only means are provided for the intralimb (brachial and crural) indices, as these are proportions and therefore make standard deviations difficult to assess (Sokal and Rohlf, 1995). The shortest and least massive males and females are generally both found amongst the Southwestern sites, though the narrowest males and females are notably found in Southeastern sites (e.g. Windover and Indian Knoll). Almost all of the sampled sites have moderate to high brachial (>77, males; >76, females) and crural (>83.5, both sexes) indices, which is expected given the warm temperate environments sampled.

Coefficients of variation (CV) are calculated from the means and standard deviations for these dimensions within each site, and are also presented in Table 9.2. A brief examination of the CVs reveals body mass to generally demonstrate much greater variation compared with stature or bi-iliac breadth; often body mass has two to three times the value of the other two dimensions. It is interesting that these latter two dimensions, which are linked in determining body mass (Ruff, 1994), also have considerably less variation overall when compared with estimated body masses. Males amongst the Southeastern subregion also tend to have higher variation in all three of these dimensions than females within the same site; Windover, Indian Knoll and the Ouchita River sites are exceptions to this pattern. Sites sampled from the Southwest and from the western Prairie demonstrate many more exceptions to this trend.

A MANCOVA examining broad subsistence categories and sex within the Southeastern subregion demonstrates no significant interaction term for these categories for stature ($F=0.001$, $p=0.97$), body mass ($F=2.727$, $p=0.10$) or bi-iliac breadth ($F=3.403$, $p=0.07$). Main effects of subsistence and sex, however, are significant for all of these dimensions ($p < 0.01$), and so all analyses are conducted by sex in comparing variation between the Archaic 'forager' and Mississippian 'agricultural' samples. A Quick-Test on an RMA regression of humerus length on radius length is significant between the males and females, but the results of femoral length against tibial length are not significant. Therefore, sexes may be pooled in examining crural index, but not for brachial index.

An examination of allometry in raw dimensions reveals that most measurements do not exhibit scaling with body size. As shown in previous studies (Holliday, 1997), femoral head size exhibits slight positive allometry with overall body size, both amongst males ($r=0.285$; $p < 0.01$) and females ($r=0.168$; $p < 0.01$). This pattern is found both in analyses of the total sample, as well as analyses within each broad subsistence category and within regions. Interestingly, there is no allometry observed in tibial length, as has been reported elsewhere using different samples (Jantz and Jantz, 1999; Sylvester, Kramer and Jungers, 2008). This suggests that any variation observed in crural indices is not related to samples with tall mean statures. Indeed, none of the limbs demonstrate allometry amongst sex, regional and subsistence divisions, with one notable exception: humeral length demonstrates negative allometry amongst Southwestern males ($r=-0.236$, $p < 0.01$), but not Southwestern females ($r=-0.087$, $p=0.21$) or members of either sex from Eastern region samples.

The overall comparison of femoral head diameter and femoral bicondylar length reveals a significant correlation between the two dimensions. Amongst males, the regressions yield a Pearson's correlation coefficient of 0.726 and a slope of 0.836, which is significantly less than a slope of 1 and corroborates the allometry analysis results. Female regressions have similar

Table 9.2 Descriptive statistics for derived morphologies used in analyses

Site name	Region	Dimension[a]	Mean	Standard Deviation	Coefficient of Variation[b]
Males					
Windover	Southeastern subregion	Stature	165.00	5.21	3.16
		Body mass	65.00	5.40	8.31
		Bi-iliac breadth	265.42	10.93	4.12
		Brachial index	79.47		
		Crural index	85.50		
Palmer and Bayshore	Southeastern subregion	Stature	162.54	5.52	3.40
		Body mass	64.79	4.25	6.56
		Bi-iliac breadth	267.50	14.11	5.27
		Brachial index	76.16		
		Crural index	84.50		
Indian Knoll	Southeastern subregion	Stature	160.37	6.72	4.19
		Body mass	60.32	4.80	7.96
		Bi-iliac breadth	258.68	10.80	4.18
		Brachial index	77.14		
		Crural index	84.51		
Eva	Southeastern subregion	Stature	161.23	6.71	4.16
		Body mass	61.90	6.01	9.71
		Bi-iliac breadth	252.67	11.71	4.63
		Brachial index	78.98		
		Crural index	84.06		
Cherry	Southeastern subregion	Stature	162.73	7.28	4.47
		Body mass	62.24	4.84	7.78
		Bi-iliac breadth	255.58	11.01	4.31
		Brachial index	77.90		
		Crural index	85.18		
Ledbetter Landing	Southeastern subregion	Stature	164.51	4.30	2.61
		Body mass	64.57	3.60	5.58
		Bi-iliac breadth	275.50	11.30	4.10
		Brachial index	77.12		
		Crural index	85.56		
St Francis and Black River Sites	Southeastern subregion	Stature	170.69	6.04	3.54
		Body mass	69.96	6.80	9.72
		Bi-iliac breadth	284.32	16.18	5.69
		Brachial index	77.55		
		Crural index	85.05		
Ouachita River Sites	Southeastern subregion	Stature	163.44	7.37	4.51
		Body mass	68.73	4.19	6.10
		Bi-iliac breadth	273.69	18.61	6.80

(*continued*)

Table 9.2 (*Continued*)

Site name	Region	Dimension[a]	Mean	Standard Deviation	Coefficient of Variation[b]
		Brachial index	78.40		
		Crural index	83.73		
Irene Mound	Southeastern subregion	Stature	165.20	6.88	4.16
		Body mass	64.29	8.42	13.10
		Bi-iliac breadth	268.48	17.42	6.49
		Brachial index	78.16		
		Crural index	85.55		
Averbuch	Southeastern subregion	Stature	165.28	8.10	4.90
		Body mass	68.09	5.51	8.09
		Bi-iliac breadth	275.40	15.94	5.79
		Brachial index	78.04		
		Crural index	84.03		
Hiwassee Island	Southeastern subregion	Stature	165.11	7.08	4.29
		Body mass	66.40	5.35	8.06
		Bi-iliac breadth	271.90	13.78	5.07
		Brachial index	77.90		
		Crural index	83.87		
Ledford Island	Southeastern subregion	Stature	166.13	4.41	2.65
		Body mass	63.86	4.47	7.00
		Bi-iliac breadth	269.73	12.95	4.80
		Brachial index	78.15		
		Crural index	84.33		
Thompson Village	Southeastern subregion	Stature	162.48	4.77	2.94
		Body mass	65.59	5.87	8.95
		Bi-iliac breadth	274.57	12.84	4.68
		Brachial index	76.32		
		Crural index	84.14		
Toqua	Southeastern subregion	Stature	163.83	5.06	3.09
		Body mass	64.57	6.42	9.94
		Bi-iliac breadth	271.91	12.94	4.76
		Brachial index	76.06		
		Crural index	83.61		
Canyon del Muerto	Southwest	Stature	161.43	5.62	3.48
		Body mass	62.79	5.70	9.08
		Bi-iliac breadth	265.59	16.35	9.87
		Brachial index	79.21		
		Crural index	86.67		
Carter Ranch	Southwest	Stature	157.85	10.23	6.48
		Body mass	63.99	3.83	5.99
		Bi-iliac breadth	270.64	12.56	4.64

Table 9.2 (*Continued*)

Site name	Region	Dimension[a]	Mean	Standard Deviation	Coefficient of Variation[b]
		Brachial index	78.55		
		Crural index	86.60		
Grasshopper	Southwest	Stature	162.57	6.20	3.81
		Body mass	61.49	5.82	9.46
		Bi-iliac breadth	264.69	9.97	3.77
		Brachial index	78.09		
		Crural index	86.30		
Ma'ip'ovi	Southwest	Stature	162.30	8.13	5.01
		Body mass	60.14	7.31	12.15
		Bi-iliac breadth	269.00	12.73	4.73
		Brachial index	78.76		
		Crural index	86.19		
Point of Pines	Southwest	Stature	161.94	5.91	3.65
		Body mass	60.77	5.72	9.41
		Bi-iliac breadth	258.71	13.99	5.41
		Brachial index	77.69		
		Crural index	86.17		
Pueblo Bonito	Southwest	Stature	160.05	8.63	5.39
		Body mass	61.77	7.21	11.67
		Bi-iliac breadth	272.50	13.22	4.85
		Brachial index	77.65		
		Crural index	85.07		
Paa-Ko	Southwest	Stature	157.96	9.72	6.15
		Body mass	62.08	5.00	8.05
		Bi-iliac breadth	267.89	14.01	5.23
		Brachial index	77.26		
		Crural index	85.25		
Hawikuh	Southwest	Stature	158.78	7.19	4.53
		Body mass	59.08	4.66	7.89
		Bi-iliac breadth	263.96	10.74	4.07
		Brachial index	77.92		
		Crural index	84.89		
Pottery Mound	Southwest	Stature	157.97	5.90	3.73
		Body mass	59.59	4.65	4.80
		Bi-iliac breadth	263.53	9.95	3.78
		Brachial index	78.21		
		Crural index	85.13		
Puye	Southwest	Stature	155.17	6.67	4.30
		Body mass	57.29	6.11	10.67
		Bi-iliac breadth	264.82	16.57	6.26

(*continued*)

Table 9.2 (*Continued*)

Site name	Region	Dimension[a]	Mean	Standard Deviation	Coefficient of Variation[b]
		Brachial index	77.06		
		Crural index	84.54		
Glen Canyon Sites	Southwest	Stature	163.89	3.92	2.39
		Body mass	63.48	4.48	7.06
		Bi-iliac breadth	266.98	11.97	4.48
		Brachial index	79.17		
		Crural index	86.81		
Pecos Pueblo	Southwest	Stature	157.07	7.27	4.63
		Body mass	62.69	4.98	7.94
		Bi-iliac breadth	262.78	11.08	4.22
		Brachial index	77.19		
		Crural index	85.31		
Dickson Mound	Western Prarie subregion	Stature	167.62	6.67	3.98
		Body mass	67.87	4.71	6.94
		Bi-iliac breadth	283.00	14.82	5.24
		Brachial index	78.05		
		Crural index	84.43		
Kuhlman Mound	Western Prarie subregion	Stature	162.82	8.71	5.36
		Body mass	63.27	3.58	5.66
		Bi-iliac breadth	270.25	8.87	3.28
		Brachial index	77.54		
		Crural index	85.77		
Middle Woodland Hopewell Sites	Western Prarie subregion	Stature	164.47	7.86	4.78
		Body mass	65.30	5.65	8.65
		Bi-iliac breadth	274.24	15.22	5.55
		Brachial index	78.42		
		Crural index	85.49		
Females					
Windover	Southeastern subregion	Stature	155.82	6.28	4.03
		Body mass	54.49	4.62	8.48
		Bi-iliac breadth	250.94	17.77	7.08
		Brachial index	77.59		
		Crural index	85.23		
Palmer and Bayshore	Southeastern subregion	Stature	157.27	4.55	2.89
		Body mass	57.41	3.28	5.71
		Bi-iliac breadth	271.08	14.48	5.34
		Brachial index	75.73		
		Crural index	84.69		
Indian Knoll	Southeastern subregion	Stature	151.93	6.17	4.06
		Body mass	50.28	4.24	8.43

Table 9.2 (*Continued*)

Site name	Region	Dimension[a]	Mean	Standard Deviation	Coefficient of Variation[b]
		Bi-iliac breadth	249.84	11.67	4.67
		Brachial index	75.00		
		Crural index	83.93		
Eva	Southeastern subregion	Stature	152.60	6.15	4.03
		Body mass	52.72	4.15	7.87
		Bi-iliac breadth	259.40	11.48	4.43
		Brachial index	76.99		
		Crural index	84.27		
Cherry	Southeastern subregion	Stature	152.27	4.58	3.01
		Body mass	50.74	2.90	5.72
		Bi-iliac breadth	247.83	5.06	2.04
		Brachial index	75.45		
		Crural index	83.37		
Ledbetter Landing	Southeastern subregion	Stature	154.46	2.58	1.67
		Body mass	55.35	1.77	3.20
		Bi-iliac breadth	259.25	2.35	0.91
		Brachial index	76.53		
		Crural index	85.21		
St Francis and Black River Sites	Southeastern subregion	Stature	158.44	3.52	2.22
		Body mass	57.13	3.79	6.63
		Bi-iliac breadth	266.75	9.77	3.66
		Brachial index	76.92		
		Crural index	83.93		
Ouachita River Sites	Southeastern subregion	Stature	163.03	6.26	3.84
		Body mass	56.73	6.43	11.33
		Bi-iliac breadth	266.69	9.92	3.72
		Brachial index	76.30		
		Crural index	84.41		
Irene Mound	Southeastern subregion	Stature	155.58	5.08	3.27
		Body mass	53.15	5.73	10.78
		Bi-iliac breadth	253.12	9.56	3.78
		Brachial index	76.16		
		Crural index	84.22		
Averbuch	Southeastern subregion	Stature	158.41	5.40	3.41
		Body mass	56.62	6.12	10.81
		Bi-iliac breadth	266.11	15.05	5.66
		Brachial index	76.11		
		Crural index	83.41		
Hiwassee Island	Southeastern subregion	Stature	155.53	4.77	3.07
		Body mass	54.71	3.99	7.29

(*continued*)

Table 9.2 (*Continued*)

Site name	Region	Dimension[a]	Mean	Standard Deviation	Coefficient of Variation[b]
		Bi-iliac breadth	264.82	13.56	5.12
		Brachial index	76.08		
		Crural index	83.01		
Ledford Island	Southeastern subregion	Stature	153.92	3.21	2.09
		Body mass	53.30	4.06	7.62
		Bi-iliac breadth	259.32	14.90	5.75
		Brachial index	76.17		
		Crural index	83.92		
Thompson Village	Southeastern subregion	Stature	152.18	5.13	3.37
		Body mass	53.07	3.63	6.84
		Bi-iliac breadth	255.00	13.17	5.16
		Brachial index	75.89		
		Crural index	83.69		
Toqua	Southeastern subregion	Stature	156.12	6.33	4.05
		Body mass	55.46	4.84	8.73
		Bi-iliac breadth	261.54	16.22	6.20
		Brachial index	75.93		
		Crural index	83.02		
Canyon del Muerto	Southwest	Stature	149.99	4.26	2.84
		Body mass	51.57	2.89	5.60
		Bi-iliac breadth	262.54	11.80	4.49
		Brachial index	78.08		
		Crural index	85.66		
Carter Ranch	Southwest	Stature	152.36	9.67	6.35
		Body mass	47.57	5.00	10.51
		Bi-iliac breadth	254.33	19.35	7.61
		Brachial index	76.92		
		Crural index	83.77		
Grasshopper	Southwest	Stature	151.56	5.72	3.77
		Body mass	52.52	4.38	8.34
		Bi-iliac breadth	256.63	16.99	6.62
		Brachial index	77.23		
		Crural index	85.28		
Ma'ip'ovi	Southwest	Stature	151.16	4.17	2.76
		Body mass	51.89	3.63	7.00
		Bi-iliac breadth	262.17	20.18	7.70
		Brachial index	77.50		
		Crural index	86.22		
Point of Pines	Southwest	Stature	150.06	5.60	3.73
		Body mass	51.78	4.49	8.67
		Bi-iliac breadth	253.75	15.65	6.17

Table 9.2 (*Continued*)

Site name	Region	Dimension[a]	Mean	Standard Deviation	Coefficient of Variation[b]
		Brachial index	77.87		
		Crural index	85.86		
Pueblo Bonito	Southwest	Stature	156.49	6.55	4.19
		Body mass	49.28	4.37	8.87
		Bi-iliac breadth	260.34	15.31	5.88
		Brachial index	77.84		
		Crural index	84.13		
Paa-Ko	Southwest	Stature	152.52	5.66	3.71
		Body mass	50.38	4.14	8.22
		Bi-iliac breadth	262.29	10.45	3.98
		Brachial index	75.62		
		Crural index	83.82		
Hawikuh	Southwest	Stature	150.13	4.53	3.02
		Body mass	50.01	4.08	8.16
		Bi-iliac breadth	258.65	12.79	4.94
		Brachial index	77.12		
		Crural index	84.39		
Pottery Mound	Southwest	Stature	151.24	4.87	3.22
		Body mass	50.87	2.99	5.88
		Bi-iliac breadth	257.23	12.20	4.74
		Brachial index	77.16		
		Crural index	84.46		
Puye	Southwest	Stature	149.21	4.98	3.34
		Body mass	48.03	3.34	6.95
		Bi-iliac breadth	251.98	10.54	4.18
		Brachial index	76.16		
		Crural index	84.43		
Glen Canyon Sites	Southwest	Stature	153.51	3.24	2.11
		Body mass	51.73	3.83	7.40
		Bi-iliac breadth	259.44	13.16	5.07
		Brachial index	77.55		
		Crural index	85.43		
Pecos Pueblo	Southwest	Stature	154.35	7.33	4.75
		Body mass	51.59	4.80	9.30
		Bi-iliac breadth	259.88	11.93	4.59
		Brachial index	75.30		
		Crural index	84.22		
Dickson Mound	Western Prarie subregion	Stature	159.97	5.67	3.54
		Body mass	57.76	4.73	8.19
		Bi-iliac breadth	268.20	12.52	4.67

(*continued*)

Table 9.2 (*Continued*)

Site name	Region	Dimension[a]	Mean	Standard Deviation	Coefficient of Variation[b]
		Brachial index	76.56		
		Crural index	84.33		
Kuhlman Mound	Western Prarie subregion	Stature	161.97	7.26	4.48
		Body mass	56.29	6.63	11.78
		Bi-iliac breadth	256.17	10.07	3.93
		Brachial index	75.80		
		Crural index	84.47		
Middle Woodland Hopewell Sites	Western Prarie subregion	Stature	159.92	6.70	4.19
		Body mass	56.02	6.45	11.51
		Bi-iliac breadth	266.11	11.55	4.34
		Brachial index	76.57		
		Crural index	84.74		

[a] Units: stature = centimetres; body mass = kilograms; bi-iliac breadth = millimetres.
[b] Standard deviation ÷ mean × 100.

results, with a Pearson's *r* of 0.703, and a slope of 0.788. These results are especially interesting in light of correlations between stature and body mass within each sex for the same samples. The correlation between body mass and stature amongst males ($r = 0.382$) and females ($r = 0.312$) are substantially lower than the raw dimensions used to calculate these derived dimensions.

9.3.2 Comparisons within Subsistence Categories in the Southeast

Both amongst males and females, body mass, bi-iliac breadth and brachial indices significantly differ amongst the samples from forager sites in the Southeast. ANOVA results indicate that males from Indian Knoll, Eva and Cherry sites are significantly less massive ($F = 3.351$, $p < 0.01$) and had narrower bi-iliac breadths ($F = 4.688$, $p < 0.01$) than the samples from the Florida sites (Windover and Palmer/Bayshore), as well as the western Tennessee Ledbetter Landing site. Contrastingly, a Quick-Test indicates that the Windover, Cherry and Eva sites' males have significantly higher brachial indices than the other sites from the Southeast. The same pattern is found in comparisons of body mass amongst females ($F = 7.676$, $p < 0.01$), but not in bi-iliac breadth, where the females at Cherry, Indian Knoll and Windover are significantly ($F = 4.445$, $p < 0.01$) narrower than females from other samples. Again, Windover and Eva, but not Cherry, have significantly higher brachial indices. Thus, despite the temporal range represented by the forager samples, the patterns of variation amongst them do not suggest diachronic change. However, with the notable exception of Ledbetter Landing, the Florida site males are the most massive and widest-bodied, and so there may be a slight geographic pattern amongst males but not females.

Interestingly, the agriculturalist samples do not demonstrate any significant differences in any dimensions, except amongst female statures. ANOVA results demonstrate that females from the more western sites – Averbuch, Saint Francis and Black River sites, and Ouachita

River sites – are significantly taller than any of the more eastern sites from the Southeastern subregion. Males demonstrate the same pattern, though it does not reach statistical significance ($F = 1.615$, $p = 0.14$).

9.3.3 Comparisons between Southeastern 'Foragers' and 'Agriculturalists'

Despite the overall lack of significant variation amongst the agricultural samples from the Southeast, given the heterogeneity amongst the forager groups, analyses comparing subsistence groups are conducted amongst sites instead of treating all sites within a subsistence category together in a combined sample. This also minimizes the potential swamping effects of the largest samples within each broad category (e.g. Windover vs. Cherry). Given the unknown variation in subsistence practices within each broad group, this approach may also be the most archaeologically (in addition to statistically) conservative. Box plots comparing stature and body mass amongst the Southeast sampled sites are presented in Figures 9.2 and 9.3, respectively.

Examining results for males, a comparison of the morphological dimensions amongst all Southeastern subregion sites distinguishes many of the foragers from the agriculturalists. The males with the highest mean statures are found amongst the agriculturalists (Figure 9.2); St Francis and Black River sites, Ledford Island, Averbuch, Irene Mound and Hiwassee are all significantly taller ($F = 2.274$, $p < 0.01$) than all of the forager site samples. Body masses demonstrate a more mixed variation amongst sites. The St Francis and Black River sites, Averbuch and Hiwassee are, on average, amongst the most massive males. This parallels the

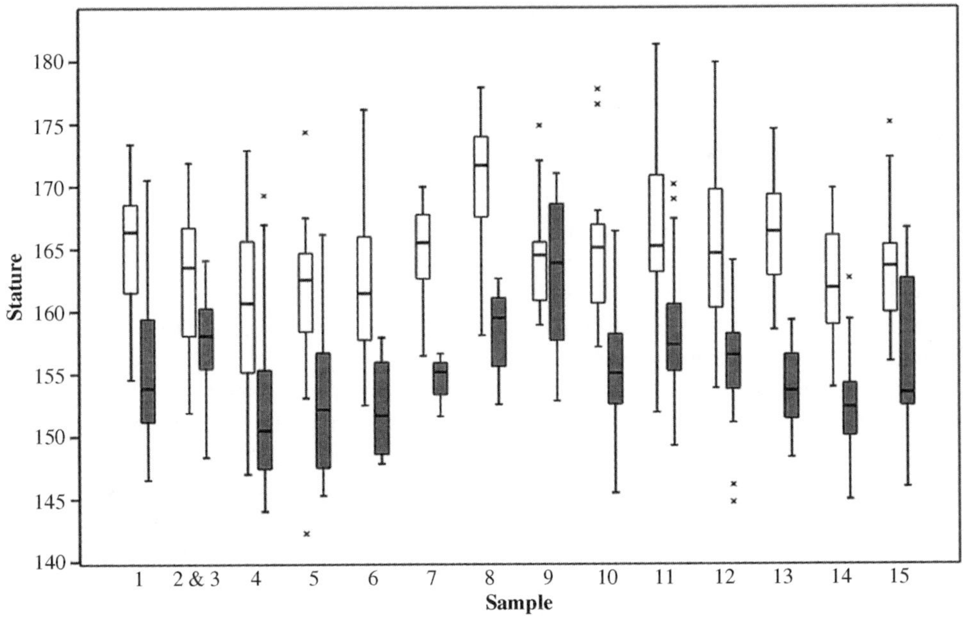

Figure 9.2 Box plots for stature amongst Southeastern sites. Numbers correspond to site names listed in Table 9.1: 1–7, foragers; 8–15, agriculturalists. White boxes, males; grey boxes, females

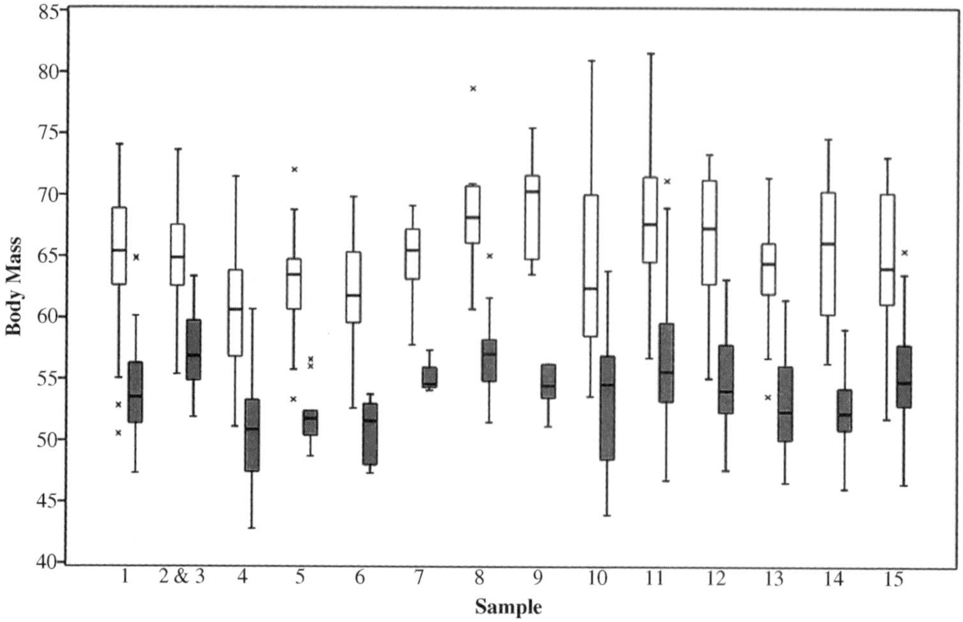

Figure 9.3 Box plots for body mass amongst Southeastern sites. Numbers correspond to site names listed in Table 9.1: 1–7, foragers; 8–15, agriculturalists. White boxes, males; grey boxes, females

high statures from these sites. In addition, males from the Ouachita River sites and Thompson Village samples, who were not significantly different from most of the foragers in statures, are also more massive than all of the foragers, except Windover ($F = 4.311$, $p < 0.01$) (Figure 9.3). Contrastingly, the males from Irene Mound, who are the fourth tallest group, are significantly less massive than all of the other agriculturalists. It is interesting that bi-iliac breadth most clearly separates out the foragers from the agriculturalists ($F = 5.010$, $p < 0.01$), where all of the forager samples are significantly narrower *except* for Ledbetter Landing. The males from the Late Archaic Ledbetter Landing site have pelves with widths second only to the St Francis and Black River sites. Importantly, no significant difference exists in crural indices amongst any of the male samples, and the only significant difference in brachial indices is found in comparisons of the Windover site and all other samples.

Like males, female samples from sites of the two subsistence categories are generally distinguished in stature, body mass and bi-iliac breadth. No significant differences exist in intralimb indices amongst the females. The tallest females are amongst the agriculturalist sites, except that, unlike the males, Ledford Island females (along with Thompson Village) are significantly shorter than the other agricultural sites, and not significantly different from most of the foragers (except Windover) ($F = 4.084$, $p < 0.01$). Body masses again do not precisely match the results for stature comparisons, though the western agriculturalist sites are more massive overall than the eastern agriculturalists (with the notably exceptional massive Palmer forager females). Similar to the males, the majority of the female forager female samples are significantly less massive, except for the noted Palmer females, as well as Ledbetter Landing. Both of these sites' females, along with Eva, distinguish themselves from the remainder of the foragers in bi-iliac breadth as well, with significantly wider body breadths than Cherry,

Indian Knoll or Windover. The agriculturalist females are collectively wider than these latter forager sites.

One caveat is that variation in body mass reported above may largely be driven by the observed variation in bi-iliac breadth. Indeed, Pearson's correlations for bi-iliac breadth and body mass amongst all of samples for the Southeast are significant (males, $r = 0.623$; females, $r = 0.552$) and considerably higher than correlations between body mass and stature (males, $r = 0.320$; females, $r = 0.229$). ANCOVAs comparing body mass amongst the sampled sites by sex, using bi-iliac breadth as a covariate, furthermore result in non-significant variation amongst sites (males, $F = 1.736$, $p = 0.06$; females, $F = 1.423$, $p = 0.15$).

Finally, sexual dimorphism for stature and body mass are reported for the Southeastern sites in Table 9.3. It is apparent that, in general, sexual dimorphism remained unchanged or even slightly increased amongst the agriculturalist samples compared to the foragers. Amongst the foragers, the Florida Late Archaic Palmer and Bayshore sites have considerably lower sexual

Table 9.3 Sexual dimorphism for selected morphological dimensions amongst Southeastern samples

Site	Dimension	Percent sexual dimorphism[a]
Windover	Stature	5.56
	Body Mass	16.17
Palmer and Bayshore	Stature	3.24
	Body Mass	11.39
Indian Knoll	Stature	5.26
	Body Mass	16.64
Eva	Stature	5.35
	Body Mass	14.83
Cherry	Stature	6.43
	Body Mass	18.48
Ledbetter Landing	Stature	6.11
	Body Mass	14.28
St Francis and Black River sites	Stature	7.18
	Body Mass	18.34
Ouachita River sites	Stature	0.25
	Body Mass	17.46
Irene Mound	Stature	5.82
	Body Mass	17.33
Averbuch	Stature	4.16
	Body Mass	16.84
Hiwassee Island	Stature	5.80
	Body Mass	17.61
Ledford Island	Stature	7.35
	Body Mass	16.54
Thompson Village	Stature	3.91
	Body Mass	19.09
Toqua	Stature	4.71
	Body Mass	14.11

[a] (male mean – female mean) ÷ (male mean) × 100.

dimorphism, which may indicate that these populations were either more stressed or that males were not genetically predisposed to larger size. Likewise, the agricultural Ouachita River sites and Thompson Village demonstrate greatly reduced sexual dimorphism in stature, but not body mass.

9.3.4 Variation amongst Agriculturalists

In order to make comparisons easier to interpret, the agriculturalist samples within each region are combined for the following analyses. This is justified for the Southeastern subregion, as comparisons amongst the agriculturalist samples do not show any significant differences in any of the morphological dimensions under examination. An ANOVA likewise does not indicate any significant differences in any of these dimensions for males or females from the western Prairie subregion. In the Southwest, however, significant variation exists in both stature and body mass amongst the sampled sites (results not shown). These differences do not present a consistent pattern, even when including geographical location or archaeological tradition association as covariates. Thus, though there is significant variation within the Southwest, these samples are combined in order to compare total variation within the region to the other areas.

ANOVA and Quick-Test results show that the agriculturalists in the Eastern region are significantly different from Southwestern sampled sites in all dimensions. Stature and body mass comparisons are presented as boxplots for stature and body mass by sex and region in Figures 9.4 and 9.5, respectively. Males and females in the Prairie subregion have the highest

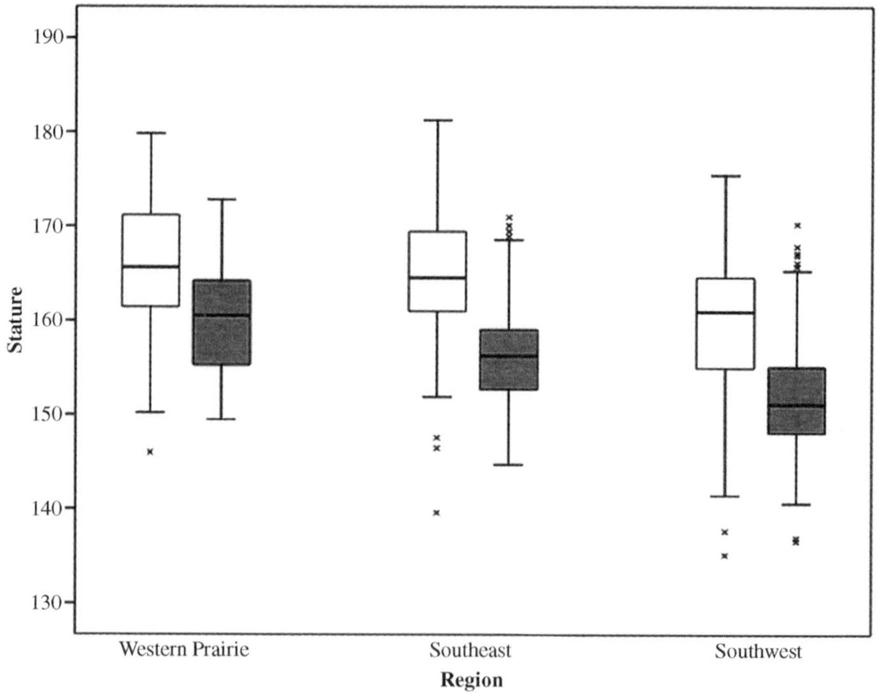

Figure 9.4 Box plots of stature amongst agriculturalists by region. White boxes, males; grey boxes, females

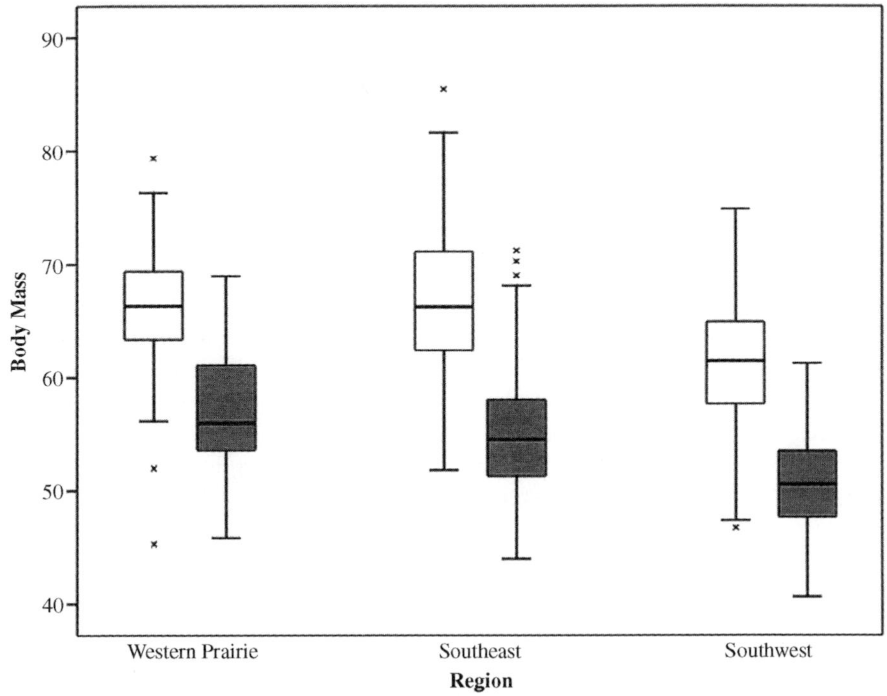

Figure 9.5 Box plots of body mass amongst agriculturalists by region. White boxes, males; grey boxes, females

mean statures, and together with the Southeastern subregion samples are significantly taller than groups from the Southwest. The same pattern occurs amongst female body masses, though Southeast males are, collectively, slightly but not significantly more massive than the Prairie males; again the Southwestern males are significantly less massive. Bi-iliac breadth also follows the same pattern as stature, and Southwest groups have the narrowest body breadths. In contrast to these results, both sexes in the Southeast have significantly lower crural indices than the Prairie or Southwest samples, which are not significantly different from each other. Brachial indices do not demonstrate significant differences amongst the regions.

9.4 DISCUSSION

Overall, the analyses indicate that significant variation is found amongst humans buried at the sampled Southeastern sites in stature, body mass and body breadth. Within the forager category, there is a significant distinction between the sites and Florida and the northern sites from modern Tennessee and Kentucky, with the exception of the Late Archaic individuals from Ledbetter Landing. The more recent Mississippian period agriculturalists from the Southeast are relatively more homogeneous, with no significant differences amongst any of the sampled sites in any of the dimensions examined.

Between the two subsistence groups in the Eastern region, there is a similar trend amongst both males and females: the agriculturalists are taller and more massive, on average. This is identical to patterns of diachronic change in stature documented using different samples from the Southeast (Boyd and Boyd, 1989; Danforth *et al.*, 2007). There is also a coincident slight increase in sexual dimorphism amongst the agriculturalist samples, accompanied by a slight increase in overall variance in stature, body mass and bi-iliac breadth (Table 9.2; Figures 9.2 and 9.3). It is noteworthy that these trends are not universal for all agriculturalist samples, and with the inclusion of the forager samples significant variation amongst the more recent sites is made more evident. Nevertheless, despite these interesting (though inconsistent) exceptions, the differences between the earlier foragers and Mississippian agriculturalists in stature parallel those in body mass. This, in turn, supports the main hypothesis stated at the end of the Introduction. In short, the long temporal perspective on the development of agriculture in the Southeast may be characterized by significant overall *increases* in body size for both males and females.

The changes in body mass may be driven more by bi-iliac breadth, as explained in the Results section above. Generally narrower body breadths of the foragers contrast markedly with the wider-bodied agriculturalists. Although bi-iliac breadth has been argued to be stable over long periods of time (Auerbach, 2007), this shift in mean body breadth may be indicative of changes correlated with subsistence economy. As crural indices do not significantly vary between these groups, despite a broad range of temporal and geographical sampling, there is some argument for genetic continuity (or at least not unmistakable signs of population replacement).

A caveat should be noted here, however. The Archaic Eva, Cherry and Indian Knoll peoples appear to be driving much of the observed significant variation in body breadth within the Southeast, both within the foragers and in comparison with the agriculturalists. These individuals are significantly narrower, especially amongst males. Yet, there is potential evidence for gene flow or population replacement into the region at the terminus of the Archaic: the individuals from Ledbetter Landing – a site occurring at the terminus of the Archaic near to Eva and Cherry – have significantly wider bodies, more like those observed amongst later Mississippians. The Archaic-Woodland transition is characterized by major climatic shifts and, as a result, likely large population movements (Anderson, Russo and Sassaman, 2007). Indeed, the temporally more recent Palmer and Bayshore samples also demonstrate some difference from earlier foragers in their higher body masses without taller statures, especially females, though the interpretation of the reasons for this change, given a lack of significantly wider pelves (which in turn means absolutely larger femoral head diameters without increases in body breadth or stature), remains equivocal.

Placed into a wider context, it is apparent that the tall statures and higher body masses observed amongst groups in the Eastern region were not occurring simultaneously in the Southwest. It is arguable that any such comparison is undermined by a large number of assumptions, primarily that Southwestern and Eastern groups shared similar genetic potentials for stature and body mass. Yet, the significantly narrower bi-iliac breadths found amongst Southwestern samples strongly implies that individuals from these sites had substantially different physiques than agriculturalists to the east: Southwestern individuals were shorter, narrower and therefore less massive than Eastern individuals, especially those from the western Prairie. Given the heterogeneity of the samples from the Southwest and the descriptive statistics from Table 9.2, it is likely that this is an over-generalization, and requires more detailed examination in future studies. Furthermore, without a comparative sample from preagricultural time periods, it cannot be ascertained if the smaller overall size amongst the Southwestern groups represents diachronic decreases or increases with agricultural intensification.

Even so, the results argue that body morphology changes are contingent on local circumstances when populations experience a shift in subsistence economy. Both Southeastern and Southwestern groups did not adopt identical forms of agriculture amongst themselves, and the planting techniques and crops utilized by Southwestern groups were not the same as those grown by populations living to the (distant) east. In addition, all of these groups supplemented their cultivars with non-domesticated flora and fauna that differed considerably by location. The great diversity in food sources and utilization was paralleled by broad differences (and occasional fluctuations) in population structure and habitation, and therefore the amounts of non-nutritional stress encountered during primary growth. In addition, coupled with differences in juvenile provisioning and the masking effects of catch-up growth, this further argues against the use of nonmetric traits associated with stress (e.g. linear enamel hypoplasia, dental caries or skeletal pathologies) as a corroborator for body size variation.

As discussed in the Introduction, multiple studies have maintained that the transition to agriculture was generally accompanied by decreases in health. Although a few papers have supported an opposite trend, the increase in stature and body mass in the Southeast – if indeed these are good indicators for overall health – is still unexpected. It is possible, as noted above, that the agriculturalist samples represent replacement populations in the Southeast compared to the Archaic foragers, or that significant gene flow into the region altered morphologies considerably with the advent of early agriculture. This cannot be supported or refuted given the available evidence presented about intralimb proportions and bi-iliac breadth, though disruptions in the archaeological record at the end of the Archaic lend some support to this possibility. Given the particularities of diachronic change between low- and high-food production within populations, it is also possible that agriculturalists experiencing severe stress or nutritional deprivation were not sampled for this study. For example, it has been well documented that the colonization of the Americas by Europeans, and the disruptions and population changes incurred with this event, led to drastically declining health amongst indigenous peoples (Hutchinson *et al.*, 2000; Larsen, 1995). It is therefore important that no post-contact sites were included in this analysis; their inclusion likely would have changed the reported patterns considerably.

One final, interesting caveat concerns the use of limb dimensions to extrapolate morphological dimensions. Body mass and stature estimations inherently incur error, though this is most often non-systematic (i.e. randomly distributed about the mean) (Auerbach and Ruff, 2004, 2010). There is no evidence to argue that the comparisons made in this chapter were inherently biased by examining estimated dimensions, rather than compare the raw skeletal measurements. Furthermore, the latter practice, as shown in the Results section above, is problematic. Possible integration between femoral head size and femoral length may lead to spurious results. Moreover, the use of femoral length alone as a proxy for stature, while acceptable in samples with identical body and limb proportions, is problematic when comparing populations with significantly different crural or cormic indices (Auerbach and Ruff, 2010), and is not advised.

9.5 CONCLUSIONS

This study has addressed many of the questions and hypotheses set out in the Introduction. Within the Eastern region, the following conclusions may be drawn about differences occurring between foragers and agriculturalists:

- Southeastern subregion foragers exhibited significant variation in body mass, body breadth and brachial indices. Stature and crural indices, however, did not significantly vary amongst the sites sampled, despite these other morphological differences and a broad temporal and geographical span in sampling.
- In contrast to the foragers, Southeastern agriculturalists did not exhibit significant variation in any morphology. However, the Southwestern agriculturalists did present significant variation in stature and body mass.
- In the Eastern region, there is a general temporal trend for humans to be more massive and taller during or after the incorporation of agriculture. This conclusion must be made with caution, however, as the genetic continuity of sampled populations in the Eastern region between the Archaic and Mississippian periods cannot be argued. Indeed, the morphological differences of the Ledbetter Landing individuals when compared with other Archaic groups could be interpreted as population replacement in parts of the region during the late Archaic.
- The observed differences in body mass appear to be driven more by differences in body breadth than stature. Generally, though, the amount of variance in stature and in bi-iliac breadth amongst all of the samples is similar.

In conclusion, the results of this study strongly caution against the 'universalizing' of patterns of morphological change occurring with the development of agriculture, as has been argued by others in recent years (Bridges, Blitz and Solano, 2000; Larsen, 1995; Rose, Marks and Tieszen, 1991). The adoption of the complex group of subsistence economies collectively termed 'agriculture' did not inherently lead to declining health, as measured by stature and by body mass. Given the confounding effects of population history and genetics, however, the relationship of these dimensions to health is not explicit. Researchers are encouraged to incorporate more morphological dimensions into their future studies of the effects of subsistence shifts on physique and health in past human populations.

ACKNOWLEDGEMENTS

I am grateful to Ron Pinhasi and Jay Stock for including this chapter in their volume, and for their useful feedback during its preparation. A special thanks is given to Christopher Ruff for generously sharing his Pecos Pueblo data, and for many beneficial conversations concerning preliminary results from this study. A number of anonymous reviewers greatly improved the arguments made in this study. I continue to be appreciative to the many institutions that have continued to grant skeletal collection access to me. A National Science Foundation Doctoral Dissertation Improvement Grant, #0550673, helped to support this research.

NOTES

1. For the purposes of this paper, the Eastern region encompasses all areas generally east of the Mississippi River, which includes parts of the Prairie, Eastern Woodlands and the Southeastern United States. See the *Handbook of North American Indians* for reference on cultural regional designations.

2. The Poverty Point tradition was one of many Archaic cultures recognized in the Eastern Woodland (Doran, 2007), and may represent the first large-scale ritual or habitation centre in North America.

REFERENCES

Anderson, D.G., Russo, M. and Sassaman, K.E. (2007) Mid-Holocene cultural dynamics in southeastern North America, in *Climate Change and Cultural Dynamics: A Global Perspective on Mid-Holocene Transitions* (eds D.G. Anderson, K.A. Maasch and D.H. Sandweiss), Academic Press, New York, pp. 457–489.

Anderson, D.G., Stahle, D.W. and Cleaveland, M.K. (1995) Paleoclimate and the potential food reserves of Mississippian societies: a case study from the Savannah River Valley. *AM. Antiquity*, **60**, 258–286.

Auerbach, B.M. (2010) Giants among us? Morphological variation and migration on the Great Plains, in *Human Variation in the Americas: the integration of archaeology and biological anthropology* (ed. B.M. Auerbach), Center for Archaeological Investigations, Carbondale, IL, pp. 172–214.

Auerbach, B.M. (2007) Human Skeletal Variation in the New World During the Holocene: Effects of Climate and Subsistence Across Geography and Time. Ph.D. Dissertation, Johns Hopkins University.

Auerbach, B.M. and Ruff, C.B. (2004) Human body mass estimation: a comparison of 'morphometric' and 'mechanical' methods. *Am. J. Phys. Anthropol.*, **125**, 331–342.

Auerbach, B.M. and Ruff, C.B. (2006) Limb bone bilateral asymmetry: variability and commonality among modern humans. *J. Hum. Evol.*, **50**, 203–218.

Auerbach, B.M. and Ruff, C.B. (2010) Stature estimation formulae for indigenous North American populations. *Am. J. Phys. Anthropol.*, **141**, 190–207.

Blake, M., Chisholm, B.S., Clark, J.E. *et al.* (1992) Prehistoric subsistence in the Soconusco region. *Curr. Anthropol.*, **33**, 83–94.

Blakely, R.L. (ed.) (1977) Biocultural Adaptation in Prehistoric America, in *Proceedings No. 11*, Southern Anthropological Society, University of Georgia, Athens, GA.

Bogin, B. (1999) *Patterns of Human Growth*, Cambridge University Press, Cambridge.

Boyd, D.C. and Boyd, C.C. (1989) A comparison of Tennessee Archaic and Mississippian maximum femoral lengths and midshaft diameters: subsistence change and postcranial variability. *Southeastern Archaeology*, **8**, 107–116.

Bridges, P.S. (1991) Skeletal evidence of changes in subsistence activities between the Archaic and Mississippian time periods in northwestern Alabama, in *What Mean These Bones? Studies in Southeastern Bioarchaeology* (eds M.L. Powell, P.S. Bridges and A.M.W. Mires), University of Alabama Press, Tuscaloosa, AL, pp. 89–101.

Bridges, P.S. (1992) Prehistoric arthritis in the Americas. *Ann. Rev. Anthropol.*, **21**, 67–91.

Bridges, P.S. (1996) Skeletal biology and behavior in ancient humans. *Evol. Anthropol.*, **4**, 112–120.

Bridges, P.S., Blitz, J.H. and Solano, M.C. (2000) Changes in long bone diaphyseal strength with horticultural intensification in west-central Illinois. *Am. J. Phys. Anthropol.*, **112**, 217–238.

Buikstra, J.E. (1991) Out of the appendix and into the dirt: comments on thirteen years of bioarchaeological research, in *What Mean These Bones? Studies in Southeastern Bioarchaeology* (eds M.L. Powell, P.S. Bridges and A.M.W. Mires), University of Alabama Press, Tuscaloosa, AL, pp. 172–188.

Buikstra, J.E., Konigsberg, L.W. and Bullington, J. (1986) Fertility and the development of agriculture in the prehistoric Midwest. *AM. Antiquity*, **51**, 528–546.

Carter, G.F. (1946) Origins of American Indian agriculture. *American Anthropologist*, **48**, 1–21.

Casas, A., Otero-Arnaiz, A., Pérez-Negrón, E. and Valiente-Banuet, A. (2007) In situ management and domestication of plants in Mesoamerica. *Annals of Botany*, **100**, 1101–1115.

Cassidy, C.M. (1984) Skeletal evidence for prehistoric subsistence adaptation in the central Ohio River Valley, in *Paleopathology at the Origins of Agriculture* (eds M.N. Cohen and G.J. Armelagos), Academic Press, New York, pp. 307–346.

Cohen, M.N. (2007) Introduction, in *Ancient Health: Skeletal Indicators of Agricultural and Economic Intensification* (eds M.N. Cohen and G.M.M. Crane-Kramer), University Press of Florida, Gainesville, pp. 1–9.

Cohen, M.N. and Armelagos, G.J. (eds) (1984a) *Paleopathology at the Origins of Agriculture*, Academic Press, New York.

Cohen, M.N. and Armelagos, G.J. (1984b) Paleopathology at the origins of agriculture: editors' summation, in *Paleopathology at the Origins of Agriculture* (eds M.N. Cohen and G.J. Armelagos), Academic Press, New York, pp. 585–601.

Cohen, M.N. and Crane-Kramer, G.M.M. (eds) (2007) *Ancient Health: Skeletal Indicators of Agricultural and Economic Intensification*, University Press of Florida, Gainesville.

Cook, D.C. (1984) Subsistence and health in the lower Illinois Valley: osteological evidence, in *Paleopathology at the Origins of Agriculture* (eds M.N. Cohen and G.J. Armelagos), Academic Press, New York, pp. 237–270.

Cook, D.C. (2007) Maize and Mississippians in the American Midwest: twenty years later, in *Ancient Health: Skeletal Indicators of Agricultural and Economic Intensification* (eds M.N. Cohen and G.M.M. Crane-Kramer), University Press of Florida, Gainesville, pp. 10–19.

Danforth, M.E., Cook, D.C. and Knick, S.G. (1994) The human remains from Carter Ranch Pueblo, Arizona: Health in isolation. *AM. Antiquity*, **59**, 88–101.

Danforth, M.E. (1999a) Nutrition and politics in prehistory. *Ann. Rev. Anthropol.*, **28**, 1–25.

Danforth, M.E. (1999b) Coming up short: Stature and nutrition among the ancient Maya of the southern lowlands, in *Reconstructing Maya Diet* (ed. C.D. White), The University of Utah Press, Salt Lake City, pp. 103–118.

Danforth, M.E., Jacobi, K.P., Wrobel, G.D. and Glassman, S. (2007) Health and the transistion to horticulture in the south-central United States, in *Ancient Health: Skeletal Indicators of Agricultural and Economic Intensification* (eds M.N. Cohen and G.M.M. Crane-Kramer), University Press of Florida, Gainesville, pp. 65–79.

DeLeon, V.B. and Auerbach, B.M. (2007) Morphological integration in human long bones. *American Journal Physical Anthropology* S44:96.

Diamond, J. and Bellwood, P. (2003) Farmers and their languages: the first expansions. *Science*, **300**, 597–603.

Diehl, M.W. (2005) Morphological observations on recently recovered early agricultural period maize cob fragments from southern Arizona. *AM. Antiquity*, **70**, 361–375.

Dillehay, T.D., Rossen, J., Andres, T.C. and Williams, D.E. (2007) Preceramic adoption of peanut, squash, and cotton in northern Peru. *Science*, **316**, 1890–1893.

Doran, G.H. (2007) A brief continental view from Windover, in *Ancient Health: Skeletal Indicators of Agricultural and Economic Intensification* (eds M.N. Cohen and G.M.M. Crane-Kramer), University Press of Florida, Gainesville, pp. 35–51.

Eveleth, P.B. and Tanner, J.M. (1990) *Worldwide Variation in Human Growth*, 2nd edn, Cambridge University Press, Cambridge.

Frishancho, A.R. (1993) *Human Adaptation and Accommodation*, University of Michigan Press, Ann Arbor.

Frishancho, A.R. and Housh, C.H. (1988) The relationship of maturity rate to body size and body proportions in children and adults. *Human Biology*, **60**, 759–770.

Fritz, G.J. (1997) A three-thousand-year-old cache of crop seeds from Marble Bluff, Arkansas, in *People, Plants, and Landscapes: Studies in Paleoethnobotany* (ed. K.J. Gremillion), University of Alabama Press, Tuscaloosa, AL, pp. 42–62.

Fritz, G.J. (2007) Keepers of Louisiana's levees: early mound builders and forest managers, in *Rethinking Agriculture: Archaeological and Ethnoarchaeological Perspectives, One World Archaeology Series No. 51* (eds T. Denham, J. Iriarte and L. Vrydaghs), Left Coast Press, Walnut Creek CA, pp. 189–209.

Gibson, J.L. (2000) *The Ancient Mounds of Poverty Point: Place of Rings*, University Press of Florida, Gainesville.

Golden, M.H. (1994) Is complete catch-up growth possible for stunted malnourished children? *Eur. J. Clin. Nutr.*, **48**, S58–S71.

Goodman, A.H., Lallo, J., Armelagos, G.J. and Rose, J.C. (1984) Health changes at Dickson Mounds, Illinois (AD 950–1300), in *Paleopathology at the Origins of Agriculture* (eds M.N. Cohen and G.J. Armelagos), Academic Press, New York, pp. 271–206.

Grine, F.E., Jungers, W.L., Tobias, P.V. and Pearson, O.M. (1995) Fossil Homo femur from Berg Aukas, northern Namibia. *Am. J. Phys. Anthropol.*, **97**, 151–185.

Hart, J.P. (1999) Maize agriculture evolution in the Eastern Woodlands of North America: a Darwinian perspective. *J. Archaeol. Meth. Theor.*, **6**, 137–179.

Hart, J.P. and Sidell, N.A. (1997) Additional evidence for early cucurbit use in the northern Eastern Woodlands of the Allegheny Front. *AM. Antiquity*, **62**, 523–537.

Holliday, T.W. (1995) Body Size and Proportions in the Late Pleistocene Western Old World and the Origins of Modern Humans Ph.D. Dissertation, University of New Mexico.

Holliday, T.W. (1997) Postcranial evidence of cold adaptation in European Neandertals. *Am. J. Phys. Anthropol.*, **104**, 245–258.

Holliday, T.W. (1999) Brachial and crural indices of European Late Upper Paleolithic and Mesolithic humans. *J. Hum. Evol.*, **36**, 549–566.

Holliday, T.W. and Ruff, C.B. (2001) Relative variation in human proximal and distal limb segment lengths. *Am. J. Phys. Anthropol.*, **116**, 26–33.

Hutchinson, D.L., Larsen, C.S., Norr, L. and Schoeninger, M.J. (2000) Agricultural melodies and alternative harmonies in Florida and Georgia, in *Biological Studies of Life in the Age of Agriculture: A View from the Southeast* (ed. P.M. Lambert), University of Alabama Press, Tuscaloosa, AL, pp. 96–115.

Hutchinson, D.L., Larsen, C.S., Schoeninger, M.J. and Norr, L. (1998) Regional variation in the pattern of maize adoption and use in Florida and Georgia. *AM. Antiquity*, **63**, 397–416.

Iriarte, J. (2007) New perspectives on plant domestication and the development of agriculture in the New World, in *Rethinking Agriculture: Archaeological and Ethnoarchaeological Perspectives* (eds T. Denham, J. Iriarte and L. Vrydaghs), One World Archaeology Series No. 51, Left Coast Press, Walnut Creek, CA, pp. 167–188.

Jantz, L.M. and Jantz, R.L. (1999) Secular change in long bone length and proportion in the United States, 1800–1970. *Am. J. Phys. Anthropol.*, **119**, 57–67.

Lambert, P.M. (ed.) (2000) *Biological Studies of Life in the Age of Agriculture: A View from the Southeast*, University of Alabama Press, Tuscaloosa.

Larsen, C.S. (1981) Skeletal and dental adaptations to the shift to agriculture on the Georgia Coast. *Curr. Anthropol.*, **22**, 422–423.

Larsen, C.S. (1995) Biological changes in human populations with agriculture. *Ann. Rev. Anthropol.*, **24**, 185–213.

Larsen, C.S., Hutchinson, D.L., Stokanowski, C.M. *et al.* (2007) Health and lifestyle in Georgia and Florida: agricultural origins and intensification in regional perspective, in *Ancient Health: Skeletal Indicators of Agricultural and Economic Intensification* (eds M.N. Cohen and G.M.M. Crane-Kramer), University Press of Florida, Gainesville, pp. 20–34.

Larsen, C.S., Shavit, R. and Griffin, M.C. (1991) Dental caries evidence for dietary change: an archaeological context, in *Advances in Dental Anthropology* (eds M.A. Kelley and C.S. Larsen), Wiley-Liss, New York, pp. 179–202.

Malina, R.M., Reyes, M.E.P., Tan, S.K. *et al.* (2004) Secular change in height, sitting height and leg length in rural Oaxaca, southern Mexico: 1968–2000. *Ann. Hum. Biol.*, **31**, 615–633.

Martorell, R. and Habicht, J.P. (1986) Growth in early childhood in developing countries, in *Human Growth: A Comprehensive Treatise*, vol. **3** (eds F. Falkner and J.M. Tanner), Plenum Press, New York, pp. 241–262.

Martorell, R., Kahn, L.K. and Schroeder, D.G. (1994) Reversibility of stunting: epidemiological findings in children from developing countries. *Eur. J. Clin. Nutr.*, **48**, S45–S57.

McLauchlan, K. (2003) Plant cultivation and forest clearance by prehistoric North Americans: pollen evidence from Fort Ancient, Ohio, USA. *Holocene*, **13**, 557–566.

Newman, M.T. (1962) Nutritional stress in man. *Am. Anthropol.*, **64**, 22–34.

Pauketat, T.R. (2004) *Ancient Cahokia and the Mississippians*, Cambridge University Press, Cambridge.

Perzigian, A.J., Tench, P.A. and Braun, D.J. (1984) Prehistoric health in the Ohio River Valley, in *Paleopathology at the Origins of Agriculture* (eds M.N. Cohen and G.J. Armelagos), Academic Press, New York, pp. 347–366.

Pickersgill, B. (2007) Domestication of plants in the Americas: insights from Mendelian and molecular genetics. *Annals of Botany*, **100**, 925–940.

Piperno, D.R. and Dillehay, T.D. (2008) Starch grains on human teeth reveal early broad crop diet in northern Peru. *PNAS*, **105**, 19622–19627.

Pohl, M.D., Pope, K.O., Jones, J.G. *et al.* (1996) Early agriculture in the Maya lowlands. *AM. Antiquity*, **7**, 355–372.

Powell, M.L., Bridges, P.S. and Mires, A.M.W. (eds) (1991) *What Mean These Bones? Studies in Southeastern Bioarchaeology*, University of Alabama Press, Tuscaloosa, Al.

Raxter, M.H., Auerbach, B.M. and Ruff, C.B. (2006) Revision of the Fully technique for estimating statures. *Am. J. Phys. Anthropol.*, **130**, 374–384.

Rose, J.C., Burnett, B.A., Nassaney, M.S. and Blauer, M.W. (1984) Paleopathology and the origins of maize agriculture in the lower Mississippi Valley and Caddoan culture areas, in *Paleopathology at the Origins of Agriculture* (eds M.N. Cohen and G.J. Armelagos), Academic Press, New York, pp. 393–424.

Rose, J.C., Marks, M.K. and Tieszen, L.L. (1991) Bioarchaeology and subsistence in the central and lower portions of the Mississippi Valley, in *What Mean These Bones? Studies in Southeastern Bioarchaeology* (eds M.L. Powell, P.S. Bridges and A.M.W. Mires), University of Alabama Press, Tuscaloosa AL pp. 7–21.

Ruff, C.B. (1991) Climate and body shape in hominid evolution. *J. Hum. Evol.*, **21**, 81–105.

Ruff, C.B. (1994) Morphological adaptation to climate in modern and fossil hominids. *Yrbk Phys. Anthropol.*, **37**, 65–107.

Ruff, C.B. (2007) Body size prediction from juvenile skeletal remains. *Am. J. Phys. Anthropol.*, **133**, 698–716.

Ruff, C.B., Larsen, C.S. and Hayes, W.C. (1984) Structural changes in the femur with the transition to agriculture on the Georgia Coast. *Am. J. Phys. Anthropol.*, **64**, 125–136.

Ruff, C.B., Scott, W.W. and Liu, A.Y.C. (1991) Articular and diaphyseal remodeling of the proximal femur with changes in body mass in adults. *Am. J. Phys. Anthropol.*, **86**, 397–413.

Russo, M. (1996) Southeastern preceramic Archaic ceremonial mounds, in *Archaeology of the Mid-Holocene Southeast* (eds K.E. Sassaman and D.G. Anderson), University Press of Florida, Gainesville, pp. 259–287.

Sassaman, K.E. (2010) The Eastern Archaic, Historicized, Alta Mira Press, New York.

Scheinsohn, V. (2003) Hunter-gatherer archaeology in South America. *Ann. Rev. Anthropol.*, **32**, 339–361.

Smith, B.D. (1992) Prehistoric plant husbandry in eastern North America, in *The Origins of Agriculture: An International Perspective* (eds C.W. Cowan and P.J. Watson), Smithsonian Institution Press, Washington, D.C., pp. 101–109.

Smith, B.D. (2001) Low-level food production. *J. Archaeol. Res.*, **9**, 1–43.

Sokal, R.R. and Rohlf, F.J. (1995) *Biometry*, 3rd edn, WH Freeman and Company, New York.

Sylvester, A.D., Kramer, P.A. and Jungers, W.L. (2008) Modern humans are not (quite) isometric. *Am. J. Phys. Anthropol.*, **137**, 371–383.

Tanner, J.M. (1986) Growth as a mirror for the conditions of society: secular trends and class distinctions, in *Human Growth: A Multidisciplinary Review* (ed. A. Demirjian), Taylor and Francis, London, pp. 3–34.

Temple, D.H., Auerbach, B.M., Nakatsukasa, M. and Larsen, C.S. (2008) Variation in limb proportions between Jomon foragers and Yayoi agriculturalists from prehistoric Japan. *Am. J. Phys. Anthropol.*, **137**, 164–174.

Tsutakawa, R.K. and Hewett, J.E. (1977) Quick test for comparing two populations with bivariate data. *Biometrics*, **33**, 215–219.

10

Evolution of Postcranial Morphology during the Agricultural Transition in Prehistoric Japan

Daniel H. Temple

Department of Anthropology, University of North Carolina at Wilmington, Wilmington, NC, USA

10.1 INTRODUCTION

The purpose of this chapter is to document and interpret patterns of variation in the postcranial skeleton of prehistoric Jomon (pre-agricultural foragers) and Yayoi (wet rice farmers) people during the agricultural transition in the Japanese Islands. In prehistoric Japan, the agricultural transition is associated with major population migrations from the East Asian continent (Brace and Nagai, 1982; Hanihara, 1991; Turner, 1992; Nakahashi, 1993; Omoto and Saitou, 1997; Hammer *et al.*, 2006). These migrations biologically and culturally subsumed the majority of indigenous foragers occupying the Japanese Islands and introduced wet rice agriculture to the region (Imamura, 1996a,b). In this sense, scholars involved in bioarchaeological analyses of phenotypic variability during the agricultural transition in prehistoric Japan are presented with the task of parsing out environmental and genetic influences on shifts in morphology. Here, environmental and genetic contributions to variability in postcranial morphology between Jomon and Yayoi people is addressed through comparisons of relative body mass, brachial and crural indices, lower limb lengths and femoral growth rates.

Geographical studies of body size amongst mammals suggest ecogeographical patterning in mass: polytypic species from cold environments are larger than conspecifics from warmer areas (Bergmann, 1847; Mayr, 1963). Enlarged body size improves heat retention in cold environments (Futuyma, 1998). Relative body mass and latitude are correlated in human groups sampled from diverse environments (Ruff, Scott and Liu, 1991; Ruff, 1994; Auerbach and Ruff, 2004).

Human Bioarchaeology of the Transition to Agriculture Edited by Ron Pinhasi and Jay T. Stock

Organisms from warmer environments have elongated limbs compared to conspecifics from colder environments (Allen, 1877). Elongated appendages increase body surface area for improved convective or evaporative cooling in warmer environments, while foreshortened appendages decrease surface area and improve heat retention in colder regions (Futuyma, 1998). In humans, distal relative to proximal limb segment length and limb length relative to skeletal trunk height are associated with a general pattern of long-term adaptation to climate (Trinkaus, 1981; Ruff, Scott and Liu, 1991, 1994, 2002; Holliday, 1997a,b 1999). Distal relative to proximal limb length (brachial and crural index) is associated with thermoregulatory adaptation because this feature influences body surface area (Trinkaus, 1981; Holliday, 1995).

Studies of lower limb length relative to sitting height found significant changes amongst Mayan migrants in the United States compared to parental generations (Bogin, 1995, 2002). Similarly, pre- and post-war Japanese express significant changes in relative lower limb length following greater availability in nutrients (Tanner *et al.*, 1982, 1988). These studies indicate that leg length is associated with nutritional status in a variety of contexts. In addition, individuals experiencing greater systemic stress loads, particularly infectious disease and malnutrition, will express reduced percentages of achieved growth velocity in long bones (Mensforth, 1985; Lovejoy, Russell and Harrison, 1990; Okazaki, 2004).

Jomon period (13 000–2300 BP) cultures were part of a 10 000 year foraging tradition in the Japanese Islands (Imamura, 1996a). Jomon foragers were the descendents of Pleistocene nomads who migrated to Japan around 20 000 BP and subsumed pre-existing knife-blade cultures (Kobayashi, 2004). One set of hypotheses surrounding the earliest migrations to the Japanese Islands suggest that the ancestors of Jomon people migrated from Sundaland (Turner, 1990, 1992; Hanihara, 1991). Other multivariate analyses of cranial and dental traits suggest a Northeast Asian 'point of origin' for the Pleistocene ancestors of Jomon foragers (Doi *et al.*, 1997; Dodo *et al.*, 1998; Pietrusewsky, 1999, 2005; Seguchi *et al.*, 2007; Hanihara and Ishida, 2009; and others).

The last glacial maximum in Japan (25 000–10 000 YBP) is characterized by glacial spread only on the mountain peaks of Honshu and Hokkaido and coniferous trees adapted to warm, moist environments in Honshu and Northern Kyushu (Tsukada, 1986). Postglacial warming is recorded from 10 000 through 4300 BP (Tsukada, 1986) in Japan, suggesting that the ancestors of Jomon foragers migrated to a relatively warm environment.

Broad reliance on cariogenic cultigens is reported during the Jomon period, despite variation in resource availability (Turner, 1979; Fujita, 1995; Todaka *et al.*, 2003; Temple, 2007a). Spikes in carious tooth frequencies are observed following climatic oscillations around 4300 BP, indicating a shift in diet across eastern and western Japan (Fujita, 1995; Temple, 2007a), with exceptions reported on Hokkaido Island (Oxenham and Matsumura, 2008). This dietary shift is not associated with an agricultural transition, as the level of carious lesions and changes in energy expended on plant care during the subsequent Yayoi period are more consistent with an agricultural economy (Sanui, 1960; Inoue *et al.*, 1986; Imamura, 1996a, 1996b; Oyamada *et al.*, 1996; Tsude, 2001; Todaka *et al.*, 2003; Temple and Larsen, 2007).

Cranial and dental size and shape varied between Jomon and Yayoi people in association with environment and gene flow (Brace and Nagai, 1982; Mizoguchi, 1986; Hanihara, 1991; Turner, 1992; Nakahashi, 1993; Pietrusewsky, 1999, 2006). Yayoi period (2500–1700 BP) agriculturalists were the descendents of people from modern-day Korea or northern China who migrated to Japan and interbred to varying degrees with indigenous Jomon foragers around 2500 BP (Brace and Nagai, 1982; Hanihara, 1991; Nakahashi, 1993; Omoto and Saitou, 1997; Pietrusewsky, 1999, 2006; Hammer *et al.*, 2006). Palynological studies of

Holocene Northeastern China indicate vegetation zones consisting of deciduous broad leafed and mixed coniferous forests similar to those found in colder, high latitude environments (Yafeng *et al.*, 1993). Migrations from these regions during the Yayoi period introduced wet rice agriculture to the Japanese Islands (Imamura, 1996a, 1996b; Hudson, 1999; Tsude, 2001).

Studies of systemic stress between Jomon and Yayoi people suggest an improvement in health following the transition to agriculture in prehistoric Japan (Koga, 2003; Okazaki, 2007; Temple, 2010). Enamel hypoplasia prevalence declines between the Jomon and Yayoi periods, likely because the introduction of wet rice agriculture provided a renewable source of food and wet rice dependence was supplemented by protein rich maritime and terrestrial resources (Temple, 2010). Cribra orbitalia prevalence was static between the two groups due to similar exposure to infectious bacteria and parasites (Temple, 2010).

With the results of previous research in mind, four hypotheses are developed for this study:

1. Jomon people are hypothesized to express similar relative body mass when compared to Yayoi people. This similarity will reflect cold adaptation in the ancestors of the two groups.
2. Jomon people will express higher distal relative to proximal intralimb indices (brachial and crural) compared to Yayoi people. This difference will reflect exposure to the climatically mild Japanese Islands in the ancestors of Late/Final Jomon people and recent migration to this region by the ancestors of Yayoi people.
3. Greater variability in limb lengths are expected amongst prehistoric people of the Jomon compared to Yayoi period; these data are expected to express positive skewness indicative of a greater distribution of individuals below median values. Greater variability in stature will reflect exposure to environmental perturbations amongst Jomon people.
4. This study predicts reduced long bone growth rates amongst prehistoric Jomon compared to Yayoi people. This will further reflect greater exposure to systemic stress amongst Jomon hunter-gatherers.

10.2 SUMMARY OF SKELETAL REMAINS

All skeletal materials were excavated from archaeological sites on Honshu and Kyushu Islands (Figure 10.1; Table 10.1). These samples are derived from four Late to Final Jomon period sites and three Yayoi period sites. Late to Final Jomon period sites date between 4300 and 2300 BP, while Yayoi period sites date between 2500 and 1700 BP. These dates were established on the basis of pottery chronologies and radiocarbon dates. Some overlap between the Jomon and Yayoi periods is noted. This overlap occurs because wet rice agriculture began in western Japan around 2500 BP, but is not found in eastern regions until approximately 2300 BP (Imamura, 1996a). Pottery chronology is an accurate way to date Jomon and Yayoi sites given repeated correspondence between these relative methods and radiocarbon analysis (Watanabe, 1966; Habu, 2004; Tanaka *et al.*, 2005). Sample sizes are listed independently for each analysis. In general, these sites were chosen because of large sample sizes and temporal proximity of each site to the agricultural transition, either before or after this economic shift.

A number of comparative samples are used by this study to model ecogeographical variation. These samples combined with approximate latitude and references are listed in Table 10.2. Bivariate plots of relative body mass specifically use Sadlermiut and Ugandan samples as a basis for comparison with Jomon and Yayoi groups. The Sadlermiut are a protohistoric

Figure 10.1 Map of sites yielding human skeletal remains utilized by this chapter. Yayoi period sites are listed by number. Jomon sites are listed by letter. Yayoi: (1) Kanenokuma, (2) Doigahama, (3) Koura; Jomon: (A) Tsukumo, (B) Yoshigo, (C) Inariyama; (D) Hobi

population (300–100 BP) from the southern point of Southampton Island, who express intralimb indices associated with long-term exposure to a high-latitude, cold climate (Auerbach, 2007). Ugandan samples are dated to approximately 50 BP and express intralimb indices consistent with long-term exposure to a low-latitude, warm climate (Ruff and Walker, 1993).

Use of all other comparative samples is justified based on the observation that limb proportions remain consistent between groups from similar climates, regardless of ancestry, assuming these groups occupied their respective environments for sufficient time (Holliday and Ruff, 1997). All samples have limb proportions that conform to ecogeographical expectations (Ruff and Walker, 1993; Holliday, 1995, 1997a; Auerbach, 2007; Temple *et al.*, 2008).

Table 10.1 Sites yielding human remains utilized by this study

Site	Period	Dates	Institution
Hobi	Late/Final Jomon	4000–2500 BP	University Museum, Tokyo
Inariyama	Late/Final Jomon	4000–2500 BP	Kyoto University
Tsukumo	Late/Final Jomon	4000–2500 BP	Kyoto University
Yoshigo	Late/Final Jomon	4000–2500 BP	Kyoto University
Doigahama	Yayoi	2500–1400 BP	Kyushu University
Kanenokuma	Yayoi	2500–1400 BP	Kyushu University
Koura	Yayoi	2500–1400 BP	Kyushu University

Table 10.2 Comparative samples for multivariate analysis of limb shape

Group	N ♂:♀	Institution[c]	Latitude	Climate	Reference
Inupiat, Kodiak Island	18 10:8	NMNH	58.0° N	Polar	Holliday, 1999
Sadlermiut, Southampton Island	54 26:28	CMC	64.1° N	Polar	Auerbach, 2007
19th/20th CE Ainu, Hokkaido	11 7:4	Sapp. Med. Univ.	43.0° N	Continental Microthermal	Temple *et al.*, 2008
Late Iron Age Poundbury, UK	50 26:24	NHM	50° N	Temperate	Auerbach and Ruff, 2004
Early Archaic, Windover Pond, FL	40 17:23	FSU	28.6° N	Temperate	Auerbach, 2007
Early Mediaeval, Straβkirchen Germany	39 13:26	SfAP	48.9° N	Temperate	Auerbach and Ruff, 2004
Edo Period, Japan	21 12:9	Kyoto Univ.	40.0° N	Temperate	Auerbach and Ruff, 2004
Pueblo IV, Cliff Dwellings, Puye, NM	37 20:17	NMNH	35.9° N	Temperate	Auerbach, 2007
Shi San Hang, Taiwan	11 4:7	Acad. Sinica; SMA	25° N	Temperate	Pietrusewsky and Tsang, 2003
Late Pleistocene, Okinawa Island, Minatowgawa I	1 1:0	UTOK	26.0° N	Temperate	Baba and Endo, 1982
Late 19th CE Aboriginal Australians	26 8:18	AMNH; NMNH; DC; MdH; MNdAE; NM; NHM	27.0° S	Tropical	Auerbach and Ruff, 2004

(*continued*)

Table 10.2 (*Continued*)

Group	N ♂ : ♀	Institution[c]	Latitude	Climate	Reference
'Negrito' Philippines	27				
	6 : 21	MdH	13.0° N	Tropical	Auerbach and Ruff, 2004
African Pygmy	10				
	3 : 7	IRSN; Univ. Gen	2.0° N	Tropical	Holliday, 1999
19th CE South	19				
African San	8 : 11	SAM	25.0° N	Tropical	Slome, 1929
Ban Chiang, Non	15				
Nok Tha, Thailand[a,b]	14 : 1	Univ. Hawai'i; UNLV; MUMS	16.5° N	Tropical	Pietrusewsky and Douglas, 2002; Douglas, 1996
Uganda	46				
	19 : 27	MU; KNM	0.0° N	Tropical	Ruff and Walker, 1993

[a] Ban Chiang postcranial measurements are available via the World Wide Web: http://seasia.museum.upenn.edu/skeletal/banchiang_data.htm.

[b] Non Nok Tha postcranial measurements are available via the World Wide Web: http://seasia.museum.upenn.edu/skeletal/nonnoktha_data.htm.

[c] *Acad. Sinica*: Academia Sinica, Taipei, Taiwan; *AMNH*: American Museum of Natural History, New York, NY; *DC*: Duckworth Osteological Collection, Cambridge, UK; *IRSN*: Institut Royale des Sciences Naturelles, Brussels, Belgium; *KNM*: Kenya National Museum, Nairobi, Kenya; *Kyoto Univ*: Kyoto University, Kyoto, Japan; *MdH*: Musée de l'Homme, Paris, France; *MNdAE*: Muzeo Nationale di Antropologia e Etnologia, Florence, Italy; *MU*: Makerere University, Kampala, Uganda; *MUMS*: Maidol University Medical School, Siriraj Hospital, Bangkok, Thailand; *NM*: Naturhistorishes Museum, Vienna, Austria; *NHM*: Natural History Museum, London, UK; *NMNH*: National Museum of Natural History, Washington, DC; *Sapp. Med. Univ*: Sapporo Medical University, Sapporo, Japan; *SAM*: South African Museum, Capetown, South Africa; *SfAP*: Staatssammlung für Anthropologie und Palaeoanatomie, Munich, Germany *SMA*: Shihsanhang Museum of Anthropology, Tapei, Taiwan; *Univ. Gen*: University of Geneva, Geneva Switzerland; *Univ. Hawai'i*: University of Hawai'i, Manoa, HI; *UNLV*: University of Nevada at Las Vegas, Las Vegas, NV; *UTOK*: University of Tokyo, Tokyo, Japan.

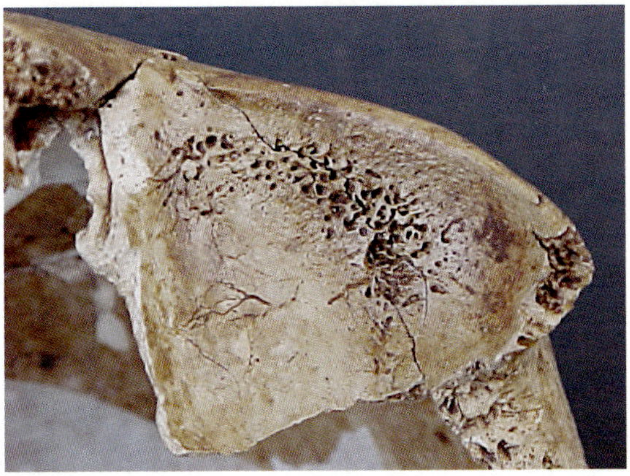

Plate 5.3 *Cribra orbitalia* from Alepotrypa Cave

Plate 5.5 Healed depressed cranial fracture from Alepotrypa Cave

Plate 13.1 Example of Middle Neolithic burial in the Finale Ligure area: the individual is laid down in a stone cist, crouched on the left side of the body

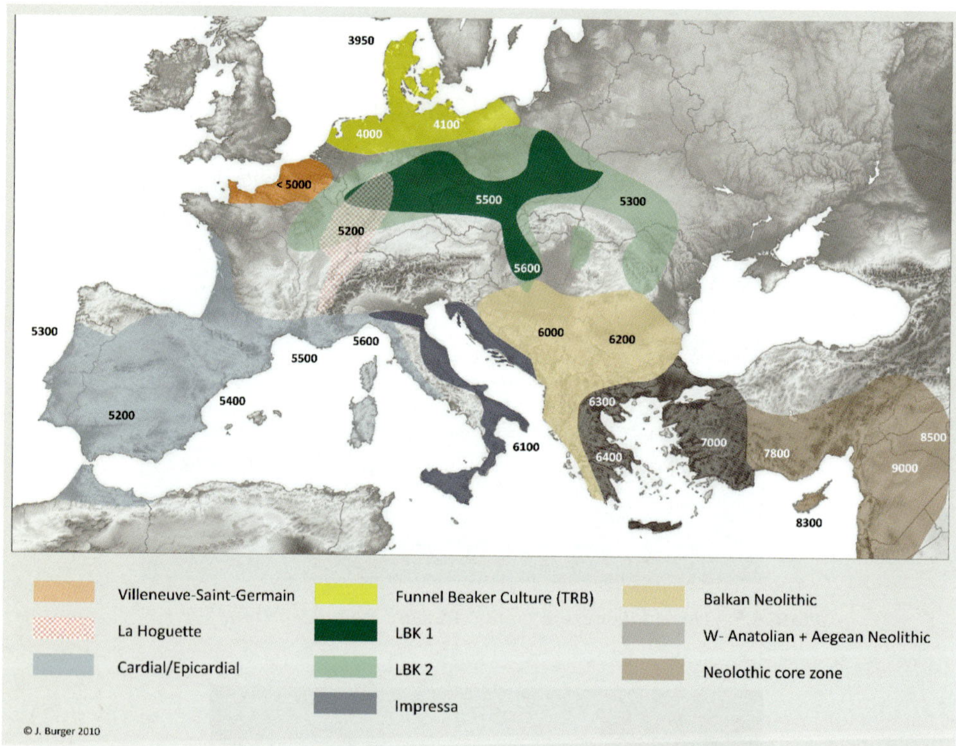

Plate 15.1 Chronological spread of the Neolithic. Numbers give the approximate earliest dates of the Neolithic in years before Christ (calBC)

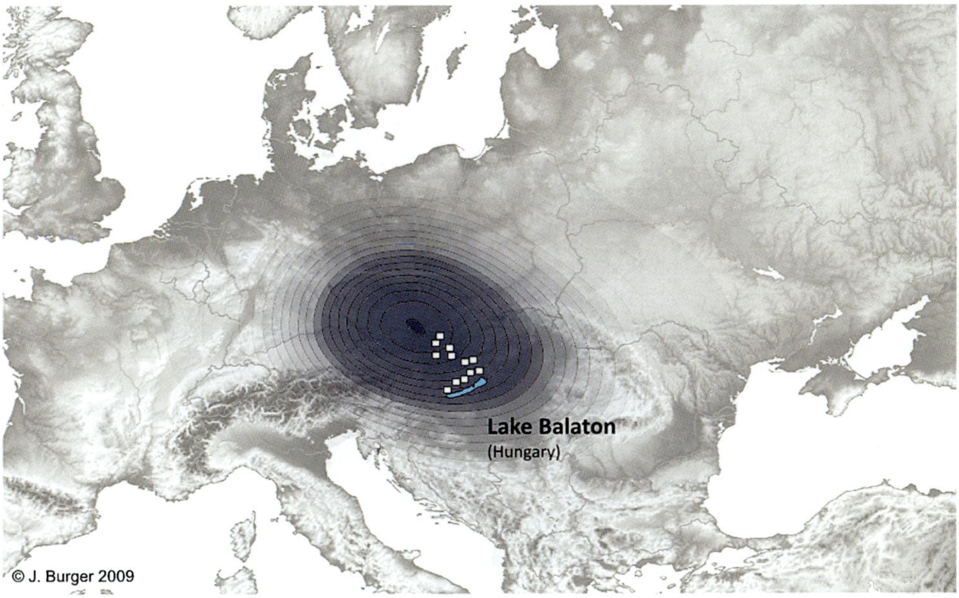

Plate 15.2 Earliest known LBK sites (5,700–5,500 calBC; white squares) north of Lake Balaton after Pavuk (2005) and geographical origin of selection of lactase persistence, after Itan *et al.*, 2009 (about 4,310–6,730 calBC, concentric blue ellipses)

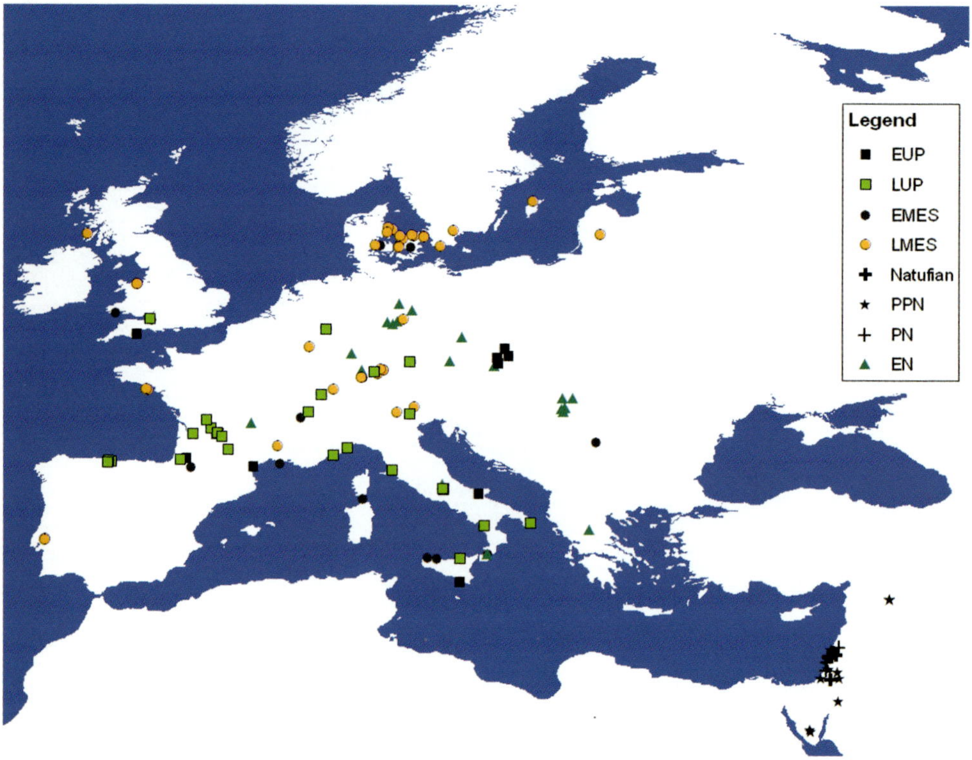

Plate 19.1 Location of sites with dental remains analysed in this study (by archaeological period)

10.3 METHODS

10.3.1 Climatic Adaptation

Climatic adaptation in Jomon and Yayoi people is explored using comparisons of relative body mass and intralimb indices. Relative body mass compares ln bicondylar femoral length (BFL) relative to ln body mass. This comparison explores body mass relative to height, where high values are commonly found amongst cold-adapted people (Ruff, 1994). Disproportionate contributions of BFL to height do, however, occur between many groups, including the Jomon and Yayoi (Wada and Motomura, 2000). Analysis of relative body mass using all limbs is, therefore, a preferred method (Holliday, 1997b). Due to space constraints, this study only uses BFL in comparisons in relative body mass and urges caution in the interpretation of results.

Body mass was estimated using superior-inferior measurements of femoral head diameter. Femoral head breadth and BFL were measured according to standard protocols (Buikstra and Ubelaker, 1994). Femoral head measurements were used in favour of morphometric methods to maximize sample sizes. Currently, the most accurate mechanically based equations used to estimate body mass are those reported by Ruff *et al.* (1991) for males and females, when the sample does not lie at the large or small size extremes for contemporary humans (Auerbach and Ruff, 2004). Under circumstances where BM is associated with large body size, equations reported by Grine *et al.* (1995) may be more appropriately applied (Auerbach and Ruff, 2004). Comparisons of Jomon and Yayoi body mass to other groups in East Asia suggest that these samples do not express a 'large bodied' phenotype (Shackelford, 2005; Temple, 2007b). This study, therefore, estimates BM amongst all samples using the sex-specific equations reported by Ruff *et al.* (1991): $BM♀ = (2.246 \cdot FHB - 35.1) \cdot 0.90$; $BM♂ = (2.741 \cdot FHB - 54.9) \cdot 0.90$. Relative body mass was compared between the groups using bivariate scatter plots. Sample sizes for these comparisons are listed in Table 10.3.

Relative limb lengths were calculated on the basis of maximum radial length (RL) relative to maximum humeral length (HL), as well as tibial length (TL) relative to bicondylar femoral length (BFL), respectively. These measurements were collected according to standard protocols; tibial length was measured as the maximum distance between the lateral condyle and medial malleolus (Buikstra and Ubelaker, 1994). All data were log transformed. Differences in limb lengths were compared using RMA regression combined with the Quick-Test method described by Tsutakawa and Hewett (1977). RMA regression is used because these comparisons do not include dependent and independent variables and were measured with a component of error. Quick-Tests are used to test the null hypothesis that no significant

Table 10.3 Sample sizes used to analyse relative body mass and relative limb length

Group	N BM[a] ♂:♀[b]	N RL rel. HL ♂:♀	N TL rel. BFL ♂:♀	N Multivariate ♂:♀
Jomon	46 : 28	39 : 32	40 : 28	29 : 14
Yayoi	37 : 32	33 : 29	28 : 27	20 : 21
Uganda	27 : 19	—	—	see Table 10.2
Sadlermiut	22 : 20	—	—	see Table 10.2

[a] Number of individuals used in relative body mass comparisons.
[b] Number of male relative to female specimens.

difference between the number of individuals above and below the regeression line will be observed between the Jomon and Yayoi samples. A significantly greater number of positive residuals will be associated with elongated distal relative to proximal limbs lengths, while a significantly greater number of negative residuals will be associated with reduced distal relative to proximal limb lengths. Sample sizes for these comparisons are listed in Table 10.3.

Before multivariate data analyses were performed, all limb measurements were size standardized using the geometric mean derived for each individual with all four limb measurements (HL, RL, BFL, TL). Geometric means were calculated using methods described by Darroch and Mosimann (1985). Multivariate calculations including cluster and principal components analysis were then performed on Jomon, Yayoi, and all comparative samples. Jomon and Yayoi sample sizes used in these comparisons are listed in Table 10.3, while comparative samples are listed in Table 10.2.

10.3.2 Systemic Stress Variability

This section compares limb lengths between Jomon and Yayoi people, to better understand how environmental factors associated with the agricultural transition contributed to variation in postcranial morphology between the two groups. Many studies use stature estimations to compare patterns of systemic stress between skeletal samples (Larsen, 1997). Accurate estimations of stature do, however, require predictive equations drawn from samples with similar body proportions to offset differences in morphology associated with ecogeographical adaptation (Ruff and Walker, 1993; Holliday and Ruff, 1997). These types of equations are not available for Jomon or Yayoi samples. As a consequence, this study compares leg length between Jomon and Yayoi groups. These methods are appropriate given the fact that changes in leg length are associated with nutritional status (Tanner *et al.*, 1982; Takamura, 1984; Bogin, 1995, 2002).

For this chapter, leg length is calculated as bicondylar femoral length (BFL) + tibial length (TL). Variation in the anatomical components of leg length between Jomon and Yayoi people is also evaluated: BFL and TL are individually compared. Jomon samples were derived from eastern and western Japan and compared as a single group. Differences in enamel hypoplasia frequencies are observed between eastern and western Jomon people (Temple, 2007a). Differences in stature are not, however, reported between these regions (Temple, 2008), suggesting that it is appropriate to combine eastern and western Jomon samples. Student's t-tests are used to test the significance of difference in mean leg length, BFL and TL between Jomon and Yayoi people. These traits are compared separately for males and females.

Genetic differences between the Jomon and Yayoi people pose a problem in understanding the environmental influence on postcranial morphology during the agricultural transition in prehistoric Japan. Each group experienced variable migratory histories (see above). As a consequence, direct comparisons of limb measurements may reflect variation in baseline metrics for each sample. However, limb lengths are quantitative traits and the distribution of these traits should follow the central limit theorem. The central limit theorem states that the distribution of quantitative traits within a sample follows a bell shaped curve. Models of quantitative traits suggest deviations from bell shaped distribution will increase with greater additive environmental variance (Mielke, Konigsberg and Relethford, 2006). Limb length *distributions* between Jomon and Yayoi people are, therefore, compared. This will further explain environmental influences on limb lengths between the two samples.

Box plots were used to graphically depict the distribution of leg length, BFL and TL for each sample. F-tests were then applied to this data to test for significant differences in variance, while skewness was calculated for each group to understand if limb lengths are symmetrically distributed in each sample. Where asymmetrical distributions of data are observed, skewness will depart from zero (0) in either positive or negative space. Negative skewness indicates a disproportionate distribution of individuals above the sample median, while positive skewness indicates a disproportionate number of individuals distributed below the sample median. Student's t-tests were utilized to calculate significant skewness. The formula chosen to calculate this statistic follows Sokal and Rohlf (1995: 174).

10.3.3 Long Bone Growth Velocity

Long bone growth patterns amongst subadults were estimated for the Jomon and Yayoi people to further investigate how environment impacted postcranial morphology during the agricultural transition. This study utilizes percentages of femoral growth and femoral growth velocity comparisons within age specific context. Tibial length was not utilized owing to poor preservation of this skeletal element. Age was estimated for each individual using standard phases of tooth development and eruption (Buikstra and Ubelaker, 1994). Ages estimated for teeth were then subjected to the moving average method described by Lovejoy *et al.* (1990). This method adjusts estimations of age based on tooth development and eruption of appropriate reference samples. Here, the reference sample is one derived from 20th CE Japanese.

Percentages of long bone growth are calculated as the maximum diaphyseal length of a femur (left side, where possible) for a given subadult, divided by the mean femoral length for all adults in a given sample. Growth velocity was calculated according to the following equation:

$$\mathrm{V_{xi}} = \mathrm{b_x} + (\mathrm{cln_x}/\mathrm{t_{xi}})$$

where b_x is the slope of the regression analysis for long bone length relative to age for each sample, cln_x is the y-intercept of the regression analysis of femoral length on age + ln age for a sample and t_{xi} represents age of the individual. Jomon samples from eastern and western Japan were again considered as a homogenous sample, this time owing to poor preservation of subadults.

10.4 RESULTS AND DISCUSSION

Bivariate plots indicate that the relative body mass of Jomon and Yayoi people were more similar to the Sadlermiut than Ungandan samples (Figure 10.2), though use of long bones such as the radius indicate some similarity between the Jomon and Ugandan samples in relative body mass (Temple and Matsumura, 2010). This result implies that the ancestors of both Jomon and Yayoi people had larger relative body sizes, possibly indicative of ancestral adaptation to high latitude, cold climates. However, it is important to note that Late Pleistocene humans express a general pattern of enlarged body mass in comparison with Holocene people (Ruff, 1994; Ruff et al., 1997).

Studies of cranial morphology indicate that the Pleistocene ancestors of Jomon people migrated to the Japanese Islands via Northeast Asia (Hanihara and Ishida, 2009; and others). Y-chromosome analysis and traditional haplotyping amongst Ainu groups find

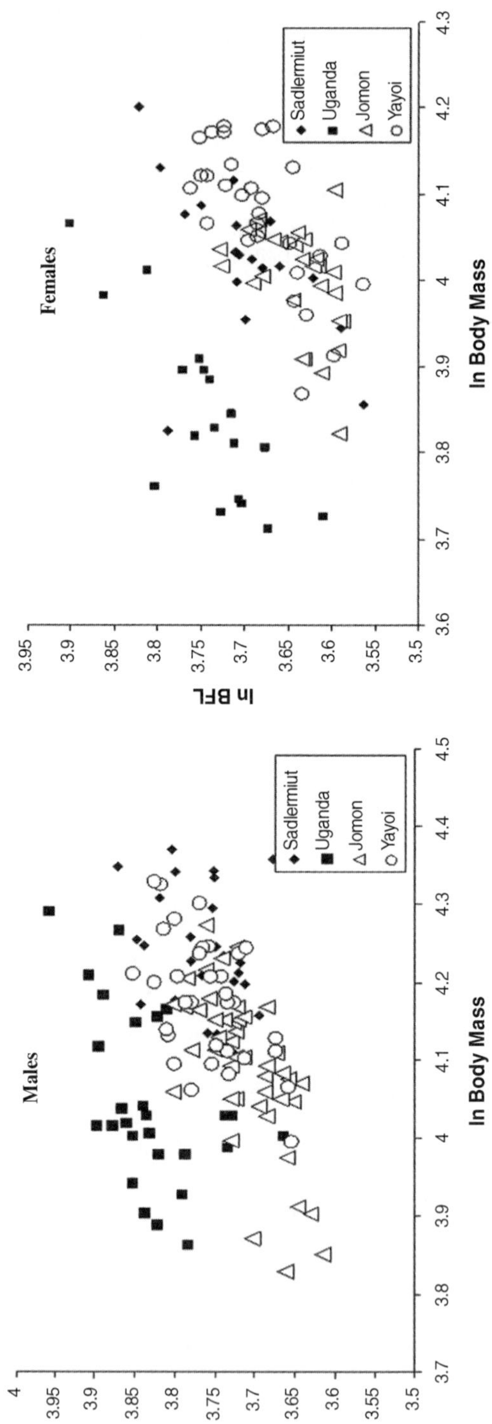

Figure 10.2 Male and female scatter plots depicting relative body mass (in kilograms) for Jomon, Yayoi, Saderlmiut, and Ugandan samples

common ancestral alleles with Central and Northeast Asian groups dating to approximately 20 000 BP (Omoto and Saitou, 1997; Hammer *et al.*, 2006). These findings are consistent with recent aDNA studies of human skeletal material from the Late Jomon Funadomori site in Hokkaido: Haplogroups M9b, M7a and D1 are shared between Jomon and indigenous foraging groups from Southeastern Siberia (Adachi *et al.*, 2009). It is further suggested that Jomon people express craniometric similarity to samples from Eurasia (Hanihara and Ishida, 2009). Pleistocene migrations from Eurasia, Southern Siberia or regions surrounding the Altai Mountains into Northeast Asia, and finally, Japan are supported by genetic studies of contemporary Ainu people (Hammer *et al.*, 2006). Long term evolution in these climates may partially explain the enlarged relative body mass of the Jomon samples.

Skeletal remains dated to the Yayoi period from Northern Kyushu and Southern Honshu represent the descendents of migrants from continental Asia (Brace and Nagai, 1982; Hanihara, 1991; Nakahashi, 1993; Hammer *et al.*, 2006). Palynological studies of Holocene Northeastern China indicate vegetation zones similar to those found in colder, high latitude environments (Yafeng *et al.*, 1993). These findings suggest that the body mass of the Yayoi people is also associated with long-term evolution in a colder climate prior to migrating to the Japanese Islands.

Mean brachial and crural indices are listed for the Jomon samples in Table 10.4. RMA regression analysis combined with Quick-Test results suggests that Jomon people have significantly elongated distal relative to proximal appendages compared to Yayoi people (Figures 10.3 and 10.4; Table 10.5). Cluster and principal components analysis of size-standardized limb lengths suggests Jomon people expressed limb proportions similar to warm-adapted groups, while Yayoi limb shape is similar to cold-adapted groups (Figures 10.5 and 10.6). These differences are associated with distal relative to proximal limb segment lengths and upper relative to lower limb lengths (Table 10.6), which may be associated with negative allometry of the forelimb and positive allometry of the hindlimb.

Differences in limb proportions between Jomon and Yayoi people are associated with varying evolutionary histories. Distal limbs, and specifically the tibia, have the greatest level of intrinsic variability in global comparisons of relative limb length (Holliday and Ruff, 2001). This variability has been experimentally demonstrated in animal models, and is likely a response to climate (Serrat *et al.*, 2008). Cold exposure is associated with vasoconstriction in the distal regions of the limbs and quite likely, disruption of chondrocyte profileration (Serrat *et al.*, 2008). The ancestors of Jomon people migrated to the Japanese Islands around 20 000 BP from Northeast Asia, possibly originally departing from Late Pleistocene Eurasia (Omoto and Saitou, 1997; Hammer *et al.*, 2006; Adachi *et al.*, 2008; Hanihara and Ishida, 2009). Early Holocene Japan was substantially warmer than these environments, with Post-Glacial warming recorded around 12 000 BP (Tsukada, 1986). This suggests that the relative elongation of

Table 10.4 Mean brachial and crural indices for Jomon, Yayoi, Sadlermiut and Ugandan samples[a]

Groups	Brachial	Crural
Jomon Males	80.2	82.1
Yayoi Males	77.4	80.8
Jomon Females	81.1	82.2
Yayoi Females	78.1	80.7

[a] Sample sizes are listed in Table 10.3.

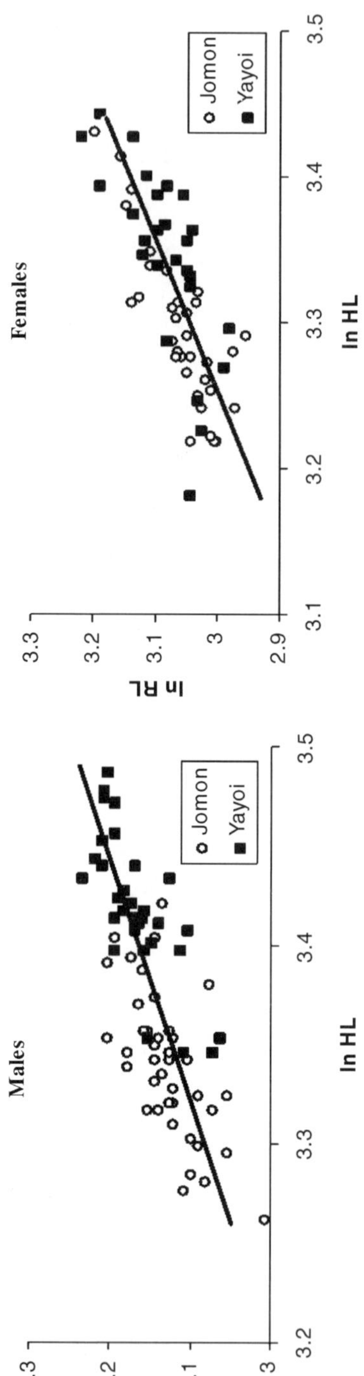

Figure 10.3 Radial relative to humeral length for male/female Jomon and Yayoi samples

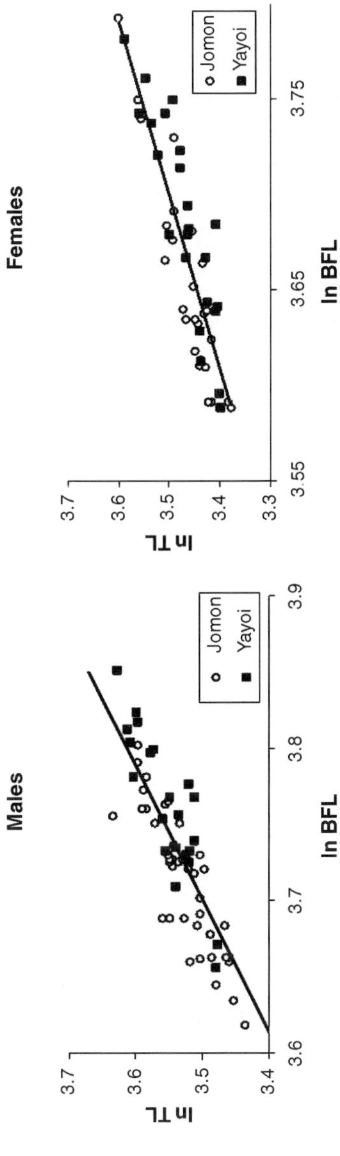

Figure 10.4 Tibial relative to bicondylar femoral length for male/female Jomon and Yayoi samples

Table 10.5 Quick-Test results for relative limb lengths

Index	Males	Females
Tibia rel. Femur	Jomon > Yayoi ($P \leq 0.004$)	Jomon > Yayoi ($P \leq 0.002$)
Radius rel. Humerus	Jomon > Yayoi ($P \leq 0.004$)	Jomon > Yayoi ($P \leq 0.05$)

Jomon distal limbs may represent a response to the more climatically mild environment of Holocene Japan. However, it is equally possible that this morphology is associated with neutral mutation and isolation. Developmental studies of limb proportions amongst Jomon samples are necessary to further test these competing hypotheses.

Yayoi people are characterized by foreshortened distal relative to proximal limb segments and cluster with groups from colder environments at higher latitudes. This result is not surprising given evidence that indicate Yayoi people were the descendents of Northeast Asian migrants to the Japanese Islands (Brace and Nagai, 1982; Hanihara, 1991; Nakahashi, 1993; Hammer *et al.*, 2006). Depth-of-time is an important consideration regarding the retention of cold-derived limb proportions amongst the Yayoi people, especially in comparison to the elongated distal appendages expressed by Jomon foragers. Previous studies suggest that long periods of evolution within a particular climate are necessary before relative limb lengths

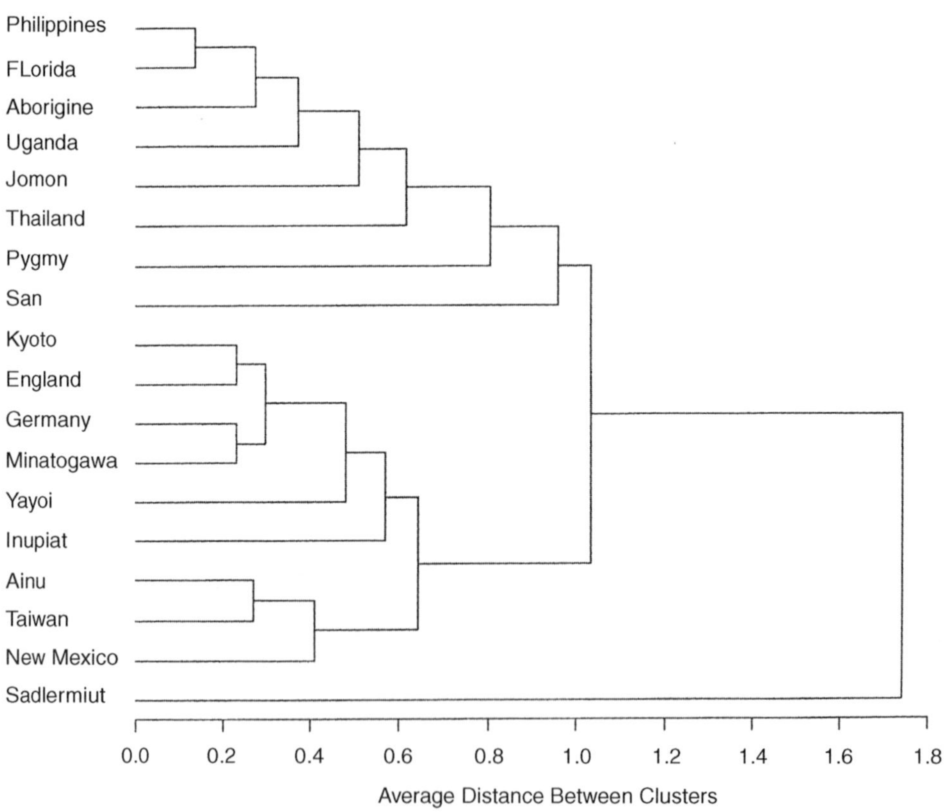

Figure 10.5 Cluster analysis of size-standardized limb lengths

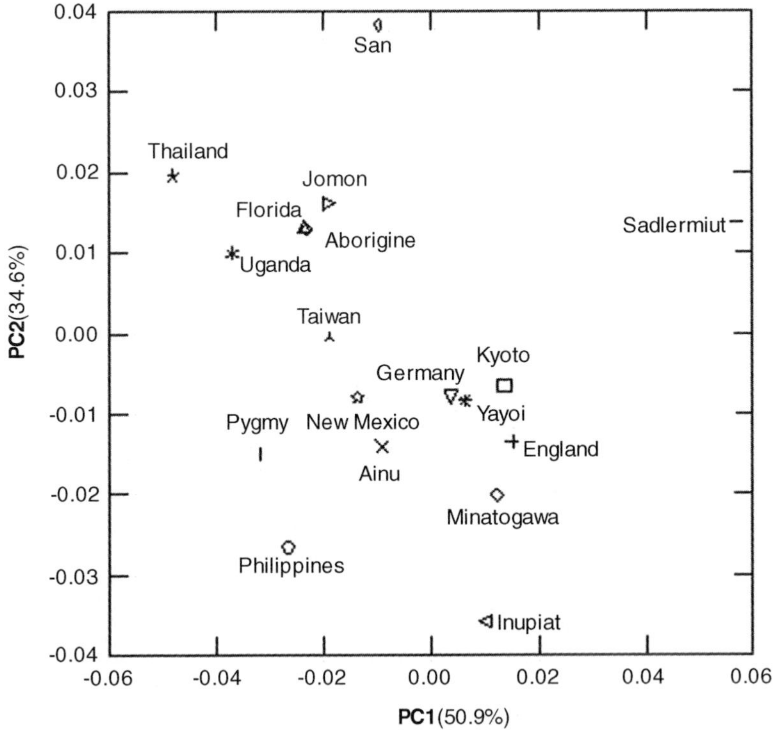

Figure 10.6 Principal components analysis of size-standardized limb lengths

adjust to climate (Ruff, 1994; Holliday, 1997, 1999; Auerbach, 2007). All Yayoi sites in this study are dated to points very near the first arrival of Northeast Asian migrant groups to Honshu and Northern Kyushu (2500 through 1700 BP). It is, therefore, likely that Yayoi people experienced little change in limb shape, given a relatively brief period of colonization of the Japanese Islands.

It is also worth noting that a climatic shift is recorded around 4300 BP and characterized by cooling temperatures (Tsukada, 1986). These cooling temperatures are, quite possibly, associated with a slight temporal decline in intralimb indices between Middle and Late/Final Jomon people (Temple, unpublished data). Intralimb indices also declined between Yayoi and historic Japanese (Yamaguchi, 1989). Under such circumstances, it is unreasonable to expect that Yayoi limb proportions would resemble that observed in Jomon people, even after prolonged exposure to the Japanese environment.

Table 10.6 Principal components of size-standardized limb lengths

Principle Components	BFL	TL	HL	RL	% VAR
Eigenvectors of PC1	0.673	–0.407	0.442	–0.431	50.9
Eigenvectors of PC2	0.497	0.640	–0.487	–0.327	34.6

Table 10.7 Sample sizes and mean limb lengths for Jomon and Yayoi groups

Group	Leg Length	Femur Length	Tibia Length
Jomon ♂	42[a] (75.8)[b]	61 (41.7)	45 (34.2)
Yayoi ♂	22 (78.3)	39 (43.5)	24 (34.8)
Jomon ♀	22 (71.2)	37 (38.9)	27 (32.3)
Yayoi ♀	25 (72.4)	35 (40.1)	26 (32.3)

[a] Sample size.
[b] Mean limb length in centimeters.

10.4.1 Long Bone Length

Sex-specific means and sample sizes for lower limb measurements are listed in Table 10.7. Significantly greater leg length is observed amongst Yayoi compared to Jomon people for both males ($t=-2.65$; $p\leq 0.01$) and females ($t=-2.32$; $p\leq 0.02$), which is primarily based upon differences in femoral length (Figure 10.7). Significantly longer femora are observed amongst Yayoi compared to Jomon males ($t=-4.14$; $p\leq 0.001$) and females ($t=-2.21$; $p\leq 0.03$); however, there are no significant differences in tibial length between Jomon and Yayoi males ($t=-1.60$; $p\leq 0.102$) or females ($t=-0.02$; $p\leq 0.98$). Box plots representing leg, femoral and tibial lengths are presented for comparisons of males (Figure 10.7) and females (Figure 10.8). Variance and skewness for these measurements are listed in Tables 10.8 and 10.9. Statistical significance for these values is included in the corresponding tables. Leg length box plots for Yayoi males suggests that a greater percentage of individuals are distributed in the lower quartiles of the distribution, while Jomon male leg length is more evenly dispersed (Figure 10.7). However, there are only slight, insignificant differences in variance (Table 10.8) and skewness (Table 10.9) between the two groups, suggesting fairly equivalent variability and symmetry of data. In contrast, larger though non-significant variance is reported for Jomon females compared to Yayoi (Table 10.8). Significantly positive skewness is reported for Jomon compared to Yayoi samples (Table 10.9). This indicates that a greater majority of Jomon female leg lengths are dispersed in the lower quartile.

Bicondylar femoral length (BFL) appears to follow an even distribution for both Jomon and Yayoi males (Figure 10.7). This observation is substantiated by a similar variance (Table 10.8) and skewness (Table 10.9) between the two groups. The BFL of Jomon females features a greater number of individuals in the lower quartiles, while Yayoi femoral lengths are more evenly distributed (Figure 10.8); however, variability is non-significant between Jomon and Yayoi females (Table 10.8). Jomon females do, however, express significant positive skewing, indicating a greater distribution of individuals in the lower quartiles (Table 10.9). This contrasts with the more evenly distributed Yayoi data (Table 10.9).

Jomon males express a relatively even distribution of tibia length (TL), while the Yayoi data is more heavily distributed below the median (Figure 10.7). The distribution of data between the two groups does, however, express approximately equal variance (Table 10.8) and non-significant divergence in skewness (Table 10.9). For females, Jomon samples express greater distributions of TL in the lower quartiles, while Yayoi samples were more evenly distributed (Figure 10.8). This is further illustrated by greater variance for Jomon compared to Yayoi samples (Table 10.8). Skewness for Jomon people is significantly positive, while Yayoi samples do not differ from zero (Table 10.9). This suggests a strong distribution of Jomon

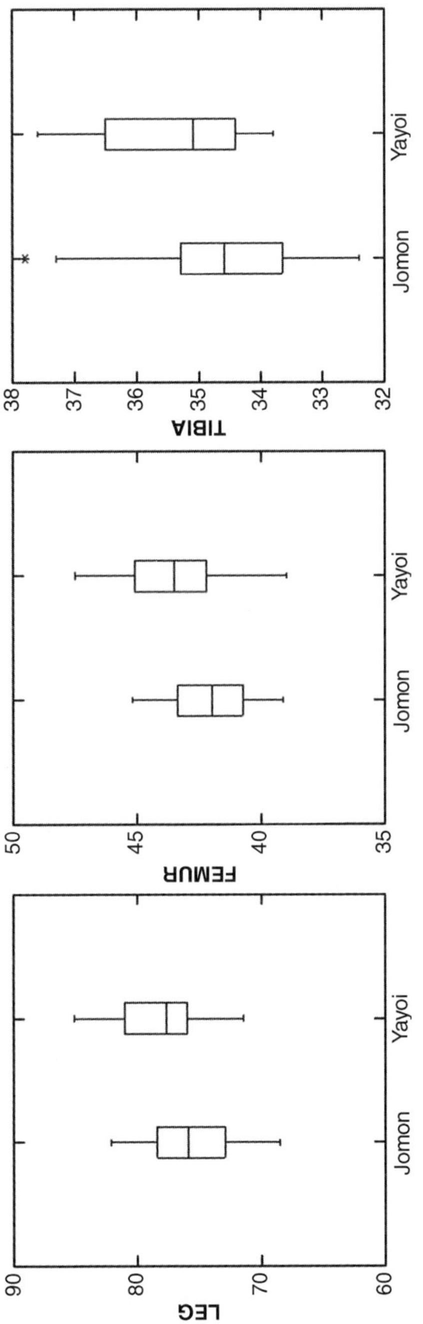

Figure 10.7 Box plots depicting male Jomon and Yayoi femur, tibia and limb length variability (cm)

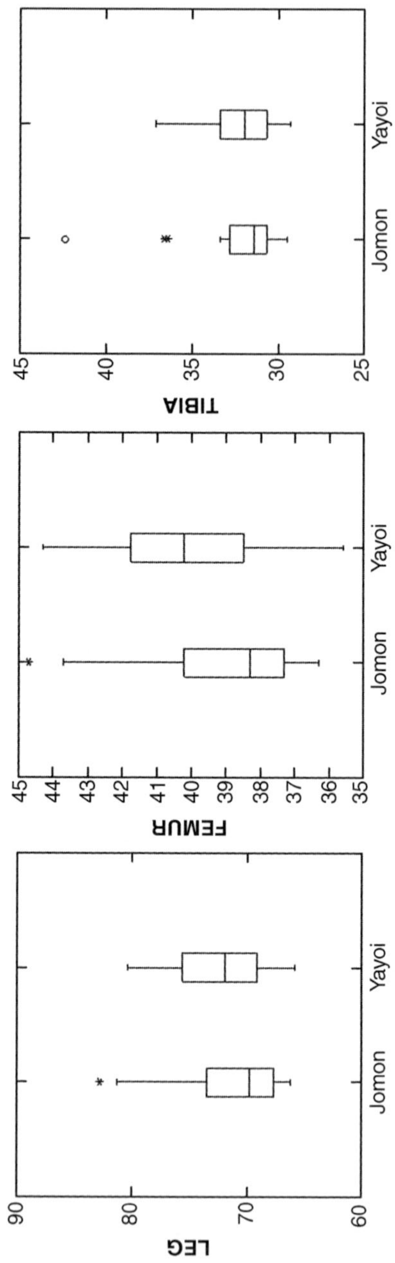

Figure 10.8 Box plots depicting female Jomon and Yayoi femur, tibia and limb length variability (cm)

Table 10.8 Variance and related F-values for limb length comparisons

Group	Leg Length	Femoral Length	Tibial Length
Jomon ♂	12.64	3.8	2.59
Yayoi ♂	12.21	4.36	2.01
F	1.04 (NS)	1.53 (NS)	1.36 (NS)
Jomon ♀	22.13	2.184	7.057
Yayoi ♀	15.68	2.177	4.067
F	1.41 (NS)	1.01 (NS)	1.89 (NS)

females in the lower quartiles of tibial length, while Yayoi samples are more evenly dispersed (Figure 10.8). These results suggest that greater numbers of leg lengths were distributed below the median in Jomon compared to Yayoi females, a trend suggesting growth suppression in these pre-agricultural foragers.

10.4.2 Long Bone Growth

Age specific sample sizes for achieved femoral growth for age, and femoral growth velocity, are listed in Table 10.10. Jomon and Yayoi people express approximately equal percentages of achieved growth throughout infancy and early childhood (Figure 10.9). Jomon people do, however, express significantly reduced percentages of achieved femoral growth between ages 5.5 and 7.5 years (Figure 10.9). Jomon foragers also experienced repressed femoral growth *velocity* compared to Yayoi people between 5.5 and 7.5 years; however, the greatest

Table 10.9 Skewness and related t-values for limb length comparisons

Leg Length ♂	Skewness	SE	T_s	Significance
Jomon	−0.031	0.365	−0.084 931 507	NS
Yayoi	−0.017	0.491	−0.034 623 218	NS
Femoral Length ♂				
Jomon	−0.201	0.306	−0.656 862 745	NS
Yayoi	−0.366	0.378	−0.968 253 968	NS
Tibial Length ♂				
Jomon	0.09	0.354	0.254 237 288	NS
Yayoi	0.125	0.472	0.264 830 508	NS
Leg Length ♀	Skewness	SE	T_s	
Jomon	1.155	0.472	2.447 033 898	0.05
Yayoi	0.161	0.464	0.346 982 759	NS
Femoral Length ♀				
Jomon	0.948	0.388	2.443 298 969	0.05
Yayoi	−0.164	0.398	−0.412 060 302	NS
Tibial Length ♀				
Jomon	2.411	0.448	5.381 696 429	0.05
Yayoi	0.703	0.456	1.541 666 667	NS

Table 10.10 Growth velocity comparisons between Jomon and Yayoi samples

Age in Years	N Jomon	Jomon Velocity	N Yayoi	Yayoi Velocity
0–1.4	6	7.7	1	15.2
1.5–2.4	1	4.8	0	—
2.5–3.4	2	3.2	1	3.4
3.5–4.4	1	2.7	3	2.7
4.5–5.4	1	2.3	0	0
5.5–6.4	2	2	1	2.2
6.5–7.4	2	1.8	1	2
7.5–8.4	1	1.7	2	1.9
8.5–9.4	2	1.6	0	—
9.5–10.4	1	1.6	1	1.8
10.5–11.4	—	—	1	1.7
11.5–12.4	1	1.5	1	1.7
Σ[a]	20		12	

[a] The number of Jomon and Yayoi observed in each group and total number within each sample are identical to the number of individuals used in the analysis of femoral growth percentages

differences in femoral growth velocity are observed during infancy. (Table 10.10) These trends are, to some extent, consistent with chronological studies of enamel hypoplasia amongst Jomon people that suggest elevated stress in later childhood (Yamamoto, 1992), though no previous studies have reported stress in early infancy amongst prehistoric Jomon people.

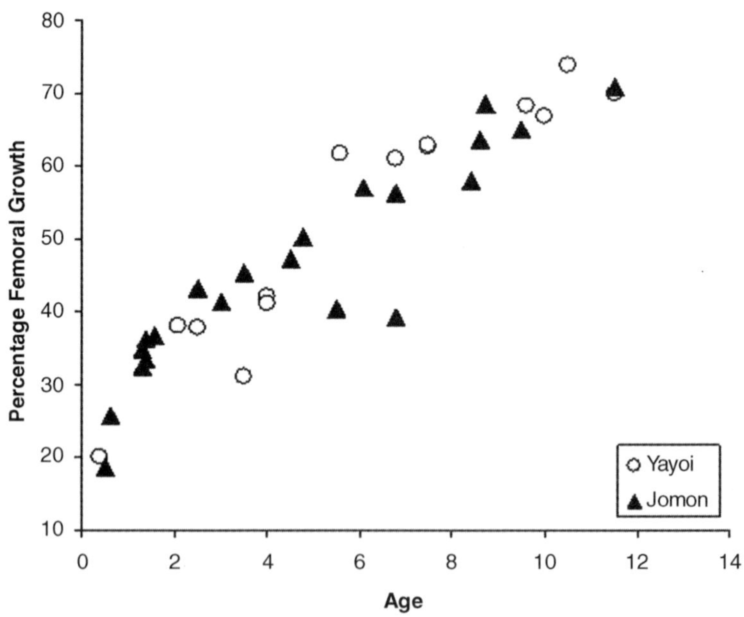

Figure 10.9 Percentages of achieved femoral growth for Jomon and Yayoi samples

These results provide support for the hypothesis that differences in Jomon and Yayoi leg length were associated, in part, with environmental variability. Stress experienced at earlier ages has a much greater impact on longitudinal growth compared to stress associated with later developmental stages (Ulijaszek, 1994; Stinson, 2000; Floyd and Littleton, 2006). Environmental variance in longitudinal growth is greater at younger ages (Ooki and Asaka, 1993) and thus, more susceptible to stress during these times. Within the context of these findings, Jomon foragers may express reduced femoral length in association with growth stunting experienced during early infancy and later childhood.

Differences in stress associated with the agricultural transition most likely contributed to the variable patterns of limb growth between Jomon and Yayoi people. These stressors include patterns of both chronic infection and malnutrition. Infectious disease and growth are linked because immunostimulation diverts essential growth factors such as IL-1, IL-6 and TNFα during the acute phase response to invading pathogens (Solomons *et al.*, 1993). This may explain patterns of growth in prehistoric samples, for example, chronic infection is associated with long bone growth retardation amongst infants from the Libben site (Lovejoy, Russell and Harrison, 1990). In particular, Lovejoy, Russell and Harrison (1990) suggest that population crowding and density likely contributed to a pattern of increased chronic infection amongst subadults and that this pattern of disease significantly disrupted long bone growth.

The Jomon people lived in densely populated environments. Eastern Japan experienced population decline during the Late/Final Jomon period, but still retained significant population density (Koyama, 1978). Western Japan experienced a population increase during the Late/Final Jomon period (Koyama, 1978). Greater frequencies of periostitis are observed amongst Late/Final compared to Middle Jomon period people, indicating an increase in chronic infectious disease over time (Temple, 2007b). Increased diversity of infectious diseases is observed during the Yayoi period as the earliest cases of confirmed tuberculosis date to this time (Suzuki and Inoue, 2006) and are associated with migratory behaviour (Suzuki, Fujita and Choi, 2008). Following the agricultural transition, significantly fewer cases of periostitis are, however, reported for Yayoi adults, while the same trend is observed between Jomon and Yayoi subadults aged 0 to 15 years (Temple, 2007b). It is, therefore, likely that the greater *frequency* of *chronic* infection, experienced at younger ages amongst Jomon compared to Yayoi people accounts for the discrepancies in femoral growth between the two samples.

It is also important to note that variation in nutritional quality and quantity is reported between Jomon and Yayoi communities. The Jomon people were reliant upon many seasonally available resources, while Yayoi groups subsisted on both seasonally available goods and agricultural products (Imamura, 1996; Chisholm and Koike, 1999; Tsude, 2001). Resource density also significantly declined during the Late/Final Jomon period, though this was not catastrophic (Koike, 1986a, 1986b). In contrast, agricultural economies were extraordinarily well developed in prehistoric Japan as they arrived as fully functioning economic systems (Tsude, 2001). In this sense, Yayoi people controlled resource yield to a greater extent than Jomon foragers and likely experienced reduced nutritional stress. Reduction of nutritional stress should then also be associated with improved percentages of achieved femoral growth and femoral growth velocities amongst Yayoi compared to Jomon people.

These trends contrast with those reported for the agricultural transition in many regions, where reduction in the quality of life is the rule rather than exception (Larsen, 1987, 1995, 2002). This process differs in Southeast and East Asia, where increases in systemic stress following the transition to wet rice economies are not uniformly observed (Douglas and Pietrusewsky, 2007; Domett and Tayles, 2007), though increased diversity of chronic infectious diseases is reported in various environments (Oxenham, Nguyen and Nguyen, 2005; Suzuki and Inoue, 2006). The introduction of wet rice agriculture and preceding environmental conditions should be considered when interpreting these unique results, particularly for Japan. First, wet rice was introduced to the Japanese Islands as a fully functioning subsistence system (Tsude, 2001). The earliest wet rice fields in Japan best resemble those from China dated to around 6000 BP (Imamura, 1996; Tsude, 2001). These fields postdate the earliest development of wet rice paddies in Japan by approximately 3000 years (Tsude, 2001). This suggests Yayoi food production may not have been exacerbated by the types of environmental fluctuations that reduced resource yield for many incipient agriculturalists (Harris, 1989). In addition, Yayoi people supplemented this diet with maritime resources (Chisholm and Koike, 1999) and even cultivated various species of freshwater fish (Nakajima *et al.*, 2010). Maritime resources provide an excellent source of nutrition, particularly when mixed with cereal products (Layrisse, Martinez-Torres and Roche, 1968; Rivera *et al.*, 2003). This contrasts with agricultural economies in other regions, where dietary breadth decreased and focus on nutritionally inadequate plants was common (Larsen, 1987, 1995, 2002). Overall, these findings suggest that Yayoi economies were reliant on predictable, renewable food sources that complemented the transition to a carbohydrate rich diet, and abated systemic stress levels during the transition to wet rice economies.

10.5 CONCLUDING REMARKS

This chapter explored variability in postcranial morphology associated with the migratory and the agricultural transition amongst the Jomon and Yayoi people. While there is similarity in body mass between the two groups, the Jomon and Yayoi people express significantly different intralimb indices, which may reflect unique migratory histories that exposed Jomon and Yayoi people to different climates. Jomon people and their ancestors evolved in the temperate Japanese Islands over approximately 15 000 years, while the ancestors of Yayoi people evolved in colder regions of continental East Asia. Exposure to the warmer environment of Japan and the colder climates of continental East Asia likely resulted in differing intralimb indices between the two groups. These unique evolutionary trajectories underscore an important component of the agricultural transition and postcranial morphological variability in prehistoric Japan. Variation in intralimb indices between Jomon and Yayoi people reflects the introduction of new alleles to the Japanese Islands and suggests that variability in postcranial morphology between the two groups is associated with migrations experienced during the agricultural transition.

Postcranial morphological variation also reflects the direct influence of environmental change concurrent with the introduction of agricultural economies to the Japanese Islands. Jomon people express a considerably foreshortened leg length compared to Yayoi people, which is associated with shorter femoral lengths and greater variability in limb lengths amongst females. In addition, percentages of achieved femoral growth and femoral growth velocity are reduced amongst Jomon compared to Yayoi subadults between 5.5 and 7.0 years of age. Significantly lower growth velocity is also observed amongst Jomon infants (0–1.4 years). Jomon people had considerably greater frequencies of periostitis and enamel hypoplasia compared to Yayoi farmers in both adult and subadult age categories (Temple, 2007b). This suggests that stress duration and chronology differed between Jomon and Yayoi people, with

Jomon showing greater evidence of chronic illness during child- and adulthood. These experiences are likely associated with the reduced leg length and femoral growth rates observed amongst Jomon compared to Yayoi people.

The results of this study indicate an improvement in health outcomes following the agricultural transition in prehistoric Japan. Greater systemic stress occurred before the transition to agricultural economies, rather than following this economic and environmental shift. This differs from previous interpretations of the agricultural transition in many regions that suggest a general decline in health and well-being (Larsen, 1987, 1995, 2006). The results reported here are, however, similar to other studies from Southeast Asia, particularly those focusing on the origins of wet rice economies. Improvements in health and/or stasis in systemic stress is observed (Pietrusewsky and Douglas, 2002; Domett and Tayles, 2007). Two reasons for these patterns of systemic stress are noted. First, prehistoric wet rice agriculturalists experienced a subsistence transition that introduced a predictable, renewable source of food to economies that may have experienced resource shortages. Second, wet rice agriculturalists continued to exploit maritime resources, a nutrient rich food. Results of this nature suggest that patterns of systemic stress during the transition to agriculture were regionally variable and should be considered within local context.

ACKNOWLEDGEMENTS

I thank Masato Nakatsukasa, Kazumichi Katayama and Wataru Yano from Kyoto University for access to and assistance with Jomon skeletal materials stored at that institution. I thank Gen Suwa, Soichiro Mizushima and Aiko Saso for allowing access to and providing assistance with the Jomon materials stored at the University Museum, University of Tokyo. I am grateful to Takahiro Nakahashi, Yoshiyuke Tanaka, Shozo Iwanaga, Koji Mizoguchi and Kenji Okazaki at Kyushu University for providing access to and assistance with the Yayoi skeletal material housed in that institution. I thank Ben Auerbach (Univ. Tennessee), Chris Ruff (Johns Hopkins Univ.), Mike Pietrusewsky (Univ. Hawaii), Trent Holliday (Tulane Univ.) and Michele Toomay Douglas for providing access to comparative materials. I further acknowledge Clark Larsen, Paul Sciulli, Sam Stout, Kris Gremillion and Debbie Guatelli-Steinberg from The Ohio State University for support with many aspects of this project. This chapter greatly benefited from the comments of two anonymous reviewers, though any mistakes are my responsibility. I finally thank Ron Pinhasi and Jay Stock for inviting me to contribute to this volume.

REFERENCES

Adachi, N., Shinoda, K., Umetsu, K. and Matusmura, H. (2009) Mitochondrial DNA analysis of Jomon skeletons from the Funadomari site, Hokkaido and its implication for the origins of Native Americans. *Am. J. Phys. Anthropol.*, **138**, 255–265.

Allen, J.A. (1877) The influence of physical conditions on the genesis of species. *Rad. Rev.*, **1**, 108–140.

Auerbach, B.M. (2007) Human Skeletal Variation in the New World during the Holocene: Effects of Climate and Subsistence across Geography and Time. Ph.D. Dissertation, Johns Hopkins University.

Auerbach, B.M. and Ruff, C.B. (2004) Human body mass estimation. A comparison of mechanical and morphometric methods. *Am. J. Phys. Anthropol.*, **125**, 331–342.

Baba, H. and Endo, B. (1982) Postcranial skeleton of Minatogawa Man, in *The Minatogawa Man: The Upper Pleistocene Man from Okinawa Island* (eds H. Suzuki and K. Hanihara), University Museum, University of Tokyo, Tokyo, Bulletin 19, pp. 61–195.

Bergmann, C. (1847) Uber die verhaltniesse der warmeokonomie der thiere zu ihrer grosse. *Gottingen Studien*, **1**, 595–708.

Bogin, B. (1995) Plasticity in the growth of Mayan refugee children living in the United States, in *Human Variability and Plasticity* (eds C.G.N. Mascie-Taylor and B. Bogin), Cambridge University Press, Cambridge, pp. 46–74.

Bogin, B.P., Smith, P., Orden, A.B. *et al.* (2002) Rapid change in height and body proportions of Mayan American children. *Am. J. Hum. Biol.*, **14**, 753–761.

Brace, C.L. and Nagai, M. (1982) Japanese tooth size: past and present. *Am. J. Phys. Anthropol.*, **59**, 399–411.

Buikstra, J.E. and Ubelaker, D.H. (eds) (1994) *Standards for Data Collection from Human Skeletal Remains*, Arkansas Archaeological Survey Research Series no. 44.

Chisholm, B. and Koike, H. (1999) Reconstructing prehistoric Japanese diet using stable isotopic analysis, in *Interdisciplinary Perspectives on the Origins of the Japanese* (ed. K. Omoto), International Research Center for Japanese Studies, Kyoto, pp. 69–73.

Darroch, J.N. and Mosimann, J.E. (1985) Canonical principal components of shape. *Biometrika*, **72**, 241–252.

Dodo, Y., Ishida, H. and Kondo, O. (1998) Ainu and Ryukyuan cranial nonmetric variation: Evidence which disputes the Ainu-Ryukyu Islands common origin theory. *Anthropol. Sc.*, **106**, 99–120.

Doi, N., Dodo, Y. and Kondo, O. (1997) Amami-Okinawans as viewed from cranial measurements. *Anthropol. Sc.*, **105**, 79.

Domett, K. and Tayles, N. (2007) Population health from the Bronze to the Iron Age in the Mun River Valley, Northeastern Thailand, in *Ancient Health: Skeletal Indicators of Agricultural and Economic Intensification* (eds M.N. Cohen and G.M.M. Crane-Kramer), University Press of Florida, Gainesville, pp. 286–299.

Douglas, M.T. (1996) Paleopathology in Human Skeletal Remains from the Pre-Metal, Bronze, and Iron Ages, Northeastern Thailand. Ph.D. Dissertation, University of Hawaii.

Douglas, M.T. and Pietrusewsky, M. (2007) Biological consequences of sedentism: Agricultural intensification in Northeastern Thailand, in *Ancient Health: Skeletal Indicators of Agricultural and Economic Intensification* (eds M.N. Cohen and G.M.M. Crane-Kramer), University Press of Florida, Gainesville, pp. 300–319.

Floyd, B. and Littleton, J. (2006) Linear enamel hypoplasia and growth in an Australian Aboriginal community: not so small, but not so healthy either. *Ann. Hum. Biol.*, **33**, 424–443.

Fujita, H. (1995) Geographical and chronological differences in dental caries in the Neolithic Jomon period of Japan. *Anthropol. Sci.*, **103**, 23–37.

Futuyma, D. (1998) *Evolutionary Biology*, Sinauer and Associates, Sunderland.

Grine, F.E., Jungers, W.L., Tobias, P.V. and Pearson, O.M. (1995) Fossil Homo femur from Berg Aukas, northern Namibia. *Am. J. Phys. Anthropol.*, **97**, 151–185.

Habu, J. (2004) *Ancient Jomon of Japan*, Cambridge University Press, Cambridge.

Hammer, M.F., Karafet, T.M., Hwayong, P. *et al.* (2006) Dual origins of the Japanese: common ground for hunter-gatherer and farmer Y-chromosomes. *J. Hum. Genet.*, **51**, 47–58.

Hanihara, K. (1991) Dual structure model for the population history of Japan. *Japan Rev.*, **2**, 1–33.

Hanihara, T. and Ishida, H. (2009) Regional differences in craniofacial diversity and the population history of Japan. *Am. J. Phys. Anthropol.*, **139**, 311–322.

Harris, D.R. (1989) An evolutionary continuum of people – plant interaction, in *Foraging and Farming: The Evolution of Plant Exploitation* (eds D.R. Harris and G.C. Hillman), Unwin Hyman, London, pp. 11–26.

Holliday, T.W. (1995) Body Size and Proportions in the Late Pleistocene Western Old World and the Origins of Modern Humans. PhD dissertation, University of New Mexico.

Holliday, T.W. (1997a) Postcranial evidence of cold adaptation in European Neandertals. *Am. J. Phys. Anthropol.*, **104**, 245–258.

Holliday, T.W. (1997b) Body proportions in Late Pleistocene Europe and modern human origins. *J. Hum. Evol.*, **32**, 423–447.

Holliday, T.W. (1999) Brachial and crural indices of European Late Upper Paleolithic and Mesolithic Humans. *J. Hum. Evol.*, **36**, 549–566.

Holliday, T.W. and Ruff, C.B. (1997) Ecogeographic patterning and stature prediction in fossil hominids: Comment on M.R. Feldesman and R.L. Fountain. *Am. J. Phys. Anthropol.*, **103**, 137–140.

Hudson, M.J. (1999) *Ruins of Identity: Ethnogenesis in the Japanese Islands*, University of Hawaii Press, Honolulu.

Imamura, K. (1996a) *Prehistoric Japan: New Perspectives on Insular East Asia*, University of Hawaii Press, Honolulu.

Imamura, K. (1996b) Jomon and Yayoi: the transition to agriculture in Japanese prehistory, in *The Origins and Spread of Agriculture and Pastoralism in Eurasia* (ed. D.R. Harris), Smithsonian Institute Press, Washington DC, pp. 442–464.

Inoue, N., Ito, G. and Kamegai, T. (1986) Dental pathology of hunter-gatherees and early farmers in prehistoric Japan, in *Prehistoric hunter-gatherers in Japan: New Research Methods* (eds T. Akazawa and C.M. Aikins), University Museum, University of Tokyo Bulletin, 27, Tokyo, pp. 163–198.

Kobayashi, T. (2004) *Jomon Reflections: Forager Life and Culture in the Prehistoric Japanese Archipelago*, Oxbow Books, Oxford.

Koga, H. (2003) Stress markers in the ancient people of western Japan. 2. Cribra orbitalia, enamel hypoplasia and the relationship between three stress markers containing Harris' lines. *Anthropol. Sci.*, **111**, 51–67. In Japanese with English abstract.

Koike, H. (1986a) Prehistoric hunting pressure and paleobiomass: An environmental reconstruction and archaeozoological analysis of a Jomon shell mound area, in *Prehistoric Hunter-Gatherers in Japan: New Research Methods* (eds T. Akazawa and C.M. Aikens), Bulletin of the University Museum, University of Tokyo, Tokyo, pp. 27–53.

Koike, H. (1986b) Jomon shell mounds and growth analysis of molluscan shells, in *Windows on the Japanese Past: Studies in Archaeology and Prehistory Center for Japanese Studies* (eds R.J. Pearson, G.L. Barnes and K. Hutterer), University of Michigan, Ann Arbor, pp. 267–279.

Koyama, S. (1978) Jomon subsistence and population. *Senri Ethno. Stud.*, **2**, 1–65.

Larsen, C.S. (1987) Bioarchaeological interpretations of subsistence economy and behavior from human skeletal remains, in *Advances in Archaeological Method and Theory* (ed. M.B. Schiffer), Academic Press, New York, pp. 339–445.

Larsen, C.S. (1995) Biological changes in human populations with agriculture. *Ann. Rev. Anthropol.*, **24**, 185–213.

Larsen, C.S. (1997) *Bioarchaeology: interpreting behavior from the human skeleton*, Cambridge University Press, Cambridge.

Larsen, C.S. (2002) Post-Pleistocene human evolution: bioarchaeology of the agricultural transition, in *Human Diet: Its Origin and Evolution* (eds P.S. Ungar and M.F. Teaford), Bergin and Garvey, West Port, pp. 19–35.

Layrisse, M., Martinez-Torres, C. and Roche, M. (1968) Effect of interaction of various foods on iron absorption. *Am. J. Clin. Nutr.*, **21**, 1175–1183.

Lovejoy, C.O., Russell, K.F. and Harrison, M.L. (1990) Long bone growth velocity in the Libben population. *Am. J. Hum. Biol.*, **2**, 533–541.

Mayr, E. (1963) *Animal Species and Evolution*, The Belknap Press of Harvard University, Cambridge.

Mensforth, R.P. (1985) Relative tibia long bone growth in the Libben and Bt-5 prehistoric skeletal populations. *Am. J. Phys. Anthropol.*, **68**, 247–262.

Mielke, J.H., Konigsberg, L.W. and Relethford, J.H. (2006) *Human Biological Variation*, Oxford University Press, Oxford.

Mizoguchi, Y. (1986) Contributions of prehistoric Far East populations to the population of modern Japan: a Q-mode path analysis based on cranial measurements, in *Prehistoric Hunter-gatherers in Japan: New Research Methods* (eds C.M. Aikens and T. Akazawa), University Museum, University of Tokyo Bulletin 27, Tokyo, pp. 107–136.

Nakahashi, T. (1993) Temporal craniometric changes from the Jomon to the Modern period in western Japan. *Am. J. Phys. Anthropol.*, **90**, 409–425.

Nakajima, T., Nakajima, M. and Yamazaki, T. (2010) Evidence for fish cultivation during the Yayoi period in western Japan. *Int. J. Osteoarchaeol.*, **20** (2), 127–134.

Okazaki, K. (2004) A morphological study on the growth pattern of ancient people in the Northern Kyushu-Yamaguchi region, Japan. *Anthropol. Sci.*, **112**, 219–223.

Omoto, K. and Saitou, N. (1997) Genetic origins of the Japanese: partial support for the dual structure hypothesis. *Am. J. Phys. Anthropol.*, **102**, 437–446.

Ooki, S. and Asaka, A. (1993) Physical growth of Japanese twins. *Acta Genet. Med. Gemell.*, **42**, 275–287.

Oyamada, J., Manabe, Y., Kitagawa, Y. and Rokutanda, A. (1996) Dental morbid condition of hunter-gatherers on Okinawa Island during the middle period of the prehistoric shell midden culture and of agriculturalists in northern Kyushu during the Yayoi period. *Anthro. Sc.*, **104**, 261–280.

Oxenham, M.F. and Matsumura, H. (2008) Oral and physiological paleohealth in cold-adapted peoples: Northeast Asia, Hokkaido. *Am. J. Phys. Anthropol.*, **135**, 64–74.

Oxenham, M.F., Nguyen, K.T. and Nguyen, L.C. (2005) Skeletal evidence for the emergence of infectious disease in Bronze and Iron Age northern Vietnam. *Am. J. Phys. Anthropol.*, **126**, 359–376.

Pietrusewsky, M. (1999) Multivariate craniometic investigations of Japanese, Asians and Pacific Islanders, in *Interdisciplinary Perspectives on the Origins of the Japanese* (ed. K. Omoto), International Research Center for Japanese Studies, Kyoto, pp. 65–104.

Pietrusewsky, M. (2005) The physical anthropology of the Pacific, East Asia, and Southeast Asia: a multivariate craniometric analysis, in *The Peopling of East Asia: Putting Together Archaeology, Linguistics, and Genetics* (eds L. Sargat, R. Blench and A. Sanchez-Mazas), Routledge Curzon, London, pp. 203–231.

Pietrusewsky, M. and Douglas, M.T. (2002) *Ban Chiang, A Prehistoric Village site in Northeastern Thailand. I: The Human Skeletal Remains*, University of Pennsylvania Press, Philadelphia.

Pietrusewsky, M. and Tsang, C.H. (2003) A preliminary assessment of health and disease in human skeletal remains from Shi San Hang: A prehistoric aboriginal site on Taiwan. *Anthropol. Sci.*, **111**, 203–223.

Rivera, J.A., Hotz, C., González-Cossío, T. and García-Guerra, A. (2003) The effect of micronutrient deficiencies on child growth: a review of results from community based supplemental trials. *J. Nutr.*, **133**, 4010S–4020.

Ruff, C.B. (1994) Morphological adaptation to climate in modern and fossil hominids. *Yrbk Phys. Anthropol.*, **37**, 65–107.

Ruff, C.B. (2002) Variation in human body size and shape. *Ann. Rev. Anthropol.*, **31**, 211–232.

Ruff, C.B. and Walker, A. (1993) Body size and body shape, in *The Nariokotome Homo Erectus Skeleton* (eds A. Walker and R. Leakey), Harvard University Press, Cambridge, pp. 234–265.

Ruff, C.B., Scott, W.W. and Liu, A.Y.C. (1991) Articular and diaphyseal remodeling of the proximal femur with changes in body mass in adults. *Am. J. Phys. Anthrol.*, **86**, 397–413.

Sanui, Y. (1960) Anthropological researches on the teeth of the prehistoric Yayoi-ancients excavated from the Doigahama site, Yamaguchi Prefecture. *J. Anthro. Sc. Nipp.*, **3**, 861–864.

Seguchi, N., Umeda, H., Nelson, A.R. and Brace, C.L. (2007) Population movement into the Japanese archipelago during antiquity: a craniofacial and odontometric perspective. *Am. J. Phys. Anthro. Supp.* **544**, 213.

Shackelford, L.L. (2005) Regional Variation in the Postcranial Robusticity of Late Upper Paleolithic Humans. PhD dissertation, Washington University, St. Louis.

Slome, D. (1929) The osteology of a Bushmen tribe. *Ann. S. Afr. Mus.*, **24**, 33–60.

Sokal, R.R. and Rohlf, F.J. (1995) *Biometry*, W.H. Freeman, New York.

Solomons, N.W., Mazariegos, M., Brown, K.H. and Klasing, K. (1993) The underprivileged developing country child: Environmental contamination and growth failure revisited. *Nutr. Rev.*, **51**(11), 327–332.

Stinson, S. (2000) Growth variation: Biological and cultural factors, in *Human Biology: An Evolutionary and Biocultural Perspective* (eds S. Stinson, B. Bogin, R. Huss-Ashmore and D. O'Rourke), Wiley-Liss, New York, pp. 425–463.

Suzuki, T. and Inoue, T. (2006) Earliest evidence of spinal tuberculosis from the Aneolithic Yayoi period in Japan. *Int. J. Osteoarchaeol.*, **17**, 392–402.

Suzuki, T., Fujita, H. and Choi, J.G. (2008) Brief communication: New evidence of tuberculosis from prehistoric Korea – Population movement and early evidence of tuberculosis in far East Asia. *Am. J. Phys. Anthropol.*, **136**, 357–360.

Takamura, K., Ohyama, S., Yamada, T. and Ishinishi, N. (1988) Changes in body proportions of Japanese medical students between 1961–1986. *Am. J. Phys. Anthropol.*, **77**, 17–22.

Tanaka, Y., Mizoguchi, K., Iwanaga, S. and Higham, T. (2005) A preliminary report on the AMS dating of human and animal skeletal remains of Yayoi period Northern Kyushu, Japan. *Bull. Japan Soc. Prom. Sc.: East Asia and Japan: Interactions and Transformations*, **3**, 39–47.

Tanner, J.M., Hayashi, T., Preece, M.A. and Cameron, N. (1982) Increase in leg length relative to sitting height in Japanese children from 1957–1977: a comparison with British and with Japanese Americans. *Ann. Hum. Biol.*, **9**, 411–423.

Temple, D.H. (2007a) Dietary variation and stress among prehistoric Jomon foragers from Japan. *Am. J. Phys. Anthropol.*, **133**, 1035–1046.

Temple, D.H. (2007b) Human Biological Variation during the Agricultural Transition in Prehistoric Japan. PhD dissertation, The Ohio State University.

Temple, D.H. (2008) What can stature variation reveal about environmental differences between prehistoric Jomon foragers? Understanding the impact of systemic stress on developmental stability. *Am. J. Hum. Biol.*, **20**, 431–439.

Temple, D.H. (2010) Patterns of systemic stress during the agricultural transition in prehistoric Japan. *Am. J. Phys. Anthro.*, **142**, 112–114.

Temple, D.H. and Larsen, C.S. (2007) Dental caries prevalence as evidence for agriculture and subsistence variation during the Yayoi period in prehistoric Japan: Biocultural interpretations of an economy in transition. *Am. J. Phys. Anthro.*, **134**, 501–512.

Temple, D.H., Auerbach, B.M., Nakatsukasa, M. and Larsen, C.S. (2008) Variation in limb proportions between Jomon foragers and Yayoi agriculturalists from prehistoric Japan. *Am. J. Phys. Anthropol.*, **137**, 164–174.

Todaka, Y., Oyamada, J., Manabe, Y., Kitagawa, Y., Kato, K and Rokutanada, A. (2003) The relationship between immigration and the prevalence of dental caries in the Yayoi people. *Antrop. Sc.*, **111**, 265–292.

Trinkaus, E. (1981) Neandertal limb features and cold adaptation, in *Aspects of Human Evolution* (eds C.B. Stringer), Taylor and Francis Ltd, London, pp. 187–224.

Tsude, H. (2001) Yayoi farmers reconsidered: new perspectives on agricultural development in East Asia. *Bull. Indo-Pac. Prehist. Assoc.*, **5**, 53–59.

Tsukada, M. (1986) Vegetation in prehistoric Japan: the last 20 000 years, in *Windows of the Japanese Past: Studies in Archaeology and Prehistory* (eds R.J. Pearson, G.L. Barnes and K.L. Hutterer), Center for Japanese Studies, University of Michigan, pp. 11–56.

Tsutakawa, R.K. and Hewett, J.E. (1977) Quick test for comparing two populations with bivariate data. *Biometrics*, **33**, 215–219.

Turner, C.G. (1979) Dental anthropological indications of agriculture among the Jomon people of central Japan. *Am. J. Phys. Anthropol.*, **51**, 619–636.

Turner, C.G. (1990) Major features of sundadonty and sinodonty, including suggestions about East Asian microevolution, population history and Late Pleistocene relationships with Australian Aboriginals. *Am. J. Phys. Anthropol.*, **82**, 295–317.

Turner, C.G. (1992) Sunadonty and sinodonty in Japan: the dental basis for a dual origin hypothesis for the peopling of the Japanese Islands, in *Japanese as a Member of the Asian Pacific Populations* (ed. K. Hanihar), International Research Center for Japanese Studies, Kyoto, pp. 97–111.

Ulijaszek, S. (1994) Between-population variation in preadolescent growth. *Eur. J. Clin. Nutr.*, **48**, S5–S14.

Wada, Y. and Motomura, H. (2000) Temporal changes in stature of western Japanese based on limb characteristics. *Anthropol. Sci.*, **108**, 147–168.

Watanabe, N. (1966) Radiocarbon dates of the Jomon and Yayoi periods in Japan Jomon oyobi Yayoi jidai no C14 nendai. *Quart. Res. Daiyonki kenkyu*, **5**, 157–168. In Japanese.

Yafeng, S., Zhaozheng, K., Sumin, W. *et al.* (1993) Mid-Holocene environments and climates in China. *Global Planet. Change*, **7**, 219–233.

Yamaguchi, B. (1989) Limb segment proportions in human skeletal remains of the Jomon period. *Bull. Nat. Sc. Mus. Series D*, **18**, 41–48.

Yamamoto, M. (1992) Secular trends of enamel hypoplasia in Japanese from the prehistoric to modern times. *J. Paleopathol. Monograph.* Pub. 2, 231–238.

SECTION C

Biomechanics and Indicators of Habitual Activity

11

The Bioarchaeology of Habitual Activity and Dietary Change in the Siberian Middle Holocene

A.R. Lieverse[1], Jay T. Stock[2], M.A. Katzenberg[3] and C.M. Haverkort[4]

[1] *Department of Archaeology and Anthropology, University of Saskatchewan, Saskatoon, Canada*
[2] *Leverhulme Centre for Human Evolutionary Studies, Department of Biological Anthropology, University of Cambridge, Cambridge, UK*
[3] *Department of Archaeology, University of Calgary, Calgary, Canada*
[4] *Department of Anthropology, University of Alberta, Edmonton, Canada*

11.1 INTRODUCTION

The abundance of middle Holocene human remains from the Cis-Baikal region of Siberia provides a unique opportunity to investigate hunter-gatherer adaptation across periods of substantial cultural change. In this paper, we examine evidence of changing mobility patterns and habitual behaviour amongst the region's Neolithic and Bronze Age occupants by integrating three lines of evidence: musculoskeletal stress markers, postcranial robusticity and the isotopic chemistry of bone and teeth. In Siberia, the terms Neolithic and Bronze Age are by convention not used to characterize agriculture and sedentism, but rather refer to the appearance of pottery and ground stone technology and the introduction of copper and/or bronze objects, respectively (Weber and Bettinger, 2003). Despite this, the region's cultural history is of great interest to us, because the Cis-Baikal is one of the few places where diachronic cultural and behavioural change can be studied within hunter-gatherers prior to the adoption of agriculture. This cultural transition is accentuated by an 800 to 1000 year interruption in the region's culture history sequence dating to the seventh millennium BP, or the Middle Neolithic period (about 7000/6800–6000/5800 calBP; Weber, 1995; Weber, Link and Katzenberg, 2002). In order to explain the circumstances surrounding this hiatus, much of our efforts have focused on better understanding the biologically and culturally distinct populations lying on either side of it, namely the pre-hiatus Kitoi culture, dating to the Early Neolithic period (8000–7000/6800 calBP), and the post-hiatus Isakovo-Serovo-Glaskovo

Human Bioarchaeology of the Transition to Agriculture Edited by Ron Pinhasi and Jay T. Stock

(ISG) cultural complex, spanning the Late Neolithic and Bronze Age (6000/5800–4000 calBP). In this chapter, we attempt to synthesize data obtained from the Kitoi and ISG skeletal records pertaining to behaviour and mobility.

11.1.1 Geographical and Archaeological Context

Lake Baikal is located in the southern part of eastern Siberia (Russia), between 52 and 58 degrees North Latitude and less than 200 km north of the Mongolian border (Figure 11.1). The Baikal Mountain Region, the vast area surrounding the lake, is divided into the Cis-Baikal to the north and west and the Trans-Baikal to the south and east. Because the two regions exhibit distinctive culture histories (Khazonov, 1994: 91; Kuzmin and Orlova, 2000), our research focuses only on the Cis-Baikal. Its geographical definition, following that adopted by Michael (1958:5), includes the basin of the Angara River from its source at the lake to Ust'-Ilimsk, the drainage of the upper Lena River to Kirensk, and the west coast of Baikal itself, including its largest island, Ol'khon (Figure 11.2).

Today, the Cis-Baikal is situated within a southern boreal forest (or taiga) biome, bordering the steppe-forest transitional zone on its southern extent. It exhibits a noticeably continental climate, with long cold winters and short mild summers. The region is rich in resources, providing a broad range of potential food sources to foraging peoples. While the taiga consists predominantly of larch (*Larix*), spruce (*Picea*), pine (*Pinus*) and fir (*Abies*), a wide variety of plant foods (e.g. berries, pine nuts and mushrooms) are available seasonally. Terrestrial fauna

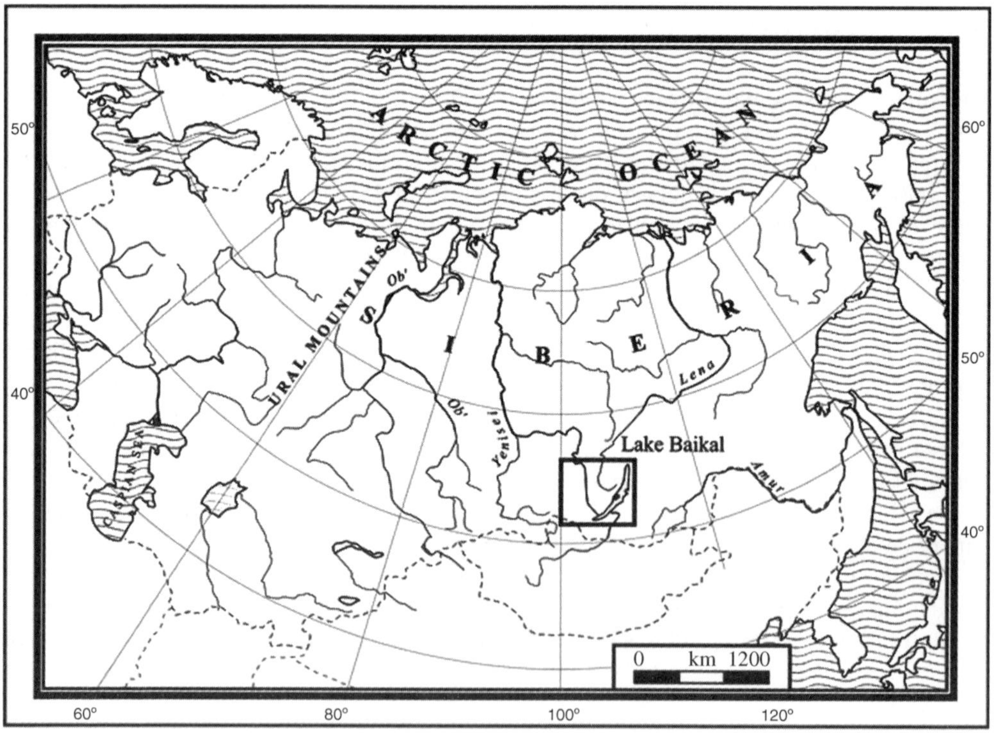

Figure 11.1 Siberia and the Cis-Baikal region

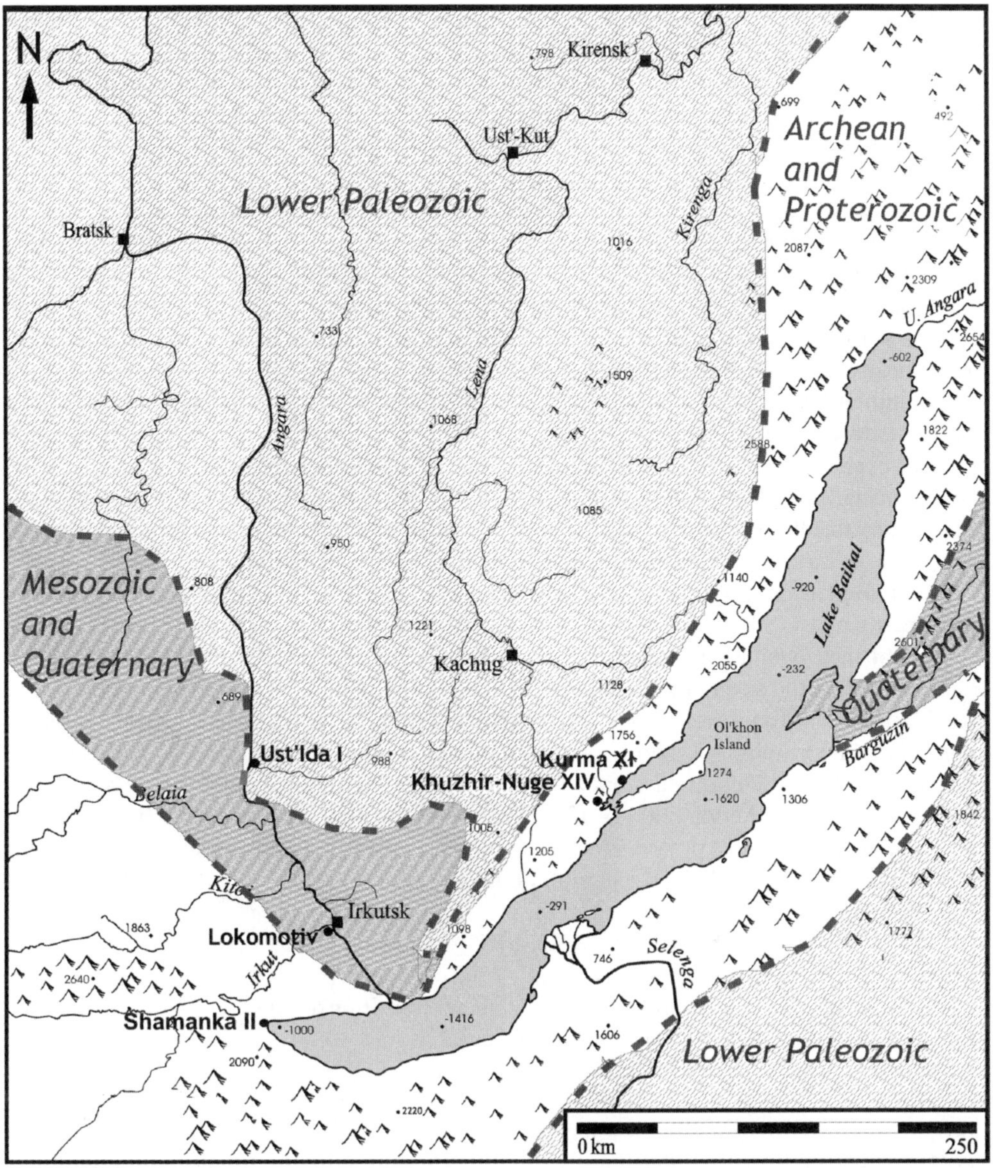

Figure 11.2 Map of the Cis-Baikal showing the location of the cemetery sites: Lokomotiv (Kitoi), Shamanka II (Kitoi), Ust'-Ida I (ISG), Khuzhir-Nuge XIV (ISG), Kurma XI (ISG). Also indicated are the main geochemical zones: Archaean and Proterozoic granites ($^{87}Sr/^{86}Sr > 0.720$); Mesozoic and Quaternary deposits ($^{87}Sr/^{86}Sr$ 0.705–0.710); Lower Palaeozoic – Cambrian and Precambrian limestone ($^{87}Sr/^{86}Sr \sim 0.709$)

are abundant and diverse in the region, with over 100 mammalian and 300 avian species identified. Of these, mammals, particularly ruminants such as red deer (*Cervus elaphus*), roe deer (*Capreolus capreolus*), elk (*Alces alces*) and reindeer (*Rangifer tarandus*), have traditionally supplied the bulk of human dietary requirements (Cordain *et al.*, 2002; Helm, 1981; Levin and Potapov, 1965; Ziker, 2002). Aquatic fauna is equally rich in the Cis-Baikal, being concentrated in its three main basins (the lake itself, the Angara and the upper Lena). In addition to fish such as grayling (*Thymallus* spp.), whitefish (*Coregonus* spp.), northern pike (*Esox lucius*), taimen (*Hucho taimen*) and lenok (*Brachymystax lenok*), freshwater seal (*Phoca sibirica*) is also available from the lake.

In comparison to the region's modern climate (also undergoing drastic change, Moore *et al.*, 2009), it is likely that environmental conditions in the Cis-Baikal were significantly different during the middle Holocene period. Global circulation models (Bush, 2007) and proxy data (Prokopenko *et al.*, 2007; White *et al.*, 2008) suggest that there has been a general trend of warming and drying in the Cis-Baikal since the end of the Pleistocene, with the most pronounced changes occurring between 7000 and 6000 years ago, coincident with the seventh millennium BP hiatus (see below). Furthermore, changes in vegetation (e.g. expansion of coniferous forests in the cooler and wetter early Holocene) appear to have migrated from south to north over time. While it is beyond the scope of this paper to discuss or speculate on environmental fluctuations and resource availability during the middle Holocene period, it seems likely that these changing conditions would have affected hunter-gatherer mobility and behaviour. At this point, these issues must remain subjects for future investigation.

The Cis-Baikal archaeological record documents more or less continuous human occupation for at least the last 20 000 years (Goebel, 1999; Weber, 1995). One notable exception to this well-established habitation sequence is a recently delineated period of substantial cultural transition during the seventh millennium BP. The current culture history model (Table 11.1), based on a mid-1990s re-evaluation of archaeological material and radiocarbon dates, reveals a considerable interruption in the region's cultural continuity (Weber, 1995; Weber, Link and Katzenberg, 2002). This hiatus, dating to the Middle Neolithic (about 7000/6800–6000/5800 calBP) and spanning 800 to 1000 years, is reflected archaeologically by both a modification to mortuary protocols (i.e. the absence of formal cemeteries) and an almost complete lack of human skeletal material. Equally noteworthy is the fact that the populations lying on either side of this discontinuity – the pre-hiatus (Early Neolithic) Kitoi peoples and the post-hiatus (Late Neolithic-Bronze Age) ISG – are biologically and culturally distinct (Weber, 1995; Weber, Link and Katzenberg, 2002).

The disclosure of a major period of hunter-gatherer transition during the Cis-Baikal's middle Holocene has prompted extensive scholarly research in a variety of disciplines, including archaeology, palaeoenvironmental studies, human osteology and ancient DNA analyses (Haverkort *et al.*, 2008; Katzenberg, Goriunova and Weber, 2009; Lieverse *et al.*, 2007a, 2007b, 2009; Losey, Nomokonova and Goriunova, 2008; Mooder *et al.*, 2005, 2006; Weber, 1995; Weber, Link and Katzenberg, 2002; Weber *et al.*, 2006; Weber, Katzenberg and Schurr, 2010; White *et al.*, 2008). Much of this research has focused on the populations pre- and

Table 11.1 Culture History Model of the Middle Holocene Cis-Baikal

Period	Culture/Mortuary Complex	Calibrated Age BP
Early Neolithic	Kitoi and Others	8000–7000/6800
Middle Neolithic	Hiatus	7000/6800–6000/5800
Late Neolithic	Isakovo-Serovo	6000/5800–5200
Bronze Age	Glaskovo	5200/5000–4000

postdating the discontinuity, particularly their biological and cultural characteristics. Genetic distinctions between the Kitoi and the ISG have been clearly demonstrated (Mooder *et al.*, 2005, 2006), as have some cultural differences such as those pertaining to mortuary ritual and material culture (Weber, 1995; Weber, Link and Katzenberg, 2002). On the other hand, possible disparities in adaptive strategies and their consequent effects (if any) of the nature of the transition remain incompletely understood. For example, while archaeological data (e.g. site location and distribution, cemetery size, grave inclusions) suggest that the Kitoi were characterized by larger, fewer and more sedentary community groups with higher (male-dominated) logistical mobility and an increased emphasis on fishing (Weber, Link and Katzenberg, 2002), this is only partially supported by human osteological evidence, whether biochemical or morphological (Haverkort *et al.*, 2008; Katzenberg, Goriunova and Weber, 2009; Lieverse *et al.*, 2007a, 2007b; Stock *et al.*, 2010). Not only do a number of behavioural characteristics appear to be held in common by all inhabitants of the region, Kitoi and ISG alike, but some of the distinctions noted amongst study populations seem to reflect geographical or environmental differences rather than cultural ones (Lieverse *et al.*, 2007a, 2007b).

Here we examine mobility patterns, diet and habitual behaviour in the middle Holocene Cis-Baikal, in order to more fully explain the adaptive strategies employed by the region's pre- and post-hiatus occupants (the Kitoi and ISG). Three lines of enquiry are investigated: musculoskeletal stress markers (MSM) of the upper limb, postcranial (femoral and humeral) robusticity and stable isotopes of carbon ($\delta^{13}C$), nitrogen ($\delta^{15}N$) and strontium ($^{87}Sr/^{86}Sr$). Human skeletal and dental remains from five Cis-Baikal cemetery populations – two predating the hiatus and three postdating it – are analysed, providing an ideal opportunity to study diachronic trends in hunter-gatherer subsistence, adaptation and cultural change prior to the adoption of agriculture.

11.1.2 Skeletal Morphology

Studies documenting the relationship between skeletal morphology and behaviour are abundant (Angel *et al.*, 1987; Churchill and Morris, 1998; Daly *et al.*, 2004; Eshed *et al.*, 2004; Hawkey and Merbs, 1995; Hawkey, 1988, 1998; Kelley and Angel, 1987; Lai and Lovell, 1992; Molnar, 2006; Ruff, Holt and Trinkaus, 2006; Robling, Burr and Turner, 2000; Warden *et al.*, 2005; Steen and Lane, 1998; Stirland, 1993; Weiss, 2003, 2004, 2007). All are based on the model of bone functional adaptation (sometimes referred to as Wolff's Law; Wolff, 1892), maintaining that bone form reflects the cumulative effects of *intra vitam* mechanical loading or strain (Ruff, Holt and Trinkaus, 2006). In the case of MSM, or activity-induced changes to muscle and ligament attachment sites on the skeleton, high levels of activity increase strain on attachment sites, stimulating osteoblastic activity and consequently affecting site morphology (i.e. increasing rugosity, furrowing, or ossification). While the effects of many other factors – including sex, body size, age at death and (in the case of upper limbs) handedness – are incompletely understood (Cook and Dougherty, 2001; Eshed *et al.*, 2004; Hawkey and Merbs, 1995; Molnar, 2006; Nagy and Hawkey, 1995; Nagy, 1998, 1999; Robb, 1998; Ruff, Holt and Trinkaus, 2006; Steen and Lane, 1998; Weiss, 2007; Wilczak, 1998), prominent MSM are typically the result of extensive muscle or ligament use, both in degree and duration, over the span of much or all of an individual's lifetime (Ruff, Holt and Trinkaus, 2006; Weiss, 2003, 2004, 2007).

Changes to skeletal robusticity, defined in long bones as the thickness of the shaft relative to its length (Martin and Saller, 1957) and more recently quantified by cross-sectional cortical geometry (Ruff, 2000), work in much the same way. Although the process of bone functional adaptation is complicated and leaves many issues still unresolved (Lanyon, 1992; Pearson and Lieberman, 2004), there remains considerable evidence for a direct link between mechanical loads and the geometric characteristics of long bone diaphyses (Ruff, Holt and Trinkaus, 2006; Robling, Burr and Turner, 2000; Warden *et al.*, 2005; Daly *et al.*, 2004). On this basis, measures of skeletal robusticity have frequently been used to interpret patterns of prehistoric behaviour (Ruff, 2008; and see Larsen and Ruff, Marchi *et al.*, and Stock *et al.*, this volume), including hunting and gathering (Stock and Pfeiffer, 2001, 2004).

Recent evidence provides a link between characteristics of long bone diaphyseal robusticity and shape, and specific patterns of habitual locomotion amongst athletes (Shaw and Stock, 2009a, 2009b). This provides strong evidence to support the interpretation of behaviour from diaphyseal robusticity in prehistoric contexts. Mobility patterns, whether terrestrial or marine through the use of watercraft, represent the greatest single characteristic of hunter-gatherer behaviour and biomechanical loading of the skeleton, and there appear to be clear differences between terrestrially- and marine-based foragers in limb biomechanics (Stock, 2006). There is evidence that the relative symmetry (circularity) of strength characteristics of the midshaft femur, quantified as the ratio of antero-posterior to medio-lateral bending strength, correlates with the general intensity of terrestrial locomotion in human populations (Ruff, 1987). A relationship between femoral midshaft shape and locomotion has been supported amongst hunter-gatherers (Stock, 2006). In addition to characteristics of diaphyseal shape, the comparison of biomechanical characteristics of the humerus and femur of the Kitoi and ISG to other hunter-gatherers with known behavioural repertoires provides a means of investigating the intensity of biomechanical loading through time. The ratio of humeral to femoral strength is of particular utility in identifying patterns of habitual mechanical loading of the upper limbs relative to the lower, and should provide a means of identifying the relative importance of watercraft use (Stock, 2006), an issue which remains unresolved amongst the middle Holocene foragers of the Cis-Baikal.

11.1.3 Stable Isotopes

Stable isotopes of carbon and nitrogen in preserved tissues (most often bone collagen) reveal information about past diet (reviewed by Katzenberg, 2008; and see other chapters in this volume by Schulting, Lillie and Budd, Papathanassiou, Grupe and Peters). Stable carbon isotopes indicate the types of plants consumed, because photosynthesis may occur by one of three different pathways that differentially discriminate against the heavier isotope, ^{13}C. Stable carbon isotopes also differentiate marine from terrestrial resource use based on differences in the $\delta^{13}C$ of the source carbon in the atmosphere and in the ocean. In freshwater ecosystems, such as Lake Baikal, source carbon is varied and fractionation varies, depending on factors such as temperature and water depth (Hecky and Hesslein, 1995). Variation in stable isotopes of nitrogen reflects foods from different trophic levels. In this region, two different food webs exist, a terrestrial food web and an aquatic one. The aquatic food web includes more steps and the high trophic level species, such as the Baikal seal and fish such as pike and sturgeon, have $\delta^{15}N$ that is more enriched in the heavier isotope than high trophic level species in the terrestrial food web. Stable isotopes of carbon and nitrogen from preserved bone collagen have been analysed from both human and faunal skeletal remains, for the purpose of reconstructing past diet in the Cis-Baikal (Katzenberg and Weber, 1999). Both prehistoric and modern mammals as well as modern fish bones have been studied in an effort to document the variation in potential foods available to the region's middle Holocene occupants (Katzenberg, Goriunova and Weber, 2009; Katzenberg *et al.*, 2010). Because plant foods in the region all follow the C_3 pathway for photosynthesis, they present little variation for stable carbon isotopes amongst plants, the terrestrial herbivores that feed on them and the human consumers of these herbivores. Furthermore, northern people generally rely more heavily on animals as sources of food than they do on plants (Cordain *et al.*, 2002) and this was undoubtedly the case amongst the ancient foragers of the Cis-Baikal. Therefore, sources of variation for stable carbon isotopes in the region largely reflect different populations of fish inhabiting various regions of the lake and its rivers, while variation in stable nitrogen isotopes represents differential reliance on terrestrial mammals vs. aquatic species (i.e. fish and seals; Katzenberg and Weber, 1999; Katzenberg *et al.*, 2010). Because the cemetery sites under consideration here vary geographically, in addition to chronologically, these differences should be reflected in the stable carbon and nitrogen ratios of the human remains recovered from them. Sites located on the shores of Baikal are expected to reflect diets with more aquatic species, including the Baikal seal (*Phoca sibirica*), while those along the Angara River should reflect diets with more terrestrial resources and riverine fish but few, if any, lake species.

Strontium (Sr) isotope ratios ($^{87}Sr/^{86}Sr$) have been used in studies of prehistoric mobility and migration for almost two decades (Bentley, 2006). The method is based on the fact that the Sr isotope ratios characterizing local bedrock can also be found in local water sources, the tissues of local flora and fauna and, via dietary intake, in the human body (Beard and Johnson, 2000; Curzon and Cutress, 1983; Ericson, 1985, 1989). Because of the similarity in chemical properties, the element Sr can replace calcium (Ca) in the hydroxyapatite lattice of hard tissues such as bone and teeth. A critical difference between tooth enamel and bone – enamel does not remodel once formed, whereas bone is remodelled throughout life – allows us to compare the Sr isotope signature characteristic of the general geographical area where a person resided during childhood and adolescence (enamel samples) with the signature characteristic of the general geographic area where a person resided during the last years of life (bone samples).

The $^{87}Sr/^{86}Sr$ method has so far been applied in studies of individual migration patterns between presumably largely sedentary agricultural communities, whereby immigrants are identified by their non-local Sr-isotope signatures (Bentley *et al.*, 2003; Bentley, Price and Stephan, 2004; Ezzo, Johnson and Price, 1997; Ezzo and Price, 2002; Grupe *et al.*, 1997; Hodell *et al.*, 2004; Knudson *et al.*, 2004; Montgomery, Budd and Evans, 2000; Price, Grupe and Schröter, 1994, 1998; Price, Manzanilla and Middleton, 2000; Price *et al.*, 1994b, 2001; Schweissing and Grupe, 2003; Tafuri *et al.*, 2006). The feasibility of using this method to study forager mobility in the Cis-Baikal region was demonstrated in a pilot study (Weber *et al.*,2003). Our use of the Sr isotope methodology in the context of hunter-gatherer mobility patterns differs in several respects from most other studies carried out so far. Most importantly, Cis-Baikal forager mobility patterns were likely considerably more complex than those for populations with an agriculture-based subsistence system, given the degree of spatial and temporal variation in resource availability in the region. While this overall complexity creates some challenges for the interpretation of the Sr isotope data, the availability of other skeletal and isotope data pertaining to subsistence-related mobility and behavioural patterns can assist with the analysis. Importantly, because Sr isotopes in hard tissues derive from dietary intake, an investigation of multiple variables pertaining to subsistence-related behaviour, such as presented here, has the potential to make valuable contributions to the relatively new application of Sr isotopes in the context of hunter-gatherer mobility. The use of watercraft in particular has important implications for dietary intake – and therefore 'Sr-catchment' – as the use of boats can dramatically increase the potential foraging range, and thereby access to different types of resources.

11.2 MATERIALS

This paper examines human skeletal remains from five Cis-Baikal cemeteries – two pre-dating the seventh millennium BP hiatus and three post-dating it – in order to examine evidence of changing mobility and behavioural patterns throughout the mid-Holocene. The pre-hiatus sites of Lokomotiv (n = 99) and Shamanka II (n = 140), located on the Angara River and at the southern tip of Lake Baikal, respectively (Figure 11.1), both represent the Early Neolithic Kitoi culture. The three post-hiatus cemeteries include the site of Ust'-Ida I (n = 67), represented by both Late Neolithic Isakovo-Serovo and Bronze Age Glaskovo components and located on the Angara River, as well as the two Bronze Age Glaskovo cemeteries of Khuzhir-Nuge XIV (n = 81) and Kurma XI (n = 20), both located in Baikal's Little Sea micro-region, across from Ol'khon Island (Figure 11.1). While the Ust'-Ida I cemetery is represented by two chronologically discrete components (Isakovo-Serovo and Glaskovo), the general consensus is that these represent two horizons of the same cultural complex, the Isakovo-Serovo-Glaskovo (or ISG) (Weber, 1995; Weber, Link and Katzenberg, 2002). Thus, in order to simplify analyses and maximize sample sizes, the two components of Ust'-Ida I are grouped together as one population here.

Sex assessment and age estimation were based largely on the standards presented by Buikstra and Ubelaker (1994:16–21), with as many techniques as possible being applied to each individual. However, even with the application of numerous metric and non-metric methods, sex assessment and age estimation were often compromised by poor preservation and incomplete element representation, particularly at the ISG sites of Khuzhir-Nuge XIV and Kurma XI. As a result, a number of individuals had to be excluded from those analyses controlling for sex and/or age at death and, in many cases, sample sizes were substantially smaller than the total number of individuals represented by the five cemeteries indicated above.

11.3 METHODS

11.3.1 Musculoskeletal Stress Markers (MSM)

Twenty-four upper limb MSM sites were documented, as presented in Table 11.2. MSM were recorded using the graded visual reference system developed by Hawkey (1988) and outlined by Hawkey and Merbs (1995) for robusticity, stress lesions and ossification. Because robusticity and stress lesions typically represent a continuum of increasing muscle and/or ligament strain, 'total' MSM for each individual were calculated by combining robusticity and stress lesion scores (as per Hawkey and Merbs, 1995). 'Total' scores were thus assigned as follows: 0 = no expression, 1 = faint robusticity, 2 = moderate robusticity, 3 = strong robusticity, 4 = faint stress lesion, 5 = moderate stress lesion and 6 = strong stress lesion. In the interest of maximizing sample sizes, MSM scores were documented for all available (left and/or right) attachment sites, being averaged if both sides were recorded. Upper limb asymmetry (i.e. reflecting differences in limb use or handedness) was found to be low for all five skeletal

Table 11.2 Summary of MSM scores and ranks

No.	MSM Site	Lokomotiv (Kitoi)						Shamanka II (Kitoi)						All Kitoi					
		Males			Females			Males			Females			Males			Females		
		x	n	r	x	n	r	x	n	r	x	n	r	x	n	r	x	n	r
1	C: Conoid ligament	2.56	18	19	2.50	22	**21**	2.51	41	**21**	2.27	13	**22**	2.53	59	**21**	2.41	35	**22**
2	C: Costoclavicular ligament	2.84	19	**24**	2.74	19	**24**	4.11	41	**24**	2.50	11	**24**	3.71	60	**24**	2.65	30	**24**
3	C: Subclavius	1.71	19	3	1.75	22	5	1.68	41	5	1.23	13	3	1.69	60	3	1.56	35	4
4	C: Trapezoid ligament	1.97	16	6	1.95	22	10	2.02	41	13	1.81	13	10	2.01	57	8	1.90	35	8
5	S: Pectoralis minor	2.43	7	15	1.86	7	6	1.65	26	4	1.58	6	5	1.82	33	5	1.73	13	6
6	S: Trapezius	2.04	12	7	1.92	6	8.5	1.63	26	3	1.92	6	12.5	1.76	38	4	1.92	12	11.5
7	H: Coracobrachialis	1.22	30	1	0.89	27	1	1.07	46	1	0.82	14	1	1.13	76	1	0.87	41	1
8	H Deltoid	2.68	31	**22**	2.37	27	19	2.58	46	**22**	2.25	14	**21**	2.62	77	**22**	2.33	41	19
9	H: Flexors (O)	2.16	29	10	2.00	23	11.5	2.01	45	12	1.92	13	12.5	2.07	74	12.5	1.97	36	14
10	H: Extensors (O)	2.52	25	17	2.35	20	18	2.27	46	17	2.15	13	19	2.36	71	17	2.27	33	18
11	H: Infraspinatus	1.82	17	5	1.74	17	4	1.86	43	7	1.56	9	4	1.85	60	6	1.67	26	5
12	H: Pectoralis major	2.76	31	**23**	2.54	27	**22**	2.59	46	**23**	2.07	14	18	2.66	77	**23**	2.38	41	**21**
13	H: Supraspinatus	2.08	18	8	1.89	18	7	2.00	42	10.5	1.73	11	9	2.03	60	10.5	1.83	29	7
14	H: Brachioradialis (O)	2.53	29	18	2.40	24	**20**	2.34	46	18	2.32	14	**23**	2.41	75	18	2.37	38	**20**
15	H: Latissimus dorsi	2.27	30	12.5	2.02	26	13.5	2.22	46	15.5	1.71	14	7.5	2.24	76	14.5	1.91	40	9.5
16	H: Teres major	2.27	30	12.5	2.02	26	13.5	2.22	46	15.5	1.71	14	7.5	2.24	76	14.5	1.91	40	9.5
17	H: Teres minor	2.13	16	9	2.00	16	11.5	1.99	40	8.5	1.89	9	11	2.03	56	10.5	1.96	25	13
18	R: Pronator teres	1.41	23	2	1.17	21	3	1.40	43	2	0.89	14	2	1.40	66	2	1.06	35	2
19	R: Biceps brachii	1.75	26	4	1.11	22	2	1.99	44	8.5	1.64	14	6	1.90	70	7	1.32	36	3
20	U: Supinator (O)	2.18	28	11	1.92	24	8.5	2.00	44	10.5	1.93	14	14	2.07	72	12.5	1.92	38	11.5
21	U: Pronator quadratus (O)	2.50	23	16	2.68	17	**23**	2.43	42	**20**	2.18	14	**20**	2.45	65	**19.5**	2.45	31	**23**
22	U: Anconeus	2.66	25	**21**	2.11	22	16	2.07	42	14	2.00	14	16	2.29	67	16	2.07	36	16
23	U: Brachialis	2.57	28	**20**	2.22	23	17	2.37	45	19	2.04	14	17	2.45	73	**19.5**	2.15	37	17
24	U: Triceps brachii	2.38	25	14	2.09	22	15	1.82	44	6	1.96	14	15	2.02	69	9	2.04	36	15

(continued)

Table 11.2 (*Continued*)

No.	MSM Site	Ust'-Ida I (ISG)						Khuzhir-Nuge XIV (ISG)						Kurma XI (ISG)						All ISG					
		Males			Females			Males			Females			Males			Females			Males			Females		
		x	n	r	x	n	r	x	n	r	x	n	r	x	n	r	x	n	r	x	n	r	x	n	r
1	C: Conoid ligament	2.19	13	19	2.30	5	**20.5**	2.33	9	18	2.00	2	/	2.00	4	**18**	2.00	5	**17.5**	2.21	26	19	2.13	12	**20**
2	C: Costoclavicular ligament	2.75	12	**24**	2.50	5	**23**	2.92	6	**23**	/	0	/	2.00	4	**18**	1.67	3	12.5	2.66	22	**24**	2.19	8	**21.5**
3	C: Subclavius	1.71	12	5	2.10	5	17	2.00	5	11	2.00	1	/	1.75	4	11.5	1.67	3	12.5	1.79	21	9	1.94	9	16.5
4	C: Trapezoid ligament	1.95	11	14	2.00	4	15.5	2.00	7	11	2.00	2	/	2.00	3	**18**	1.75	2	15	1.98	21	14	1.94	8	16.5
5	S: Pectoralis minor	1.79	7	7	1.67	3	8	2.00	1	11	/	0	/	/	0	/	/	0	/	1.81	8	11	1.67	3	9.5
6	S: Trapezius	1.50	2	4	1.88	4	14	3.00	1	**24**	/	0	/	/	0	/	/	0	/	2.00	3	15	1.88	4	14
7	H: Coracobrachialis	1.28	16	1	0.95	10	1	1.08	12	2	/	0	/	1.17	6	3	0.83	3	1	1.19	34	1	0.92	13	1
8	H Deltoid	2.41	17	**22**	2.15	10	18	2.32	14	17	2.00	2	/	2.07	7	**20**	2.00	5	**17.5**	2.32	38	**22**	2.09	17	18
9	H: Flexors (O)	1.82	11	10	1.57	7	4	1.80	5	6	/	0	/	1.50	2	8	1.50	2	9.5	1.78	18	7.5	1.56	9	5.5
10	H: Extensors (O)	2.18	11	18	1.83	6	12	2.00	2	11	/	0	/	1.00	1	1.5	1.00	1	3.5	2.07	14	16	1.71	7	13
11	H: Infraspinatus	1.83	9	11	1.60	5	5	2.00	2	11	/	0	/	1.00	3	1.5	/	0	/	1.68	14	4	1.60	5	7
12	H: Pectoralis major	2.44	17	**23**	2.30	10	**20.5**	2.58	13	**22**	2.00	1	/	1.86	7	15	2.00	5	**17.5**	2.38	37	**23**	2.19	16	**21.5**
13	H: Supraspinatus	1.89	9	13	1.75	6	10	2.00	1	11	/	0	/	1.50	2	8	1.00	2	3.5	1.83	12	12	1.56	8	5.5
14	H: Brachioradialis (O)	2.31	16	**21**	2.28	9	19	2.38	8	19	2.00	1	/	1.80	5	13	1.67	3	12.5	2.24	29	**21**	2.12	13	19
15	H: Latissimus dorsi	1.81	16	8.5	1.83	9	12	1.80	10	6	2.00	1	/	1.43	7	5.5	1.17	3	6.5	1.73	33	5.5	1.69	13	11.5
16	H: Teres major	1.81	16	8.5	1.83	9	12	1.80	10	6	2.00	1	/	1.43	7	5.5	1.17	3	6.5	1.73	33	5.5	1.69	13	11.5
17	H: Teres minor	1.88	8	12	1.67	3	8	1.00	1	1	/	0	/	/	0	/	/	0	/	1.78	9	7.5	1.67	3	9.5
18	R: Pronator teres	1.33	12	2	1.14	7	2	1.39	9	3	/	0	/	1.88	4	16	1.25	4	8	1.44	25	2	1.18	11	2
19	R: Biceps brachii	1.35	13	3	1.67	6	8	1.67	9	4	2.00	1	/	1.38	4	4	1.00	3	3.5	1.46	26	3	1.50	10	4
20	U: Supinator (O)	2.28	16	**20**	2.50	8	**23**	2.13	16	16	2.00	2	/	2.40	5	**21**	2.00	4	**17.5**	2.23	37	**20**	2.29	14	**23**
21	U: Pronator quadratus (O)	2.13	15	17	2.50	7	**23**	2.43	7	**20**	/	0	/	1.67	3	10	2.13	4	**20**	2.16	25	17	2.36	11	**24**
22	U: Anconeus	1.77	13	6	2.00	7	15.5	2.08	6	15	2.00	1	/	1.50	4	8	1.67	3	12.5	1.80	23	10	1.91	11	15
23	U: Brachialis	2.00	16	15	1.63	8	6	2.47	16	**21**	2.00	2	/	1.83	6	14	1.50	5	9.5	2.17	38	18	1.63	15	8
24	U: Triceps brachii	2.05	11	16	1.50	6	3	2.00	3	11	/	0	/	1.75	4	11.5	1.00	1	3.5	1.97	18	13	1.43	7	3

[a]MSM Site: C, clavicle; S, scapula; H, humerus; R, radius; U, ulna; all insertion sites unless otherwise indicated as origin site (O).
x, mean score (averaged for left and/or right sides).
n, number of individuals.
r, rank (the highest five ranked scores for each group are in bold font).
/, no data.

populations, warranting the use of left and/or right side MSM scores interchangeably here. In addition, the effects of body size were also found to be negligible (Lieverse *et al.*,2009).

For the purpose of statistical analyses, the seven most commonly scored MSM sites – the *deltoid, pectoralis major, latissimus dorsi, teres major, biceps brachii, supinator*, and *triceps brachii* – were aggregated into one composite variable for all observable individuals (adults of determined age and sex), following the procedures employed by Weiss (2003, 2007). These seven MSM were chosen not only because their common documentation on the Cis-Baikal skeletal remains provided the least amount of potential error (i.e. substitutions, see below), but also because they are all rather distinctive, frequently cited by other scholars, and typically associated with specific activity patterns (Kennedy, 1989; Nagy, 1999; Peterson, 1998; Robb, 1998). Absent values, being inevitable, were substituted with the average of all remaining values comprising the composite for each individual (after Weiss, 2003, 2007). While substitution rates were generally low (7.8%), individuals with more than three substitutions were excluded from analyses. In addition to statistical analyses (t-tests, Wilcoxon rank sum tests and Mann-Whitney U tests, depending on data quality), rank patterns (Table 11.2) were also examined in order to provide insight into specific muscle and/or ligament use. Ranks were assigned from the highest number (i.e. 24) down to the lowest (i.e. 1).

11.3.2 Postcranial Robusticity

Long bone diaphyseal robusticity was quantified at 35% of the length of the humerus and at the femoral midshaft using moulds of periosteal contours. This method provides data that are highly correlated with cross-sectional geometry derived from radiographic methods and provides a reliable indicator of diaphyseal strength (Stock and Shaw, 2007). The polar second moment of area (J) is commonly calculated as an estimate of the torsional rigidity of a diaphysis as it is related to the average bending moment when it is raised to the power of 0.73 (Ruff, 1995; Ruff *et al.*, 1993), and is an indicator of general levels of biomechanical loading. It is calculated as the sum of any two perpendicular second moments of area (I_{max} and I_{min}; I_x and I_y). These represent rigidity of the bone, and can be used to calculate ratios (I_{max}/I_{min}; I_x/I_y), which provide an indicator of the relative symmetry of cortical bone around the diaphyseal centroid, and hence the shape of the diaphysis. A recent study has demonstrated that total subperiosteal area (TA) is highly correlated with J ($r^2 = 0.996$), and thus provides a robust estimate of diaphyseal rigidity (Stock and Shaw, 2007). Thus, ratios of humerus to femur TA provide a size-free ratio of upper body to lower body strength. Thus, humeral 35% TA/femoral 50% TA and I_x/I_y ratios are compared to those from two groups of marine watercraft-using foragers – the Andaman Islanders of Southeast Asia and the Yahgan of Tierra del Fuego (Argentina) – and highly mobile terrestrially based hunter-gatherers, the Later Stone Age of Southern Africa. A fourth sample of Archaic hunter-gatherers from the Great Lakes region of Southern Ontario is included as a comparative group of middle Holocene hunter-gatherers, although it is at present unknown whether or not they used watercraft. Univariate ANOVA was used to test for differences in femoral shaft shape and humeral to femoral TA ratios.

11.3.3 Stable Isotopes of Carbon and Nitrogen

Collagen for carbon (C) and nitrogen (N) analyses was extracted from bone samples following the method of Sealy (1986), as previously described in Katzenberg and Weber (1999) and Katzenberg, Goriunova and Weber (2009). Collagen samples were analysed on either a

MicroMass Prism mass spectrometer, or a Finnegan Mat Delta Plus mass spectrometer in the Isotope Science Laboratory at the University of Calgary (analyses have been ongoing since the mid-1990s). Precision for $\delta^{13}C$ and $\delta^{15}N$ is $\pm 0.2‰$. Results for stable isotopes of carbon and nitrogen are reported relative to the international standards, VPDB and AIR, respectively. Only samples from older children and adults are reported and all C/N ratios are within the accepted range for well-preserved collagen (atomic C/N 2.9 to 3.6).

11.3.4 Stable Isotopes of Strontium

Strontium isotope signatures were measured in samples obtained from the three permanent molars (M1, M2, M3) and from a long bone (femur). Only individuals with these four samples available, and from three of the five cemeteries (Lokomotiv, Ust'Ida I and Khuzhir-Nuge XIV), were included in these analyses. Samples were prepared with a Dremel tool and diamond-cutting disk from standardized sampling locations on femoral shafts and on the buccal or lingual surfaces of the three molars. Various surface cleaning and decontamination procedures were included in the sample preparation and strontium purification protocols. Purified Sr samples were submitted for mass spectrometry analysis – thermal ionization mass spectrometry (TIMS) and solution mode multi-collector inductively coupled plasma mass spectrometry (MC-ICP-MS). All ratios were calculated relative to the NIST SRM987 strontium isotope standard using a value of 0.710 245. External reproducibility of the standard was 0.710 194 (SD 0.000 033) and 0.710 242 (SD 0.000 041) for the TIMS and MC-ICP-MS analyses, respectively. All samples were processed and analysed by the Radiogenic Isotope Facility, Department of Earth and Atmospheric Sciences, University of Alberta, under the joint direction of R.A. Creaser and A. Simonetti. Full details of the methodology used are outlined in Haverkort *et al.*,2008.

11.4 RESULTS AND DISCUSSION

11.4.1 Musculoskeletal Stress Markers (MSM)

Aggregate MSM data revealed some substantial differences in overall upper limb usage between the Early Neolithic (pre-hiatus) Kitoi occupants of the Cis-Baikal and the Late Neolithic-Bronze Age (post-hiatus) ISG (Lieverse *et al.*, 2009). The former appear to have been characterized by increased heterogeneity, with Kitoi males exhibiting significantly more pronounced MSM – reflecting greater or more intensive overall upper limb use – than both contemporary females and ISG males (p-values from 0.003–0.05). Furthermore, the discrepancy between Early Neolithic males and females appears to have increased with advancing age at death (Figure 11.3). Negligible sexual dimorphism was observed amongst Kitoi young adults, but statistically significant sexual differences were documented amongst middle adults (p-values from 0.000–0.05). In addition, while no age-related distinctions were documented for Kitoi females, the opposite was true for males: middle adult males exhibited significantly higher MSM scores than did their younger male counterparts (p-values from 0.005–0.05). This pattern seemed to be absent amongst ISG peoples, or at least to be much less obvious, but data were insufficient to determine this with any degree of certainty. In fact, unlike their Kitoi antecedents, the Late Neolithic-Bronze Age inhabitants of the Cis-Baikal appear to have exhibited few sex-related differences in overall upper limb usage.

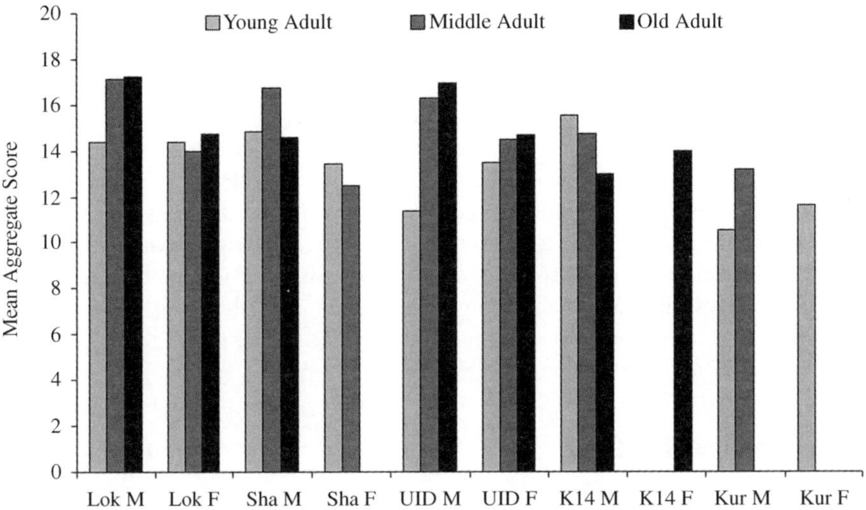

Figure 11.3 Mean aggregate scores for males (M) and females (F) across age at death categories (young adult, 20–35 years; middle adult, 35–50 years; old adult, 50+ years); Lok, Lokomotiv (Kitoi), Sha, Shamanka II (Kitoi), UID, Ust'-Ida I (ISG), K14, Khuzhir-Nuge XIV (ISG), Kur, Kurma XI (ISG)

One explanation for this discrepancy between the Kitoi and ISG populations is the lower residential, and thus higher male-dominated logistical, mobility characterizing the former group (Weber, Link and Katzenberg, 2002). Logistical forays undertaken by Kitoi males would have involved the acquisition of large terrestrial mammals such as ruminants and increased strain on most of the body's muscle and ligament attachment sites, including those of the upper limbs. Activities engaged in would not only have included the transportation of goods and resources over long distances and across difficult terrain, but also a myriad of other tasks such as manufacturing and repairing tools, harvesting and processing animals, establishing and dismantling temporary camps, and most likely, propelling and portaging watercraft. While these activities did not produce a distinctive pattern of muscle and ligament use amongst Early Neolithic males (see next section), they did increase upper limb strain overall by their higher intensity and/or longer duration.

Rank patterning of MSM provided additional information about specific upper limb activities employed in the Cis-Baikal, but only slight evidence of disparity amongst the populations or between the sexes (Lieverse *et al.*, 2009). All middle Holocene occupants of the region – Kitoi and ISG, male and female, alike – exhibited the same six high-ranked MSM attachment sites (the conoid and costoclavicular ligaments and the *pectoralis major*, *deltoid*, *pronator quadratus* and *brachioradialis* muscles) and the same three low-ranked ones (the *coracobrachialis*, *pronator teres* and *biceps brachii* muscles). All of the six high-ranked sites are compatible with activities involving watercraft propulsion and, in fact, four of them have been previously identified amongst a number of past populations as consistent with paddling (Eshed *et al.*, 2004; Hawkey and Merbs, 1995; Hawkey and Street, 1992; Lai and Lovell, 1992; Steen and Lane, 1998; Stirland, 1991). Thus, the evidence suggests that watercraft use was common in the Cis-Baikal and likely undertaken by all or most community members throughout the middle Holocene period. This observation is particularly interesting given the complete lack of archaeological evidence to date for boats or watercraft-related implements of any kind. It is also noteworthy that variation in rank patterning pertained almost exclusively to the Late Neolithic-Bronze Age ISG populations, despite the relative homogeneity in their overall upper limb use as indicated by aggregate data (see above). All ISG individuals appear to have engaged in activities involving the *supinator* muscle (e.g. overhead throwing, kayaking with a double-bladed paddle) more commonly and/or strenuously than their Early Neolithic Kitoi predecessors, while ISG females alone seem to have utilized both their forearm flexors and extensors (e.g. activities such as simple lifting) less frequently and/or intensively than did either their contemporary males or Kitoi individuals (of either sex). Furthermore, rank patterning did not differentiate Early Neolithic individuals by sex. Considering that aggregate data revealed significantly higher MSM scores amongst Kitoi males, their lack of distinctive rank patterns suggested differences in the degree (intensity and/or duration), rather than the type, of upper limb activities employed.

11.4.2 Postcranial Robusticity

When we consider long bone cross-sectional geometry, there is considerable variation in femoral midshaft I_x/I_y ratios amongst the hunter-gatherer samples (Figure 11.4). Very high

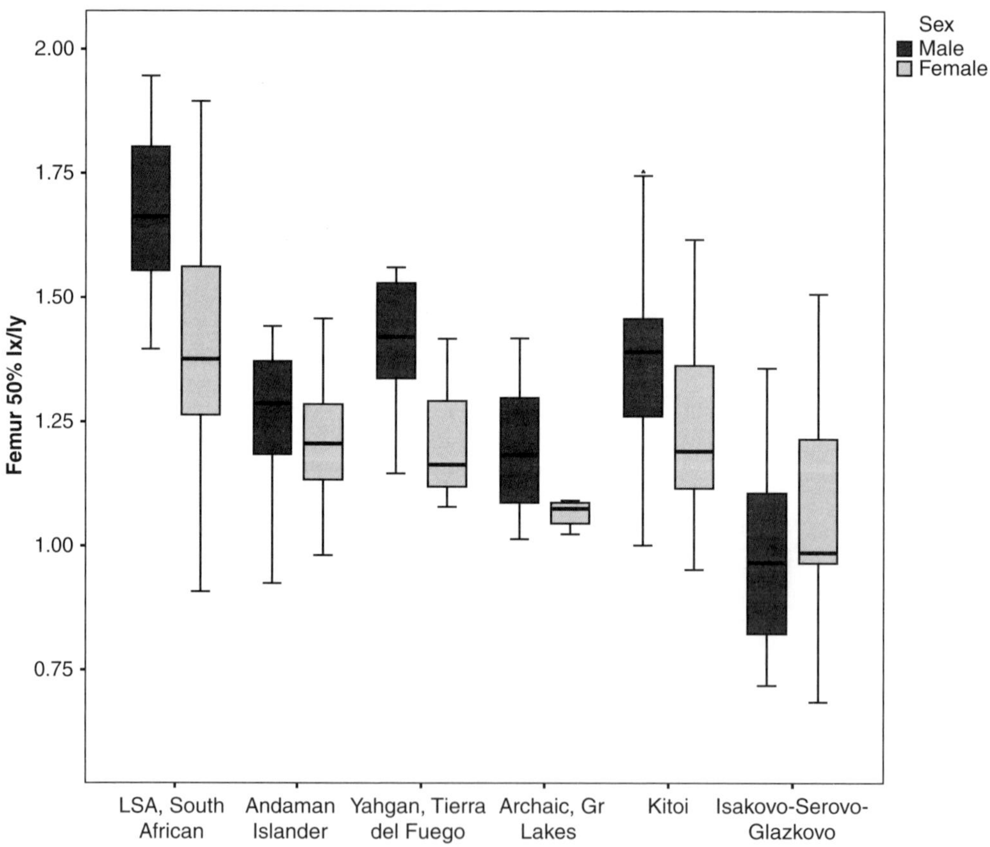

Figure 11.4 Femoral midshaft Ix/Iy ratios

Table 11.3 Summary of middle Holocene Cis-Baikal diaphyseal robusticity

	Kitoi			Isakovo-Serovo-Glazkovo			
Property	n	Mean	S.D.	n	Mean	S.D.	Sig.[a]
Male							
% Humerus 35% *TA*/Femur 50% *TA*	29	52.75	4.14	11	52.69	5.73	0.970
Femur 50% I_x/I_y ratio	31	1.368	0.196	17	0.988	0.189	>0.001
Female							
% Humerus 35% *TA*/Femur 50% *TA*	16	47.38	3.18	5	38.18	5.72	>0.001
Femur 50% I_x/I_y ratio	19	1.223	0.176	11	1.062	0.267	0.055

[a] Difference is significant based upon ANOVA with $\alpha = 0.05$.

ratios are found amongst the highly mobile Later Stone Age South African sample, indicating femoral diaphyses, which feature antero-posterior hypertrophy and relatively narrow breadths medio-laterally. This morphological pattern likely relates to a combination of significant cycles of annual and logistic mobility, as well as the biomechanical influence of rugged terrain (Stock and Pfeiffer, 2004). In contrast, the Andaman Islanders, Yahgan and Archaic Ontarians show moderate levels of femoral shaft circularity, which are similar to the Kitoi sample. In contrast to this pattern, the ISG have femoral diaphyses that are, on average, roughly circular, indicating more symmetrical loading regimes in antero-posterior and medio-lateral planes. The contrast between the antero-posteriorly elongated Kitoi and the later ISG diaphyses (Table 11.3) is statistically significant for the comparison of males ($F = 42.257$; $p < 0.001$), and nearly significant for the comparison of female subsamples ($F = 4.005$; $p = 0.055$).

The calculation of a ratio to reflect the percentage of humeral 35% total subperiosteal area (*TA*) relative to femoral midshaft *TA* provides a means of comparing the strength of the upper limb relative to the lower limb. Comparison of hunter-gatherers demonstrates a clear and significant distinction between the LSA Southern Africans, who were highly mobile terrestrial foragers lacking watercraft, and other samples known to have used watercraft extensively (Andaman Islanders, Yahgan; Figure 11.5). These contrasts suggest that there is a biomechanical signature of watercraft use in the form of humeral cortical bone hypertrophy relative to femoral strength. When the middle Holocene populations of the Cis-Baikal are interpreted in this context, both the male and female Kitoi (Early Neolithic) have robust humeri relative to femora. This biomechanical signature could result from a variety of habitual behaviours that place heavy mechanical demands on the upper limb, but the use of watercraft may be the most plausible interpretation.

The ISG (Late Neolithic-Bronze Age) males are not significantly different than the Kitoi ($F = 0.001$; $p = 0.970$) and so appear to have had a similar pattern of biomechanical loading. In contrast, there is a significant reduction in humeral *TA*/femoral *TA* between the Kitoi and ISG females ($F = 21.652$; $p < 0.001$), such that the latter have the lowest average values amongst the samples reported here. This appears to be a function of the ISG females having a

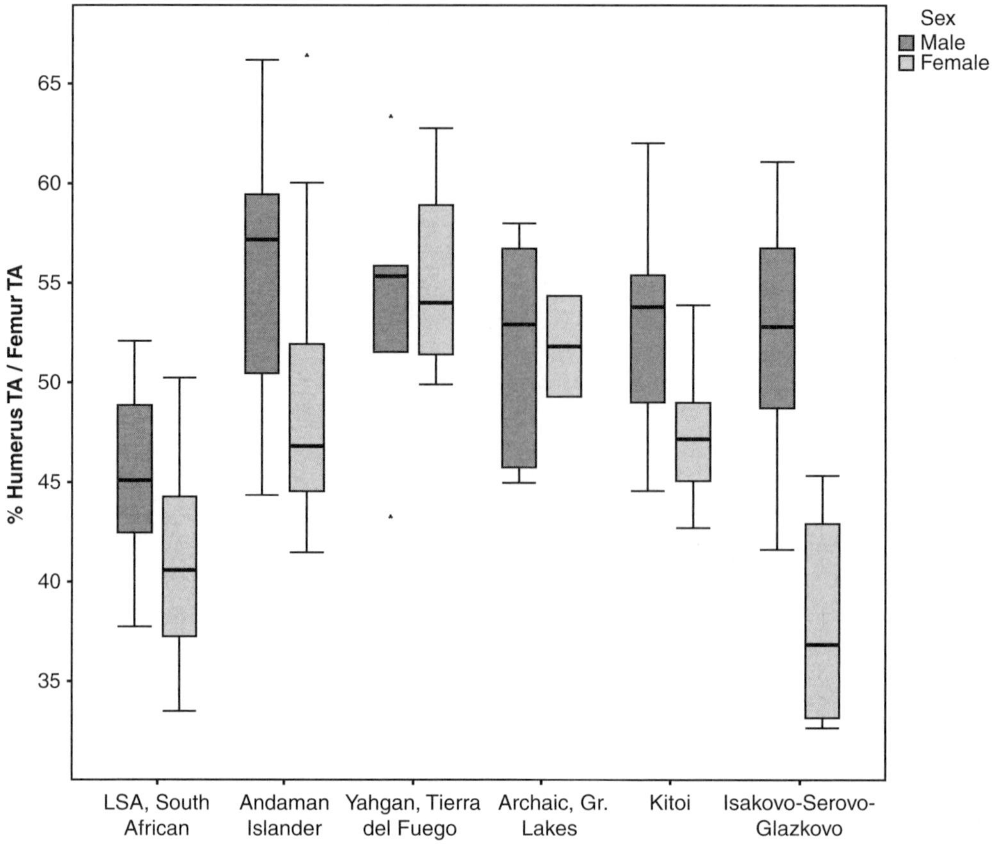

Figure 11.5 Percentage Humerus 35% TA/Femur Midshaft TA

combination of relatively robust femora and gracile humeri. Overall, the pattern of limb robusticity suggests that there was a reduction in antero-posterior loading of the lower limb, or more symmetrical/dynamic loading of the lower limb, between the Kitoi and ISG. This could reflect a decrease in terrestrial mobility, as there are only minor differences in physique between these groups (Stock *et al.*, 2010). Despite this, the pattern of humeral to femoral strength suggests continued importance of watercraft to the mid-Holocene populations of the Cis-Baikal, and particularly the males. There is a significant difference in the nature of female limb robusticity between the Early Neolithic and Late Neolithic-Bronze Age, suggesting that either Kitoi females were involved in watercraft propulsion and ISG females were not, or that there were other significant biomechanical influences on the upper limb in the Kitoi that contrast with a relatively low intensity of manual loading amongst the female ISG.

11.4.3 Stable Isotopes of Carbon and Nitrogen

Summary data for diet reconstruction using stable isotopes of carbon and nitrogen are presented in Table 11.4. The two Early Neolithic Kitoi sites of Lokomotiv and Shamanka II are statistically similar with respect to $\delta^{13}C$ but significantly different from the Late Neolithic-Bronze Age sites (one way ANOVA, $F = 42.46$; $p < 0.001$). In fact, no sites within

Table 11.4 Summary data for diet reconstruction using stable isotopes of carbon and nitrogen (modified from Katzenberg, Goriunova and Weber, 2009; Katzenberg *et al.*, 2010)

	Region	Period	Sample size	Mean $d^{13}C$	s.d.	Mean $d^{15}C$	s.d.
Lokomotiv	Angara	Early Neolithic	90	–15.6	3.4	14.0	1.2
Shamanka	Baikal	Early Neolithic	66	–16.3	0.8	14.6	1.0
Ust' Ida	Angara	ISG	53	–17.9	0.8	11.8	0.8
Khuzhir Nuge XIV	Baikal	Bronze Age	73	–18.6	0.9	13.8	1.7
Kurma XI	Baikal	Bronze Age	19	–18.6	1.1	14.3	1.7

each of the two time periods (Kitoi or ISG) are significantly different for $\delta^{13}C$. The Kitoi sites are more enriched in the heavier isotope of carbon, resulting in less negative $\delta^{13}C$ values. This might be caused by a greater reliance on fish that inhabit the littoral (shallow coastal) zones of the lake. Ecological research on Lake Baikal (Kiyashko *et al.*, 1998) and earlier analyses of fish (Katzenberg and Weber, 1999; Weber, Link and Katzenberg, 2002) have provided evidence for enrichment in the heavier isotope of carbon amongst littoral species based on variation associated with the sources of carbon, related to a number of biological and physical variables (Hecky and Hesslein, 1995). Since some of those characteristic species were found in the upper reaches of the Angara River, near the site of Lokomotiv, prior to the modern construction of a hydroelectric dam in Irkutsk (Kozhov, 1963), people residing around this site would likely have had access to them. It is also possible that those interred at Lokomotiv travelled to Baikal, located only 65 km upstream, for fishing or other subsistence activities during their lifetimes. Another possibility is that the isotope values are similar but for different reasons. For example, the outflow of Lake Baikal into the Angara is an area that attracts waterfowl. So far, only one modern grey duck (*Anas formosa*) has been analysed from this region, but the stable isotope values for carbon and nitrogen are very similar to those of the individuals buried in Lokomotiv ($\delta^{13}C = -15.2$; $\delta^{15}N = 14.8$). This source of meat and eggs, which is exploited by many northern people (Alekseenko, 1999; Ziker, 2002), could be another source of food leading to higher $\delta^{13}C$ and $\delta^{15}N$ values at this site.

The ISG sites of Khuzhir-Nuge XIV and Kurma XI are located close to one another on the western shore of Lake Baikal, so it is not surprising that the people interred at these cemeteries have very similar stable isotope values. Mean $\delta^{13}C$ values are the same and the small difference in $\delta^{15}N$ is not significant, given the much smaller number of individuals buried at Kurma XI. The site of Ust'-Ida I clusters with the other post-hiatus sites for $\delta^{13}C$, but is different from all other sites for $\delta^{15}N$. Located furthest from Lake Baikal, sources of food with higher $\delta^{15}N$, such as Baikal seals and high trophic level fish, would have been less available to the people interred at Ust'-Ida I. It is likely that they relied more on terrestrial mammals, resulting in $\delta^{15}N$ values that are lower than those of individuals consuming more fish and, potentially, seals. Lokomotiv individuals have higher mean $\delta^{15}N$ than those from Ust'-Ida I and are more similar to populations interred on Lake Baikal, suggesting that they exploited similar resources. Their higher $\delta^{13}C$ suggests that these were the littoral resources that are commonly found in the upper reaches of the Angara River or in the lake itself.

One possible interpretation of the isotope data with respect to shifting dietary resources is that ISG lake dwellers (i.e. individuals from Khuzhir-Nuge XIV and Kurma XI) had access to boats better equipped to travel further from the shore in order to fish open water species, whereas earlier people were restricted to exploiting the near-shore littoral species. This would explain the lower $\delta^{13}C$ yet high $\delta^{15}N$ characterizing Khuzhir-Nuge XIV and Kurma XI. Ust'-Ida I differences are best explained by the site's location farther from the lake, with decreased access to species with higher $\delta^{15}N$. In addition, Khuzhir-Nuge XIV and Kurma XI are located close to rivers that may have served as spawning streams for fish, thereby providing reliable resources at particular times of the year.

11.4.4 Stable Isotopes of Strontium

The $^{87}Sr/^{86}Sr$ ratios measured in four samples (3 enamel, 1 bone) per individual are shown in Figure 11.6. The patterns for each of the three populations considered here (Lokomotiv, Ust'Ida I and Khuzhir-Nuge XIV) are quite distinct, reflecting a combination of three different factors: geographical location of the sites, dietary differences between the populations and, possibly, mobility patterns (Haverkort, Bazaliiskii and Savel'ev, 2010). Our comparative environmental framework, based on geochemical data reported for the Cis-Baikal area and $^{87}Sr/^{86}Sr$ ratios measured in prehistoric and modern faunal specimens, suggests that the Cis-Baikal region is characterized by three major geochemical zones. These offer a preliminary indication of the $^{87}Sr/^{86}Sr$ ratios associated with some of the major resources available to the Kitoi and ISG hunter-gatherers (Haverkort *et al.*, 2008; Weber *et al.*, 2003). The $^{87}Sr/^{86}Sr$ ratio for Lake Baikal water is reported to be around 0.7085 (Kenison Falkner *et al.*, 1992), although samples of northern pike (*Esox lucius*) from the Little Sea region ranged from 0.70 985 to 0.72 829, suggesting the possibility of locally higher isotope ratios due to incomplete mixing of the shallow waters of the Little Sea (Haverkort *et al.*,2008). Aquatic resources from the first section of the Angara River, the only river flowing out of Lake Baikal, are expected to be characterized by $^{87}Sr/^{86}Sr$ ratios similar to that of Lake Baikal water, that is, around 0.709. The ratios for terrestrial resources from the western and eastern hinterlands along this section of the river reflect the geological differences on either side of the river, with somewhat higher ratios found on the western side and lower ratios (~0.709) found towards the east.

The area to the east of the Angara River (the Central Siberian Plateau) includes the Upper Lena watershed. This extensive region consists mostly of limestone with $^{87}Sr/^{86}Sr$ ratios around 0.709. Resources obtained from the west coast of Lake Baikal appear to be characterized by ratios of more than 0.720. For our interpretation of the human Sr isotope ratios, another important finding is the relatively high elemental Sr concentrations in fish bones (Haverkort *et al.*, 2008). Ethnographic accounts indicate that several contemporary Siberian groups do not bone their fish when preparing them for food or storage, such that fish bones are part of the diet (Argounova, 2002; Ziker, 2002). This observation is critical in that a high proportion of fish in the diet may increase the likelihood that fish bones are included as well, which may tend to skew the human $^{87}Sr/^{86}Sr$ ratios towards the values found in fish bones, that is the lower end of the range (~0.709).

Almost all of the $^{87}Sr/^{86}Sr$ ratios obtained for Ust'Ida I cluster together tightly toward the lower end of the Cis-Baikal signatures, around 0.709 to 0.710. Because these isotope ratios correspond to both aquatic and terrestrial resources from a wide area, including Lake Baikal, the Angara River and the extensive plateau to the east, it is difficult to draw any conclusions with regard to foraging range and mobility patterns. The low variability in Sr ratios for this group may indicate a large foraging range, whereby the overall variation is 'averaged' out so that all individuals end up with similar ratios. Alternatively, it may reflect a strategy whereby the whole group and all ages moved around frequently, and together as a group (residential

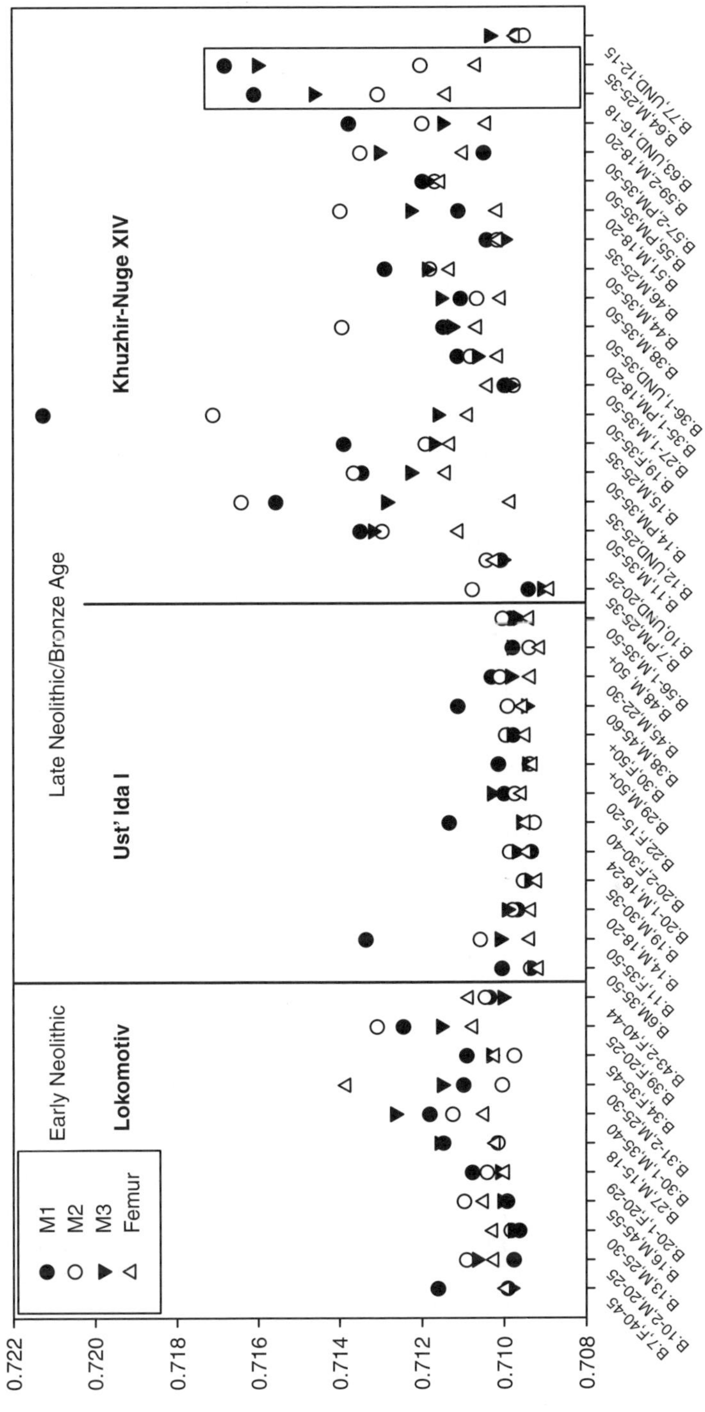

Figure 11.6 Strontium isotope ratios in human tooth and bone samples for individuals from Lokomotiv, Ust'Ida and Khuzhir-Nuge XIV

mobility), again resulting in similar ratios. A third possibility is that the foraging range falls entirely within the low-variability region of the Central Siberian Plateau, such that even considerable differences in mobility between individuals cannot be detected.

The somewhat higher $^{87}Sr/^{86}Sr$ ratios for the Kitoi from Lokomotiv correspond well with the geographical location of the Lokomotiv cemetery on the western side of the Angara River. A localized Sr isotope signature for Lokomotiv individuals would also agree well with the observation that the $\delta^{13}C$ data for individuals from cemeteries along the river appear to co-vary with local fish signatures (Katzenberg and Weber, 1999). In addition, the somewhat higher $^{87}Sr/^{86}Sr$ ratios for individuals from this site could be the result of the additional exploitation of terrestrial resources from along the west coast of the lake, where Sr ratios are generally higher. Relative to Ust'Ida I, Lokomotiv also shows a higher degree of variability both within and between individuals, possibly indicative of a more logistical pattern of mobility, though theoretically this could also partially reflect differential access to resources, resulting in differences in dietary intake.

The $^{87}Sr/^{86}Sr$ ratios in many of the samples from ISG individuals buried at Khuzhir-Nuge XIV are much higher than those observed for either Lokomotiv (Kitoi) or Ust'-Ida I (ISG). This result seems to indicate that terrestrial resources from along the west coast of Lake Baikal played an important role in the economy of these people, in contrast to that of the ISG communities at Ust'Ida I. The overall higher Sr isotope ratios for burials at Khuzhir-Nuge XIV may also partially be accounted for by the reduced emphasis on aquatic resources in comparison to the earlier Kitoi. If fish (bones) and seal constituted a smaller proportion of the overall diet, the dominant effect of fish signatures may have been less important, resulting in the overall higher $^{87}Sr/^{86}Sr$ ratios observed for individuals from Khuzhir-Nuge XIV. The high degree of variability amongst individuals from Khuzhir-Nuge XIV could indicate a pattern of logistical mobility. Furthermore, the similarity in lifetime mobility patterning for some individuals buried in rows at this cemetery (e.g. burials 63, 64) could point to the possibility of family- or lineage-specific territories. Different families or lineages might then have accessed different 'Sr-catchments' (i.e. different territories, possibly with different combinations of resources). The observed differences between individuals could then partly reflect differences in dietary intake, similar to the situation for Lokomotiv individuals, and not necessarily solely differences in mobility strategy. However, it is important to keep in mind that our use here of the term 'mobility strategy' (such as logistic or residential) tends to refer to a characteristic of a group of contemporary individuals, whereas the individuals in our samples do not represent any such group specifically, given the temporal range of use of the cemeteries.

11.5 SUMMARY

The results of upper limb MSM and long bone (femoral and humeral) robusticity analyses are complementary, together providing a more detailed picture of behaviour and mobility patterns in the middle Holocene Cis-Baikal. Both lines of enquiry identify some significant differences between the Early Neolithic Kitoi occupants of the region and the Late Neolithic-Bronze Age ISG. MSM data point to higher heterogeneity in overall upper limb activity levels amongst Early Neolithic populations: Kitoi males exhibited significantly more pronounced muscle and ligament attachment sites than contemporary females, with sexual disparity increasing substantially with advancing age at death. On the other hand, activity *patterns* seem to have varied more during the latter period, with the *supinator* muscle being ranked high amongst the all ISG individuals and the forearm flexors and extensors being ranked generally low amongst ISG females. Finally, for both Kitoi and ISG peoples, upper limb patterns, particularly the consistently high-ranked costoclavicular ligament and *deltoid* and *pectoralis major* muscles, are consistent with watercraft use. This is particularly noteworthy because, at present, there is no direct archaeological evidence for boats or associated paraphernalia during this time period.

Two MSM observations – strong evidence for watercraft use by all occupants of the region and a decrease in at least some upper limb usage amongst ISG females – are supported by long bone robusticity data, specifically humeral-femoral total subperiosteal area (*TA*) ratios. These ratios reveal generally robust humeri compared to femora amongst the middle Holocene occupants of the Cis-Baikal, a trait consistent with the biomechanical signature of watercraft use observed amongst other hunter-gatherer groups. Humeral-femoral TA ratios also distinguish ISG females as having less robust humeri compared to femora, suggesting decreased upper limb loading and/or increased lower limb loading. In light of femoral diaphyseal shape characteristics (see below), it is likely that these ratios reflect a decrease in upper limb use, rather than an increase in lower limb use. Furthermore, because MSM evidence for watercraft use amongst ISG females remains, any decrease in their upper limb loading likely involved other activities, possibly those associated with mobility.

Analyses of femoral diaphyseal shape reveal greater antero-posterior cortical hypertrophy, indicative of increased antero-posterior femoral loading amongst the Kitoi when compared to the ISG. This elongation has been found to be consistent with significant cycles of annual and logistic mobility amongst other foraging populations (Stock, 2006), often reflecting travel over rugged terrain. When interpreted in the Cis-Baikal context, then, Early Neolithic Kitoi males and females appear to have engaged in higher levels of terrestrial mobility – whether residential or logistical – than their ISG successors. This lack of sexual disparity in femoral diaphyseal shape seems to contrast with archaeological data, suggesting lower residential and higher male-dominated logistical mobility during the Early Neolithic period (Weber, Link and Katzenberg, 2002). However, biomechanical signatures presumably represent aggregate levels of mechanical loading over many years of life and are likely insensitive to whether the mobility is logistical or residential. As such, the interpretations of mobility based upon different methods may not necessarily be exclusive; for example, a pattern of Early Neolithic female mobility which included intensive local foraging would be supported by both lines of evidence.

The isotope data presented here add a valuable dimension to the behaviour and mobility insights gleaned from skeletal morphology. The combination of strontium isotope analyses with those of carbon and nitrogen allow us to further expand on our interpretations of differing adaptive strategies within the context of cultural change. Carbon and nitrogen isotope data reveal that, with the probable exception of those groups residing a substantial distance (>100 km) from Lake Baikal, aquatic resources enriched in the heavier isotope of nitrogen (i.e. high trophic level fish and seal) appear to have comprised a substantial portion of middle Holocene diets. In the Early Neolithic period, these resources seem to have been dominated by littoral (shallow water) fish species, as indicated by the higher $\delta^{13}C$ values characterizing the Kitoi sites of Shamanka II and Lokomotiv. Amongst the later ISG, on the other hand, $\delta^{13}C$ ratios (at the coastal sites of Khuzhir-Nuge XIV and Kurma XI) are substantially lower, suggesting that aquatic resource consumption likely included more open water lake species. This difference may point to technological advances, such as watercraft better equipped to travel further from the lake shore, made during the Late Neolithic-Bronze

Age. Finally, nitrogen isotope data reveal that those individuals residing considerable distances from the lake (and the upper reaches of the Angara River) appear to have consumed more terrestrial resources (e.g. herbivorous mammals) with lower $\delta^{15}N$ values relative to aquatic resources.

Strontium isotope analyses have focused on three of the five sites investigated (Lokomotiv, Ust'-Ida I and Khuzhir-Nuge XIV) and yielded a number of interesting trends applicable to the Cis-Baikal region as a whole. Strontium values at the Kitoi site of Lokomotiv show some degree of intra-site and intra-individual variability, consistent with (but not necessarily indicative of) logistical mobility. In addition, Sr signatures from this site correspond to its location on the west side of the Angara River, suggesting that individuals interred there did not frequently or regularly cross the river for the purpose of resource acquisition. This may reflect the technological limitations of Early Neolithic watercraft – restricting fish consumption to littoral species (see above) and preventing confident navigation of the Angara – or it may simply be the result of numerous other cultural factors (e.g. territoriality, religious beliefs, etc.). Alternatively, the somewhat higher $^{87}Sr/^{86}Sr$ ratios for Lokomotiv individuals could indicate that these populations accessed some terrestrial resources from along the west coast of the lake. This observation is not inconsistent with the stable carbon and nitrogen isotope data. The two ISG sites yield quite different results, at least partly resulting from their relative locations on the Angara and the lake itself. Strontium values at Ust'-Ida I, while clustering together and showing the least amount of variation, can be interpreted in a broad range of potential mobility and subsistence scenarios and require further investigation. Those from Khuzhir-Nuge XIV, on the other hand, are higher than those of the two Angara River sites, suggesting the predominant use of terrestrial resources from the west side of the lake, possibly combined with a decreased focus on aquatic fauna. However, considering nitrogen isotopic evidence for aquatic resource use at this site (see above), the latter seems less plausible.

11.6 CONCLUSIONS

This paper provides a unique synthesis of biomechanical and isotopic data for long-term changes in subsistence behaviour amongst hunter-gatherers. Collectively, the biomechanical evidence points towards a continuing importance of watercraft use throughout the Holocene, but one that is associated with a decrease in the intensity of terrestrial mobility during the Late Neolithic ISG. They also provide some evidence for changes in the sexual division of labour, with ISG women showing a reduction in humeral robusticity. The isotopic data provide evidence for the continued importance of fish in diets throughout the Holocene, although they suggest that there may have been a dietary shift from shallow water species towards more open water species from the Kitoi to ISG periods. The strontium isotopes suggest that there may have been increased use of terrestrial resources to the west of Lake Baikal or along the Angara River amongst the post-hiatus ISG samples. When taken in combination, the data are complimentary, and underscore the importance of watercraft based fishing throughout the Holocene, although they suggest a changing efficiency in the use of watercraft towards more confident deep water fishing and use of watercraft to exploit resources across the lake and along the Angara River. Our ability to interpret diachronic variation in hunter-gatherer diet, mobility and behaviour is becoming an increasingly important factor in our interpretations of long-term subsistence transitions in prehistory, particularly in understanding foraging preceding the origins of agriculture. Collectively, this research highlights the benefits of multivariate approaches to the study of prehistoric behavioural ecology and demonstrates that it is possible to use bioarchaeological methods to detect subtle changes in hunter-gatherer behaviour through time.

REFERENCES

Alekseenko, E.A. (1999) Ket, in *The Cambridge Encyclopedia of Hunters and Gatherers* (eds R.B. Lee and R. Daly), University of Cambridge Press, Cambridge, pp. 156–160.

Angel, J.L., Kelley, J.O., Parrington, M. and Pinter, S. (1987) Life stresses of the free Black community as represented by the First African Baptists Church, Philadelphia. *Am. J. Phys. Anthropol.*, **74**, 213–299.

Beard, B.L. and Johnson, C.M. (2000) Strontium isotope composition of skeletal material can determine the birth place and geographic mobility of humans and animals. *J. Forensic Sci.*, **45**, 1049–1061.

Bentley, R.A. (2006) Strontium isotopes from the earth to the archaeological skeleton: a review. *J. Archaeol. Method Th.*, **13** (3), 135–187.

Bentley, R.A., Krause, R., Price, T.D. and Kaufmann, B. (2003) Human mobility at the Early Neolithic settlement of Vaihingen, Germany: evidence from strontium isotope analysis. *Archaeometry*, **45**, 471–486.

Bentley, R.A., Price, T.D. and Stephan, E. (2004) Determining the 'local' $^{87}Sr/^{86}Sr$ range for archaeological skeletons: a case study from Neolithic Europe. *J. Archaeol. Sci.*, **31**, 365–375.

Buikstra, J.E. and Ubelaker, D.H. (eds) (1994) *Standards for Data Collection from Human Skeletal Remains*, Archaeological Survey Research Series 44, Arkansas Archaeological Society, Fayetteville, AR.

Bush, A. (2007) Holocene climate change in the Cis-Baikal region, in *Conference Proceedings, The Ethnohistory and Archaeology of Northern Eurasia: Theories, Methods, and Practice* (ed. A.V. Kharinskii), Izd-vo IrGTU, Irkutsk-Edmonton, pp. 386–388.

Churchill, S.E. and Morris, A.G. (1998) Muscle marking morphology and labour intensity among Khoisan foragers. *Int. J. Osteoarchaeol.*, **8**, 390–411.

Cook, D.C. and Dougherty, S.P. (2001) Row, row, row your boat: activity patterns and skeletal robusticity in a series from Chirikof Island, Alaska (abstract). *Am. J. Phys. Anthropol.* (Suppl. 32), 53.

Cordain, L., Miller, J.B., Eaton, S.B. *et al.* (2002) Plant-animal subsistence ratios and macronutrient energy estimations in worldwide hunter-gatherer diets. *Am. J. Clin. Nutr.*, **71** (3), 682–692.

Curzon, M.E. and Cutress, T.W. (1983) *Trace Elements and Dental Disease*, John Wright, PSG, Boston.

Daly, R.M., Saxon, L., Tumer, C.H. *et al.* (2004) The relationship between muscle size and bone geometry during growth and in response to exercise. *Bone*, **34**, 281–287.

Ericson, J.E. (1985) Strontium isotope characterization in the study of prehistoric human ecology. *J. Hum. Evol.*, **14**, 503–514.

Ericson, J.E. (1989) Some problems and potentials of strontium isotope analysis for human and animal ecology, in *Stable Isotopes in Ecological Research* (eds P.W. Rundel, J.R. Ehleringer and K.A. Nagy), Springer-Verlag, New York, pp. 252–259.

Eshed, V., Gopher, A., Galili, E. and Hershkovitz, I. (2004) Musculoskeletal stress markers in Natufian hunter-gatherers and Neolithic farmers in the Levant: the upper limb. *Am. J. Phys. Anthropol.*, **123**, 303–315.

Ezzo, J.A. and Price, T.D. (2002) Migration, regional reorganization, and spatial group composition at Grasshopper Pueblo, Arizona. *J. Archaeol. Sci.*, **29**, 499–520.

Ezzo, J.A., Johnson, C.M. and Price, T.D. (1997) Analytical perspectives on prehistoric migration: A case study from East-Central Arizona. *J. Archaeol. Sci.*, **24**, 447–466.

Goebel, T. (1999) Pleistocene human colonization of Siberia and peopling of the Americas: an ecological approach. *Evol. Anthropol.*, **8** (6), 208–227.

Grupe, G., Price, T.D., Schröter, P. *et al.* (1997) Mobility of Bell Beaker people revealed by strontium isotope ratios of tooth and bone: a study of southern Bavarian skeletal remains. *Appl. Geochem.*, **12**, 517–525.

Haverkort, C.M., Weber, A.W., Katzenberg, M.A. *et al.* (2008) Hunter-gatherer mobility strategies and resource use based on strontium ($^{87}Sr/^{86}Sr$) analysis: a case study from middle Holocene Lake Baikal, Siberia. *J. Archaeol. Sci.*, **35**, 1265–1280.

Haverkort, C.M., Bazaliiskii, V.I. and Savel'ev, N.A. (2010) Identifying hunter-gatherer mobility patterns using strontium isotopes, in *Prehistoric Hunter-Gatherers of the Baikal Region, Siberia* (eds A.W. Weber, M.A. Katzenberg and T.G. Schurr), University of Pennsylvania Museum of Archaeology and, Anthropology, University of Pennsylvania Press, Philadelphia.

Hawkey, D.E. (1988) Use of Upper Extremity Enthesopathies to Indicate Habitual Activity Patterns. MA Thesis. Tempe, AZ: Arizona State University.

Hawkey, D.E. (1998) Disability, compassion, and the skeletal record: using musculoskeletal stress markers (MSM) to construct an osteobiography from early New Mexico. *Int. J. Osteoarch.*, **8**, 326–340.

Hawkey, D.E. and Merbs, C.F. (1995) Activity-induced musculoskeletal stress markers (MSM) and subsistence strategy changes among ancient Hudson Bay Eskimos. *Int. J. Osteoarch.*, **5**, 324–338.

Hawkey, D.E. and Street, S.R. (1992) Activity-induced stress markers in prehistoric human remains from eastern Aleutian Islands (abstract). *Am. J. Phys. Anthropol.* (Suppl. 14), 89.

Hecky, R.E. and Hesslein, R.H. (1995) Contributions of benthic algae to lake food webs as revealed by stable isotope analysis. *J. N. Am. Benthol. Soc.*, **14**, 631–653.

Helm, J. (ed.) (1981) *Handbook of the North American Indians*, Subarctic, vol. **6**, Smithsonian Institution, Washington, DC.

Hodell, D.A., Quinn, R.L., Brenner, M. and Kamenov, G. (2004) Spatial variation of strontium isotopes ($^{87}Sr/^{86}Sr$) in the Maya region: a tool for tracking ancient human migration. *J. Archaeol. Sci.*, **31**, 585–601.

Katzenberg, M.A. (2008) Stable isotope analysis: A tool for studying past diet, demography and life history, in *Biological Anthropology of the Human Skeleton*, 2nd edn (eds M.A. Katzenberg and S.R. Saunders), New York, Wiley-Liss, pp. 413–441.

Katzenberg, M.A. and Weber, A.W. (1999) Stable isotope ecology and palaeodiet in the Lake Baikal region of Siberia. *J. Archaeol. Sci.*, **26**, 651–659.

Katzenberg, M.A., Goriunova, O.I. and Weber, A.W. (2009) Paleodiet reconstruction of Bronze Age. Siberians from the mortuary site of Khuzhir-Nuge XIV, Lake Baikal. *J. Archaeol. Sci.*, **36**, 663–674.

Katzenberg, M.A., Bazaliiskii, V.I., Goriunova, O.I. *et al.* (2010) Diet reconstruction of prehistoric hunter-gatherers in the Lake Baikal region, in *Prehistoric Hunter-Gatherers of the Baikal Region* (eds A.W. Weber, M.A. Katzenberg and T.G. Schurr), Bioarchaeological Studies of Past Life Ways, Philadelphia, University of Pennsylvania Museum of Archaeology and, Anthropology, University of Pennsylvania Press, Siberia.

Kelley, J.O. and Angel, J.L. (1987) Life stresses of slavery. *Am. J. Phys. Anthropol.*, **74**, 199–211.

Kenison Falkner, K., Lebaron, G., Thouron, D. *et al.* (1992) Cr, V and Sr in Lake Baikal. *EOS T. Am. Geophys. Un.*, **73** (14), 140.

Kennedy, K.A.R. (1989) Skeletal markers of occupational stress, in *Reconstruction of Life from the Skeleton* (eds M.Y. Işcan and K.A.R. Kennedy), Alan R. Liss, New York, pp. 129–160.

Khazonov, A.M. (1994) *Nomads and the Outside World*, 2nd edn, University of Wisconsin Press, Madison.

Kiyashko, S.I., Richard, P., Chandler, T. *et al.* (1998) Stable carbon isotope ratios differentiate autotrophs supporting animal diversity in Lake Baikal. *C. R. Acad. Sci. III*, **321** (6), 509–516.

Knudson, K.J., Price, T.D., Buikstra, J.E. and Blom, D.E. (2004) The use of strontium isotope analysis to investigate Tiwanaku migration and mortuary ritual in Bolivia and Peru. *Archaeometry*, **46**, 5–18.

Kozhov, M. (1963) *Lake Baikal and Its Life*, Dr. W. Junk, Publishers, The Hague.

Kuzmin, Y.V. and Orlova, L.A. (2000) The Neolithization of Siberia and the Russian Far East: Radiocarbon evidence. *Antiquity*, **74** (284), 356–364.

Lai, P. and Lovell, N.C. (1992) Skeletal markers of occupational stress in the fur trade: a case study from a Hudson's Bay Company fur trade post. *Int. J. Osteoarch.*, **2**, 221–234.

Lanyon, L.E. (1992) Control of bone architecture by functional load bearing. *J. Bone Miner. Res.*, **7** (Suppl. 2), S369–S375.

Levin, M.G. and Potapov, L.P. (1965) *Peoples of Siberia*, University of Chicago Press, Chicago.

Lieverse, A.R., Weber, A.W., Bazaliiskiy, V.I. *et al.* (2007) Osteoarthritis in the mid-Holocene Cis-Baikal: skeletal indicators of hunter-gatherer adaptation and cultural change. *Am. J. Phys. Anthropol.*, **132**, 1–16.

Lieverse, A.R., Link, D.W., Bazaliiskiy, V.I. *et al.* (2007) Dental health indicators of hunter-gatherer adaptation and cultural change in Siberia's Cis-Baikal. *Am. J. Phys. Anthropol.*, **134**, 323–339.

Lieverse, A.R., Bazaliiskii, V.B., Goriunova, O.I. and Weber, A.W. (2009) Upper limb musculoskeletal stress markers among middle Holocene foragers of Siberia's Cis-Baikal region. *Am. J. Phys. Anthropol.*, **138**, 458–472.

Losey, R.J., Nomokonova, T. and Goriunova, O.I. (2008) Fishing ancient Lake Baikal: inferences from the reconstruction of harvested perch (*Perca fluviatilis*) size at Ityrkhei, Siberia. *J. Archaeol. Sci.*, **35** (3), 577–590.

Martin, R. and Saller, K. (1957) *Lehrbuch der Antropologie*, Gustav Fischer Verlag, Stuttgart.

Michael, H.N. (1958) The Neolithic Age in Eastern Siberia. *T. Am. Philos. Soc.*, **48** (2), 1–108.

Molnar, P. (2006) Tracing prehistoric activities: musculoskeletal stress marker analysis of a Stone-Age population on the island of Gotland in the Baltic Sea. *Am. J. Phys. Anthropol.*, **129**, 12–23.

Montgomery, J., Budd, P. and Evans, J. (2000) Reconstructing the lifetime movements of ancient people: A Neolithic case study from southern England. *Eur. J. Arch.*, **3**, 370–385.

Mooder, K.P., Weber, A.W., Bamforth, F.J. *et al.* (2005) Matrilineal affinities and prehistoric Siberian mortuary practices: a case study from Neolithic Lake Baikal. *J. Archaeol. Sci.*, **32**, 619–634.

Mooder, K.P., Schurr, T.G., Bamforth, F.J. *et al.* (2006) Population affinities of Neolithic Siberians: a snapshot from prehistory Lake Baikal. *Am. J. Phys. Anthropol.*, **129**, 349–361.

Moore, M.V., Hampton, S.E., Izmest'eva, L.R. *et al.* (2009) Climate change and the World's 'Sacred Sea' – Lake Baikal, Siberia. *Bioscience*, **59**, 405–417.

Nagy, B.L.B. (1998) Age, activity, and musculoskeletal stress markers (abstract). *Am. J. Phys. Anthropol.* (Suppl. 26), 168–169.

Nagy, B.L.B. (1999) Bioarchaeological evidence for atl-atl use in prehistory (abstract). *Am. J. Phys. Anthropol.* (Suppl. 28), 208.

Nagy, B.L.B. and Hawkey, D.E. (1995) Musculoskeletal stress markers as indicators of sexual division of labour: multivariate analyses (abstract). *Am. J. Phys. Anthropol.* (Suppl. 20), 158.

Pearson, O.M. and Lieberman, D.E. (2004) The aging of Wolff's 'law': ontogeny and responses to mechanical loading of cortical bone. *Am. J. Phys. Anthropol.*, **47**, 63–99.

Peterson, J. (1998) The Natufian hunting conundrum: spears, atlatls, or bows? Musculoskeletal and armature evidence. *Int. J. Osteoarch.*, **8**, 378–389.

Price, T.D., Grupe, G. and Schröter, P. (1994) Reconstruction of migration patterns in the Bell Beaker period by stable strontium isotope analysis. *Appl. Geochem.*, **9**, 413–417.

Price, T.D., Johnson, C.M., Ezzo, J.A. *et al.* (1994) Residential mobility in the prehistoric Southwest United States: A preliminary study using strontium isotope analysis. *J. Archaeol. Sci.*, **21**, 315–330.

Price, T.D., Grupe, G. and Schröter, P. (1998) Migration in the Bell Beaker period of central Europe. *Antiquity*, **72**, 405–411.

Price, T.D., Manzanilla, L. and Middleton, W.D. (2000) Immigration and the ancient city of Teotihuacan in Mexico: A study using strontium isotope ratios in human bone and teeth. *J. Archaeol. Sci.*, **27**, 903–913.

Price, T.D., Bentley, R.A., Lüning, J. *et al.* (2001) Prehistoric human migration in the Linearbandkeramik of Central Europe. *Antiquity*, **75**, 593–603.

Prokopenko, A.A., Khursevich, G.K., Bezrukova, E.V. *et al.* (2007) Paleoenvironmental proxy records from Lake Hovsgol, Mongolia: new data on the Holocene climate in the Lake Baikal watershed. *Quaternary Res.*, **68**, 2–17.

Robb, J.E. (1998) The interpretation of skeletal muscle sites: a statistical approach. *Int. J. Osteoarchaeol.*, **8**, 363–377.

Robling, A.G., Burr, D.B. and Turner, C.H. (2000) Partitioning a daily mechanical stimulus into discrete loading bouts improves the osteogenic response to loading. *J. Bone Miner. Res.*, **15**, 1596–1602.

Ruff, C.B. (1987) Sexual dimorphism in human lower limb bone structure: relationship to subsistence strategy and sexual division of labour. *J. Hum. Evol.*, **16**, 391–416.

Ruff, C.B. (1995) Biomechanics of the hip and birth in early Homo. *Am. J. Phys. Anthropol.*, **98**, 527–547.

Ruff, C.B. (2000) Biomechanical analyses of archaeological human skeletons, in *Biological Anthropology of the Human Skeleton* (eds M.A. Katsenberg and S.R. Saunders), Wiley-Liss, Toronto, pp. 71–102.

Ruff, C.B. (2008) Biomechanical analyses of archaeological human skeletons, in *Biological Anthropology of the Human Skeleton*, 2nd edn (eds M.A. Katzenberg and S.R. Saunders), Wiley-Liss, Hoboken, pp. 183–206.

Ruff, C.B., Trinkaus, E., Walker, A. and Larsen, C.S. (1993) Postcranial robusticity in homo. I: Temporal trends and mechanical interpretation. *Am. J. Phys. Anthropol.*, **91**, 21–53.

Ruff, C., Holt, B. and Trinkaus, E. (2006) Who's afraid of the Big Bad Wolff?: 'Wolff's Law' and bone functional adaptation. *Am. J. Phys. Anthropol.*, **129**, 494–498.

Schweissing, M.M. and Grupe, G. (2003) Stable strontium isotopes in human teeth and bone: a key to migration events of the late Roman period in Bavaria. *J. Archaeol. Sci.*, **30**, 1373–1383.

Sealy, J. (1986) *Stable Carbon Isotopes and Prehistoric Diets in the South-Western Cape Province, South Africa*, BAR International Series 293, Cambridge Monographs in African Archaeology 15, Cambridge UK.

Shaw, C.N. and Stock, J.T. (2009) Habitual throwing and swimming correspond with upper limb diaphyseal strength and shape in modern human athletes. *Am. J. Phys. Anthropol.*, **140**, 160–172.

Shaw, C.N. and Stock, J.T. (2009) Intensity, repetitiveness, and directionality of habitual adolescent mobility patterns influence the tibial diaphysis morphology of athletes. *Am. J. Phys. Anthropol.*, **140**, 149–159.

Steen, S.L. and Lane, R.W. (1998) Evaluation of habitual activities among two Alaskan Eskimo populations based on musculoskeletal stress markers. *Int. J. Osteoarch.*, **8**, 341–353.

Stirland, A. (1991) Diagnosis of occupationally related pathology: can it be done? in *Human Paleopathology: Current Syntheses and Future Options* (eds D.J. Ortner and A.C. Aufderheide), Smithsonian Institution Press, Washington DC, pp. 40–47.

Stirland, A. (1993) Asymmetry and activity-related change in the male humerus. *Int. J. Osteoarch.*, **3**, 105–113.

Stock, J.T. (2006) Hunter-gatherer postcranial robusticity relative to patterns of mobility, climatic adaptation, and selection for tissue economy. *Am. J. Phys. Anthropol.*, **131** (2), 194–204.

Stock, J. and Pfeiffer, S. (2001) Linking structural variability in long bone diaphyses to habitual behaviors: Foragers from the southern African Later Stone Age and the Andaman Islands. *Am. J. Phys. Anthropol.*, **115**, 337–348.

Stock, J.T. and Pfeiffer, S.K. (2004) Long bone robusticity and subsistence behaviour among Later Stone Age foragers of the forest and fynbos biomes of South Africa. *J. Archaeol. Sci.*, **31** (7), 999–1013.

Stock, J.T. and Shaw, C.N. (2007) Which measures of skeletal robusticity are robust? A comparison of external methods of quantifying diaphyseal strength to cross-sectional geometric properties. *Am. J. Phys. Anthropol.*, **134** (3), 412–423.

Stock, J.T., Bazaliiskii, V.I., Goriunova, O.I., *et al.* (2010) Skeletal morphology, climatic adaptation and habitual behaviour among the mid-Holocene populations of the Cis-Baikal, relative to other hunter-gatherers, in *Prehistoric Hunter-Gatherers of the Baikal Region, Siberia: Bioarchaeological Studies of Past Lifeways* (eds A. Weber, M.A. Katzenberg and T. Schurr), University of Pennsylvania Museum of Archaeology and, Anthropology Pubs., Philadelphia.

Tafuri, M.A., Bentley, R.A., Manzi, G. and Lernia, S. (2006) Mobility and kinship in the prehistoric Sahara: Strontium isotope analysis of Holocene human skeletons from the Acacus Mts (southwestern Libya). *J. Anthropol. Archaeol.*, **25**, 390–402.

Warden, S.J., Hurst, J.A., Sanders, M.S. *et al.* (2005) Bone adaptation to a mechanical loading program significantly increases skeletal fatigue resistance. *J. Bone Min. Res.*, **20**, 809–816.

Weber, A.W. (1995) The Neolithic and early Bronze Age of the Lake Baikal region: a review of recent research. *J. World Prehist.*, **9** (1), 99–165.

Weber, A.W. and Bettinger, R. (2003) Current goals of mid-Holocene hunter-gatherer archaeology in the Lake Baikal region, in *Prehistoric Foragers of the Cis-Baikal, Siberia, Proceedings of the first conference of the Baikal Archaeology Project* (eds A.W. Weber and H.G. McKenzie), CCI Press, Edmonton, AB, pp. 1–14.

Weber, A.W., Link, D.W. and Katzenberg, M.A. (2002) Hunter-gatherer culture change and continuity in the Middle Holocene of the Cis-Baikal, Siberia. *J. Anthropol. Archaeol.*, **21** (2), 230–299.

Weber, A.W., Creaser, R., Goriunova, O.I. and Haverkort, C.M. (2003) Strontium isotope tracers in enamel of permanent human molars provide new insights into prehistoric hunter-gatherer procurement and mobility patterns: A pilot study of a Middle Holocene group from Cis-Baikal, in *Prehistoric Foragers of the Cis-Baikal, Siberia, Proceedings of the first conference of the Baikal Archaeology Project* (eds A.W. Weber and H.G. McKenzie), CCI Press, Edmonton, AB, pp. 133–153.

Weber, A.W., Beukins, R., Bazaliiskii, V.I. *et al.* (2006) Radiocarbon dates from Neolithic and Bronze Age hunter-gatherer cemeteries in the Cis-Baikal region of Siberia. *Radiocarbon*, **48** (3), 1–40.

Weber, A.W., Katzenberg, M.A. and Schurr, T.G. (eds) (2010) *Prehistoric Hunter-Gatherers of the Baikal Region, Siberia: Bioarchaeological Studies of Past Life Ways*, University of Pennsylvania Museum of Archaeology and Anthropology, University of Pennsylvania Press, Philadelphia.

Weiss, E. (2003) Understanding muscle markers: aggregation and construct validity. *Am. J. Phys. Anthropol.*, **121**, 230–240.

Weiss, E. (2004) Understanding muscle markers: lower limbs. *Am. J. Phys. Anthropol.*, **125**, 232–238.

Weiss, E. (2007) Muscle markers revisited: activity pattern reconstruction with controls in a central Californian Amerind population. *Am. J. Phys. Anthropol.*, **133**, 931–940.

White, D., Preece, R.C., Shchetnikov, A.A. *et al.* (2008) A Holocene molluscan succession from floodplain sediments of the upper Lena River valley (Lake Baikal region), Siberia. *Quaternary Sci. Rev.*, **27**, 962–987.

Wilczak, C.A. (1998) Consideration of sexual dimorphism, age, and asymmetry in quantitative measurements of muscle insertion sites. *Int. J. Osteoarch.*, **8**, 311–325.

Wolff, J. (1892) *Das Gesetz der Transformation der Knochen*, A. Hirchwild, Berlin.

Ziker, J.P. (2002) *Peoples of the Tundra: Northern Siberians in the Post-Communist Transition*, Waveland Press, Long Grove, IL.

12

'An External Agency of Considerable Importance': The Stresses of Agriculture in the Foraging-to-Farming Transition in Eastern North America

Clark Spencer Larsen[1] and Christopher Ruff[2]

[1] *Department of Anthropology, The Ohio State University, Columbus, OH, USA*
[2] *Center for Functional Anatomy and Evolution, Johns Hopkins University, Baltimore, MD, USA*

12.1 INTRODUCTION

Eastern North America was occupied for most of its prehistory by small-scale societies, whose lifeways and associated diets focused exclusively on gathering of wild plants, hunting and fishing. Dietary adaptation changed at two key times during the last several thousand years of prehistory (Gremillion, 2002; Lusteck, 2006; Milner, 2004; Smith, 1995; Smith and Yarnell, 2009; Hutchinson, 2002; Larsen, in press). The first took place 2000 to 4000 years ago in the midcontinental region (Ohio, Kentucky, Illinois) and involved the domestication of at least five native weedy plants as important food sources. The second and more pervasive event involved the spread and adoption of domesticated maize throughout much of the region between AD 800 and 1100, earlier in some regions (e.g. the Illinois River valley) than in others (e.g. Georgia coast, southern Canada). Although remains of maize have been found well prior to the ninth century AD (Riley *et al.*, 1994), its presence is not measureable in terms of dietary contribution until the end of the first millennium AD (Ambrose, 1987). The precise timing and importance of maize is now well documented via stable isotope analysis (Ambrose, 1987; Hutchinson *et al.*, 1998; Schoeninger, 2009; Smith, 1990; Staller, Tykot and Benz, 2006). Unlike the earlier domestication event, the adoption of maize in later prehistory led to clearly discernible changes

Human Bioarchaeology of the Transition to Agriculture Edited by Ron Pinhasi and Jay T. Stock

in health, including an increase in infectious disease, growth disruption and poorer quality of life in general (various authors in Cohen and Armelagos, 1984; Cohen and Crane-Kramer, 2007; Lambert, 2000; Powell, Bridges and Mires, 1991; Steckel and Rose, 2002).

With the introduction and intensification of maize, there was a simultaneous increase in population size, sedentism, social and political complexity, long-distance trade, adoption of belief systems shared over large regions and intensified resource extraction (Milner, 2004). With regard to subsistence intensification, the physical behaviour, activity and lifestyle must have undergone significant changes. Anthropologists are interested in answering the question: 'How did the shift from foraging to farming affect workload and activity?'

In order to address this question, this chapter uses osteoarthritis and long bone cross-sectional geometric data derived from three regional studies in eastern North America, namely the Pickwick Basin area of the Tennessee River valley (northern Alabama), the lower Illinois River valley (southern Illinois) and the Georgia Bight (Atlantic coastal Georgia and Florida). Specifically, we outline the results of these regional investigations, and then draw conclusions about behavioural and lifestyle implications of the foraging to farming transition for native populations living in eastern North America. Many other bioarchaeological investigations have recorded data pertinent to inferences about behaviour (especially osteoarthritis). However, the three regional studies are placed within the context of large skeletal series and availability of excellent and comprehensive contextual data on diet, resource acquisition and population history.

Osteoarthritis (or degenerative joint disease) is a multifactorial disorder characterized by degeneration of diarthrodial joints (e.g. elbows, knees and shoulders). The closely related degenerative disorder, osteoarthrosis, involves the amphiarthrodial joints (vertebral bodies). The degenerative changes associated with osteoarthritis and osteoarthrosis are so similar that here we refer to the two disorders collectively as osteoarthritis. Osteoarthritis has been linked with a variety of causal factors, including genetic predisposition, normal anatomical variation, climate, obesity and body mass, trauma and age (Hough, 2001; Poole, 1999; Radin, Paul and Rose, 1972; Radin *et al.*, 1991). However, the primary causal factor is the mechanical environment involving strenuous physical activity (reviewed in Larsen, 1997). The process behind articular degeneration is slow, taking as long as 15 to 20 years (Poole, 1999).

It is rarely possible to link osteoarthritis to a specific activity. Rather, it is a general indicator of joint loading useful for documenting degree of long-term stress as well as patterns of more general motions and habitual movement (Knüsel, 2000; Larsen, 1997). The pathology includes proliferative exophytic growth of new bone on joint margins ('marginal lipping'), and in more severe cases, rarefaction and thinning of the articular surface owing to degeneration of the articular cartilage. In circumstances when the cartilage is missing, there is direct bone-on-bone articulation, resulting in polishing called eburnation.

Central to reconstruction and interpretation of past behaviour via the study of human skeletal remains is the notion that bone is adapted to the mechanical environment surrounding it. That is, while genetically controlled in primary form, bone morphology is highly influenced by mechanical loading in mature form. The general concept that mechanical loading plays a key role in shaping skeletal morphology is known as Wolff's Law (Ruff, 2008). Owing to the somewhat different use of the term than what was originally intended in the late nineteenth century when it was first formulated by Julius Wolff, Ruff, Holt and Trinkaus (2006) suggest that it should be substituted with 'bone functional adaptation'.

It has been well established that long bone diaphyses can be modelled as engineering beams, and therefore can be subject to the same analyses that engineers use to design buildings and

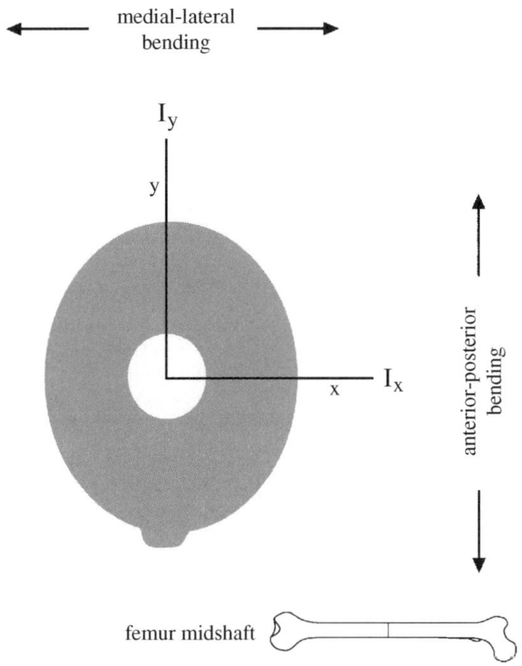

Figure 12.1 Cross-section of femur midshaft showing I_x and I_y geometric properties. I_x measures resistance strength. to bending in anterior-posterior plane and I_y measures resistance to bending in the medial-lateral plane. J, or resistance to torsional stresses, is the sum of I_x and I_y. The top of the cross-section is anterior and the bottom of the cross-section is posterior. Adapted from Larsen, 2002

other kinds of structures. That is, cross-sections of long bones – such as the midshafts of femora, tibiae and humeri – can be analysed via measurement of cross-sectional geometric properties (Figure 12.1). It is now well recognized that the strength of a bone is increased via formation of new bone to meet the demands of increased activity, and this is well documented in both animal and human studies (Macdonald, Cooper and McKay, 2009; Shaw and Stock, 2009) and applied to numerous bioarchaeological contexts (Larsen, 1997; Ruff, 2008). The bone strength is documented via measurement of cross-sectional geometric properties. The higher the value of the cross-sectional geometric property, the greater its ability to resist breakage due to bending or twisting (torsion), the two most critical forces that long bones are typically subjected to. In the lower limb bones, bending and torsional forces occur owing to body weight and muscle activity. The upper limb in humans is not involved in locomotion, therefore bending and torsion (e.g. in the humerus) is due to muscle forces used in lifting and carrying.

The cross-sectional geometric properties depend on both the *amount* and *distribution* of bone in a cross-section. The former are measured as 'areas' and the latter as 'second moments of area'. Areas include total subperiosteal area (*TA*), endosteal or medullary area (*MA*), and cortical area (*CA*). *CA* measures the amount of bone in a cross-section and is a strength indicator of a long bone diaphysis under pure axial loading (loading that is simultaneously applied to both ends of a straight bone). Second moments of area include '*I*' values, or measures of resistance to bending forces in specific planes, such as in the anteroposterior and mediolateral directions of bending (I_x, I_y) or as maximum and minimum bending (I_{max}, I_{min}).

Researchers also report on resistance to torsion, or polar moment of inertia (*J*), which is a sum of bending values in perpendicular planes and also represents an overall indicator of bone bending strength. This discussion will focus on *CA* and *J* as indicators of bone strength. Because body size has a strong influence on these variables (Ruff, 2000), it must be factored into such analyses. This can be done directly by estimating body mass (weight) from other skeletal dimensions such as femoral head size (Auerbach and Ruff, 2004), or less directly by using powers of bone length as a proxy (Ruff *et al.*, 1993). In the latter case though, similar body proportions must be assumed, or differences in body proportions must be factored into comparisons (Ruff *et al.*, 1993; Ruff, 2000).

While indicators of osteoarthritis are clearly visible on dry bone, methods of data collection and analysis tend to be diverse, resulting in studies that may have limited comparative value (Bridges, 1993a). Cross-sectional geometry requires access to the bone cross-section. This can be accomplished by either cutting the bone in the section region or by imaging technology. Because most museums will not allow destructive access (cutting and exposure of the section), imaging via radiography or computed tomography (CT) is almost always required (see Ruff, 2008 for details). Owing to the more recent application of engineering tools to skeletal analysis and the greater level of technology required in order to measure section properties, the results of such studies are fewer in number than those available from osteoarthritis (see Larsen, 1997; Ruff, 2008 for reviews).

Osteoarthritis and cross-sectional geometry clearly provide significant insight into behavioural regimes of past populations. Thus, both measures are affected by adult age-at-death (see Larsen, 1997; Ruff, 2008 for detailed treatments of the role of age and the ageing process on these variables). Moreover, it is often the case that estimates of adult age-at-death are not sufficiently precise to set aside the possibility that longevity has influenced the outcome of these variables within a skeletal series or between series. Elsewhere, we have discussed how samples presented in this study have been apportioned for age or otherwise considered in the analyses (Larsen, 1982; Larsen *et al.*, 2007; Ruff *et al.*, 1984; Ruff and Larsen, 1990, 2001). With respect to osteoarthritis, we note that patterning of the condition relating to inter-joint variation is less sensitive to demographic differences in age-at-death than prevalence.

12.1.1 Behaviour and Lifestyle: Suggestive Insights from Early Bioarchaeological Studies

For eastern North America, glimpses at prevalence and pattern of osteoarthritis as a behavioural marker of lifestyle and activity are presented in reports on collections of human remains excavated in the 1930s to 1950s, especially with regard to the large samples – numbering in the hundreds of individuals – deriving from work relief programmes and federal construction projects that required mitigation (see Larsen, n.d.; Milner and Jacobi, 2006). Study of skeletons of prehistoric foragers from Archaic period Indian Knoll, Kentucky, for example, revealed that the majority of adults, including young individuals, had lumbar osteoarthritis, some of which was severe (Cassidy, 1984; Snow, 1948). This finding implied a highly demanding, physically-active lifestyle related to resource acquisition in hunting, gathering and collecting.

Comparison of an earlier and later skeletal series in Virginia representing foragers and maize farmers, respectively, revealed an increase in osteoarthritis (Hoyme and Bass, 1962). Hoyme and Bass's study recognized the introduction of agriculture as 'an external agency of considerable importance' (p. 350), and suggested that the farming lifestyle was more physically demanding than the foraging lifestyle. Unfortunately, owing to poor preservation, the osteoarthritis data were based on a small subset of the collection, and interpretations were suggestive only. Morphological evidence indicated that the foraging-to-farming transition in this setting may have involved a decrease in mobility. That is, measurement of the degree of flatness of the tibia midshaft (platycnemia) revealed that forager male tibiae had marked transverse flattening. Contrary to other authorities who had argued for nutritional causes, Hoyme and Bass (1962) argued that transverse flattening has a functional/behavioural cause, such as from 'muscular stresses'. Although the argument was not based in the context of modern biomechanical models, nonetheless it was prescient of work done decades later (see below). Importantly, these earlier investigations set the stage for large-scale, regionally based studies of lifestyle (Larsen, n.d.).

12.1.2 Pickwick Basin of the Tennessee River Valley

Following approval of the construction of the Pickwick Landing Dam in 1934, a survey of some 75 square miles was undertaken in order to identify and excavate archaeological sites that would be flooded. An unprecedented number of skeletons – nearly 2000 – were excavated from various localities in Tennessee and Alabama. A subsample of the series was studied and reported on by Marshall Newman and Charles Snow (1942). Their investigation of these remains was one of the first regionally-based bioarchaeological studies in eastern North America, and although preliminary and descriptive it is an essential resource for some observations on and interpretations of behavioural adaptation. The sample represents remains of prehistoric foragers (about 6000–1000 BC) and prehistoric farmers (about AD 1200–1500) (Bridges, 1991a). In their preliminary comparison of the earlier and later series, Newman and Snow (1942) found some evidence for an increase in robusticity of the long bones (increase in midshaft diameters of the femur, tibia, and humerus). In their view, the record suggested some degree of biological change, perhaps associated with the shift from hunting and gathering to agriculture.

In order to test the hypothesis that dietary adaptation in this setting involved greater physical demand and workload, Patricia Bridges undertook a restudy of a subsample of the remains from Alabama studied by Newman and Snow, focusing on osteoarthritis and cross-sectional geometry (Bridges, 1989a,b,c 1990, 1991a,b, 1994) . Analysis of osteoarthritis prevalence reveals some important patterns of variation (Table 12.1). First, males tend to have a higher prevalence than females. Second, articular joints most affected include the shoulder, elbow and knee. Third, the foragers have somewhat higher prevalence of osteoarthritis than farmers. While the differences do not reach statistical significance, the pattern of higher prevalence in the forager series is consistent. On the other hand, the forager series has a somewhat older age structure in the foragers (Bridges, 1991a), explaining at least in part the temporal variation.

Analysis of cross-sectional geometry derived from CT scans of femora and humeri somewhat conflicts with the osteoarthritis results (Bridges, 1989b,c, 1991a,b). That is, bone area (*CA*) and polar second moments of area (*J*) for the humerus and femur increased in farmers relative to foragers (Table 12.2). These findings are consistent with the relatively greater external dimensions of long bones in the farmers, a characteristic that is suggestive of greater mechanical loading (Larsen, 1997). The trends are present in both males and females, but males show relatively greater strength increase in the femur only, whereas females show the change in both femur and humerus. This is consistent with Bridges's notion that there may have been a change in division of labour with the shift to maize farming, which she attributes to the greater involvement of females than males in tending crops. Moreover, females express greater

Table 12.1 Osteoarthritis for the Pickwick Basin[a]

Articular joint	Foragers % n.		Farmers % n.	
	L	R	L	R
Males				
Shoulder	39.1 (23)	45.9 (24)	34.8 (23)	34.6 (26)
Elbow	37.0 (29)	48.2 (29)	31.0 (29)	24.1 (29)
Wrist	17.9 (28)	24.0 (25)	0.0 (26)	15.4 (26)
Hip	4.3 (23)	4.3 (23)	0.0 (22)	0.0 (22)
Knee	35.8 (28)	34.4 (29)	25.9 (27)	11.1 (27)
Ankle	18.5 (27)	7.1 (28)	0.0 (28)	8.0 (25)
Females				
Shoulder	30.0 (20)	30.4 (23)	15.4 (26)	21.7 (23)
Elbow	40.0 (35)	57.6 (33)	21.5 (28)	28.6 (28)
Wrist	4.2 (24)	25.9 (27)	7.6 (26)	0.0 (20)
Hip	0.0 (20)	5.6 (18)	5.9 (17)	0.0 (22)
Knee	23.3 (30)	34.5 (29)	17.9 (28)	26.1 (23)
Ankle	6.2 (32)	13.3 (30)	4.3 (23)	0.0 (28)

[a]Adapted from Bridges, 1991a; study reported by side for individuals older than 30 years only.

increase in humerus robusticity than males, which she attributes to processing corn, such as with wooden mortars and pestles documented in the ethnographic literature for women. In this regard, the external dimensions for left and right humeri are nearly identical in the female farmers, indicating that the use of the limbs was approximately equal.

At first glance, the differences in results between the moderate decrease in osteoarthritis and increase in cross-sectional geometry would appear to be contradictory with the former, indicating a decline in activity and the later an increase in activity. While osteoarthritis is due to mechanical factors, some of the variation may be explained by the fact that short-term, unusually high mechanical loads and microtrauma may play a role in developing arthritis, whereas changes in cross-sectional geometry are more usually the result of long-term repetition of lower magnitude mechanical loading (Bridges, 1991a). Rather than being contradictory, it seems more likely that osteoarthritis and cross-sectional geometry track different *kinds* of activity. Based on the well-known link between behaviour and skeletal modelling and remodelling, along with the significant biomechanical changes in the Pickwick Basin series, it seems most likely that level of physical activity and workload increased in the foraging to farming transition here.

12.1.3 Lower Illinois River Valley

The lower Illinois River Valley has been the focus of intensive archaeological investigation since the early 1960s, as highway construction required mitigation and study of cultural resources. This setting is amongst the best archaeologically documented regions in North America, and it is represented by a rich record of village and mortuary remains. Bioarchaeological research dealing with population history (biodistance), palaeodemography, diet and population health (palaeopathology) provides an essential context for understanding the

Table 12.2 Cross-sectional geometric properties for the Pickwick Basin[a]

Property	Foragers	Farmers
Males, Femur	n = 18	17
Length	435	447
CA	401	444*
J	39940	50465*
CA_{STD}	487	497*
J_{STD}	345	377*
Females, Femur	n = 6	n = 11
Length	404	410
CA	323	336
J	26856	33463*
CA_{STD}	490	488
J_{STD}	344	397
Males, Humerus	n = 17	n = 13
Length	319	324
CA	209	203
J	12637	11949
CA_{STD}	644	597
J_{STD}	571	497
Females, Humerus	n = 6	n = 14
Length	288	299
CA	157	159*
J	6728	9279*
CA_{STD}	657	595*
J_{STD}	524	592*

[a]Adapted from Bridges, 1985. Standard error values not available.
*$p < 0.05$, t-test.
Size standardization: $CA_{STD} = CA/Length^3 \times 10^8$; $J_{STD} = J/Length^{5.33} \times 10^{12}$; see Ruff *et al.*, 1993.

groups that occupied the region, and is especially well-known for the temporal sequence of Middle Woodland (50 BC–AD 200), Late Woodland (AD 600–1050), and Mississippian (AD 1050–1250) (Buikstra, 1984, 1988; Cook, 1984). Throughout the sequence, populations ate cultivated plants to varying degrees, including a focus on native seed crops in the Middle Woodland and early Late Woodland, continuation of seed production coupled with introduction of maize in appreciable amounts in the late Late Woodland, and full-blown maize production and intensification in the Mississippian period. For the sequence as a whole, populations became increasingly reliant on crop production, culminating in a rapid shift to, and intensification of, maize agriculture in the last prehistoric period.

Although more limited in scope than Bridges's study of the Pickwick Basin series, analysis of osteoarthritis in cervical vertebrae increased in agricultural females (males show no change) in the lower Illinois River valley of west-central Illinois (Pickering, 1984 cited in Cook, 1984; and see Bridges, 1991a). The case was made that the increase in females but not in males represents their new role in food production – raising and tending crops and preparation of food relating to agricultural produce.

Table 12.3 Cross-sectional geometric properties for the lower Illinois River valley[a]

Property	Middle Woodland	Early Late Woodland	Late Late Woodland	Mississippian
Males, Femur	(n = 10)	(n = 10)	(n = 10)	(n = 10)
Length	453 (3.7)	459 (4.0)	451 (3.0)	451 (3.6)
CA	400 (18.3)	418 (14.9)	424 (11.7)	428 (18.1)
J	49066 (4558)	46773 (2605)	52665 (2979)	55602 (3766)
CA_{STD}	435 (16.4)	437 (21.8)	444 (14.6)	429 (14.0)
J_{STD}	340 (15.9)	309 (22.5)	347 (16.8)	343 (21.4)
Females, Femur	(n = 10)	(n = 10)	(n = 10)	(n = 10)
Length	421 (2.8)	421 (2.9)	418 (3.3)	418 (3.2)
CA	278 (8.4)	316 (10.9)	299 (9.5)	311 (16.9)
J	26006 (1234)	31231 (1336)	27367 (1426)	30545 (3158)
CA_{STD}	367 (18.7)	414 (14.4)	434 (11.6)*	401 (13.8)
J_{STD}	261 (18.0)	309 (11,4)	326 (17.8)*	290 (18.5)
Males, Humerus	(n = 10)	(n = 10)	(n = 10)	(n = 10)
Length	330 (3.1)	332 (2.2)	327 (2.1)	325 (2.5)
CA	168 (10.5)	163 (10.9)	155 (12.2)	160 (7.1)
J	11458 (1266)	9766 (720)	13272 (1060)	12283 (927)
CA_{STD}	171 (8.1)	163 (11.5)	153 (9.2)	159 (4.9)
J_{STD}	445 (44.8)	353 (35.8)	466 (62.0)	429 (16.1)
Females, Humerus	(n = 10)	(n = 10)	(n = 10)	(n = 10)
Length	303 (2.7)	304 (2.3)	302 (2.2)	302 (2.3)
CA	103 (4.6)	125 (5.0)	123 (4.3)	97 (7.2)**
J	5574 (307)	7379 (524)*	7495 (519)*	5742 (471)
CA_{STD}	113 (5.7)	138 (4.7)	138* (3.5)	105 (7.5)**
J_{STD}	330 (20.6)	428 (14.6)	493 (31.4)*	321 (39.6)**

[a]Adapted from Bridges, Blitz and Solano, 2000. Standard errors in parentheses following values.
*$p < 0.05$, Tukey test with Middle Woodland.
**$p < 0.05$, Tukey test with Early Late and Late Late Woodland.
Size standardization: $CA_{STD} = CA/Length^3 \times 10^8$; $J_{STD} = J/Length^{5.33} \times 10^{12}$; see Ruff *et al.*, 1993.

In order to test the hypothesis that workload increased with agriculture, Bridges, Blitz and Solano (2000) measured cross-sectional geometric properties. The skeletal series used for the analysis derives from the aforementioned archaeological research programme in the lower Illinois valley, resulting in the excavation and study of hundreds of skeletons from excellent archaeological contexts. Analysis of femoral *CA* and *J* show little evidence of change in males; none of the changes reaches statistical significance (Table 12.3). In contrast, females show increased bone strength through the end of the Late Woodland period, followed by a decrease in the Mississippian maize farmers. These changes reach statistical significance.

For the most part, external bone dimensions mirror these changes. This pattern indicates that bone strength changed prior to the adoption of maize, and suggests that the first of two shifts in diet adaptation likely influenced lifestyle and workload, at least for women. On the other hand, the adoption of maize in later prehistory also had an effect, resulting in decline in workload, again mainly in women. This decline in workload suggests that other activities besides maize production and processing may have come into play. Bridges, Blitz and Solano (2000) suggest that innovations in food preparation technology, such as soaking, boiling and other techniques, may have helped to reduce workload by replacing corn pounding (or at least supplementing it) with boiling. Indeed, the innovations in ceramic technology – a transition that took place in the Late Woodland period – argues for this possibility. Moreover, recent analysis of work required in harvesting and preparation of starchy seed (e.g. *Chenopodium*) indicates that it requires considerable time and effort, resulting in relatively low return rates (Gremillion, 2004). Thus, a high and likely increasing work load in the earlier periods is consistent with the increase in bone strength, followed by a decline that is perhaps related to a reduction in time and effort needed to process maize kernels in the lower Illinois River valley.

12.1.4 The Georgia Bight

Like the Pickwick Basin, bioarchaeological research in this region got its first big boost during the era of the Great Depression in the 1930s and early 1940s. Work relief archaeological crews excavated hundreds of skeletons and (occasionally) associated village sites in the Georgia Bight, especially along the north Georgia coast near the modern city of Savannah and on St Simons Island, one of numerous sea islands (Larsen, 1982, n.d.). Unlike the Pickwick Basin, no large published reports were completed on these remains. However, Hulse (1941) produced a descriptive summary report on a series of remains from the Irene Mound site, a large Mississippian centre. These remains were later re-studied by Clark Larsen in his analysis of Georgia coastal populations (Larsen, 1982, 1984) and expanded to a large regional study of prehistoric and mission-era bioarchaeology of the Georgia Bight generally (Larsen, 1990, 2001; Larsen *et al.*, 1992a, 2002, 2007). The latter investigation focused on the adoption of maize farming in later prehistory (post-AD 1150) and its intensification in the colonial period during the process of missionization (AD 1550–1700), investigating health, dietary adaptation and lifestyle. The forager samples derive from about 500 BC to AD 1150. However, most of the forager samples are from after about AD 700. Prehistorically, the region is dominated by a pan-Mississippian florescence after about AD 1000 (Anderson, 1994). Unlike the lower Illinois River valley, only one agricultural event took place – the adoption of maize. Moreover, the region was subject to Spanish colonialism, establishment of missions and labour exploitation, mainly for agricultural production (Worth, 1995, 2001). This transition – adoption of maize agriculture and later intensification of agricultural production and consumption – is now well documented by stable isotope analysis (Larsen *et al.*, 1992b; Hutchinson *et al.*, 1998, 2000) and dental caries prevalence (Larsen, 1982; Larsen *et al.*, 1992a, 2007).

Documentation of osteoarthritis in the series reveals a clear trend of decline in prevalence in comparison of prehistoric foragers and farmers from Georgia coastal sites (Table 12.4). For non-vertebral joints, the highest prevalence is in the shoulder, elbow and knee, and lowest in the wrist, hand, hip, ankle and foot. The highest prevalence in the skeleton is in the vertebral column, especially the cervical and lumbar vertebra. Male prevalence is largely higher than female prevalence for nearly all joints. Indication of a change in mechanical environment in the colonial period is provided by the dataset on osteoarthritis. Overall, there is a considerably higher prevalence of osteoarthritis, and strikingly so. For example, the vertebrae show very elevated frequencies. Males and females have generally the same prevalence profile, but the shoulder osteoarthritis is considerably higher for females than males. The general pattern of greater prevalence of osteoarthritis in the colonial-era populations may due to an older adult group relative to the prehistoric farmers (Larsen *et al.*, 2002). However, comparisons of

Table 12.4 Osteoarthritis for the Georgia Bight[a]

Articular joint	Foragers % n.	Farmers % n.	Late Mission Guale % n.
All Adults			
Cervical	26.0 (50)	16.4 (189)	68.3 (60)*
Thoracic	15.6 (32)	11.4 (175)	68.3 (60)*
Lumbar	41.9 (43)	24.5 (163)*	67.2 (58)*
Shoulder	9.7 (113)	5.3 (207)	15.2 (66)*
Hip	12.0 (108)	6.8 (206)	10.5 (67)
Wrist	5.9 (84)	1.1 (187)*	13.2 (68)*
Hand	5.0 (40)	3.0 (165)	5.9 (68)
Males			
Cervical	33.3 (18)	27.5 (80)	74.1 (27)*
Thoracic	16.7 (12)	19.2 (73)	74.1 (27)*
Lumbar	50.0 (14)	37.7 (69)	75.0 (28)*
Shoulder	17.7 (34)	10.1 (79)	16.7 (30)
Hip	13.2 (38)	12.5 (80)	9.4 (32)
Wrist	15.4 (26)	2.7 (75)*	16.1 (31)*
Hand	8.3 (12)	5.8 (69)	3.3 (30)
Females			
Cervical	22.2 (27)	8.4 (95)	65.6 (32)*
Thoracic	16.7 (18)	5.5 (91)	65.6 (32)*
Lumbar	38.5 (26)	15.3 (85)*	60.0 (30)*
Shoulder	5.2 (58)	2.9 (105)	13.9 (36)*
Hip	13.2 (53)	3.8 (105)*	11.4 (35)
Wrist	2.3 (43)	0.0 (96)	10.8 (37)*
Hand	4.0 (25)	1.2 (85)	7.9 (38)

[a]Adapted from Larsen *et al.*, 2002; presence per individual reported if one or other side affected, includes all adults.
*$p < 0.05$, chi-square.

osteoarthritis prevalence between late prehistoric and colonial-period farmers show significant differences in comparison of the same five-year age groups ($p < 0.05$, chi-square).

Cross-sectional analysis of prehistoric and historic populations provides an abundant record of behavioural adaptation and workload over the sequence. Specifically, four periods representing prehistoric and historic-era Guale, a tribal group on the Georgia coast, provide a temporal and regional perspective – prehistoric foragers (pre-AD 1000), prehistoric farmers (AD 1000–1450), Early Mission Guale (AD 1600–1680) and Late Mission Guale (AD 1680–1700). The Early Mission Guale sample is from Mission Santa Catalina on St Catherine's Island, Georgia, and the Late Mission Guale sample is from the descendant Mission Santa Catalina on Amelia Island, Florida (Larsen *et al.*, 2007). In addition, data for skeletal series representing the historic-era Yamasee, a tribal group contemporary with Early Mission Guale, are included in the analysis (see Ruff and Larsen, 2001 for details on the samples).

Cross-sectional properties show interesting patterns of variation, largely reflecting the patterns of inferred workload and activity based on analysis of osteoarthritis (Table 12.5). For males, there is a decline in bone strength in comparison of foragers and farmers, followed by

Table 12.5 Cross-sectional geometric properties for the Georgia Bight[a]

Property	Foragers	Farmers	Early Mission Guale	Late Mission Guale	Early Mission Yamasee
Males, Femur	(n = 6)	(n = 21)	(n = 9)	(n = 21)	(n = 8)
Length	418 (8.8)	421 (4.1)	420 (7.9)	420 (3.9)	426 (9.8)
CA	440 (23.0)	393 (9.3)	400 (13.5)	395 (8.1)	399 (14.5)
J	53336 (4549)	42863 (1898)	47216 (2252)	47634 (1270)	45803 (3317)
CA_{STD}	601 (24.8)	531 (19.4)	547 (31.9)	536 (13.3)	519 (21.3)
J_{STD}	563 (27.2)	444 (22.0)	502 (37.1)	506 (21.0)	444 (22.7)
Females, Femur	(n = 8)	(n = 24)	(n = 9)	(n = 23)	(n = 8)
Length	408 (5.5)	401 (5.3)	404 (6.5)	392 (3.2)	392 (3.2)
CA	315 (21.4)	301 (11.9)	343 (12.2)	315 (12.2)	299 (8.9)
J	32210 (1949)	29270 (2172)	32656 (2217)	31813 (1653)	27147 (1303)
CA_{STD}	470 (37.9)	469 (15.6)	521 (20.1)	524 (20.6)	500 (22.0)
J_{STD}	398 (28.9)	381 (18.1)	417 (21.5)	481 (25.4)	412 (27.0)
Males, Humerus	(n = 8)	(n = 31)	(n = 11)	(n = 24)	(n = 8)
Length	309 (6.3)	319 (2.4)	314 (4.0)	314 (2.7)	322 (6.4)
CA	215 (14.4)	202 (5.3)	201 (10.6)	212 (5.1)	206 (6.0)
J	14038 (1077)	12052 (508)	12032 (951)	13710 (470)	13543 (1064)
CA_{STD}	736 (63.2)	630 (21.2)	655 (34.9)	690 (18.5)	620 (35.5)
J_{STD}	770 (85.0)	547 (23.3)	603 (57.8)	686 (29.8)	578 (40.6)
Females, Humerus	(n = 7)	(n = 30)	(n = 8)	(n = 24)	(n = 12)
Length	288 (6.0)	305 (3.6)	301 (5.6)	299 (2.6)	293 (3.2)
CA	158 (6.1)	150 (5.7)	154 (7.3)	144 (5.8)	125 (7.2)
J	6999 (350)	7303 (392)	6623 (587)	7226 (372)	5849 (290)
CA_{STD}	670 (48.6)	537 (25.8)	566 (26.8)	541 (21.6)	497 (26.3)
J_{STD}	558 (52.8)	426 (24.3)	406 (32.6)	467 (24.9)	416 (24.0)

[a]Adapted from Ruff and Larsen, 2001; Larsen *et al.*, 2007. Standard errors in parentheses following values. Size standardization: $CA_{STD} = CA/Length^3 \times 10^8$, $J_{STD} = J/Length^{5.33} \times 10^{12}$; see Ruff, 2008.

successive increases in the mission populations. The Yamasee Mission series shows values that are relatively high, but most similar to the prehistoric farmers. In some key respects, females show a similar pattern of change and variation as males. For the femur, the reduction is very slight in the foraging to farming transition, much less so than in males. Moreover, Guale females show a strong increase in bone strength in the early mission, and especially, late mission period. For the humerus, the pattern of change in male bone strength is identical to the femur. Females are quite different from males, however, in that the bone strength for the humerus continues to decline in the Early Mission Guale, but then increases in the Late Mission Guale.

One important element of lifestyle and workload is mobility. In general, foragers are regarded as more mobile than farmers. This is in keeping with the notion that most foragers are mobile for significant parts of the year. Farmers, on the other hand, are more tethered to a specific locality, as tending, maintenance and harvesting of crops occurs in one place. Ruff (1987, 1999, 2008; and see Larsen *et al.*, 2008) observed in a sample of foragers, farmers and industrial societies that the ratio of relative anterior-posterior bending to medial-lateral bending in the femur midshaft tends to be relatively high ($I_x/I_y > 1.0$) in foragers and less so in farmers ($I_x/I_y \sim 1.0$) and in industrial societies. This reflects the fact that highly mobile

Table 12.6 I_x/I_y midshaft femur. mobility index for the Georgia Bight[a]

Sex	Foragers	Farmers	Early Mission Guale	Late Mission Guale	Early Mission Yamasee
Males	1.333	1.153	1.192	1.137	1.110
Females	1.207	1.122	0.978	0.949	1.068

[a]Adapted from Ruff and Larsen, 2001.

individuals – those engaged in frequent running, walking and long distance travel generally – tend to have higher anterior-posterior bending loads in the region of the lower limb about the knee (the region extending from the midshaft femur to the midshaft tibia) (Ruff, 1987, 1999). For the Georgia Bight, analysis of the I_x/I_y ratio shows a decline in the forager-to-farmer transition in this setting (Table 12.6). Interestingly, there is a reversal of the trend in the mission period, but for males only. The reversal is due to a number of males in the Early Mission Guale and Late Mission Guale with high I_x/I_y ratios. These ratios may reflect the fact that some native males in the mission system were recruited for movement of food and other goods over long distances and other activities requiring travel (Ruff and Larsen, 1990). Overall, the contrast in the Georgia Bight is between prehistoric foragers and all the later groups.

Compared within a wide range of populations studied from North America, including those from the Georgia Bight and Alabama, the level of sexual dimorphism in the I_x/I_y ratio is consistently higher in foragers than farmers (Figure 12.2). This reflects the fact that males are relatively more mobile amongst hunter-gatherers than agriculturalists, owing to the greater role of males in hunting and long distance travel, whereas females are usually more tethered to the home base (Ruff, 1987). The two prehistoric groups from coastal Georgia – the foragers and farmers – are consistent with this pattern. However, the two contact-era mission Guale samples deviate strongly from this pattern for the midshaft femur. This appears to be due in part to the presence of some extreme male outliers, individuals having very high I_x/I_y ratios in the two groups as well as decreased levels of mobility in females. Interestingly, the Yamasee, another Georgia Bight early historic sample who were less acculturated than the Guale, do not show this extreme sexual dimorphism, likely because males were not recruited in the same way as the acculturated Guale.

In terms of overall strength, temporal changes in mechanical loading of the lower and upper limbs in the Georgia samples – based on second moments of area for the femur and humerus – are broadly similar (except for the lack of increase in humeral strength amongst Mission period females). This indicates that loading affected both ambulatory functions and use of the legs and non-ambulatory functions and use of the arms. Overall, the pattern indicates a decline in workload with the adoption of maize agriculture, followed by a general increase in workload and activity with the arrival of Spaniards and establishment of the mission system. The increase in mechanical loading is fully consistent with the historical records, namely that native people were recruited as a source of labour, mostly for food (crop) production, but also for other physically-demanding activities (Larsen, 1990; Worth, 2001).

12.1.5 Comparisons of the Three Regions

In many ways, eastern North America is unified by some common adaptive themes for the human populations that occupied it prior to its colonization by Europeans (Milner, 2004). We

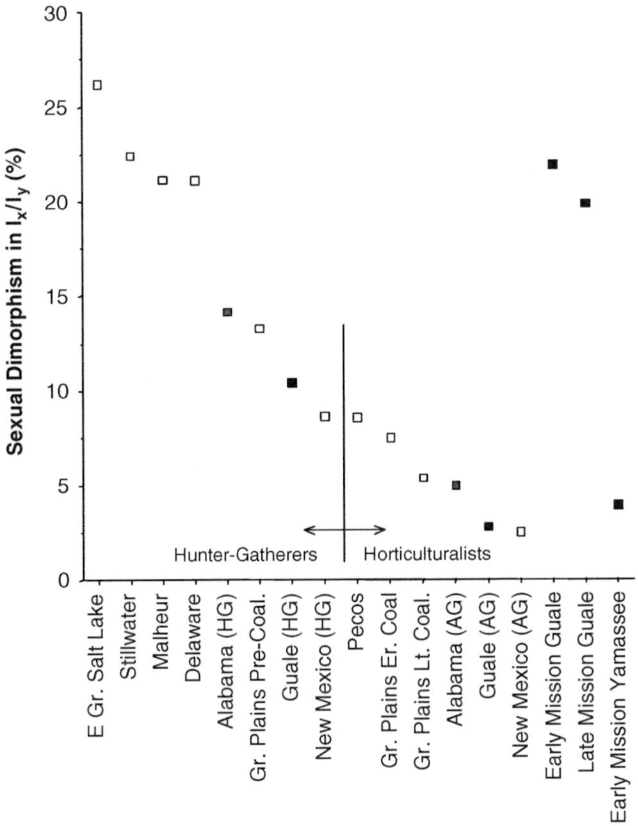

Figure 12.2 Sexual dimorphism {Male–Female./Female × 100} in femoral I_x/I_y in Georgia and Florida black boxes., Alabama grey boxes, and other North American archaeological skeletal series open boxes. The Georgia Bight prehistoric hunter-gatherers are 'Guale HG', Georgia Bight prehistoric farmers are 'Guale AG', and the three mission series are Santa Catalina on St Catherine's Island 'Early Mission Guale', Santa Catalina on Amelia Island 'Late Mission Guale', and Santa Maria de Yamassee on Amelia Island 'Early Mission Yamasee'. Note the very high degree of sexual dimorphism in the two Guale mission populations (see text for description). After Ruff and Larsen, 2001

argue that the adoption of agriculture and resulting changes in means of acquiring food is paramount. On the other hand, the character of the landscape in eastern North America is not uniform and adaptations, including those associated with activity and behaviour generally, should reflect the diversity of the region overall. Analysis of osteoarthritis and cross-sectional geometry from these three regions are consistent with this conclusion, strongly suggesting that populations adapt and modify behaviour on a local level (Ruff, 1999). In fact, given the differences in terrain and ecology, we would have been surprised to have seen a uniformity of response to the agricultural transition. The common theme here is that all three regions saw a shift from foraging to farming, resulting in a commitment to maize-based agriculture in the final centuries prior to European contact. In addition, the lower Illinois River valley saw an earlier phase of plant domestication that likely played a fundamental role in the development of increasingly complex societies during the Middle Woodland period and after, setting the stage for the remarkable increases in social complexity and commitment to an economy fuelled in part by maize agriculture.

The archaeological record suggests a high degree of success of native populations in all three regions, particularly if we evaluate degree of success by population size increase. Yet, these regions show fundamental differences in ecology that would predict different challenges and adaptive responses, at least with respect to activity and workload. For one thing, the Pickwick Basin and the lower Illinois River valley settings were focused on river systems and surrounding uplands. In sharp contrast, the Georgia Bight populations were living in a coastal setting, having significant access to marine resources during prehistory. After the arrival of Europeans, the marine component of diet declined significantly and maize production and consumption increased, essentially replacing the marine component of diet (Hutchinson *et al.*, 1998, 2000). The key point here is that while both settings saw an increasing focus on domesticated plants, the ecological settings are quite different in these key areas. The records of osteoarthritis and cross-sectional geometry described here are consistent with the notion that differences in ecology result in different behavioural adaptations.

The comparative record for osteoarthritis is limited to prehistoric foragers and farmers in the Pickwick Basin and the Georgia Bight (Figure 12.3). As is most common in archaeological settings worldwide (Bridges, 1992; Larsen, 1997), adult males in both settings express higher prevalence of osteoarthritis than adult females. This difference reflects the greater mechanical demands associated with male-based labour (e.g. hunting larger game) than female-based labour. Both regions clearly express these differences in osteoarthritis prevalence. Moreover, both regions show highest prevalence for the shoulder, elbow and knee joints (Tables 12.1 and 12.2). Bridges's (1992) overview of osteoarthritis patterning in the Americas – including

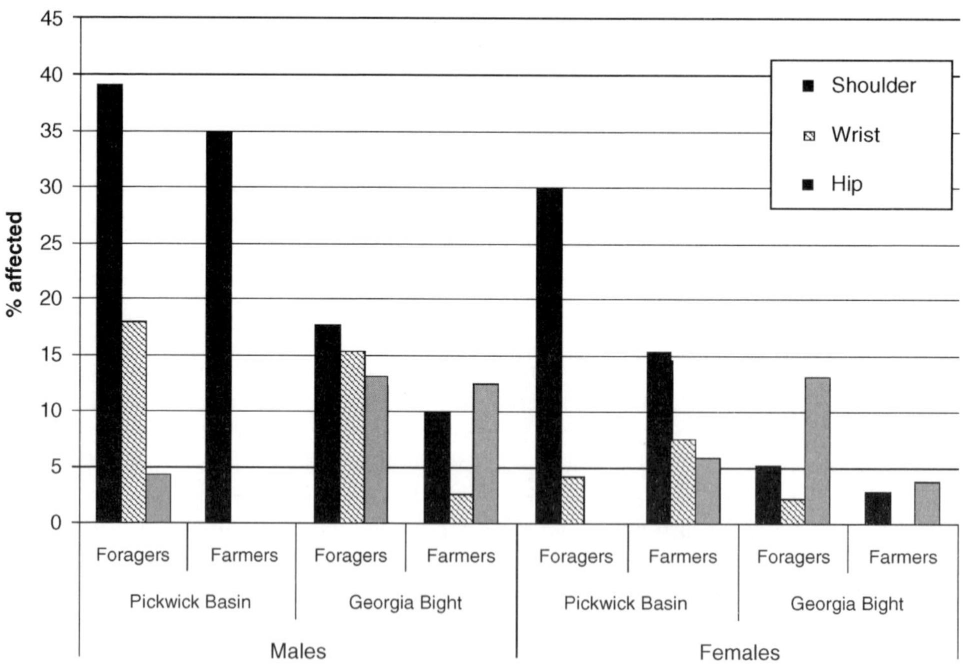

Figure 12.3 Comparisons of osteoarthritis prevalence in Pickwick Basin and Georgia Bight foragers and farmers

prehistoric Native Americans and individuals of European and African descent – reveal that this pattern is not universal. For example, modern Euroamericans and African Americans express the highest prevalence in the hip and lowest in the elbow, and there is a high amount of variation in prehistoric societies.

There are some important patterns that suggest differences in behavioural adaptation in adults when comparing these two settings. For example, there is a modest increase in osteoarthritis prevalence in the Pickwick Basin setting, which may reflect some change in behaviour, at least as it affects osteoarthritis. In contrast, there is a significant reduction in osteoarthritis in comparison of foragers and farmers in the Georgia Bight. The degree of change reveals something very different about behaviour in this setting, involving decline in mechanical loading of the articular joints. This pattern is in sharp contrast to the historic-era mission populations showing a dramatic increase in osteoarthritis. Clearly then, this change reflects a change in activity, which most likely reflects the exploitive nature of labour conscription in the Spanish Mission system (Hann, 1988; Worth, 1995, 2001). Thus, social and economic forces generated by a more dominant society profoundly affected diet and labour.

Direct comparisons between the regional series are not possible, owing to the limitation of the studies themselves and variation in how researchers collect osteoarthritis and cross-sectional geometric data. However, the comparison of periods within each of the three series is informative about activity and workload variation. Here, we use *J* for comparing cross-sectional geometry for assessing overall bone strength for the femoral (Figure 12.4a) and humeral (Figure 12.4b) midshaft regions. This comparison reveals striking variation in the three groups. In males, femoral midshaft strength greatly increases in the Pickwick Basin, shows no change in the lower Illinois River valley and greatly decreases in the prehistoric Georgia Bight. In females, femoral midshaft strength increases in the Pickwick Basin, increases in the lower Illinois River valley in the first three periods, but then decreases in the maize agriculturalists and decreases in the Georgia Bight prior to the colonial period. Comparisons of humerus cross-sectional geometry show broadly similar patterns across the three regions. However, male humeri show no change in the Pickwick Basin, no change in the lower Illinois River valley and decrease in the Georgia Bight. Females show increase in the Pickwick Basin, increase in the Late Woodland in the lower Illinois River valley, followed by profound decrease in the Mississippian period for the region and decrease in the Georgia Bight. In the Georgia Bight, like the femur, the humerus shows a marked increase in the colonial period.

Overall, when considered collectively, the changes in the three regions are complex. However, they show several trends. Amongst the most interesting are the different patterns of change in males and females in the foraging-to-farming transition. The extent of change is relatively greater in females in the Pickwick Basin and Illinois River valley, but greater in males in the Georgia Bight. This suggests that the foraging-to-farming transition had a greater effect on women in the terrestrial, inland setting and on men in the marine, coastal setting. In two of the regions (Pickwick Basin, Georgia Bight), one agricultural phase occurred, whereas in one region (Illinois River valley), there were two phases of agriculture. The latter involved activity changes in both phases, with different results – an increase in activity in the shift to oily starches, followed by a decrease following the adoption of maize. The implications of biomechanical analysis are that the transition had behavioural significance well before maize became important in diet. These results also show that the agricultural transition was not a monolithic event having the same result. Indeed, the factors of terrain and ecology – sharply

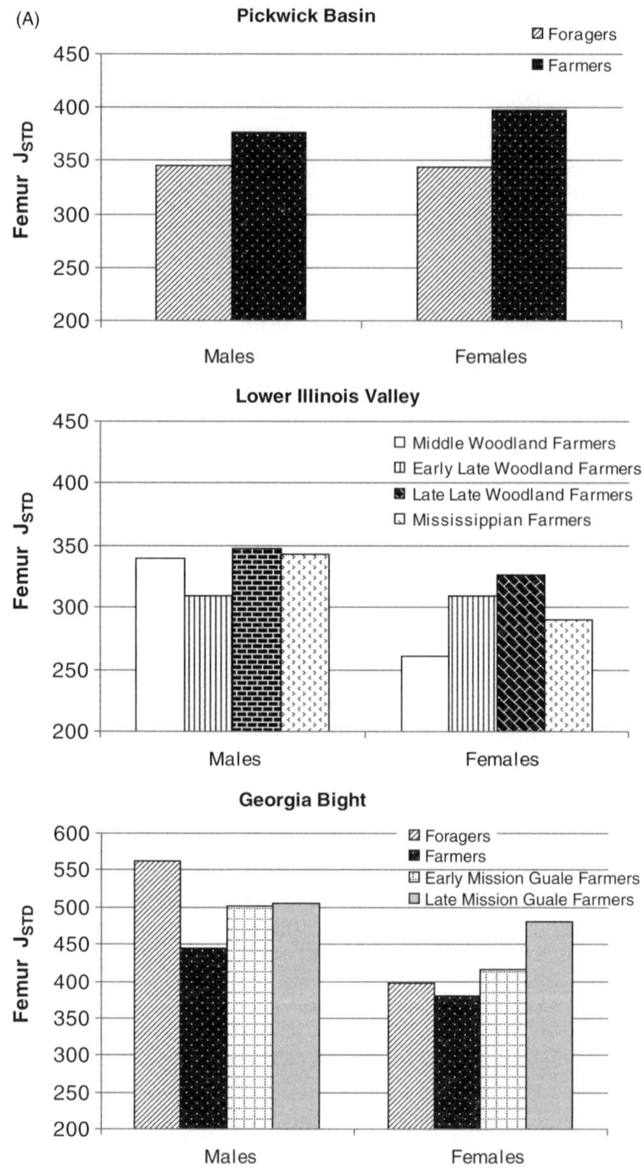

Figure 12.4 (a) Mean comparisons of femoral J for Pickwick Basin, Lower Illinois Valley, and Georgia Bight. (b) Mean comparisons of humeral J for Pickwick Basin, Lower Illinois Valley and Georgia Bight. Pickwick Basin farmers focused on maize production. Lower Illinois River valley populations were incipient, weedy plant farmers in the Middle Woodland period, followed by introduction of and increasing dependence on maize farming beginning in the Early Late Woodland period culminating during the Mississippian period. Georgia Bight prehistoric farmers focused on maize, followed by intensive maize production in the contact-era Early Mission and Late Mission periods

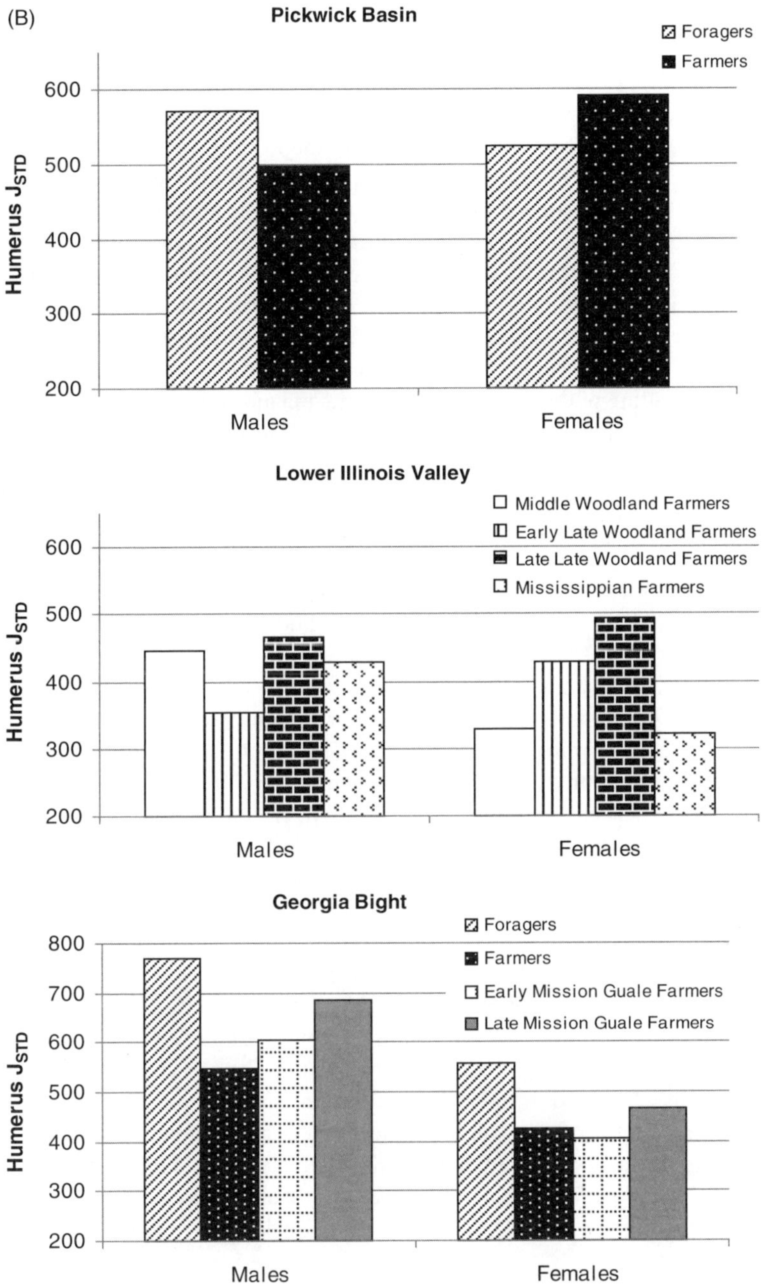

Figure 12.4 (*Continued*)

contrasting when comparing the inland and coastal settings – likely influenced the kinds and levels of activity that have affected skeletal growth, development and adult morphology.

These comparisons of three regions of eastern North America clearly emphasize the variation in response to changes in living circumstances, especially in regard to the farming transition, even with regard to the arrival of the single crop of maize. Even in relatively restricted areas, there appear to be behavioural variations. Williamson (2000), for example, has compared osteoarthritis prevalence of late prehistoric, maize agriculturalists from uplands of north Georgia with contemporary populations from the coast. Like the coastal setting, males exceeded females in osteoarthritis for nearly every articular joint. However, upland females and males have consistently greater prevalence of osteoarthritis than coastal females and males, suggesting greater mechanical demand in the uplands setting than the lowlands setting. Williamson (2000) suggests that greater terrain relief created circumstances conducive to more mechanical loading of the articular joints in the uplands setting. For example, carrying of loads up and down hills may have contributed to greater prevalence of osteoarthritis in general and in the vertebral column and knees. This is a nearly identical outcome to what has been reported in comparisons of upland and lowland populations in Iran and Iraq (cf. Rathbun, 1984). The greater mechanical demand in more difficult terrain is also clearly expressed in comparison of cross-sectional geometry in a variety of populations in North America (Ruff, 1999; Larsen *et al.*, 2008). With few exceptions, populations living in mountainous, upland settings express greater lower limb bone strength relative to body size than populations living in less mountainous settings (Ruff, 1999; Larsen *et al.*, 2008).

Variation in activity is consistent with other comparisons of health and quality of life amongst populations in eastern North America. For example, while health generally declined with the rise of maize farming (e.g. various in Cohen and Armelagos, 1984; Steckel and Rose, 2002), growing evidence indicates that with respect to some health indicators (e.g. dental caries), health outcomes show considerable variation, which is almost certainly due to local circumstances (e.g. specific diets and lifestyles associated with resource acquisition). It is this variability in physical activity and health that provides such an important platform for investigation of outcomes of adaptive transitions and consequences for the human condition.

Finally, study of osteoarthritis and cross-sectional geometry shows consistency of results in some settings but not in other settings. In this regard, the different results in the transition to maize farming in comparison of the Pickwick Basin and Georgia Bight show that osteoarthritis and cross-sectional geometry temporal changes are somewhat different in the Pickwick Basin, whereas in the Georgia Bight they are the same. It may be the case that osteoarthritis and cross-sectional bone distribution track different kinds of mechanical stresses. Alternatively, they may track different times at which the mechanical demand begins. For example, cross-sectional geometry reflects workload that begins relatively early in life, whereas osteoarthritis is found predominantly in older adults. Thus, the two variables may well reflect behaviours that are emphasized at different periods of life history (Knüsel, 1993; Bridges, 1993b).

12.2 CONCLUSIONS

The study of activity and workload as documented by osteoarthritis and cross-sectional geometry provides important insight into lifestyle in past societies. This allows inferences about the difficulty of the transition from foraging to farming lifeway, variation in the response of humans to new challenges and other adaptive shifts. While only three regions have been studied in eastern North America, the studies nonetheless provide perspective on variation in human responses to broadly the same economic changes. Specifically, males and females show relatively different responses to the adoption of agriculture, which reflect their different roles in the acquisition of resources, mostly related to food. In addition, these studies reveal different patterns of osteoarthritis change and cross-sectional geometry change. In this regard, the populations from the Pickwick Basin of north-western Alabama show different results in osteoarthritis and cross-sectional geometry, whereas the results are similar in the Georgia Bight. Finally, it has long been assumed that maize agriculture was the only adaptive shift involving plant cultigens that affected human biology. However, comparisons within these three regional contexts indicate that the shift from foraging to farming, as well as farming intensification, involved significant behavioural transformations. Importantly, the analysis of Woodland period populations from the Illinois River valley shows a shift in behaviour and its biological impact well in advance of the adoption of maize agriculture in late prehistory. Future studies should investigate why the variation exists, explaining local adaptation and incorporation of factors such as terrain, climate and ecological variation, as we continue to develop and refine our growing understanding of the foraging to farming transition, its causes and its outcomes for human societies in eastern North America and elsewhere.

ACKNOWLEDGEMENTS

This chapter is a contribution to the La Florida Bioarchaeology Project. We thank Ron Pinhasi and Jay Stock for their invitation to contribute to this book. Leslie Williams, the editors, and two anonymous reviewers provided helpful comments on earlier drafts. We acknowledge the US National Science Foundation for support of bioarchaeological research in the Georgia Bight over the last three decades. We dedicate this chapter to the late Patricia S. Bridges, a pioneer in the application of biomechanical analysis to bioarchaeological enquiry.

REFERENCES

Ambrose, S.H. (1987) Chemical and isotopic techniques of diet reconstruction in eastern North America, in *Emergent Horticultural Economies of the Eastern Woodlands* (ed. W.F. Keegan), Southern Illinois University at Carbondale, Center for, Archaeological Investigations, Occasional Paper No. 7., pp. 78–107.

Anderson, D.G. (1994) *The Savannah River Chiefdoms: Political Change in the Late Prehistoric Southeast*, University of Alabama Press, Tuscaloosa.

Auerbach, B.M. and Ruff, C.B. (2004) Human body mass estimation: a comparison of 'morphometric' and 'mechanical' methods. *Am. J. Phys. Anthropol.*, **125**, 331–342.

Bridges, P.S. (1989a) Changes in activities with the shift to agriculture in the southeastern United States. *Curr. Anthropol.*, **30**, 385–394.

Bridges, P.S. (1989b) Spondylolysis and its relationship to degenerative joint disease in the prehistoric southeastern United States. *Am. J. Phys. Anthropol.*, **79**, 321–329.

Bridges, P.S. (1989c) Bone cortical area in the evaluation of nutrition and activity levels. *Am. J. Hum. Biol.*, **1**, 785–792.

Bridges, P.S. (1990) Osteological correlates of weapon use, in *A Life in Science: Papers in Honor of J. Lawrence Angel* (ed. J.E. Buikstra), Center for American Archeology, Scientific Papers No. 6., pp. 87–98.

Bridges, P.S. (1991a) Degenerative joint disease in hunter-gatherers and agriculturalists from the southeastern United States. *Am. J. Phys. Anthropol.*, **85**, 379–391.

Bridges, P.S. (1991b) Skeletal evidence of changes in subsistence activities between the Archaic and Mississippian time periods in northwestern Alabama, in *What Mean These Bones? Studies in Southeastern Bioarchaeology* (eds M.L. Powell, P.S. Bridges and A.M.W. Mires), University of Alabama Press, Tuscaloosa, pp. 89–101.

Bridges, P.S. (1992) Prehistoric arthritis in the Americas. *Ann. Rev. Anth.*, **21**, 67–91.

Bridges, P.S. (1993a) The effect of variation in methodology on the outcome of osteoarthritic studies. *Int. J. Osteoarch.*, **3**, 289–295.

Bridges, P.S. (1993b) Reply to Dr Knüsel. *Am. J. Phys. Anthropol.*, **91**, 526–527.

Bridges, P.S. (1994) Vertebral arthritis and physical activities in the prehistoric southeastern United States. *Am. J. Phys. Anthropol.*, **93**, 83–93.

Bridges, P.S., Blitz, J.H. and Solano, M.C. (2000) Changes in long bone diaphyseal strength with horticultural intensification in west-central Illinois. *Am. J. Phys. Anthropol.*, **112**, 217–238.

Buikstra, J.E. (1984) The lower Illinois River region: a prehistoric context for the study of ancient diet and health, in *Paleopathology at the Origins of Agriculture* (eds M.N. Cohen and G.J. Armelagos), Academic Press, Orlando, Florida, pp. 215–234.

Buikstra, J.E. (1988) *The Mound-Builders of Eastern North America: A Regional Perspective*, Elfde Kroon-Voordracht, Amsterdam, The Netherlands.

Cassidy, C.M. (1984) Skeletal evidence for prehistoric subsistence adaptation in the central Ohio River valley, in *Paleopathology at the Origins of Agriculture* (eds M.N. Cohen and G.J. Armelagos), Academic Press, Orlando, Florida, pp. 307–345.

Cohen, M.N. and Armelagos, G.J. (eds) (1984) *Paleopathology at the Origins of Agriculture*, Academic Press, Orlando, Florida.

Cohen, M.N. and Crane-Kramer, G.M.M. (eds) (2007) *Ancient Health: Skeletal Indicators of Agricultural and Economic Intensification*, University Press of Florida, Gainesville.

Cook, D.C. (1984) Subsistence and health in the lower Illinois Valley: osteological evidence, in *Paleopathology at the Origins of Agriculture* (eds M.N. Cohen and G.J. Armelagos), Academic Press, Orlando, pp. 235–269.

Gremillion, K.J. (2002) The development and dispersal of agricultural systems in the Woodland period Southeast, in *The Woodland Southeast* (eds D.G. Anderson and R.C. Mainfort Jr), University of Alabama Press, Tuscaloosa, pp. 483–501.

Gremillion, K.J. (2004) Seed processing and the origins of food production in eastern North America. *AM Antiquity*, **69**, 215–233.

Hann, J.H. (1988) *Apalachee: The Land Between the Rivers*, University Presses of Florida, Gainesville.

Hough, A.J. Jr (2001) Pathology of osteoarthritis, in *Arthritis and Allied Conditions* (ed. W.J. Koopman), Lippincott, Williams and Wilkins, Philadelphia.

Hoyme, L.E. and Bass, W.M. (1962) Human skeletal remains from the Tollifero Ha6. and Clarksville Mc14. sites, John H. Kerr Reservoir Basin, Virginia, in *Archeology of the John H. Kerr Reservoir Basin, Roanoke River, Virginia-North Carolina* (eds C.F. Miller), Smithsonian Institution Bureau of American Ethnology, Bulletin No. 182., pp. 329–400.

Hulse, F.S. (1941) The people who lived at Irene: physical anthropology, in *Irene Mound Site, Chatham County, Georgia* (eds J. Caldwell and C. McCann), University of Georgia Press, Athens, pp. 57–68.

Hutchinson, D.L. (2002) *Foraging, Farming, and Coastal Biocultural Adaptation in Late Prehistoric North Carolina*, University Press of Florida, Gainesville.

Hutchinson, D.L., Larsen, C.S., Norr, L. and Schoeninger, M.J. (2000) Agricultural melodies and alternative harmonies in Florida and Georgia, in *Bioarchaeological Studies of Life in the Age of Agriculture* (ed. P.M. Lambert), University of Alabama Press, Tuscaloosa, pp. 116–133.

Hutchinson, D.L., Larsen, C.S., Schoeninger, M.J. and Norr, L. (1998) Regional variation in the pattern of maize adoption and use in Florida and Georgia. *AM Antiquity*, **63**, 397–416.

Knüsel, C. (1993) On the biomechanical and osteoarthritic differences between hunter-gatherers and agriculturalists. *Am. J. Phys. Anthropol.*, **91**, 523–527.

Knüsel, C. (2000) Bone adaptation and its relationship to physical activity in the past, in *Human Osteology in Archaeology and Forensic Science* (eds M. Cox and S. Mays), Greenwich Media, London, United Kingdom, pp. 381–401.

Lambert, P.M. (ed.) (2000) *Bioarchaeologial Studies of Life in the Age of Agriculture: A View from the Southeast*, University of Alabama Press, Tuscaloosa.

Larsen, C.S. (1982) *The Anthropology of St Catherine's Island: 3. Prehistoric Human Biological Adaptation*, Anthropological Papers of the American Museum of Natural History 57, part 3, American Museum of Natural History, New York.

Larsen, C.S.(1984) Health and disease in prehistoric Georgia: the transition to agriculture, in *Paleopathology at the Origins of Agriculture* (eds M.N. Cohen and G.J. Armelagos), Academic Press, Orlando, FL, pp. 367–392.

Larsen, C.S. (ed.) (1990) *The Archaeology of Mission Santa Catalina de Guale: 2. Biocultural Interpretations of a Population in Transition*, Anthropological Papers of the American Museum of Natural History No. 68, American Museum of Natural History, New York.

Larsen, C.S. (1997) *Bioarchaeology: Interpreting Behavior from the Human Skeleton*, Cambridge University Press, Cambridge, UK.

Larsen, C.S. (ed.) (2001) Bioarchaeology of Spanish Florida, in *Bioarchaeology of Spanish Florida: The Impact of Colonialism* (ed. C.S. Larsen), University Press of Florida, Gainesville, pp. 22–51.

Larsen, C.S. (in press) History of paleopathology in the American Southeast: from pox to population, in *The History of Paleopathology* (eds J.E. Buikstra and C.A. Roberts), Oxford University Press, New York.

Larsen, C.S., Crosby, A.W., Griffin, M.C. *et al.* (2002) A biohistory of health and behavior in the Georgia Bight: the agricultural transition and the impact of European contact, in *The Backbone of History: Health and Nutrition in the Western Hemisphere* (eds R.H. Steckel J.C. Rose), Cambridge University Press, New York, pp. 406–439.

Larsen, C.S., Hutchinson, D.L., Stojanowski, C.M. *et al.* (2007) Health and lifestyle in Georgia and Florida: agricultural origins and intensification in regional perspective, in *Ancient Health: Skeletal Indicators of Agricultural and Economic Intensification* (eds M.N. Cohen and G.M.M. Crane-Kramer), University Press of Florida, Gainesville, pp. 20–34.

Larsen, C.S., Kelly, R.L., Ruff, C.B. *et al.* (2008) Living on the margins: biobehavioral adaptations in the western Great Basin, in *Case Studies in Environmental Archaeology* (eds E. Reitz, C.M. Scarry and S.J. Scudder), Springer, New York, pp. 161–189.

Larsen, C.S., Ruff, C.B., Schoeninger, M.J. and Hutchinson, D.L. (1992a) Population decline and extinction in La Florida, in *Disease and Demography in the Americas* (eds J.W. Verano and D.H. Ubelaker), Smithsonian Institution Press, Washington, pp. 25–39.

Larsen, C.S., Schoeninger, M.J., van der Merwe, N.J., *et al.* (1992b) Carbon and nitrogen stable isotopic signatures of human dietary change in the Georgia Bight. *Am. J. Phys. Anthropol.*, **89**, 197–214.

Lusteck, R. (2006) The migrations of maize into the southeastern United States, in *Histories of Maize: Multidisciplinary Approaches to the Prehistory, Linguistics, Biogeography, Domestication, and Evolution of Maize* (eds J. Staller, R. Tykot and B. Benz), Academic Press, Burlington, Massachusetts, pp. 521–528.

Macdonald, H.M., Cooper, D.M.L. and McKay, H.A. (2009) Anterior-posterior bending strength at the tibial midshaft increases with physical activity in boys: evidence for non-uniform geometric adaptation. *Osteoporosis Int.*, **20**, 61–70.

Milner, G.R. (2004) *The Moundbuilders: Ancient Peoples of Eastern North America*, Thames & Hudson, London, UK.

Milner, G.R. and Jacobi, K.P. (2006) A New Deal for human osteology, in *Bioarchaeology: The Contextual Analysis of Human Remains* (eds J.E. Buikstra and L.A. Beck), Elsevier, Burlington, MA, pp. 113–129.

Newman, M.T. and Snow, C.E. (1942) Preliminary report on the skeletal material from Pickwick Basin, Alabama, in *An Archaeological Survey of Pickwick Basin in the Adjacent Portions of the States of Alabama, Mississippi and Tennessee* (eds W.S. Webb and D.L. DeJarnette), Bulletin No. 129, Bureau of American Ethnology, Washington, DC, pp. 393–507.

Pickering, R.B. (1984) *Patterns of Degenerative Joint Disease in Middle Woodland, Late Woodland, and Mississippian Skeletal Series from the Lower Illinois Valley*. PhD Dissertation, Northwestern University, Evanston, Illinois.

Poole, A.R. (1999) An introduction to the pathophysiology of osteoarthritis. *Front. Biosci.*, **4**, 662–670.

Powell, M.L., Bridges, P.S. and Mires, A.M.W. (eds) (1991) *What Mean These Bones? Studies in Southeastern Bioarchaeology*, University of Alabama Press, Tuscaloosa.

Radin, E.L., Burr, D.B., Caterson, B., *et al.* (1991) Mechanical determinants of osteoarthrosis. *Semin. Arthritis. Rheu.*, **21**(3), (Suppl. 2), 12–21.

Radin, E.L., Paul, I.L. and Rose, R.M. (1972) Role of mechanical factors in pathogenesis of primary osteoarthritis. *The Lancet*, **1**, 519–522.

Rathbun, T.A. (1984) Skeletal pathology from the Paleolithic through the Metal Ages in Iran and Iraq, in *Paleopathology at the Origins of Agriculture* (eds M.N. Cohen and G.J. Armelagos), Academic Press, Orlando, Florida, pp. 137–167.

Riley, T.J., Walz, G.R., Bareis, C.J. *et al.* (1994) Accelerator mass spectrometry AMS dates confirm early Zea mays in the Mississippi River valley. *AM Antiquity*, **59**, 490–498.

Ruff, C.B. (1987) Sexual dimorphism in human lower limb bone structure: relationship to subsistence strategy and sexual division of labor. *J. Human Evol.*, **16**, 391–416.

Ruff, C.B. (1999) Skeletal structure and behavioral patterns of prehistoric Great Basin populations, in *Understanding Prehistoric Lifeways in the Great Basin Wetlands: Bioarchaeological Reconstruction and Interpretation* (eds B.E. Hemphill and C.S. Larsen), University of Utah Press, Salt Lake City, pp. 290–320.

Ruff, C.B. (2000) Body size, body shape, and long bone strength in modern humans. *J. Human Evol.*, **38**, 269–290.

Ruff, C.B. (2008) Biomechanical analyses of archaeological human skeletons, in *Biological Anthropology of the Human Skeleton*, 2nd edn (eds M.A. Katzenberg and S.R. Saunders), Wiley-Liss, Hoboken, pp. 183–206.

Ruff, C.B., Holt, B.H. and Trinkaus, E. (2006) Who's afraid of the big bad Wolff? 'Wolff's law' and bone functional adaptation. *Am. J. Phys. Anthropol.*, **129**, 484–498.

Ruff, C.B. and Larsen, C.S. (1990) Postcranial biomechanical adaptations to subsistence strategy changes on the Georgia coast, in *The Archaeology of Mission Santa Catalina de Guale: 2. Biocultural Interpretations of a Population in Transition* (ed. C.S. Larsen), Anthropological Papers of the American Museum of Natural History No. 68, American Museum of Natural History, New York, pp. 94–120.

Ruff, C.B. and Larsen, C.S. (2001) Reconstructing behavior in Spanish Florida: the biomechanical evidence, in *Bioarchaeology of Spanish Florida: The Impact of Colonialism* (ed. C.S. Larsen), University Press of Florida, Gainesville, FL, pp. 113–145.

Ruff, C.B., Larsen, C.S. and Hayes, W.C. (1984) Structural changes in the femur with the transition to agriculture on the Georgia coast. *Am. J. Phys. Anthrop.*, **64**, 125–136.

Ruff, C.B., Trinkaus, E., Walker, A. and Larsen, C.S. (1993) Postcranial robusticity in Homo, I: Temporal trends and mechanical interpretation. *Am. J. Phys. Anthrop.*, **91**, 21–53.

Schoeninger, M.J. (2009) Stable isotope evidence for the adoption of maize agriculture. *Curr. Anthropol.*, **50**, 633–640.

Shaw, C.N. and Stock, J.T. (2009) Intensity, repetitiveness, and directionality of habitual adolescent mobility patterns influence the tibial diaphysis morphology of athletes. *Am. J. Phys. Anthropol.*, **140**, 149–159.

Smith, B.D. (1995) *The Emergence of Agriculture*, Scientific American Library, New York.

Smith, B.D. (1990) Origins of agriculture in eastern North America. *Science*, **246**, 1566–1571.

Smith, B.D. and Yarnell, R.A. (2009) Initial formation of an indigenous crop complex in eastern North America at 3800 BP, *PNAS*, **106**, 6561–6566.

Snow, C.E. (1948) *Indian Knoll Skeletons of Site Oh 2, Ohio County, Kentucky*, University of Kentucky, Reports in Anthropology 4 no. 3, part 2.

Staller, J., Tykot, R. and Benz, B. (eds) (2006) *Histories of Maize: Multidisciplinary Approaches to the Prehistory, Linguistics, Biogeography, Domestication, and Evolution of Maize. Burlington,* Academic Press, Massachusetts.

Steckel, R.H. and Rose, J.C. (eds) (2002) *The Backbone of History: Long-Term Trends in Health and Nutrition in the Americas*, Cambridge University Press, New York.

Williamson, M.A. (2000) A comparison of degenerative joint disease between upland and coastal prehistoric agriculturalists from Georgia, in *Bioarchaeological Studies of Life in the Age of Agriculture: A View from the Southeast* (ed. P.M. Lambert), University of Alabama Press, Tuscaloosa, pp. 134–147.

Worth, J.E. (1995) The Struggle for the Georgia Coast: An Eighteenth-Century Spanish Retrospective on Guale and Mocama. Anthropological Papers of the American Museum of Natural History No. 75.

Worth, J.E. (2001) The ethnohistorical context of bioarchaeology in Spanish Florida, in *Bioarchaeology of Spanish Florida: The Impact of Colonialism* (ed. C.S. Larsen), University Press of Florida, Gainesville, pp. 1–21.

13

Mobility and Lower Limb Robusticity of a Pastoralist Neolithic Population from North-Western Italy

Damiano Marchi[1], Vitale Sparacello[2] and Colin Shaw[3]

[1] *Department of Evolutionary Anthropology, Duke University, Durham, NC, USA*
[2] *Department of Anthropology, University of New Mexico, Albuquerque, USA*
[3] *Department of Anthropology, Pennsylvania State University, State College, PA, USA*

13.1 INTRODUCTION

Western Liguria, a coastal region located in Northwestern Italy, has provided a wealth of palaeoanthropological material. Excavations conducted since the end of the nineteenth century have unearthed several skeletons at the Grimaldi and Arene Candide caves, which date to the Early Upper Palaeolithic and Late Upper Palaeolithic. In addition, many Neolithic burials have been unearthed at Arene Candide and other nearby caves in the Finale Ligure area. This material has provided the rare opportunity to study regional diachronic variations in biological parameters, including health status, body size and skeletal robusticity, within the framework of changing subsistence patterns, from foraging to early farming.

Comparisons between Ligurian Late Upper Palaeolithic and Neolithic skeletal samples show a diachronic reduction in stature and decline in general health indicators, trends known to accompany the transition to agriculture elsewhere (Cohen and Armelagos, 1984). However, osteometric analyses have shown that while lower limb robusticity decreases, there is an increase in upper limb robusticity in the Ligurian Neolithic (LIG) sample (Formicola, 1997). The observed changes in robusticity suggest that the Neolithic transition in Liguria did not result in wholesale decline in bone strength, as documented elsewhere, and underscores the

Human Bioarchaeology of the Transition to Agriculture Edited by Ron Pinhasi and Jay T. Stock

need for measures of robusticity that reflect adaptation to specific mechanical loads and physical activity.

Though traditional osteometric parameters provide a baseline measure of robusticity, long bone diaphyses behave like beams when they are mechanically loaded, therefore the diaphyses can be analysed using the same theoretical principles used by engineers when designing structures. The ability of a beam to resist stresses resulting from externally applied loads can be calculated given the cross-sectional geometric properties. By accounting for the internal architecture of the diaphysis we can better quantify robusticity and its relationship with specific activity patterns (Ruff, Holt and Trinkaus, 2006, and references therein).

Archaeological and bioarchaeological studies suggest that the change in subsistence patterns that took place from the Upper Palaeolithic through the Neolithic in Europe involved a significant decrease in mobility (Holt and Formicola, 2008). In general, variation in lower limb long bone diaphyseal morphology can be used to infer changes in habitual mobility patterns. Specifically, variation in mobility can alter loading patterns of the midshaft femur and tibia (Ruff and Hayes, 1983), resulting in changes in diaphyseal robusticity and shape. In particular, it has been demonstrated that variation in bone cross-sectional size more accurately reflects adaptation to the magnitude of mechanical load, while variation in bone cross-sectional shape more accurately reflects adaptation to the 'type' of mechanical load (Trinkaus *et al.*, 1991). Comparisons of the skeletal remains of hunter-gatherer and agricultural populations indicate a pattern of decreased lower limb diaphyseal robusticity and increased cross-sectional circularity, thought to reflect decreased levels of mobility and physical demand in agricultural populations (Larsen, 1995). A similar reduction in lower limb cross-sectional dimensions was found in a European sample following the transition from the Mesolithic to the Neolithic, and may also reflect a reduction in mobility (Ruff *et al.*, 2006).

Previous studies (Marchi *et al.*, 2006; Sparacello and Marchi, 2008) on the Ligurian Neolithic population have suggested that this group did not follow the trend of reduced lower limb robusticity that is characteristic of the diachronic studies cited above. In fact, compared with various European hunter-gatherer populations, the lower limb robusticity of the Ligurian Neolithic is similar or higher. These results have been explained as a consequence of the subsistence strategy (pastoralism, on uneven terrain) practised by this population (Marchi, 2008). This finding indicates that mobility does not necessarily decrease with the onset of the Neolithic, and that general activity patterns may have been region specific. Overall, variability in subsistence practices may be part of the complex mosaic of tempo, modes and agents influencing the spread of the European Neolithic (Lahr, Foley and Pinhasi, 2000).

The primary purpose of this chapter is to investigate the morphological trends in Liguria, throughout the Neolithic period, and to broaden the comparative context by including cave burials from caves located within the same pastoral system (Maggi and Nisbet, 1990), as well as modern humans with known habitual activity patterns. The secondary purpose is to investigate behavioural implications for variation in the tibia/fibula complex associated with the transition to agriculture. Previous work has suggested that variation in this complex may be associated with locomotor patterns that involve pronounced loading in the mediolateral plane (Marchi, 2007). Mediolateral forces generated through travel on uneven terrain may produce well defined stresses on the fibula. Here we explore the contribution of the fibula to the biomechanics of the human leg to assess whether this bone may enhance our understanding of the mobility patterns of past populations.

13.1.1 Archaeological Background

Archaeological evidences suggest that the subsistence economy of the Ligurian Neolithic primarily involved agricultural and herding related activities. Agricultural practices can be inferred from the presence of cereal grinding tools in the Middle Neolithic levels (from 4900–4820 calBC to 4450–4250 calBC; Maggi, 1997a) at Arene Candide (Starnini and Voytek, 1997), and from carbonized remains of cereal, as well as their impressions on clay (Arobba, Deferrari and Nisbet, 1997). The frequent use of bimanual grinding tools is suggested by previous biomechanical analyses that have reported low levels of bilateral asymmetry in the upper limbs of Ligurian Neolithic females (Marchi *et al.*, 2006).

There is no evidence, however, that Neolithic agricultural practices in Western Liguria were well developed. The dense forests of holm oak groves, the ruggedness of the territory and the absence of a coastal plain prevented the development of intensive cereal agriculture in the area throughout history, forcing Ligurian populations to rely on other resources. It appears that, despite the abundance of small eclogite axes (used in woodworking) in Neolithic layers (Starnini and Voytek, 1997), modern humans had little impact on forest coverage until the Late Neolithic (Nisbet, 1997a). Furthermore, land clearing through the extensive use of fire was not introduced in Liguria until the end of the Neolithic (Maggi, 1997b). It has been suggested that Neolithic small-scale agriculture was performed through planting small cereal fields in the woods, leaving *in situ* pruned tree stumps (Pessina and Tiné, 2008).

Stable isotope analyses of skeletal samples from Arene Candide indicate that the ratio of meat to vegetable consumption was similar to that seen in the Late Epigravettian, as indicated by the analyses of skeletal remains found in the same cave. This provides further evidence to suggest that although domesticated animals and plants replaced gathered products, vegetable consumption was less important than for other Neolithic communities and a significant part of the diet was derived from meat and dairy products (Le Bras-Goude *et al.*, 2006).

Several lines of evidence point towards a mainly pastoral economy, indicating that agriculture was not the focus of Ligurian Neolithic subsistence. Faunal and micromorphological soil analyses at Arene Candide suggest that, from the beginning, Ligurian Neolithic subsistence focused on sheep breeding, integrated with deer and boar hunting (Rowley-Conwy, 1997). Faunal analyses indicate that by the Early Middle Neolithic there occurs a drop in hunting activities and an intensification of pastoral practices through the introduction of cattle, and especially goats. Goats appear in the sediments at the beginning of the Middle Neolithic and increase steadily until they constitute one-third of the total faunal assemblage (Rowley-Conwy, 2000).

Pastoralism in Western Liguria was likely to have been favoured due to the availability of numerous karstic caves and rock shelters that would have provided natural stables for the herds (Maggi and Nisbet, 2000). At Arene Candide, occupational layers are indicative of seasonal human occupation and stabling; this suggests that the cave may have been an out-station of a larger short-range transhumance pastoral system (Rowley-Conwy, 1997), likely involving the accessible caves of the Finale Ligure and other nearby areas (Maggi and Nisbet, 1990).

Maggi and Nisbet (2000) proposed that because the Ligurian landscape lacked pastures, pollarding (recurrent tree pruning) was an important strategy amongst Ligurian Neolithic people, in order to procure alternative and additional sources of animal fodder in the form of foliage and brushwood. Micromorphological analyses of Arene Candide sediments and coprolites support this hypothesis, showing that a significant part of animal fodder was derived from thin branches, twigs and leaves (Macphail *et al.*, 1997; Nisbet, 1997b). Evidence supporting the pollarding hypothesis has also been derived from biomechanical analyses:

previous studies on the Ligurian Neolithic sample have found an appreciable degree of bilateral asymmetry in male humeral cross-sectional robusticity, compatible with the habitual use of small axes in tree pruning and other unimanual activities (Sparacello and Marchi, 2008). The practice of pollarding may have influenced the mobility of Ligurian Neolithic people: since the same tree can only be used once every 3 to 4 years, pollarding requires the exploitation of extensive wooded areas.

13.2 MATERIALS AND METHODS

The Ligurian Neolithic skeletal material was unearthed from five caves and rock shelters (Arene Candide, Arma dell'Aquila, Bergeggi, Boragni and Pollera) located near Finale Ligure (Savona, Italy) (Table 13.1). All the caves lie within a radius of 10 km, and during the

Table 13.1 Ligurian Neolithic sample composition

Individual	Sex	Notes
Arene Candide 2 Tiné[a]	M	Previous studies[b,c]
Arene Candide 7	M	Previous studies[b,c,d]
Arene Candide 8	M	Previous studies[b,c,d]
Arene Candide IX	M	Previous studies[b,c]
Arene Candide XIII	M	Present study
Arene Candide EVI	M	Previous studies[b,c,d]
Arma dell'Aquila 2	M	Present study
Bergeggi 2	M	Present study
Bergeggi 3	M	Present study
Bergeggi 4[e]	M	Present study
Pollera 10	M	Previous studies[b,c,d]
Pollera 13	M	Previous studies[b,c,d]
Pollera 30	M	Previous studies[b,c,d]
Pollera 32[f]	M	Present study
Pollera 6246	M	Previous studies[b,c,d]
Cavernetta Boragni 2	M	Present study
Arene Candide XII[a]	F	Previous studies[b,c]
Arene Candide EIV	F	Previous studies[b,c,d]
Arma dell'Aquila 1	F	Previous studies[b,c,d]
Arma dell'Aquila V[a]	F	Previous studies[b,c,d]
Bergeggi 5[a]	F	Present study
Cavernetta Boragni 1	F	Present study
Pollera 1Tiné	F	Previous studies[b,c,d]
Pollera 12	F	Previous studies[b,c,d]
Pollera 14	F	Previous studies[b,c,d]
Pollera 33	F	Previous studies[b,c,d]

[a] Fibula not present.
[b] Marchi *et al.*, 2006.
[c] Sparacello and Marchi, 2008.
[d] Marchi, 2008.
[e] Only femur present.
[f] Femur not present.

Figure 13.1 Example of Middle Neolithic burial in the Finale Ligure area: the individual is laid down in a stone cist, crouched on the left side of the body. (See Plate 13.1 for a colour version of this image)

Neolithic were likely part of the same pastoral system described for the Arene Candide cave (Rowley-Conwy, 1997).

Neolithic burials in Liguria show a consistent pattern not found in other periods of the region: adults are laid down in a stone cist, crouched on the left side of the body (Figure 13.1). Although most of the individuals can be ascribed to the Middle Neolithic Square-Mouthed Pottery culture (Parenti and Messeri, 1962), precise information about stratigraphic position is lacking for some burials. In this study, we assigned the skeletal material to the Neolithic if available information on the deposition of the body in the burial was consistent with Neolithic practices, or if the skeletal material was found in association with eclogite axes or Neolithic ceramics.

The comparative sample includes 220 individuals (Table 13.2). Late Upper Palaeolithic, Mesolithic and Eneolithic comparative data were generously provided by C.B. Ruff, B.M. Holt and V. Sládek (Ruff *et al.*, 2006). All are European and have a temporal range of 19000 BP (Late Upper Palaeolithic) through to 4000 BP (Eneolithic). The Late Upper Palaeolithic and Mesolithic samples are derived from sites throughout Europe (Holt, 2003) and ethnographic and archaeological evidence suggest that both groups were characterized by elevated levels of residential mobility (Montet-White, 1991; Kelly, 1983). The Eneolithic sample comes mostly from sites in Central Europe. Previous biomechanical analyses (Sládek, Berner and Sailer, 2006a, 2006b) indicated similar femoral and tibial diaphyseal robusticity between the Eneolithic sample and an agricultural Early Bronze Age sample from the same area, suggesting a low level of mobility. Iron Age data were collected by one of the authors (VSS) on Samnite burials laid down in the fifth–sixth century BC in the Alfedena necropolis (Abruzzo, Central Italy). This Samnite population undertook agro-pastoral subsistence practices in the mountainous region of Abruzzo (Italy), with a greater focus on agriculture in the small fertile areas, such as Alfedena (Sparacello, Pearson and Coppa, 2009). The medieval data (Marchi, 2007) were collected by one of the authors (DM) and come from a German necropolis located south of the Neuburg hill, on the southern bank of the river Danube. The necropolis has been used from the end of the seventh to the beginning of the eighth century AD. According to recent investigations, even though a few burials show higher than average grave goods indicative of

Table 13.2 Comparative sample

Period	Years BP	All bones[a]		Femora[a]		Tibiae[a]		Fibulae[a]	
		Men	Women	Men	Women	Men	Women	Men	Women
LUP[b]	19 000–10 000	16	7	16	7	13	5	—	—
MESO[c]	9000–5300	28	12	28	11	23	6	—	—
ENEOL[d]	4800–4200	31	22	29	21	28	17	—	—
IRONAGE[e]	2400–2600	35	17	27	15	32	14	21	8
MED[f]	1300–1400	14	12	—	—	14	12	14	12
MODERN[g]	0	21	—	—	—	21	—	21	—

[a] Number of bones included in cross-sectional geometric analyses for each sex. Only one side was taken for each individual, usually the right. The left side was taken when the right was not present.
[b] Comparative Late Upper Palaeolithic sample (Holt, 2003; Ruff *et al.*, 2006).
[c] Comparative Mesolithic sample (Holt, 2003; Ruff *et al.*, 2006).
[d] Comparative Eneolithic sample (Sládek, Berner and Sailer, 2006a, 2006b; Ruff *et al.*, 2006).
[e] Comparative Iron Age sample (Sparacello, Pearson and Coppa, 2009).
[f] Comparative Medieval sample (Marchi, 2007).
[g] Comparative modern sample (Shaw and Stock, 2009).

higher social status, the population represented in the graves was generally rural and lived in a relatively flat terrain (Benjamin Hoeke, personal communication). The modern data were collected by one of the authors (CNS) and represent living adult males between the ages of 19 and 30 who led sedentary lifestyles, undertaking less than 1 hour of strenuous physical activity per week (Shaw and Stock, 2009). Sample sizes differ depending on the variable analysed, on account of missing data for some individuals.

13.2.1 Cross-Sectional Geometric Properties

Cross-sectional properties were calculated at midshaft (50% bone length) (bone lengths of femur and tibia as defined by Ruff, 2002; bone length of fibula as defined in Marchi, 2004), and estimates of endosteal contours were obtained from measurements of biplanar radiographs of the diaphysis. In absence of a CT scanner, this method yields reasonably accurate results (Stock, 2002; O'Neill and Ruff, 2004). Cross-sectional properties of the living sample were calculated using peripheral quantitative computed tomography (pQCT).

For the Iron Age and Ligurian samples (Table 13.1), cross-sectional geometric (CSG) properties were calculated from a 'solid bone' section (i.e. only the external contour was sampled), without reconstructing the contour of the medullary cavity. Actual CSG properties were estimated from 'solid bone' results using regression equations. The method is derived from a recent study (Stock and Shaw, 2007) predicting polar second moment of area (J) and diaphyseal shape indices (I_{max}/I_{min} and I_x/I_y, see below) from the external dimension of cross-sections. We developed similar equations based on actual CSG properties (hollow bone) and periosteal contour (solid bone) data, and used these to estimate hollow bone CSG properties from periosteal contour data. We used CSG data from the Ligurian sample (Marchi *et al.*, 2006), a medieval sample from Liguria (Sparacello and Marchi, 2008, and unpublished data) and from 30 femora of the Iron Age sample that were sectioned at

Table 13.3 Regression equations used to predict actual cross-sectional geometric properties from the 'solid bone' section

	Model parameters and coefficients					Error	
Properties	n	r^2	b_0	b_1	P	SE Estimate	Mean PPE[a]
I_{max}[b]	204	0.999	0.995	−0.021	0.000	0.041	2.94%
I_{min}[c]	204	0.999	0.991	−0.012	0.000	0.041	2.86%
J[d]	204	0.999	0.994	−0.002	0.000	0.04	2.85%
I_{max}/I_{min}[e]	174	0.996	1.012	−0.007	0.000	0.017	1.26%

[a] PPE = (observed y − predicted y)/predicted y × 100. Based on natural logarithm back-transformed predicted and observed values.
[b] ln I_{max} vs. ln solid I_{max}.
[c] ln I_{min} vs. ln solid I_{min}.
[d] ln J vs. ln solid J.
[e] ln I_{max}/I_{min} vs. ln solid I_{max}/I_{min}.

midshaft for previous investigations (Bondioli, Corruccini and Macchiarelli, 1986). In addition, we included in the regression CSG data on 37 fibulae of the medieval sample used in this comparison.

Even though the sample used to create the regression equations used here is smaller (n = 202) than that used in Stock and Shaw (2007) (n = 1431), regression equations hold high predictive power (Table 13.3). In fact, mean percent prediction error (observed value − predicted value)/predicted value × 100, obtained for the regression equations fall within the range of error generally accepted using the methods involving external moulds and biplanar X-rays (Stock, 2002; Stock and Shaw, 2007).

Cross-sectional properties calculated include second moments of area. Maximum (I_{max}) and minimum (I_{min}) second moments of area reflect the maximum and minimum bending rigidity of the bone, respectively. Polar second moment of area (J) represents torsional as well as (twice) average bending rigidity. Diaphyseal shape ratios (I_{max}/I_{min}) were used to evaluate variability in diaphyseal shape. Finally, tibio-fibular proportions were quantified using an index (property of fibula/property of tibia) × 100 for J, I_{max} and I_{min}, which takes into account the diaphyseal robusticity of both bones.

Cross-sectional properties were standardized using the product of body mass estimates and bone length.[2] Body mass for the Ligurian Neolithic and the comparative samples was calculated using two different but complementary techniques: one based on estimated stature and bi-iliac (maximum pelvic) breadth and one based on the mean of three regression equations using femoral head breadth (Auerbach and Ruff, 2004). Stature (Table 13.4) was estimated from maximum femoral length using Sjovold's (1990) 'Caucasian' formula, based on modern Euroamericans (see Ruff *et al.*, 2006 for discussion and justification).

It has been proposed that femoral diaphyseal shape ratios may be sensitive to differences in body shape (Ruff *et al.*, 2006 but see Pearson, Cordero and Bubsy, 2006). Specifically, as pelvic interacetabular distance relative to femoral length increases, cross-sections may become more mediolaterally buttressed, causing a decrease in diaphyseal shape ratios not connected to mobility levels. We analysed body shape variation in our samples to determine if this factor may have influenced our results. Body shape was expressed as the index bi-iliac breadth/femoral maximum length. Bi-iliac breadth was used as a proxy for interacetabular distance

Table 13.4 Temporal differences and sexual dimorphism in body proportions

Bi-iliac breadth/ Femoral length[a]				Multiple comparisons with Hommel's sequential Bonferroni correction		
Males	Mean (n)	SD	Sexual Dimorphism[b]	MESO	LIG	ENEOL
LUP[c]	0.63 (15)	0.03	NS	NS	*	*
MESO	0.62 (28)	0.02	NS		*	*
LIG[d]	0.66 (10)	0.02	NS			*
ENEOL	0.59 (29)	0.02	*			
Females	Mean (n)	SD		MESO	LIG	ENEOL
LUP	0.65 (7)	0.05		NS	NS	NS
MESO	0.63 (11)	0.02			°	NS
LIG	0.66 (8)	0.03				°
ENEOL	0.63 (21)	0.04				

[a] All the variables are explained in the text.
[b] Mann-Whitney U-test.
[c] For abbreviations, see Table 13.2.
[d] Ligurian Neolithic sample.
[e] *$p < 0.05$; ° $0.10 < p < 0.05$; NS = non significant difference. Results of Mann-Whitney U test corrected for multiple comparisons using Hommel's sequential Bonferroni correction.

(Ruff *et al.*, 2006), and femoral maximum length was used as a proxy for stature. Because of the limited availability of bi-iliac breadth data for both sexes, this analysis was carried out only on the Late Upper Palaeolithic, the Mesolithic, the Ligurian Neolithic and the Eneolithic samples.

13.2.2 Statistical Analysis

Certain variables included in these comparisons do not show a normal distribution (results not shown here). Therefore statistical evaluation of differences between temporal periods was carried out using the non-parametric Mann-Whitney U-Test corrected for multiple comparisons using Hommel's sequential Bonferroni procedure (Hommel, 1988). The nonparametric Mann-Whitney U-Test was used to evaluate differences in sexual dimorphism within each temporal period. Factorial ANOVA bivariate plots for both sexes are used to illustrate diachronic trends and to visualize the differences between sexes and temporal periods.

13.3 RESULTS

13.3.1 Body Shape

The comparisons of 'body shape' amongst all groups, calculated as the ratio of bi-iliac breadth/ femur length, indicate that the Ligurian Neolithic sample has greater body breadth than all groups (Table 13.4). Nevertheless, the difference is significant only amongst males. In addition, the Eneolithic sample is the only one showing significant sexual dimorphism in body proportions.

13.3.2 Femur

The results for the comparison of femur torsional rigidity (*J*) show a slight decrease from the Late Upper Palaeolithic through to the Eneolithic, for males and females (Figure 13.2; Table 13.5). The only significant differences are found amongst males, and reveal that the Eneolithic sample is especially gracile. No significant differences were found for minimum femoral rigidity (I_{min}) between Ligurian males and all other groups, suggesting that variation in I_{max} is driving the differences in torsional rigidity (Table 13.5). Interestingly, the Iron Age sample which, like the Ligurian sample had settled in a mountainous area, displays an average femoral bending rigidity similar to the more mobile Late Upper Palaeolithic and Mesolithic populations (Figure 13.2). Females show the same general pattern observed in males (Figure 13.2), even though differences amongst groups do not reach significance.

Ligurian Neolithic males have amongst the most elliptical shape at the femoral midshaft (I_{max}/I_{min}), and are more similar to Late Upper Palaeolithic males than to any other group (Figure 13.3; Table 13.5). Ligurian Neolithic females do not show significant differences in femoral midshaft shape when compared to all the other temporal groups (Figure 13.3; Table 13.5).

Sexual dimorphism in diaphyseal strength is generally high in all samples, with males displaying higher values than females for most cross-sectional properties in all time periods

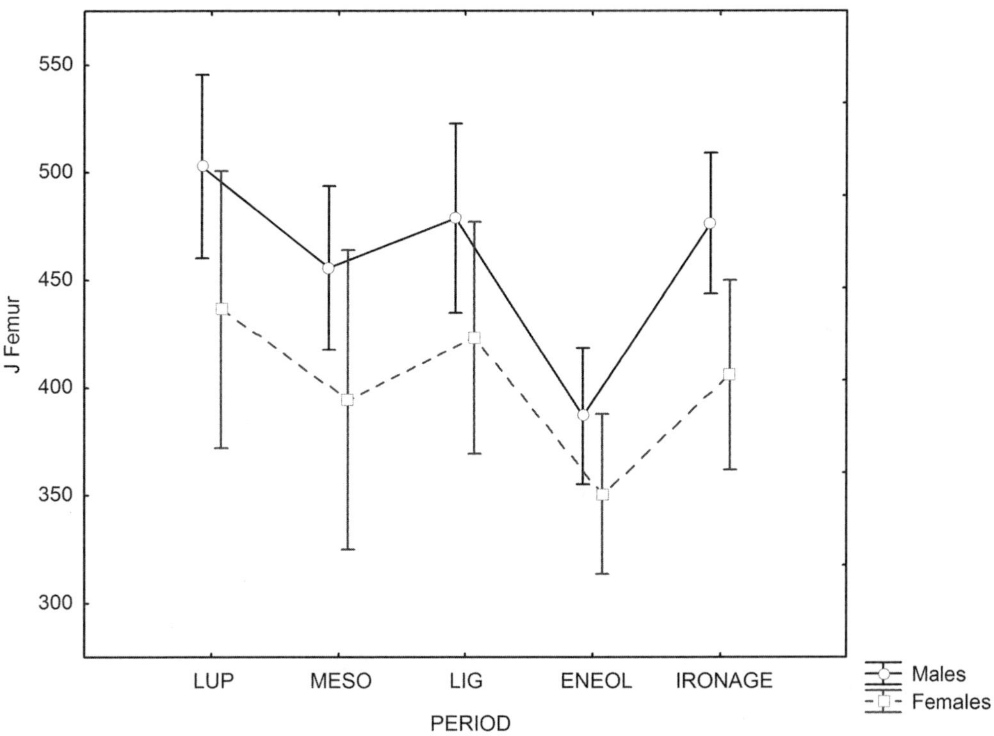

Figure 13.2 Two-factors ANOVA plot for midshaft femoral polar second moment of area (*J*). Factors: temporal period (*x*-axis); sex (line). Vertical bars denote 95% confidence intervals. LUP: Late Upper Palaeolithic; MESO: Mesolithic; LIG: Ligurian Neolithic; ENEOL: Eneolithic; IRONAGE: Iron Age

Table 13.5 Temporal differences in diaphyseal geometric properties of the femur in males and females

	Males		Multiple comparisons with Hommel's sequential Bonferroni correction				Females		Multiple comparisons with Hommel's sequential Bonferroni correction			
I_{max}[a]	Mean (n)	SD.	MESO	LIG	ENEOL	IRONAGE	Mean (n)	SD	MESO	LIG	ENEOL	IRONAGE
LUP[b]	300.95 (16)	63.86	NS	NS	*	NS	255.95 (7)	67.85	NS	NS	°	NS
MESO	254.00 (21)	51.17		NS	°	NS	219.73 (6)	40.94		NS	NS	NS
LIG	284.47 (15)	75.14			*	NS	235.58 (10)	56.84			NS	NS
ENEOL	216.95 (29)	36.51				*	191.39 (21)	43.34				NS
IRONAGE	261.17 (27)	44.18					220.15 (15)	54.58				
I_{min}			MESO	LIG	ENEOL	IRONAGE			MESO	LIG	ENEOL	IRONAGE
LUP	201.81 (16)	41.29	NS	NS	*	NS	180.50 (7)	32.94	NS	NS	NS	NS
MESO	205.09 (21)	42.74		NS	*	NS	174.77 (6)	29.48		NS	NS	NS
LIG	195.54 (15)	50.17			NS	NS	188.27 (10)	33.28			NS	NS
ENEOL	170.01 (29)	26.98				*	159.46 (21)	34.85				NS
IRONAGE	215.15 (27)	35.22					185.82 (15)	49.7				
J			MESO	LIG	ENEOL	IRONAGE			MESO	LIG	ENEOL	IRONAGE
LUP	502.76 (16)	97.62	NS	NS	*	NS	436.45 (7)	97.51	NS	NS	NS	NS
MESO	474.19 (28)	86.77		NS	*	NS	409.27 (11)	70.86		NS	NS	NS
LIG	478.72 (15)	120.48			*	NS	423.22 (10)	85.84			NS	NS
ENEOL	386.95 (29)	59.73				*	350.85 (21)	76.95				NS
IRONAGE	476.43 (27)	75.66					406.06 (15)	102.3				
I_{max}/I_{min}			MESO	LIG	ENEOL	IRONAGE			MESO	LIG	ENEOL	IRONAGE
LUP	1.51 (16)	0.23	*	NS	*	*	1.40 (7)	0.19	NS	NS	°	*
MESO	1.24 (20)	0.14		*	NS	NS	1.26 (6)	0.10		NS	NS	NS
LIG	1.47 (15)	0.22			*	*	1.25 (10)	0.16			NS	NS
ENEOL	1.29 (29)	0.17				NS	1.20 (21)	0.10				NS
IRONAGE	1.20 (27)	0.13					1.18 (15)	0.14				

[a] All the variables are explained in the text. I_{max}, I_{min} and J are divided by body mass × bone length^2.
[b] For abbreviations, see Tables 13.2 and 13.4.
*$p < 0.05$; ° $0.10 < p < 0.05$; NS = non significant difference. Results of Mann-Whitney. U-test corrected for multiple comparisons using Hommel's sequential Bonferroni correction.

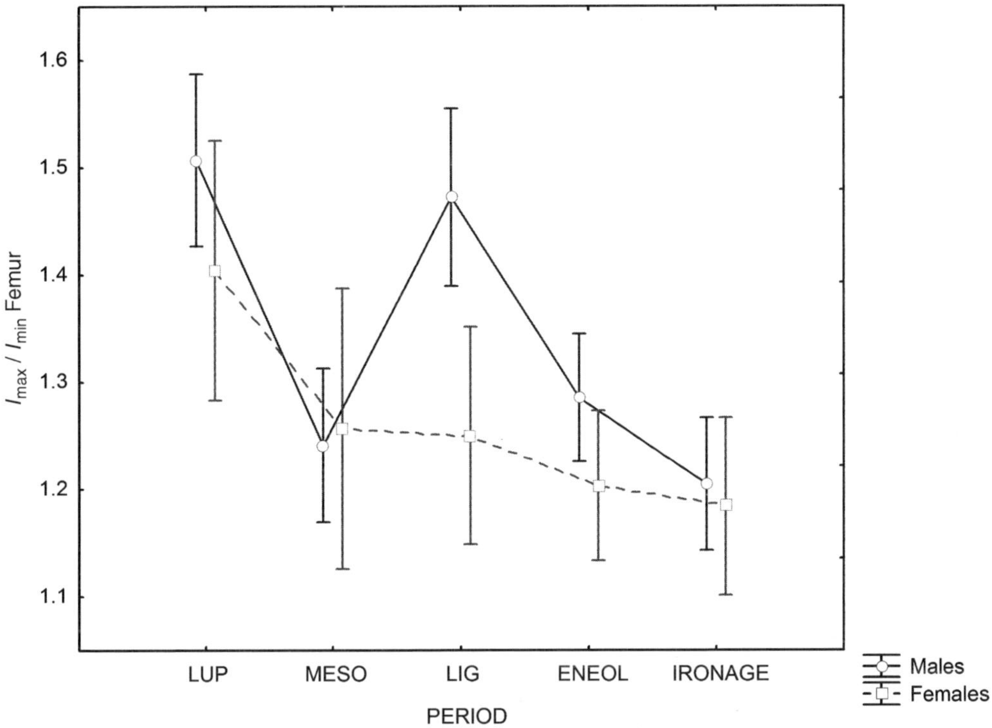

Figure 13.3 Two-factors ANOVA plot for midshaft femoral diaphyseal shape ratio (I_{max}/I_{min}). Factors: temporal period (x-axis); sex (line). Vertical bars denote 95% confidence intervals. LUP: Late Upper Palaeolithic; MESO: Mesolithic; LIG: Ligurian Neolithic; ENEOL: Eneolithic; IRONAGE: Iron Age

(Table 13.6). The Ligurian Neolithic group presents the highest level of sexual dimorphism for I_{max}, and the lowest for I_{min}. Although there were no significant differences in sexual dimorphism for these two cross-sectional properties, the Ligurian Neolithic sample shows the highest level of sexual dimorphism for femoral midshaft shape (I_{max}/I_{min}) ($p < 0.05$), with males displaying more elliptical femoral midshafts than females.

13.3.3 Tibia

The comparison of tibial diaphyseal rigidity (J, I_{max} and I_{min}) shows a general decrease from the Late Upper Palaeolithic through to the modern sample (Figure 13.4; Tables 13.7 and 13.8), for both males and females. However, for almost all indicators of diaphyseal rigidity, Ligurian Neolithic males do not significantly differ from the groups from earlier periods (Late Upper Palaeolithic and Mesolithic) that were characterized by high residential mobility. The only exception to this is minimum bending rigidity (I_{min}), where Eneolithic and modern males are not significantly different than Ligurian Neolithic males. This means that the higher value of J shown by the Ligurian Neolithic sample – when compared to later samples – is mainly due to higher I_{max} values. Females generally show the same trend seen in males (Figure 13.4), but with fewer significant differences (Table 13.8).

Table 13.6 Sexual dimorphism in diaphyseal geometric properties of the femur, tibia and fibula across temporal periods

Femur	LUP[a]	MESO	LIG	ENEOL	IRONAGE	MED
I_{max}[b,c,d]	17.58	15.59	20.75°	13.35*	18.64*	—
I_{min}	11.80	17.35°	3.86	6.62	15.78°	—
J	15.19	15.86*	13.11	10.29°	17.33°	—
I_{max}/I_{min}	7.31	–1.25	18.6*	6.80*	1.76	—
Tibia						
I_{max}	15.76	27.91*	34.83*	20.87*	22.61*	20.62*
I_{min}	–5.4	24.22	16.70*	14.81*	8.59	10.29
J	9.15	16.19°	29.10*	18.99*	17.85*	16.88°
I_{max}/I_{min}	25.16*	8.47	15.29°	5.94	15.46*	8.52°
Fibula						
I_{max}	—	—	19.14	—	18.49	3.71
I_{min}	—	—	48.09*	—	28.39°	2.58
J	—	—	28.23°	—	21.65	3.28
I_{max}/I_{min}	—	—	–15.24	—	–9.5	2.45

[a] For abbreviations, see Tables 13.2 and 13.4.
[b] All the variables are explained in the text.
[c] Sexual dimorphism calculated as: ((males – females)/females) * 100.
[d] I_{max}, I_{min}, and J are divided by body mass × bone length^2.
*$p < 0.05$; ° $0.10 < p < 0.05$; NS = non significant difference. Significance levels are determined from Mann–Whitney U-test between sexes within temporal groups.

Ligurian Neolithic male tibial midshaft shape is amongst the most elliptical of the samples analysed here (Table 13.7). However, all temporal periods feature tibiae with accentuated elliptical shapes. Differences are only significant when the more elliptical Ligurian sample is compared to the more circular medieval sample. Associated with this is a trend of generally increasing circularity of the tibial diaphysis, from the Late Upper Palaeolithic to the modern sample, although the majority of this variation is not significantly different (Figure 13.5). For the female sample, diaphyseal shape shows little change from the Late Upper Palaeolithic through to the Eneolithic, while more tibial diaphysis circularity is observable from the Iron Age through to the medieval period (Figure 13.5). The only finding of significant differences is that of more elliptical diaphyses amongst the Ligurian Neolithic females when compared to the medieval females (Table 13.8).

Males generally display more robust and elliptically shaped tibial midshafts than females across all groups (Table 13.6). This is particularly pronounced amongst the Ligurian Neolithic sample, where the highest levels of sexual dimorphism are found. In summary, males are significantly more robust than females (J, I_{max} and I_{min}), and present more elliptically shaped cross-sections compared with females (although not significant at the alpha = 0.5 level, Table 13.6).

13.3.4 Fibula

Fibular data for the Late Upper Palaeolithic, Mesolithic and Eneolithic were not available; therefore the comparative sample for the fibular analysis is composed of Iron Age, medieval

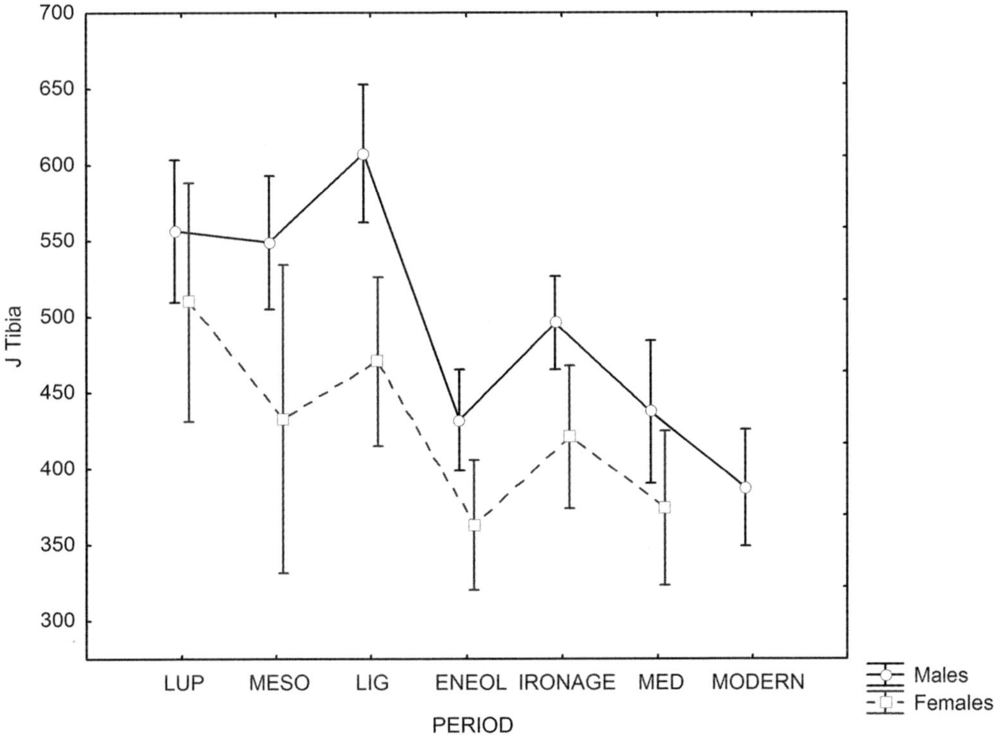

Figure 13.4 Two-factors ANOVA plot for midshaft tibial polar second moment of area (J). Factors: temporal period (x-axis); sex (line). Vertical bars denote 95% confidence intervals. LUP: Late Upper Palaeolithic; MESO: Mesolithic; LIG: Ligurian Neolithic; ENEOL: Eneolithic; IRONAGE: Iron Age; MED: medieval; MODERN: modern

and modern samples. In addition, the modern sample did not include measurements taken on females.

Comparisons of fibular rigidity (J, I_{max} and I_{min}) indicate a general decrease through time, similar to that reported for the tibia (Figure 13.6). Ligurian Neolithic male J is the highest amongst all periods, but does not differ significantly compared with the rugged terrain dwelling Iron Age individuals (Table 13.9). The same trend is apparent for I_{max} and I_{min}. Therefore, high levels of fibular J amongst the Ligurian Neolithic males are the result of high values for both I_{max} and I_{min}. Ligurian Neolithic females show higher fibular torsional rigidity (J) than all other groups; however, none of these differences were statistically significant (Table 13.9). The Ligurian Neolithic females also show higher I_{max} and I_{min} than other groups, though only I_{max} showed any significant pairwise comparisons. The lack of significant results for the other variables is likely due to the small female sample size, which restricts our ability for definitive comparisons.

Ligurian Neolithic male fibular midshaft shapes are not significantly different from any of the comparative samples (Figure 13.7; Table 13.9). All groups display a fibula with an elliptical shape (I_{max}/I_{min} ratios considerably higher than 1). Interestingly, Ligurian Neolithic and Iron Age females show a more elliptical fibula than males, a result of relatively lower I_{min} values (Table 13.9). Both Ligurian Neolithic and Iron Age females display significantly more

Table 13.7 Temporal differences in diaphyseal geometric properties of the tibia: males

			Multiple comparisons with Hommel's sequential Bonferroni correction					
I_{max}[a]	Mean (n)	SD	MESO	LIG	ENEOL	IRONAGE	MED	MODERN
LUP[b]	406.15 (14)	77.4	NS	NS	*	*	*	*
MESO	388.72 (16)	83.67		NS	*	NS	*	*
LIG	435.31 (15)	112.81			*	*	*	*
ENEOL	302.44 (28)	58.86				*	NS	NS
IRONAGE	342.43 (33)	49.16					*	*
MED	287.75 (14)	56.29						NS
MODERN	268.39 (21)	46.31						
I_{min}			MESO	LIG	ENEOL	IRONAGE	MED	MODERN
LUP	150.28 (14)	32.86	NS	NS	NS	NS	NS	*
MESO	160.33 (16)	44.20		NS	°	NS	NS	*
LIG	172.52 (15)	36.55			*	NS	NS	*
ENEOL	129.68 (28)	23.07				*	NS	NS
IRONAGE	153.97 (33)	29.03					NS	*
MED	149.5 (14)	26.38						*
MODERN	119.06 (21)	18.95						
J			MESO	LIG	ENEOL	IRONAGE	MED	MODERN
LUP	556.48 (14)	93.83	NS	NS	*	NS	*	*
MESO	545.07 (19)	115.51		NS	*	NS	*	*
LIG	607.55 (15)	143.13			*	*	*	*
ENEOL	432.11 (28)	73.33				*	NS	NS
IRONAGE	495.94 (33)	69.17					NS	*
MED	437.25 (14)	78.16						NS

MODERN	387.45 (21)	60.61						
I_{max}/I_{min}			MESO	LIG	ENEOL	IRONAGE	MED	MODERN
LUP	2.77 (14)	0.52	NS	NS	°	NS	*	*
MESO	2.5 (19)	0.34		NS	NS	NS	*	NS
LIG	2.54 (15)	0.44		NS	NS	NS	*	NS
ENEOL	2.37 (28)	0.45				NS	*	NS
IRONAGE	2.27 (33)	0.37					*	NS
MED	1.93 (14)	0.29						*
MODERN	2.27 (21)	0.30						

[a] All the variables are explained in the text. I_{max}, I_{min} and J are divided by the product of body mass and the second power of bone length.
[b] For abbreviations, see Tables 13.2 and 13.4.
*$p < 0.05$; ° $0.10 < p < 0.05$; NS = non significant difference. Results of Mann-Whitney U-test corrected for multiple comparisons using Hommel's sequential Bonferroni correction.

Table 13.8 Temporal differences in diaphyseal geometric properties of the tibia: females

			Multiple comparisons with Hommel's sequential Bonferroni correction				
I_{max} [a]	Mean (n)	SD	MESO	LIG	ENEOL	IRONAGE	MED
LUP[b]	350.85 (5)	62.54	NS	NS	*	NS	°
MESO	303.9 (3)	1.59		NS	NS	NS	NS
LIG	322.85 (10)	60.32			*	NS	*
ENEOL	250.21 (17)	51.76				NS	NS
IRONAGE	279.29 (14)	82.77					NS
MED	238.57 (12)	58.56					
I_{min}			MESO	LIG	ENEOL	IRONAGE	MED
LUP	158.85 (5)	18.19	NS	NS	*	NS	NS
MESO	129.07 (3)	12.36		NS	NS	NS	NS
LIG	147.83 (10)	33.1			*	NS	NS
ENEOL	112.95 (17)	25.11				NS	NS
IRONAGE	141.79 (14)	32.66					NS
MED	135.55 (12)	30.64					
J			MESO	LIG	ENEOL	IRONAGE	MED
LUP	509.82 (5)	75.18	NS	NS	*	NS	*
MESO	469.14 (6)	145.84		NS	NS	NS	NS
LIG	470.59 (10)	91.46			*	NS	°
ENEOL	363.16 (17)	74.33				NS	NS
IRONAGE	420.80 (14)	108.81					NS
MED	374.11 (12)	83.12					
I_{max}/I_{min}			MESO	LIG	ENEOL	IRONAGE	MED
LUP	2.21 (5)	0.30	NS	NS	NS	NS	°
MESO	2.30 (4)	0.23		NS	NS	NS	NS
LIG	2.21 (10)	0.24			NS	NS	*
ENEOL	2.24 (17)	0.3				°	*
IRONAGE	1.96 (14)	0.42					NS
MED	1.78 (12)	0.34					

[a] All the variables are explained in the text. I_{max}, I_{min} and J are divided by the product of body mass and the second power of bone length.
[b] For abbreviations, see Tables 13.2 and 13.4.
*$p < 0.05$; ° $0.10 < p < 0.05$; NS = non significant difference. Results of Mann-Whitney U-test corrected for multiple comparisons using Hommel's sequential Bonferroni correction.

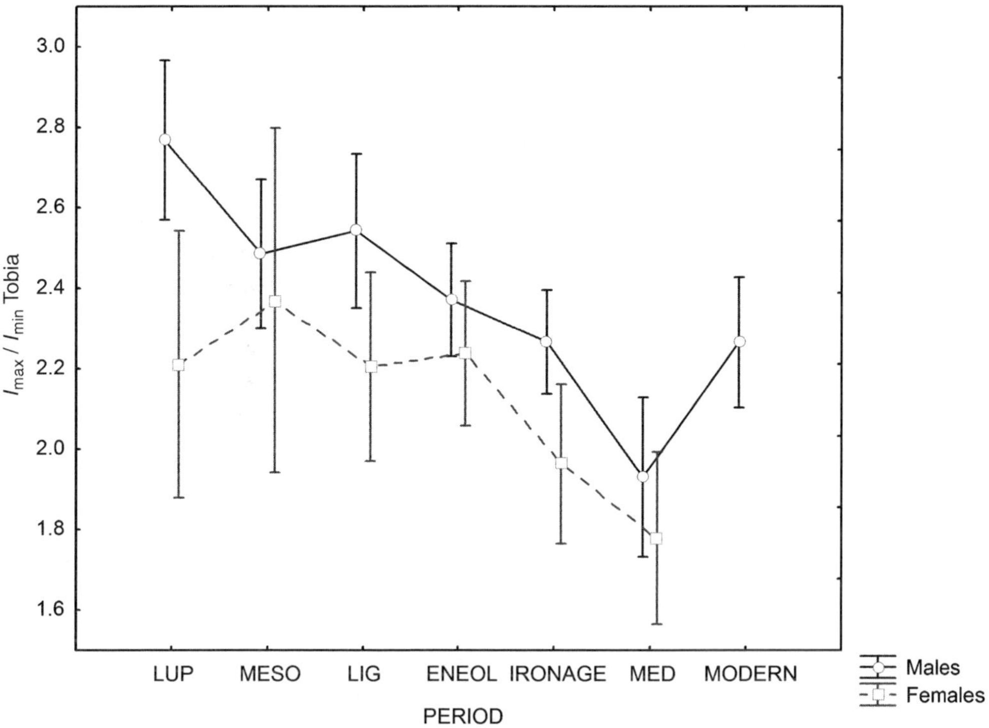

Figure 13.5 Two-factors ANOVA plot for midshaft tibial diaphyseal shape ratio (I_{max}/I_{min}). Factors: temporal period (x-axis); sex (line). Vertical bars denote 95% confidence intervals. LUP: Late Upper Palaeolithic; MESO: Mesolithic; LIG: Ligurian Neolithic; ENEOL: Eneolithic; IRONAGE: Iron Age; MED: medieval; MODERN: modern

elliptical fibular diaphyses compared with medieval females. Although a trend of increasing fibular circularity is present from the Ligurian Neolithic sample through to the medieval female sample, the Ligurian and Iron Age females are not significantly different. Again, this could be the result of sample size, and indicates that larger samples are necessary to properly assess the validity of this trend.

Males generally display more robust fibular midshafts compared with females across all groups, although significant differences are only found in the Ligurian Neolithic sample for I_{min} (Table 13.6). The Ligurian Neolithic and Iron Age female samples show more elliptical midshaft diaphyses than males. Even though sexual dimorphism is not significant for fibular I_{max} and I_{max}/I_{min}, the difference between the sexes is higher for the Ligurian Neolithic sample compared to all other groups. These differences decrease consistently from the Ligurian Neolithic through to the medieval sample for all variables.

13.3.5 Between-Bone Comparison: Tibiofibular Complex

Relative (to the tibia) fibular diaphyseal rigidity (J, I_{max} and I_{min}) amongst Ligurian Neolithic males is amongst the highest of all periods (Figure 13.8; Table 13.10). Interestingly, Iron Age

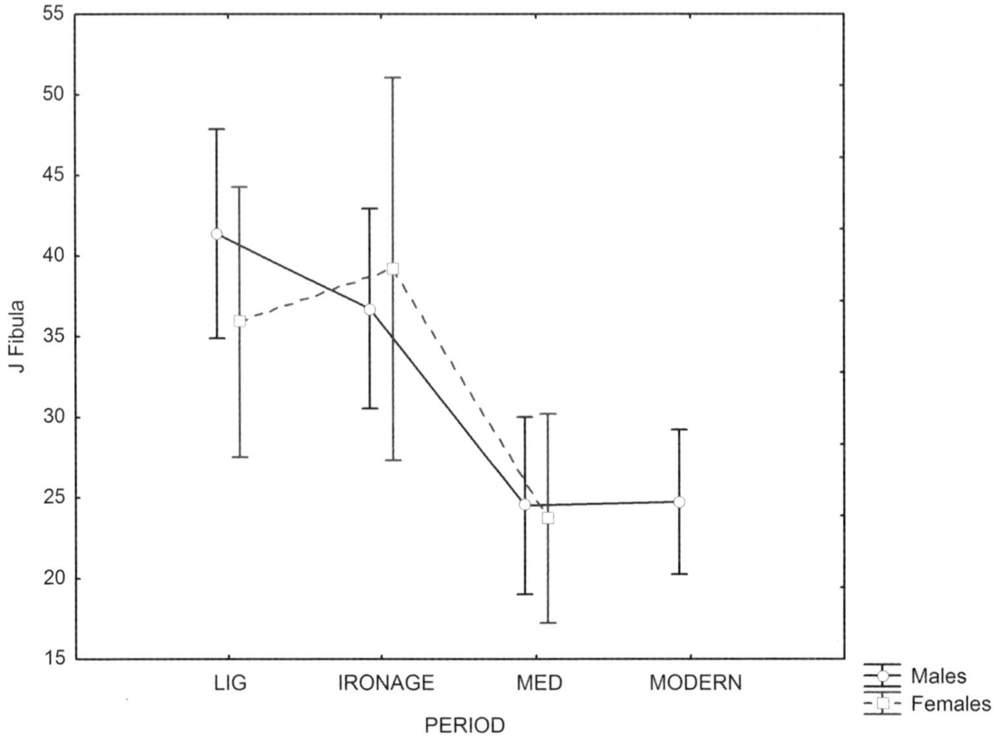

Figure 13.6 Two-factors ANOVA plot for midshaft fibular polar second moment of area (J). Factors: temporal period (x-axis); sex (line). Vertical bars denote 95% confidence intervals. LIG: Ligurian Neolithic; IRONAGE: Iron Age; MED: medieval; MODERN: modern

males show similar rigidity to Ligurian Neolithic males. However, though Figure 13.8 shows a sharp decline in relative torsional rigidity of the fibula from Iron Age to medieval, the differences are not significant after correction for multiple comparisons. Significant differences are found for comparisons involving I_{min} (Table 13.10).

Ligurian Neolithic females display relatively more rigid fibulae (high I_{max}, I_{min} and J) than found amongst other periods; nevertheless, the differences are non-significant after correction for multiple comparisons (Table 13.10). A larger sample size is necessary to properly ascertain whether this lack of significance is the result of similar biomechanical loading patterns amongst periods, or due to sampling error associated with small sample size. In summary, a non-significant trend is found involving a decrease in relative fibular robusticity from the Ligurian Neolithic through to the Iron Age, the medieval and modern samples.

13.4 DISCUSSION

13.4.1 Femur and Tibia

Change in the lower limb robusticity in the genus *Homo* shows a consistent decline in relative bone strength during the Pleistocene, and a further decrease during the Holocene (Ruff, 2008). This is likely related to diachronic changes in subsistence and to technological advancements

Table 13.9 Temporal differences in diaphyseal geometric properties of the fibula of males and females

	Males		Multiple comparisons with Hommel's sequential Bonferroni correction			Females		Multiple comparisons with Hommel's sequential Bonferroni correction	
I_{max}[a]	Mean (n)	SD	IRONAGE[b]	MED	MODERN	Mean (n)	SD	IRONAGE	MED
LIG	27.87 (15)	7.73	NS	*	*	23.39 (7)	9.06	NS	*
IRONAGE	24.49 (21)	9.06		*	*	20.67 (8)	8.55		NS
MED	15.3 (14)	5			NS	14.75 (12)	6.11		
MODERN	16.56 (21)	5.86							
I_{min}			IRONAGE	MED	MODERN			IRONAGE	MED
LIG	15.97 (15)	5.71	°	*	*	10.78 (7)	3.3	NS	NS
IRONAGE	12.6 (21)	3.92		*	*	9.81 (8)	4.76		NS
MED	9.23 (14)	2.39			NS	9 (12)	3.66		
MODERN	8.19 (21)	2.99							
J			IRONAGE	MED	MODERN			IRONAGE	MED
LIG	43.84 (15)	12.21	NS	*	*	34.19 (7)	12.13	NS	NS
IRONAGE	37.09 (21)	12.09		*	*	30.49 (8)	12.87		NS
MED	24.53 (14)	6.8			NS	23.75 (12)	9.67		
MODERN	24.75 (21)	8.31							
I_{max}/I_{min}			IRONAGE	MED	MODERN			IRONAGE	MED
LIG	1.85 (15)	0.53	NS	NS	NS	2.19 (7)	0.45	NS	*
IRONAGE	1.99 (21)	0.55		NS	NS	2.2 (8)	0.59		*
MED	1.69 (14)	0.43			NS	1.65 (12)	0.24		
MODERN	2.18 (21)	0.96							

[a] All the variables are explained in the text. I_{max}, I_{min} and J are divided by body mass × bone length^2.

[b] For abbreviations, see Tables 13.2 and 13.4.

*$p < 0.05$; ° $0.10 < p < 0.05$; NS = non significant difference. Results of Mann-Whitney U-test corrected for multiple comparisons using Hommel's sequential Bonferroni correction.

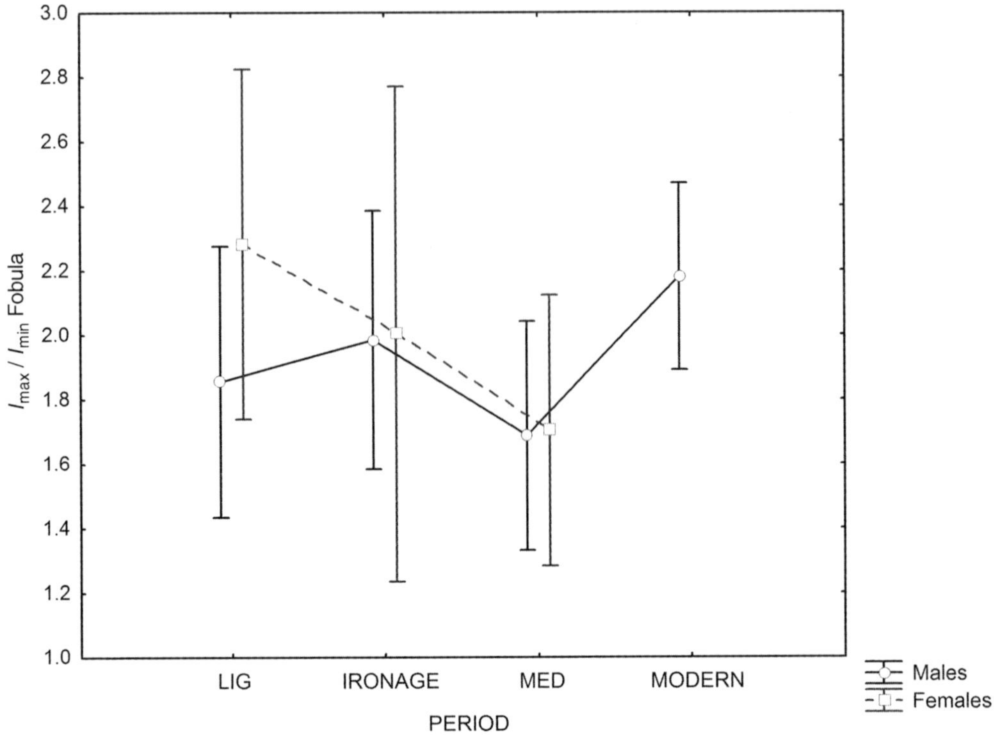

Figure 13.7 Two-factors ANOVA plot for midshaft fibular diaphyseal shape ratio (I_{max}/I_{min}). Factors: temporal period (x-axis); sex (line). Vertical bars denote 95% confidence intervals. LIG: Ligurian Neolithic; IRONAGE: Iron Age; MED: medieval; MODERN: modern

that have progressively shielded the body from physical demands of habitual labour (Ruff *et al.*, 1993). The temporal changes observed in this study from the Late Upper Palaeolithic through to the Eneolithic, for most of the variables related to diaphyseal robusticity and shape, are similar to the results of other studies. This suggests a real diachronic trend in post-Palaeolithic times of decreasing lower limb robusticity, especially at the Mesolithic/Neolithic boundary (Holt, 2003; Ruff *et al.*, 2006; Holt and Formicola, 2008).

However, these results also show that Ligurian Neolithic individuals (particularly males) do not show the reduction in lower limb mechanical properties that characterizes the Eneolithic sample, and are instead more similar to populations from earlier time periods. The Eneolithic comparative sample has been described as relatively sedentary, performing mixed agriculture subsistence patterns (Sládek, Berner and Sailer, 2006a, 2006b), while the Late Upper Palaeolithic and Mesolithic samples come from populations of hunter-gatherers characterized by high residential mobility (Holt, 2003). Although archaeological and biological evidence suggest that mobility levels decreased through the Late Upper Palaeolithic and Mesolithic, with respect to the Middle Upper Palaeolithic (Holt and Formicola, 2008), it is generally accepted that an even more marked change in mobility accompanied the shift to agriculture (Ruff *et al.*, 2006). Ligurian Neolithic people do not fit into this trend: their similarity to earlier hunter-gatherers indicates a higher degree of mobility, at least for males.

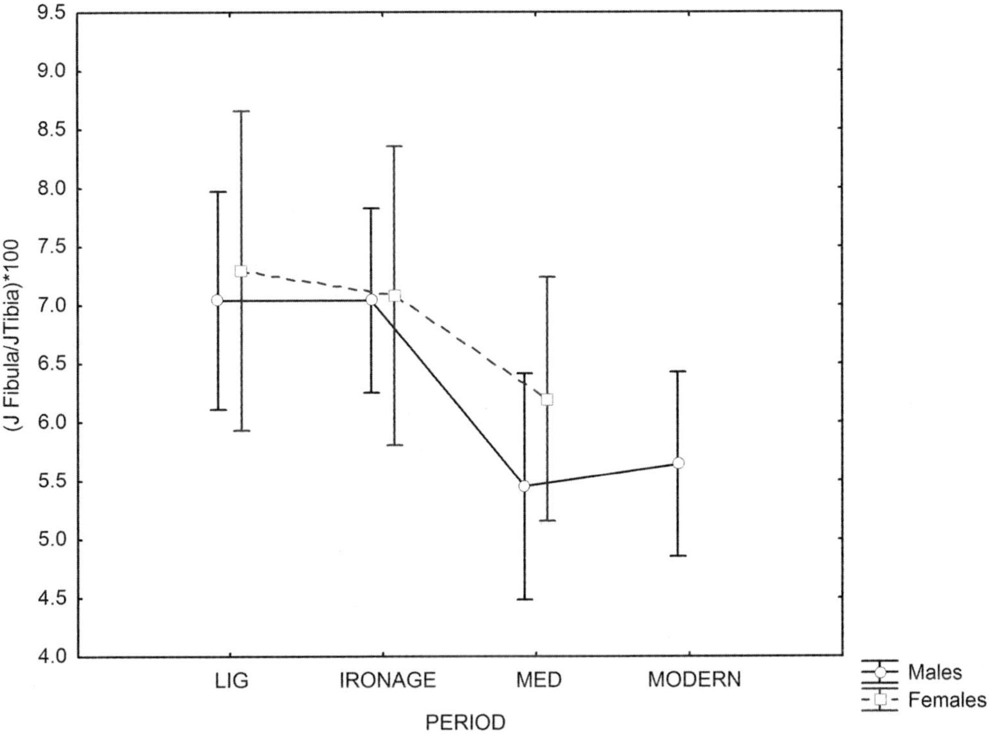

Figure 13.8 Two-factors ANOVA plot for the tibio-fibular robusticity index (*J* fibula/*J* tibia)*100. Factors: temporal period (*x*-axis); sex (line). Vertical bars denote 95% confidence intervals. LIG: Ligurian Neolithic; IRONAGE: Iron Age; MED: medieval; MODERN: modern

Shape index results are consistent with this scenario. Previous studies (Ruff, 1987; Holt, 2003) have provided evidence of a link between mobility and femoral A-P bending rigidity, supporting the use of diaphyseal shape ratios as indicators of mobility levels. The results reported here describe a significant decrease in femoral diaphyseal shape index (increased diaphyseal circularity) from the Late Upper Palaeolithic through to the modern sample, which is most pronounced in males and is consistent with a scenario of decreasing mobility from hunting and gathering to more sedentary subsistence activities. Again, the Ligurian Neolithic male sample shows femoral diaphyseal shape similar to the highly-mobile Late Upper Palaeolithic.

Differences in tibial shape have been associated with differences in terrestrial mobility levels (Ruff and Hayes, 1983; Holt, 2003; Stock, 2006); however, the behavioural interpretations are somewhat more complex than is the case for analyses of the femur. Stock (2006) found more modest correlations between tibial diaphyseal shape and mobility than for femoral midshafts shape. However, a recent study (Shaw and Stock, 2009) reported high levels of tibial rigidity and elliptical diaphyseal shapes (anteroposterior strengthening) for a sample of varsity cross-country runners. In contrast, varsity field hockey players displayed similarly high levels of tibial rigidity, yet significantly more circular tibial shapes (greater mediolateral strengthening). These differences were attributed to variation in loading orientation patterns associated with

Table 13.10 Temporal differences in the relative fibular diaphyseal rigidity of males and females

	Males		Multiple comparisons with Hommel's sequential Bonferroni correction			Females		Multiple comparisons with Hommel's sequential Bonferroni correction	
% J Fib[a]	Mean (n)	SD	IRONAGE	MED	MODERN	Mean (n)	SD	IRONAGE	MED
LIG[b]	7.04 (15)	1.77	NS	°	°	7.29 (7)	1.99	NS	NS
IRONAGE	7.04 (21)	2.00		°	°	7.08 (8)	1.56		NS
MED	5.45 (14)	1.32			NS	6.20 (12)	2.22		
MODERN	5.64 (21)	1.72							
% I_{max} Fib[a]			IRONAGE	MED	MODERN			IRONAGE	MED
LIG	6.36 (15)	1.91	NS	NS	NS	7.21 (7)	2.26	NS	NS
IRONAGE	6.74 (21)	2.34		°	*	7.14 (8)	1.49		NS
MED	5.20 (14)	1.52			NS	5.98 (12)	1.97		
MODERN	5.43 (21)	1.68							
% I_{min} Fib[a]			IRONAGE	MED	MODERN			IRONAGE	MED
LIG	8.83 (15)	2.14	NS	*	*	7.52 (7)	1.67	NS	NS
IRONAGE	7.82 (21)	2.11		*	*	7.24 (8)	3.08		NS
MED	6.01 (14)	1.47			NS	6.70 (12)	2.94		
MODERN	6.16 (21)	2.36							

[a] (J fibula/J tibia) × 100; (I_{max} fibula/I_{max} tibia) × 100; (I_{min} fibula/I_{min} tibia) × 100.
[b] For abbreviations, see Tables 13.2 and 13.4.
*$p < 0.05$; ° $0.10 < p < 0.05$; NS = non significant difference. Results of Mann-Whitney U-test corrected for multiple comparisons using Hommel's sequential Bonferroni correction.

each sport. Therefore, tibial diaphyseal shape ratios taken alone do not always correspond with mobility levels, but it is important to interpret them in association with diaphyseal robusticity levels. Low diaphyseal shape ratios associated with high levels of tibial robusticity may be indicative of high mobility with frequent changes of direction.

The shape of Ligurian Neolithic male tibial midshafts falls between the Late Upper Palaeolithic and Eneolithic samples and is not significantly different from either. Ligurian Neolithic tibial rigidity, however, is higher than that of both the Late Upper Palaeolithic and Eneolithic samples, with the latter comparison reaching significance. Therefore, the diaphyseal shape of Ligurian Neolithic people, when viewed in association with high tibial and femoral rigidity, suggests a high level of mobility and varied loading regimes, possibly caused by the uneven terrain where Ligurian Neolithic people dwelled.

It has been proposed that femoral diaphyseal shape ratios may be sensitive to differences in body shape (Ruff *et al.*, 2006). Specifically, as pelvic interacetabular distance relative to femoral length increases, cross-sections may become more mediolaterally buttressed, causing a decrease in diaphyseal shape ratios not connected to mobility levels. However, our results suggest that body breadth is not a significant factor in this analysis. Although the Ligurian Neolithic males show greater relative body breadth than the other samples (Table 13.4), they have significantly less circular midshaft diaphysis than the more slender Eneolithic males, mainly due to an increase in I_{max}. Therefore, in the Ligurian Neolithic sample, differences in femoral and tibial diaphysis shape may reflect adaptation to mobility patterns, the primary difference being pronounced anteroposterior bending strength. The same conclusion holds for the female samples, given that they show a similar pattern of variation in body proportions, although the differences between temporal periods are not significant.

13.4.2 Sexual Dimorphism

Sex-related differences in long bone cross-sectional properties have been associated with variation in sex-specific mobility patterns related to subsistence strategy (Ruff, 1987, 1999). The Ligurian Neolithic males displayed more robust and less circular femoral and tibial midshafts than females, a relationship that is greater than all other periods. Within the Ligurian Neolithic sample, body proportions are comparable between the sexes (Table 13.4), which suggests that variation in body breadth has not inordinately influenced tibial and femoral diaphyseal shape. Thus, the results presented here support the assertion of greater mobility in Ligurian Neolithic males when compared to females, which may reflects marked differences in gender roles, with males potentially performing the majority of the tasks requiring long-distance travel. This is compatible with the sexual division of labour present in contemporary herding societies, where men perform most of the herding duties, and other related activities (such as pollarding), while females are often involved in agricultural activities and food processing (Murdock and Provost, 1973; Dyson-Hudson and Dyson-Hudson, 1980; Marchi, 2008, and references therein).

Archaeological evidence, such as virtually identical burial patterns in both sexes, suggests that gender ideology was not well developed in the Neolithic (Robb, 1994b, 2007). However, the interpretation of Neolithic rock engravings suggests a gender based partitioning of occupational space, with male images associated with more marginal areas, while female focus appear to be closer to the villages. This led Robb (1994b, 2007) to hypothesize that males performed mobile 'outdoor activities', such as herding and hunting, and females performed activities involving greater proximity to the village, such as agriculture-based tasks and gathering. This scenario of a sex-based division of labour is consistent with the morphological results presented here.

13.4.3 Fibula and Tibia/Fibula Complex

Another factor that must be considered when studying lower limb bone rigidity is that the tibia is paired with the fibula. The fibula has only rarely been considered in anthropological studies of the lower limb (Marchi, 2004, 2007). However, several studies performed *in vivo* (Takebe *et al.*, 1984; Goh *et al.*, 1992; Funk *et al.*, 2004) found that in the human leg the fibula has a weight-bearing function, carrying a varying percentage of the biomechanical load (6–19%).

The temporal changes observed from the Ligurian Neolithic to the modern sample in fibular robusticity (J, I_{max} and I_{min}) show trends similar to those observed for the tibia. A trend of decreasing robusticity for the fibula (both standardized and relative to the tibia) from the Ligurian Neolithic to the Iron Age to the medieval and modern samples, suggest that, similar to variation in tibial robusticity, fibular robusticity may correlate with mobility levels.

As previously noted for the tibia, terrain attributes may influence fibular robusticity. High levels of mobility, performed on extremely uneven terrain, as is the case for the Ligurian Neolithic population, requires greater eversion and inversion loading of the foot than produced while locomoting on even terrain. In a study conducted on a hominoid skeletal sample, Marchi (2007) found that fibular robusticity (relative to the tibia) is correlated with the degree of mobility of the ankle joint, particularly with regard to plantarflexion/dorsiflexion and inversion/eversion movements. In addition, a recent study that assessed tibial deformation during axial compression of the leg (Funk *et al.*, 2007) found that peak bending moments migrated medially when the foot was inverted, and laterally when the foot was everted. Therefore, bending moments produced in association with inversion and/or eversion loading of the foot may increase mediolateral bending rigidity in the tibia and, as a consequence, osseous adaptations to these strains. Support for this hypothesis is given by the high tibial I_{min} values found in the present study amongst groups settled in mountainous areas; both the Ligurian Neolithic and Iron Age populations.

Given that the fibula is lateral to the tibia, it may be expected that as a consequence of its position it would be subject to more M-L orientated loads, and therefore increase its bending rigidity. In agreement with this scenario, Ligurian Neolithic male fibulae were significantly more robust than those of medieval and Modern males, but not significantly different from those of Iron Age males. This pattern is apparent for both standardized fibular robusticity and relative fibular robusticity.

The preliminary results of an ongoing project (Marchi and Show, 2010) carried out on the relative (to the tibia) robusticity of the fibula of varsity athletes seem to indicate a relatively more robust fibula for field hockey players than for cross-country runners. Generally, cross-country running requires the athlete to proceed in a relatively straight line the majority of the time. Movement patterns associated with field hockey instead necessitate repeated changes of direction. These changes in direction may be responsible for more M-L orientated loading on the leg that eventually would lead to the relatively more robust fibula of the field hockey players than that of the cross-country runners. This result provides further support to the interpretation of the results obtained in this study.

There is now preliminary evidence to suggest that the fibula may provide particularly important biomechanical information in some circumstances, specifically when there is considerable terrain relief. This is the first time that the fibula has been included in a bioarchaeological study and even though results are not always significant, evaluation of the mechanical properties of the fibula (particularly when in association with the tibia) have aided the assessment of the mobility patterns of the Ligurian Neolithic population.

13.4.4 The Agricultural Transition in Liguria

Femoral, tibial and fibular biomechanical properties depict a coherent scenario of high mobility in Ligurian Neolithic males. These results confirm that the shift to Neolithic subsistence patterns based on agriculture and animal husbandry did not result in a uniform decrease in mobility throughout Europe. At a regional level, more complex trends in habitual behaviour may have appeared. In the case of the Ligurian Neolithic people, the high level of male mobility can be attributed to an emphasis on pastoral activities, for which there is abundant archaeological evidence. The degree of reliance on herding, hypothesized from the inferred mobility patterns of Ligurian Neolithic males, was unexpected for Neolithic times. Robb (1994a) hypothesized that pastoral groups may have developed skeletal properties similar to hunter-gatherers, due to habitual travel through rough and marginal areas, and anticipated the results of mechanical studies on prehistoric Italian populations from the Neolithic to the Iron Age. For the Neolithic, it was hypothesized that the integration of small herds in an agriculture-based village society would result in a relatively low average level of mobility. Our results are not consistent with Robb's expectations, and correspond better with the high mobility levels associated with intense pastoralism expected from peoples of the Italian Copper Age (Robb, 1994a,b). Moreover, isotopic evidence suggests that a significant part of the Ligurian Neolithic diet was composed of meat and dairy products (Le Bras-Goude *et al.*, 2006). This provides complimentary evidence that the people of the Ligurian Neolithic may have had a degree of reliance on pastoralism and secondary products (cheese, milk and wool). Interestingly, this focus on pastoralism would have happened before the Copper Age socio-economic transformation, which led to a system of exchange 'regulated by the circulation of generalized value via prestige competition' (Robb, 1994b: 30).

Neolithic resource exploitation in the Italian peninsula involved marked variability within a broad agropastoral framework. This included intensive agricultural exploitation in the Tavoliere (Apulia), shifting agriculture in the Po plain, fresh water resources exploitation at the fringes of the Alps, and subsistence integration with hunting and gathering (Guidi and Piperno, 1992; Robb, 2007; Pessina and Tiné, 2008). This variability reflects a certain degree of plasticity required by early production economies, which were able to adapt to the specific environmental condition of each region. Ligurian Neolithic subsistence would therefore be one part of a broader mosaic of varying subsistence strategies. It has been suggested that the variability in Neolithic resource exploitation observed in the Italian peninsula did not develop out of a need to compensate for scarce agricultural output. Given the low population density, the availability of cultivable land was never a limiting factor for the degree of reliance on agriculture, even in the most mountainous areas (Jarman and Webley, 1975; Robb, 2007). Accordingly, many authors suggest that nomadism never replaced the permanent village as the dominant form of settlement, and that agriculture rather than secondary animal products was the basis of the economy throughout Italian prehistory (Guidi and Piperno, 1992; Robb, 1994a, 2007). Given this perspective, the mobility patterns of the Ligurian Neolithic people, and their reliance on pastoralism is peculiar within the context of the Italian Neolithic landscape. This emphasis on pastoral activities may be due to the fact that the geographical makeup of Western Liguria includes a rich variety of easily defendable caves that would have provided appropriate shelter to herds. However, other explanations are possible and require further investigation, also taking into account the possible influence of cultural factors and social choice on the development of this subsistence economy.

Behavioural changes associated with subsistence patterns are likely to have been amongst the primary factors influencing long-term changes in lower limb robusticity. Nevertheless, the effect of terrain relief on long bone structure adds another level of complexity to the interpretation of changes in lower limb diaphyseal robusticity (Ruff, 1999, 2008). Liguria has an extremely rugged terrain, which may have increased the biomechanical load associated with locomotion, and therefore had a significant impact on the relatively high lower limb robusticity found in the Ligurian Neolithic sample. This presumption is supported by the finding of comparable results for Iron Age samples. These individuals were settled in the mountainous area of Abruzzo (Central Italy) and display higher maximum and minimum second moments of area (for both males and females) compared to the Eneolithic sample, which are associated with less rugged plain settlements. Nevertheless, it has been suggested that diaphyseal shape ratios are not significantly influenced by terrain (Ruff, 1999, 2008), thus the discussion of mobility levels (that can be inferred from diaphyseal shape alone) may not be appreciably influenced by differences in topographic relief present amongst the groups compared here.

Genetic continuity amongst Italian prehistoric groups may be a factor to take into account in the analysis of robusticity, especially when we try to untangle the possible effect of terrain. The possibility that the Iron Age sample might constitute a more robust immigrant population from another region may confound our argument. However, the idea of biological continuity within the Italian peninsula, from Neolithic times to early historic times, is supported by more recent analyses of genetic markers from living populations, and cranial, dental, metric and nonmetric traits analyses (Coppa *et al.*, 1998, 2007).

13.5 CONCLUSION

In this study of the activity patterns of the Ligurian Neolithic people, we have included several new burials, expanded the comparative samples, and included – for the first time in a bioarchaeological study – the fibula amongst the bones analysed.

Our results provide further evidence to indicate that Ligurian Neolithic femora and tibiae do not show the reduction in diaphyseal rigidity characteristic of agriculturalist samples from the same time period. Rather, the Ligurian Neolithic sample is more similar to mobile European Late Upper Palaeolithic and Mesolithic populations. Furthermore, Ligurian Neolithic peoples display remarkable levels of sexual dimorphism for diaphyseal shape, suggesting that males were more mobile than females. These results support the assertion that the Ligurian Neolithic people were not sedentary, and were in fact highly mobile, travelling across very rugged terrain, and relied mainly on pastoralism rather than agriculture. A secondary purpose of this study was to explore the contribution of the fibula to the biomechanics of the human leg in past populations. The results presented here indicate that this bone does contribute to the diaphyseal rigidity of the lower leg, and its inclusion provides a better understanding of past population mobility especially when considered in association with the tibia.

Overall, these results do not invalidate the general biological trend of reduction in lower limb robusticity observed from the Pleistocene through the Holocene. Rather, they suggest that when the transition to agriculture is analysed at the regional level, a more complex mosaic of habitual behaviours may appear. This study demonstrates that a biological approach to the reconstruction of past subsistence activities provides the opportunity to gain a more subtle understanding of broad behavioural changes. Through biomechanical analysis it is possible to view the Neolithic transition as a dynamic process involving various degrees of reliance to herding within a broad agropastoral framework. This highlights the importance of directly assessing regional and temporal biological variation in relation to multiple environmental factors, including subsistence strategy and terrain relief.

ACKNOWLEDGEMENTS

We express our gratitude to A. Del Lucchese and the Soprintendenza Archeologica della Liguria for the permission to examine the material. Thanks are also due to G. Rossi and P. Garibaldi of the Museo di Archeologia Ligure in Genova Pegli, and to D. Arobba, G. Vicino and A. De Pascale of the Museo Civico di Archeologia in Finale Ligure for the active collaboration and assistance when working in the Museums, and to C.B. Ruff, B.M. Holt and V. Sládek for allowing us to use their cross-sectional data for comparison. We also thank Benjamin Hoeke for providing unpublished information of the German medieval comparative sample. Finally, we wish to thank B.M. Holt and V. Formicola for precious comments and suggestions during the years that we have studied the Ligurian Neolithic material.

REFERENCES

Arobba, D., Deferrari, G. and Nisbet, R. (1997) *Carpological Investigation at the Cave of Arene Candide: An Impression of Caryopsis of Barley on a Ceramic Sherd*, Memorie dell'Istituto Italiano di Paleontologia Umana 5, pp. 119–123.

Auerbach, B.M. and Ruff, C.B. (2004) Human body mass estimation: a comparison of 'morphometric' and 'mechanical' methods. *Am. J. Phys. Anthropol.*, **125**, 331–342.

Bondioli, L., Corruccini, R.S. and Macchiarelli, R. (1986) Familial segregation in the Iron age community of Alfedena, Abruzzo, Italy, based on osteodental trait analysis. *Am. J. Phys. Anthropol.*, **71**, 393–400.

Cohen, M.N. and Armelagos, G.J. (1984) *Paleopathology at the Origins of Agriculture*, Academic Press, Orlando.

Coppa, A., Cucina, A., Mancinelli, D. *et al.* (1998) Dental anthropology of central-southern. Iron age italy: the evidence of metric versus nonmetric traits. *Am. J. Phys. Anthropol.*, **107**, 371–386.

Coppa, A., Cucina, A., Lucci, M. *et al.* (2007) Origins and spread of agriculture in Italy: A nonmetric dental analysis. *Am. J. Phys. Anthropol.*, **133**, 918–930.

Dyson-Hudson, R. and Dyson-Hudson, N. (1980) Nomadic pastoralism. *Ann. Rev. Anthropol.*, **9**, 15–61.

Formicola, V. (1997) The Neolithic transition in Western Liguria: the current status of the anthropological research, in *Arene Candide: A Functional and Environmental Assessment of the Holocene Sequence Excavations Bernabò Brea-Cardini 1940–1950* (ed. R. Maggi), Memorie dell'Istituto Italiano di Paleontologia Umana 5, pp. 599–604.

Funk, J.R., Rudd, R.W., Kerrigan, J.R. and Crandall, J.R. (2004) The effect of tibial curvature and fibular loading on the tibia index. *Traffic Injury Prevention*, **5**, 164–172.

Funk, J.R., Rudd, R.W., Kerrigan, J.R. and Crandall, J.R. (2007) The line of action in the tibia during axial compression of the leg. *J. Biomech.*, **40**, 2277–2282.

Goh, J.C.H., Lee, H.H., Ang, E.J. *et al.* (1992) Biomechanical study on the load-bearing characteristics of the fibula and the effects of fibular resection. *Clinical Orthopaedics*, **279**, 223–228.

Guidi, A. and Piperno, M. (1992) *Italia Preistorica*, Laterza, Roma.

Holt, B.M. (2003) Mobility in Upper Palaeolithic and Mesolithic Europe: evidence from the lower limb. *Am. J. Phys. Anthropol.*, **122**, 200–215.

Holt, B.M. and Formicola, V. (2008) Hunters of the Ice Age: the biology of Upper Palaeolithic people. *Yrbk Phys. Anthropol.*, **51**, 70–99.

Hommel, G. (1988) A stagewise rejective multiple test procedure on a modified Bonferroni test. *Biometrika.*, **75**, 383–386.

Jarman, M. and Webley, D. (1975) Settlement and land use in Capitanata, Italy, in *Palaeoeconomy* (ed. E. Higgs), Cambridge University Press, Cambridge, pp. 177–231.

Kelly, R.L. (1983) Hunter-gatherer mobility strategies. *J. Anthropol. Res.*, **39**, 277–306.

Lahr, M., Foley, R. and Pinhasi, R. (2000) Expected regional patterns of Mesolithic-Neolithic human population admixture in Europe based on archaeological evidence, in *Archaeogenetics: DNA and the Population Prehistory of Europe* (eds C. Renfrew and K. Boyle), McDonald Institute for Archaeological Research, Monographs, Cambridge, pp. 81–88.

Larsen, C.S. (1995) Biological changes in human populations with agriculture. *Ann. Rev. Anthropol.*, **24**, 185–213.

Le Bras-Goude, G., Binder, D., Formicola, V. *et al.* (2006) Stratégies de subsistance et analyse culturelle de populations néolithiques de Ligurie: approche par l'étude isotopique δ^{13}C et δ^{15}N. des restes osseux. Bulletins et Mémoires de la **société** d'Anthropologie de Paris 18: 43–53.

Macphail, R.I., Courty, M.A., Hather, J. and Wattez, J. (1997) The soil micromorphological evidence of domestic occupation and stabling activities, in *Arene Candide: A Functional and Environmental Assessment of the Holocene Sequence Excavations Bernabó Brea-Cardini 1940–1950* (ed. R. Maggi), Memorie dell'Istituto Italiano di Paleontologia Umana 5, pp. 53–88.

Maggi, R. (1997a) The radiocarbon chronology, in *Arene Candide: A Functional and Environmental Assessment of the Holocene Sequence Excavations Bernabó Brea-Cardini 1940–1950* (ed. R. Maggi), Memorie dell'Istituto Italiano di Paleontologia Umana, 5, pp. 31–52.

Maggi, R. (1997b) *Aspetti di Archeologia del Territorio in Liguria: La Formazione Del Paesaggio Dal Neolitico all'Età del Bronzo, Annali dell'Istituto 'Alcide Cervi' 19*, Edizioni Dedalo, Bari.

Maggi, R. and Nisbet, R. (1990) Prehistoric pastoralism in Liguria. *Rivista di Studi Liguri*, 56, 26BAR International Series 207, 5–296.

Maggi, R. and Nisbet, R. (2000) Alberi da foraggio e scalvatura neolitica: nuovi dati dalle Arene Candide, in *La Neolitizzazione Tra Oriente e Occidente, Atti Convegno Di Studi* (eds A. Pessina and G. Muscio), Edizioni del Museo Friulano di Storia Naturale, Udine, pp. 289–308.

Marchi, D. (2004) Cross-sectional geometry of the limb bones of the Hominoidea: its relationships with locomotion and posture. Ph.D. Dissertation, University of Pisa, Pisa.

Marchi, D. (2007) Relative strength of the tibia and fibula and locomotor behaviour in hominoids. *J. Human. Evol.*, **53**, 647–655.

Marchi, D. (2008) Relationships between lower limb cross-sectional geometry and mobility: the case of a Neolithic sample from Italy. *Am. J. Phys. Anthropol.*, **137**, 188–200.

Marchi, D. and Show, C.N. (2010) Does variation in fibular robusticity reflect variation in mobility patterns? *Am. J. Phys. Anthropol.*, Suppl 50, 163.

Marchi, D., Sparacello, V.S., Holt, B.M. and Formicola, V. (2006) Biomechanical approach to the reconstruction of activity patterns in Neolithic Western Liguria, Italy. *Am. J. Phys. Anthropol.*, **131**, 447–455.

Montet-White, A. (1991) Lithic acquisition, settlements and territory in the Epigravettian of Central Europe, in *Raw Material Economies Among Prehistoric Hunter-Gatherers* (eds A. Montet-White and S. Holen), University of Kansas Publications in Anthropology 19, Lawrence, pp. 205–219.

Murdock, G.P. and Provost, C. (1973) Factors in the division of labors by sex: a cross-cultural analysis. *Ethnology*, **12**, 203–225.

Nisbet, R. (1997a) Arene Candide: charcoal remains and prehistoric woodland use, in *Arene Candide: A Functional and Environmental Assessment of the Holocene Sequence Excavations Bernabò Brea-Cardini 1940–1950* (ed. R. Maggi), Memorie dell'Istituto Italiano di Paleontologia Umana 5, pp. 103–112.

Nisbet, R. (1997b) The phytoliths from the Neolithic levels of Arene Candide, in *Arene Candide: A Functional and Environmental Assessment of the Holocene Sequence Excavations Bernabò Brea-Cardini 1940–1950* (ed. R. Maggi), Memorie dell'Istituto Italiano di Paleontologia Umana 5, pp. 113–118.

O'Neill, M.C. and Ruff, C.B. (2004) Estimating human long bone cross-sectional geometric properties: a comparison of noninvasive methods. *J. Human. Evol.*, **47**, 221–235.

Parenti, R. and Messeri, P. (1962) *I Resti Scheletrici Umani Del Neolitico Ligure, Paleontographia Italica 50*, Tipografia Moderna, Pisa.

Pearson, O.M., Cordero, R.M. and Busby, A.M. (2006) How different were Neanderthals' habitual activities? A comparative analysis with diverse groups of recent humans, in *Neanderthals Revisited: New Approaches and Perspectives* (eds K. Harvati and T. Harrison), Springer, New York, pp. 135–156.

Pessina, A. and Tiné, V. (2008) *Archeologia del Neolitico – l'Italia tra VI e IV Millennio a.C*, Carocci editore, Roma.

Robb, J. (1994a) Skeletal signs of activity in the Italian Metal Ages: methodological and interpretative notes. *Human Evol.*, **3**, 215–229.

Robb, J. (1994b) Gender contradictions: moral coalitions and inequality in prehistoric Italy. *J. Eur. Archaeol.*, **2**, 20–49.

Robb, J. (2007) *The Early Mediterranean Village – Agency, Material Culture, and Social Change in Neolithic Italy*, Cambridge University Press, Cambridge.

Rowley-Conwy, P. (1997) The animal bones from Arene Candide Holocene sequence: Final report, in *Arene Candide: A Functional and Environmental Assessment of the Holocene Sequence Excavations Bernabò Brea-Cardini 1940–1950* (ed. R. Maggi), Memorie dell'Istituto Italiano di Paleontologia Umana 5, pp. 153–195.

Rowley-Conwy, P. (2000) Milking caprines, hunting pigs: the Neolithic economy of Arene Candide in its West Mediterranean context, in *Animal Bones, Human Societies* (ed. P. Rowley-Conwy), Oxbow Press, Oxford, pp. 124–132.

Ruff, C.B. (1987) Sexual dimorphism in human lower limb bone structure: relationship to subsistence strategy and sexual division of labor. *J. Human. Evol.*, **16**, 391–416.

Ruff, C.B. (1999) Skeletal structure and behavioural patterns of prehistoric Great Basin populations, in *Prehistoric Lifeways in the Great Basin Wetlands: Bioarchaeological Reconstruction and Interpretation* (eds B.E. Hemphill and C.S. Larsen), University of Utah Press, Salt Lake City, pp. 290–320.

Ruff, C.B. (2002) Long bone articular and diaphyseal structure in Old World monkeys and apes. I. Locomotor effects. *Am. J. Phys. Anthropol.*, **119**, 305–342.

Ruff, C.B. (2008) Biomechanical analyses of archeological human skeletons, in *Biological Anthropology of the Human Skeleton*, 2nd edn (eds M.A. Katzenberg and S.R. Saunders), John Wiley & Sons, New York, pp. 183–206.

Ruff, C.B. and Hayes, W.C. (1983) Cross-sectional geometry of the Pecos Pueblo femora and tibiae-a biomechanical investigation: I. Methods and general pattern of variation. *Am. J. Phys. Anthropol.*, **60**, 359–381.

Ruff, C.B., Holt, B.M., Sládek, V. *et al.* (2006) Body size, body proportions, and mobility in the Tyrolean 'Iceman'. *J. Human Evol.*, **51**, 91–101.

Ruff, C.B., Holt, B.M. and Trinkaus, E. (2006) Who's afraid of the big bad Wolff?: 'Wolff's law' and bone functional adaptation. *Am. J. Phys. Anthropol.*, **129**, 484–498.

Ruff, C.B., Trinkaus, E., Walker, A. and Larsen, C.S. (1993) Postcranial robusticity in Homo. I. Temporal trends and mechanical interpretation. *Am. J. Phys. Anthropol.*, **91**, 21–54.

Shaw, C.N. and Stock, J.T. (2009) Intensity, repetitiveness, and directionality of habitual adolescent mobility patterns influence the tibial diaphysis morphology of athletes. *Am. J. Phys. Anthropol.*, **140**, 149–159.

Sjovold, T. (1990) Estimation of stature from long bones utilizing the line of organic correlation. *Human Evol.*, **5**, 431–447.

Sládek, V., Berner, M. and Sailer, R. (2006a) Mobility in central European late Eneolithic and early Bronze Age: tibial cross-sectional geometry. *J. Archaeol. Sci.*, **33**, 470–482.

Sládek, V., Berner, M. and Sailer, R. (2006b) Mobility in central European late Eneolithic and early Bronze Age: femoral cross-sectional geometry. *Am. J. Phys. Anthropol.*, **130**, 320–332.

Sparacello, V.S. and Marchi, D. (2008) Mobility and subsistence economy: a diachronic comparison between two groups settled in the same geographical area Liguria, Italy. *Am. J. Phys. Anthropol.*, **136**, 485–495.

Sparacello, V.S., Pearson, O.M. and Coppa, A. (2009) Cross-sectional geometry of a warlike Samnite sample from the Alfedena necropolis Italy. *Am. J. Phys. Anthropol.*, (Suppl 48), 244.

Starnini, A. and Voytek, B. (1997) New lights on old stones, in *Arene Candide: A Functional and Environmental Assessment of the Holocene Sequence Excavations Bernabó Brea-Cardini 1940–50* (ed. R. Maggi), Memorie dell'Istituto Italiano di Paleontologia Umana, 5, pp. 427–511.

Stock, J.T. (2002) A test of two methods of radiographically deriving long bone cross-sectional properties compared to direct sectioning of the diaphysis. *Int. J. Osteoarch.*, **12**, 335–342.

Stock, J.T. (2006) Hunter-gatherer postcranial robusticity relative to patterns of mobility, climatic adaptation, and selection for tissue economy. *Am. J. Phys. Anthropol.*, **131**, 194–204.

Stock, J.T. and Shaw, C.N. (2007) Which measures of skeletal robusticity are robust? A comparison of external methods of quantifying diaphyseal strength to cross-sectional geometric properties. *Am. J. Phys. Anthropol.*, **134**, 412–423.

Takebe, K., Nakagawa, A., Minami, H. *et al.* (1984) Role of the fibula in weight-bearing. *Clinical Orthopaedics*, **184**, 289–192.

Trinkaus, E., Churchill, S.E., Villemeur, I. *et al.* (1991) Robusticity versus shape: the functional interpretation of neandertal appendicular morphology. *J. Anthrop. Soc. Nippon.*, **99**, 257–278.

14

Body Size, Skeletal Biomechanics, Mobility and Habitual Activity from the Late Palaeolithic to the Mid-Dynastic Nile Valley

Jay T. Stock[1], Matthew C. O'Neill[2], Christopher B. Ruff[2], Melissa Zabecki[3], Laura Shackelford[4] and Jerome C. Rose[5]

[1] *Leverhulme Centre for Human Evolutionary Studies, Department of Biological Anthropology, University of Cambridge, Cambridge, UK*
[2] *Centre for Functional Anatomy and Evolution, Johns Hopkins University, Baltimore, MD, USA*
[3] *Department of Behavioural Sciences, University of Arkansas, Fort Smith*
[4] *Department of Anthropology, University of Illinois at Urbana-Champaign, Urbana, IL, USA*
[5] *Department of Anthropology, University of Arkansas, Fayetteville, AR, USA*

14.1 INTRODUCTION

The ancient history of the Nile Valley is of particular interest for those studying long-term trends in human cultural and biological adaptation, due to the presence of an excellent series of human remains and associated material culture dating from the Late Palaeolithic to the present. The late Pleistocene of this region is represented by the archaeological record of hunter-gatherers at sites such as Jebel Sahaba. The subsequent Predynastic, or Neolithic Period (5500–3100 BC), in current Egypt and Nubia is characterized by a transition from nomadic pastoralism to agricultural proto-city-states (Kemp, 1989). It has been argued that this period forms the basis for the Pharaonic state (Wilkinson, 1999), which developed in the Early Dynastic Period. The first and second Dynasties mark the onset of the first monumental architecture, a developed system of writing, and a combination of agricultural and individual

Human Bioarchaeology of the Transition to Agriculture Edited by Ron Pinhasi and Jay T. Stock

task specialization (Bard, 2003), which led to the further development and expansion of the Egyptian empire with its hierarchically-organized bureaucratic structure. By the 12th Dynasty, in the Middle Kingdom, the Egyptian empire had expanded considerably and ruled over a large population along the Nile. The temporal span of the Late Palaeolithic through the 12th Dynasty extends from Late Pleistocene hunting and gathering populations (about 13 000–9000 BC) through the earliest agriculturalists (about 5000–4000 BC), to the formation and expansion of the Egyptian empire (about 3100–1500 BC; Starling and Stock, 2007). As such, it spans a period of subsistence transition but also follows the origins of agriculture with state formation and associated changes in infrastructure, influence and trade. These transitions have been well-documented in the Nile Valley, and provide a unique opportunity to study associated changes in human biology and skeletal morphology (Zakrzewski, 2002, 2007). The current study examines the body size and biomechanical correlates of these transitions in Egypt and Nubia through the analysis of variation in long bone cross-sectional geometry. Four populations are compared, which span the temporal range of 13 000 to 1500 BC, a period which includes the origins of agriculture and the formation and expansion of the Egyptian state (Starling and Stock, 2007).

The transition from hunting and gathering to an agricultural subsistence strategy is amongst the most profound changes in the relationship between humans and the natural environment, as it is often viewed as marking the most significant shift towards property ownership, social complexity and hierarchy (Childe, 1936; Diamond, 1999). In addition, agriculture can be viewed as a way of colonizing a new niche, which leads to reduced interbirth intervals, facilitates birth stacking, and leads to dramatic population growth (Wells and Stock, 2007). However, it is also possible that demographic shifts underpin this cultural change (Boserup, 1965), making returns from a hunting and gathering lifestyle insufficient to support an increased population size (Cohen, 1977). Whether population size was an important catalyst for, or a consequence of, the transition to agriculture, the positive feedback between demography and culture certainly underpinned subsequent urbanization and state formation. In this context, the origins of agriculture can be seen as, arguably, the most significant social and biological transition in human history.

The origin of domesticated animals and plants in the Nile Valley is commonly attributed to cultural diffusion from the Near East rather than *in situ* domestication (Smith, 1995). However, recent evidence suggests that there were multiple regions of agricultural origin with subsequent localized diffusions (Jones and Brown, 2000), which may have influenced cultural change in the region. For example, centres of domestication of both cattle and grains in the Sahara may have influenced the Neolithic of the Nile Valley (Warfe, 2003; Hanotte *et al.*, 2002). The complexity of the cultural history of the Nile Valley throughout the Holocene may have also influenced the development of the Neolithic and subsequent cultural change in a variety of ways (Holmes, 1993; Warfe, 2003).

The earliest documentation of food production in the Nile Valley comes in the form of evidence for barley, cattle, and either sheep or goats in various Neolithic sites of el-Badari, Maadi, Merimbd, Omari and the Fayum at approximately 5000 BC (Arkell and Ucko, 1965; Hassan, 1984). There were no permanent dwellings at the site of el-Badari, which suggests that while exploiting barley and domesticated animals, the Badarians may have remained semi-nomadic pastoralists rather than committed farmers (Hassan, 1988). While herding may have been a component of the Badarian culture, it is clear that the intensification of agriculture in the Nile Valley occurred in this early Predynastic Period, and by the beginning of the first Dynasty there were a number of large sedentary communities, which were politically united under the first pharaohs (Kemp, 1989). The subsequent intensification of agriculture in the region can be characterized by increasing population size and density, increasing agricultural production, and more efficient distribution of resources through the administration of the Dynastic states (Malek, 2003).

The population history of the Nile has been of considerable recent interest and focuses on two competing hypotheses. The first suggests that the Egyptian dynasties developed *in situ* from the earlier Predynastic and Neolithic populations represented at sites such as el-Badari. The second scenario suggests that migration of people from western Asia led to the development of the Egyptian state (Petrie, 1920, 1939; Kantor, 1965). In general, the archaeological evidence suggests that the Egyptian state had an indigenous origin (Hassan, 1988). Two recent studies provide evidence for population dynamics in the Nile Valley throughout the Holocene. Zakrzewski (2007) demonstrates evidence for broad population continuity through time on the basis of craniometric variation, with some level of population movement. Several recent analyses of dental variation come to essentially the same conclusion (Irish, 2005, 2006; Schillaci *et al.*, 2009). Thus, in the most general terms, there is strong evidence for population continuity along the Nile from the late Palaeolithic through the Egyptian Empire. However, the diffusion of agricultural technologies into the Nile from other regions, and the subsequent trade networks of the Egyptian empire, would have undoubtedly brought with it people and genes from other regions to varying extents through time and space.

Our understanding of the biological impact of the transition to agriculture in the Nile Valley is known from a few key studies. Analyses of the human remains from Wadi Halfa (9000–6000 BC) in Sudanese Nubia (Armelagos, 1969; Armelagos *et al.*, 1972; Hummert, 1983; Hummert and Van Gerven, 1983; Martin *et al.*, 1984; Vagn Nielsen, 1970; Van Gerven *et al.*, 1981) provide critical insights, including the recognition of significant morphological similarities in the crania to individuals from Jebel Sahaba and a general description of the postcranial skeletons as robust and muscular based on the heaviness of the bones and marked muscle attachments, but this work does not include Neolithic material (6000–3600 BC). More recently, Shackelford (2007) has analysed the postcranial skeletons of Jebel Sahaba and Wadi Halfa within a temporal context. In an analysis of these Palaeolithic samples relative to a Predynastic Egyptian sample (5000–4000 BC), the lower limb bones show a decrease in diaphyseal rigidity (as reflected by J, polar second moment of area) over time, as well as a change in the distribution of cortical bone (I_x/I_y). When the right and left humeri were evaluated separately, there was a decrease in axial rigidity (CA) through time, as well as a change in distal humerus diaphyseal shape (I_x/I_y), although there was no significant decrease in torsional rigidity (J) through time on either side. Zakrzewski (2003) documented changes in human body size from the Neolithic through the Middle Kingdom (1800 BC) and demonstrated a general increase in body size across this period. This trend is interpreted as reflective of increasing development of social hierarchy, while sexual differences in stature are interpreted as evidence of preferential provisioning of males over females.

Starling and Stock (2007) have also investigated changes in health across the transition to agriculture in the Nile Valley, finding a considerable increase in the frequency of linear-enamel hypoplasia (LEH) between the Upper Palaeolithic and Neolithic. The highest frequency of LEH was found amongst the population from el-Badari, which suggests that the transition from hunting and gathering to herding and the cultivation of grains was initially stressful. This stress may have stemmed from either a reduction in nutritional quality or an increase in the impact of other pathologies. However, the study also noted a subsequent decrease in LEH from the first through 12th Dynasties. This provides the first evidence for a recovery in health following the Neolithic, which appears to be associated with the formation of the Egyptian state. In this case, improvements in trade and infrastructure may have provided Dynastic populations with better and more consistent access to resources and greater cultural buffering from nutritional stress and disease.

At present, it is uncertain how the shift from hunting and gathering to a predominantly agricultural subsistence strategy may have altered the activity levels and mobility of human populations in Northeast Africa. Of particular relevance for understanding trends in activity levels and mobility through time are the size and shapes of the humeral and femoral diaphyses. Long bone diaphyseal shape and rigidity, most typically quantified through cross-sectional geometric properties, has frequently been used to interpret the habitual behaviour of prehistoric human populations (Larsen, 1995, 1997; Holt, 2003; Stock and Pfeiffer, 2004; Marchi *et al.*, 2006; Ruff, Holt and Trinkaus, 2006; Ruff *et al.*, 2006; Ruff, 2008 and see Larsen and Ruff, Marchi *et al.*, and Lieverse *et al.*, this volume). Activity level and mobility are fundamental behavioural characteristics of human populations and, based on both experimental and theoretical grounds, can be expected to engender the most repetitive forms of loading on the upper and lower limbs. These loading patterns should be reflected in the shape and rigidity of the humeral and femoral diaphyses (Ruff, 1987; Stock and Pfeiffer, 2001; Weiss, 2003; Holt, 2003; Marchi *et al.*, 2006; Ruff, Holt and Trinkaus, 2006; Ruff *et al.*, 2006; Stock, 2006; Shaw and Stock, 2009).

The most commonly calculated measures of long bone rigidity include the second moments of area (I_x, I_y, I_{min}, I_{max}) and polar moment of area (J) (Stock, 2002; O'Neill and Ruff, 2004). Indices of diaphyseal shape are commonly calculated as ratios of second moments of area (I_x/I_y, I_{min}/I_{max}). Collectively, these variables quantify the amount and distribution of cortical bone within the diaphysis, and have previously been used to investigate temporal trends in activity levels (upper limb) and mobility (lower limb) from other archaeological skeletal series (Holt, 2003; Larsen *et al.*, 1995; Ruff, 1987, 1994, 1999; Stock and Pfeiffer, 2004; Stock, 2006). While polar second moments of area (J) have been commonly used to estimate the biomechanical rigidity of long-bone diaphyses, indices of diaphyseal shape based on second moments of area (I_x/I_y or I_{min}/I_{max} ratios) have been most commonly applied to femoral midshafts to interpret mobility from archaeological skeletal series (Holt, 2003; Larsen *et al.*, 1995; Ruff, 1987, 1994, 1999; Stock and Pfeiffer, 2004; Stock, 2006). This is based on the theoretical prediction that repetitive antero-posterior loading of the lower limb associated with locomotion will favour structural hypertrophy along the antero-posterior plane, producing cross-sectional morphology with higher I_x/I_y ratios. While it is well known that this relationship is complex and influenced by other factors (Ruff, Holt and Trinkaus, 2006), recent evidence supports a general correspondence between lower limb diaphyseal shapes and habitual locomotion amongst human athletes (Shaw and Stock, 2009).

These biomechanical approaches have been applied to archaeological remains from a number of regions to investigate variation in habitual activity with the transition to agriculture. For example, changes in patterns of upper and lower limb structure have been previously documented across the transition from hunting and gathering to agriculture in the US (Larsen *et al.*, 2002; Ruff, 1984, 1987; Bridges, 1989), with some studies showing an increase in upper limb rigidity and a decrease in lower limb rigidity with adoption of a more sedentary lifestyle (Ruff and Larsen, 1990; Ruff, 1984, 1987). However, other studies have demonstrated opposite trends in some populations (Bridges, 1989; Marchi *et al.*, 2006; Marchi, 2008), suggesting that the biomechanical impact of the transition to agriculture on skeletal biomechanics is regionally specific (Ruff, 1999, 2008).

This study examines evidence for changes in human growth and habitual activity in the Nile Valley from the Late Pleistocene through to the Mid-Dynastic period (the Middle Kingdom) in the second half of the Holocene, using morphological variation in human skeletal remains from the sites of: Jebel Sahaba (Upper Palaeolithic, 13 000–9000 BC), el-Badari (Neolithic/Predynastic, 5000–4000 BC), Hierakonpolis (Predynastic, 4000–3000 BC) and Kerma (12th Dynasty, 2100–1500 BC). The skeletal remains from these sites (Figure 14.1) provide a range

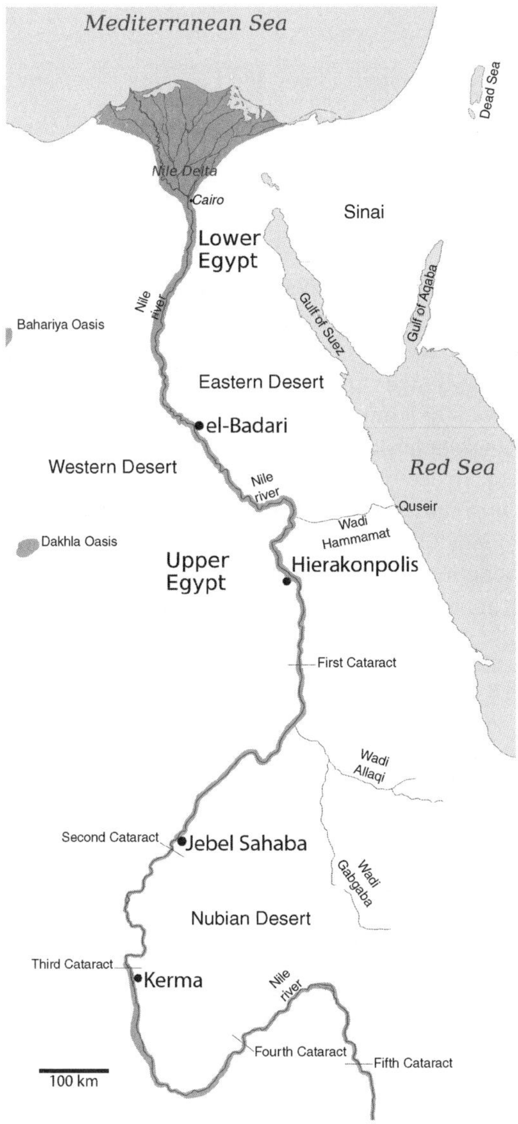

Figure 14.1 Map of the Nile River showing location of sites used in this study

of variation in the population history of the Nile Valley, but perhaps most importantly include two key contrasts, between:

1. the hunter-gatherers of the late Pleistocene (Jebel Sahaba) and the earliest Neolithic remains of the site of el-Badari; and
2. the el-Badari and somewhat later Hierakonpolis and Kerma sites, which span the Predynastic Period of the initial adoption and intensification of agriculture through the formation of the Egyptian State.

We use osteometric data to investigate long-term variation in several key morphological dimensions:

1. body size, as reflected by estimated stature and body mass;
2. upper limb strength, as reflected by humerus torsional rigidity (J) and shape index (I_x/I_y); and
3. lower limb strength, as reflected by femoral midshaft torsional rigidity (J) and shape index I_x/I_y.

The terms strength and rigidity have often been used interchangeably in the literature. Here, we use strength as a generic term to describe biomechanical characteristics of bone, while we refer to second moments of area as reflecting diaphyseal rigidity, thus rigidity is a specific quantification of strength. Body size variation in populations is often correlated with nutrition and stress (see other chapters by Ginter, Mieklejohn and Babb, Auerbach, and Temple, this volume). Based on previous studies and other lines of evidence (Zakrzewski, 2003; Starling and Stock, 2007), we predict a decrease in body size with the transition to agriculture in North eastern Africa, followed by a subsequent increase. Given the variation in biomechanical trends observed elsewhere, we predict a general increase in upper limb strength, and a decrease in lower limb diaphyseal strength and a trend towards greater circularity of the femoral diaphyses (i.e. an I_x/I_y index closer to 1.0) with the shift to agriculture and increased sedentism.

14.2 MATERIALS AND METHODS

The skeletal remains used in this study were derived from archaeological contexts along the Nile in Egypt and Nubia, dating between 13 000 and 1500 BC (Table 14.1; Figure 14.1). While two of these sites fall within present-day Egypt (el-Badari, Hierakonpolis), the others (Jebel

Table 14.1 Samples used in this study

Population	Time Period	♂	♀	Indet.	N
Jebel Sahaba (Late Palaeolithic)	13 000–9000 BC	11	8	0	19
el-Badari (Predynastic)	5000–4000 BC	3	5	2	10
Hierakonpolis	4000–3000 BC	32	47	7	86
Kerma (Dynastic)	2100–1500 BC	13	14	1	28
Total		59	74	10	143

Note: All dates are approximate, based primarily on Kemp (1989) and Zakrzewski (2003).

Sahaba, Kerma) come from the upper Nile in Sudanese Nubia. The latter samples are included here for several reasons:

1. the upper Nile in Nubia has always formed part of ecological zone of the Nile, and part of the associated cultural sphere through the development of the Egyptian empire;
2. the Jebel Sahaba remains are the best preserved late Palaeolithic hunter-gatherers from Northeast Africa, and hence represent our best means of quantifying morphological variation prior to the origins of the agriculture in the region; and
3. the remains from Kerma can be used to provide a preliminary characterization of habitual activity of one community following considerable development of the Egyptian empire.

Only skeletally mature adults, with all long bone epiphyses fused, were included in this analysis. Sex determination was made on the basis of sexually dimorphic characteristics of the pelvis wherever possible, using standard indicators. Sex could not be determined for a small subset of individuals ($N = 10$), so these were included in only general sample comparisons. Femoral lengths were used to calculate stature, on the basis of stature regression equations developed for Egyptians (Raxter *et al.*, 2008), and to compare general trends in body size through time. Femoral head diameters were used to estimate body mass, using the mean of three regression equations (Ruff *et al.*, 1997). Some specimens could not be included in the body size-standardized comparisons because of missing femoral heads, which led to some variation in sample and comparison sizes between analyses. The comparison of long-bone strength focuses on variation in the humeri and femora as an indicator of the mechanical adaptation in the upper and lower limbs. Cross-sectional properties were calculated at the mid-distal location of the humeral diaphyses (35% of length from the distal end), and at the midshaft of the femoral diaphyses. Cross-sectional contours were collected using a method that combines silicone moulds of the periosteal dimensions and bi-planar radiographs (Trinkaus and Ruff, 1989), and which has been demonstrated to provide accurate and consistent estimates of bone rigidity (Stock, 2002; O'Neill and Ruff, 2004).

Cross-sectional properties calculated include second moments of area, which represent the resistance of a bone to bending loads. Maximum (I_{max}) and minimum (I_{min}) second moments of area were calculated for each section, as well as second moments of area in the x and y planes (I_x, I_y). Circularity indices (I_{max}/I_{min}, I_x/I_y) were used to characterize the relative symmetry or asymmetry of rigidity in perpendicular planes. Finally, the polar moment of area (J) was calculated as a representation of torsional and (twice) average bending rigidity. Cross-sectional area measurements were standardized to body size using body mass estimates, while second moments of area were standardized using the product of body mass and moment arm (bone) length^2 (Ruff, 2008). Statistical comparisons were made using two-way ANOVAs with sex and site as factors; the Hochberg GT2 *post-hoc* test, which is well suited for use with unequal sample sizes, was used to test for pairwise, population-level differences. All individuals, including those of indeterminate sex, were used in the ANOVAs of population comparisons, while only those individuals of determined sex were included in sex-based analyses. While parametric statistics were used for quantitative comparisons between groups, the data were also plotted using box plots. This may be more appropriate for the smaller and variably-sized samples. In most cases, the medians in the box plots, and the resulting sample positions, correspond closely with the means presented in the tables.

14.3 RESULTS

The results of comparisons of estimated stature and body mass are found in Table 14.2. Overall, 63% of the variance in stature can be explained by the factors of population and sex. Much of this diversity is accounted for by a relatively consistent sexual dimorphism, accounting for 46% of variation in stature. In pairwise *post-hoc* comparisons between populations, Jebel Sahaba and Kerma had the greatest stature and can be viewed as a homogenous subset; however, these samples were significantly different from the el-Badari and Hierakonpolis results. These differences account for an overall trend of a reduction of stature from the earliest hunter-gatherers to the Neolithic/Predynastic, and a subsequent increase by the Middle Kingdom (Figure 14.2). The comparison of estimated body mass illustrates a very similar trend (Figure 14.3), with the highest values found amongst the Jebel Sahaba and Kerma samples. The strength of the model in this case is somewhat weaker, accounting for 41% of the variation in body mass (21% Sex, 9% Population), and while there were fewer significant pairwise contrasts, the Hierakonpolis sample was significantly different from both the Jebel Sahaba and Kerma samples.

The comparison of long-bone diaphyseal rigidity, represented by polar second moments of area of the humerus and femur, illustrates some interesting trends (Table 14.3; Figures 14.4 and 14.5). The Jebel Sahaba sample features the highest levels of diaphyseal rigidity amongst males, for both the humerus and femur. The few females of the Jebel Sahaba sample are more gracile, resulting in a high level of sexual dimorphism amongst these hunter-gatherers. *Post-hoc* comparisons demonstrate that Jebel Sahaba humeral rigidity was significantly higher than in the el-Badari, Hierakonpolis and Kerma samples, both in the combined sample and amongst the males. Femoral rigidity of the Jebel Sahaba males was also significantly higher than all other groups, while at the group level, the femora of Jebel Sahaba were significantly stronger than the Hierakonpolis and Kerma samples. In this context, sexual dimorphism seems to be generally higher amongst the Jebel Sahaba hunter-gatherers than the other groups; however, this trend is most pronounced for humeral strength and may be influenced by sampling bias due to the small size of the el-Badari sample. As such, it should be interpreted as a preliminary observation.

Overall, this pattern suggests a significant reduction in the intensity of mechanical loading amongst males, from the hunting and gathering Jebel Sahaba population to the earliest Neolithic. Temporal trends in male humeral strengths following the adoption of agriculture (el-Badari) were generally homogeneous. The pattern amongst women suggests that there was a general, but non-significant decrease in humeral rigidity from the Badarian period to the populations at Hierakonpolis and Kerma. In this case, a slight, non-significant increase in both humeral and femoral rigidity was found between the Jebel Sahaba and el-Badari groups, and increasingly lower values found amongst the Hierakonpolis and Kerma females.

The shapes of the humeral diaphyses, as represented by I_x/I_y ratios, showed interesting trends through time (Table 14.4; Figure 14.6). In general, humeral mid-distal diaphyseal shapes were highly variable, with means clustering between 0.95 (representing relatively wide diaphyses amongst the Jebel Sahaba males) and 1.13. In contrast the sex-pooled el-Badari sample had a mean circularity index of 1.22, illustrating that diaphyses have relatively greater rigidity in the antero-posterior dimension (I_x) than the other groups. However, this relationship only reached statistical significance in the pairwise comparison of the Jebel Sahaba and el-Badari samples, both amongst the males and when sexes were combined.

Table 14.2 Estimated stature and body mass

		Stature (cm)[a]			Sig *post-hoc* comparisons[b]	Body Mass (kg)[b]			Sig *post-hoc* comparisons[b]
		N	mean	s.d.		N	mean	s.d.	
Jebel Sahaba	Male	11	169.5	4.60		10	63.3	5.90	
	Female	8	158.7	3.97		4	55.6	6.93	
	All	19	164.9	6.91	el-Badari, Hierakonpolis	14	61.1	6.93	Hierakonpolis
el-Badari	Male	3	164.9	1.72		3	57.4	7.91	
	Female	5	152.5	0.94		5	53.2	1.87	
	All	8	157.1	6.49	Jebel Sahaba, Kerma	8	54.8	4.96	
Hierakonpolis	Male	32	165.1	5.15	Kerma	20	60.1	6.74	
	Female	47	156.0	4.60		26	52.5	5.18	
	All	79	159.7	6.58	Jebel Sahaba, Kerma	46	55.8	6.97	Jebel Sahaba, Kerma
Kerma	Male	13	170.8	4.95		13	64.7	5.91	
	Female	14	157.0	4.20		14	55.0	4.75	
	All	27	163.6	8.37	el-Badari, Hierakonpolis	27	59.7	7.20	Hierakonpolis

ANOVA	d.f.	F	Sig	Var
Model	1	29.845	0	0.626
Population	3	6.315	0.001	0.132
Sex	1	105.375	0	0.457
Pop × Sex	3	1.898	0.133	0.044

ANOVA	d.f.	F	Sig.	Var
Model	1	8.719	0	0.412
Population	3	2.993	0.035	0.094
Sex	1	23.29	0	0.211
Pop × Sex	3	0.503	0.681	0.017

[a] Stature estimated using the regression equation of Raxter *et al.* (2008).
[b] *Post-hoc* tests using Hochberg GT2 test, $\alpha = 0.05$.
[c] Body mass estimated using the mean of three regression equations found in Ruff *et al.* (1997).

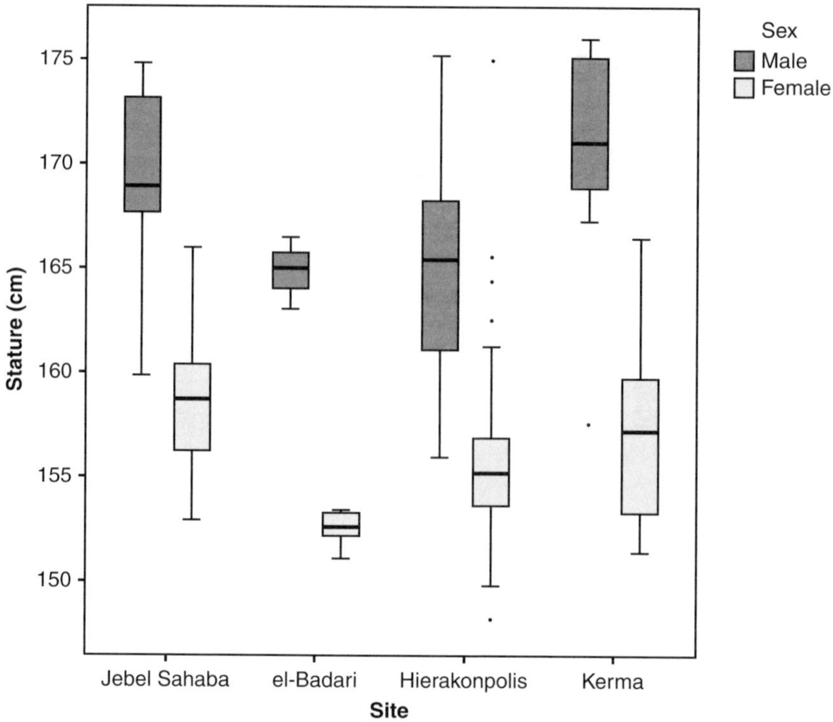

Figure 14.2 Variation in estimated stature

For the femur, there was considerable variation in midshaft shape amongst the males through time, which contrasts with a relative homogeneity amongst females (Table 14.4; Figure 14.7). The only pairwise comparison that achieved statistical significance was between the pooled-sex samples of el-Badari and Hierakonpolis. Some additional individuals of 'indeterminate' sex were included in this analysis, where shaft dimensions were available without body mass estimates or complete bone lengths.

14.4 DISCUSSION

In this study, skeletal measures of body size were analysed to evaluate the long-term impact of the transition to agriculture in the Nile Valley. It has previously been noted that the transition to agriculture in the Nile, Valley is associated with a deterioration and subsequent improvement in health, as reflected by a dramatic increase in the frequency of linear enamel hypoplasia between the Jebel Sahaba and el-Badari samples, followed by a reduction in frequencies in subsequent populations of the Nile, including the Kerma sample (Starling and Stock, 2007). Here, we demonstrate that this transition is also associated with a modest reduction and subsequent improvement in stature and body mass. This trend could be broadly interpreted in the context of models of a relationship between body size and nutrition. In this case, the greater body size of early hunter-gatherers may reflect the benefit of broadly-based hunting and gathering subsistence. With the onset of the Neolithic, the dietary diversity of hunter-gatherers

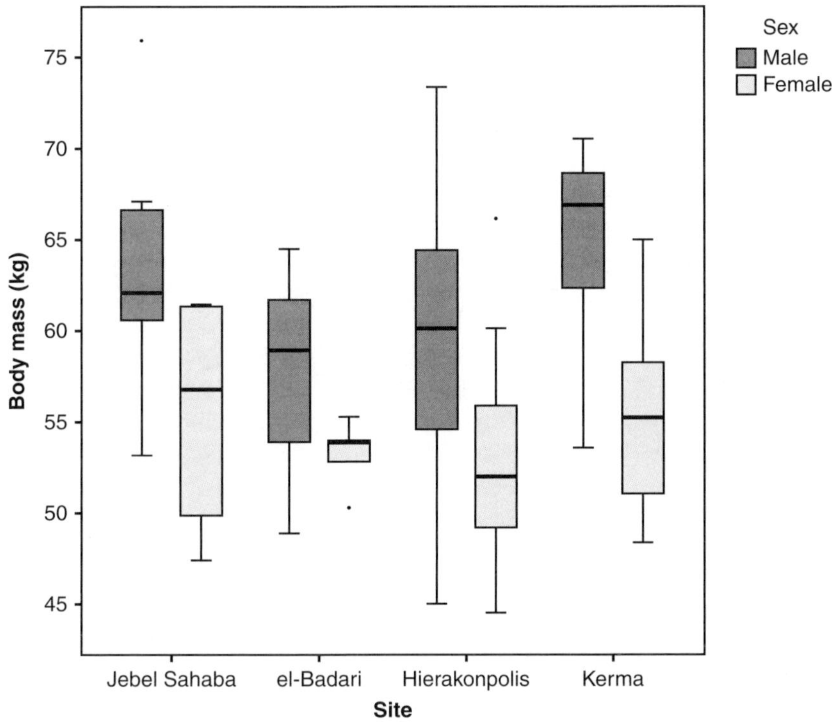

Figure 14.3 Variation in estimated body mass

is replaced with dietary specialization on one or a few cereal crops and the products of domestic animals. The potential nutritional implications of this are further compounded by the potential transmission of zoonotic diseases associated with living in close proximity to domestic animals, as well as related increases in population density and poor hygiene. Increasing sedentism and population density are almost universally associated with increases in infectious disease (Cohen, 1989; Steckel and Rose, 2002; Stuart-Macadam and Kent, 1992) and may underpin the reduction in stature in the Predynastic Period. Archaeological evidence suggests that the Badarian civilization had higher population density than did other contemporaneous civilizations (Gabriel, 1987; Hassan, 1988). The increase in stature found by the 12th Dynasty at Kerma suggests that the development of resources and infrastructure associated with the Egyptian empire may have had a positive influence on the health of the population. Of course, this general trend may be specific to the Kerma sample, whether due to direct cultural influence or population migration, as it is at odds with previous research showing a general decline in stature following the sixth Dynasty (about 2160 BC) in Egypt (Zakrzewski, 2003). It is also possible that the broader range of values and sexual dimorphism in the Kerma sample may reflect greater variation in social status in this period or differences in underlying genetic diversity.

Long-term trends in long-bone rigidity, as reflected by polar second moments of area of the mid-distal humerus and femoral midshaft, provide evidence of changes in habitual behaviour through time. The strength of the Jebel Sahaba male humeri and femora provide evidence of high levels of mechanical loading amongst the Late Pleistocene hunter-gatherers of the Nile.

Table 14.3 Humerus and femur polar second moments of area (J)

		Humerus 35%[a]			Sig *post-hoc* comparisons[b]	Femur 50%[a]			Sig *post-hoc* comparisons[b]
		N	mean	s.d.		N	mean	s.d.	
Jebel Sahaba	Male	7	225.4	39.3	el-Badari, Hierakonpolis, Kerma	7	460.9	58.8	el-Badari, Hierakonpolis, Kerma
	Female	2	142.1	49.1		2	366.9	30.2	
	All	9	206.9	53.0	el-Badari, Hierakonpolis, Kerma	9	440.0	66.5	Hierakonpolis, Kerma
el-Badari	Male	3	128.1	31.7	Jebel Sahaba	2	262.1	2.1	Jebel Sahaba
	Female	4	155.5	43.8		5	404.6	136.2	Hierakonpolis, Kerma
	All	7	143.7	38.8	Jebel Sahaba	7	362.9	131.2	Kerma
Hierakonpolis	Male	14	149.9	30.0	Jebel Sahaba	17	342.2	72.4	Jebel Sahaba
	Female	18	142.4	34.0		23	273.3	46.9	el-Badari
	All	32	145.6	32.0	Jebel Sahaba	40	302.6	67.7	Jebel Sahaba
Kerma	Male	11	147.5	24.6	Jebel Sahaba	12	316.7	38.7	Jebel Sahaba
	Female	12	114.0	30.0		13	253.2	39.2	el-Badari
	All	23	130.0	31.9	Jebel Sahaba	25	283.7	50.0	Jebel Sahaba, el-Badari

ANOVA	d.f.	F	Sig	Var	ANOVA	d.f.	F	Sig.	Var
Model	1	154.2	0.051	0.994	Model	1	11.96	0.020	0.999
Population	3	1.12	0.460	0.531	Population	3	7.96	0.389	0.588
Sex	1	1.89	0.248	0.342	Sex	1	1.21	0.611	0.078
Pop × Sex	3	3.90	0.013	0.156	Pop × Sex	3	5.59	0.002	0.164

[a] Standardized to body size, using the formula = J/(estimated body mass × bone length^2) × 100 000.
[b] *Post-hoc* tests using Hochberg GT2 test, $\alpha = 0.05$.

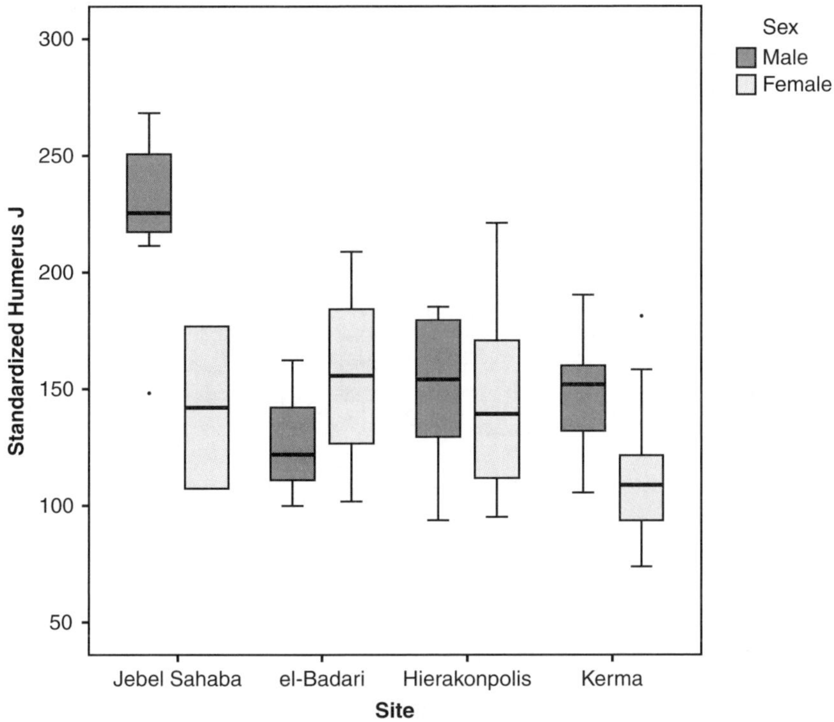

Figure 14.4 Body size standardized humerus mid-distal (35%) rigidity, *J*

The Jebel Sahaba sample exhibits the highest levels of sexual dimorphism in humeral rigidity, but morphological trends from the earliest Neolithic at el-Badari through to the 12th Dynasty at Kerma are variable. In general, humeral strength amongst males drops considerably between the Jebel Sahaba and el-Badari samples, but remains relatively consistent afterwards. Femoral midshaft rigidity shows a similar trend amongst males, where the Jebel Sahaba sample was found to be significantly more robust than the three other groups. This illustrates a reduction in femoral strength with the transition to agriculture amongst men, followed by relative homogeneity in diaphyseal strengths amongst men. The lower levels of humeral and femoral rigidity amongst the males of each of the agricultural samples suggest that there was a general reduction in habitual loading of the upper and lower limbs with the transition to agriculture in this region.

A different pattern of bone strength was found amongst the female subsamples. Humeral strength was relatively homogenous between the Jebel Sahaba and el-Badari females, followed by an incremental reduction in strength from the el-Badari to Kerma samples. While femoral strength amongst the females showed no significant differences between the Jebel Sahaba and Badari samples, a significant decrease in midshaft rigidity between el-Badari and both Hierakonpolis and Kerma samples is evident. The morphological pattern amongst women is suggestive of relatively consistent mechanical loading of the skeleton amongst the Jebel Sahaba hunter-gatherers and the Badarian Neolithic, followed by reductions in the intensity of habitual activity with the development of the Egyptian Empire.

In these analyses, the small sample sizes from el-Badari limit the extent to which we can interpret these trends. In particular, the low values for Badarian male femoral *Js* may be an

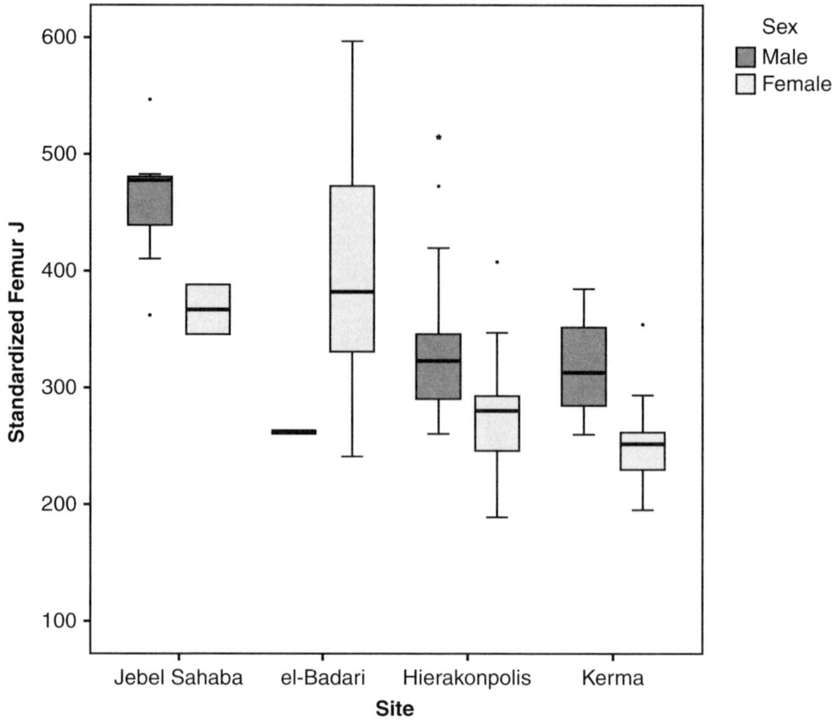

Figure 14.5 Body size standardized femoral midshaft (50%) rigidity, *J*

artefact of sampling (N = 2). The comparison of the humeral mid-distal shape suggests that there is a unique pattern of morphology amongst the el-Badari sample, with higher I_x/I_y values indicating greater rigidity of these diaphyses in the anterio-posterior plane. While our ability to interpret this trend is limited by the relatively small sample sizes, the differences are suggestive of a shift to mechanical loading that emphasized repetitive antero-posterior strains. A plausible explanation for this could be the use of groundstone technology in food preparation; however, it is possible that the shape of the humerus may be influenced by variation in physique or general biomechanical loading. Similarly high levels of humeral antero-posterior rigidity (I_x) have been identified in populations from Spanish Florida (Ruff and Larsen, 2001) and amongst female Later Stone Age foragers from Southern Africa (Stock and Pfeiffer, 2004). In the latter case, the morphological trend was interpreted as a possible correlate of digging stick use; however, a common trend of each of these studies is that they are a feature of relatively gracile skeletons. It is possible that lower humeral loading may lead to less developed flexor and extensor ridges, leading to lower values of medio-lateral rigidity (I_y).

In contrast, comparisons of femoral midshaft shape create a relatively complex, sexually-dimorphic pattern. While females showed very little variation in femoral shape across the time span of the samples compared here, the males showed considerable variation, both within and between samples. The highest values were found amongst the el-Badari males, which were significantly different from the low indices amongst the Hierakonpolis males. These could be interpreted as evidence for a reduction in antero-posterior loading between these groups, if we interpret these data using proposed and supported relationships between shape and

Table 14.4 Humerus and femur diaphyseal shape, I_x/I_y

		Humerus 35%			Sig *post-hoc* comparisons[a]			Femur 50%			Sig *post-hoc* comparisons[a]
		N	mean	s.d.				N	mean	s.d.	
Jebel Sahaba	Male	10	0.957	0.181				7	1.282	0.325	
	Female	9	1.132	0.255				8	1.126	0.240	
	All	19	1.039	0.231	el-Badari			15	1.199	0.284	
el-Badari	Male	3	1.262	0.036				2	1.556	0.469	
	Female	4	1.261	0.231				5	1.175	0.175	
	All	9	1.219	0.185	Jebel Sahaba			9	1.320	0.292	Hierakonpolis
Hierakonpolis	Male	29	1.010	0.111				17	1.049	0.280	
	Female	40	1.117	0.200				25	1.141	0.205	
	All	76	1.072	0.169				48	1.093	0.230	el-Badari
Kerma	Male	11	1.074	0.109				12	1.315	0.235	
	Female	12	1.112	0.068				13	1.071	0.162	
	All	24	1.092	0.088				26	1.197	0.233	
ANOVA	d.f.	F	Sig	Var		ANOVA	d.f.	F	Sig.	Var	
Model	1	901.72	0.000	0.997		Model	1	11.96	0.020	0.999	
Population	3	2.603	0.115	0.458		Population	3	7.96	0.389	0.588	
Sex	2	2.617	0.109	0.274		Sex	1	1.21	0.611	0.078	
Pop × Sex	5	0.851	0.851	0.035		Pop × Sex	3	5.59	0.002	0.164	

[a] *Post-hoc* significant comparisons using Hochberg GT2 test, $\alpha = 0.05$.

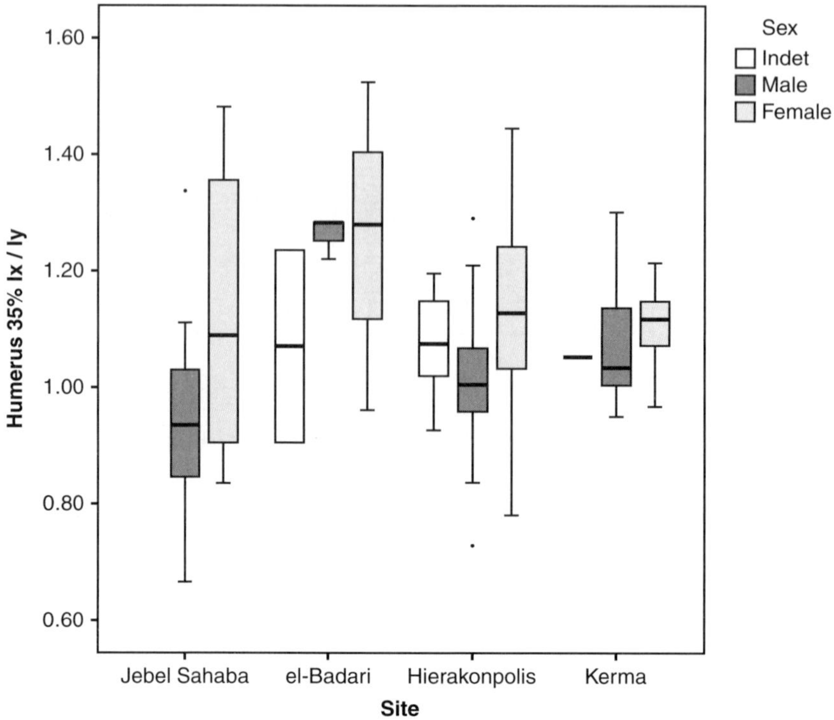

Figure 14.6 Humerus mid-distal (35%) shape, I_x/I_y

mobility (Ruff, 1987; Stock, 2006; Shaw and Stock, 2009). However, this interpretive model would predict that we would find the highest I_x/I_y ratios amongst the sample from Jebel Sahaba, and this is not the case. At present, the most likely explanation of this trend is potential sampling bias within the el-Badari sample. This is due to 1 male (el-Badari 11.5.15) having a shape index of 1.89, which is at the high end of the range found amongst the rest of our sample. Clearly, the small sample sizes of some groups may have been particularly relevant in interpreting the results of some comparisons, especially those involving males from el-Badari. The divergence of values between the two Badarian males suggests that sampling bias may be driving the values here, since there is no evidence of pathology in either individual. Femoral diaphyseal shape is known to be influenced by variation in physique as well as mobility (Ruff, 1995; Weaver, 2003; Ruff, Holt and Trinkaus, 2006; Ruff *et al.*, 2006; Stock, 2006), which may be a factor in explaining individual variation in these small samples, and may be driving some of the variation we see between samples, such as Hierakonpolis and Kerma.

The population from Kerma shows an unexpected increase in sexual dimorphism in body size and both humeral and femoral strength. This could be related to changes in nutrition and habitual behaviour, or may be a consequence of greater heterogeneity amongst this population due to migration along the Nile Valley (Thompson *et al.*, 2008). The increased biomechanical variation seen in the Kerma sample may be a reflection of this heterogeneity. The decrease and subsequent increase in body size and sexual dimorphism, and general decrease in diaphyseal strength observed in this study, may be interpreted in the context of the transition to agriculture and following development of the Egyptian state. However, they may also be influenced by

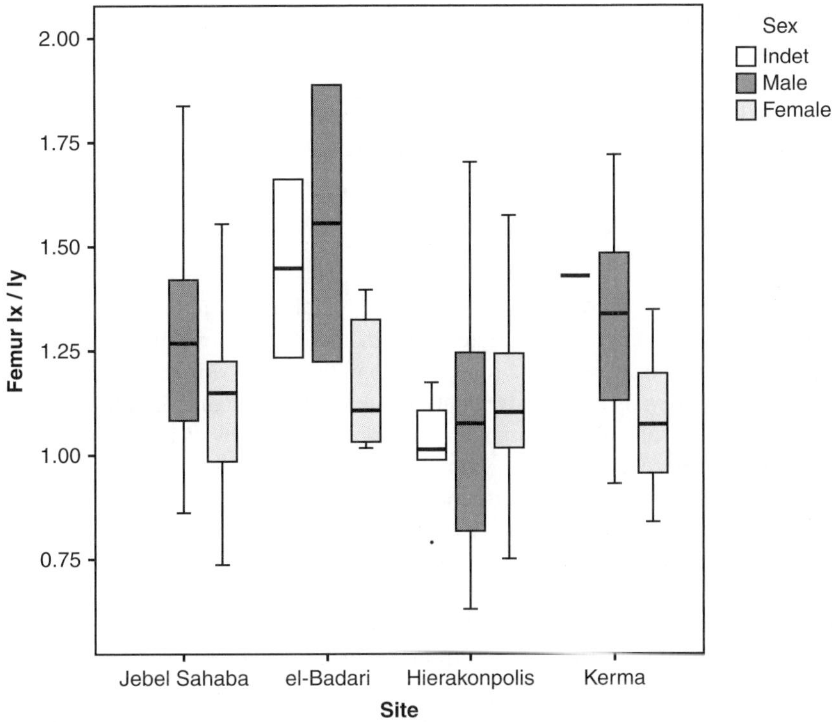

Figure 14.7 Femur midshaft (50%) shape, I_x/I_y

the development of social hierarchy. In particular, the relatively higher levels of sexual dimorphism in long-bone strength in the Jebel Sahaba sample, and body mass and humeral rigidity of the Kerma sample, may reflect a social hierarchy in relation to sex as well as the division of labour.

While the current study identifies interesting trends in body size and skeletal strength through time in the Nile Valley, the results should be considered as preliminary for a number of reasons. There is a fairly wide geographical range amongst the samples, and there were regional differences in the history of the transition to agriculture. In addition, while the samples share general characteristics of their regional ecology, they span nearly 10 000 years of history. The inclusion of both early and late samples from Nubia was a necessity of the current study, which may have some implications for interpretation. Most early work considered Upper and Lower Egyptians to be genetically-distinct populations; however, more recent analyses suggest that these populations are not sufficiently distinct to consider either non-indigenous (Zakrzewski, 2007). A craniometric study found the Kerma population to be morphologically similar to a Lower Egyptian Predynastic population (Keita, 1990). Thus, while they may have existed on the margins of the growing Egyptian empire, their inclusion in comparisons of earlier Nubian and Nile Valley Predynastic samples appears warranted.

The results presented here suggest a complex pattern of morphological change spanning 10 000 years of prehistory along the Nile River. A decrease and subsequent increase in body size is noted from the Late Palaeolithic to Neolithic and then Dynastic periods. A general reduction in bone strength is noted, but this occurs between the Late Palaeolithic and Neolithic for males, and between the Neolithic and Dynastic periods for females. The broad geographical and temporal span of the populations, when combined with the modest sample sizes at some sites, requires that these trends be interpreted as preliminary observations. Future research should investigate the body size and skeletal strength of other populations of the Late Palaeolithic, Neolithic and Dynastic to confirm the results reported here.

14.5 CONCLUSIONS

This study has identified broad trends in body size and long bone rigidity amongst hunter-gatherer, nomadic pastoralist and agricultural dynastic populations of the Nile Valley. These trends provide preliminary evidence for a decline in body size with the origins of agriculture, followed by a recovery, as previously identified using linear enamel hypoplasia (Starling and Stock, 2007). The results also identify a reduction in the levels of habitual activity with the transition to agriculture, although they emphasize differences in the timing of this transition amongst men and women. The reduction in male humeral and femoral strengths occurs between the Jebel Sahaba and el-Badari samples, with the earliest expression of the Neolithic culture. Female activity levels appear to have decreased later or not at all, with a modest non-significant reduction in humeral strength between the Hierakonpolis and Kerma samples, and a significant reduction in femur strength between the el-Badari and both Hierakonpolis and Kerma samples. In this context, the current analysis provides preliminary evidence for a complex pattern in the sexual division of labour through time, and different patterns of change amongst men and women. Mechanical loading of the skeleton appears to have been significantly reduced for males with the initial transition to agriculture, while female mechanical loading shows only modest decreases that occur only much later, with the formation and development of the Egyptian state. Future research should focus on further delineation of these trends in relation to social change in the Nile. This could be achieved through more finely resolved temporal and spatial analyses within the region. This could be achieved through detailed study of the temporal depth of Sudanese Nubian remains, or by adding further resolution to the skeletal series within Upper and Lower Egypt.

ACKNOWLEDGEMENTS

This research was made possible with financial support from the Leverhulme Trust, the Natural Environment Research Council, UK, and the Arts and Humanities Research Council, UK, Johns Hopkins University School of Medicine, National Science Foundation (#BCS-0119754, #BCS-0314002), Sigma Xi and Washington University.

REFERENCES

Arkell, A.J. and Ucko, P.J. (1965) Review of Predynastic development in the Nile Valley. *Curr. Anthropol.*, **6**(2), 145–166.

Armelagos, G.J. (1969) Disease in ancient Nubia. *Science*, **163**, 255–259.

Armelagos, G.J., Mielke, J.H., Owen, K.H. *et al.* (1972) Bone growth and development in prehistoric populations from Sudanese Nubia. *J. Human Evol.*, **1**, 89–119.

Bard, K.A. (2003) The emergence of the Egyptian State (c. 3200–2686 BC), in *The Oxford History of Ancient Egypt* (ed. I. Shaw), Oxford University Press, Oxford, pp. 57–82.

Raxter, M.H., Ruff, C.B., Azab, A. *et al.* (2008) Stature estimation in ancient Egyptians: A new technique based on anatomical reconstruction of stature. *Am. J. Phys. Anthropol.*, **136**(2), 147–155.

Bridges, P.S. (1989) Changes in activities with the shift to agriculture in the Southeastern United States. *Curr. Anthropol.*, **30**(3), 385–394.

Boserup, E. (1965) *The Conditions of Agricultural Growth*, George Allen and Urwin, London.

Childe, V.G. (1936) *Man Makes Himself*, Watts and Co., London.

Cohen, M.N. (1977) *The Food Crisis in Prehistory: Overpopulation and the Origins of Agriculture*, Yale University Press, New Haven.

Cohen, M.N. (1989) *Health and the Rise of Civilization*, Yale University Press, New Haven.

Diamond, J. (1999) *Guns, Germs, and Steel*, W.W. Norton & Co., New York.

Gabriel, B. (1987) Palaeoecological evidence from Neolithic fireplaces in the Sahara. *Afr. Archaeol. Rev.*, **5**, 93–103.

Hassan, F.A. (1984) Environment and subsistence in Predynastic Egypt, in *From Hunters to Farmers: The Causes and Consequences of Food Production in Africa* (eds J.D. Clark and S.A. Brandt), University of California Press, Berkeley, pp. 57–64.

Hanotte, O., Bradley, D.G., Ochieng, J.W. *et al.* (2002) African pastoralism: Genetic imprints of origins and migrations. *Science*, **296**, 336–339.

Hassan, F.A. (1988) The Predynastic of Egypt. *J. World Prehis.*, **2**(2), 137–185.

Holmes, D.L. (1993) Rise of the Nile delta. *Nature*, **363**, 402–403.

Holt, B.M. (2003) Mobility in upper Paleolithic and Mesolithic Europe: Evidence from the lower limb. *Am. J. Phys. Anthropol.*, **122**, 200–215.

Hummert, J.R. (1983) Cortical bone growth and dietary stress among subadults from Nubia's Batn El Hajar. *Am. J. Phys. Anthropol.*, **62**, 167–176.

Hummert, J.R. and Van Gerven, D.P. (1983) Skeletal growth in a medieval population from Sudanese Nubia. *Am. J. Phys. Anthropol.*, **60**, 471–478.

Irish, J.D. (2005) Population continuity vs. discontinuity revisited: Dental affinities among late Paleolithic through Christian-Era Nubians. *Am. J. Phys. Anthropol.*, **128**, 520–535.

Irish, J.D. (2006) Who were the ancient Egyptians? Dental affinities among Neolithic through Post-dynastic peoples. *Am. J. Phys. Anthropol.*, **129**, 529–543.

Jones, M. and Brown, T. (2000) Agriculture origins: the evidence of modern and ancient DNA. *The Holocene*, **10**(6), 769–776.

Kantor, H.J. (1965) The relative chronology of Egypt and its foreign correlations before the Late Bronze Age, in *Chronologies in Old World Archaeology* (ed. R.W. Ehrich), University of Chicago Press, Chicago, pp. 1–46.

Keita, S.O.Y. (1990) Studies of ancient crania from northern Africa. *Am. J. Phys. Anthropol.*, **83**, 35–48.

Kemp, B.J. (1989) *Ancient Egypt: Anatomy of a Civilization*, Routledge, London.

Larsen, C.S. (1995) Biological changes in human populations with agriculture. *Annu. Rev. Anthropol.*, **24**, 185–213.

Larsen, C.S., Ruff, C.B. and Kelly, R.L. (1995) Structural analysis of the Stillwater postcranial human remains: Behavioural implications of articular joint pathology and long bone diaphyseal morphology, in *Bioarchaeology of the Stillwater Marsh: Prehistoric Human Adaptation in the Western Great Basin*, (eds C.S. Larsen and R.L. Kelly), Anthropological Papers of the American Museum of Natural History, New York, pp. 107–133.

Larsen, C.S. (1997) *Bioarchaeology: Interpreting Behavior from the Human Skeleton*, Cambridge University Press, Cambridge.

Larsen, C.S., Crosby, A.W., Griffin, M.C. *et al.* (2002) A biohistory of health and behavior in the Georgia Bight: the agricultural transition and the impact of contact, in *The Backbone of History: Health and*

Nutrition in the Western Hemisphere (eds R.H. Steckel and J.C. Rose), Cambridge University Press, New York, pp. 406–439.
Malek, J. (2003) The Old Kingdom (c. 2686–2160 BC), in *The Oxford History of Ancient Egypt* (ed. I. Shaw), Oxford University Press, Oxford, pp. 83–107.
Marchi, D., Sparacello, V.S., Holt, B.M. and Formicola, V. (2006) A biomechanical approach to the reconstruction of activity patterns in Neolithic Western Liguria (Italy). *Am. J. Phys. Anthropol.*, **131**, 447–455.
Marchi, D. (2008) Relationships between lower limb cross-sectional geometry and mobility: the case of a Neolithic sample from Italy. *Am. J. Phys. Anthropol.*, **137**, 188–200.
Martin, D.L., Armelagos, G.J., Goodman, A.H. and Van Gerven, D.P., (1984) The effects of socioeconomic change in prehistoric Africa: Sudanese Nubia as a case study, in *Paleopathology at the Origins of Agriculture* (eds M.N. Cohen and G.J. Armelagos), Academic Press, Orlando, Florida, pp. 193–214.
O'Neill, M.C. and Ruff, C.B. (2004) Estimating human long bone cross-sectional geometric properties: a comparison of noninvasive methods. *J. Human Evol.*, **47**, 221–235.
Petrie, W.M.F. (1920) *Prehistoric Egypt*. British School of Archaeology and Egyptian Research Account Publication 31, London.
Petrie, W.M.F. (1939) *The Making of Egypt*, Sheldon Press, London.
Ruff, C.B. (1984) Allometry between length and cross-sectional dimensions of the femur and tibia in *Homo sapiens sapiens*. *Am. J. Phys. Anthropol.*, **65**, 347–358.
Ruff, C.B. (1987) Sexual dimorphism in human lower limb bone structure: relationship to subsistence strategy and sexual division of labour. *J. Human Evol.*, **16**, 391–416.
Ruff, C.B. (1994) Morphological adaptation to climate in modern and fossil hominids. *Yearb. Phys. Anthropol.*, **37**, 65–107.
Ruff, C.B. (1995) Biomechanics of the hip and birth in early *Homo*. *Am. J. Phys. Anthropol.*, **98**, 527–574.
Ruff, C.B. (1999) Skeletal structure and behavioral patterns of prehistoric Great Basin populations, in *Prehistoric Lifeways in the Great Basin Wetlands: Bioarchaeological Reconstruction and Interpretation* (eds B.E. Hemphill and C.S. Larsen), Univ. Utah Press, Salt Lake City, pp. 290–320.
Ruff, C.B., Trinkaus, E. and Holliday, T.W. (1997) Body mass and encephalization in Pleistocene Homo. *Nature*, **387**, 173–176.
Ruff, C.B. and Larsen, C.S. (1990) Postcranial biomechanical adaptations to subsistence strategy changes on the Georgia coast. *Anthro. Papers Amer. Mus. Nat. Hist.*, **68**, 94–120.
Ruff, C.B. and Larsen, C.S. (2001) Reconstructing behavior in Spanish Florida: The biomechanical evidence, in *Bioarchaeology of Spanish Florida: The Impact of Colonialism* (ed. C.S. Larsen), Univ. Press of Florida, Gainesville, pp. 113–145.
Ruff, C.B. (2008) Biomechanical analyses of archaeological human skeletal samples, in *Biological Anthropology of the Human Skeleton*, 2nd edn (eds M.A. Katzenburg and S.R. Saunders), John Wiley and Sons, Inc., New York, pp. 183–206.
Ruff, C.B., Holt, B. and Trinkaus, E. (2006) Who's afraid of the big bad Wolff?: 'Wolff's Law' and bone functional adaptation. *Am. J. Phys. Anthropol.*, **129**, 484–498.
Ruff, C.B., Holt, B.M., Sládek, V. *et al.* (2006) Body size, body shape, and long bone strength of the Tyrolean 'Iceman'. *J. Human Evol.*, **51**, 91–101.
Schillaci, M.A., Irish, J.D. and Wood, C.C.E. (2009) Further analysis of the population history of ancient Egyptians. *Am. J. Phys. Anthropol.*, **139**(2), 235–243.
Shackelford, L.L. (2007) Postcranial evidence for regional variation in the Late Pleistocene. *Am. J. Phys. Anthropol.*, **133**, 655–668.
Shaw, C.N. and Stock, J.T. (2009) Intensity, frequency, and directionality of habitual adolescent mobility patterns: Influences tibial diaphysis morphology in modern humans. *Am. J. Phys. Anthropol.*, **140**, 149–159.
Smith, B.D. (1995) *The Emergence of Agriculture*, Scientific American Library, New York.

Starling, A. and Stock, J.T. (2007) Dental indicators of health and stress in early Egyptian and Nubian agriculturalists: A difficult transition and gradual recovery. *Am. J. Phys. Anthropol.*, **134**, 28–38.

Steckel, R.H. and Rose, J.C. (2002) *The Backbone of History: Health and Nutrition in the Western Hemisphere*, Cambridge University Press, Cambridge.

Stock, J. (2002) A test of two methods of radiographically deriving long bone cross-sectional properties compared to direct sectioning of the diaphysis. *Int. J. Osteoarchaeol.*, **12**, 335–342.

Stock, J.T. (2006) Hunter-gatherer postcranial robusticity relative to patterns of mobility, climatic adaptation, and selection for tissue economy. *Am. J. Phys. Anthropol.*, **131**(2), 194–204.

Stock, J. and Pfeiffer, S. (2001) Linking structural variability in long bone diaphyses to habitual behaviors: Foragers from the southern African Later Stone Age and the Andaman Islands. *Am. J. Phys. Anthropol.*, **115**, 337–348.

Stock, J.T. and Pfeiffer, S.K. (2004) Long bone robusticity and subsistence behaviour among Later Stone Age foragers of the forest and fynbos biomes of South Africa. *J. Archaeol. Sci.*, **31**(7), 999–1013.

Thompson, A.H., Chaix, L. and Richards, M.P. (2008) Stable isotopes and diet at Ancient Kerma, Upper Nubia (Sudan). *J. Archaeol. Sci.*, **35**(2), 376–387.

Trinkaus, E. and Ruff, C. (1989) Diaphyseal cross-sectional morphology and biomechanics of the Fond-de-Foret 1 femur and the Spy 2 femur and tibia. *Bulletin de la Société Royale Belge d'Anthropologie et de Préhistoire*, **100**, 33–42.

Stuart-Macadam, P. and Kent, S. (1992) *Diet, Demography, and Disease: Changing Perspectives on Anemia*, Aldine de Gruyter, New York.

Vagn Nielsen, O. (1970) *The Nubian Skeleton Through 4,000 Years*, Andelsbogtrykkeriet i Odense, Copenhagen.

Van Gerven, D.P., Sandford, M.K. and Hummert, J.R. (1981) Mortality and culture change in Nubia's Batn el Hajar. *J. Human Evol.*, **10**, 395–408.

Warfe, A.R. (2003) Cultural origins of the Egyptian Neolithic and Predynastic: An evaluation of the evidence from the Dakhleh oasis (South Central Egypt). *Afr. Archaeol. Rev.*, **20**(4), 175–202.

Weaver, T.D. (2003) The shape of the Neandertal femur is primarily the consequence of a hyperpolar body form. *Proc. Natl. Acad. Sci. USA*, **100**, 6926–6929.

Wells, J.C.K. and Stock, J.T. (2007) The biology of the colonizing ape. *Yearb. Phys. Anthropol.*, **50**, 191–222.

Weiss, E. (2003) Effects of rowing on humeral strength. *Am. J. Phys. Anthropol.*, **121**(4), 293–302.

Wilkinson, T.A.H. (1999) *Early Dynastic Egypt*, Routledge, London.

Zakrzewski, S.R. (2002) Continuity and change: a biological history of ancient Egypt. University of Cambridge. (Unpublished PhD dissertation).

Zakrzewski, S.R. (2003) Variation in ancient Egyptian stature and body proportions. *Am. J. Phys. Anthropol.*, **121**(3), 219–229.

Zakrzewski, S.R. (2007) Population continuity or population change: Formation of the ancient Egyptian state. *Am. J. Phys. Anthropol.*, **132**, 501–509.

SECTION D

Archaeogenetics, Palaeodemography, Cranial and Dental Morphology

15

The Palaeopopulationgenetics of Humans, Cattle and Dairying in Neolithic Europe

Joachim Burger[1] and Mark G. Thomas[2]

[1] *Institut für Anthropologie, AG Palaeogenetik, Johannes Gutenberg-Universität, Germany*
[2] *Research Department of Genetics, Evolution and Environment, University College London, London, UK*

15.1 INTRODUCTION

The term 'Neolithic' most often refers to a package of features including permanent settlements, agriculture, pottery and animal husbandry. This definition is valid for an area between Western Anatolia and Central Europe, but approaches the limits of its utility in many other parts of the world. For example, pottery has been found, for example, in pre-Neolithic Africa, and signs of the 'Neolithic package' are found in Mesolithic cultures of the Baltic region. The term 'Neolithic' is also frequently used purely chronologically and so has been applied to parallel hunter-gatherer societies who are culturally 'Mesolithic', even if chronologically they live in a Neolithic phase. Such variability in definitions and nomenclature means that the term 'Neolithic' does not describe a universal and uniform process, a point that should be taken into consideration not only in archaeology, but also in studies of population biology.

Humans first started managing goat sheep, pig and cattle some 11 000 to 10 000 years ago in a region between the Levant, the Zagros mountains and Central Anatolia (Zeder, 2008). After 6500 calBC, archaeological signatures of a Neolithic package first appeared outside of the core region (Figure 15.1), in archaeological sites in southern and western Anatolia, and seem to have spread there from central Anatolia (Özdogan and Basgelen, 1999). By around 6400 calBC, a Neolithic package appeared in the Greek Aegean (monochrome phase in Thessaly and the Peleponnes and proto-Sesklo in other regions), and slightly later around the Marmara Sea, at the north-west tip of Anatolia and in Eastern Thracia.

Human Bioarchaeology of the Transition to Agriculture Edited by Ron Pinhasi and Jay T. Stock

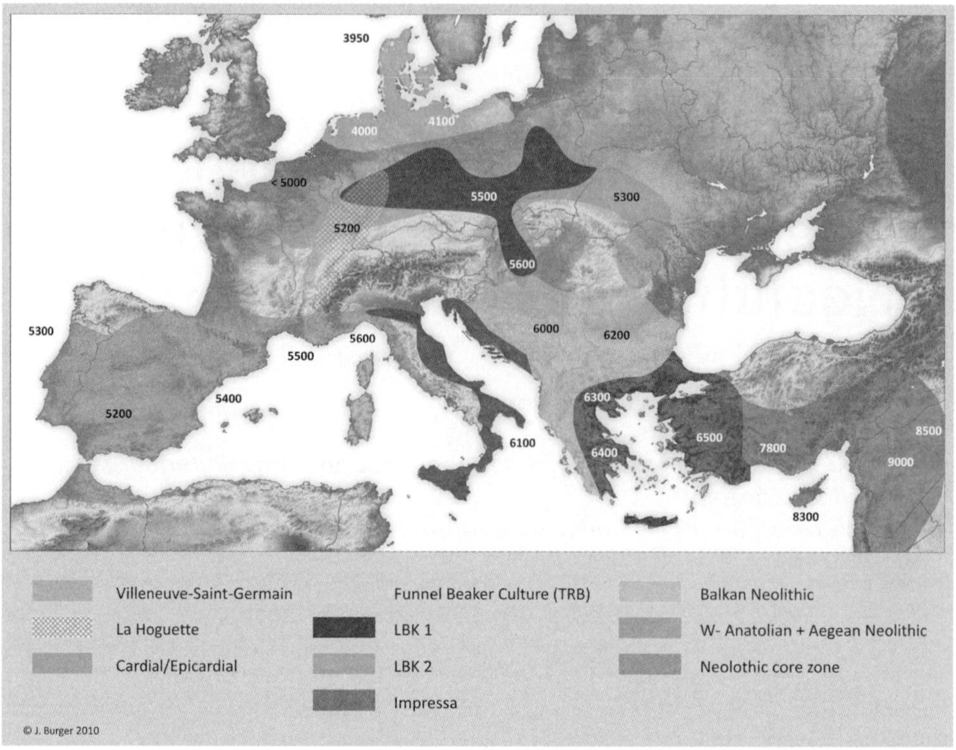

Figure 15.1 Chronological spread of the Neolithic. Numbers give the approximate earliest dates of the Neolithic in years before Christ (calBC). (See Plate 15.1 for a colour version of this image)

At this point, pottery style analysis indicates two culturally distinct but geographically overlapping trajectories for the spread of Neolithic culture in Europe: a so-called 'Mediterranean Route' and a 'Danubian/Balkan Route'. The former spread via southern Italy and the Adriatic (so-called Adriatic Impressa) to the islands of the Tuscan archipelago and later on to the south of France and the Iberian Peninsula (the so-called Cardial Ware culture). In the following, we focus on the so-called 'Danubian/Balkan Route', which is better studied from a palaeogenetic point of view.

After an initial phase of monochrome ceramics (proto-Sesklo and proto-Starčevo), a multitude of archaeologically defined cultures develop on the Balkan Peninsula and in the Carpathian Basin. These cultures can be broadly subdivided into the Karanovo I complex and the Starčevo-Körös-Çris complex. The Linearbandkeramik culture (LBK) spreads rapidly from what is believed to be its Starčevo forerunners in Central Europe from about 5500 calBC (Pavuk, 2005). Prior to 4100 calBC, Neolithic culture remains confined to these limits and only later expands to the North German lowlands and other regions of Northern Europe (Figure 15.1).

In this chapter, we compile the findings from the major palaeogenetic studies of the European Neolithic, gathered in recent years. We take into account studies of humans as well as of domestic animals, which could act as proxies for human population dynamics. First, we outline the differences between modern model based approaches and classical phylogeographic interpretation of palaeogenetic data in population history inference. We then summarize and discuss how palaeogenetics has been used to examine key questions on the demographic and evolutionary histories of Neolithic cattle and humans, including recent studies on the spread of lactase persistence.

15.2 THE AIM OF PALAEOGENETIC STUDIES IN THE CONTEXT OF THE NEOLITHIC TRANSITION IN EUROPE

Following the pioneering work of Luca Cavalli-Sforza in the 1970s and 1980s (Cavalli-Sforza *et al.*, 1986; Bowcock *et al.*, 1987), a series of studies have been performed over the last 20 years that attempt to infer human demographic history through patterns of genetic variability in modern populations. In the early days of research, classical genetic markers and later mitochondrial DNA (mtDNA) were the main contributors to this field, particularly through the interpretation of summary patterns of variation, such as Principle Components Analysis (Cavalli-Sforza, Menozzi and Piazza, 1996) and phylogeographic (Avise, 2000) analysis of hypervariable region I (HVR I) sequences (Richards *et al.*, 2000).

A central tenant of phylogeographic inference is that the nodes of major clades on an inferred phylogenetic tree, or a network, correlate with demographic events in time and space. However, this approach has been criticized (Goldstein and Chikhi, 2002; Nielsen and Beaumont, 2009, Beaumont *et al.*, 2010, but also see Templeton, 2009, 2010) and studies based on computer simulations (Knowles and Maddison, 2002; Panchal and Beaumont, 2007) have highlighted major problems with the more systematic incarnations of phylogeographic inference (Templeton, 1998). While phylogeographic inference has dominated palaeogenetics over the last 15 years, explicit model-based statistical inference approaches – particularly those employing coalescent theory (Griffiths and Tavare, 1994) – have developed apace in the field of population genetics. Coalescent theory permits the joining of probabilities of lineages (when viewed backwards through time) based on calculations of a range of different demographic conditions (e.g. population constancy, growth, etc.). These approaches emphasize the stochastic nature of the genealogical process in populations and recognize that very different 'gene trees' can arise from very similar demographic histories, and vice-versa. In other words, the gene tree and the demographic history of a population are, to a greater or lesser extent, decoupled and, in a sense, the gene tree can be thought of as a nuisance parameter. Lineage sorting – the phenomenon whereby lineages first coalesce not with lineages in the same population, but rather with lineages in a related population, illustrates the dangers of misinterpretation if the history of these two populations is derived from a phylogenetic tree/network. Similar problems arise when interpreting summary patterns of genetic variation, such as Principal Components Analysis (Cavalli-Sforza, Menozzi and Piazza, 1996). For example, different gradients of Principle Components in space have been interpreted as signatures of past migration events. Recent simulation studies have shown that components may form predictable patterns in space, but that these do not necessarily represent demographic episodes, if indeed they have any sensible interpretation at all (Novembre and Stephens, 2008).

The challenge in population genetic inference is to understand, in a statistical framework, what historical scenarios could have given rise to that tree/network, and in relation to the geographical location of samples. The solution to this problem is to explore (by simulation) different historical scenarios and search for the conditions under which the data, or some description of the data, has the highest probability of arising. This approach not only allows parameters of an evolutionary model to be estimated but also allows formal statistical comparison of different models of demographic history. Unfortunately this is not a trivial task, because appropriate models need to be used that are sufficiently complex to reflect the important processes that shape patterns of genetic variation, but are not so complex that they are difficult to implement and, when implemented, can explain any data. However, important developments such as Approximate Bayesian Computation (Beaumont, Zhang and Balding, 2002) make this task much easier as it allows estimation of parameters of a model, and comparison of different models, without the need to compute exact probabilities of data under specified models.

Because of problems of inference when using DNA data from modern individuals, as outlined above, the most direct way to examine demographic and evolutionary history is to obtain ancient DNA data directly from archaeological skeletal populations in the timeframe under study. This does not mean that the outlined inferential challenges disappear altogether. But for many of the questions archaeologists are interested in, such as population continuity vs. admixture, the task becomes somewhat easier. Accordingly, by applying palaeogenetic methods, archaeological hypotheses concerning the spread of people, raw materials, artefacts and pottery styles (and thereby of 'cultures') can be tested. Although palaeodemography is often taken as representing the biological analogue of culture, the extent of any correlation between genetic and material culture data may actually be widely different for different episodes in prehistory.

15.3 THE IMPORT AND MOVEMENT OF DOMESTIC ANIMALS

The early Neolithic of the Near East and Anatolia is characterized by the management of animals including goats, sheep, pigs and cattle, at the latest by 8500 calBC (Zeder, 2008). The transformation of wild forms of these four species into domestic animals is, alongside sedentism and crop (cereals and legumes) agriculture, a characteristic of early Neolithic cultures. This transformation was most likely a long process – taking at least two millennia – and appears to have taken place in the core regions of the Near East and Anatolia, long before the spread of agriculture into western Anatolia and southeast Europe. As both the morphological and the population-genetic consequences of the management of wild animals are unlikely to differ significantly from those of full domestication, uncertainty persists on where and when domestication in the full zoological sense actually took place. Foreseeably, palaeogenetic analyses of gene variants associated with the 'domesticated' phenotype will provide crucial information on this question in the future. The information we have so far is mainly from non-coding mitochondrial DNA, which does not tell us much about the phenotype of an animal but should contain information on its demographic history. The best investigated species in Neolithic times is cattle. Initial studies on the mitochondrial variability of present-day cattle in the Near East, Anatolia, North Africa and Europe showed that the variability in the vicinity of the postulated Neolithic domestication process is higher than that in Europe and/or North Africa (Troy *et al.*, 2001). The prevalence of T and T3 lineages in Europe and of T1 lineages in North Africa is consistent with the idea of an export of early domestic animals from the Neolithic core zone to Europe and North Africa. Nevertheless, this dataset can be interpreted in a variety of ways (e.g. the reduced variability in Europe could be the result of historical breeding practices). In order to further examine this question, our group has examined skeletal remains from the Neolithic period in central, north and south-east Europe.

They found that the genetic variability of today's cattle is similar to that of early Neolithic cattle (Bollongino *et al.*, 2006). But a deeper understanding of these results requires assessment of whether European domesticated cattle were introduced into Europe from the Near East or were descended wholly or in part from wild European cattle.

Morphologically, European wild cattle (*Bos primigenius*), otherwise known as *auroch*, differ from domestic cattle (*Bos taurus*), mainly in terms of the size of skeletal elements. This represents a methodical problem as individual male taurine and individual female aurochs have a significant overlap in size of all skeletal elements. Therefore, in order to be able to study known European aurochs, Pleistocene or Mesolithic Holocene specimens must be analysed. So far, all known aurochs identified carry mtDNA haplotypes that are classified phylogenetically as P or E. These lineages are distinct from those found in Anatolian and Middle Eastern taurine cattle (which are mostly T- and Q-types) (Scheu and Bollongino, unpublished data). As all known wild cattle in central and northern Europe carry P-types and to some extent E-types (Edwards *et al.*, 2007), the T- and Q-types have been proposed to originate from outside Europe, probably from early Neolithic domesticates in Anatolia or the Near East. Using coalescent simulations conditioned on both ancient and modern DNA sequences, it should be possible to estimate key parameters of the domestication history of these earliest cattle, such as the effective size of the population at the time of domestication.

Admixture between imported domestic cattle and indigenous European aurochs can, to a first-order approximation, be excluded in the case of central Europe (Bollongino *et al.*, 2006; Scheu *et al.*, 2008; Edwards *et al.*, 2007), but this remains a plausible scenario for other parts of Europe (Beja-Pereira *et al.*, 2006). In south-eastern Europe, most auroch bones from Early Neolithic sites are of relatively small-sized specimens and hence the differentiation between *Bos taurus* and *Bos primigenius* on the basis of archaeozoological criteria is problematic.

With respect to pigs, extensive ancient and modern mtDNA data have been analysed and interpreted in a phylogeographic framework to indicate multiple independent centres of domestication. (Larson *et al.*, 2005). We consider the latter scenario unlikely, and given that wild boar is very common in Europe, it is more likely that progressive admixture between local wild boar and imported (domesticated) Anatolian pigs is responsible for the pattern observed (Larson *et al.*, 2007). However, as with the other domestic species – sheep and goat – more extensive coalescent modelling studies need to be performed to fully and explicitly examine models of the spread of these animals in Europe. Thus, there is evidence that both pig and cattle have been imported to south-east Europe and then spread over the continent within three millennia, with pigs being mixed with local wild boar and cattle remaining maternally unmixed with local aurochs. With regards to sheep and goat, there is little doubt that they were imported from the Near East and/or Anatolia into Europe, as there are no wild progenitors of these species in Europe.

The fact that four different species were disseminated over such a long time and moved into such a large area previously free of agriculture, and also considering the difficulties of breeding and transport in a new ecological niche, the question arises as to which cultural and technical abilities were necessary to establish the basis for the spread of stock farming. Certainly, the rapid dissemination of all four species over the continent is unthinkable without favourable social and economic conditions and it is tempting to speculate that the Mesolithic-Neolithic transition in Europe involved some professional trade in domestic animals. Furthermore, it is clear that the four domestic species did not migrate without humans. But the extent to which humans moved into Europe continues to be debated (Barbujani, Bertorelle and Chikhi, 1998; Torroni *et al.*, 2000, Chikhi *et al.*, 2002, Dupanloup *et al.*, 2004; Belle, Landry and Barbujani, 2006). As mass migration of Neolithic farmers seems at first glance unlikely, the predominant view of continental archaeologists over the last few decades was that a few pioneers were responsible for the dissemination of Neolithic agriculture (Lüning, 2007) and that Neolithic economy and lifestyle were adopted by local Mesolithic hunter-gatherer populations (Kind, 1998; Gronenborn, 1997). As we discuss in the following section, recent palaeogenetic evidence challenges this traditional view, at least for Central and Northern Europe.

15.4 THE DEMOGRAPHY OF LATE HUNTER-GATHERERS AND EARLY FARMERS

Little is known about the palaeogenetics of late Glacial and early Holocene human populations in Europe. The first molecular genetic study of early LBK skeletons from Central Europe – all dating to between 5500 to 5000 calBC – did little to reveal the origin of Europe's first agriculturists, but did produce a surprising result. Although the LBK was the pioneer Neolithic farming culture of central Europe, their mitochondrial lineages appear to have been severely diluted over the subsequent millennia and some of them are rare in modern Europeans (Haak *et al.*, 2005). Two main explanations have been offered to account for this observation. The first, which is consistent with some archaeological theories (Lüning, 2007), is that the early LBK farmers were small groups of pioneers whose lineages were later replaced by surrounding hunter-gatherers. The second scenario is that these pioneer LBK farmers were to a large extent replaced by subsequent waves of farmers carrying different mtDNA lineages. However, currently neither of these scenarios receives support either from archaeology or palaeogenetics. In a more recent study and using coalescent simulations, Bramanti and colleagues (2009) further confirmed the observation that the LBK farmers were not the direct ancestors of modern Central Europeans, but to date, the missing continuity between the early farmers of central Europe and modern Europeans remains to be explained. It is likely that a combination of a series of demographic processes, which post-date the initial colonization of central Europe by the pioneer LBK farmers, is responsible for this lack of genetic continuity. Future analysis of palaeogenetic data from later periods, including spatially explicit serial modelling of heterochronous aDNA data, will allow us to assess the effects of these demographic processes in more detail.

As part of the same study, Bramanti and colleagues (2009) also examined mitochondrial DNA from skeletons of European hunters-gatherers from Upper Palaeolithic, Mesolithic and Neolithic periods. The individuals examined came from different locations in southwest Germany, northern Germany and Scandinavia. Specimens from various archaeological periods were considered as a single group of 'hunter-gatherers' because of their assumed similar subsistence strategies, even though it is not certain if they belong to any single population in a narrower biological sense. As a group, they represent the descendents of the hunters-gatherers that re-colonized central and northern Europe from southern European refugia after the end of the last Glacial maximum, around 20 000 BP. The following major climatic/environmental changes during deglaciation, and corresponding re-colonization of the biota in northern Europe, meant that large parts of this region were also re-occupied by human populations. These hunter-gatherers are probably the descendents of the first anatomically modern humans to settle in Europe around 45 000 years ago. The authors have shown that, in genetic terms, the LBK sample and the hunter-gatherer sample were indeed significantly different, as expressed by a comparatively high F_{ST} value (0.163). F_{ST} is a measure of genetic difference between populations and quantifies the proportion of genetic variation in a set of population samples that is attributable to differences between those samples. For comparison, the observed F_{ST} (0.163) between the two prehistoric groups was higher than the highest values found between similarly sized samples from a set of global populations ($F_{ST} = 0.133$). Serial coalescent simulations – that is coalescent simulations where a simulated lineage can be sampled at different points in time – were carried out in order to determine whether or not the difference between the LBK sample and the hunter-gatherer sample, as measured by F_{ST}, could be explained under the null-hypothesis of population continuity. Because genetic differences between populations in space and time are determined by a variety of factors, including ancestral population sizes, it was necessary to perform the simulations under a wide range of assumed combinations of population size at the start of the Neolithic and the Upper Palaeolithic colonization of Europe some 45 000 years ago. These simulations revealed no continuity scenarios in which a significant proportion of the simulated F_{ST} values were the same or higher than those actually observed. This allowed direct continuity between the hunter-gatherer sample and the LBK sample – the null-hypothesis – to be rejected. In another palaeogenetic study on the genetic contribution of late hunter-gatherers to modern Europeans, Malmström and colleagues (Malmström *et al.*, 2009) showed that the peoples of the last major hunter-gatherer complex in Europe, the Pitted Ware culture of southern Scandinavia, were unlikely to be the direct ancestors of modern Swedes, Norwegians or Saami. They did, however, show that direct continuity between the Pitted Ware peoples and those of the eastern Baltic region was possible. This does not mean that they actually were the ancestors of modern Eastern Baltic populations, as it can also – and more plausibly – be explained by common ancestry between the sampled populations. It is intriguing to note that the predominant lineage found in both the central European and the Scandinavia, the Pitted Ware hunter-gatherers' haplogroup U (Bramanti *et al.*, 2009; Malmström *et al.*, 2009) is the same as that observed in a single Upper Palaeolithic Russian burial from the site of Kostenkii 14 (Markina Gora), which is indirectly dated to approximately 30 000 year ago (Krause *et al.*, 2010), although in our opinion it is hard to exclude contamination in such a study (Pääbo *et al.*, 2004) and the pit contained no associated artefacts and hence without direct dating of the specimen, it is difficult possible that the pit is of a younger age (Pinhasi, personal communication).

The study by Bramanti and colleagues (2009) was not the first genetic study to address the question of hunter-gatherer contribution to early European farming populations or to modern Europeans (Barbujani, Sokal and Oden, 1995; Barbujani, Bertorelle and Chikhi, 1998; Chikhi *et al.*, 1998; Richards *et al.*, 2000; Semino *et al.*, 2000; Chikhi *et al.*, 2002; Dupanloup *et al.*, 2004; Currat and Excoffier, 2005). However, these previous studies were all based on modern genetic data – where inference requires many explicit, and in some cases implicit assumptions. Also, most were aimed at obtaining estimates of admixture between hunter-gatherers and incoming farmers, which were analysed by looking at genetic markers of modern Europeans, and using questionable proxies for ancestral source populations (specific lineages in the case of those studies based on phylogeographic inference, modern population samples in the case of those studies that used explicit modelling). Therefore, given the uncertainties of the inference process (see above), the null-hypothesis of population continuity remained to be formally rejected. Because Bramanti and colleagues (2009) sampled directly from ancient populations living around the time of the transition from hunter-gathering to farming, they – for the first time – were able formally reject continuity. It is likely that modern Europeans are a product of admixture between these two groups – Central European hunter-gatherers and

LBK farmers – and probably other ancestral source populations. But those other ancestral source populations remain to be identified and represented with ancient DNA data. Without such data, any estimates of admixture proportions in early farmers, late Neolithic populations and any subsequent European peoples up to the present day, will remain unreliable.

Since the first LBK farmers of central Europe were clearly not the direct descendents of local hunter-gatherers, they must have migrated in from another region. There are, as yet, no palaeogenetic data to indicate the most probable region of origin of the early LBK farmers. From an archaeological perspective, the most plausible region of origin is around the area of Lake Balaton in present-day Hungary (Pavuk, 2005; Figure 15.2), since this is the region (along with south-west Slovakia) where the LBK first developed around 5700 calBC from the predecessor Starčevo culture. From 5500 calBC until the end of its initial expansion phase around 5300 calBC, the LBK culture spread rapidly across central Europe, with settlements very soon reaching from the Rhine valley to east Poland. By around 5000 calBC, the LBK culture had expanded further, stretching from the Paris Basin through to the Ukraine. Since the LBK culture spread from a clearly restricted region, and because at least the early LBK farmers investigated were such highly mobile immigrants, sampling ancient skeletal material from the Lake Balaton core region will be essential to better understand the genetic origins of Central Europe's first farmers. Unfortunately, there is currently an almost complete absence of skeletons from the oldest phase of the LBK, that is between 5600 and 5400 calBC (Lüning, 2005).

If the scenario of a migration from the north-western part of the Balkans westwards to central Europe during the Neolithic could be examined using ancient DNA data, then the question that

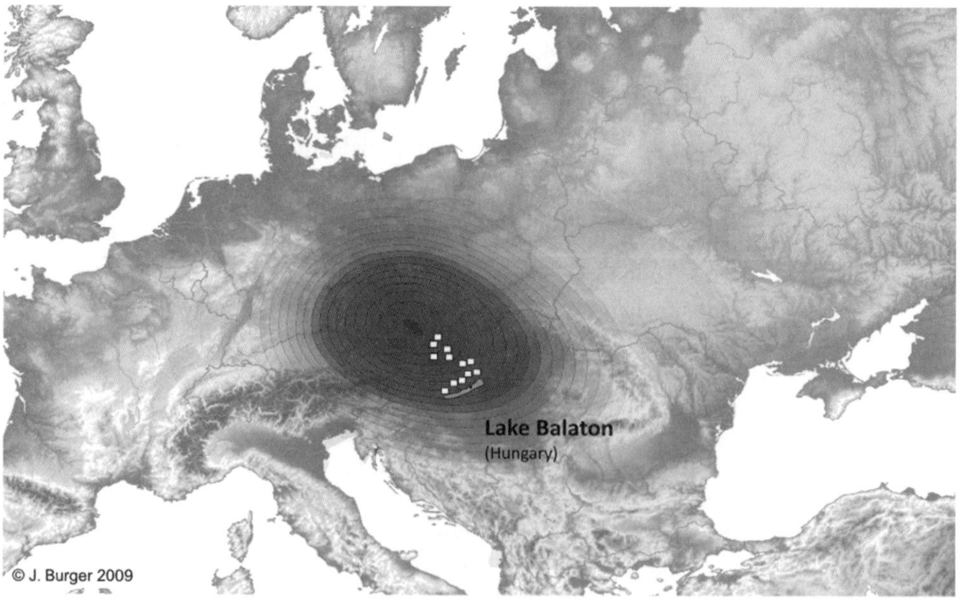

Figure 15.2 Earliest known LBK sites (5700–5500 calBC; white squares) north of Lake Balaton after Pavuk (2005) and geographical origin of selection of lactase persistence after Itan *et al.*, 2009 (about 4310–6730 calBC, concentric blue ellipses). (See Plate 15.2 for a colour version of this image)

immediately arises is whether it was a single event (perhaps of a mass migration) or perhaps only one of series of Neolithic expansions. It is possible that populations of the Starčevo-Körös-Cris complex were themselves the direct descendents of the earliest farmers to enter Europe, presumably from Western Anatolia or even from in the western foothills of the secondary Neolithic core zone in Central Anatolia. If we consider the new picture that is now emerging for Central Europe on the basis of palaeogenetic data from domestic animals and humans, then an ultimate Anatolian (or even Near-Eastern) origin no longer seems as implausible as it once was.

We also need to consider the vexing question of hunter-farmer contacts in Europe. It is. however, necessary to point out that cultural and economic contacts may or may not lead to biological admixture. There is clear archaeological evidence of cultural exchange between hunter-gatherers and central Europe's first farmers, as witnessed, for example, in the Mesolithic silex artefacts found in LBK settlements (Gronenborn, 1999). Because the exchange of underlying silex technologies (e.g. sourcing, tool production methods) primarily involved hunting weapons, it is not unreasonable to assume that contact was primarily mediated by men. So far, there has been no palaeogenetic evidence of any admixture at all, if only because ancient Y-chromosome data has yet to be generated. The analysis of ancient Y-chromosome data is more challenging than for mtDNA, because of the reduced survival of nuclear DNA over time, but recent developments in sequencing technologies (Mardis, 2008) do offer a solution to this problem. However, where data does exist, there appears to be little if any exchange of mitochondria in the early centuries following the Neolithic transition in central Europe. Perhaps the most plausible scenario is that two distinct societies existed during this early Neolithic period, which interacted with each other culturally, but maintained strict female marriage restrictions and fundamentally differed in their economic and cultural attributes. There are numerous present-day parallels for this kind, where ethnic groups live side by side but have different economic systems. To give one example, in Mali the semi-nomadic Fulbe are cattle breeders and pastoralists, the Dogon are cropping farmers but also keep some goats and sheep, and the Bozos are fishers of the Niger river. These three linguistically separated groups exchange the products of their work, for example the Fulbe exchange milk for millet with the Dogon but keep a strict marriage ban (Huysecom, personal communication). Although the precise nature of co-existence in Neolithic Europe will probably remain unknown, to infer the progressive degree of admixture between the two groups during the fourth and third millennia BC is an upcoming task of palaeogenetic research. Besides purely demographic phenomena and culturally defined barriers, selection of certain genetic traits is likely to have played an important role in the central European Neolithic transition. One such example is the rise of lactase persistence in Europe, a process on which we focus in the following section.

15.5 THE ROLE OF, AND ADAPTATIONS TO DAIRYING

Lactose is the main carbohydrate in milk and is a major energy source for most young mammals. The enzyme responsible for hydrolysis of lactose into glucose and galactose is lactase. Without this enzyme, mammals are unable to break down and thus utilize lactose. After the weaning period is over, lactase production usually declines. However, some humans continue to produce lactase throughout adult life, and are thus able to digest the lactose found in fresh milk; a trait which is called lactase persistence (LP). Genetic studies on modern populations have shown that a common variant of the lactase gene (LCT) known as –13 910*T, is strongly associated with, and may be causative of LP in Europe (Enattah *et al.*, 2002) and is one of the most strongly selected alleles in the human genome in Europeans in the last 30 000 years (Tishkoff *et al.*, 2007; Bersaglieri *et al.*, 2004; Coelho *et al.*, 2005). In Europe, clines are seen with LP found at frequencies of 10 to 30% in the south-eastern part of the continent, 50 to 0% in Central Europe, rising to 70 to 95% in north-western continental Europe and the British Isles. A similar trend is seen for the –13 910*T allele (Ingram *et al.*, 2009). Palaeogenetic typing of this variant site in 9 early and middle Neolithic skeletons revealed an absence of the –13 910*T allele in all cases (Burger *et al.*, 2007). Although this is a relatively small sample, statistical analysis demonstrated significantly lower frequencies of LP when compared to modern Europeans from the same region. This is consistent with previous studies, inferring high selection coefficients acting on this trait. It is unlikely that natural selection would have driven the –13 910*T allele to high frequencies without a supply of fresh milk and this ties the biological evolution of LP to the culture of dairying through a gene-culture co-evolution process. Clear evidence of milk production can be seen in South-east Europe as early as 6200 calBC, using lipid analysis on pot sherds (Evershed *et al.*, 2008). In order to better understand the co-evolution of LP and dairying in Europe, Itan and colleagues (Itan *et al.*, 2009) developed a demic computer simulation in order to examine how demographic and evolutionary parameters could have shaped both the modern distribution of LP in Europe and the timing of the arrival of farming at different locations throughout Europe. The study modelled the spread of dairying and non-dairying farmers into a Europe that was previously occupied by hunter-gatherers, under the plausible assumption that an LP-associated allele would only be selected in dairying farmers. Values for a number of different parameters must have shaped this process, including:

1. the extent of sporadic unidirectional migration;
2. the extent of gene flow between different cultural groups;
3. the extent of gene flow between neighbouring demes;
4. the extent to which people take up the culture of their neighbours;
5. the strength of selection favouring LP; and
6. the origin time and location of LP-dairying co-evolution.

By choosing random values for these parameters (within reasonable ranges), performing the simulation, and then comparing outcomes to observed data using Approximate Bayesian Computation (Beaumont, Zhang and Balding, 2002), the authors were able to identify parameter values that best explained the modern distribution of LP in Europe and the timing of the arrival of farming at different locations throughout Europe. Although the LP allele is most frequent in Northern Europe today, the simulations that best explained the observed data (on the distribution of LP and the arrival time of farming at different locations) required LP-dairying co-evolution to start in an area between the Carpathian Basin and Central Europe (Figure 15.2) between 6260 and 8680 years ago. The LP selection coefficient had inferred values between 0.0518 and 0.159 (in dairying farmers only). The inferred location and dates for the co-evolution of LP and dairying correspond well with the origins of the LBK culture in the Lake Balaton region. This again is in accordance with Bramanti *et al.* (2009) who inferred a massive immigration of the LBK farmers from this region.

15.6 CONCLUSIONS

By integrating evidence from ancient and modern DNA, archaeozoology and archaeology, we now envisage the following plausible scenario: whereas in the early Neolithic there are already archaeometric indications of milk usage in West Anatolia (Evershed *et al.*, 2008) and the Balkans (Craig *et al.*, 2005), these early farmers (who imported their cattle ultimately from the Neolithic Core zone in South-east Anatolia and the Northern Levant), were not able to consume significant quantities of fresh milk. Instead, they would have extracted the nutritional benefits of dairying by processing milk to make yoghurt and cheese. In this process much or most of the lactose in the milk is converted to fatty acids, rendering the product consumable by lactase non-persistent individuals.

Starting in early Neolithic south-eastern Central Europe, LBK people began to settle in Central Europe from 5500 calBC, without mixing significantly with the local hunters and gatherers. Although there are attested cultural contacts between the two groups, exchange of females must have been limited. An initially rare –13 910*T allele began to rise in frequency as fresh milk became more readily available and the ability to drink it became more advantageous. We are not yet sure why LP provided such a big advantage at the specific time. The fact that it takes longer to ferment milk in cold climates may be a factor, but some buffering of the food supply, when compared to the boom-and-bust of seasonal crops, is most likely to have been involved, especially amongst pioneering farmers in uncharted territories. Significantly, the reduced mortality of post-weaning period lactase persistent children would have had a major demographic effect over the centuries. Already by the Middle Neolithic this co-evolutionary process had resulted in a specialized dairying economy something like we see in sites of the Rössen culture (Benecke, 1994). From 4100 calBC, the Neolithic spread across the central low mountain regions to the North German lowlands and Northern Europe, would have been aided by the constancy of milk supply. Accordingly, a further rising in the frequency of the LP occurred, driven in part by selection and in part by the process of allele surfing (Edmonds, Lillie and Cavalli-Sforza, 2004; Klopfstein, Currat and Excoffier, 2006). Following further demographic expansion, the farmers that reached the north-western reaches of Europe were now predominately LP dairyers.

Such scenarios are easy to envisage and eminently plausible. But to develop any real confidence in them, it is necessary to show by quantitative simulation modelling that the relevant processes fit well with ancient and modern DNA data, and also with the archaeological data. Only by this route, we believe, can the different data sources satisfactorily be integrated into a true understanding of the human past.

ACKNOWLEDGEMENT

We thank Jens Lüning, Norbert Benecke, Pascale Gerbault, Adam Powell and Yuval Itan for useful comments, Bernhard Weninger for critical reading of the manuscript. MT was supported by the AHRC Centre for the Evolution of Cultural Diversity.

REFERENCES

Avise, J.C. (2000) *Phylogeography. The History and Formation of Species*, Harvard University Press, Cambridge, London.

Barbujani, G., Bertorelle, G. and Chikhi, L. (1998) Evidence for Paleolithic and Neolithic gene flow in Europe. *Am. J. Hum. Genet.*, **62**, 488–492.

Barbujani, G., Sokal, R.R. and Oden, N.L. (1995) Indo-European origins: a computer-simulation test of five hypotheses. *Am. J. Phys. Anthropol.*, **96**, 109–132.

Beaumont, M.A., Nielsen, R., Robert, C. *et al.* (2010) In defence of model-based inference in phylogeography. *Mol. Ecol.*, **19**, 436–446.

Beaumont, M.A., Zhang, W. and Balding, D.J. (2002) Approximate Bayesian computation in population genetics. *Genetics*, **162**, 2025–2035.

Beja-Pereira, A., Caramelli, D., Lalueza-Fox, C. *et al.* (2006) The origin of European cattle: evidence from modern and ancient DNA. *Proc. Natl. Acad. Sci. U. S. A.*, **103**, 8113–8118.

Belle, E.M., Landry, P.A. and Barbujani, G. (2006) Origins and evolution of the Europeans' genome: evidence from multiple microsatellite loci. *Proc. Biol. Sci.*, **273**, 1595–1602.

Benecke, N. (1994) *Der Mensch und seine Haustiere. Die Geschichte Einer Jahrtausendealten Beziehung*, Konrad Theiss Verlag, Stuttgart.

Bersaglieri, T., Sabeti, P.C., Patterson, N. *et al.* (2004) Genetic signatures of strong recent positive selection at the lactase gene. *Am. J. Hum. Genet.*, **74**, 1111–1120.

Bollongino, R., Edwards, C.J., Alt, K.W. *et al.* (2006) Early history of European domestic cattle as revealed by ancient DNA. *Biol. Lett.*, **2**, 155–159.

Bowcock, A.M., Bucci, C., Hebert, J.M. *et al.* (1987) Study of 47 DNA markers in five populations from four continents. *Gene Geogr.*, **1**, 47–64.

Bramanti, B., Thomas, M.G., Haak, W. *et al.* (2009) Genetic discontinuity between local hunter-gatherers and central Europe's first farmers. *Science*, **326**, 137–140.

Burger, J., Kirchner, M., Bramanti, B. *et al.* (2007) Absence of the lactase-persistence-associated allele in early Neolithic Europeans. *Proc. Natl. Acad. Sci. U. S. A.*, **104**, 3736–3741.

Cavalli-Sforza, L.L., Kidd, J.R., Kidd, K.K. *et al.* (1986) DNA markers and genetic variation in the human species. *Cold Spring Harb. Symp. Quant. Biol.*, **51** (Pt 1), 411–417.

Cavalli-Sforza, L.L., Menozzi, P. and Piazza, A. (1996) *The History and Geography of Human Genes*, Princeton University Press, Princeton.

Chikhi, L., Destro-Bisol, G., Bertorelle, G. *et al.* (1998) Clines of nuclear DNA markers suggest a largely Neolithic ancestry of the European gene pool. *Proc. Natl. Acad. Sci. U. S. A.*, **95**, 9053–9058.

Chikhi, L., Nichols, R.A., Barbujani, G. and Beaumont, M.A. (2002) Y genetic data support the Neolithic demic diffusion model. *Proc. Natl. Acad. Sci. U. S. A.*, **99**, 11008–11013.

Coelho, M., Luiselli, D., Bertorelle, G. *et al.* (2005) Microsatellite variation and evolution of human lactase persistence. *Hum. Genet.*, **117**, 329–339.

Craig, O.E., Chapman, J., Heron, C. *et al.* (2005) Did the first farmers of central and eastern Europe produce dairy foods? *Antiquity*, **79**, 882–894.

Currat, M. and Excoffier, L. (2005) The effect of the Neolithic expansion on European molecular diversity. *Proc. Biol. Sci.*, **272**, 679–688.

Dupanloup, I., Bertorelle, G., Chikhi, L. and Barbujani, G. (2004) Estimating the impact of prehistoric admixture on the genome of Europeans. *Mol. Biol. Evol.*, **21**, 1361–1372.

Edmonds, C.A., Lillie, A.S. and Cavalli-Sforza, L.L. (2004) Mutations arising in the wave front of an expanding population. *Proc. Natl. Acad. Sci. U.S.A.*, **101**, 975–979.

Edwards, C.J., Bollongino, R., Scheu, A. *et al.* (2007) Mitochondrial DNA analysis shows a Near Eastern Neolithic origin for domestic cattle and no indication of domestication of European aurochs. *Proc. Biol. Sci.*, **274**, 1377–1385.

Enattah, N.S., Sahi, T., Savilahti, E. *et al.* (2002) Identification of a variant associated with adult-type hypolactasia. *Nat. Genet.*, **30**, 233–237.

Evershed, R.P., Payne, S., Sherratt, A.G. *et al.* (2008) Earliest date for milk use in the Near East and southeastern Europe linked to cattle herding. *Nature*, **455**, 528–531.

Goldstein, D.B. and Chikhi, L. (2002) Human migrations and population structure: what we know and why it matters. *Annu. Rev. Genomics Hum. Genet.*, **3**, 129–152.

Griffiths, R.C. and Tavare, S. (1994) Sampling theory for neutral alleles in a varying environment. *Philos. Trans. R. Soc. Lond. B Biol. Sci.*, **344**, 403–410.

Gronenborn, D. (1999) A variation on a basic theme: the transition to farming in southern central Europe. *J. World Prehistory*, **13**, 123–210.

Gronenborn, D. (1997) *Silexartefakte der ältestbandkeramischen Kultur. Mit einem Beitrag von Jean-Paul Caspar*, Universitätsforschungen zur prähistorischen Archäologie, 37, Habelt Bonn.

Haak, W., Forster, P., Bramanti, B. *et al.* (2005) Ancient DNA from the first European farmers in 7500-year-old Neolithic sites. *Science*, **310**, 1016–1018.

Ingram, C.J., Mulcare, C.A., Itan, Y. *et al.* (2009) Lactose digestion and the evolutionary genetics of lactase persistence. *Hum. Genet.*, **124**, 579–591.

Itan, Y., Powell, A., Beaumont, M.A. *et al.* (2009) The origins of lactase persistence in Europe. *PLoS Comput. Biol.*, **5**, e1000491.

Kind, C.-J. (1998) Komplexe Wildbeuter und frühe Ackerbauern. Bemerkungen zur Ausbreitung der Linearbandkeramik im südlichen Mitteleuropa. *Germania*, **76/1**, 1–24.

Klopfstein, S., Currat, M. and Excoffier, L. (2006) The fate of mutations surfing on the wave of a range expansion. *Mol. Biol. Evol.*, **23**, 482–490.

Knowles, L.L. and Maddison, W.P. (2002) Statistical phylogeography. *Mol. Ecol.*, **11**, 2623–2635.

Krause, J., Briggs, A.W., Kircher, M. *et al.* (2010) A complete mtDNA genome of an early modern human from Kostenki. *Russia. Curr. Biol.*, **20**(3), 231–236.

Larson, G., Dobney, K., Albarella, U. *et al.* (2005) Worldwide phylogeography of wild boar reveals multiple centers of pig domestication. *Science*, **307**, 1618–1621.

Larson, G., Albarella, U., Dobney, K. *et al.* (2007) Ancient DNA, pig domestication, and the spread of the Neolithic into Europe. *Proc. Natl. Acad. Sci. U. S. A.*, **104**, 15276–15281.

Lüning, J. (2005) Bandkeramische Hofplätze und absolute Chronologie der Bandkeramik. (eds J. Lüning, C. Friedrich and A. Zimmermann), Die Bandkeramik im 21. Jahrhundert: Symposium in der Abteil Brauweiler bei Köln, pp. 49–74.

Lüning, J. (2007) Bandkeramiker und Vor-Bandkeramiker – Die Entstehung des Neolithikums in Mitteleuropa, in *Vor 12 000 Jahren in Anatolien – Die ältesten Monumente der Menschheit* (ed. Badisches Landesmuseum Karlsruhe), Konrad Theiss Verlag, Stuttgart, pp. 177–189.

Malmström, H., Gilbert, M.T., Thomas, M.G. *et al.* (2009) Ancient DNA reveals lack of continuity between Neolithic hunter-gatherers and contemporary Scandinavians. *Curr. Biol.*, **19**, 1758–1762.

Mardis, E.R. (2008) The impact of next-generation sequencing technology on genetics. *Trends Genet.*, **24**, 133–145.

Nielsen, R. and Beaumont, M.A. (2009) Statistical inferences in phylogeography. *Mol. Ecol.*, **18**, 1034–1047.

Novembre, J. and Stephens, M. (2008) Interpreting principal component analyses of spatial population genetic variation. *Nat. Genet.*, **40**, 646–649.

Özdogan, M. and Basgelen, N. (1999) *Neolithic in Turkey. The Cradle of Civilization*, Ancient Anatolians Civilizations Series 3, Istanbul.

Pääbo, S., Poinar, H., Serre, D. *et al.* (2004) Genetic analyses from ancient DNA. *Annu. Rev. Genet.*, **38**, 645–679.

Panchal, M. and Beaumont, M.A. (2007) The automation and evaluation of nested clade phylogeographic analysis. *Evolution*, **61**, 1466–1480.

Pavuk, J. (2005) Typologische Geschichte der Linearbandkeramik. (eds J. Lüning, C. Friedrich, A. Zimmermann), Die Bandkeramik im 21. Jahrhundert: Symposium in der Abteil Brauweiler bei Köln, pp. 17–39.

Richards, M., Macaulay, V., Hickey, E. *et al.* (2000) Tracing European founder lineages in the Near Eastern mtDNA pool. *Am. J. Hum. Genet.*, **67**, 1251–1276.

Scheu, A., Hartz, S., Schmölcke, U. *et al.* (2008) Ancient DNA provides no evidence for independent domestication of cattle in Mesolithic Rosenhof, Northern Germany. *J. Arch. Sci.*, **35**, 1257–1264.

Semino, O., Passarino, G., Quintana-Murci, L. *et al.* (2000) MtDNA and Y-chromosome polymorphisms in Hungary: inferences from the Palaeolithic, Neolithic and Uralic influences on the modern Hungarian gene pool. *Eur. J. Hum. Genet.*, **8**, 339–346.

Templeton, A.R. (2009) Why does a method that fails continue to be used: the answer. *Evolution*, **63**, 807–812.

Templeton, A.R. (2010) Coalescent-based, maximum likelihood inference in phylogeography. *Mol. Ecol. Early View*. doi:10.1111/j.1365-294X.2009.04514.x.

Templeton, A.R. (1998) Nested clade analyses of phylogeographic data: testing hypotheses about gene flow and population history. *Mol. Ecol.*, **7**, 381–397.

Tishkoff, S.A., Reed, F.A., Ranciaro, A. *et al.* (2007) Convergent adaptation of human lactase persistence in Africa and Europe. *Nat. Genet.*, **39**, 31–40.

Torroni, A., Richards, M., Macaulay, V. *et al.* (2000) mtDNA Haplogroups and Frequency Patterns in Europe. *Am. J. Hum. Genet.*, **66**, 1173–1177.

Troy, C.S., MacHugh, D.E., Bailey, J.F. *et al.* (2001) Genetic evidence for Near-Eastern origins of European cattle. *Nature*, **410**, 1088–1091.

Zeder, M. (2008) Domestication and early agriculture in the Mediterranean basin: origins, diffusion, and impact. *Proc. Natl. Acad. Sci. USA*, **105**, 11597–11604.

16

The Genetics of the Neolithic Transition: New Light on Differences Between Hunter-Gatherers and Farmers in Southern Sweden

Anna Linderholm

The Archaeological Research Laboratory, Stockholm University, Stockholm, Sweden

16.1 INTRODUCTION

During the last 150 years, many theories have been proposed to explain the nature and timing of the Neolithic transition in Europe. Considerable attention has been placed on the analysis of changes in economy, diet and lifestyle that characterized this transition. Southern Scandinavia is one of the 'frontier' regions in which the transition occurred approximately two millennia after the appearance of the first Neolithic settlements in south-east Europe (cf. Price, 2000). There are several competing hypotheses on the nature of the Neolithization process in Southern Sweden and neighbouring regions. One theory is that in this region, farming was adopted as a response to environmental changes (Zvelebi and Rowley-Conwy, 1984), while another theory proposes that the Neolithization process was followed by a return to hunter-gatherer subsistence, during the middle Neolithic (Carlsson, 1998).

Agriculture was introduced to Scandinavia by the Funnel Beaker Culture (TRB after the German Trichterrandbecherkultur) and the Swedish TRB was the most northern outpost of this culture in Scandinavia (Malmer, 2002). A contemporaneous hunter-gatherer population – the Pitted Ware culture (PWC) – existed in Southern Sweden (Malmer, 2002; Eriksson *et al.*, 2008).

Genetic research on the origin of the Europeans began in the late 1970s by Cavalli-Sforza and colleagues and was originally based on non-molecular (or 'classical') markers (Cavalli-

Human Bioarchaeology of the Transition to Agriculture Edited by Ron Pinhasi and Jay T. Stock

Sforza, 1994). Several decades later, there is still no consensus amongst archaeologists/geneticists regarding the extent to which the transition to agriculture in Europe involved the indigenous transition of local hunter-gatherer populations who adopted farming, the arrival of exogenous farmers from the Near East/Anatolia, or a combination of both. Analysis of modern European populations using non-molecular and molecular markers has yielded variable and often contrasting results (Cavalli-Sforza, 1994; Chikhi *et al.*, 1998; Richards and Macaulay, 2000; Underhill *et al.*,2000; Torroni *et al.*, 2001; Richards *et al.*, 2002; Tambets *et al.*, 2004; Pereira *et al.*, 2005). These studies are all based on the assumption that it is possible to detect and reconstruct the nature and timing of past demographic processes by studying genetic variation amongst modern Europeans. This assumption is problematic since, as will be discussed below, it appears that in at least some regions the genetic 'signal' of Neolithic farmers left no traces, probably due to post-Neolithic demographic processes (e.g. major population migrations from Central Asia).

In recent years. a small number of studies have examined mitochondrial DNA of the actual Neolithic and pre-Neolithic human skeletons from Europe (see Burger and Thomas, this volume). Advances in ancient DNA (aDNA) methods and next generation sequencing have now paved the way for new studies, which can directly assess the genetic structure of past European populations. Furthermore, the ability to combine ancient DNA data with other archaeological data offers the potential to shed new light on past societies. This chapter will focus on genetic, isotopic and archaeological studies of Mesolithic/Neolithic populations from Southern Sweden, with a particular focus on three aDNA case studies that illustrate the potential and limitations in the application of such methods to the analysis of archaeological populations.

16.2 THE NEOLITHIC TRANSITION IN SWEDEN

The Mesolithic marks the establishment of human residence in Sweden and it stretches from 8200 BC until the introduction of agriculture in the Neolithic around 4000 BC (Larsson, Lindgren and Nordqvist, 1997; Malmer, 2002). The Neolithic in Sweden is traditionally divided into four periods: Early Neolithic (EN) about 4000 BC, Middle Neolithic A (MN A) about 3300 to 2700 BC, Middle Neolithic B (MN B) about 2700 to 2300 BC and Late Neolithic (LN) about 2300 to 1800 BC. During these periods different cultural groups emerged and disappeared. The extent to which inter-group contacts led to genetic admixture, replacement or a combination of both is currently unknown (cf. Papmehl-Dufay, 2006).

In Scandinavia and neighbouring regions, the EN is associated with the Funnel Beaker Culture (also known as TRB) and the formation of large megalithic stone monuments, dolmens and passage graves. This practice continues during later phases (Sherratt, 1990; Tilley, 1996; Price, 2000). It seems that across northern and western Europe the constructions of the first megalithic monuments are closely linked to the introduction of agriculture (Renfrew, 1981; Hodder, 1990). The ritualistic focus of these megalithic monuments is revealed through a constant reuse of the chambers for human burials, sometimes spanning several centuries as well as the scattered pottery depositions made by the local communities (Tilley, 1996). In Sweden the highest concentration of these monuments is found in the provinces of Skåne, Halland, Bohuslän and in the Falbygden area, all situated in the south. However, some megalithic monuments and burials also appear in low numbers elsewhere, for example on Öland (Persson and Sjögren, 2001) (Figure 16.1).

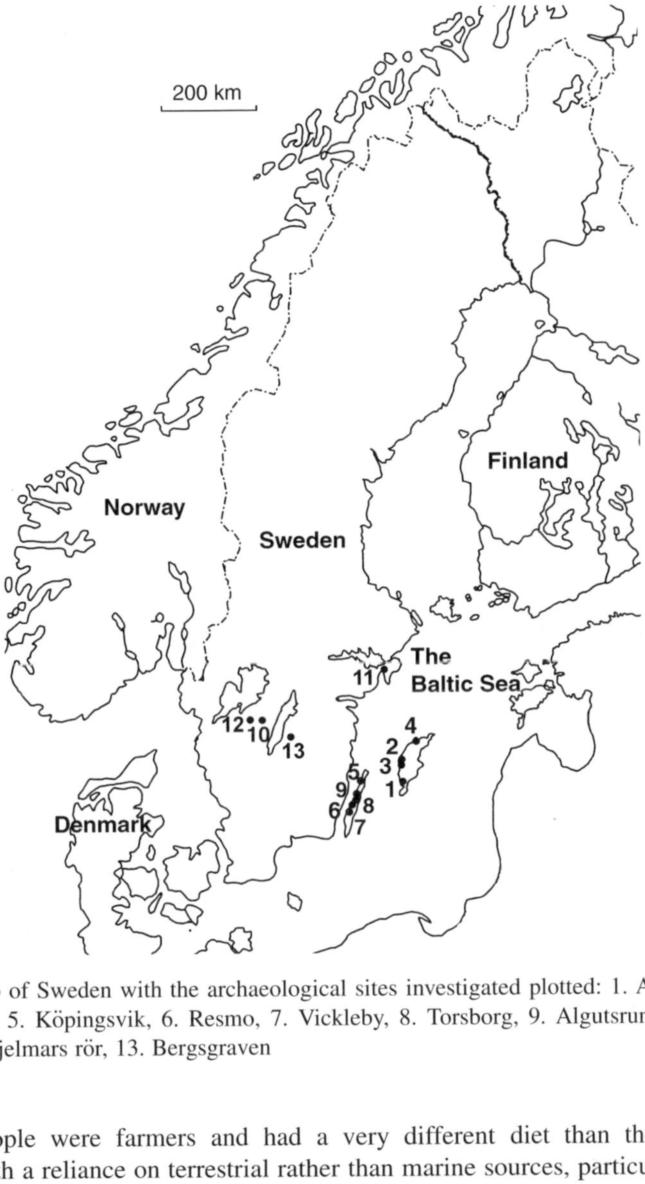

Figure 16.1 Map of Sweden with the archaeological sites investigated plotted: 1. Ajvide, 2. Visby, 3. Fridtorp, 4. Ire, 5. Köpingsvik, 6. Resmo, 7. Vickleby, 8. Torsborg, 9. Algutsrum, 10. Gökhem, 11. Korsnäs, 12. Hjelmars rör, 13. Bergsgraven

The TRB people were farmers and had a very different diet than their Mesolithic predecessors, with a reliance on terrestrial rather than marine sources, particularly domesticated crops, meat from a range of domesticated animals, and milk products (Lidén, 1995; Eriksson *et al.*, 2008). In Scandinavia the TRB period had its peak population density and distribution during the Middle Neolithic (Malmer, 2002). The Pitted Ware Culture (PWC) appears in Scandinavia during the late fourth millennium and early third millennium BC. In contrast with the TRB, who predominantly occupied inland locales (cf. Malmer, 2002), the PWC were coastal hunter-gatherers and foragers whose subsistence spectrum mostly relied on fishing and gathering of marine resources. In Southern Scandinavia these two cultures co-existed during a period that spans several centuries,while on Öland they co-exist during the period between 3500 and 2500 BC (Eriksson *et al.*, 2008). In the second half of the Middle Neolithic, the Battle Axe Culture (BAC) appears in the same region. The BAC, like the TRB, was predominantly an agricultural-based society. It is clearly distinguished from both TRB and PWC on the basis of its material culture (Malmer, 1975). At the end of the Neolithic there seems to have been some sort of cultural unification as there is no archaeological evidence for the co-existence of different cultures with different subsistence strategies. This cultural homogeneity seems to be associated with the adoption of agricultural subsistence practices by all populations who inhabited this region (Carlsson, 2001).

In most regions of Europe the Neolithic economy consisted of domesticated livestock (sheep, goat, cattle, pig) and plants (mainly cereals and legumes). Unlike most regions in south-east Europe (with the exception of the Danube Gorges) and Central Europe, in southern Scandinavia the transition to a fully Neolithic economy took more than a millennium to accomplish (Tilley, 1996; Eriksson *et al.*, 2008). The TRB subsistence was based on mixed farming practices, cereals (wheat and barley) and legumes (peas, beans and flax) (Larsson, 2007). Barley was a crop that was well suited to the cooler climactic conditions of Southern Scandinavia. The domesticated animals that were reared comprised a mixture of cattle, pig and sheep/goat, which provided important sources of food in the form of meat and dairy products, as well as manure for soil fertilization and a number of essential raw materials, such as hides and wool (Sherratt, 1990). While only a small percentage of wild animals are found in the archaeological record of TRB sites, it is clear that some traditional exploitation of wild resources such as hunting of game, fishing and fowling prevailed alongside farming (Larsson, 1985; Midgley, 1992).

16.3 ANCIENT DNA AND THE NEOLITHIC TRANSITION IN EUROPE

Ancient DNA research started in the 1980s, but has only recently begun to gain momentum following major advances in extraction, amplification and sequencing of ancient DNA strains, and procedures to establish the authenticity of aDNA on the basis of a set of criteria (cf. Cooper and Poinar, 2000; Hofreiter *et al.*, 2001; Gilbert *et al.*, 2005). The study of aDNA allows for the direct analysis of genes from individuals of known location, date and archaeological provenance. However, most surviving ancient human nuclear or mitochondrial DNA is fragmentary, present in small amounts (and often low copy numbers), and can be easily contaminated from modern human DNA (Hofreiter *et al.*, 2001). Because of the low concentration of aDNA in bone and its usually degraded state, researchers strive to gain as much information as possible from very short DNA fragments (Lindahl, 1993; Smith *et al.*, 2003). Ancient DNA studies are now beginning to provide critical data about the timing and past distribution of genetic variation, allowing tests of the relative roles of selection, mutation, migration and/or genetic drift. The main source of aDNA has traditionally been mitochondrial DNA (mtDNA), due to the high copy number within cells (up to 1000s in comparison to the single copy of nuclear DNA per cell). Working with ancient human DNA has been challenging due to the risk of contamination from the archaeologists, laboratory workers and any other humans that were in contact with the samples. Consequently, in the analysis of human aDNA, extra care has to be taken to produce authentic results and eliminate and/or explain any possibility of contamination (Hofreiter *et al.*, 2001; Pääbo *et al.*, 2004; Gilbert *et al.*, 2005; Willerslev and Cooper, 2005; Malmström *et al.*, 2005, 2007; Linderholm *et al.*, 2008). Due to contamination issues palaeogenetic studies of human aDNA have only gained momentum during the last decade (Haak *et al.*, 2005; Kuch *et al.*, 2007; Burger *et al.*, 2007; Bramanti *et al.*, 2009;

Malmström *et al.*, 2009, 2010;) following improvements in the prevention control, detection and decontamination of exogenous DNA. Increasingly powerful methods, including genomic technology, have started to remove this constraint and gradually more nuclear DNA is being seen as a suitable target.

16.3.1 Ancient Mitochondrial DNA Studies

The complete mitochondrial genome is transmitted from mother to offspring without any genetic modifications such as meiosis and recombination. This means that studies based on mtDNA will investigate the maternal lineage of the population (Helgason *et al.*, 2000). Human mtDNA is thought to be a non-recombining molecule (Pakendorf and Stoneking, 2005) and its large sequence diversity, high copy number and high mutation rate make it ideal for ancient DNA studies and particularly those that aim to investigate past populations phylogenies and phylogeography (Gilbert *et al.*, 2003; Ho and Gilbert, 2010). However, only recently has the study of ancient mtDNA markers been applied to the investigation of hunter-farmer contact, and the nature of the Neolithic transition in Europe (Bramanti *et al.*, 2009, Malmström *et al.*, 2009, 1010).

Haak and colleagues (2005) investigated 24 central European Neolithic specimens from the Linearbandkeramik (LBK) culture, in order to assess the extent of the genetic contribution from Neolithic female lineages (mtDNA) to present-day Europeans. The sequencing of the HVS1 region of the mitochondrial DNA showed that 25% of the LBK samples belonged to a haplotype N1a that today is very rare in Europe (prevalence of ~0.2%). This indicates a 150-fold decline in the frequency of this allele over a period of about 7500 years. This result suggests that the first Neolithic farmers did not leave a strong genetic signature on modern European lineages (Haak *et al.*, 2005). A more recent follow-up study by the same group of researchers further investigated the relationship between hunter-gatherer and farmers in central Europe, by comparing 20 mtDNA sequences from central and northern Mesolithic European hunter-gatherers sequences from 25 Neolithic farmers (some of which were available from previous studies). In addition, mtDNA sequences were compared to modern sequences in order to identify any genetic affinities between the two ancient population and modern Europeans. Their study showed genetic differences between the hunter-gatherers, farmers and modern European populations, with 82% of hunter-gatherers having mtDNA sequences that are rare amongst present-day central Europeans (Bramanti *et al.*, 2009; Burger and Thomas, this volume). This indicates that the genetic markers of Mesolithic hunter-gatherers did not become part of the modern European genome to any large extent. This result also indicates that the first farmers in Europe were not descendants of local hunter-gatherers and that the genetic legacy of these Neolithic farmers must have been wiped out by subsequent migrations, drift and other demographic processes.

A recent study investigated the origin of local hunter-gatherer population in Sweden (the Pitted Ware culture, PWC) using 22 samples of both the hunter-gatherers and farmers from Southern Sweden (Malmström *et al.*, 2009). A 316 bp long mtDNA D-loop fragment was sequenced to assign a haplotype to each individual. When comparing the obtained sequences to those of modern Swedes, no evidence for genetic continuity was detected. This result is in accord with those obtained by Bramanti *et al.* (2009) that suggests that the genetic legacies of both the LBK and TRB cultures did not survive to the modern day. However, the results also indicate, as in the case of the study by Bramanti *et al.* (Ibid.), that the PWC hunter-gatherers did not contribute to the modern Swedish gene pool (Malmström *et al.*, 2009). These recent aDNA studies provide results that question the validity of inferences about past demographic processes, which are based on genetic data of modern Europeans and stress the importance of aDNA studies that directly sample past populations.

16.3.2 Ancient Nuclear DNA Studies

There are several ways to investigate differences in nuclear DNA (nuDNA). Some of the best targets are Single Nucleotide Polymorphisms (SNPs), which are single base pair positions in the DNA at which different sequence alternatives (alleles) exist (Brookes, 1999). SNPs are slow-evolving markers when compared to fast-evolving short tandem repeats (STRs), which are commonly used in forensic studies. Each SNP is a single base mutation, such as a shift from a C to a T at one specific location of the DNA sequence (Jobling Hurles and Tyler-Smith, 2004). Recently ancient DNA research began to focus on the study of nuDNA and particularly SNPs – the biggest advantage being that only a short sequence is needed for analysis, which is useful as most aDNA is in a fragmented state. In recent years more attention has been placed on ancient nuDNA studies of genes associated with different diseases, or genes expressing specific morphological or physiological characters (Mishmar *et al.*, 2003; Lalueza-Fox *et al.*, 2007; Ludwig *et al.*, 2009).

16.3.3 Next-Generation Sequencing

Major technological progress has been made in the last few years in the development of new commercially available DNA sequencing methods. The new next-generation sequencing technologies (NGS) seek to amplify single DNA strands from fragmented samples, thus avoiding the need for cloning, and perform sequencing reactions on the amplified strands (for a review, cf. Mardis, 2008; Shendure and Ji, 2008; Metzker, 2010). The three main methods/sequencers currently in use are the 454 GenomeSequencer FLX (Roche Applied Science), the Illumina (Solexa) Genome Analyser and the Applied Biosystems ABI SOLiD system. The 454 GS FLX system is based on pyrosequencing (Ronaghi *et al.*, 1996) and utilizes library fragments that have been ligated with specific 454 adaptors that bind to a complementary sequences on an agorose bead, resulting in one single fragment per bead. Each bead will then be enclosed by oil in an emulsion PCR reaction. The result of this PCR is that each single fragment is independently amplified *en masse* and this fragment (in multiples, simulating the cloning step) can then be sequenced using the pyrosequencing method. The average read-length is 330 bases (the read length range lies between 400 and 500 bases) and the reads yield on average over 100 Mb of data per run (Mardis, 2008; Metzker, 2010). With the Illumina/Solexa technology (Metzker, 2010), DNA fragments in the library are ligated at both ends with specific adaptors that ensures their immobilization. After immobilization, each fragment is amplified using bridge PCR, creating clusters of amplicons (simulating the cloning step). Once the fragment has been amplified, the sequence-by-synthesis takes place using a hybridization sequencing method. The average read-length is 75 or 100 bases and the reads yield between 1 and 3 Gb of data per run on average (Mardis, 2008, 2010). The ABI SOLiD sequencing system also relies on short adaptor-ligated fragmented libraries. After the fragments have been captured to the beads, an emulsion PCR amplifies each fragment individually (simulating the cloning step). The beads are then deposited on to a glass surface where the fragments are being sequenced using a ligation method for the sequence-by-synthesis step. The average read-length is 30 or 50 bases and the reads yield between 3 and 10 Gb of data per run on average (Mardis, 2008;

Metzker, 2010). To further develop the sequencing methods, we have to be able to sequence single molecules without any prior amplification and methods for this step are already underway (Metzker, 2010).

The use of these NGS techniques produces large numbers of relatively short reads, which are useful for many studies. In the aDNA field, they have been used for sequencing nuclear genomes of extinct cave bears (Stiller *et al.*, 2009), mammoths (Poinar *et al.*, 2006), Neanderthals (Green *et al.*, 2006; Noonan *et al.*, 2006; Briggs *et al.*, 2009) and of prehistoric humans (Linderholm, 2008; Malmström *et al.*, 2009; Rasmussen *et al.*, 2010; Olivieri *et al.*, 2010) to mention a few (for a review, cf. Hofreiter, 2008; Ho and Gilbert, 2010).

One of the main issues in the NGS-era is the processing and analysis of the vast amount of genetic information that is obtained in a relatively short time and at diminishing costs. Consequently, there is a growing need for new methods in the fields of bioinformatics and statistics, which can assist the sorting, aligning and assemblage of all the reads into workable information packages (Flicek and Birney, 2009). At the same time, there is a need to ensure high data quality control, in order to verify authenticity and assure the correct interpretations of the obtained datasets.

16.4 ANCIENT DNA STUDIES OF THE MESOLITHIC–NEOLITHIC TRANSITION IN SWEDEN

Hypotheses about the origin and affinities of PWC people have varied between those that view them as relict late Mesolithic populations, while others view them as being biologically affiliated with the TRB (farming) populations and believe that the economic and cultural difference between these groups were due to the fact that they inhabited different geographical regions and consequently were engaged in different economic activities. Other major questions are the place of origin of the TRB people, and the manner by which their Neolithic culture spread in Scandinavia (Larsson, Lindgren and Nordqvist, 1997; Malmer, 2002; Eriksson *et al.*, 2008). The discussion so far has been based on the archaeological record, namely the study of artefacts (mainly pottery), burial practices, dwellings and diet. In the three cases described below, DNA was used for the first time on prehistoric Swedish human skeletal material in order to shed new light on the issue of the Neolithic transition in this region and on the admixture and migrations of past populations. DNA was extracted from individuals from both (Mesolithic/Neolithic) hunter-gatherer and (Neolithic) farming populations. In these studies, both mtDNA and nuDNA were examined in order to look at both haplotype affiliation as well as some more specific phenotypic traits that provide adaptive advantages, such as lactose tolerance or the case of the CCR5 Δ32-deliton.

16.4.1 Case Study 1: Hunter/Gatherers vs. Farmers, Where did they Come from?

The aim of this study was to investigate the origin of two populations: the TRB and PWC, their genetic relationship and its potential impact on the genomes of modern Europeans from Southern Scandinavia and neighbouring regions. In this case study, 111 individuals were investigated from the period spanning from the middle Neolithic to the early Bronze Age (3300–1100 calBC). The assignment of the skeletal remains was based on their material culture as well as on the dietary signals (Erikson *et al.*, 2008). The HSV1 was sequenced in order to assign haplogroup status to each individual examined (Torroni *et al.*, 2001; Richards and Macaulay, 2000; Haak *et al.*, 2005; Burger *et al.*, 2007). In order to retrieve the 341 bp sequenced needed for this, a large-scale sequencing approach comprising of 111 individuals, 40 animals and 126 control samples were analysed, using aDNA samples obtained from teeth and/or long bones of the archaeological skeletons that were retrieved from 14 sites from Southern Sweden and the Baltic Sea islands of Öland and Gotland (Figure 16.1). Ancient DNA from these human samples was extracted at two especially dedicated aDNA laboratories at the Archaeological Research Laboratory in Stockholm or at the National Board of Forensic Medicine in Linköping, Sweden, and samples were treated and extracted as previously described (cf. Yang *et al.*, 1998; Malmström *et al.*, 2007). The duplicate extracts were then amplified and analysed using a method based on a PCR amplification of HVSI using 5′tetranucleotide-tagged primers (Binladen *et al.*, 2007; Malmström *et al.*, 2009; Linderholm, unpublished). The strength of this method lies in the fact that it makes it possible to retrieve massive amounts of amplicon clone sequences, and the authentication of the sequenced data.

Each individual was assigned to a given haplogroup based on those reported in genetic studies of various modern European and Asian populations (Torroni *et al.*, 2001; Richards and Macaulay, 2000; Tamberts *et al.*, 2004; Pereira *et al.*, 2005; Lappalainen *et al.*, 2008). From this large dataset, 53 individuals (31 Pitted Ware, 7 Funnel Beaker and 15 Bronze Age) were chosen on the basis of aDNA authenticity criteria and their data was used for the generation of a reduced median network. The animal samples, and extraction blanks in some cases, contained traces of human DNA, although it could be easily identified as modern contaminants. None of the contaminations could be reproduced. The obtained sequences from the archaeological specimens was then compared to three modern datasets (Swedish, Sami and Latvian, cf. Helgason *et al.*, 2001; Sajantila *et al.*, 1996; Lappalainen *et al.*, 2008), using a reduced median network (threshold $r = 1$) (Figure 16.2) (Linderholm, 2008; Linderholm, unpublished). All datasets were analysed using F-statistics and analysis of molecular variance (AMOVA), in order to test for differences between and within populations.

The mtDNA results point to the fact that the TRB and PWC are two genetically distinct populations ($p < 0.001$) (Linderholm, 2008; Linderholm, unpublished). Furthermore, the PWC population appears to be more closely related to the modern Latvian population than the contemporaneous TRB population or any of the other modern population examined. This could imply that the PWC population may have had an eastern/central European origin, whereas the TRB population may have had a continental European origin (Linderholm, 2008; Malmström *et al.*, 2009; Linderholm *unpublished*). These results could also imply that in Sweden the PWC were part of a large hunter-gatherer complex that spanned vast areas of the central and eastern parts of Europe.

The results suggest that in northern Europe the Neolithic transition involved the replacement of local Mesolithic hunter-gatherer groups by incoming farmers. This model supports the one made by both Stenbäck (2003) and Wyszomirska (1984), which suggests close contact between the PWC and the north-east European Comb Ceramic Culture (CCC, also known as the Pit-Comb Ware culture), which was panecontemporaneous with the PWC (4200–2000 calBC). Artefacts belonging to the CCC have been found in northern parts of Norway, Sweden and Finland, and were also found as far south as Poland and as far east as Russia (Werbart, 1998). Neither of the two populations (PWC and TRB) shows any genetic affinities to the Sami population when compared on the basis of mtDNA sequence. It therefore appears that the ancient mtDNA analyses provide genetic evidence that the TRB and the PWC cultures were

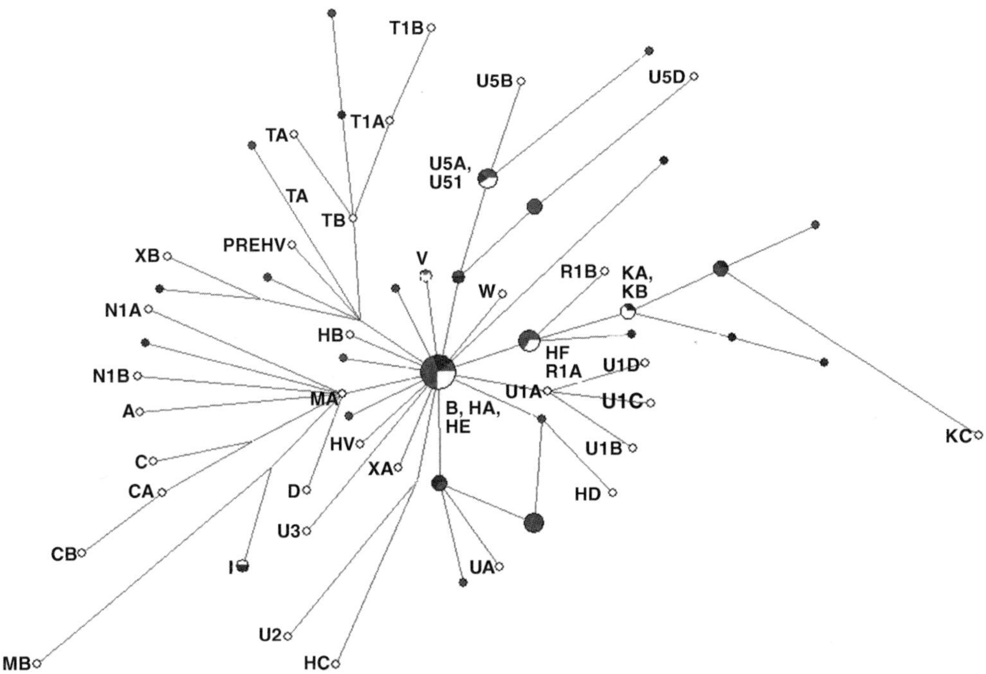

Figure 16.2 A reduced median network based on the HVS1 region for the Neolithic to Bronze Age Swedish human samples (Network 4.1, www.fluxusengineering.com) together with known haplogroups. Colour/population cod: Blue = TRB (Funnel Beaker Culture), Red = PWC (Pitted Ware Culture), Green = BA (Bronze Age), White = Haplogroups

two separate populations. This view is also in agreement with the data published by Malmström *et al.* (2009). These populations lived side by side for a few hundred years, with different lifestyles, economic strategies and other cultural characteristics (Erikson *et al.*, 2008). By the onset of the Bronze Age, the farming population that prevails in Sweden shows close genetic resemblance to the modern-day Swedish population.

16.4.2 Case Study 2: Lactose Intolerance/Tolerance in Prehistory

The so-called secondary Neolithic revolution involved the utilization of new products from domesticated animals, such as wool and milk (Sherratt, 1990). Unfermented milk contains lactose. Lactose intolerance refers to the inability to digest milk, which is a common condition amongst the great majority of adult populations across the world and is caused by a deactivation of the lactase-phlorinzin hydrolas enzyme after weaning (also see Burger and Thomas, this volume). The ability to digest lactose (i.e. lactose tolerance) is the result of a single mutation upstream the lactase gene at the 13 910 bp position. Lactose tolerance is thought to have evolved within a relatively short period of time in association with the emergence of farming and particularly the consumption of unfermented milk (Bellwood, 2004; Beja-Pereira *et al.*, 2006; Tishkoff *et al.*, 2007; Itan *et al.*, 2009; Ingram *et al.*, 2009). In Europe, a C to T substitution had occurred, thus making it possible for adults to digest unfermented milk (Enattah *et al.*, 2002). Lactose tolerance must have provided an advantage to farmers, as it allowed them to include milk in their diet and benefit from its nutritious value (Beja-Pereira *et al.*, 2003; Burger *et al.*, 2007).

Lactose tolerance has a very clear distribution pattern in modern-day Europe (Kuokkanen *et al.*, 2005). Two forms of lactose tolerance more than likely evolved in parallel in Europe and in Africa, as they were caused by separate mutations (Swallow, 2003; Tishkoff *et al.*, 2007; Ingram *et al.*, 2009). This pattern may be explained to some extent by migration (Beja-Pereira *et al.*, 2003; Gerbault *et al.*, 2009). The origin and timing of lactase persistence in Europe is the focus of a paper by Itan *et al.* (2009). The paper's starting assumption is that the lactase persistence can only provide a selective advantage if there is a supply of fresh milk. Itan *et al.* then examined the probable place of origin of this mutation by using a series of spatial simulations. The combination of the –13 910*T allele frequency and the dates for the introduction of farming across Europe suggest that the –13 910*T allele underwent selection for the first time around 7500 years ago. The authors could also infer that this event most likely took place in a region located between the central Balkans and central Europe (Itan *et al.*, 2009). The estimated age of the –13 910*T allele fits well with the estimated dates for the arrival of farming in Europe.

Another study examined the impact of selection and demography on the somewhat peculiar world distribution of lactase persistence (Gerbault *et al.*, 2009). The worldwide distribution of the lactase persistence trait can only be explained by positive selection and so far there have been two hypotheses to explain the distribution pattern. One hypothesis is of gene-culture co-evolution (gcc), as fresh milk gave an evolutionary advantage to people with the –13 910*T mutation. Another hypothesis suggests that the distribution of lactose tolerance is associated with the need for vitamin-D (which is present in milk), in high latitude regions with limited sunlight (which triggers metabolic pathways that replenish vitamin D deficiencies). The lactose would then be a substitute for the vitamin-D, calcium assimilation (cal). Using computer simulations of all the known –13 910*T frequencies across the world (in 115 populations), Gerbault and his team found that the gcc hypothesis seemed to fit the data from Africa, whereas the cal hypothesis is compatible with the data from Europe (Gerbault *et al.*, 2009). These findings could explain the high –13 910*T frequency found presently in Sweden (74%).

A recent study carried out at three aDNA laboratories (Stockholm, Linkoping and Uppsala) investigated the frequency of the gene-variant responsible for lactase persistence in a Swedish Neolithic population (Linderholm, 2008; Malmström *et al.*, 2010; Linderholm, unpublished). The aim was to assess two aspects:

1. the ability to digest milk amongst adults from a PWC hunter-gatherer population; and
2. the genetic relationship of the PWC population to a modern-day Swedish population based on the study of this genetic marker.

The study used duplicate samples (teeth and/or ulna, femur or fibula) from 14 PWC individuals from 4 archaeological sites on Gotland in the Baltic Sea, which are dated to the Middle Neolithic (2800–2200 calBC) and thus post-date the first appearance of agriculture in Sweden by more than a millennium. Ancient DNA was extracted and amplified using standard methods (Yang *et al.*, 1998; Malmström *et al.*, 2007). The C/T polymorphism, at position –13 910 in the lactase gene, was amplified by two different fragments: one of 53 bp and a second of 168 bp. These amplicons were then pyrosequenced for allele identification, as described in Anderung *et al.* (2005). All of the 14 samples yielded replicable results and 10 were successfully SNP-typed more than 4 times, thus ensuring the authenticity of the typing. A few positive results

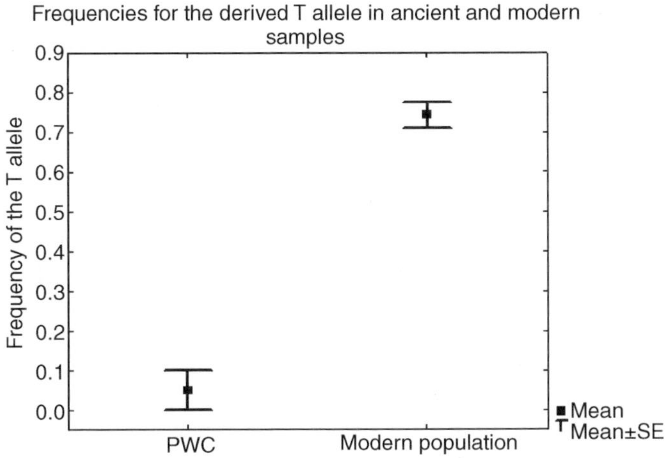

Figure 16.3 Frequency of the –13 910*T polymorphic site in the lactase gene in two datasets: an extant Swedish population and a Swedish Neolithic hunter-gatherer population (PWC)

were detected in the negative controls, but they could not be reproduced (Malmström *et al.*, 2010).

Only one out of the ten samples analysed, a heterozygote, revealed the presence of the T-allele. Given these results, the T-allele frequency in the analysed PWC population is 0.05, which is significantly different from the T-allele frequency of 0.74 amongst a modern Swedish population ($n = 97$, chi-square test, $p < 0.0001$) (Figure 16.3). In this case study we have relied on several supporting arguments as proof of the obtained results' authenticity (Richards, Sykes and Hedges, 1995; Kolman and Tuross, 2000; Malmström *et al.*, 2005). Not only do we see a difference in the allele frequency between the prehistoric samples and the negative controls, but also this difference is present in the success rate of the sample categories. To further strengthen our case, the material has been pre-screened for contamination, as reported in Linderholm *et al.* (2008).

The results obtained suggest that the PWC population, with its low T-allele frequency, has either undergone some extreme positive selection, which led to the observed modern-day frequency of 0.74 or alternatively did not contribute its genetic legacy to modern-day Swedish populations (Malmström *et al.*, 2010). Since the frequency of the derived T-allele is strongly linked to the ability to consume milk as an adult, it seems likely that the increase in T-allele frequency in Sweden is closely linked to the Neolithization process.

Analysis has also been performed on individuals representing the TRB population; the outcome of these analyses has not yet been totally verified. In the dataset we can detect a large rise in T-allele frequency in Sweden amongst the farming community (unpublished data).

16.4.3 Case Study 3: The Timing of the CCR5-Δ32 Deletion

The last case is a study of a 32 bp deletion in the CCR5 gene that provides protection against HIV I in individuals homozygous for the deletion (Biti *et al.*, 1996; Liu *et al.*, 1996; Paxton *et al.*, 1996). This deletion has its highest frequency in northern Europe and its frequency decreases towards the southwest; a clinical pattern similar to the lactose tolerance allele distribution (Tishkoff *et al.*, 2007). The aim was to try to date the mutation and see whether it is associated with the Neolithization process. It was also of interest to investigate whether it is possible to detect different frequencies between the TRB and the PWC. The CCR5 32 bp deletion seems to be restricted to Caucasian populations. Coalescence analysis of this mutation has suggested two different timeframes for the emergence of this mutation: fourteenth century AD or 3000 BC. The geographical distribution of the 32 bp deletion in Europe has been explained by either genetic drift (Martinson *et al.*, 1997), or positive selection due to resistance to a major disease (Dean *et al.*, 1996). While the timing of this 32 bp deletion and the factors that caused it remain elusive, it is evident that it originated from a single mutation event.

The aDNA analysis included bones and teeth of human skeletons from two Swedish Mesolithic sites and six Swedish Neolithic sites dating from 7000 BC to 1700 BC. Ancient DNA was successfully extracted from 17 out of 46 individuals at the Archaeological Research Laboratory in Stockholm. The deletion was found in 11 out of the 17 individuals (7 heterozygots and 4 homozygots); the CCR-Δ32 mutation frequency of 17.1%, which does not differ significantly from the frequency amongst modern Swedes (14.3%). The samples were reproduced in a laboratory in Madrid, Centro UCM-ISCIII de Investigación sobre Evolución y Comportamiento Humanos. We further tried to see if there was a difference between the TRB and the PWC cultures. However, due to the low number of individuals that gave any positive results, no unambiguous conclusions can be drawn concerning the difference between the two cultures and the only conclusions that could be drawn are that the deletion already existed during the Mesolithic and that its frequency seems to have increased in the Neolithic.

16.5 STABLE ISOTOPE STUDIES OF THE NEOLITHIC TRANSITION IN SWEDEN

The diet of individuals from the PWC and TRB cultures was investigated by analysing bulk carbon and nitrogen isotopic ratios (for more information regarding the methods, see contributions by Schulting, Lillie and Budd, Papathanassiou, Grupe and Peters, Lieverse *et al.*, this volume). The analysis included 123 human samples (plus 27 faunal samples) from 9 archaeological sites on the island of Öland and from the TRB and PWC cultures that co-existed on the island during several centuries (Erikson *et al.*, 2008). The main aim was to try to date the change in diet connected with the introduction of the Neolithic (by the TRB), by analysing stable isotopes of carbon (δ^{13}C) and nitrogen (δ^{15}N). The overall results from the δ^{13}C and δ^{15}N analyses gave a large range of values: for carbon δ^{13}C –21.3 to –13.1‰ and for nitrogen δ^{15}N 8.2 to 18.9‰. A typical PWC range for carbon was δ^{13}C –15.8 to –13.1‰ and for nitrogen δ^{15}N 15.2 to 18.9‰ (these are the values from the site of Köpingsvik). For a typical TRB population, the range for carbon was δ^{13}C –21.0 to –17.0‰ and for nitrogen δ^{15}N 8.8 to 14.9‰ (these are the values from the site of Resmo). The results show a gradual transition from a hunter-gatherer diet to a diet based on agriculture took place by the TRB people (which is thought to be associated with the introduction of farming on the island) only by the end of the Neolithic and hence much later than had been previously thought. The stable isotope data were also compared to climatic data from this time. No change in climate can be perceived in connection with this shift, which makes the notion that it was a culturally induced process more plausible (Linderholm, 2008; Erikson *et al.*, 2008).

16.6 DISCUSSION AND CONCLUSIONS

The introduction of farming in Sweden is attributed to the Funnel Beaker culture, an event that took place over 6000 years ago. Three case studies were presented in an attempt to answer the question of the origin of the earliest farmers in Sweden, their potential admixture with local hunter-gatherers and the contribution of both to modern Scandinavian populations. In the first case study, attempts were made to profile over 100 individuals to ascertain their mitochondrial haplogroup. Over 50 individuals were successfully typed and based on these results it has been shown that the PWC and TRB cultures were genetically distinct populations. The genetic legacy of the TRB continued into the Bronze Age and further in time, suggesting that they made a genetic contribution to modern Swedish populations. In contrast, the PWC seem to have left no genetic legacy to the modern Swedish population, but their genetic signals can be detected in modern north-eastern European populations.

These results are in agreement with those obtained from the second case study. The fact that lactose tolerance is very low in the PWC population indicates that they were more than likely not the founding population that contributed its genetic legacy to the modern Swedish population. None of these results should come as a big surprise; as the archaeological record indicates that the differences between the two culture groups are substantial. Not only did they have different dietary patterns, but they also occupied different ecotones, had different pottery styles and burial customs (cf. Itan *et al.*, 2009).

In the third case study, the focus was placed on the detection of a specific deletion in order to date its emergence and to examine if there are any differences in frequency of the deletion between the TRB and PWC population. The frequency of this deletion amongst modern Swedish subjects is very high and it can be traced back to the Mesolithic, but in a much lower frequency. But no conclusive differences in deletion frequency can be seen between the TRB and the PWC.

As discussed above, aDNA analysis of cases from central Europe indicate that LBK farmers differed in various genetic markers from both contemporary Mesolithic hunter-gatherers and modern European populations (Haak *et al.*, 2005; Bramanti *et al.*, 2009). A rather different picture emerges from Spain, as Sampietro and colleagues (2007) showed that in the case of 11 Spanish Neolithic, the ancient mtDNA haplotypes still prevail amongst modern European and Near Eastern populations. The different results could reflect variations in the nature and timing of the Neolithic transition in different parts of Europe, whereas in some regions, such as south-east Europe, Central Europe and Southern Scandinavia, the transition did not seem to entail any genetic contribution of local hunter-gatherers to the Neolithic gene pool. However, there is also no indication that the Neolithic genetic legacy contributed to the modern European gene pool. In Spain, and possibly in other parts of Mediterranean and Atlantic Europe, there is no discontinuity, as evident from the study of mtDNA, between the Neolithic populations and modern-day Europeans and it therefore suggests the possibility that their genetic legacy did not disappear, although we must not rule out new re-introduction of various markers by migrants, for example, from eastern Europe and Asia, during subsequent post-Neolithic periods. These studies highlight the complexity and some of the challenges involved in any studies, which attempt to reconstruct the demographic history of past populations.

The cases presented here, taken together with other genetic studies concerning the Neolithization process in Europe, indicate that the process was far more complex and variable

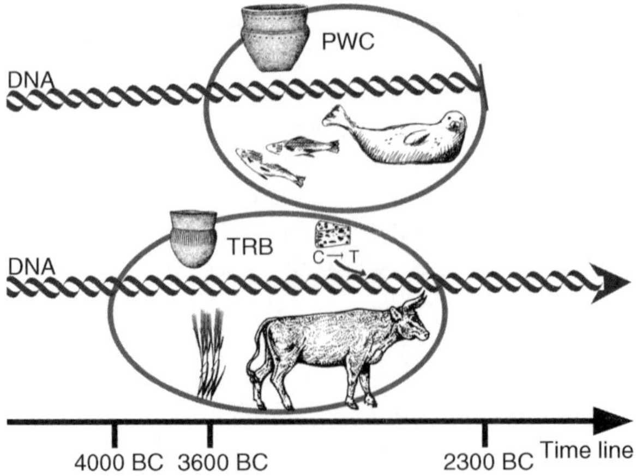

Figure 16.4 A generalization of the relationship between PWC and TRB, showing their main food source and ceramic style over a timeline (made by L. Bergstrom)

than was first thought. They establish the fact that the PWC and the TRB are two genetically different populations (Figure 16.4). The PWC population probably originated from the north-east, and they left no genetic impact on modern Scandinavian populations. In contrast, the Neolithic TRB populations of Sweden originated in continental Europe and show some genetic similarities to modern Scandinavian populations. We still do not know what happened to the Mesolithic hunter-gatherer populations from southern Sweden, as no studies have been carried out on early Holocene Mesolithic samples. The data presented here does not resolve all of the issues, but it does provide a clear indication of the great potential of future ancient DNA studies and the analysis of molecular, archaeological and isotopic data in order to shed more light on the Neolithic transition.

REFERENCES

Anderung, C., Bouwman, A., Persson, P. *et al.* (2005) Prehistoric contacts over the Straits of Gibraltar indicated by genetic analysis of Iberian Bronze Age cattle. *Proc. Natl. Acad. Sci. U.S.A*, **102**, 8431–8435.

Beja-Pereira, A., Luikart, G., England, P.R. *et al.* (2003) Gene-culture co-evolution between cattle milk protein genes and human lactase genes. *Nat. Genet.*, **35**, 311–313.

Beja-Pereira, A., Caramelli, D., Lalueza-Fox, C. *et al.* (2006) The origin of European cattle: Evidence from modern and ancient DNA. *Proc. Natl. Acad. Sci. U.S.A*, **103**, 8113–8118.

Bellwood, P. (2004) *First Farmers: The Origins of Agricultural Societies*, Wiley-Blackwell, Malden, MA.

Binladen, J., Gilbert, M.T.P., Bollback, J.P. *et al.* (2007) The use of coded PCR primers enables high-throughput sequencing of multiple homolog amplification products by 454 parallel sequencing. *PLoS ONE*, **2**, e197.

Biti, R., French, R., Young, J. *et al.* (1996) HIV-1 infection in an individual homozygous for the CCR5 deletion allele. *Nat. Med.*, **2**, 1240–1243.

Bramanti, B., Thomas, M.G., Haak, W. *et al.* (2009) Genetic discontinuity between local hunter-gatherers and central Europe's first farmers. *Science*, **326**, 137–140.

Briggs, A.W., Good, J.M., Green, R.E. *et al.* (2009) Targeted retrieval and analysis of five Neandertal mtDNA genomes. *Science*, **325**, 318–321.

Brookes, A.J. (1999) The essence of SNPs. *Gene*, **234**, 177–186.

Burger, J., Kirchner, M., Bramanti, B. *et al.* (2007) Absence of the lactase-persistence-associated allele in early Neolithic Europeans. *Proc. Natl. Acad. Sci. U S A*, **104**, 3736–3741.

Carlsson, A. (1998) Tolkande arkeologi och svensk forntidshistoria. Stenåldern. Stockholm studies in archaeology, 17. Stockholm, Sweden.

Carlsson, A. (2001) Tolkande arkeologi och svensk forntidshistoria. Bronsåldern (med senneolitikum och förromersk järnålder). Stockholm studies in archaeology, 22. Stockholm, Sweden.

Cavalli-Sforza, L.L. (1994) *The History and Geography of Human Genes*, Princeton University Press, Princeton.

Chikhi, L., Destro-Bisol, G., Bertorelle, G. *et al.* (1998) Clines of nuclear DNA markers suggest a largely Neolithic ancestry of the European gene pool. *Proc. Natl. Acad. Sci. U.S.A*, **21**, 9053–9058.

Cooper, A. and Poinar, H. (2000) Ancient DNA: Do it right or not at all. *Science*, **289**, 1139.

Dean, M., Carrington, M., Winkler, C. *et al.* (1996) Genetic restriction of HIV-1 infection and progression to AIDS by a deletion allele of the CCR5 structural gene. *Science*, **273**, 1856–1862.

Enattah, N.S., Sahi, T., Savilahti, E. *et al.* (2002) Identification of a variant associated with adult-type hypolactasia. *Nat. Genet.*, **30**, 233–237.

Eriksson, G., Linderholm, A., Fornander, E. *et al.* (2008) Same island, different diet: Cultural evolution of food practice on Öland, Sweden, from the Mesolithic to the Roman Period. *J. Anthropol. Archaeol.*, **27**, 520–543.

Flicek, P. and Birney, E. (2009) Sense from sequence reads: methods for alignment and assembly. *Nat. Methods*, **11**, S6–S12.

Gerbault, P., Moret, C., Currat, M. and Sanchez-Mazas, A. (2009) Impact of selection and demography on the diffusion of lactase persistence. *PLoS One*, **4**, e6369.

Gilbert, M.T.P., Willerslev, E., Hansen, A.J. *et al.* (2003) Distribution patterns of post mortem damage in human mitochondrial DNA. *Am. J. Hum. Genet.*, **72**, 32–47.

Gilbert, M.T.P., Bandelt, H.-J., Hofreiter, M. and Barnes, I. (2005) Assessing ancient DNA studies. *Trends Ecol. Evol.*, **20**, 541–544.

Green, R.E., Krause, J., Ptak, S.E. *et al.* (2006) Analysis of one million base pairs of Neanderthal DNA. *Nature*, **444**, 330–36.

Haak, W., Forster, P., Bramanti, B. *et al.* (2005) Ancient DNA from the first European farmers in 7500-year-old Neolithic sites. *Science*, **310**, 1016–1018.

Helgason, A., Sigurdadóttir, S., Gulcher, J.R. *et al.* (2000) mtDNA and the origin of the Icelanders: deciphering signals of recent population history. *Am. J. Hum. Genet.*, **66**, 999–1016.

Helgason, A., Hickey, E., Goodacre, S. *et al.* (2001) mtDNA and the islands of the North Atlantic: estimating the proportions of Norse and Gaelic ancestry. *Am. J. Hum. Genet.*, **68**, 723–737.

Ho, S.Y. and Gilbert, M.T. (2010) Ancient mitogenomics. *Mitochondrion*, **10**(1), 1–11, Epub 2009 Sep27.

Hodder, I. (1990) *The Domestication of Europe*, Basil Blackwell, Oxford.

Hofreiter, M. (2008) Palaeogenomics. *Comptes Rendus Palevol*, **7**, 113–124.

Hofreiter, M., Serre, D., Poinar, H.N. *et al.* (2001) Ancient DNA. *Nat. Rev. Genet.*, **2**, 353–358.

Ingram, C.J., Mulcare, C.A., Itan, Y. *et al.* (2009) Lactose digestion and the evolutionary genetics of lactase persistence. *Hum. Genet.*, **124**, 579–591.

Itan, Y., Powell, A., Beaumont, M.A. *et al.* (2009) The origins of lactase persistence in Europe. *PLoS Comput. Biol.*, **5**, e1000491.

Jobling, M.A., Hurles, M.E. and Tyler-Smith, C. (2004) *Human Evolutionary Genetics. Origins, Peoples and Disease*, Garland Publishing, UK.

Kolman, C.J. and Tuross, N. (2000) Ancient DNA analysis of human populations. *Am. J. Phys. Anthropol.*, **111**, 5–23.

Kuch, M., Gröcke, D.R., Knyf, M.C. *et al.* (2007) A preliminary analysis of the DNA and diet of the extinct Beothuk: A systematic approach to ancient human DNA. *Am. J. Phys. Anthropol.*, **132**, 594–604.

Kuokkanen, M., Butzow, R., Rasinperä, H. *et al.* (2005) Lactase persistence and ovarian carcinoma risk in Finland, Poland and Sweden. *Intl. J. Cancer*, **117**, 90–94.

Lalueza-Fox, C., Rompler, H., Caramelli, D. *et al.* (2007) A melanocortin 1 receptor allele suggests varying pigmentation among neanderthals. *Science*, **318**, 1453–1455.

Lappalainen, T., Laitinen, V., Salmela, E. *et al.* (2008) Migration waves to the Baltic Sea region. *Ann. Hum. Genet.*, **72**, 337–48.

Larsson, M. (1985) *The early Neolithic Funnel-Beaker Culture in South-West Scania, Sweden*, BAR, Oxford.

Larsson, M. (2007) Mistrust traditions, consider innovations? The Mesolithic-Neolithic transition in southern Scandinavia. *Proc. Br. Acad.*, **144**, 595–616.

Larsson, M., Lindgren, C. and Nordqvist, B. (1997) Regionalitet under mesolitikum. Från senglacial tid till senatlatisk tid i Syd- och Mellansverige, in *Regionalt och inetrregionalt. Stenåldersundersökningar i Syd- och Mellansverige* (eds M. Larsson and E. Olsson), Riksantikvarieämbetet Arkeologiska Undersökningar skrift, nr 23, Stockholm.

Lindahl, T. (1993) Instability and decay of the primary structure of DNA. *Nature*, **362**, 709–715.

Lidén, K. (1995) Megaliths, agriculture, and social complexity: A diet study of two Swedish Megaliths populations. *J. Anthropl. Archaeol.*, **14**, 404–417.

Linderholm, A. (2008) Migration in Prehistory. DNA and stable isotope analyses of Swedish skeletal material. (Theses and Papers in Scientific Archaeology 10). Stockholm.

Linderholm, A., Malmström, H., Lidén, K. *et al.* (2008) Cryptic contamination and phylogenetic nonsense. *PLoS One*, **3**, e2316.

Liu, R., Paxton, W.A., Choe, S. *et al.* (1996) Homozygous defect in HIV-1 coreceptor accounts for resistance of some multiply-exposed individuals to HIV-1 infection. *Cell*, **86**, 367–377.

Ludwig, A., Pruvost, M., Reissmann, M. *et al.* (2009) Coat color variation at the beginning of horse domestication. *Science*, **324**, 485.

Malmer, M. (1975) *Stridsyxekulturen I Sverige och Norge*, Libertryck, Stockholm.

Malmer, M. (2002) *The Neolithic of south Sweden: TRB, GRK, and STR. Royal Swedish Academy of Letters, History and Antiquities*, Almqvist & Wiksell International, Stockholm.

Malmström, H., Linderholm, A., Lidén, K. *et al.* (2010) High frequency of lactose intolerance in a prehistoric hunter-gatherer population in northern Europe. *BMC Evol. Biol.*, **10**, 89.

Malmström, H., Stora, J., Dalen, L. *et al.* (2005) Extensive human DNA contamination in extracts from ancient dog bones and teeth. *Mol. Biol. Evol.*, **22**, 2040–2047.

Malmström, H., Svensson, E.M., Gilbert, M.T.P. *et al.* (2007) More on contamination: the use of asymmetric molecular behaviour to identify authentic ancient human DNA. *Mol. Biol. Evol.*, **24**, 998–1004.

Malmström, H., Gilbert, M., Thomas, M. *et al.* (2009) Ancient DNA reveals lack of continuity between neolithic hunter-gatherers and contemporary Scandinavians. *Curr. Biol.*, **19**, 1758–1762.

Mardis, E.R. (2008) Next-generation DNA sequencing methods. *Annu. Rev. Genomics Hum. Genet.*, **9**, 387–402.

Martinson, J.J., Chapman, N.H., Reeds, D.C. *et al.* (1997) Global distribution of the CCR5 gene 32-base pair deletion. *Nat. Genet.*, **16**, 100–102.

Metzker, M.L. (2010) Sequencing technologies – the next generation. *Nat. Rev. Genet.*, **11**, 31–46.

Midgley, M.S. (1992) *TRB Culture: The First Farmers of the North European Plain*, Edinburgh U.P, Edinburgh.

Mishmar, D., Ruiz-Pesini, E., Golik, P. *et al.* (2003) Natural selection shaped regional mtDNA variation in humans. *Proc. Natl. Acad. Sci.*, **100**, 171–176.

Noonan, J.P., Coop, G., Kudaravalli, S. *et al.* (2006) Sequencing and analysis of Neanderthal genomic DNA. *Science*, **314**, 1113–1118.

Olivieri, C., Ermini, L., Rizzi, E. *et al.* (2010) Characterization of nucleotide misincorporation patterns in the iceman's mitochondrial DNA. *PLoS One*, **5**, e8629.

Pääbo, S., Poinar, H., Serre, D. *et al.* (2004) Genetic analyses from ancient DNA. *Ann. Rev. Genet.*, **38**, 645–679.

Pakendorf, B. and Stoneking, M. (2005) Mitochondrial DNA and human evolution. *Annu. Rev. Genomics Hum. Genet.*, **6**, 165–183.

Papmehl-Dufay, L. (2006) Shaping an identity. Pitted Ware pottery and potters in southeast Sweden. (Theses and Papers in Scientific Archaeology 7). Stockholm.

Paxton, W.A., Martin, S.R., Tse, D. *et al.* (1996) Relative resistance to HIV-1 infection of CD4 lymphocytes from persons who remain uninfected despite multiple high-risk sexual exposure. *Nat. Med.*, **2**, 412–417.

Pereira, L., Richards, M., Goios, A. *et al.* (2005) High-resolution mtDNA evidence for the late-glacial resettlement of Europe from an Iberian refugium. *Genome Res.*, **15**, 19–24.

Persson, P. and Sjögren, K.-G. (2001) Falbygdens gånggrifter. Del 1. Undersökningar 1985-1998. GOTARC Ser. C. Nr 34. Göteborg.

Poinar, H.N., Schwarz, C., Qi, J. *et al.* (2006) Metagenomics to paleogenomics: large-scale sequencing of mammoth DNA. *Science*, **311**, 392–394.

Price, T.D. (2000) The introduction of farming in northern Europe, in *Europe's First Farmers* (ed. T.D. Price), Cambridge University Press, Cambridge, pp. 260–300.

Rasmussen, M., Li, Y., Lindgreen, S. *et al.* (2010) Ancient human genome sequence of an extinct Palaeo-Eskimo. *Nature*, **463** 757–762.

Renfrew, C. (1981) The Megalith Builders of Western Europe, in *The Megalithic Monuments of Western Europe* (ed. C. Renfrew), Thames and Hudson Ltd, London, pp. 8–17.

Richards, M. and Macaulay, V. (2000) Genetic data and the colonization of Europe: Genealogies and founders, in *Archaeogenetics: DNA and the Population Prehistory of Europe* (eds C. Renfrew and K. Boyle), McDonald Institute for Archaeological Research, Cambridge, pp. 139–151.

Richards, M.B., Sykes, B.C. and Hedges, R.E.M. (1995) Authenticating DNA extracted from ancient skeletal remains. *J. Archaeol. Sci.*, **22**, 291–299.

Richards, M., Macaulay, V., Torroni, A. and Bandelt, H.J. (2002) In search of geographical patterns in European mitochondrial DNA. *Am. J. Hum. Genet.*, **71**, 1168–1174.

Ronaghi, M., Karamohamed, S., Pettersson, B. *et al.* (1996) Real-time DNA sequencing using detection of pyrophosphate release. *Anal. Biochem.*, **242**, 84–89.

Sajantila, A., Salem, A.H., Savolainen, P. *et al.* (1996) Paternal and maternal DNA lineages reveal a bottleneck in the founding of the Finnish population. *Proc. Natl. Acad. Sci.*, **93**, 12035–12039.

Shendure, J. and Ji, H. (2008) Next-generation DNA sequencing. *Nat. Biotechnol.*, **10**, 1135–1145.

Sherratt, A. (1990) The genesis of megaliths: Monumentality, ethnicity and social complexity in Neolithic north-west Europe. *World Archaeol.*, **22**, 147–167.

Smith, C.I., Chamberlain, A.T., Riley, M.S. *et al.* (2003) The thermal history of human fossils and the likelihood of successful DNA amplification. *J. Hum. Evol.*, **45**, 203–217.

Stenbäck, N. (2003) Människor vid havet. Platser och keramik på ålandsöarna perioden 3500-2000 f.Kr. (Theses and Papers in Stockholm Studies in Archaeology 28). Stockholm.

Stiller, M., Knapp, M., Stenzel, U. *et al.* (2009) Direct multiplex sequencing (DMPS) – a novel method for targeted high-throughput sequencing of ancient and highly degraded DNA. *Genome Res.*, **10**, 1843–1843.

Swallow, D.M. (2003) Genetics of lactase persistence and lactose intolerance. *Ann. Rev. Genet.*, **37**, 197–219.

Tambets, K., Rootsi, S., Kivisild, T. *et al.* (2004) The western and eastern roots of the Saami – the story of genetic 'outliers' told by mitochondrial DNA and Y chromosomes. *Am. J. Hum. Genet.*, **74**, 661–682.

Tilley, C. (1996) *An Ethnography of the Neolithic. Early Prehistoric Societies in Southern Scandinavia*, Cambridge University Press, Cambridge.

Tishkoff, S.A., Reed, F.A., Ranciaro, A. *et al.* (2007) Convergent adaptation of human lactase persistence in Africa and Europe. *Nat. Genet.*, **39**, 31–40.

Torroni, A., Bandelt, H.J., Macaulay, V. *et al.* (2001) A signal, from human mtDNA, of postglacial recolonization in Europe. *Am. J. Hum. Genet.*, **69**, 844–852.

Underhill, P.A., Shen, P., Lin, A.A. *et al.* (2000) Y-chromosome sequence variation and the history of human populations. *Nat. Genet.*, **26**, 358–361.

Werbart, B. (1998) Subneolithic: What is it? – 'Subneolithic' societies and the conservative economies of the circum-Baltic region, in *Harvesting the Sea, Farming the Forest: The Emergence of Neolithic Societies in the Baltic Region. Imprint Sheffield* (eds. M. Zvelebil, L. Domanska and R. Dennell), Sheffield Academic Press, England, pp. 37–44.

Willerslev, E. and Cooper, A. (2005) Ancient DNA. *Proc. Roy. Soc. B*, **272**, 3–16.

Wyszomirska, B. (1984) *Figurplastik och gravskick hos Nord- och Nordösteuropas neolitiska fångstkulturer*, Acta Archaeologica Lundensia 18, Lund.

Yang, D.Y., Eng, B., Waye, J.S. *et al.* (1998) Technical note: Improved DNA extraction from ancient bones using silica-based spin columns. *Am. J. Phys. Anthropol.*, **105**, 539–543.

Zvelebi, M. and Rowley-Conwy, P. (1984) Transition to farming in northern Europe: a hunter-gatherer perspective. *Norwegian Archaeological Rev.*, **17**, 104–27.

17

Palaeodemography of Southern Levantine Pre-Pottery Neolithic Populations: Regional and Temporal Perspectives

Vered Eshed[1] and Ehud Galili[2]

[1] *Israel Antiquities Authority, Tel Aviv, Israel*
[2] *Israel Antiquities Authority, Atlit, Israel and Zinman Institute of Archaeology, University of Haifa*

17.1 INTRODUCTION

The Neolithic transition involved profound changes in mobility, social organization, subsistence and technology (cf. Kuijt, 2000; Byrd, 2005). In the Levant, the old lifestyle previously practised (hunting and gathering) for thousands of years, was abandoned about 10 000 years ago (Bar-Yosef, 1995, 2001; Belfer-Cohen, 1991). In the southern Levant, the transition is associated with Pre-Pottery Neolithic populations, who were the first to establish farming communities (Bar-Yosef and Meadow, 1995; Kuijt and Goring-Morris, 2002; Simmons, 2007). This transition had an impact on all aspects of human life, including demography, as greater control over food resources lead to greater life expectancy and increasing fertility (Cohen and Armelagos, 1984; Eshed *et al.*, 2004b; Cohen and Crane-Kramer, 2007; Smith and Kolska-Horwitz, 2007). Pre-Pottery Neolithic sites (PPNB 8550–6750 calBC and PPNC 6750–6250 calBC) reveal the earliest evidence of plant domestication in the late 10th Millennium BC (Bar-Yosef and Belfer-Cohen, 1989). Domesticated animals appeared in the southern Levant later, during the ninth Millennium BC, yet hunting continued throughout the Pre Pottery Neolithic (PPN) period.

PPN sites are found throughout the Levant in a variety of environments. Neolithic sites vary largely in size. There are small temporary sites of a few huts as well as 'mega-size' sites of 10 to 15 hectares, usually built of stone and mud bricks, creating a whole new anthropogenic landscape (Eshed, 2001; Eshed *et al.*, 2004a,b). Burials in the PPN are on-site and are located in houses (under floors) or in courtyards. These include both primary and secondary burials,

Human Bioarchaeology of the Transition to Agriculture Edited by Ron Pinhasi and Jay T. Stock

mostly of single individuals, and group burials (Kuijt and Goring–Morris, 2002; Byrd, 2005; Galili *et al.*, 2005; Simmons, 2007; Eshed, Hershkovitz and Goring-Morris, 2008; Eshed and Nadel, in preparation).

Palaeodemographic studies share a range of limitations, and hence reconstruction of demographic patterns of ancient populations is a difficult and complicated task. Skeletal series recovered from archaeological sites often display demographic patterns, which differ from those of living or historically-documented populations (Sattenspiel and Harpending, 1983; Paine and Boldsen, 2002; Eshed *et al.*, 2004b; Eshed, Hershkovitz and Goring-Morris, 2008). Palaeodemographic studies must consider a range of biases that can affect the reconstruction of demographic patterns of past populations. Estimated sex and age determinations that are based on various methods introduce bias to the reconstructing of life tables, to the sex ratios and child/adult ratios (Hassan, 1981; Konigsberg and Frankenberg, 1992; Hoppa and Vaupel, 2002; Lovejoy, Meindl and Mensforth, 1985b; Paine and Harpending, 1998; Bocquet-Appel and Masset, 1996; Jackes, 1993, 2003). There are also problems resulting from sampling and preservation. An archaeological 'population' represents only a sample of the entire population, which may or may not be representative of the actual population. Furthermore, the final representation of the age groups and sexes in a given skeletal assemblage (i.e. 'population') is affected by various biases such as different burial customs of infants and children (Goring-Morris, 2005; Kirkbride, 1966; Rollefson, 2000). Burial practices (i.e. primary/secondary burials, solitary/multiple burials), may vary widely within the same general population and may affect the total Minimum Number of Individuals (MNI) (Goring-Morris, 2005; Kirkbride, 1966; Rollefson, 2000). Differential preservation through the studied sites can significantly affect the skeletal remains recovered at both the intra- and inter-sample levels, that is infant and children's bones are especially susceptible to poor preservation relative to adult remains (Simmons, Horwitz and Goring-Morris, 2007; Pinhasi and Bourbou, 2008).

These difficulties do not necessarily mean that we should give up our efforts to reconstruct the demographic profile of past populations. These criticisms of palaeodemographic studies, however, were the driving force for the search for new, better and more reliable methods (Hoppa and Vaupel, 2002; Paine, 1989, 2000; Konigsberg and Frankenberg, 1992; Gage, 1988, 1990). It is hoped that the present study can provide an overview of the palaeodemographic profile of Middle and Late PPN period in the Levant – a period of great importance to our understanding of the effect of the Neolithic transition on the demography and other social and biological aspects of these early farming populations.

Farming creates crowded, dense, permanent settlements, which are often associated with and increased prevalence of infectious diseases. Daily contact with livestock could have led to human infections by zoonoses, and changes in diet could have adversely affected human health (Cohen, 1989; Armelagos, Goodman and Jacobs, 1991; Cohen and Armelagos, 1984). The effect of the transition to agriculture on the demography and health of human populations is still an unanswered question, over which opinions remain strongly divided (Cohen and Armelagos, 1984; Bocquet-Appel and Naji, 2006; Bocquet-Appel, 2009; Cohen and Crane-Kramer, 2007). Cohen and Armelagos (1984) concluded that the major changes following the transition to agriculture in various parts of the world were higher rates of infection, a decline in overall quality of nutrition, a reduction in physical stress, and a decline in mean age at death (and/or life expectancy at various ages).

Not all studies agree with these conclusions (cf. Papathanasiou, this volume). For example, lower mean age at death amongst agriculturalists, compared to hunter-gathers, was interpreted as a reflection of an increase in mortality and declining life expectancy with the shift to agriculture (Cohen and Armelagos, 1984; Kennedy, 1984). However, mean age at death is related to fertility and birthrate, and not mortality (Sattenspiel and Harpending, 1983; Cashdan, 1985). Human populations experiencing growth will have a greater number of younger individuals, resulting in a larger proportion of juvenile relative to adult skeletons. The Neolithic Demographic Transition (NDT) model takes into consideration the variation in the birth rate (fertility) and mortality as the main factors that are responsible for the drastic population increase following the transition to agriculture in most world regions (Bocquet-Appel, 2008, 2009; Bocquet-Appel and Naji, 2006). Indeed, in many parts of the world, birth rate increased during the adoption to agriculture, while mortality showed no change over the same period. For example, Buikstra, Konigsberg and Bullington (1986) and Johansson and Horowitz (1986) claimed that when considering the wide variety of ecological conditions under which local populations adopted agriculture, it is reasonable to expect a wide variety of demographic responses to the adoption of agriculture and to its intensification. Gage and DeWitte (2009) stated that the 'demographic models' of the agricultural transition are largely based on unsubstantiated assumptions. Though generally population growth accompanied the agricultural transition, it is not clear how fertility and mortality changed (if at all) consequent to the transition, or whether the transition caused a decline in health. If there is a negative relationship between mortality and health, then when health declines, mortality increases (Bocquet-Appel, 2009; Coale, 1972). Therefore, researchers should be aware of the underlying assumptions and be open to the testing of future empirical evidence (Gage and DeWitte (2009)).

A previous study of the demographic changes following the transition from a hunting-gathering way of life (Natufians) to an agricultural, food producing economy (Neolithic) in the southern Levant indicated no increase in mortality with the advent of agriculture (Eshed, 2001; Eshed *et al.*, 2004b). On the contrary, both life expectancy at birth (24.6 vs. 25.5 years) and adults' mean age at death (31.2 vs. 32.1 years) increased slightly from the Natufian to the Neolithic period (assuming stationary populations). Yet, the transition to agriculture affected males and females differently. While females experienced a decrease in the mean age at death in the Neolithic period, males experienced an increase (Eshed, 2001; Eshed *et al.*, 2004b). Natufian women lived longer, probably because of less frequent births, while Natufian males were more likely to die earlier, perhaps because of dangerous hunting activities and inter-social conflicts (Eshed *et al.*, 2004b). It was shown that in the Levant, the mean age at death for females declined with the introduction of agriculture and fertility increased (Eshed *et al.*, 2004b). These finds suggested that the population increased (about $r = 0.5$–1%) following the transition to an agriculture way of life (Eshed *et al.*, 2004b). Hershkovitz and Gopher (1990) estimated that the mean age at death for the PPNB populations (based on combined skeletal samples of several small Neolithic groups) was 18.4 years. The number of people older than 45 years was small (ranging from 0–15%). According to Kurt and Rohrer-Ertl (1981) and Smith, Bar-Yosef and Sillen (1984), mortality rates of PPN populations in the southern Levant probably remained high (50% before the age of 20 years). A combined sample of 303 individuals from PPNB sites contained 32.3% children aged 0 to 14 years, whereas only 23.1% of a combined Natufian sample of 368 individuals were children. In the PPNB of the Levant a sample, 16.8% of adolescents were reported, and the ratio of children/adolescents/adults was 98/51/154. Hershkovitz and Gopher (1990) concluded that there was a rapid population growth in the PPNB period compared to the previous Natufian period.

The demographic characteristics of the southern Levant populations before (Natufian) and after the transition to agriculture (Pre Pottery Neolithic) were previously studied by Eshed *et al.*, 2004b. The aim of the present study is to reconstruct the demographic profile of the Middle-Late Pre pottery Neolithic populations of the southern Levant, by focusing on differences in the demographic characteristics of three large skeletal populations from the southern Levantine Neolithic sites of Atlit Yam (AY), Kfar HaHoresh (KHH) and 'Ain Ghazal (AG) (Figure 17.1, and see below). These are also compared to the demographic patterns of the pooled PPNB populations based on a sample of 15 sites. Subsequently, we address possible associations of the demographic profiles with the emergence of agro-pastoral-marine communities, as well as with additional differences in subsistence strategies and environmental conditions.

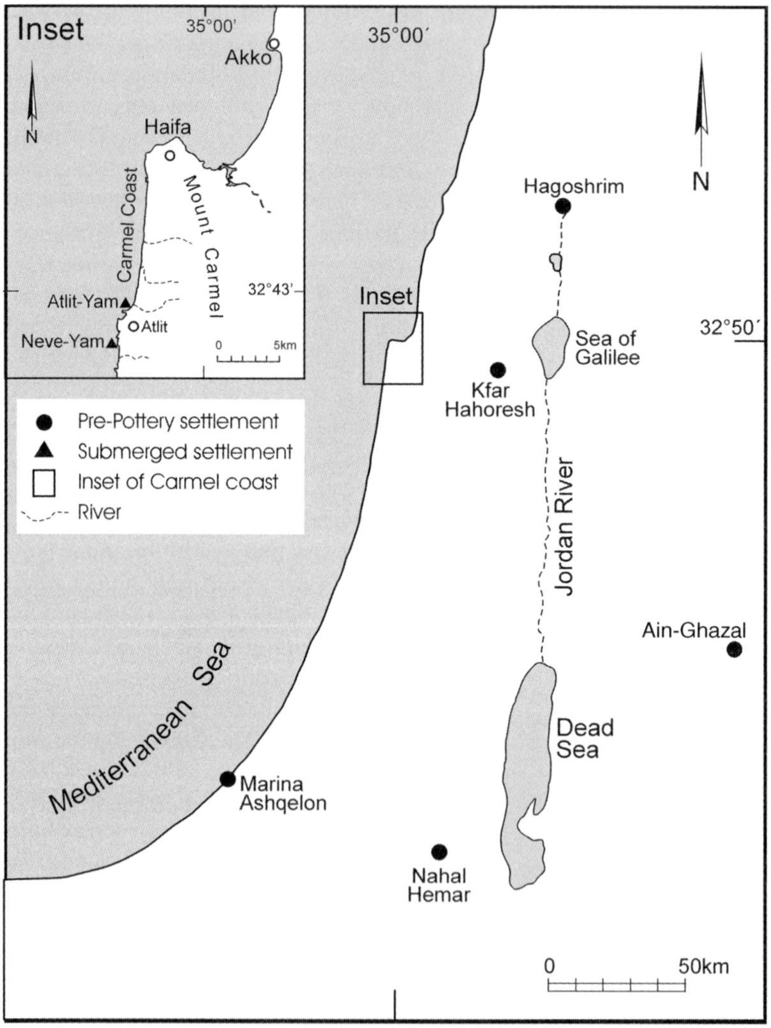

Figure 17.1 Map of middle and Late PPN sites in the south Levant, of the study

17.2 SAMPLES AND ARCHAEOLOGICAL CONTEXT

Excavations in the sites of Atlit-Yam (AY), Kfar HaHoresh (KHH) and 'Ain Ghazal (AG) (Figure 17.1) yielded numerous skeletal remains and revealed many new aspects about human lifestyle health and activity during the transition to agriculture. Below, we provide a brief summary of the human bioarchaeological context of each site.

17.2.1 Kfar HaHoresh

The site of Kfar HaHoresh is situated on Nazareth Hills, Lower Galilee, Israel (Figure 17.1), and revealed strata that span the early-middle and late PPNB (about 8550–6750 calBC). On the basis of its location and the contextual associations of the material culture recovered, the site was interpreted as a long-lived but sporadically used cult and mortuary centre (Goring-Morris, 2000, 2005; Goring-Morris *et al.*, 1994/5, 1998, Goring-Morris, Boaretto and Weiner, 2001; Goring-Morris and Horwitz, 2007; Eshed, Hershkovitz and Goring-Morris, 2008). To date, a total 54 human skeletons from this site were studied (Eshed, Hershkovitz and Goring-Morris, 2008). The skeletal sample was derived from the systematic excavations of 425 m^2. Grave contexts comprised of both primary and secondary burials included both individual and multiple interments (Horwitz and Goring-Morris, 2004; Goring-Morris and Horwitz, 2007; Eshed, Hershkovitz and Goring-Morris, 2008).

The KHH burials included some unique elements, in addition to the more common Levantine PPN burial practices such as the use of stone slabs, stelae or other 'megalithic stones', which were embedded in plaster surfaces for the purpose of marking the skeletons and graves associated with funerary structures, as well as specific features such as post-holes accompanying the graves. Grave offerings were found in a considerable number of cases. Clear association of human burials with animal bones was also noted, and in many cases it included complete animals (e.g. gazelle) and/or articulated parts of animals (cattle). In addition, secondary group burials appeared in organized patterns. In one of the secondary burials, a number of human mandibles were placed together in one location (Goring-Morris *et al.*, 1994/5, 1998; Goring-Morris, Boaretto and Weiner, 2001; Goring-Morris, 2000, 2005; Eshed, Hershkovitz and Goring-Morris, 2008). Whether these new practices were unique to the specialized, regional, funerary centre at KHH, or may have prevailed in other contemporaneous locations, is yet to be revealed.

17.2.2 'Ain Ghazal

The site of 'Ain Ghazal is located near the modern city of Amman, Jordan (Figure 17.1). The site was continuously inhabited from about 8200 to 5800 calBC. Changes were identified in the archaeological record and the two millennia of occupation were divided into several periods: Middle Pre-Pottery Neolithic B (MPPNB) 8200 to 7500 calBC, Late Pre-Pottery Neolithic B (LPPNB) 7500 to 6750 calBC, Pre-Pottery Neolithic C (PPNC) 6750 to 6250 calBC and Yarmoukian Pottery Neolithic 6250 to 5800 calBC. During the seventh and sixth millennia BC, the inhabitants lived at the site all year round and subsisted on farming, hunting and sheep herding (in the LPPNB), pigs (PPNC) and cattle (by the end of the PPNC or Yarmoukian, if not earlier) (Rollefson, 1985, 1989, 2000, 2001; Rollefson and Kohler-Rollefson, 1993; Rollefson, Simmons and Kafafi, 1992; Rollefson *et al.*, 1995). Their subsistence spectrum also involved the cultivation of domestic wheat, barley, lentils, peas and chickpeas. Two caches of plaster statues intentionally buried beneath the floor of long-abandoned houses were discovered at these periods.

Eighty middle PPN burials from 'Ain Ghazal are discussed here (data used are based on Rolston in Rollefson *et al.*, 1995). In most adult burials, the skeletons were found in a flexed position below the floors of the houses or in a courtyard and the skulls were removed after the initial burial. About one-third of the adolescent and adult burials occurred in trash deposits in a variety of postures and always with the skull present and articulated to the body. Differences in these burials may have suggested varied treatments and reverence of individuals, and therefore social distinctions (Rollefson, Simmons and Kafafi, 1992: 461). Many of the buried skulls were treated with plaster moulded into naturalistic facial features, and some were painted. Caches of skulls disarticulated from their bodies were found intentionally buried in groups, orientated in the same direction (Rollefson, Simmons and Kafafi, 1992; Rollefson *et al.*, 1995). Many of the skulls were buried beneath floors, a practice that may indicate an ancestor cult. About half of the PPNC buried individuals were in secondary burials in, or near, corridor houses, believed to be used as storage for the property of semi-nomadic herders during long periods of site desertion (Rollefson, Simmons and Kafafi, 1992; Rollefson and Kohler-Rollefson, 1993; Kuijt, 2008).

17.2.3 Atlit-Yam

The submerged PPNC site of Atlit-Yam (about 7000–6200 calBC) lies at 8 to 12 m below the present sea level, off the northern Carmel coast, 10 km south of Haifa, Israel (Figure 17.1). It is the largest and best preserved submerged PPNC site and the only known coastal PPNC site with *in-situ* human remains (Galili *et al.*, 1993, 2005; Eshed, 2001; Eshed *et al.*, 2004a,b).

The archaeological remains included rectangular dwellings (Figure 17.2) with long walls, ritual structures built of megaliths (Figure 17.3), water wells and paved areas. In addition, stone, bone, wood and flint artefacts were recovered, as well as well preserved faunal and floral remains. The economy of the village was complex, based on farming, grazing, hunting, gathering and fishing. AY is the earliest example of a Mediterranean fishing village, subsisting simultaneously on farming, grazing, hunting gathering and fishing (Galili, Gopher and Horwitz, 2002; Galili *et al.*, 2005).

Sixty-five human skeletons from the site were studied. Burials were on-site, and most were concentrated in its north-west section (Areas K and L; Figure 17.2). Most burials were located near or within structures, mainly stone-built walls. Graves were usually simple, showing minimal investment. The deceased were interred in simple pits dug in the clay, sometimes marked by a few stones on top or around the grave. Most burials were primary, although evidence for secondary burials does exist (Figures 17.4 and 17.5). Group burials comprise one-third of all primary burials. Isolated bones were found throughout the site. These probably represented post-depositional exposure and scattering by sea erosion. Most of the deceased were found in flexed or semi-flexed positions, with no clear orientation pattern (Figure 17.6). Skulls were usually intact. Offerings were modest, but certainly existed in a considerable number of graves (Galili *et al.*, 2005).

The combined PPNB and PPNC samples presented in this study include the skeletal remains of 271 individuals, including 100 subadults and 171 adults from 15 PPN sites excavated in the Mediterranean zone of the southern Levant (Table 17.1). The excavations discovered largely sedentary communities subsisting on combined farming, hunting and foraging. The sites analysed are PPNB Nahal Oren (11), Abu Gosh (30), Beisamoun (19), Horvat Galil (4), Yiftahel (8), KHH (54), AG (80) and the PPNC AY (65). The skeletons were studied by V.E. with the exception of the material from AG (published by Rolston in Rollefson *et al.*, 1995).

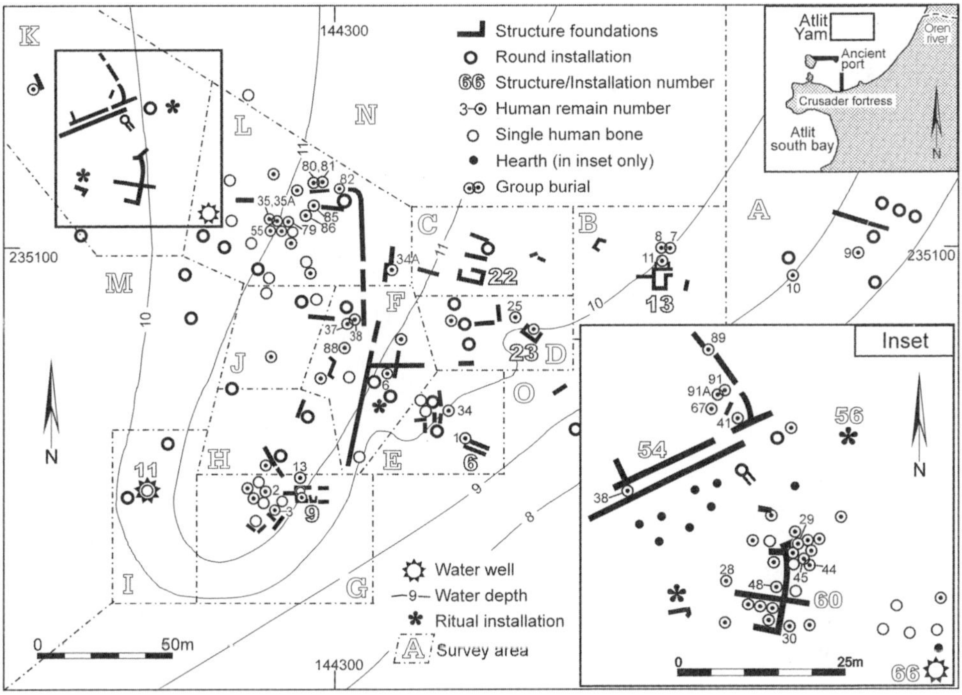

Figure 17.2 Plan of Atlit-Yam: Structures and human remains

17.3 METHODS

The Number of Individuals (NI): This parameter was calculated using the greatest number of same age-range related diagnostic and indicative bones. This included all bones of individuals that were recovered from what has been interpreted in the field as either intended graves (primary or secondary) or unintended burial contexts. Post depositional processes in prehistoric sites differentially affect the representation of certain skeletal elements and the overall number

Figure 17.3 Megalithic structure in the submerged PPNC site of Atlit-Yam

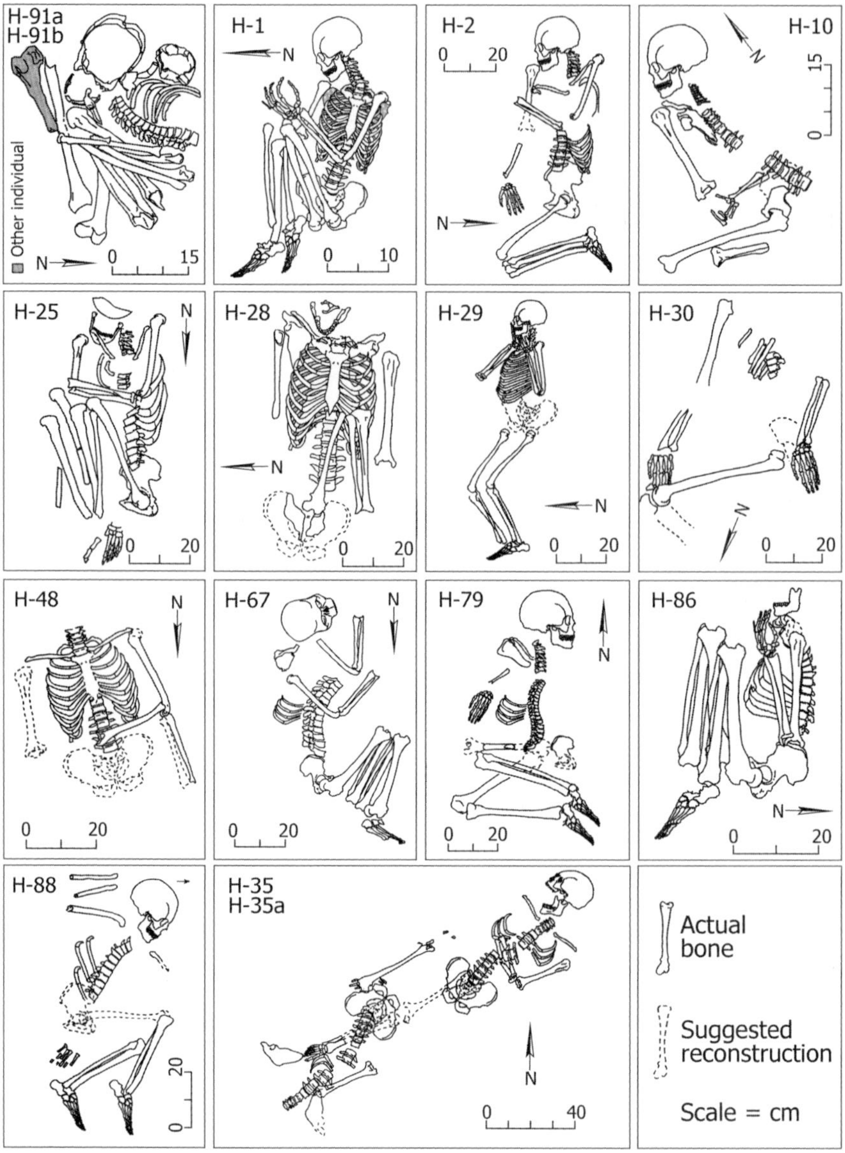

Figure 17.4 Typical articulated skeletons from Atlit-Yam (adults)

of recovered bones. Thus the NI usually represented a considerably lower number of individuals compared to the actual number originally deposited in a given site.

Minimum number of individuals (MNI): This parameter was calculated for each defined grave separately. The sample included all individuals recovered from what were interpreted as intended burial contexts, whether primary or secondary. Isolated skulls were not included in the count, since the postcranial bones of the same individuals may have been recovered (from other parts of the site). Scattered bones of children were taken into consideration, only if their ages were not represented in the sample.

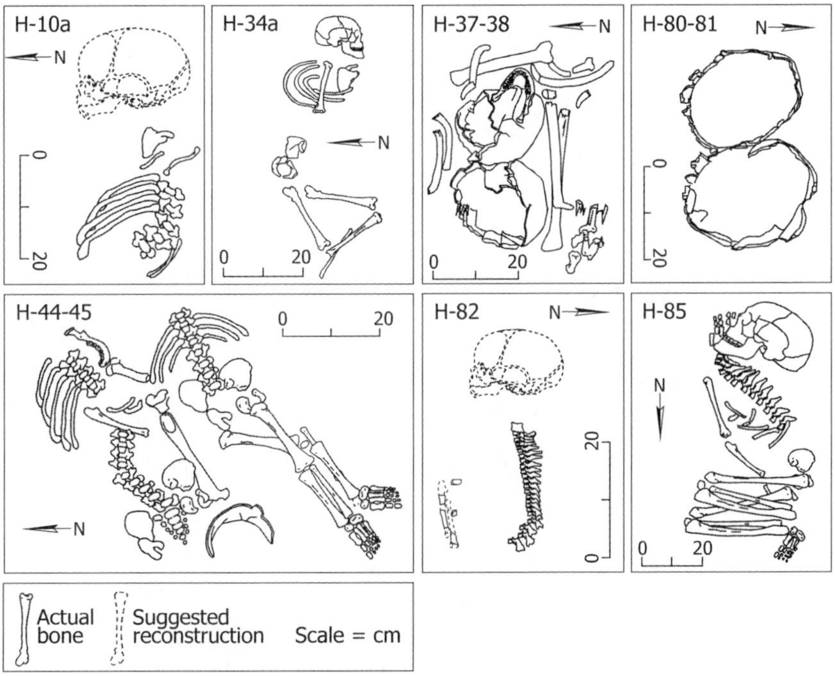

Figure 17.5 Typical primary and secondary burials from Atlit-Yam (children)

Estimated Number of Individuals (ENI) (also reported as the Maximum number of Individual: Max. NI): This parameter comprises the total number of individuals recovered at a given site. ENI was calculated by referring to the skeletal material found on the site while considering the archaeological context. It included:

1. the greatest number of the same age-range related diagnostic and indicative bones found at the site;
2. all individuals that were recovered from intentional burials;
3. isolated post cranial bones or skulls that did not belong to the *in-situ* burials and were not already represented in the sample.

ENI is thus greater in magnitude than MNI and provides a more realistic estimation of the skeletal population. In this study we calculated ENI for AY and KHH (Eshed, Hershkovitz and Goring-Morris, 2008).

17.3.1 Sex Determination for the Adult Population

Sex determination of all individuals over 15 years of age was based on standard criteria: morphology of skulls and long bones (Bass, 1987), morphology of the innominate bones (Bass, 1987; Segebarth-Orban, 1980; Washburn, 1948) and the vertical diameter of the femoral heads (Bass, 1987). The sex ratio was calculated as (males/females) $\times$ 100.

Figure 17.6 Primary flexed burials of an adult individual from the site of Atlit-Yam

17.3.1.1 Age Determination

Adult age determinations were based on morphological changes of:

- the auricular surface of the ilium (Lovejoy *et al.*, 1985a);
- the pubic symphysis (Brooks and Suchey, 1990);
- stages of tooth attrition (using standards modified from Hillson, 1986 and Lovejoy, 1985);
- morphological changes of the sternal end of the ribs (Loth and Iscan, 1989); and
- the presence of osteophytes and arthritic lesions on the vertebra (Nathan, 1962).

For young adults (15–25 years old). the additional criteria were the fusion of long bone epiphyses (Johnston and Zimmer, 1989), and of the sternal ends of the clavicles (Szilvassy, 1980). In the case of subadults (0–15 years of age), the criteria also included the long bone diaphyseal lengths (Bass, 1987), fusion of long bone epiphyses (Johnston and Zimmer, 1989) and stages of tooth eruption and development (Ubelaker, 1989). An estimated average age

Table 17.1 The Neolithic sample used in this study

Pre-Pottery Neolithic B/C sites	Subadults	Adults	Total
Nahal Oren[a]	1	10	11
Abu Gosh[b]	3	27	30
Beisamoun[b]	7	12	19
Horvat Galil[c]	1	3	4
Yiftahel[a,d]	3	5	8
Kfar HaHoresh[1,1*]	19	35	54
'Ain Ghazal[e]	42	38	80
Atlit Yam[a]	24	41	65
Total	100	171	271

[a] Eshed, 2001, Eshed, Hershkovitz and Goring-Morris, 2008*.
[b] Arensburg *et al.*, 1978.
[c] Hershkovitz and Gopher, 1988.
[d] Hershkovitz *et al.*, 1986.
[e] Rolston in Rollefson *et al.*, 1995.

for each skeleton was computed by calculating a simple arithmetic mean of the ages obtained by the different ageing methods.

17.3.1.2 Palaeodemographic Methods

Life tables were constructed using a 10-year cohort, following Ubelaker (1974). All adult individuals for whom age could not be determined were added to the adult age categories according to the relative frequency of individuals found in each age group. This addition was based on the assumption that the observed percentages of the age groups indeed represent the real mortality pattern of the population, assuming stationary populations (Johansson and Horowitz, 1986). It was suggested elsewhere that population growth rates in the Levant at the advent of agriculture were approximately 0.5 to 1% (Eshed *et al.*, 2004b; Caldwell and Caldwell, 2003). Bonneuil (2005) argued that it may be wrong to use the assumption of stability for the purpose of examining palaeodemographic assemblages. Averaging a series of palaeodemographic life tables might provide more reasonable results than using a single table. However, this method has not been practically applied (Gage and DeWitte (2009)).

17.3.1.3 Population Growth

This is an important issue in the interpretation of palaeodemographic trends. A growing population will appear to have a lower expectation of life and lower Mean Age at Death (MAAD), while a declining population will appear to have a higher expectation of life and a higher MAAD. Statistical analyses were carried out using Chi-square tests.

17.4 RESULTS

The main demographic characteristics and demographic (mortality) curves were constructed for the sites of KHH (54 individuals), AG (80 individuals) and AY (65 individuals). All the sites

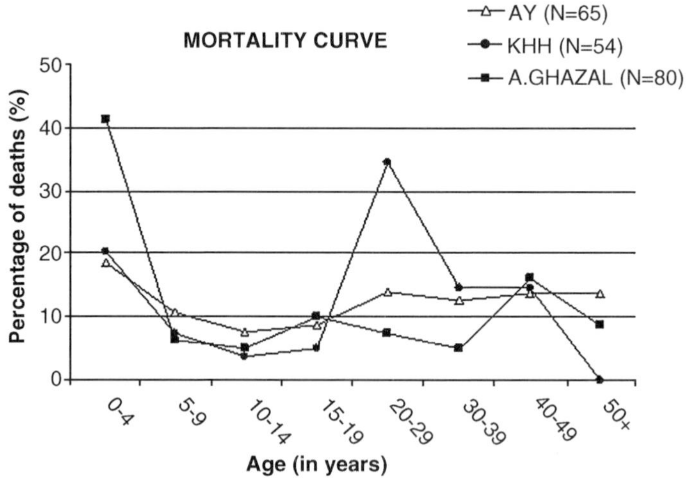

Figure 17.7 Mortality curve of three Neolithic populations (AY, KHH and AG)

in the study (in addition to KHH, AG and AY) were then combined together to the general Neolithic sample (a total of 271 individuals. Figure 17.7).

17.4.1 The Demographic Characteristics of Atlit Yam, Kfar HaHoresh and 'Ain Ghazal

17.4.1.1 Atlit Yam

Sixty-five skeletons were identified, including 24 aubadults (36.9%) and 41 adult (63.1%) individuals (Tables 17.1 and 17.2). Amongst the adults, 20 males and 15 females were identified. The life expectancy at birth (e^0x) for the AY population was calculated to be 25.7 years. The mean age at death for the adults was 36.9 years.

17.4.1.2 Kfar HaHoresh

Fifty-four skeletons were identified, including 19 subadults (35.1%) and 35 adult (64.8%) individuals (Tables 17.1 and 17.2). Amongst the sexed adults, 19 males and 5 females were identified. The life expectancy at birth (e^0x) was calculated to be 23 years. The mean age at death for the adults was 30.8 years (Eshed, Hershkovitz and Goring-Morris, 2008).

17.4.1.3 'Ain Ghazal

Eighty skeletons were identified, including 42 subadults (52.5%) and 38 adult (47.5%) individuals (Tables 17.1 and 17.2). Amongst the adults, 20 males and 15 females were identified. The life expectancy at birth (e^0x) was calculated to be 20.4 years. The mean age at death for adults was 36.8 years.

17.4.2 Mortality Patterns

The mortality pattern of the entire population (the three sites) is provided in Figure 17.7. The main differences between the AY and AG populations, according to the mortality curves (Figure 17.7), were:

Table 17.2 Age and sex distribution in the Neolithic populations used in the study

Site	N	Sex				Age								
		M	F	?	C	0–4	5–9	10–14	15–19	20–29	30–39	40–49	50+	15<
Kfar HaHoresh	54	19	5	13	17	11	4	2	2	13	5	5	0	12[a]
Abu Gosh	30	7	3	17	3		3	4	4		2			17
Beisamoun	19	3	4	5	7	6	1							12
Nahal Oren	11	1	1	8	1		1		1	1	4		2	2[b]
Horvat Galil	4	1	2	0	1	0	1	0	0	1	0	0	0	2
Yiftahel	8	2	3		3	3			2	1		1		1
‘Ain Ghazal	80	20	15		42	33	5	4	8	6	4	13	7	
Atlit-Yam	65	20	15	6	24	12	7	5	5	7	5	5	5	14[c]

[a] 1 individual is >40 years; 4 individuals are >30 years; and 4 individuals are >20 years.
[b] 1 individual from this group is >20 years old.
[c] 1 individual are >30 years; and 2 individuals are >20 years.

1. relatively high death rates in the 20 to 39 cohort at AY population; and
2. more people reached the advanced age of 50+ years, in AY (13% vs. 9%, respectively, $\chi2 = 0.9487$, df = 1, p = 0.330).

The KHH mortality curve is different from both the AY and AG in the high mortality rates amongst the young adults group of 20 to 29 years and in the total absence of old individuals (50+ years).

The demographic characteristics of the combined Neolithic sample are presented in Tables 17.1 and 17.2. In total, 271 skeletons were identified, including 100 subadults (36.9%) and 171 adults (63.1%). Amongst the adults, 73 males and 48 females were identified. The life expectancy at birth in the general Neolithic population was 22.9 years.

The mortality pattern of the combined Neolithic sample and the AY site is presented in Figure 17.8. Both curves show similar values for most age cohorts; however, several interesting differences were revealed. In the combined Neolithic sample, more individuals died in the adult age groups of 15 to 40 vs. the AY population (42.1% vs. 23%; $\chi2 = 10.3462$, df = 1, $p = 0.001$). Consequently, more individuals reach the advance age of 50 years in the AY compared to combined Neolithic population (14% vs. 8%, respectively, Figure 17.7; significant at $p = 0.001$).

The comparison of sex ratios at AY and AG (1.33 males to 1 female) and the combined Neolithic sample (1.52 males to 1 female) were similar. The KHH population has a different ratio of 3.8 males to 1 female (Tables 17.1 and 17.4). Given that KHH was interpreted as a long-lived but sporadically used cult and mortuary locality (Goring-Morris, 2000, 2005; Goring-Morris *et al.*, 1994/5, 1998), the sex ratio may be biased and does not reflect the true sex ratio in the population (Eshed, Hershkovitz and Goring-Morris, 2008; Koerner and Blakely, 1985). Thus, in order to identify the patterns of male/female sex ratio and the associated cultural and economic aspects, we compared the populations of AY and AG, as both sites were residential villages located in two different ecological niches. AY was a PPNC coastal site and its inhabitants mainly relied on farming, herding and fishing, while AG was an inland site and its inhabitants mainly relied on herding and farming. The mortality distribution diagram demonstrated considerable differences between the populations of the two sites.

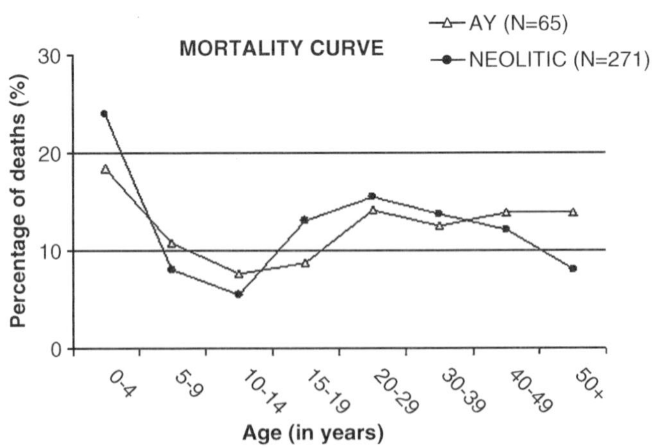

Figure 17.8 Mortality curve of AY population compared to the general Neolithic populations

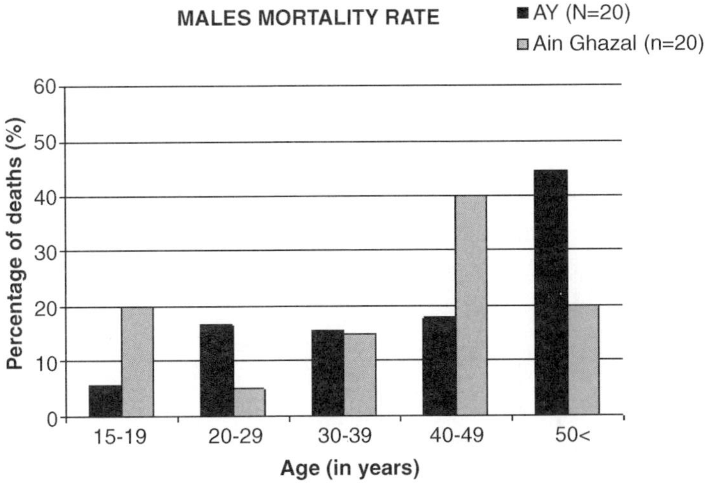

Figure 17.9 Male's mortality rates of Atlit-Yam and Ain-Ghazal populations

17.4.2.1 Males

The examination of the mortality curve of both sexes (Figure 17.9) shows in the case of AY, that more adults reached the advanced age of 50+ years than in the case of the AG population. This pattern is more pronounced when only males are considered: 44.4% of the males in AY lived beyond the age of 50 years, in comparison to only 20% in the AG population (Figure 17.9, $\chi 2 = 2.849$, df $= 1, p = 0.091$). The AY male population displayed a higher mean age at death, compared to that of AG (43 vs. 39 years, respectively, Table 17.4). Life expectancy ($e^{o}x$) at the age of 40 years was therefore lower for the AG population compared to the AY: 8.5 years vs. 10 years, respectively (Table 17.3). In the 30 to 49 cohorts, male mortality rates in both populations were similar (15.6% AY vs. 15% AG). In the younger-adult age category, more males died in the 15 to 19 year cohorts in the AG population compared to AY population (20% vs. 5.6%, respectively, ($\chi 2 = 1.889$, df $= 1$, $p = 0.169$); the trend reversed with more deaths in the 20 to 29 year cohort in the AY population (16.7 vs. 5%, Figure 17.9).

17.4.2.2 Females

Females of AY had a similar mean age at death to that of AG females (33.2 y and 31.0, respectively, Table 17.4), although the mortality pattern according to age was different (Figure 17.10). The AY population had low values of female mortality between the ages of 15 to 19 years and high rates of death between the ages of 30 to 39 years, when compared to the AG females. The pattern of females' death in the AG is one of the high mortality rates for the 15 to 19 years cohort and low mortality rates in the 30 to 39 years cohort. The death rates at the age cohorts of 20 to 29, and 40 to 49 years were similar in both sites.

17.4.3 The Subadult Populations

The subadult/adult ratio in AY is similar to that of the combined Neolithic sample, (0.59 vs. 0.58, respectively) and the KHH ratio is lower (0.54). The only site in which the number of subadults exceeds the number of adults was AG, with a ratio of 1.11:1 (Table 17.1; AG vs.

Table 17.3 Life tables for the Atlit-Yam population and the combined Neolithic sample

	All Neolithic							Atlit Yam Population	
Age group (X)	Number of deaths (Dx)	Percentage of deaths (dx)	Survivors entering (Ix)	Probability of death (qx)	Total year lived between X and X + 5 (Lx)	Total year lived between X and X + 5 (Tx)	Life expectancy (e0x)	Number of deaths (Dx)	Life expectancy (e0x)
0–9	19	29.2	100.0	0.29	853.9	2568.0	25.7	87	22.9
10–19	11	16.9	70.8	0.24	625.4	1714.2	24.2	50	21.4
20–29	9	13.8	54.3	0.26	472.7	1088.7	20.0	42	17.6
30–39	8	12.3	40.2	0.31	339.4	616.0	15.3	37	13.4
40–49	9	13.8	27.7	0.50	207.4	276.6	10.0	33	9.0
50 <	9	13.8	13.8	1.00	69.1	69.1	5.0	22	5.0
Sum	65	100.0	0.0	0.0				271	

Table 17.4 Demographic parameters of the three key Neolithic sites of the study and for the combined Neolithic sample

Demographic parameters	Life expectancy at birth (e^0x)	Mean Age at Death (MAAD) adult	Mean Age at Death males	Mean Age at Death females	Sex Ratio (males/females) × 100
Sites					
Atlit Yam	25.7	36.9	43.0	33.2	133
Ain Ghazal	20.4	36.8	39.0	31.0	133
Kfar HaHoresh	23.0	30.8	31.5	25.6	380
Combined Neolithic PPNB/C (271)	22.9	33.4			152

combined Neolithic sample, $\chi^2 = 6.239$, df $= 1$, $p = 0.012$). The mortality pattern in all three sites and the combined Neolithic sample indicates high mortality values amongst infants and young children (0–4 years) in all the populations (Figures 17.7 and 17.8, Table 17.3). However, a comparably high mortality value for the young age cohort of 0 to 4 years was observed in the case of the AG population (41.3% compared to 18.5%, 21% and 24% for the AY, KHH and the combined Neolithic sample, respectively) (Figure 17.7 and 17.8). After the age of 5 years, mortality decreased steadily until the age of 15 years.

17.5 DISCUSSION

The demographic patterns described above indicate that there are considerable differences between the demographic profiles of the MPPNB and LPPNB populations (AG, KHH), that of the AY population, and that of the general Neolithic population. The significant patterns found

Figure 17.10 Female's mortality rates of Atlit-Yam and Ain-Ghazal populations

are discussed below, in an attempt to identify economic and environmental reasons for the differences observed.

High life expectancy was observed in the case of AY when compared to AG and the combined Neolithic sample. As seen from the assessment of the mortality curves (Figures 17.7 and 17.8), more individuals of the AY population reached the advanced age of 50+ years compared to the AG population or the general Neolithic sample. This pattern was more pronounced when only males were considered. The males at AY displayed a higher mean age at death, compared to that of AG (43 vs. 39 years of age). Life expectancy ($e^{o}x$) at the age of 40 years was therefore lower for the AG males compared to the AY (8.5 years vs. 10 years, respectively, Table 17.3). Thus, 44.4% of the AY males lived beyond the age of 50 years, in comparison to only 20% in the AG population (Figure 17.9, p significant at $= 0.09$). The higher life expectancy of AY males relative to AG males may have been associated with dietary and environmental differences between the two populations. The combined agro-pastoral-marine subsistence and sedentary way of life in AY may have provided more secure and stable food supplies and better balanced nutrition than in AG (Galili, Gopher and Horwitz, 2002; Galili, 2004). The regular and frequent consumption of marine resources would have provided a rich supply of protein, fat and micronutrients. Cereal, vegetables, fruits, meat, and later, other animal products as well as seafood, especially fish, are the core elements of the present-day Mediterranean diet, which is known to be nutritious and healthy (Galili *et al.*, 2005).

The lower life expectancy of AY females when compared to males deserves some attention. Assuming that males and females shared the same diet and environmental conditions, the lower life expectancy of AY women may be attributed to complications associated with pregnancy and birth. The increase in adult female mortality following the onset of the Neolithic period could have been due to an increase in the number of births and earlier onset of pregnancy (Eshed *et al.*, 2004b). Boldsen and P'Aine (1995) found a similar trend in adult mortality when comparing Mesolithic and Neolithic populations of Europe. They demonstrated that during the Neolithic and subsequent periods, female conditional survivorship through the reproductive years declined *vis-à-vis* male survivorship of the same ages. More direct evidence is provided in the profile of archaeological cemetery populations, which Bocquet-Appel (2002); and Bocquet-Appel and Naji, 2006 believe to represent a 'signature' of increasing fertility after the origins of agriculture, followed somewhat later by an increase in mortality (Cohen, 2009). Death risks for pregnant women rise with the number of births (Friedlander, 1996). During pregnancy and birth, the risk for older women is higher than for younger women, since the former are more prone to illnesses and complications (Friedlander, 1996). Hershkovitz *et al.* (1991) suggested that the AY population adapted to malaria, which was common in the marshy coastal environment, by being infected with anaemia thalassaemia. The influence of this syndrome on females vs. males may have induced female vulnerability during pregnancy and birth and may have been one of the reasons why females did not reach the age of 50 in AY.

Comprehensive demographic studies of the PPNB, AG population and the AY PPNC population enable a comparison of the two populations. That demographic comparison was made because yet little is known on the demographic patterns of AG PPNC population. The observed high infant/child mortality at AG relative to AY at the ages of birth to 5 years of age is not a product of adverse taphonomic processes in the latter, as the preservation of fauna (Horwitz, Lernau and Galili, 2007; Greenfield, Galili and Horwitz, 2007) and human bones (Galili *et al.*, 2005) at AY is remarkably good. Although different burial customs of children in these two sites may have contributed to these differences, it is unlikely that such differences could have induced such significant patterns.

Assuming that the reproductive potential of the Neolithic populations was similar (average number of pregnancies per female), it is probable that the differences between the abilities to produce and preserve sufficient amounts of food resources between these two societies had a major impact on the survivorship of infants/children and on the rate of population growth. Environmental conditions and the prevalence of diseases probably also influenced subadult survival. The AY site was a permanent settlement, subsisting on balanced, diverse agro-pastoral marine resources (Galili, Gopher and Horwitz, 2002, Galili *et al.*, 2005). In the PPNC period at AG, a considerable portion of the population subsisted on herding and transhumance nomadic way of life away from the site (Rollefson and Kohler-Rollefson, 1989, 1993). Therefore a significant segment of the population left the site for the great part of the year. It was suggested that during the middle and late PPN, the productivity of the environment started to deteriorate due to over-exploitation of natural resources. This interpretation is supported by Smith and Kolska-Horwitz's report (2007), which asserts that the people of the MPPNB were healthier than those of the previous periods (Late/Final Natufian and PPNA). They demonstrated the existence of two major low points in the health status of skeletal samples from the southern Levant. The first occurred during the Late Natufian-PPNA transition, while the second appeared in the end of the PPN period (LPPNB-PPNC transition). Both periods are characterized by environmental deterioration and were associated with changes in patterns of human settlement (Goring-Morris and Belfer-Cohen, 1998, 2008). There is also evidence of a decline in food resources during the Late and Final Natufian, which is associated with the onset of the Younger Dryas cold and dry climatic phase (Belfer-Cohen, 1991; Bar-Yosef, 1998, 2001). At the end of the PPNA, there was an increase in the exploitation of wild cereals together with intensification of the reliance on small game and reptiles as food resources. This resulted in increased energy consumption for smaller returns (Stiner and Monroe, 2002; Munro, 2004; Smith and Kolska-Horwitz, 2007).

An improvement in the overall health of southern Levantine populations is evident during the MPPNB. It followed the onset of a fully agricultural economy (Eshed, Gopher and Hershkovitz, 2006; Eshed *et al.*, in press) with the appearance of a subsistence system comprising of domestic sheep and goats, domestic cereals and legumes and possibly milk products (Smith and Kolska-Horwitz, 2007). Eshed *et al.* (2010) reported a significant increase in the number of individuals with inflammatory bone-induced lesions following the transition from the Natufian to the Neolithic. A previous palaeodemographic study (Eshed *et al.*, 2004b) on the Natufian and Neolithic populations showed no evidence of reduction in life expectancy, which seems to run against evidence for an increase in inflammatory disease during the Pre Pottery Neolithic period. We thus speculate that an increase in resistance to infectious diseases began at the onset of the MPPNB (Eshed *et al.,* 2010).

In AG, an increase in the population density was observed towards the end of the PPN (Rollefson and Kohler-Rollefson, 1993). The higher numbers of surviving children in AY may reflect more favourable living conditions at this site when compared to AG. This may demonstrate a local expression of the benefits that a fully sedentary mode of life in a combination with better diet conferred on some Neolithic populations (Price and Gebauer, 1995). Such mode of life would have resulted in an overall increase population size and density (Hassan, 1973; Johnson and Earle, 2000; Bocquet-Appel, 2002), and may be associated with the high percent of surviving children in AY. Pearsall (2009) questioned the possibility of the NTD appearance, if one of the two principal factors of fertility energetics – the energy gain provided by sedentism – is absent. This would mean that the energy gain from food, despite the energy loss ensuing from mobility, would in itself lead to high fertility (birth rate). The

NDT is, in fact, the effect of a relatively abrupt change in maternal energetics on fertility (Valeggia and Ellison, 2004, with the relative metabolic-load model) following the shift to farming economy. In such situations, both the physical activity involved in mobility and the maternal stress of child transportation decrease and a subsequent increase in fertility energetics is to be expected (Bocquet-Appel and Naji, 2006; Bocquet-Appel, 2008).

Permanent settlers have more efficient methods of food production and storage than herders and thus food surplus may have promoted higher rates of reproduction relative to less efficient nomadic populations (Locay, 1983, 1989; Kramer and Boone, 2002; Eshed *et al.*, 2004b; Eshed, Gopher and Hershkovitz, 2006; Kuijt, 2008). In addition, environmental stress of a nomadic way of life may have induced higher rates of child mortality in these societies relative to the sedentary life at AY. It is assumed that inhabitation in stone built permanent houses was more favourable for infants than inhabitation in ephemeral shelters. In this sense, it is worth mentioning that over 50% of the mortality events amongst traditional nomadic herders and villagers who inhabited high altitude mountainous regions in Anatolia until the early 1960s were due to complications resulted from pneumonia (Yakar J. personal communication, 2009). Also, in pastoral nomadic societies, a major indicator of wealth and prestige is herd size, which depends mainly on the ability to obtain pasture and water for the animals. In agricultural societies subsisting on farming, man power, including the availability of child labour, is a most important factor in deriving maximum benefit from the environment (Kramer and Boone, 2002). This is probably true for societies subsisting on farming and fishing. Thus investing in resources which will keep children alive and in good health could have been crucial for agricultural societies (Kramer and Boone, 2002), but less so for herding societies. This may have resulted in more children reaching adulthood in permanent settlements, and higher mortality of children in nomadic societies.

It is important to point out that the KHH mortality curve is different from both the AY and AG in the high mortality rates amongst the young adults of 20 to 29 years of age, and in the total absence of older individuals (>50 years age). This may be the result of bias in mortality patterns, which in the case of KHH could be associated with its function as a cultic mortuary centre rather than a settlement (Eshed, Hershkovitz and Goring-Morris, 2008; Goring-Morris, 2000, 2005; Goring-Morris *et al.*, 2001).

17.6 CONCLUSIONS

The demographic differences between the MPPNB and LPPNB populations (KHH and AY) and between AY and the general Neolithic population can be summarized as follows.

1. The males at AY had a higher mean age at death than in the AG. This may be associated with dietary and environmental differences between the two populations, including more secure and stable food supplies and better balanced nutrition in AY than in the AG site. The sedentary way of life in the AY vs. the pastoral life of a portion of the AG population may have also contributed to these differences in life expectancy. In AY more people reached a greater age amongst the combined Neolithic sample.
2. In AY we report a lower life expectancy of females compared to males. This could have been the outcome of risks associated with increasing number of births and earlier onset of pregnancy in sedentary Neolithic communities. In AY these may have counteracted the benefits gained from a balanced diet and better environmental condition. Also the influence

of thalassaemia on females vs. males may have induced female vulnerability during to pregnancy and delivery because of blood loss during menstruation and birth.

3. The observed high infant/child mortality at AG relative to AY is not likely to be the product of differential taphonomic processes. AY was a permanent year-round settlement, and its inhabitants subsisted on a retentively balanced, diverse agropastoral and marine resources. In contrast, a portion of the PPN AG population may have subsisted on herding and a nomadic way of life. It is possible that the differences between the abilities of the two populations to produce and preserve sufficient amounts of food resources affected the number of surviving children. The higher numbers of surviving children in AY possibly reflects better living conditions provided by a sedentary, more sheltering mode of life in combination with better diet. Also in agricultural societies man power is a most important factor in deriving maximum benefit from the environment. Thus investing resources in keeping children alive could have been more worthwhile in an agrarian society than in herding society.
4. The observed high mortality rates amongst the young adults group, the high mortality rates in the 20 to 29 years cohort and the absence of old individuals in KHH may be the result of bias in burial patterns. This bias may be the result of the specialized function of the site (a burial cultic mortuary centre serving surrounding communities), as claimed by the excavator.

In summary, this study highlights differences in the mortality profile and life expectancy of different Neolithic populations from the southern Levant. Perhaps the main conclusion is that the palaeodemographic patterns of Neolithic populations were not uniform and that differences both between populations and between sexes reflect variations in an array of economic, cultural and social elements that defined each culture. These provide us with insights into the complexities and heterogeneities that underline past societies and the need to further investigate regional variations before we can detect and define any pan-regional changes in demography, which were associated with the transition to agriculture in the Near East, Anatolia and other world regions.

REFERENCES

Armelagos, G.J., Goodman, A.H. and Jacobs, K.H. (1991) The origin of Agriculture: Population growth during a period of declining health. *Popul. Environ.*, **13**, 9–22.

Arensburg B, Smith P, Yakar R. 1978. The human remains from Abou Gosh. In: Lechevallier M, editor. Abou Gosh et Beisamoun. Deux gisements du VIIe millénaire avant l'ère Chrétienne en Israël. Mémoires et Travaux du Centre de Recherches Préhistoriques Français de Jerusalem, **2**:107–120, Paris: Association Paléorient.

Bar-Yosef, O. (1995) Earliest food producers: Pre-Pottery Neolithic (8000–5500), in *The Archaeology of Society in the Holy Land* (ed. T.E. Levy), Leicester University Press, London, pp. 190–204.

Bar-Yosef, O. (1998) The Natufian culture in the Levant, threshold to the origins of agriculture. *Evol. Anthropol.*, **6**, 159–177.

Bar-Yosef, O. (2001) From Sedentary Foragers to Village Hierarchies: the Emergence of Social Institutions: The Origins of human social institutions. Proceedings of the British Academy, 110 (ed. G. Runciman), pp. 1–38.

Bar-Yosef, O. and Meadow, R. (1995) The origins of agriculture in the Near East, in *Last Hunters, First Farmers: New Perspectives on the Prehistoric Transition to Agriculture* (eds T.D. Price and A.B. Gebauer), School of American Research Press, Santa Fe, pp. 39–94.

Bar-Yosef, O. and Belfer-Cohen, A. (1989) The Levantine 'PPNB' interaction sphere, in *People and Culture in Change* (ed. I. Hershkovitz), BAR International Series 508, Oxford, pp. 59–72.

Bass, W.M. (1987) *Human Osteology. A Laboratory and Field Manual*, 3rd edn, Missouri Archaeological Society, Columbia.

Belfer-Cohen, A. (1991) The Natufian in the Levant. *Ann. Rev. Anthropol.*, **20**, 167–186.

Bocquet-Appel, J.P. (2002) Paleoanthropological traces of a Neolithic demographic transition. *Curr. Anthropol.*, **43**, 637–650.

Bocquet-Appel, J.P. (2008) Explaining the Neolithic demographic transition, in *The Neolithic Demographic Transition and its Consequences* (eds J.P. Bocquet-Appel and O. Bar-Yosef), Springer, Berlin, pp. 35–56.

Bocquet-Appel, J.P. (2009) Rethinking the origins of agriculture: The demographic impact of the agricultural system in human history. *Curr. Anthropol.*, **50**, 657–660.

Bocquet-Appel, J.P. and Masset, C. (1996) Paleodemography: Expectancy and false hope. *Am. J. Phys. Anthropol.*, **99**, 571–583.

Bocquet-Appel, J.P. and Naji, S. (2006) Testing the hypothesis of a worldwide Neolithic demographic transition: Corroboration from American cemeteries. *Currt Anthropol.*, **47**, 341–351.

Boldsen, J.L. and P'Aine, R.R. (1995) Defining extreme longevity from the Mesolithic to the Middle Ages estimates based on skeletal data, in *Exceptional Longevity from Prehistory to Present* (eds B. Jeune and J.W. Vaupel), Odense University Press, Odense, pp. 25–36.

Bonneuil, N. (2005) Fitting to a distribution of deaths by age with application to paleodemography: the route closest to a stable population. *Currt Anthropol.*, **46**, S29–S45.

Brooks, S. and Suchey, J.M. (1990) Skeletal age determination based on the os pubis: a comparison of Acsadi – Nemeskeri and Suchey – Brooks methods. *Hum. Evol.*, **5**, 227–238.

Buikstra, J.E., Konigsberg, L. and Bullington, J. (1986) Fertility and the development of agriculture in the prehistoric midwest. *Am. Antiquity*, **51**, 528–546.

Byrd, B.F. (2005) Reassessing the emergence of village life in the Near East. *J. Archaeol. Res.*, **13**, 231–289.

Caldwell, J.C. and Caldwell, B.K. (2003) Was there a Neolithic mortality crisis? *J. Population Res.*, **20**, 153–168.

Cashdan, E.A. (1985) Natural fertility, birth spacing, and the first demographic transition. *Am. Anthropol.*, **87**, 650–653.

Coale, A. (1972) *The Growth and Structure of Human Populations*, Princeton University Press, Princeton, NJ.

Cohen, M.N. and Armelagos, G.J. (1984) *Paleopathology at the Origins of Agriculture*, Academic Press, Orlando.

Cohen, M.N. (1989) *Health and the Rise of Civilization*, Yale University Press, New Haven.

Cohen, M.N. (2009) Introduction: Rethinking the origins of agriculture. *Curr. Anthropol.*, **50**, 591–595.

Cohen, M.N. and Crane-Kramer, M.M. (eds) (2007) *Ancient Health: Skeletal Indicators of Agricultural and Economic Intensification*, University of Florida Press, Gainesville, FL.

Eshed, V. (2001) From foraging to farming in the Holocene (the Pre–Pottery Neolithic period 8300–6000 BC) in the southern Levant: the skeletal evidence. Unpublished PhD. thesis, Tel Aviv University (In Hebrew).

Eshed, V., Gopher, A., Galili, E. and Hershkovitz, I. (2004a) Musculoskeletal Stress Markers (MSM) in Natufian hunter-gatherers and Neolithic farmers in the Levant: The upper limb. *Am. J. Phys. Anthropol.*, **123**, 303–315.

Eshed, V., Gopher, A., Gage, T.B. and Hershkovitz, I. (2004b) Has the transition to agriculture reshaped the demographic structure of prehistoric populations? New evidence from the Levant. *Am. J. Phys. Anthropol.*, **124**, 315–329.

Eshed, V., Gopher, A. and Hershkovitz, I. (2006) Tooth wear and dental pathology at the advent of agriculture: New evidence from the Levant. *Am. J. Phys. Anthropol.*, **130**, 145–159.

Eshed, V., Hershkovitz, I. and Goring-Morris, A.N. (2008) A re-evaluation of burial customs in the Pre-Pottery Neolithic B in light of paleodemographic analysis of the human remains from Kfar HaHoresh, Israel. *Paléorient*, **34**, 91–103.

Eshed, V., Gopher, A., Pinhasi, R. and Hershkovitz, I. (2010) Paleopathology and the origin of agriculture in the Levant. *Am. J. Phys. Anthropol.* (in press)

Eshed, V. and Nadel, D. (in preparation) Changes in burial practices from the Pre-Pottery to the Pottery Neolithic periods in the Levant: The case-study of Tel Roim West, Northern Israel.

Friedlander, N.J. (1996) The relation of lifetime reproduction to survivorship in women and men: A retrospective study. *Am. J. Hum. Biol.*, **8**, 771–783.

Gage, T.B. (1988) Mathematical hazard models of mortality: An alternative to model life tables. *Am. J. Phys. Antropol.*, **76**, 429–441.

Gage, T.B. (1990) Variation and classification of human age patterns of mortality. Analysis using competing hazards models. *Hum. Biol.*, **62**, 589–617.

Gage, T.B. and DeWitte, S. (2009) Rethinking the origins of agriculture: What do we know about the agricultural demographic transition? *Curr. Anthropol.*, **50**, 649–655.

Galili, E. (2004) Submerged settlements of the Ninth to Seventh Millennia BP off the Carmel Coast. Unpublished PhD thesis, Tel Aviv University.

Galili, E., Weinstein–Evron, M., Hershkovitz, I. *et al.* (1993) Atlit-Yam: A Prehistoric site on the sea floor off the Israeli Coast. *J. Field Archaeol.*, **20**, 133–157.

Galili, E., Gopher, A. and Horwitz, L.K. (2002) The emergence and dispersion of the Mediterranean fishing village: Evidence from submerged Neolithic settlements off the Carmel Coast, Israel. *J. Med. Archaeol.*, **15**, 167–198.

Galili, E., Eshed, V., Gopher, A. and Hershkovitz, I. (2005) Burial practices at the submerged PPNC site of Atlit-Yam, Northern coast of Israel: What do they tell us about the final phase of the Pre-Pottery Neolithic culture. *BASOR*, **339**, 1–19.

Goring-Morris, A.N. (2000) The quick and the dead: the social context of Aceramic Neolithic mortuary practices as seen from Kfar HaHoresh, in *Life in Neolithic Farming Communities: Social Organization, Identity, and Differentiation* (ed. I. Kuijt), Kluwer Academic, New York, pp. 103–135.

Goring-Morris, A.N. (2005) Life, death and the emergence of differential status in the Near Eastern Neolithic: Evidence from Kfar HaHoresh, Lower Galilee, Israel, in *Archaeological Perspectives on the Transmission and Transformation of Culture in the Eastern Mediterranean* (ed. J. Clark), Council for British Research in the Levant and Oxbow, Books, Oxford, pp. 89–105.

Goring-Morris, A.N., Goren, Y., Horowitz, L.K., *et al.* ((1994) –95) The 1992 season of excavations at the Pre-Pottery Neolithic B settlement of Kfar HaHoresh, *Mitekufat Haeven*, **26**, 74–121.

Goring-Morris, A.N. and Belfer-Cohen, A. (1998) The articulation of cultural processes and Late Quaternary environmental changes in Cisjordan. *Paléorient*, **23**, 71–93.

Goring-Morris, A.N. and Belfer-Cohen, A. (2008) A roof over one's head: Developments in Near Eastern residential architecture across the Epipalaeolithic-Neolithic transition, in *The Neolithic Demographic Transition and its Consequences* (eds J.P. Bocquet-Appel and O. Bar-Yosef), Springer, Berlin, pp. 239–286.

Goring-Morris, A.N., Burns, R., Davidzon, A. *et al.* (1998) The 1997 season of excavations at the mortuary site of Kfar HaHoresh, Galilee, Israel. *Neo-Lithics*, **3**, 1–4.

Goring-Morris, A.N., Boaretto, E. and Weiner, S. (2001) Radiometric dating the PPNB mortuary site of Kfar HaHoresh, lower Galilee, Israel: Problems and preliminary results. *Mitekufat Haeven*, **31**, 213–218.

Goring-Morris, A.N. and Horwitz, L.K. (2007) Funerals and feasts in the Near Eastern Pre-Pottery Neolithic B. *Antiquity*, **81**, 902–919.

Greenfield, H., Galili, E. and Horwitz, L.K. (2007) The butchered animal bones from Newe-Yam, A submerged Pottery Neolithic site off the Carmel Coast. *Mitekufat Haeven*, **36**, 173–200.

Hassan, F.A. (1973) On methods of population growth during the Neolithic. *Curr. Anthropol.*, **14**, 535–542.

Hassan, F.A. (1981) *Demographic Archaeology*, Academic Press, New York.

Hershkovitz I, Garfinkel Y, Arensburg B. 1986. Neolithic Skeletal Remains at Yiftahel, Area C, Israel. Paléorient **12**(1), 73–81.

Hershkovitz, I. and Gopher, A. (1988) Human remains from Horvat Galil. A Pre-Pottery Neolithic site in the Upper Galilee, Israel. *Paléorient*, **14**(1), 119–123.

Hershkovitz, I. and Gopher, A. (1990) Paleodemography, burial customs, and food-producing economy at the beginning of the Holocene: A perspective from the southern Levant. *Mitekufat Haeven*, **23**, 9–47.

Hershkovitz, I., Ring, B., Spears, M., *et al.* (1991) Possible congenital hemolytic anemia in prehistoric coastal inhabitants of Israel. *Am. J. Phys. Anthropol.*, **85**, 7–13.

Hillson, S. (1986) *Teeth*, Cambridge University Press, Cambridge.

Hoppa, R.D. and Vaupel, J.W. (2002) The Rostock manifesto for paleodemography, in *Paleodemography: Age Distributions from Skeletal Samples* (eds R.D. Hoppa and J.W. Vaupel), Cambridge University Press, Cambridge, pp. 1–8.

Horwitz, L.K. and Goring-Morris, A.N. (2004) Animals and ritual during the Levantine PPNB: a case study from the site of Kfar HaHoresh, Israel. *Anthropozoologica*, **39**, 165–178.

Horwitz, L.K., Lernau, O. and Galili, E. (2007) Fauna from the submerged Pottery Neolithic site off Newe-Yam, Northern Israel. *Journal of the Israel Prehistoric Society*, **36**, 139–171.

Jackes, M. (1993) On paradox and osteology. *Curr. Anthropol.*, **34**, 435–438.

Jackes, M. (2003) Testing a method: Paleodemography and proportional fitting. *Am. J. Phys. Antropol.*, **121**, 385–386.

Johnston, F.E. and Zimmer, L.O. (1989) Assessment of growth and age in the subadults skeleton, in *Reconstruction of Life from the Skeleton* (eds M.Y. Iscan and A.R. Kenneth), Alan R. Liss, New York, pp. 11–21.

Johansson, R. and Horowitz, B. (1986) Estimating mortality in skeletal populations: Influence of the growth rate on the interpretation of levels and trends during the transition to agriculture. *Am. J. Phys. Anthropol.*, **71**, 233–250.

Johnson, A.W. and Earle, T.K. (2000) *The Evolution of Human Societies: From Foraging Group to Agrarian State*, Stanford University Press, Stanford, CA.

Kennedy, K.A.R. (1984) Growth, nutrition, and pathology in changing paleodemographic settings in South Asia, in *Paleopathology at the Origins of Agriculture* (eds M.N. Cohen and G.J. Armelagos), Academic Press, Orlando, pp. 169–192.

Kirkbride, D. (1966) Five seasons at the pre-pottery Neolithic village of Beidha in Jordan. *Palestine Exp. Quart.*, **98**, 8–72.

Koerner, B.D. and Blakely, R.L. (1985) DJD, subsistence and sex roles in the protohistoric King Site in Georgia. *Am. J. Phys. Anthropol.*, **66**, 190.

Konigsberg, L.W. and Frankenberg, S.R. (1992) Estimation of age structure in anthropological demography. *Am. J. Phys. Antropol.*, **89**, 235–256.

Kramer, K.L. and Boone, J.L. (2002) Why intensive agriculturalists have higher fertility: a household energy budget approach. *Curr. Anthropol.*, **43**, 511–517.

Kuijt, I. (ed.) (2000) *Life in Neolithic Farming Communities: Social Organization, Identity, and Differentiation*, Kluwer Academic, New York.

Kuijt, I. (2008) Demography and storage systems during the southern Levantine Neolithic demographic transition, in *The Neolithic Demographic Transition and its Consequences* (eds J.P. Bocquet-Appel and O. Bar-Yosef), Springer, Dordrecht, pp. 287–313.

Kuijt, I. and Goring–Morris, N. (2002) Foraging, farming, and social complexity in the Pre-Pottery Neolithic of the southern Levant: a review and synthesis. *J. World Prehist.*, **16**, 361–440.

Kurt, G. and Rohrer-Ertl, O. (1981) On the anthropology of Mesolithic to Chalcolithic human remains from the Tell es-Sultan in Jericho, Jordan, in *Excavations at Jericho III* (ed. K.M. Kenyon), British School of Archaeology Jerusalem, London, pp. 407–499.

Locay, L. (1983) Aboriginal Population Density of North American Indians, Department of Economics, University of Chicago, Unpublished PhD dissertation.

Locay, L. (1989) From hunting and gathering to agriculture. *Econ. Dev. Cult. Change*, **37**, 737–756.

Loth, S.R. and Iscan, M.Y. (1989) Morphological assessment of age in the adult: The thoracic region, in *Age Markers in the Human Skeleton* (ed. M.Y. Iscan), Charles C. Thomas, Illinois, pp. 105–136.

Lovejoy, C.O. (1985) Dental wear in the Libben population: Its functional pattern and role in the determination of adult skeletal age at death. *Am. J. Phys. Antropol.*, **68**, 47–56.

Lovejoy, C.O., Meindl, R.S., Prysbeck, T.R. and Mensforth, R.P. (1985a) Chronological metamorphosis of the auricular surface of the ilium: A new method for the determination of adult skeletal age at death. *Am. J. Phys. Antropol.*, **68**, 15–28.

Lovejoy, C.O., Meindl, R.S. and Mensforth, R.P. (1985b) Multifactorial determination of skeletal age at death: A method and blind tests of its accuracy. *Am. J. Phys. Antropol.*, **68**, 1–14.

Munro, N.D. (2004) Zooarchaeological measures of hunting pressure and occupation intensity in the Natufian – Implications for agricultural origins. *Curr. Anthropol.*, **45**, S5–S33.

Nathan, H. (1962) Osteophytes of the vertebral column. An anatomical study of their development according to age, races and sex with considerations as to their etiology and significance. *J. Bone Joint Surg.*, **44**, 243–268.

Paine, R.R. (1989) Model life table fitting by maximum likelihood estimation; a procedure to reconstruct paleodemogrphic characteristics from skeletal age distribution. *Am. J. Phys. Antropol.*, **80**, 49–58.

Paine, R.R. (2000) Population crashes in prehistory, and there is no paleodemographer there to hear it, does it make a sound? *Am. J. Phys. Anthropol.*, **112**, 181–190.

Paine, R.R. and Harpending, H.C. (1998) The effect of sample bias on paleodemographic fertility estimates. *Am. J. Phys. Anthropol.*, **105**, 231–240.

Paine, R.R. and Boldsen, J.L. (2002) Linking age-at-death distributions and ancient population dynamics: a case study, in *Paleodemography: Age Distributions from Skeletal Samples* (eds R.D. Hoppa and J.W. Vaupel), Cambridge University Press, Cambridge, pp. 167–180.

Pearsall, D.M. (2009) Rethinking the origins of agriculture: investigating the transition to agriculture. *Curr. Anthropol.*, **50**, 609–613.

Pinhasi, R. and Bourbou, C. (2008) How representative are human skeletal assemblages for population analysis and interpretation? Implications for palaeopathological and palaeoepidemiological investigations, in *Advances in Human Palaeopathology* (eds R. Pinhasi and S. Mays), Wiley-Liss, New York, pp. 31–34.

Price, T.D. and Gebauer, A. (1995) New perspectives on the transition to agriculture, in *Last Hunters, First Farmers: New Perspectives on the Prehistoric Transition to Agriculture* (eds T.D. Price and A. Gebauer), School of American Research Press, Santa Fe, NM, pp. 132–139.

Rollefson, G.O. (1985) The 1983 season at the early Neolithic of 'Ain Ghazal. *Nat. Geog. Res.*, **1**, 44–62.

Rollefson, G.O. (1989) The late aceramic Neolithic of the Levant: a synthesis. *Paléorient*, **15**, 168–173.

Rollefson, G.O. (2000) Ritual and Social Structure at Neolithic 'Ain Ghazal, in *Life in Neolithic Farming Communities. Social Organization, Identity, and Differentiation* (ed. I. Kuijt), Kluwer Academic/Plenum, New York, pp. 163–190.

Rollefson, G.O. (2001) The Neolithic period, in *The Archaeology of Jordan* (eds B. MacDonal, R., Adams and P. Bienkowski), Sheffield Academic Press, Sheffield, pp. 67–105.

Rollefson, G.O., Simmons, A.H. and Kafafi, Z. (1992) Neolithic cultures at 'Ain Ghazal, Jordan. *J. Field Archaeol.*, **19**, 443–470.

Rollefson, G.O., Simmons, A.H., Donaldson, M.L. *et al.* (1995) Excavation at the Pre-Pottery Neolithic B Village of 'Ain Ghazal (Jordan), 1983. *Mitteilungen der Deutschen Orient Gesellschäft*, **117**, 69–116.

Rollefson, G.O. and Kohler-Rollefson, I. (1989) The Collapse of Early Neolithic Settlements in the Southern Levant. People and culture in change: proceedings of the 2nd symposium on Upper Paleolithic, Mesolithic and Neolithic populations of Europe and the Mediterranean Basin, Oxford: British Archaeology Reports, International Series 508 (i), Archaeopress (ed. I. Hershkovitz), pp. 73–89.

Rollefson, G.O. and Kohler-Rollefson, I. (1993) PPNC adaptation in the first half of the 6th millennium B.C. *Paléorient*, **19**, 33–42.

Sattenspiel, L. and Harpending, H. (1983) Stable populations and skeletal age. *Am. Antiquity*, **48**, 489–498.

Segebarth-Orban, R. (1980) An evaluation of the skeletal dimorphism of the human innominate bone. *J. Hum. Evol.*, **9**, 601–607.

Simmons, A. (2007) *The Neolithic Revolution in the Near East: Transforming the Human Landscape*, The University of Arizona Press, Tucson.

Simmons, T., Horwitz, L.K. and Goring-Morris, A.N. (2007) 'What ceremony else?' Taphonomy and the ritual treatment of the dead in the Pre-Pottery Neolithic B mortuary complex at Kfar HaHoresh, Israel, in *Faces from the Past: Diachronic Patterns in the Biology and Health Status of Human Populations from the Eastern Mediterranean*, Papers in Honour of Patricia Smith (eds M. Faerman, L.K. Horwitz, T. Kahan and U. Zilberan), BAR Internayional Series 1603, Oxford, pp. 1–27.

Smith, P., Bar-Yosef, O. and Sillen, A. (1984) Archaeological and skeletal evidence for dietary change during the late Pleistocene/early Holocene in the Levant, in *Paleopathology at the Origins of Agriculture* (eds M.N. Cohen and G.J. Armelagos), Academic Press, New York, pp. 101–136.

Smith, P. and Kolska-Horwitz, L.K. (2007) Ancestors and inheritors: A bioarchaeological perspective on the transition to agropastoralism in the southern Levant, in *Ancient Health: Skeletal Indicators of Agricultural and Economic Intensification* (eds M.N. Cohen and M.M. Crane-Kramer), University of Florida Press, G'Ainesville, FL, pp. 207–222.

Stiner, M.C. and Monroe, N.D. (2002) Approaches to prehistoric diet breadth, demography, and prey ranking systems in time and space. *J. Archaeol. Method Th.*, **9**, 181–214.

Szilvassy, J. (1980) Age determination on the sternal articular faces of the clavicula. *J. Hum. Evol.*, **9**, 609–610.

Ubelaker, D.H. (1989) *Human Skeletal Remains: Excavation, Analysis, Interpretation*, 2nd edn, Taraxacum, Washington DC.

Ubelaker, D.H. (1974) *Reconstruction of Demographic Profiles from Ossuary Skeletal Samples. A Case Study from the Tidewater Pormac*, Smithsonian Institution Press, Washington. DC.

Valeggia, C. and Ellison, P.T. (2004) Lactational amenorrhoea in well-nourished Toba women of Formosa, Argentina. *J. Biosco. Sci.*, **36**, 573–595.

Washburn, S.L. (1948) Sex differences in the pubic bone. *Am. J. Phys. Antropol.*, **6**, 199–208.

18

Skeletal Differentiation at the Southernmost Frontier of Andean Agriculture

Marina L. Sardi and Marien Béguelin

División Antropología, Museo de La Plata, Universidad Nacional de La Plata

18.1 INTRODUCTION

The transition from foraging to food production in South America developed independently and simultaneously in many places. Peru was the most important centre of domestication in this continent. Earliest evidences of crop plants are dated around 8200 calBC, although greater reliance on domesticates began around 3800 to 2500 calBC (Fiedel, 1996; Hastorf, 1999).

The factors that led to the domestication of plants and animals remain unclear. Evidence suggests that the low resource availability and/or demographic growth were not the main causes of change in subsistence strategy (Fiedel, 1996) since in the Peruvian coastal areas, domestication appeared in areas rich in marine resources (Hastorf, 1999), while in the Central Andes, it appeared in areas of low population density (Fiedel, 1996). For instance, in the Supe Valley of the Peruvian coast, preceramic complex societies lived between 4600 and 1900 calBC (Solís, Haas and Creamer, 2001). These populations cultivated cotton that was used for fishing nets because they obtained animal proteins entirely from the sea, complementing their subsistence by contacts with Andean populations (Sandweiss *et al.*, 2008; Solís, Haas and Creamer, 2001). After 1900 calBC, food consumption was based on plant and animal domesticates.

Domestication expanded southwards throughout the Andes and the Pacific coast, through contacts amongst populations. Prior to the adoption of agriculture, with populations on both sides of the South Central Andes there are no signs of subsistence change until approximately 3800 calBC (Castro and Tarragó, 1992; Gambier, 1993; Aschero, 1994; Planella and Tagle, 2004; Núñez *et al.*, 2006). These populations based their subsistence on hunting of wild camelids and became increasingly mobile, diversifying their technology for food acquisition and for exploiting a wide variety of resources.

Human Bioarchaeology of the Transition to Agriculture Edited by Ron Pinhasi and Jay T. Stock

The earliest evidence of plant and animal domestication in the north of Chile and in the northwest of Argentina are dated between 2500 and 1200 calBC (Castro and Tarragó, 1992; Aschero, 1994; Yacobaccio *et al.*, 1997/98). Populations of the South-Central Andes adopted pottery around 1400 to 1200 calBC and became more settled (Raffino, 2007). The transition to full agricultural economy was gradual. It was not until 50 years calAD that agriculture, pastoralism and sedentism became the main subsistence strategy (Olivera and Podestá, 1993).

18.2 AGRICULTURE EXPANSION AT THE ARGENTINE CENTRE-WEST AND ITS BIOLOGICAL CONSEQUENCES

The southernmost frontier of Andean agriculture was the Diamante River at the Argentine Centre-West (ACW) (Gambier, 1993; Lagiglia, 2001; Gil, 2003; Gil and Neme, 2008). The ACW is an archaeological area comprised of San Juan and Mendoza provinces (Figure 18.1), which is divided into southern and northern regions according to the subsistence strategies displayed by populations at the arrival of Spaniards: part of the Mendoza province located to the south of the Diamante River was inhabited by hunter-gatherers, whereas north to the Diamante River – San Juan and north of Mendoza provinces – was inhabited by farmers. In both regions, human occupation has been detected as early as 10900 to 10500 calBC (Gambier, 1976; García, 1997; Lagiglia, 2002). Populations relied on hunting guanaco (*Lama guanicoe*), ñandú (*Rhea americana*), and other small mammals, and on the gathering of wild plants by seasonal movements between lowlands and highlands (Gambier, 1993; Lagiglia, 2002).

The earliest evidence of agriculture in the northern region is dated to about 3000 calBC (Bárcena, 1985), although some researchers propose that it is not older than 630 calBC (García, 1992; Lagiglia, 2001). Local groups gradually adopted agriculture and pastoralism as complementary resources to hunting, gathering and fishing. Some cultigens included in the diet were potato (*Solanum tuberosum*), manioc (*Manihot esculenta*), beans (*Phaseolus vulgaris*) and maize (*Zea mays*). Domesticated camelids were mainly reared for wool production, and for transport. Although populations of the northern region occasionally consumed domesticated camelids (as *charqui*, a dry meat that can be stored), their most important source of animal protein was derived from game (Rusconi, 1961; Gambier, 1993).

Northern ACW farmers maintained a broad-spectrum diet, but around 600 calAD, agriculture began to be of primary importance to their economy (Gambier, 1993). As in other Andean populations, the introduction of grinding tools for food preparation (Lagiglia, 1997) and pottery use for cooking (Raffino, 2007) suggest the consumption of softer diets. Changes in

Figure 18.1 Argentine Centre-West (after Sardi, Novellino and Pucciarelli, 2006). The farmer sample was recovered at: 1. Uspallata, 2. Mendoza, 3. Capiz, 33. Calingasta, 34. San Juan, 35. Caliningasta Barrealito, 36. Angualasto, 37. Pachimoco, 38. Huaco. The hunter-gatherer sample was recovered at: 4. Campo Las Julias, 5. Arroyo Imperial, 6. Arroyo El Tigre, 7. Dique Villa 25 (Los Reyunos), 8. Dique Villa 25 (La Hedionda), 9. Dique Villa 25 de Mayo, 10. Los Coroneles, 11. Médano Puesto Díaz, 12. Arroyo Los Jilgueros, 13. Rincón del Atuel, 14. Loma del Eje, 15. Cañada Seca, 16. Jaime Prats, 17. Cerro Negro, 18. Agua del Médano, 19. El Nihuil, 20. Agua del Zapallo, 21. Cerro Meson, 22. Respolar, 23. La Herradura, 24. Puesto Aisol, 25. El Sosneado, 26. Puesto Tierras Blancas, 27. Cerro Mesa, 28. El Manzano, 29. La Matancilla, 30. El Chacay, 31. La Cañada, 32. Sur de Malargüe. Reproduced, with permission, from Sardi, M. L., Novellino, P. S., Pucciarelli, H. M. (2006) Craniofacial Morphology in the Argentine Center-West: Consequences of the Transition to Food Production, American Journal of Physical Anthropology 130:333–343 © Wiley

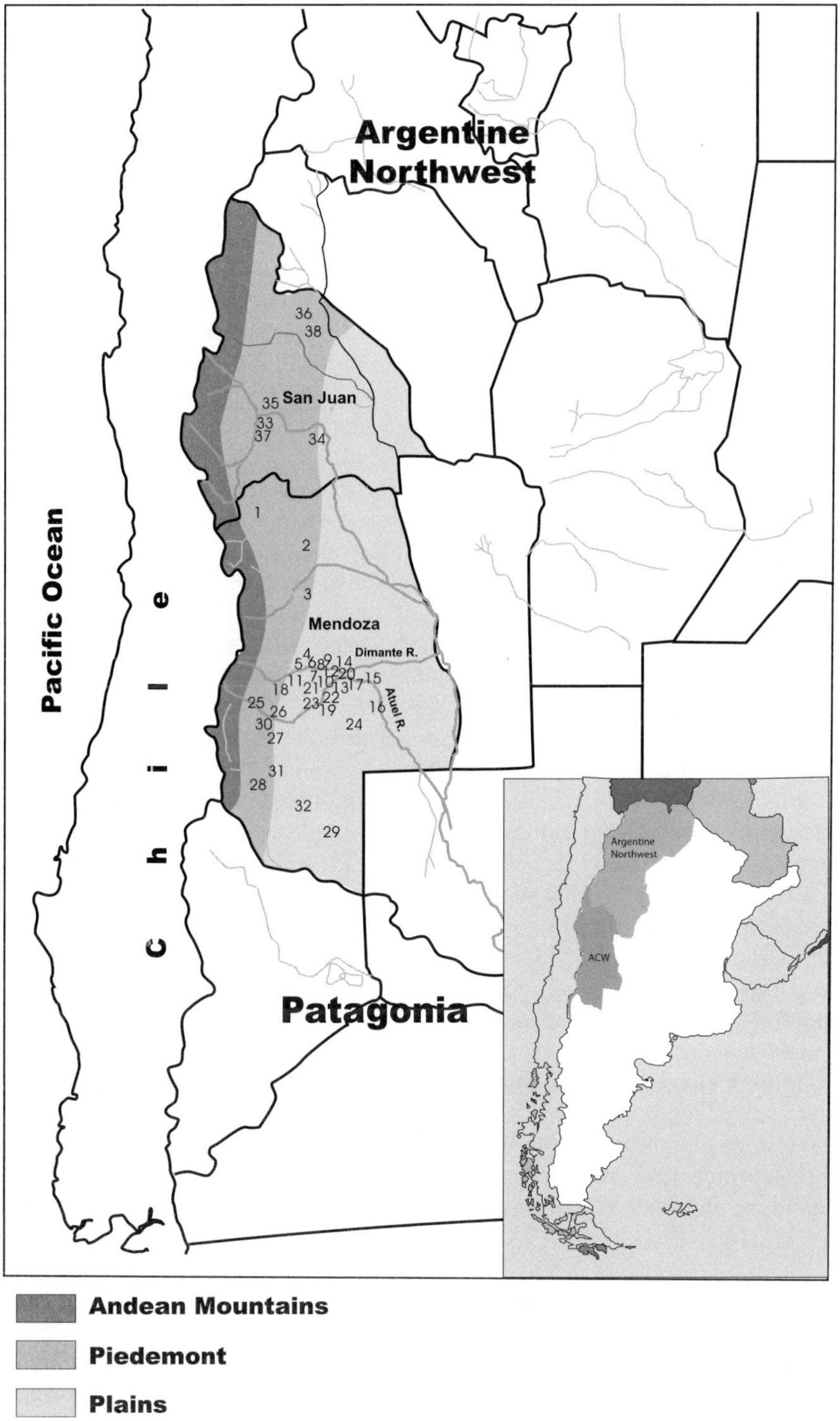
Argentine Northwest
San Juan
Mendoza
Dimante R.
Atuel R.
Pacific Ocean
Chile
Patagonia
Argentine Northwest
ACW
Andean Mountains
Piedemont
Plains

settlement patterns indicate the emergence of permanent and semi-permanent villages. In the northern Mendoza province, the concentration of archaeological assemblages and their association with collective mills suggest an increase in sedentism (Lagiglia, 2002). In San Juan province, the remains show small villages were located close to river valleys, with many kinds of irrigation systems and dwellings for animals (Gambier, 1993; Raffino, 2007).

Unlike northern ACW groups, southern groups of the ACW never adopted plant and animal domestication (Gil, 2002; Neme and Gil, 2008). Southern ACW aborigines had a hunter-gatherer subsistence strategy until ethno-historical times, moving on a seasonal basis to hunt and acquire other resources. From 600 calBC to 0, these populations began to occupy seasonal residential campsites located at an altitude of over 3000 m above sea level (Neme, 2002; Durán *et al.*, 2006). Scrapers (used for skin preparation) are a diagnostic element of habitat occupation and since they are broadly dispersed they suggest high mobility and short-term occupations (Lagiglia, 2002). Some cultural shifts were also introduced. Southern ACW groups began to exploit less diverse animal species, focusing instead on big game (Neme and Gil, 2008). They incorporated pottery (Lagiglia, 2002) and small mills into their economy, apparently for processing pigments rather than wild plants for food (Castro and Tarragó, 1992; Lagiglia, 1997).

The study of carbon and nitrogen stable-isotope ratios of human populations from the southern region (Gil, 2003; Novellino *et al.*, 2004; Gil *et al.*, 2009) indicates the consumption of a small proportion of cultivable plants, which were probably obtained by exchanges with northern farmers (Gil, 2003). In contrast, stable isotope analyses of populations from the northern region indicate a marked increase in the consumption of C4 resources, for example maize, after 1000 calAD (Gil *et al.*, 2006, 2009).

As in the case of many regional case studies (cf. Cohen and Armelagos, 1984; Larsen, 1995), the transition to food production in the ACW was associated with a decline in health and skeletal variation due to changes in both the nutritional content of food and mobility patterns. The skeletons of hunter-gatherers from the southern regions have low prevalence of caries (Novellino, Guichón and Lagiglia, 1996; Novellino, 2002) and dental hypoplasias (Novellino and Gil, 2007), which suggest that these populations did not suffer from nutritional deficiencies. Northern farmers had higher frequencies of caries than southern hunter-gatherers (Bernal *et al.*, 2007). Hunter-gatherers also show a higher prevalence of dental wear than in the case of farmers, which indicates greater masticatory loads in the case of the former (Bernal *et al.*, 2007). With respect to cranial morphology (Sardi, Novellino and Pucciarelli, 2006), farmers are characterized by smaller crania and a relative reduction in size of the structures that support masticatory and neck muscles. The pattern of differentiation was interpreted as the result of a reduction in muscular loadings due to the softer consistency of the diet consumed after the transition to food production. Nevertheless, systemic factors, such as the consumption of a low-protein diet or the decreased muscular loadings associated with sedentism, could also contribute to the observed pattern.

18.3 SKELETAL MORPHOGENESIS

Bones support and protect soft tissues and resist internal and external forces. The fundamental form of an endochondral bone is outlined by the 3D organization of the cartilaginous model (Hall, 2005). In contrast, minor morphogenetic features are much more dependent on epigenetic factors, such as hormones, mechanical loads and bioelectrical and biophysical events, acting at different structural levels (e.g. cells, tissues, organs) (Atchley and Hall, 1991). As Moss (1973, 1997) proposed for the skull in the Functional Matrix hypothesis, the rate and direction of bone and cartilage growth are not regulated by means of their own genetic control; instead, they are epigenetically modified by the growth of the functional matrices (soft tissues and functional spaces) associated with them. Even though different skeletal components (e.g. skull, trunk, limbs) present their own growth pattern, those that are integrated across development due to common embryological origins, function or epigenetic interactions tend to co-vary (Cheverud, 1996). For instance, a given factor, either developmental or environmental, may affect the adult morphology of one or more skeletal units, depending on its local or systemic influence. However, not all bones react equally under the same stimuli.

Mechanical loads are important non-heritable factors that affect bone morphology. Usual physical activities and body mass exert different demands on the skeleton and modify those structures more sensitive to loading. While the long bone diaphysis is subject to functional adaptation (Ruff, 1987, 2005), the epiphysis does not remodel to the same extent as the diaphyses (Ruff, Scott and Liu, 1991; Lieberman, Devlin and Pearson, 2001). As a general trend, muscular exertions upon the skeleton decreased after the transition to food production (Ruff, Larsen and Hayes, 1984; Larsen, 1995), promoting gracilization of the skull and lower limbs. However, some evidence shows a different pattern (Bridges, Blitz and Solano, 2000); thus, trends in long bone strength and dimensions may depend on local activity pattern.

18.4 OBJECTIVE AND HYPOTHESES

This study focuses on skeletal variation of Late Holocene hunter-gatherers and farmers of the Argentine Centre-West. The purpose is to evaluate morphological differentiation in the face and proximal bones of the upper and lower limbs. The ACW was inhabited by populations that may have been genetically related, and occupied a similar landscape with similar climatic conditions. It is anticipated that inter-population variations in the morphology of these morphological structures are not the outcome of variable ecological and climatic conditions, but rather the outcome of variations in subsistence. The general hypothesis to be tested is therefore that different subsistence strategies are associated with changes in skeletal morphology. This hypothesis is broken down into specific hypotheses in order to assess some potential sources of variation.

First, the nutritional content of diet likely changed following increased reliance on crops (Gil *et al.*, 2006, 2009). Diets became less diverse and poorer in nutritional content, as suggested by the higher frequency of dental caries (Novellino, 2002; Bernal *et al.*, 2007). Inadequate food intake and nutrient-deficient food, as in the case of low protein diets, lead to growth retardation and consequently a reduction in final stature (Stini, 1969; Tanner, 1988; Bogin, 1999). Experimental studies indicate that under-nutrition is associated with a reduction in cranial and post-cranial size (Pucciarelli, 1981; Pucciarelli and Goya, 1983; Pucciarelli, Dressino and Niveiro, 1990; Miller and German, 1999; Reichling and German, 2000), with eyes and brains proportionally larger because they receive nutritional priority during growth (Pucciarelli, 1981; Reichling and German, 2000). Another consequence is the reduction of sexual dimorphism, because males are often more affected than females (Tanner, 1962). If the populations of the ACW were experiencing nutritional stress, we might expect a reduction in sexual dimorphism and size variation in several skeletal components, especially in structures with extended growth period such as long bone shafts.

A second source of variation is related to biomechanical loadings of the skeleton as a result of habitual behaviour. Several archaeological indicators (Castro and Tarragó, 1992; Gambier, 1993; Lagiglia, 2002; Babot, 2006) suggest that farmers of the ACW consumed more processed diets than hunter-gatherers. ACW farming populations, in turn, also have smaller masticatory structures than their hunter-gatherer counterparts (Sardi, Novellino and Pucciarelli, 2006). Because farmers had more settled lifestyles, they might have experienced less loading on their lower limbs, but probably greater loading on the upper limbs as a result of activities linked to cultivation, such as harvesting and the grinding of grains. Skeletal loadings and physical exercise might have changed with the transition to agriculture and sedentism and they might have exerted influence on one or several skeletal components. If morphological differentiation between foraging and farming populations was affected by biomechanical loadings, then skeletal structures known to be responsive to different levels of habitual activity (long bones shafts) are expected to differ. Physical activities could also have systemic effects due to variation in growth hormone circulation, which is usually increased after exercise (Kalu, Banu and Wang, 2000; Weltman *et al.*, 2001; Parks, 2002).

18.5 SAMPLES

The analysed samples comprise skulls, humeri and femora of adult individuals derived from different localities of both regions of ACW (Table 18.1, Figure 18.1). All skeletal elements present ossified growth centres. When the pelvis and skull were available, the standards described by Buikstra and Ubelaker (1994) were followed to assess sex. Many individuals were recovered from secondary burials for whom sex determination could not be carried out. The sex of these individuals was assessed by applying discriminant functions derived from metric characteristics of long bones of those individuals whose sex could be estimated.

The sample from the southern region is represented by 66 adult skulls, 49 humeri and 41 femora (Table 18.1), recovered from different sites located in lowlands and highlands (Figure 18.1). Only 10 individuals are represented by all three skeletal components. Chronologically, the sample includes individuals from the Early Late (580 calBC–620 years calAD) to Final Late (< 620 calAD) Holocene, as indicated by direct radiocarbon dates of some of these specimens (cf. Novellino and Guichón, 1999; Novellino *et al.*, 2004; Béguelin *et al.*, 2006; Bernal *et al.*, 2007). In a previous study (Béguelin *et al.*, 2006), no significant variation in dental and postcranial morphology and in cranial epigenetic traits was observed between the sample from the Early Late Holocene and that from the Final Late Holocene.

Table 18.1 Sample composition

		Skulls	Humeri		Femora	
hunter-gatherers (HG)			*right*	*left*	*right*	*left*
	males	43	14	12	6	8
	females	23	10	9	12	13
farmers (F)						
	males	40	13	15	16	16
	females	43	9	9	8	8
total		149	46	45	42	45

Archaeological assemblages associated with the whole sample suggest that these individuals practised a mobile hunter-gatherer economy. For at least half of the sample, assemblages do not include pottery (Novellino, Guichón and Lagiglia, 1996).

The sample derived from the northern region includes 25 individuals represented by all 3 skeletal components and another 58 individuals only consisting of cranial material (Figure 18.1, Table 18.1). Radiocarbon dates are not available for this skeletal material, but 40 individuals were recovered from sites located in Calingasta and Angualasto, which were peopled around calAD 700 to 1500 (Michieli, 2000). For other remains, archaeological indicators suggest that they can be assigned to the Late Agriculturalist period (calAD 1300–1500) (Gambier, 1993) and to the ethno-historical agricultural populations of the area (Lehmann Nitsche, 1910). All material is housed at the Museo de Historia Natural (San Rafael), Museo Etnográfico (Buenos Aires) and Museo de La Plata (La Plata) in Argentina.

18.6 SKELETAL COMPONENTS AND STATISTICAL ANALYSES

The morphological assessment was based on the Functional Matrix hypothesis (Moss, 1973), as applied in previous studies (Pucciarelli, Dressino and Niveiro, 1990; Sardi, Ramírez Rozzi and Pucciarelli, 2004; Sardi, Novellino and Pucciarelli, 2006). Because half of the sample features artificial cranial deformation, only facial components, which were not affected by deformation (Sardi, Novellino and Pucciarelli, 2006), were analysed. Length, breadth and height were measured on the optic, respiratory, masticatory and alveolar components, and volumetric indices were calculated to estimate size (Table 18.2). Although cranial components are morphologically integrated, they encompass different functional matrices and present quite independent growth patterns (Sardi and Ramírez Rozzi, 2005). The ocular globe is the functional matrix of the optic component and can be considered as a relative proxy of brain size, since it has a common embryological origin with the brain and it is associated with neurocranial size (Sardi, Ramírez Rozzi and Pucciarelli, 2004; Sardi and Ramírez Rozzi, 2005). The respiratory component involves the cavity for respiration and olfaction, while the masticatory component involves the temporal and part of the masseter muscles, both components being relative proxies of body size (Enlow and Hans, 1996). The alveolar component involves the teeth and oral tissues and is the last cranial region to attain adult size (Enlow and Hans, 1996).

The main functional matrices of the limb bones are the muscles. Length, epicondylar breadth and midshaft diameters were measured on humeri and femora after Buikstra and Ubelaker (1994) (Table 18.2). These measures encompass some of the dimensionality of the limb bones and their shape is influenced by genetic factors and phenotypic plasticity. Femoral length is a proxy for stature, which in turn is an important indicator of health and nutrition (Bogin, 1999; Mieklejohn and Baab, this volume), sexual dimorphism and general trends in body size (Stini, 1969; Tanner, 1988; Auerbach and Ruff, 2004). Femoral head diameter is a proxy of body mass (Ruff, Scott and Liu, 1991). Four indices were calculated (Table 18.2). The ratio of femoral head diameter – highly correlated to body mass (Ruff, Scott and Liu, 1991), to femoral length – highly correlated to stature (Feldesman and Fountain, 1996), is employed here as a proxy for body linearity (Auerbach and Ruff, 2004). Femoral and humeral robusticity indices were calculated as a ratio between diaphyseal breadths and bone length (Table 18.2). We

Table 18.2 Skeletal components, measurements and indices

Skeletal component	Abbreviation	Measurement
optic	OL	length: dacryon-optic foramen
	OB	breadth: dacryon-ectoconchion
	OH	height: supraorbitary-infraorbitary
respiratory	RL	length: subspinale-posterior nasal spine
	RB	breadth: widest extension of anterior nasal aperture
	RH	height: nasion-subspinale
masticatory	ML	length: zygomaxillare-posterior border of glenoid cavity
	MB	breadth: anterior sulcus of sphenotemporal crest-lower point of zygotemporal suture
	MH	height: lower border of zygotemporal suture-upper temporal line at coronal intersection
alveolar	AL	length: external prosthion-posterior alveolar border
	AB	breadth: from left to right alveolar borders, at unions between second and third molars
	AH	height: intermaxillary synchondrosis-alveolar border, at unions between second and third molars
cranial volumetric index	VI	$\sqrt[3]{(\text{length}^* \text{ breadth}^* \text{ height})}$
humeral	HML	humeral maximum length
	HEB	humeral epicondylar breadth
	HVDH	humeral vertical diameter of the head
	HMDM	humeral maximum diameter at midshaft
	HmDM	humeral minimum diameter at midshaft
humeral robusticity index	HRI	(HMDM + HmDM)/HML* 100
femoral	FBL	femoral bicondylar length
	FEB	femoral epicondylar breadth
	FMHD	femoral maximum head diameter
	FSD	femoral anterior-posterior (sagittal) midshaft diameter
	FTD	femoral medial-lateral (transverse) midshaft diameter
femoral robusticity index	FRI	(FSD + FTD)/FBL* 100
linearity index	LI	FMHD/FBL
femoral midshaft shape index	FMSSI	FSD/FTD
differences between means	DM	(Mean of hunter-gatherers − mean of farmers)/[(mean of hunter-gatherers + mean of farmers)/2]* 100

assume that robusticity is strongly related to activity patterns (Pearson, 2000; Ruff, 2005; Stock and Shaw, 2007). For instance, mobility was associated with femoral cross-sectional geometry, and according to Ruff (1987), femoral external dimensions are highly correlated with cross-sectional geometry (Ruff, 1987; Stock and Shaw, 2007). Femoral midshaft shape was therefore assessed (Table 18.2) as a ratio between sagittal and transversal diameters.

Sexual proportions vary between both samples and almost all variables are sexually dimorphic; thus, comparisons were done separately by sex. All variables are normally distributed (based on a Kolmogorov Smirnov test). Since the hunter-gatherer sample is composed by individuals derived from greater number of sites with a longer time span than in the case of the farmer sample, variances were compared using a Levene's test and the result indicate homogeneity in the case of all comparisons.

T-tests were performed in order to look for differences in mean values between hunter-gatherers and farmers. Analyses were done separating bones of the right side from those of the left side. Differences between means of size measurements were calculated (Table 18.2) to observe the direction and the magnitude of variation.

18.7 RESULTS AND DISCUSSION

Tables 18.3 and 18.4 and Figure 18.2 summarize the comparisons of the cranial and post-cranial measurements. Differences between mean dimensions are shown in Figure 18.3. The most important result is the observed reduction in the size of the face and limbs in the farmer sample. While the pattern of facial variation was similar for both sexes (Table 18.3), limb variation was different between males and females (Table 18.4). Most of the facial measurements were larger amongst hunter-gatherers, but only masticatory length and breadth and optic length and height presented significant variation (Table 18.3). In the comparison of limb variation, bones of both sides present the same pattern; males differ in both humeral and femoral epicondylar breadths, whereas females differ in humeral diaphysis breadths, humeral head diameter, and femoral length and breadth dimensions (Table 18.4). Post-cranial indices were mostly non-significant, with the exception of the linearity index and left femoral robusticity, which are greater amongst hunter-gatherer females and the femoral midshaft shape index, which was greater amongst farmer females (Table 18.4).

In order to interpret variation, it is necessary to place it in relevant archaeological and ethno-historical contexts. When information obtained from past populations of the ACW was not sufficient, some inferences were made based on information derived from populations of the Argentine Northwest (Figure 18.1). This is an archaeological area located to the north of the ACW that underwent similar cultural shifts after the adoption of plant and animal domestication (Castro and Tarragó, 1992; Gambier, 1993; Lagiglia, 2002; Raffino, 2007).

Changes in facial variation observed were similar to those reported previously (Sardi, Novellino and Pucciarelli, 2006), despite the larger sample size analysed here. The optic and the respiratory components were the least variable (Figure 18.3). Only the masticatory volumetric index was greater amongst hunter-gatherers (Table 18.3). Muscular thickness and facial measurements are positively correlated with the magnitude of masticatory forces (Kiliaridis, 1995; Raadsheer *et al.*, 1999; Herring *et al.*, 2001). Experimental studies suggest that the mastication of soft diets contributes to the reduction of cortical bone thickness (Bresin, Kiliaridis and Strid, 1999), the shortening of maxillary arches (Beecher, Corrucini and Freeman, 1983), reduction of muscular size (Ciochon, Nisbett and Corruccini, 1997) and

Table 18.3 T-test for comparing means in facial measurements and volumetric indices (hunter-gatherers – farmers)

males	× HG ± sd	× F ± sd	t	p	
OL	40.90 ± 2.62	39.69 ± 2.38	2.21	0.030	a
OB	40.49 ± 2.02	41.04 ± 1.85	–1.28	0.203	
OH	35.11 ± 1.74	36.04 ± 1.99	–2.29	0.025	a
RL	51.48 ± 3.44	50.60 ± 2.57	1.31	0.195	
RB	24.53 ± 1.68	24.89 ± 1.82	–0.94	0.352	
RH	54.52 ± 3.33	54.78 ± 2.94	–0.37	0.713	
ML	65.55 ± 3.99	62.74 ± 3.38	3.46	0.001	b
MB	28.04 ± 2.87	26.87 ± 2.94	1.83	0.070	
MH	99.73 ± 8.60	98.91 ± 6.23	0.50	0.622	
AL	54.00 ± 3.06	52.88 ± 3.59	1.54	0.127	
AB	62.54 ± 3.70	63.25 ± 3.49	–0.89	0.376	
AH	12.15 ± 3.20	11.93 ± 2.62	0.34	0.738	
OVI	38.56 ± 1.31	38.69 ± 1.36	–0.44	0.663	
RVI	40.79 ± 1.96	40.82 ± 1.62	–0.06	0.950	
MVI	56.47 ± 3.20	54.74 ± 3.09	2.50	0.014	a
AVI	34.19 ± 3.65	33.88 ± 3.16	0.42	0.677	
females					
OL	41.14 ± 1.85	38.58 ± 1.91	5.25	0.000	b
OB	38.97 ± 1.69	39.37 ± 1.32	–1.06	0.292	
OH	34.83 ± 1.72	35.52 ± 1.92	–1.45	0.151	
RL	48.67 ± 3.25	47.40 ± 2.57	1.73	0.089	
RB	24.55 ± 1.62	24.68 ± 1.62	–0.30	0.762	
RH	52.54 ± 2.78	52.12 ± 2.52	0.64	0.527	
ML	63.57 ± 3.56	59.42 ± 2.80	5.22	0.000	b
MB	26.86 ± 3.52	24.62 ± 2.35	3.10	0.003	b
MH	93.59 ± 6.06	92.14 ± 5.70	0.96	0.340	
AL	51.63 ± 3.52	50.25 ± 2.17	1.98	0.053	
AB	61.76 ± 4.25	60.08 ± 4.42	1.49	0.141	
AH	9.94 ± 1.93	10.02 ± 2.27	–0.14	0.891	
OVI	38.06 ± 0.99	37.62 ± 1.09	1.58	0.119	
RVI	39.57 ± 1.94	39.18 ± 1.40	0.95	0.346	
MVI	53.95 ± 3.43	51.00 ± 2.51	3.99	0.000	b
AVI	31.42 ± 2.82	30.86 ± 2.75	0.79	0.434	

[a] $p < 0.05$.
[b] $p < 0.01$.

decreased growth, particularly in transverse dimensions and in posterior portions of the skull (Lieberman *et al.*, 2004). Size reduction of masticatory and dental components associated with the acquisition of plant and animal domestication has also been suggested in other studies (cf. studies by Carlson and Van Gerven (1979) for Lower Nubia, Sardi, Ramírez Rozzi and Pucciarelli (2004) for Europe and North Africa, and Pinhasi, Eshed and Shaw (2008) for the southern Levant).

In the ACW and adjacent areas (e.g. Argentine Northwest, North of Chile, North of Patagonia), grinding stone tools were introduced during the Middle Holocene by foragers (Lagiglia, 2002; Babot, 2006; Núñez *et al.*, 2006) and were used initially for the processing of

Table 18.4 T-test for comparing means in limb measurements and indices (hunter-gatherers – farmers)

males	*right bones* × HG ± sd	× F ± sd	t	p		*left bones* × HG ± sd	× F ± sd	t	p	
HML	315.86 ± 14.7	317.92 ± 13.3	–0.38	0.706		315.67 ± 16.1	314.17 ± 13.0	0.27	0.792	
HEB	61.71 ± 2.87	59.08 ± 3.43	2.18	0.039	[a]	60.50 ± 3.21	58.71 ± 3.15	1.45	0.159	
HVDH	46.22 ± 1.71	45.46 ± 2.93	0.83	0.414		46.11 ± 1.74	45.01 ± 3.06	1.11	0.277	
HMDM	21.80 ± 1.50	21.70 ± 2.60	0.12	0.907		20.70 ± 1.24	21.01 ± 1.90	–0.49	0.630	
HmDM	17.40 ± 1.16	16.91 ± 2.31	0.71	0.486		17.10 ± 1.26	16.84 ± 1.51	0.48	0.636	
FBL	453.38 ± 17.4	442.69 ± 18.8	1.21	0.241		449.93 ± 13.8	441.57 ± 17.6	1.17	0.254	
FEB	83.27 ± 3.91	79.06 ± 3.09	2.65	0.015	[a]	81.81 ± 4.07	78.34 ± 3.38	2.22	0.037	[a]
FMHD	46.64 ± 2.56	46.12 ± 2.86	0.39	0.699		46.98 ± 1.76	45.64 ± 2.33	1.42	0.169	
FSD	32.02 ± 3.04	29.58 ± 2.61	1.87	0.077		31.67 ± 3.13	29.55 ± 2.62	1.75	0.093	
FTD	26.74 ± 1.47	25.66 ± 2.00	1.20	0.243		26.47 ± 1.59	25.67 ± 2.37	0.86	0.402	
HRI	12.42 ± 0.78	12.14 ± 1.33	0.67	0.508		11.99 ± 0.83	12.06 ± 1.02	–0.18	0.859	
FRI	12.96 ± 0.75	12.49 ± 0.83	1.22	0.238		12.92 ± 0.92	12.51 ± 0.74	1.19	0.247	
LI	0.103 ± 0.01	0.104 ± 0.01	–0.59	0.561		0.104 ± 0.01	0.103 ± 0.01	0.65	0.520	
FMSSI	1.19 ± 0.09	1.16 ± 0.11	0.79	0.024		1.19 ± 0.08	1.16 ± 0.13	0.81	0.428	
females										
HML	300.70 ± 14.3	296.33 ± 7.26	0.82	0.422		296.42 ± 8.00	292.74 ± 8.07	0.97	0.346	
HEB	54.80 ± 2.39	51.33 ± 2.87	2.87	0.011	[a]	55.13 ± 2.09	51.22 ± 3.11	3.13	0.006	[b]
HVDH	40.87 ± 2.43	38.25 ± 1.40	2.83	0.011	[a]	41.60 ± 2.09	37.96 ± 1.57	4.19	0.001	[b]
HMDM	19.77 ± 1.61	19.25 ± 2.13	0.61	0.550		19.10 ± 1.34	18.62 ± 1.63	0.69	0.499	
HmDM	16.04 ± 1.73	14.28 ± 1.00	2.68	0.016	[a]	16.31 ± 1.83	14.69 ± 1.25	2.20	0.043	[a]
FBL	422.42 ± 19.4	403.00 ± 13.7	2.44	0.025	[a]	421.75 ± 19.1	402.75 ± 12.5	2.49	0.022	[a]
FEB	77.01 ± 4.67	69.75 ± 1.67	4.19	0.001	[b]	76.23 ± 4.09	69.38 ± 1.92	4.41	0.000	[b]
FMHD	44.11 ± 2.54	39.56 ± 1.91	4.31	0.000	[b]	43.21 ± 2.55	38.86 ± 1.97	4.11	0.001	[b]
FSD	26.79 ± 1.79	25.59 ± 1.24	1.64	0.118		27.33 ± 2.33	24.78 ± 1.20	2.85	0.010	[a]
FTD	26.16 ± 1.77	23.51 ± 1.72	3.31	0.004	[b]	26.57 ± 1.80	24.30 ± 1.34	3.06	0.006	[b]
HRI	11.92 ± 1.07	11.31 ± 0.86	1.36	0.192		11.95 ± 1.01	11.38 ± 0.86	1.30	0.211	
FRI	12.54 ± 0.66	12.19 ± 0.70	1.13	0.271		12.78 ± 0.63	12.19 ± 0.55	2.16	0.044	[a]
LI	0.105 ± 0.01	0.098 ± 0.01	2.70	0.015	[a]	0.102 ± 0.01	0.097 ± 0.01	2.95	0.008	[b]
FMSSI	1.03 ± 0.06	1.09 ± 0.05	–2.46	0.024	[a]	1.03 ± 0.06	1.02 ± 0.05	0.31	0.763	

[a] $p < 0.05$.
[b] $p < 0.01$.

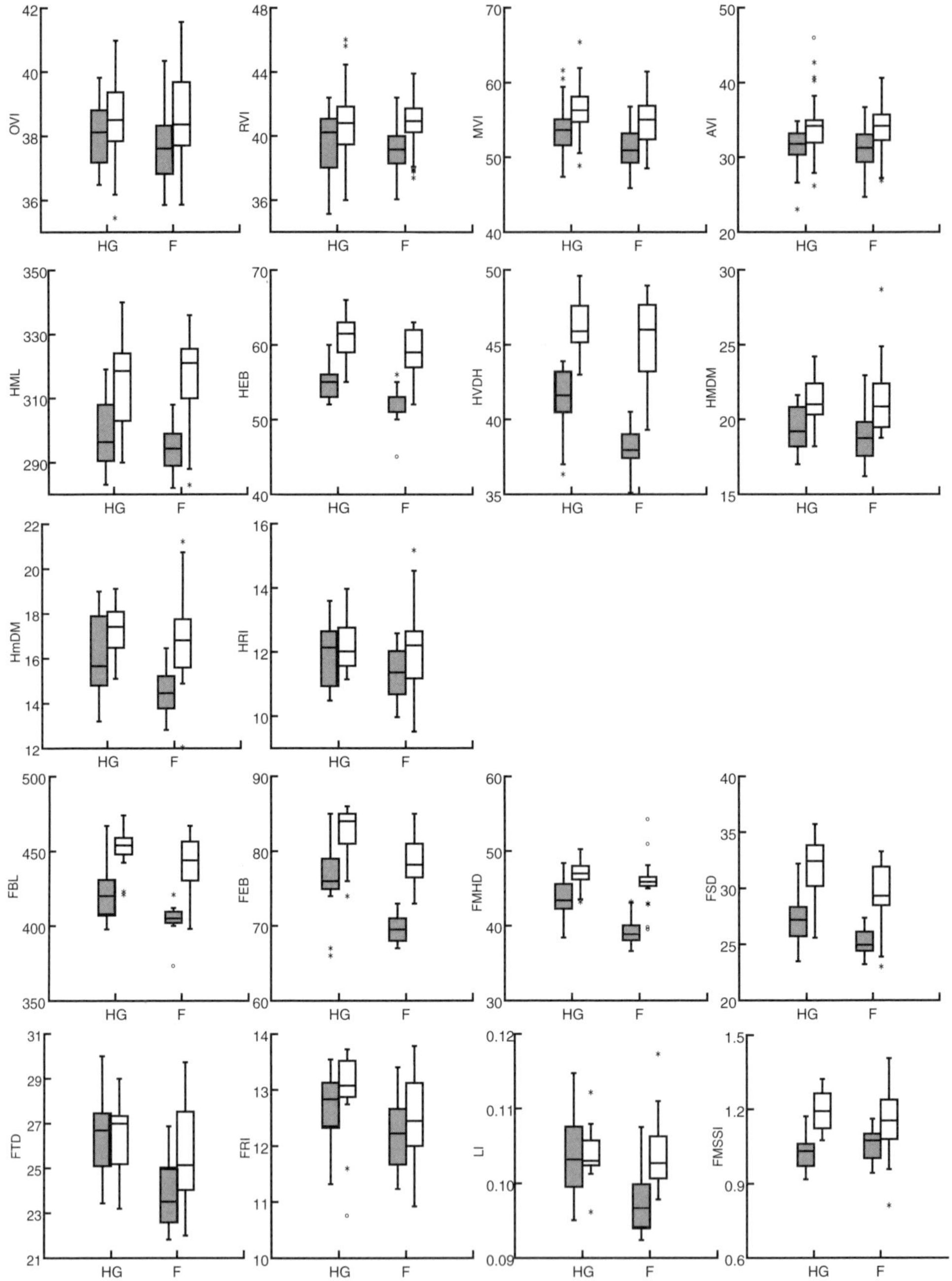

Figure 18.2 Box plots of facial indices and limb bone measurements and indices in hunter-gatherers (HG) and farmers (F), males (white boxes) and females (grey boxes). Bones of both sides were pooled together to look for differences in the post-cranial comparisons

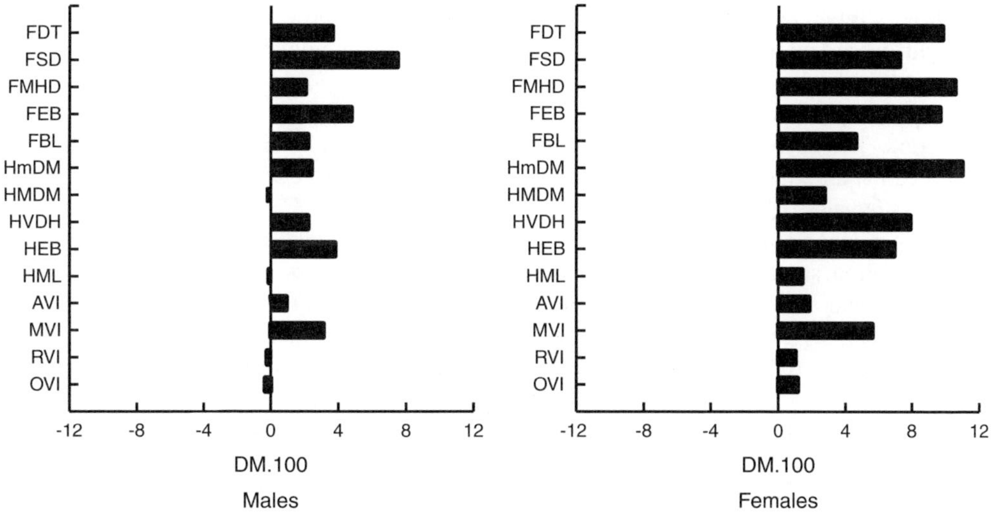

Figure 18.3 Percent differences between means (DM) of measurements and indices between hunter-gatherers and farmers (Table 18.2). Positive values indicate a greater mean value in hunter-gatherers. Bones of both sides were pooled together to look for differences in the post-cranial comparisons

wild plants for food as well as for pigments. The use of mills and mortars intensified and by ca. calAD 50 were fully utilized by farmers for processing food (Babot, 2006) which, together with the use of pottery for cooking, resulted in softer diets. It is possible therefore that reduction in loadings of the masticatory complex due to differences in food consistency between hunter-gatherers and farmers could promote facial gracilization amongst the ACW.

With respect to limb bones, a decrease in size with the transition to food production was observed in females, being the reduction in the femoral dimensions greater than in the humeral dimensions (Table 18.4, Figure 18.3). Variation in limb length and thus stature may be attributed to one or more systemic factors such as changes in the quality of nutrition and health. However, if a given factor influenced growing structures, cumulative effects produced significant variation only in the femur because femoral length was on average shorter in female farmers; humeral length did not show any variation between the two populations.

Variation in other limb bones measurements (e.g. midshaft diameters) between hunter-gatherers and farmers of the ACW might be affected by daily activities (Stock and Pfeiffer, 2004; Ruff, 2005). It has been proposed that there is an association between lower limb robusticity, diaphyseal shape and terrestrial mobility (Ruff, 1987, 2005; Stock and Pfeiffer, 2001; Wescott, 2006). The basic shape of the femur is circular, yet bending loads due to running, climbing and all contractions on the quadriceps cause antero-posterior bending and therefore the antero-posterior cross-section is elongated (Ruff, 1987). The antero-posterior elongation of femoral midshaft has been observed throughout populations of different subsistence patterns and further interpreted as a result of changes in mobility (Ruff, 1987, 1994, 1999; Larsen, 1995; Stock and Pfeiffer, 2004).

According to our results (Table 18.4), hunter-gatherer females had the most robust femora, but only in the case of the left side, probably as a result of sample composition. On the other hand, highly significant variation between female hunter-gatherers and farmers was observed in the case of femoral transversal midshaft diameter and no variation in the sagittal diameter (Table 18.4). Results on midshaft shape did not show a clear pattern and, thus, straightforward inferences about levels of terrestrial mobility amongst ACW females cannot be made on the basis of this index.

Femoral epiphyseal breadth (FMHD and FEB) was on average greater amongst hunter-gatherers. The epicondylar diameter of the femur was the only measurement that differed significantly amongst males in both sides (Table 18.4) and showed highly significant variation amongst females.

Comparisons of the size of femoral epicondyles can be used as an estimation of body mass variation, since epicondyles support a great part of body mass and attach ligaments of the knee joint and many muscles of the thigh and leg: gastrocnemius, plantaris, popliteus and abductor magnus (Scheuer and Black, 2000). On other hand, the linearity index varied amongst females because ACW farmers underwent greater reduction in the femoral head than femoral length diameter (Table 18.4, Figure 18.3). Therefore, it is possible to infer that female farmers of ACW had lower body mass than female hunter-gatherers, while the latter were stockier.

Body mass is strongly associated with many other components, such as muscle, bone and fat. Articular surfaces are assumed to be less responsive to environmental factors (Ruff, 1991; Lieberman, Devlin and Pearson, 2001); however, relatively few studies investigated potential variables that affect articular surfaces. Pfeiffer and Sealy (2006) interpreted variation in limb head diameters (i.e. body mass) amongst foragers of South Africa as a result of changes in diet. Even though a low-protein diet could promote body mass reduction in farmers of ACW, it is difficult to explain why there are no differences between males from the two populations. Since no ethnohistorical or archaeological data suggest gender-specific differences in access to dietary resources, it is possible that the reduction of loadings upon muscles of the leg amongst farmers influenced on the epicondylar reduction.

Humeral measurements show less variation than femoral measurements. Although the humerus has been much less studied, studies of upper limb robusticity and bilateral asymmetry provide some degree of insight about habitual behaviours (Stock and Pfeiffer, 2001; Ruff, 2005). Because the main part of the hunter-gatherer sample was recovered from secondary burials, we cannot associate left and right skeletal parts and study lateralization. With respect to the humeral robusticity index, no differences were obtained between hunter-gatherers and farmers in either sex, despite variations in the minimum diameter at midshaft. However, epicondylar breadth was smaller in farmers (Table 18.4). Humeral epicondyles are sites of attachment of muscles of the forearm: the superficial flexor muscles of the forearm and the long flexors of the digits (Scheuer and Black, 2000). It is probable that the humeral epicondylar reduction amongst farmers was a consequence of the lesser muscular loadings on the forearm and the hand.

The findings of this study may contrast with results obtained elsewhere (Eshed *et al.*, 2004). However, information derived from the archaeological record of the Argentine Northwest indicates that despite regional variations, changes to technological strategies associated with the adoption and intensification of agriculture and sedentism (after calAD 0) reduced physical demands. Lithic tools of daily use were manufactured with very low investments of time and skills (Hocsman, 2006; Elías, 2007). Only those tools devoted to agricultural activities (e.g. shovels, mortars, hoes) denote greater time investment in their manufacture. According to Babot (2006), as grinding activities intensified across the Late Holocene, grinding stone tools were manufactured in order to improve efficiency of mills as well as to reduce the amount of time invested in grinding activities needed to obtain greater amounts of processed food.

Despite evidence that local loadings could independently affect different skeletal structures, it is likely that all size measurements are correlated (Corruccini, 1987). The overall pattern of variation amongst populations of the ACW expresses size reduction in farmers, specifically a reduction in stature, body mass, femoral robusticity and masticatory volume. This pattern may partially reflect a decrease in muscle mass. A previous study (Sardi, Novellino and Pucciarelli, 2006) found similar results with the occipital (i.e. posteroneural) component, where neck muscles attach, being smaller in farmers than in hunter-gatherers.

According to Figure 18.2, females were smaller than males in the case of all comparisons. Farmers, on the other hand, seem more dimorphic than hunter-gatherers. The different patterns of variation between males and females can therefore be interpreted as a change in sexual dimorphism.

An association between sexual dimorphism and subsistence strategy has been explored by many scholars. Frayer (1980) stated that technological changes during the Mesolithic-Neolithic transition produced a reduction in male body size and, therefore, a decrease in sexual dimorphism. Ruff (1987) and Cole (1994) observed that sexual dimorphism in lower limb bone cross-sectional geometry and external breadths declined from hunting-gathering through agricultural to industrial groups. Collier (1993) found similar results amongst populations of different economies, but he proposed that economy does not influence bone strength in a simple manner, but is rather the total set of activities that influence variation and dimorphism. Wescott (2006) found that sexual dimorphism in femoral midshaft and robusticity is greater in more mobile populations.

The potential to expand interpretations about the different patterns of variation between males and females of the ACW depends upon additional information derived from archaeological and ethnographical descriptions. Amongst hunter-gatherers, it is probable that both sexes engaged in foraging activities, which may have started during childhood, and that may explain the lesser sexual dimorphism. Foraging activities might require great physical exertion, with less task specialization and long distance seasonal movements by the whole group (Neme, 2002; Durán *et al.*, 2006).

Some studies suggest that there was a marked sexual division of labour amongst farmer of the ACW and the Argentine Northwest. Following the transition to agriculture (which entailed crop and livestock domestication and sedentism) muscular loadings decreased, and this trend affected females more than males. It is unclear which sex was more engaged in agricultural activities. But it is clear that females were engaged in household activities in the village. Weaving and textile production was an important task in the northern region of the ACW (Rusconi, 1961; Michieli, 2000) that, together with grinding of crops and wild plants for food as well as the control of animals in their dwellings, are activities known to be carried out by women in historical times (Rusconi, 1961; Göbel, 1998, 2002).

Male farmers showed fewer morphological metric changes than females with respect to ACW hunter-gatherers, mainly because they maintained higher levels of physical exertions associated with activities outside the domestic domain. Hunting was an important male task until historical times (Rusconi, 1961). Another task was caravan trade, which was an important economic task performed exclusively by men and children from 6 to 7 years old (Göbel, 1998, 2002), in order to interchange crops, textiles, amongst other products, with populations throughout the Andean mountains (Nielsen, 2001). Even if men and children used animals for the transport of products, they were often engaged in long distance walks, around 40 km a day, as has been reported in various ethnographic descriptions (Göbel, 1998, 2002).

The overall pattern of skeletal variation may express multiple and confounding factors. The hypothesis that a low-protein diet, which is associated with farming, exerted an influence on skeletal morphology is neither rejected nor strongly supported. Only females showed the expected pattern of variation – reduction in cranial and postcranial size – while postcranial reduction amongst males was mostly non-significant. This contrasts with previous statements (Tanner, 1962) that nutritional deficiencies affect males more than females, promoting a reduction in sexual dimorphism. The comparison of males did not fit the expected pattern and, moreover, farmers were more sexually dimorphic than hunter-gatherers. Thus, it does not appear as though morphological changes associated with the transition to agriculture in the ACW were driven by the consumption of low-protein diets. The pattern of skeletal variation is better explained as being a consequence of a reduction in muscular mass in some body components. The reduction in masticatory muscles amongst farmers of the ACW suggests a corresponding decrease in masticatory forces following the reliance on softer foodstuffs. It is also likely that lower limb morphology was affected to some degree by the reduction in physical activities. Loadings are not the only factor affecting skeletal morphology directly/locally; and other factors might also act upon facial and post-cranial dimensions.

Changes in physical activities may be associated with changes in growth hormone (GH). Growth hormone plays a very important role in overall skeletal growth, promoting proliferation and differentiation of growth plates, which affects bone size and muscle mass (Vogl *et al.*, 1993; Barr and McKay, 1998; Banu *et al.*, 2001; Forwood *et al.*, 2001; Frago and Chowen, 2005). Increase in circulating growth hormone has been reported to occur following moderate to intense exercise (Kalu, Banu and Wang, 2000; Weltman *et al.*, 2001; Parks, 2002). Conversely, decreased loadings of bones result in bone loss. Variation in GH circulation throughout ontogeny is linked to variation in limb lengths (Ohlsson *et al.*, 1998; Frago and Chowen, 2005). During adulthood, GH circulation is associated with variation in muscle mass (Fernández and LeRoith, 2005), which in turn affects bone remodelling. As shown the experimental study of Lieberman (1996), animals with greater physical activity have thicker cortical bone mass, even in bones that are not under direct biomechanical loading during exercise, such as the skull. A reduction of GH circulation may thus account for the differentiation observed in female farmers of ACW.

18.8 CONCLUSIONS

Late Holocene native populations of the ACW differed in the morphology of facial and limb skeletal components. The general hypothesis that was tested – that different subsistence strategies are associated with changes in skeletal morphology – cannot be rejected. Farmers show smaller size of all components, but this reduction was greater and more significant amongst females than amongst males. Female farmers are characterized by shorter limb length and smaller areas of muscle insertion than female hunter-gatherers and a more linear body shape (body mass relative to stature).

The factors that best explain the overall morphological patterns are related to gender and population-specific variation in a range of physical activities, either by direct muscular loading on skeletal components or by systemic influence of growth hormone. The increase in sedentism amongst farmers and particularly amongst females, the improvement in technologies, the consumption of softer diets, and a gender-based division of labour (i.e. females less mobile than males) can be some of the factors that account for the overall pattern of variation in populations of the ACW. Future studies on these populations that focus on biomechanical analysis of their skeletal structures can shed more light on this issue.

ACKNOWLEDGEMENTS

We thank the editors of this book for the invitation to contribute to this chapter; two anonymous reviewers for important corrections on earlier versions of the manuscript; the curators of the collections for access to specimens in their care; Marco Giovannetti for significant contributions and suggestions; Paula Novellino, for useful information about collections. Marta Roa, Cecilia Páez, Darío Hermo, Valeria Bernal and Paula González also contributed to different aspects of this chapter. The Consejo Nacional de Investigaciones Científicas y Técnicas (CONICET, Argentina) supports our research.

REFERENCES

Aschero, C. (1994) Reflexiones desde el Arcaico Tardío (6000–3000 AP). *Rumitacana. Revista de Antropología*, **1**, 13–17.

Atchley, W.R. and Hall, B.K. (1991) A model for development and evolution of complex morphological structures. *Biol. Rev. Camb. Philos. Soc.*, **66**, 101–157.

Auerbach, B.M. and Ruff, C.B. (2004) Human body mass estimation: a comparison of 'Morphometric' and' 'mechanical' methods. *Am. J. Phys. Anthropol.*, **125**, 331–342.

Babot, P. (2006) El papel de la molienda en la transición hacia la producción agropastoril: Un análisis desde la Puna Meridional Argentina. *Estudios Atacameños*, **32**, 75–92.

Banu, J., Orhii, P.B., Okafor, M.C. *et al.* (2001) Analysis of the effects of growth hormone, exercise and food restriction on cancellous bone in different bone sites in middle-aged female rats. *Mech. Ageing. Dev.*, **122**, 849–864.

Bárcena, R. (1985) Agricultores y alfareros tempranos del noroeste de Mendoza según la excavación arqueológica de varios abrigos rocosos. Resúmenes de las IX Jornadas de Investigaciones de la Universidad Nacional de Córdoba, pp. 154.

Barr, S.I. and McKay, H.A. (1998) Nutrition, exercise, and bone status in youth. *Int. J. Sport Nutr.*, **8**, 124–142.

Beecher, R.M., Corrucini, R.S. and Freeman, M. (1983) Craniofacial correlates of dietary consistency in a nonhuman primate. *J. Craniofac. Genet. Dev. Biol.*, **3**, 193–202.

Béguelin, M. *et al.* (2006) El poblamiento humano tardío del sur de Mendoza y su relación con el norte de Patagonia: una discusión bioarqueológica, in *Arqueología y ambiente de áreas naturales protegidas de la provincia de Mendoza. Volumen Especial N. 61 Anales de Arqueología y Etnología* (eds V. Durán and V. Cortegoso), UNC, pp. 11–32.

Bernal, V., Novellino, P., Gonzalez, P.N. and Perez, S.I. (2007) Role of wild plant foods among late Holocene hunter-gatherers from central and north Patagonia (South America): an approach from dental evidence. *Am. J. Phys. Anthropol.*, **133**, 1047–1059.

Bogin, B. (1999) *Patterns of Human Growth*, Cambridge University Press, Cambridge.

Bresin, A., Kiliaridis, S. and Strid, K.G. (1999) Effect of masticatory function on the internal bone structure in the mandible of the growing rat. *Eur. J. Oral Sci.*, **107**, 35–44.

Bridges, P.S., Blitz, J.H. and Solano, M.C. (2000) Changes in long bone diaphyseal strength with horticultural intensification in west-central Illinois. *Am. J. Phys. Anthropol.*, **112**, 217–238.

Buikstra, J.E. and Ubelaker, D.H. (1994) Standards for data collection from human skeletal remains. Arkansas: Archaeological Survey Research 44.

Carlson, D.S. and Van Gerven, D.P. (1979) Masticatory function and Post-Pleistocene in Nubia. *Am. J. Phys. Anthropol.*, **46**, 495–506.

Castro, V.R. and Tarragó, M.N. (1992) Los inicios de la producción de alimentos en el cono sur de América. *Revista de Arqueología Americana*, **6**, 91–124.

Cheverud, J.M. (1996) Developmental integration and the evolution of pleiotropy. *Am. Zool.*, **36**, 44–50.

Ciochon, R.L., Nisbett, R.A. and Corruccini, R.S. (1997) Dietary consistency and craniofacial development related to masticatory function in minipigs. *J. Craniofac. Genet. Dev. Biol.*, **17**, 96–102.

Cohen, M.N. and Armelagos, G.J. (1984) *Paleopathology and the Origins of Agriculture*, Academic Press, Orlando.

Cole, T.M. III (1994) Size and shape of the femur and tibia in northern Plains Indians, in *Skeletal Biology in the Great Plains: Migration, Warfare, Health, and Subsistence* (eds D.W. Owsley and R.L. Jantz), Smithsonian Institution Press, Washington, DC, pp. 219–234.

Collier, S. (1993) Sexual dimorphism in relation to big-game hunting and economy in modern human populations. *Am. J. Phys. Anthropol.*, **91**, 485–504.

Corruccini, R. (1987) Shapes in morphometrics: Comparative analyses. *Am. J. Phys. Anthropol.*, **73**, 289–303.

Durán, V., Neme, G., Cortegoso, V. and Gil, A. (2006) Arqueología del área natural protegida Laguna del Diamante (Mendoza, Argentina), in *Arqueología y ambiente de áreas naturales protegidas de la provincia de Mendoza. Volumen Especial N. 61 Anales de Arqueología y Etnología* (eds V. Durán and V. Cortegoso), UNC, pp. 91–143.

Elías, A.M. (2007) Tecnología lítica en las sociedades tardías de Antofagasta de la Sierra (Puna Meridional Argentina). *Estudios Atacameños*, **33**, 50–85.

Enlow, D.H. and Hans, M.G. (1996) *Crecimiento Facial*, McGraw-Hill Interamericana, Mexico DF.

Eshed, V., Gopher, A., Galili, E. and Hershkovitz, I. (2004) Musculoskeletal stress markers in Natufian hunter-gatherers and Neolithic farmers in the Levant: the upper limb. *Am. J. Phys. Anthropol.*, **123**, 303–315.

Feldesman, M.R. and Fountain, R.L. (1996) Race specificity and the femur/stature ratio. *Am. J. Phys. Anthropol.*, **100**, 207–224.

Fernández, A.M. and LeRoith, D. (2005) Skeletal muscle, in *The Growth Hormone/Insulin-Like Growth Factor Axis During Development* (eds I. Varela-Nieto and J.A. Chowen), Springer, New York, pp. 117–147.

Fiedel, S.A. (1996) *Prehistoria de América*, Crítica, Barcelona.

Forwood, M.R., Li, L., Kelly, W.L. and Bennett, M.B. (2001) Growth hormone is permissive for skeletal adaptation to mechanical loading. *J. Bone. Miner. Res.*, **16**, 2284–2290.

Frago, L.M. and Chowen, J.A. (2005) Basic physiology of the growth hormone/insulin-like growth factor axis, in *The Growth Hormone/Insulin-Like Growth Factor Axis During Development* (eds I. Varela–Nieto and J.A. Chowen), Springer, New York, pp. 1–25.

Frayer, D.W. (1980) Sexual dimorphism and cultural evolution in the late Pleistocene and Holocene of Europe. *J. Hum. Evol.*, **9**, 399–415.

Gambier, M. (1976) Ecología y arqueología de los Andes Centrales Argentino-Chilenos. *Memorias del IV Congreso Nacional de Arqueología Argentina*, **3**, 185–199.

Gambier, M. (1993) *Prehistoria de San Juan*, Editorial Universidad Nacional de San Juan, San Juan.

García, A. (1992) Hacia un ordenamiento preliminar de las ocupaciones prehistóricas agrícolas precerámicas y agroalfareras en el N.O. de Mendoza. *Revista de Estudios Regionales*, **10**, 7–34.

García, A. (1997) La Ocupación humana del Centro Oeste argentino hacia el límite Pleistoceno-Holoceno: El componente Paleoindio del sitio Agua de la Cueva, sector Sur. Ph. D. Universidad Nacional de Cuyo, Argentina.

Gil, A. (2002) El registro arqueológico y la ocupación humana de La Payunia, in *Entre montañas y desiertos: arqueología del sur de Mendoza* (eds A. Gil and G. Neme), Sociedad Argentina de Antropología, Buenos Aires, pp. 103–118.

Gil, A. (2003) Zea *mays* on the South American periphery: Chronology and dietary importance. *Curr. Anthropol.*, **44**, 295–300.

Gil, A. and Neme, G. (2008) Faunal exploitation and agricultural transitions in the South American Agricultural Limit. *Int. J. Osteoarchaeol.*, **18**, 293–306.

Gil, A., Shelnut, N., Neme, G. *et al.* (2006) Isótopos estables y dieta humana en el centro oeste: Datos muestras de San Juan. *Revista Cazadores y Recolectores del Cono Sur*, **1**, 149–161.

Gil, A., Neme, G.A., Tykot, R.H. *et al.* (2009) Stable isotopes and maize consumption in Central Western Argentina. *Int. J. Osteoarchaeol.*, **19**, 215–236.

Göbel, B. (1998) Salir de viaje. Producción partoril e intercambio económico en el noroeste argentino. *Bonner Amerikanistische Studien*, **30**, 867–891.

Göbel, B. (2002) Identidades socials y medio ambiente: la multiplicidad de los significados del espacio en la Puna de Atacama. *Cuadernos del Instituto Nacional de Antropología y Pensamiento Latinoamericano*, **19**, 267–296.

Hall, B.K. (2005) *Bone and Cartilage: Developmental and Evolutionary Skeletal Biology*, Elsevier Academic Press, London.

Hastorf, C.A. (1999) Cultural implications of crop introductions in Andean prehistory, in *The Prehistory of Food: Appetites for Change* (ed. C. Gosden and J. Hather), Routledge, London, pp. 35–58.

Herring, S.W., Rafferty, K.L., Liu, Z.J. and Marshall, C.D. (2001) Jaw muscles and the skull in mammals: the biomechanics of mastication. *Comp. Biochem. Physiol. A*, **131**, 207–219.

Hocsman, S. (2006) Tecnología lítica en la transición de cazadores recolectores a sociedades agropastoriles en la porción meridional de los Andes Centro Sur. *Estudios Atacameños*, **32**, 59–73.

Kalu, D.N., Banu, J. and Wang, L. (2000) How cancellous and cortical bones adapt to loading and growth hormone. *J. Musculoskel Neuron Interact*, **1**, 19–23.

Kiliaridis, S. (1995) Masticatory muscle influence on craniofacial growth. *Acta Odontol. Scand.*, **53**, 196–202.

Lagiglia, H. (1997) *Arqueología de Cazadores-Recolectores Cordilleranos de Altura*, Ediciones Ciencia y Arte, San Rafael.

Lagiglia, H. (2001) *Los Orígenes de la agricultura en Argentina* (eds E. Berberián and A. Nielsen), Argentina Prehistórica, Córdoba, pp. 41–81.

Lagiglia, H. (2002) Arqueología prehistórica del sur mendocino y sus relaciones con el centro oeste argentino, in *Entre montañas y desiertos: arqueología del sur de Mendoza* (eds A. Gil and G. Neme), Sociedad Argentina de Antropología, Buenos Aires, pp. 43–64.

Larsen, C.S. (1995) Biological changes in human populations with agriculture. *Ann. Rev. Anthropol.*, **24**, 185–213.

Lehmann Nitsche, R. (1910) *Catálogo de la Sección Antropología del Museo de La Plata*, Ed. Coni., Buenos Aires.

Lieberman, D.E. (1996) How and why humans grow thin skulls: Experimental evidence for systemic cortical robusticity. *Am. J. Phys. Anthropol.*, **101**, 217–236.

Lieberman, D.E., Devlin, M.J. and Pearson, O.M. (2001) Articular area responses to mechanical loading: effects of exercise, age, and skeletal location. *Am. J. Phys. Anthropol.*, **116**, 266–277.

Lieberman, D.E., Krovitz, G.E., Yates, F.W. *et al.* (2004) Effects of food processing on masticatory strain and craniofacial growth in a retrognathic face. *J. Hum. Evol.*, **46**, 655–677.

Michieli, C. (2000) Telas rectangulares decoradas: Piezas de vestimenta del Periodo Tardío Preincaico (San Juan, Argentina). *Estudios Atacameños*, **20**, 77–90.

Miller, J.P. and German, R.Z. (1999) Protein malnutrition affects the growth trajectories of the craniofacial skeleton in rats. *Nutr. Requirements*, **129**, 2061–2069.

Moss, M.L. (1973) A functional cranial analysis of primate craniofacial growth. *Symp. IVth Int. Congr. Primat.*, **3**, 191–208.

Moss, M.L. (1997) The functional matrix hypothesis revisited. 4. The epigenetic antithesis and the resolving synthesis. *Am. J. Orthod. Dentofac. Orthop.*, **112**, 410–417.

Neme, G. (2002) Arqueología del alto valle del Río Atuel: modelos, problemas y perspectivas en el estudio de las regiones de altura del sur de Mendoza, in *Entre montañas y desiertos: arqueología del sur de Mendoza* (eds A. Gil and G. Neme), Sociedad Argentina de Antropología, Buenos Aires, pp. 65–84.

Neme, G. and Gil, A. (2008) Faunal exploitation and agricultural transitions in the South American Agricultural Limit. *Int. J. Osteoarchaeol.*, **18**, 293–306.

Nielsen, A. (2001) Evolución social en quebrada de Humahuaca (AD 700-1536), in *Historia Argentina Prehispánica*, vol. **1** (eds E. Berberián and A. Nielsen), Editorial Brujas, Córdoba, pp. 171–264.

Novellino, P.S. (2002) Bioarqueología del Sur de Mendoza, in *Entre montañas y desiertos: arqueología del sur de Mendoza* (eds A. Gil and G. Neme), Sociedad Argentina de Antropología, Buenos Aires, pp. 119–139.

Novellino, P.S., Gil, A., Neme, G. and Durán, V. (2004) El consumo de maíz en el Holoceno tardío del oeste argentino: isótopos estables y caries. *Revista Española de Antropología Americana*, **34**, 85–110.

Novellino, P.S. and Gil, A. (2007) Estrés nutricional, hipoplasia y explotación de recursos en el centro sur de Mendoza (Argentina). *Intersecciones en Antropología*, **8**, 17–29.

Novellino, P.S. and Guichón, R.A. (1999) Primeros resultados de isótopos estables para el sur mendocino. *Revista Argentina de Antropología Biológica*, **2**, 323–334.

Novellino, P.S., Guichón, R.A. and Lagiglia, H. (1996) Indicadores biológicos en restos humanos del sur de Mendoza: Sitio Jaime Prats. *Revista de Arqueología y Prehistoria del Instituto de Ciencias Antropológicas*, **6**, 69–82.

Núñez, A.L., Cartajena, F.I., Carrasco, G.C. *et al.* (2006) Emergencia de comunidades pastoralistas formativas en el sureste de la Puna de Atacama. *Estudios Atacameños*, **32**, 91–117.

Ohlsson, C., Bengtsson, B.A., Isaksson, O.G. *et al.* (1998) Growth hormone and bone. *Endocr. Rev.*, **19**, 55–79.

Olivera, D. and Podestá, M. (1993) Los recursos del arte: Arte rupestre y sistemas de asentamiento-subsistencia formativos en la Puna Meridional Argentina. *Arqueología*, **3**, 93–141.

Parks, J.S. (2002) Endocrine disorders of growth, in *Human Growth and Development* (ed. N. Cameron), Elsevier, Amsterdam, pp. 213–236.

Pearson, O.M. (2000) Activity, climate, and postcranial robusticity. *Curr. Anthropol.*, **41**, 569–607.

Pfeiffer, S. and Sealy, J.C. (2006) Body size among holocene foragers of the cape ecozone, Southern Africa. *Am. J. Phys. Anthropol.*, **129**, 1–11.

Pinhasi, R., Eshed, V. and Shaw, P. (2008) Evolutionary changes in the masticatory complex following the transition to farming in the southern Levant. *Am. J. Phys. Anthropol.*, **135**, 136–148.

Planella, M.T. and Tagle, B.A. (2004) Inicios de presencia de cultívenos en la zona central de Chile, períodos Arcaico y Alfarero temprano. *Chungará*, **36**, 387–399.

Pucciarelli, H.M. (1981) Growth of the functional components of the rat skull and its alteration by nutritional effects. *Am. J. Phys. Anthropol.*, **56**, 33–41.

Pucciarelli, H.M., Dressino, V. and Niveiro, M.H. (1990) Changes in skull components of the squirrel monkey evoked by growth and nutrition: an experimental study. *Am. J. Phys. Anthropol.*, **81**, 35–543.

Pucciarelli, H.M. and Goya, R.G. (1983) Effects of post-weaning malnutrition on the weight of the head components in rats. *Acta Anat.*, **115**, 231–237.

Raadsheer, M.C., van Eijden, T.M., van Ginkel, F.C. and Prahl-Andersen, B. (1999) Contribution of jaw muscle size and craniofacial morphology to human bite force magnitude. *J. Dent. Res.*, **78**, 31–42.

Raffino, R.A. (2007) *Poblaciones indígenas en Argentina. Urbanismo y proceso social precolombino*, Emecé, Buenos Aires.

Reichling, T.D. and German, R.Z. (2000) Bones, muscles and visceral organs of protein-malnourished rats (*Rattus norvegicus*) grow more slowly but for longer durations to reach normal final size. *Nut. Requirements*, **130**, 2326–2332.

Ruff, C.B. (1987) Sexual dimorphism in human lower limb bone structure: Relationship to subsistence strategy and sexual division of labor. *J. Hum. Evol.*, **16**, 391–416.

Ruff, C.B. (1991) Climate and body shape in hominid evolution. *J. Hum. Evol.*, **21**, 81–105.

Ruff, C.B. (1994) Morphological adaptation to climate in modern and fossil hominids. *Yrbk Phys. Anthropol.*, **37**, 65–107.

Ruff, C.B. (1999) Skeletal structure and behavioral patterns of prehistoric Great Basin populations, in *Understanding Prehistoric Lifeways in the Great Basin Wetlands: Bioarchaeological Reconstruction and Interpretation* (eds B.E. Hemphill and C.S. Larsen), University of Utah Press, Salt Lake City, pp. 290–320.

Ruff, C.B. (2005) Mechanical determinants of bone form: Insights from skeletal remains. *J. Musculoskelet Neuronal Interact*, **5**, 202–212.

Ruff, C.B., Larsen, C.S. and Hayes, W.C. (1984) Structural changes in the femur with the transition to agriculture on the Georgia Coast. *Am. J. Phys. Anthropol.*, **64**, 125–136.

Ruff, C.B., Scott, W.W. and Liu, A.Y.C. (1991) Articular and diaphyseal remodeling of the proximal femur with changes in body mass in adults. *Am. J. Phys. Anthropol.*, **86**, 397–413.

Sandweiss, D.H., Solís, R.S., Moseley, M.E. *et al.* (2008) Environmental change and economic development in coastal Peru between 5800 and 3600 years ago. *Proc. Natl. Acad. Sci. U. S. A.*, **106**, 1359–1363.

Rusconi, C. (1961) *Poblaciones pre y posthispánicas de Mendoza*, Imprenta Oficial de Mendoza, Mendoza.

Sardi, M.L., Ramírez Rozzi, F. and Pucciarelli, H.M. (2004) The Neolithic transition in Europe and North Africa. The functional craniology contribution. *Anthrop. Anz.*, **62**, 129–145.

Sardi, M.L. and Ramírez Rozzi, F.V. (2005) Cross-sectional study of human craniofacial growth. *Ann. Hum. Biol.*, **32**, 390–396.

Sardi, M.L., Novellino, P. and Pucciarelli, H.M. (2006) Craniofacial morphology in the Argentine Center-West: consequences of the transition to food production. *Am. J. Phys. Anthropol.*, **130**, 33–343.

Scheuer, L. and Black, S. (2000) *Developmental Juvenile Osteology*, Academic Press, San Diego.

Solís, R.S., Haas, J. and Creamer, W. (2001) Dating Caral, a preceramic site in the Supe Valley on the central coast of Peru. *Science*, **292**, 723–726.

Stini, W.A. (1969) Nutritional stress and growth: Sex difference in adaptive response. *Am. J. Phys. Anthropol.*, **31**, 417–426.

Stock, J. and Pfeiffer, S. (2001) Linking structural variability in long bone diaphyses to habitual behaviors: Foragers from the Southern African Later Stone Age and the Andaman Islands. *Am. J. Phys. Anthropol.*, **115**, 337–348.

Stock, J. and Pfeiffer, S. (2004) Long bone robusticity and subsistence behaviour among Later Stone Age foragers of the forest and fynbos biomes of South Africa. *J. Archaeol. Sci.*, **31**, 999–1013.

Stock, J.T. and Shaw, C.N. (2007) Which measures of diaphyseal robusticity are robust? A comparison of external methods of quantifying the strength of long bone diaphyses to cross-sectional geometric properties. *Am. J. Phys. Anthropol.*, **134**, 412–423.

Tanner, J.M. (1962) *Growth at Adolescence*, Blackwell, Oxford.

Tanner, J.M. (1988) Human growth and constitution, in *Human Biology: An Introduction to Human Evolution, Variation, Growth and Adaptability* (eds G.A. Harrison, J.M. Tanner, D.R. Pilbeam and P.T. Baker), Oxford Science Press, Oxford, pp. 339–432.

Vogl, C., Atchley, W.R., Cowley, D.E. *et al.* (1993) The epigenetic influence of growth hormone on skeletal development. *Growth Dev. Aging*, **57**, 163–182.

Weltman, A., Weltman, J.Y., Veldhuis, J.D. and Hartman, M.L. (2001) Body composition, physical exercise, growth hormone and obesity. *Eat Weight Disord.*, **6**, 28–37.

Wescott, D.J. (2006) Effect of mobility on femur midshaft external shape and robusticity. *Am. J. Phys. Anthropol.*, **130**, 201–213.

Yacobaccio, H.D., Madero, C.M., Malmierca, M.P. and Reigadas, M.C. (1997–1998) Caza, domesticación y pastoreo de camélidos en la Puna argentina. *Relaciones de la Sociedad Argentina de Antropología*, **22–23**, 389–418.

19

Dental Reduction and the Transition to Agriculture in Europe

Ron Pinhasi[1] and Christopher Meiklejohn[2]

[1] *Department of Archaeology, University College Cork, Cork, Ireland*
[2] *Department of Anthropology, University of Winnipeg, Winnipeg, Canada*

19.1 INTRODUCTION

Diachronic dental crown size reduction has been observed in recent hominines and modern human populations (Brace, 1966; 1976; 1979; 1980; Brace and Mahler, 1971; Brace, Rosenberg and Hunt, 1987; Brose and Wolpoff, 1971; Calcagno and Gibson, 1988a,b; Chamla, 1980; Frayer, 1977, 1978; Kieser, 1990; Meiklejohn and Schentag, 1988; Pinhasi, 1998; Pinhasi, Eshed and Shaw, 2008; Smith, 1977; y'Edynak, 1983; 1989). However, the magnitude of the trend varies by tooth type and dimension (Brace, Rosenberg and Hunt, 1987; Frayer, 1978; 1984; Wolpoff, 1971) and is more pronounced and uniform in maxillary than in mandibular teeth (Calcagno, 1989). Most of these studies claim a universal reduction trend associated with the transition to agriculture, particularly associated with diet changes and the prevalence of periodontal disease. However, there is a need to assess the universality of such a reduction trend. The limitations of most of these studies is that they report regional trends that may not necessarily indicate universal processes and, in the case of work on Late Pleistocene and early Holocene populations, include data from specimens with poor chronological provenance. Dating and re-dating of some supposedly Upper Palaeolithic and Mesolithic fossils show that many are Neolithic or later (for example, direct dating of the 'Gravettian' Balla child from Hungary; Tillier *et al.*, 2009), and reports including such material are of questionable reliability.

Crown dimensions are complex phenotypic traits controlled and shaped by genetic, ontogenetic and environmental factors. Studies of twins (Dempsey and Townsend, 2001; Potter *et al.*, 1976; Townsend and Brown, 1978) and of parents and offspring (Goose, 1971) indicate that while heritability values for dental crown size varies, most exceed 60% (Dempsey and Townsend, 2001; Harzer, 1987; Townsend, 1992). Crown breadth (buccolingual) is less

Human Bioarchaeology of the Transition to Agriculture Edited by Ron Pinhasi and Jay T. Stock

influenced by environment and has a higher heritability component than (mesiodistal) crown length (Calcagno, 1986), suggesting different genetic control for each dimension (Potter and Nance, 1976; Potter *et al.*, 1976). While population and sex-specific variability influence diachronic trends in crown reduction, their interaction remains unclear.

Dahlberg's (1963) observation of considerable population-specific variability in tooth size and form led him to hypothesize those changes in human dentition result from relaxation of certain environmental pressures, proposing that European populations have a smaller tooth mass than populations in 'less favoured environments'. However, the specific environmental factors leading to reduction were unclear as was whether some predated the Neolithic transition (and hence were not associated with changes in plant processing and cooking technology). Frayer (1978) summarized three key points regarding dental reduction in Late Pleistocene Europe:

1. a trend to marked differential reduction in tooth size, continuing into the Late Mesolithic with reduction in the anterior dentition most pronounced in the mandible;
2. a sex-specific dimorphic differentiated pattern of reduction; and
3. a decrease in variation in crown dimensions from the Early to Late Upper Palaeolithic and Mesolithic.

He saw these emerging as independent responses to the evolution of a more sophisticated Late Palaeolithic techno-cultural complex, with these innovations the main factors responsible for observed variation in dental reduction.

This chapter examines dental reduction trends in European Late Pleistocene hunter-gatherers and early Holocene farmers. A previous study (Pinhasi, Eshed and Shaw, 2008) detected a Levantine dental reduction trend mostly affecting buccolingual dimensions and on average 12-fold greater than that reported by Frayer (1978) and Brace, Rosenberg and Hunt (1987) for Late Pleistocene and Early Holocene Europe. In order to overcome the heterogeneity in assessment of chronology of European Palaeolithic and Mesolithic specimens, we applied a quality control protocol (see below) to a large odontometric dataset in order to:

1. test whether we could discern long-term diachronic reduction in buccolingual and/or mesiodistal crown dimensions.
2. examine whether reduction was associated with a corresponding reduction in variability in dental dimensions.
3. assess sex-specific variability during the main archaeological phases.

The first aim is addressed by examining size-specific dental crown reduction trends in Central European populations. We narrowed the focus to this region since any reduction trend may be a regional rather than a continental phenomenon, and this is the only region in Europe with a sufficient number of specimens spanning the chronological interval from early Upper Palaeolithic to Early Neolithic. For the second and third aims, we included all European specimens (following application of the quality control protocol, see below). Changes in variability were then examined per dimension. In order to address the third aim, we examined changes in size-free mesiodistal/buccolingual ratios.

19.2 MATERIALS AND METHODS

19.2.1 Quality Control, Datasets and the Temporal Division

A key issue in the analysis of dental trends is quality control of the chronology and accuracy of archaeological association of the human remains. The inclusion of specimens without reliable chronological attributes (which are based on radiometric dating of specimens and/or context) and the association of human remains with key archaeological sequences (in the case of indirect dating of human remains by dating of charcoal, one or other associated archaeological finds) can lead to erroneous results. Odontometric data were collected for 607 individuals ranging from Early Upper Palaeolithic to the late Neolithic (Figure 19.1). In our dataset, the association of a given specimen with a chronological period depends on reliability of both its archaeological contextualization and chronological attribution (details of the process are outlined in detail below). Archaeological control depends on quality of excavation, recovery methods, stratigraphic positioning, site formation processes and so on. Chronological attribution refers to quality of chronology with direct radiocarbon dating of a find preferable to 'indirect' dating of the associated layer/unit. The lowest level of accuracy would determine that a find is Upper Palaeolithic, but without the possibility to be certain in regards to its sub-period. In the latter situation, a possible error margin of about 25 000 years is involved. Mid-range accuracy might further identify a find as Magdalenian, reducing the error margin to approximately 7000 years.

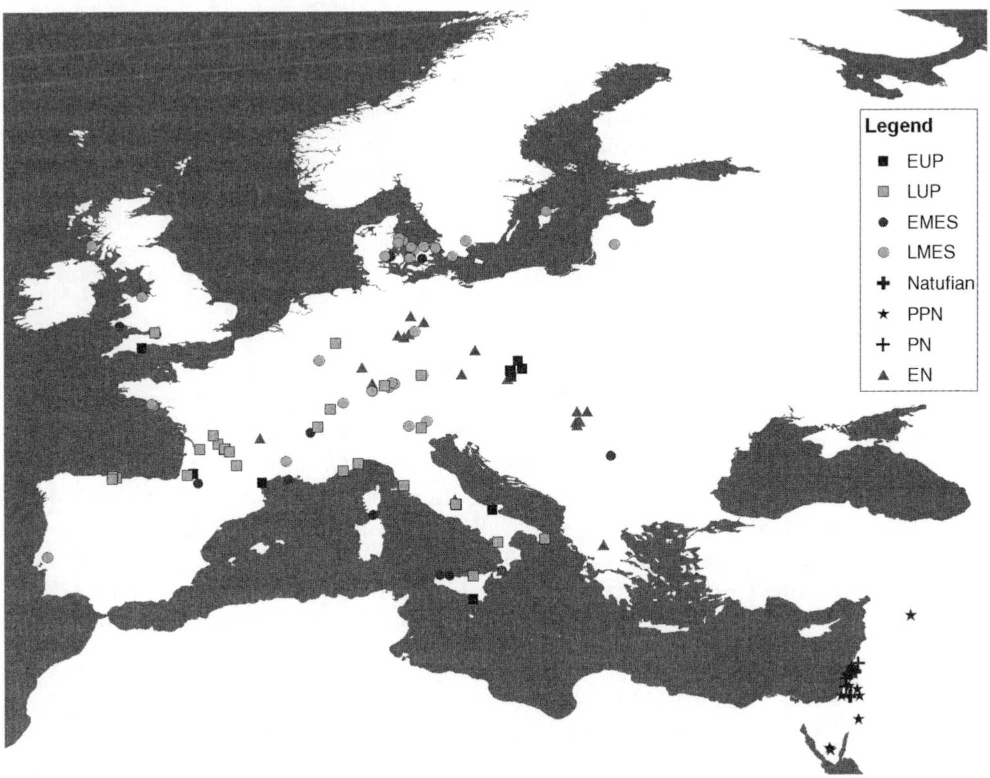

Figure 19.1 Location of sites with dental remains analysed in this study (by archaeological period). (See Plate 19.1 for a colour version of this image)

Further identification to Magdalenian VI would reduce this margin to approximately 2000 years or less. Highest accuracy would involve direct radiocarbon dating of the fossil, followed by bone or wood charcoal clearly associated with the find. The error margin would be further defined by the standard error of the date.

Although the dataset used here is largely the same as that used by Frayer (1978), the quality control measures implemented in this study allow more accurate chronological control, thus enabling finer diachronic dental dimension trends to be explored. As a result, a second issue of quality control, inter-observer error, is accounted for. We can be reasonably certain that all teeth were measured using a common protocol. Though small parts of the dataset were derived from the literature, such studies usually included work accounting for the inter-observer issue.

Over the past 30 years, the direct dating of human skeletal material has increased by leaps and bounds, especially for Upper Palaeolithic and Mesolithic material (the change is less clear for the Neolithic and later periods; see also Meiklejohn and Babb, this volume; Meiklejohn, 2009; Meiklejohn *et al.*, 1984, 2009). In addition, the dating of the underlying archaeological sequence has been incredibly augmented. We now have much clearer chronological control over data derived from skeletal material. In addition, direct dating has shown that generalized archaeological assessment of skeletal provenience can be in major error. Perhaps most obvious here is the reassignment of a number of specimens from Stetten ob Lontal (Vogelherd), until recently attributed to the Aurignacian (Conard and Bolus, 2003). Direct dating has shown that the finds are Neolithic intrusions, undetected at the time of excavation (Conard, Grootes and Smith, 2004).

In this chapter, a significant part of the work centres on the anthropological database of Upper Palaeolithic through Neolithic human remains created and maintained by one of us (Meiklejohn, nd). This contains chronological, anthropological and archaeological information for each human site or specimen, and began in the early 1990s with the need to keep current the information on Mesolithic remains published by Newell, Constandse-Westermann and Meiklejohn (1979), to which was added the information collected for the survey of Meiklejohn *et al.* (1984). This was then augmented with the information from Oakley, Campbell and Molleson (1971), and has been kept current since then. The main focus has been assessment of absolute or relative chronology and information on stable isotopes for each entry, with each assigned to one of the following six criteria:

1. *Level A*: 'good association', involves cases with direct ^{14}C dates on skeletal material, and falls into two groups. The first involves directly dated single burials (e.g. Brno – Francouzská Street [Brno 2], directly dated to 23 680 ± 200; OxA-8293 [Pettitt and Trinkaus, 2000]). The second involves sites with multiple burials with a *prima facie* case for a short-lived burial process including all individuals. Very rarely are all individuals dated. Those with direct dates are associated with these dates, while others are assigned the mean date for all obtained for the site. As an example, the Epigravettian burials from Arene Candide have six small sample AMS dates (two earlier dates are excluded) (Formicola *et al.*, 2005). Direct dates are used for two specimens, at 9925 ± 50 (OxA-10099) and 10 655 ± 55 BP (OxA-11001). The average date (10 450 BP) is assigned to the other seven individuals.
2. *Level B*: 'reasonable association', involves cases with clear association; there are one or more ^{14}C dates for associated archaeological material but no direct dates of the human remains. Placement of some cases here or in level A is sometimes a judgement call, dependent on the degree to which dated material is associated with the burials. For example,

at Cro Magnon the dated shells are from the burial context, but there is a possibility that they are significantly earlier than the age of the associated skeletons.

3. *Level C*: comprises skeletal material linked to a date or archaeological level but only loosely (e.g. at Abri Pataud the general archaeology of the site is clear, but the potential error for any given specimen may be >5000 years).
4. *Level D*: comprises material that may be archaeologically sound but with no dates for either skeletal material or site. Examples include Cap Blanc and Chancelade, almost certainly Magdalenian burials but without direct confirmation.
5. *Level E*: involves material that may be well dated but to later than the Early Neolithic. It lies beyond the focus of this paper.
6. *Level F*: comprises material for which reliable chronological assessment is currently unavailable. A good example would Combe Capelle, originally viewed as Chatelperronian, but with arguments for being Aurignacian, Gravettian, or even younger.

Based on the application of the above approach, we removed 96 specimens from the original sample, of which 56 were originally attributed to the Early Upper Palaeolithic, the rest to later Upper Palaeolithic and Mesolithic contexts. We then selected specimens attributed to levels A and B only and did not include specimens from levels C through F (103 in total).

The sample was then divided into the following temporal groups (Table 19.1):

1. *Early Upper Palaeolithic (pre-Glacial Maximum)*: specimens from the Aurignacian and Gravettian contexts.
2. *Late Upper Palaeolithic*: specimens from the Solutrean, Magdalenian, Epigravettian and Azilian contexts.
3. *Early Mesolithic (post-glacial)*: specimens from Mesolithic contexts older than 8000 BP (uncalibrated).
4. *Late Mesolithic*: specimens from Mesolithic contexts younger than 8000 BP (uncalibrated).
5. *Early Neolithic*: specimens from Early Neolithic contexts (e.g. Körös-Çris-Starčevo complex (SKC), Linienbankeramik (LBK), Cardial, and Karanovo).

19.2.2 Methods

Dental dimensions consisting of mesiodistal crown length and buccolingual breadth were taken following Frayer's (1978) procedure; MD length was maximum length unrelated to the

Table 19.1 A chronological table of the odontometric samples analysed in the study

Period	N(fem)	N(males)	N (tot)	Code
Early Upper Palaeolithic	10	15	73	1-EUP
Late Upper Palaeolithic	16	22	85	2- LUP
Early Mesolithic	20	33	119	3-EMES
Late Mesolithic	82	93	254	4-LMES
Early Neolithic SE and Central Europe	29	44	76	5- NEOL

midpoint of the tooth, BL breadth maximum breadth perpendicular to the length. Only dimensions of left teeth were recorded, unless missing, in which case tooth dimensions from the right antimere were used.

Any study of prehistoric dental trends must consider occlusal wear. In posterior teeth, the mesiodistal dimensions are more affected than the buccolingual. In the case of the anterior dentition, the mesiodistal dimensions of the incisors are most likely to change, particularly for Upper Palaeolithic and Mesolithic populations, due to severe interproximal wear of the anterior dentition (Frayer, 1978). Thus, we should be especially cautious about trends in the anterior dentition. The dataset measured by RP excludes teeth with pronounced interproximal wear. The data for most Upper Palaeolithic and Mesolithic European specimens used here are from Frayer (1978) who also eliminated teeth with severe wear. A scale was not provided (Frayer, 1978, 24) but rather a measure on the degree to which wear would impede measurement. For incisors, only unworn or slightly worn teeth were measured, for canines all but heavily worn teeth were measured, while on posterior teeth those with slight to moderate wear were included, but special attention was paid to the amount of interproximal wear.

In order to assess dental reduction trends, we regressed each of the 32 tooth dimensions against its age (radiocarbon or estimated). Linear regressions of changed dimension over time were calculated for Central Europe, the only region with a good diachronic sequence from the Early Upper Palaeolithic to Late Neolithic. Analysis of Variance (ANOVA) tests were performed on each regression slope, in order to reject the null hypothesis that a given slope is not significantly different from zero.

We analysed 32 linear regressions of mesiodistal and buccolingual dimensions, which represent the complete adult dentition. For each dimension, a statistically significant value (set at $p \leq 0.05$) for the regression slope's coefficient indicates a significant diachronic trend. These were then examined to assess the magnitude and direction of change over time. In addition, scatter plots for all significant dimensions were visually examined to support the linearity in the trends. The Coefficient of Variation (CV) provides a standardized way to compare the magnitude of morphological variation in the dentition and other morphological structures (Sokal and Braumann, 1980):

$$CV = \frac{100 \times S}{X} \tag{19.1}$$

where S is the standard deviation and X is the arithmetic mean.

Since samples were generally small, we applied the following correction:

$$CV_c = CV\left(1 + \left(\frac{1}{4n}\right)\right) \tag{19.2}$$

where $n =$ sample size (Plavcan and Cope, 2001).

Low CV values suggest that a given dimension is subject to directional selection (Leamy and Bader, 1970; Mayr, 1963). While there are cases where the amount of variation in the phenotype has increased with directional selection (Guthrie, 1963), the majority of studies indicate that directional selection is responsible for reduction in phenotypic variability (Frayer, 1978). CVs were calculated for the dataset to detect diachronic changes in variability of mesiodistal vs. buccolingual dimensions for upper and lower dentition by tooth dimension and period. Sex-specific variation is examined by comparing percent dimorphism (%D) values, which are calculated as follows:

$$\%D\frac{X_{\text{males}}-X_{\text{females}}}{X_{\text{males}}}\times 100 \tag{19.3}$$

19.3 RESULTS

19.3.1 Dental Dimensions

Results of the regression analyses of tooth dimensions for Central European specimens are provided in Table 19.2. All slopes are significant at $p \leq 0.05$ (both upper and lower dentition), and are positive, implying a general diachronic trend of dental reduction in central Europe from the Late Pleistocene to Middle Holocene (Upper Palaeolithic/Mesolithic/Early Neolithic). The average slope magnitude is 0.046 mm/ka (millimetre per thousand years) for the upper dentition, 0.036 mm/ka for the lower.

19.3.2 Coefficient of Variation

Results of the analysis of CV by period for upper and lower dentition are provided in Tables 19.3 and 19.4, respectively. A T-test for the difference in mean corrected CV values for mesiodistal and buccolingual dimensions indicates that the former are significantly more variable in dimensions than the latter ($p = 0.02$). This is also apparent from visual inspection of box plots by dimension type and period (Figure 19.2). The analysis was repeated for each period. Differences in corrected CVs between mesiodistal and buccolingual dimensions were

Table 19.2 Linear regressions by tooth dimensions – Central Europe Upper Palaeolithic to Mesolithic

Upper	N	r^2	Slope (mm/1000y)	Sig.	Lower	N	r^2	Slope (mm/1000y)	Sig.
I1MD	47	0.204	**0.037**	**	I1MD	57	0.091	**0.023**	*
I1BL	56	0.209	**0.033**	**	I1BL	62	0.084	**0.014**	*
I2MD	45	0.207	**0.004**	**	I2MD	65	0.004	**0.027**	**
I2BL	52	0.239	**0.039**	**	I2BL	70	0.241	**0.003**	**
CMD	73	0.224	**0.039**	**	CMD	72	0.221	**0.004**	**
CBL	74	0.244	**0.056**	**	CBL	73	0.367	**0.062**	**
P3MD	73	0.041	**0.046**	**	P3MD	74	0.102	**0.017**	**
P3BL	74	0.321	**0.049**	**	P3BL	75	0.359	**0.051**	**
P4MD	71	0.232	**0.031**	**	P4MD	78	0.267	**0.035**	**
P4BL	73	0.197	**0.035**	**	P4BL	78	0.154	**0.003**	**
M1MD	86	0.031	**0.005**	**	M1MD	98	0.227	**0.044**	**
M1BL	87	0.226	**0.038**	**	M1BL	97	0.097	**0.019**	**
M2MD	80	0.041	**0.062**	**	M2MD	85	0.276	**0.052**	**
M2BL	82	0.271	**0.054**	**	M2BL	85	0.246	**0.004**	**
M3MD	66	0.343	**0.065**	**	M3MD	66	0.157	**0.047**	**
M3BL	67	0.382	**0.007**	**	M3BL	67	0.288	**0.051**	**

[a] Values that are significant ($p \leq 0.05$) are in bold.

Table 19.3 Descriptive statistics and corrected coefficient of variations, upper dentition

Period	Dimension	N	Mean	SD	CV	Cor_CV	Dimension	N	Mean	SD	CV	Cor_CV
EUP	UI1MD	12	9.19	0.58	6.32	6.45	UP4MD	16	7.11	0.38	5.36	5.45
LUP		9	8.62	0.87	10.04	10.32		25	6.65	0.40	6.06	6.12
EMES		10	8.85	0.41	4.61	4.73		36	6.55	0.40	6.13	6.17
LMES		82	9.12	0.70	7.70	7.72		123	6.67	0.39	5.89	5.90
ENEOL		31	8.22	0.51	6.21	6.26		64	6.36	0.47	7.39	7.42
EUP	UI1BL	17	7.72	0.36	4.61	4.68	UP4BL	17	9.78	0.67	6.83	6.93
LUP		17	7.58	0.46	6.02	6.11		26	9.72	0.52	5.32	5.37
EMES		23	7.30	0.55	7.60	7.69		41	9.56	0.63	6.56	6.60
LMES		105	7.33	0.53	7.16	7.18		129	9.53	0.51	5.39	5.40
ENEOL		31	7.09	0.49	6.92	6.98		64	9.02	0.59	6.49	6.52
EUP	UI2MD	10	7.30	0.60	8.19	8.40	UM1MD	26	10.85	0.56	5.17	5.22
LUP		13	6.82	0.63	9.31	9.48		30	10.55	0.49	4.65	4.68
EMES		4	7.01	0.47	6.70	7.12		48	10.32	0.48	4.68	4.71
LMES		70	7.02	0.61	8.67	8.70		141	10.45	0.57	5.50	5.51
ENEOL		34	6.24	0.57	9.13	9.20		64	9.57	0.67	7.02	7.04
EUP	UI2BL	13	7.08	0.46	6.50	6.63	UM1BL	27	12.24	0.72	5.91	5.96
LUP		21	6.42	0.38	5.86	5.93		33	12.24	0.57	4.66	4.69
EMES		17	6.47	0.60	9.33	9.47		55	11.88	0.58	4.92	4.94
LMES		95	6.38	0.50	7.86	7.88		151	11.95	0.56	4.66	4.67
ENEOL		34	6.32	0.63	10.01	10.09		64	11.29	0.56	4.94	4.96

EUP	UCMD	15	8.13	0.49	6.00	6.10	UM2MD	22	10.52	0.74	7.03	7.11
LUP		21	7.88	0.41	5.24	5.31		31	9.73	0.74	7.57	7.63
EMES		29	7.82	0.43	5.53	5.58		52	9.48	0.68	7.20	7.23
LMES		124	7.83	0.54	6.84	6.86		158	9.73	0.65	6.72	6.73
ENEOL		57	7.28	0.52	7.11	7.14		64	9.14	0.70	7.70	7.73
EUP	UCBL	15	9.27	0.82	8.81	8.96	UM2BL	24	12.52	0.79	6.34	6.41
LUP		23	8.78	0.66	7.55	7.63		32	12.28	0.72	5.86	5.90
EMES		32	8.52	0.73	8.63	8.69		52	11.80	0.74	6.29	6.32
LMES		129	8.59	0.64	7.46	7.47		155	11.80	0.78	6.63	6.64
ENEOL		58	8.15	0.67	8.18	8.21		64	11.32	0.89	7.84	7.87
EUP	UP3MD	15	7.43	0.44	5.97	6.07	UM3MD	18	9.73	0.95	9.77	9.90
LUP		25	6.84	0.50	7.26	7.34		30	8.63	0.80	9.24	9.31
EMES		31	6.72	0.46	6.91	6.97		56	8.77	0.89	10.13	10.17
LMES		113	6.95	0.47	6.69	6.70		127	8.78	0.66	7.56	7.57
ENEOL		59	6.49	0.47	7.21	7.24		49	8.61	0.80	9.28	9.33
EUP	UP3BL	16	9.79	0.66	6.78	6.89	UM3BL	19	11.89	0.89	7.45	7.55
LUP		24	9.61	0.71	7.40	7.48		32	11.19	1.15	10.31	10.39
EMES		36	9.56	0.59	6.14	6.18		56	11.02	1.02	9.28	9.33
LMES		122	9.42	0.59	6.28	6.29		126	11.24	0.85	7.58	7.59
ENEOL		60	8.72	0.48	5.56	5.58		49	10.54	0.87	8.20	8.25

Table 19.4 Descriptive statistics and corrected coefficient of variations, lower dentition

Period	Dimension	N	Mean	SD	CV	Cr_CV	Dimension	N	Mean	SD	CV	Cr_CV
EUP	LI1MD	9	5.57	0.86	15.35	15.78	lp4md	15	7.36	0.54	7.29	7.41
LUP		13	5.39	0.55	10.23	10.42		28	6.97	0.44	6.34	6.40
EMES		7	5.40	0.54	10.08	10.44		40	6.90	0.43	6.29	6.33
LMES		74	5.48	0.42	7.66	7.69		142	6.95	0.50	7.15	7.17
ENEOL		37	5.06	0.37	7.23	7.28		59	6.68	0.48	7.17	7.20
EUP	LI1BL	16	6.15	0.33	5.31	5.39	lp4bl	15	8.66	0.52	5.97	6.06
LUP		27	6.25	0.45	7.12	7.19		28	8.52	0.50	5.82	5.87
EMES		29	6.14	0.39	6.41	6.46		39	8.47	0.56	6.59	6.64
LMES		111	6.12	0.39	6.38	6.40		152	8.36	0.49	5.83	5.84
ENEOL		38	5.93	0.43	7.18	7.23		59	8.10	0.59	7.26	7.29
EUP	li2md	13	6.32	0.61	9.71	9.90	lm1md	28	11.66	0.80	6.85	6.91
LUP		15	6.02	0.41	6.77	6.89		38	11.45	0.65	5.65	5.69
EMES		10	5.88	0.30	5.13	5.25		38	11.34	0.54	4.74	4.77
LMES		95	6.08	0.42	6.85	6.87		146	11.33	0.67	5.91	5.92
ENEOL		49	5.71	0.48	8.39	8.44		61	10.49	0.70	6.71	6.74
EUP	li2bl	17	6.89	0.43	6.20	6.29	lm1bl	28	10.95	0.55	5.02	5.06
LUP		28	6.65	0.39	5.91	5.97		39	11.07	0.66	5.93	5.97
EMES		40	6.42	0.37	5.82	5.86		47	10.92	0.48	4.42	4.44
LMES		134	6.45	0.41	6.35	6.36		165	10.92	0.48	4.41	4.42
ENEOL		49	6.31	0.45	7.07	7.10		61	10.40	0.47	4.53	4.55

EUP	lcmd	14	7.40	0.63	8.57	8.73	lm2md	27	11.21	0.98	8.72	8.80
LUP		35	6.97	0.59	8.53	8.59		37	10.89	0.71	6.51	6.55
EMES		41	6.77	0.48	7.10	7.14		54	10.62	0.69	6.48	6.51
LMES		154	6.88	0.47	6.90	6.91		146	10.62	0.68	6.44	6.46
ENEOL		53	6.51	0.56	8.66	8.70		60	10.15	0.62	6.12	6.15
EUP	lcbl	14	8.76	0.70	8.00	8.14	lm2bl	27	10.80	0.79	7.34	7.41
LUP		36	8.25	0.76	9.22	9.28		39	10.76	0.81	7.56	7.60
EMES		47	7.81	0.55	7.09	7.13		56	10.58	0.57	5.39	5.41
LMES		155	7.74	0.57	7.33	7.35		154	10.48	0.60	5.75	5.76
ENEOL		52	7.64	0.64	8.41	8.45		60	10.00	0.55	5.48	5.50
EUP	lp3md	17	7.04	0.37	5.22	5.29	lm3md	13	11.20	1.20	10.68	10.88
LUP		28	6.92	0.45	6.55	6.61		29	10.59	0.76	7.16	7.23
EMES		46	6.87	0.41	5.99	6.02		48	10.30	0.79	7.65	7.69
LMES		146	6.94	0.42	6.04	6.05		132	10.53	0.70	6.68	6.69
ENEOL		54	6.60	0.45	6.79	6.82		44	10.28	0.77	7.46	7.50
EUP	lp3bl	17	8.43	0.61	7.28	7.38	lm3bl	14	10.82	0.86	7.98	8.12
LUP		29	8.13	0.55	6.79	6.85		30	10.53	0.53	5.08	5.12
EMES		48	7.93	0.54	6.86	6.89		48	10.06	0.67	6.65	6.68
LMES		154	7.88	0.50	6.29	6.30		135	10.27	0.64	6.24	6.25
ENEOL		56	7.49	0.53	7.10	7.13		44	9.70	0.61	6.28	6.31

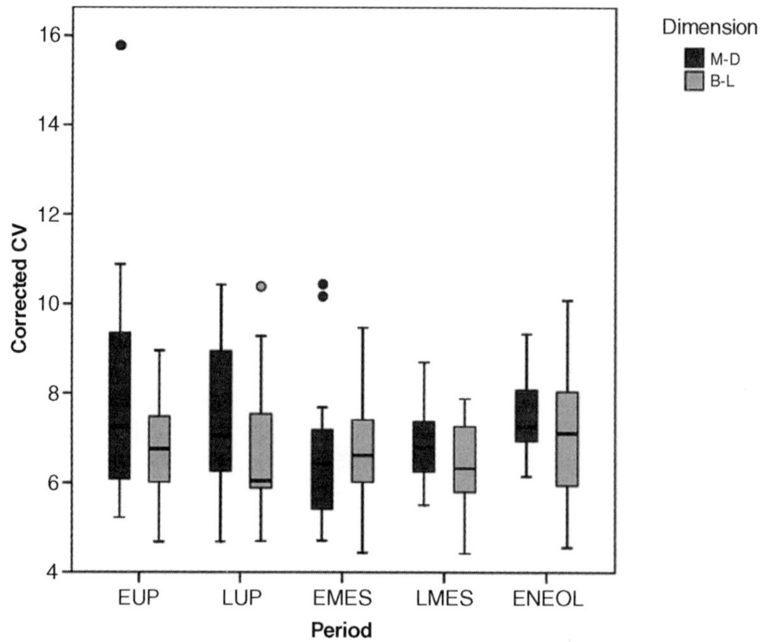

Figure 19.2 Box plots of pairs of corrected coefficient of variation values by period

non-significant. A T-test for difference in mean corrected CV values for upper and lower dentition indicates no significant difference overall or by group. Next, a T-test was performed to assess the difference in mean corrected CV values for the anterior (incisors and canines) and posterior (premolars and molars) dentition. Results indicate a significant difference ($p \leq 0.0001$) for the total sample. When the analysis was repeated for each group, only the Late Mesolithic ($p = 0.001$) and Early Neolithic ($p = 0.01$) were significant and in both cases variability was larger for anterior dimensions.

19.3.3 ANOVA and Post-Hoc Tests

Anova tests were performed in order to assess whether there are significant differences in mean dental dimensions amongst periods. Results indicate that all dimensions are significantly different at $p \leq 0.05$. Next, we examined which groups differ, using Tamhane's T2 (which is a conservative test that is appropriate when group sizes are unequal and/or when homogeneity of variances may be violated)) *post-hoc* test (which is not sensitive to unequal group sizes) and setting the significance level to $p \leq 0.05$.

Results (Table 19.5) show that Early Upper Palaeolithic (group 1) and Early Neolithic (group 5) groups have the highest number of paired differences from other periods. The Early Palaeolithic group differs from all others in the upper buccolingual dimension of the central incisor and mesiodistal dimensions of first premolar and third molar. This group also differs from one or more of the others in nearly all dimensions. The Early Neolithic group differs from all others in the upper and lower buccolingual dimensions of first and second premolars, the first molar and the mesiodistal dimension of first molar. In addition, the Early Neolithic group differs from all others in lower mesiodistal and buccolingual dimensions of the second molar, the mesiodistal dimensions of the canine and third molar, and the upper buccolingual

Table 19.5 Results of the *post-hoc* tests, ANOVA total set, by period

Dimension	Significant groups	Dimension	Significant groups
Upper		Lower	
I1MD	5 and 1,3,4; 1 and 4	I1MD	4 and 5
I1BL	1 and 4,5	I1BL	—
I2MD	5 and 1,4	I2MD	5 and 1,4
I2BL	1 and 2,3,4,5	I2BL	5 and 1,2; 1 and 3,4
CMD	5 and 1,2,3,4	CMD	5 and 1,2,4; 1 and 3
CBL	5 and 1,2,4	CBL	1 and 3,4,5; 2 and 3,4,5
P3MD	1 and 2,3,4,5; 5 and 2,4	P3MD	5 and 1,2,3,4
P3BL	5 and 1,2,3,4	P3BL	5 and 1,2,3,4; 4 and 1
P4MD	1 and 2,3,4,5; 5 and 4	P4MD	5 and 1,4
P4BL	5 and 1,2,3,4	P4BL	5 and 1,2,3,4
M1MD	5 and 1,2,3,4; 1 and 3,4	M1MD	5 and 1,2,3,4
M1BL	5 and 1,2,3,4	M1BL	5 and 1,2,3,4
M2MD	5 and 1,2,4; 1 and 2,3,4	M2MD	5 and 1,2,3,4
M2BL	1 and 3,4,5; 2 and 4; 5 and 2,3,4; 3 and 2	M2BL	5 and 1,2,3,4
M3MD	1 and 2,3,4,5	M3MD	—
M3BL	1 and 3,5; 5 and 4	M3BL	5 and 1,2,4; 3 and 2

dimension of the second molar. The most prevailing differences are between the Early Neolithic and both the Upper Palaeolithic (group 1) and late Mesolithic (group 4). This group differs from the rest in buccolingual dimensions of the upper canine, both premolars, first and second molars and mesiodistal dimensions of all molars. There is therefore some correspondence between the upper and lower dentition, as the outlier in both cases is the Early Neolithic. In both cases, differences are generally more pronounced in the posterior dentition and particularly the first and second molars.

19.3.4 Sexual Dimorphism

Kruskal-Wallis tests were carried out to test differences in percent dimorphism values for the five groups, by dimension (buccolingual vs. mesiodistal), arcade (lower vs. upper), tooth position (anterior-incisors and canines, posterior-premolars and molars) and by period. The significance level was set at $p \leq 0.01$ and all entries with a sample of less than five were excluded. The results show a significant difference in dimorphism in only the lower posterior mesiodistal dimensions ($p = 0.009$). Visual inspection of box plots by period dimension, and type (lower vs. upper (mesiodistal vs, buccolingual, Figures 19.3a and b) and anterior vs. posterior (Figures 19.4a and b) indicates a reduction in percent dimorphism during the Early Neolithic, which is particularly pronounced in the upper anterior mesiodistal dimensions. In addition, the Late Upper Palaeolithic group shows a high degree of variability in percent dimorphism values for the anterior dentition.

19.4 DISCUSSION

Our results show period-specific patterns of dental reduction amongst Late Pleistocene and Early Holocene humans. The analysis of diachronic trends in upper and lower crown

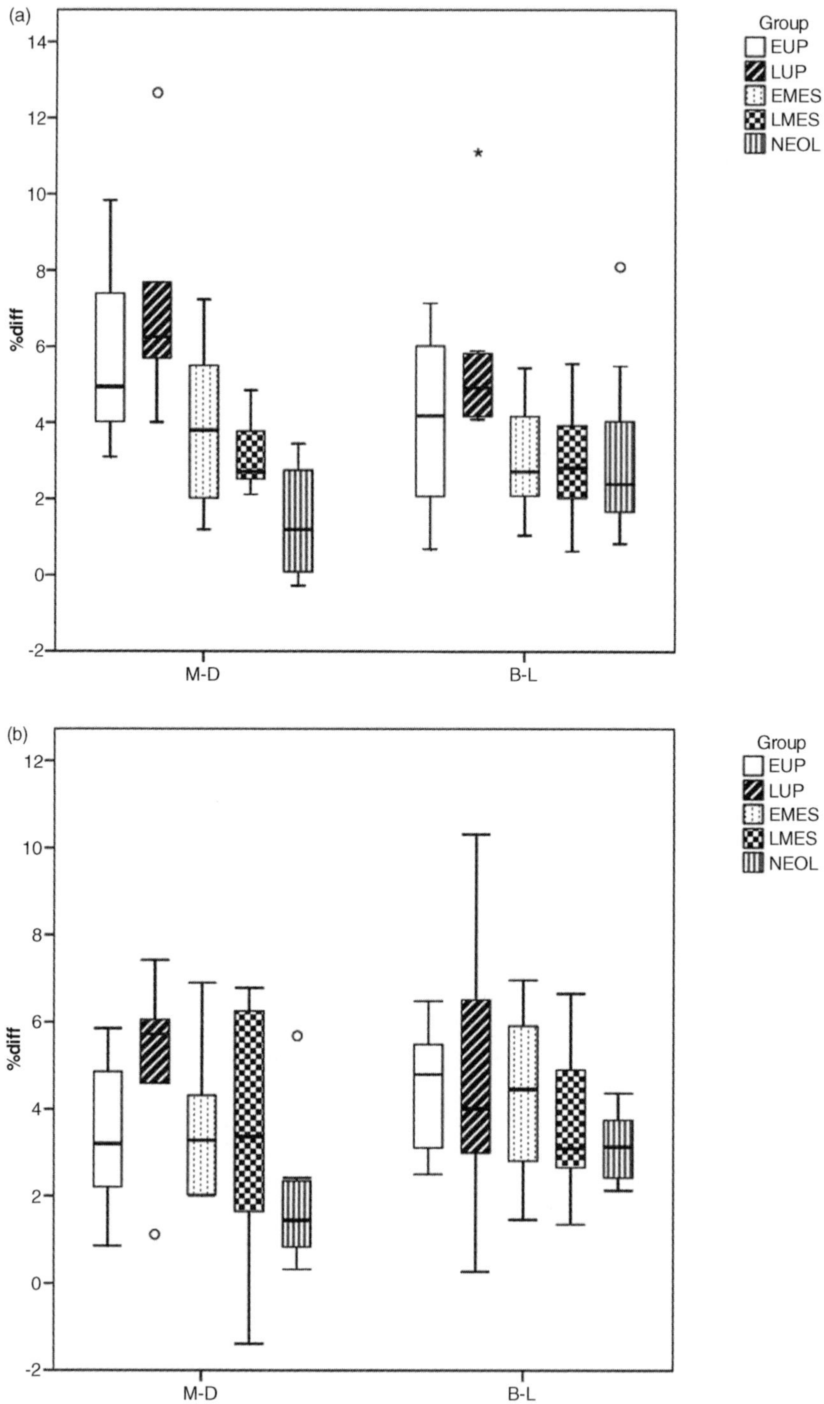

Figure 19.3 Box plots of percent dimorphism by dimension type of (a) the lower arcade, (b) the upper arcade (open circles: extreme values, stars: outliers)

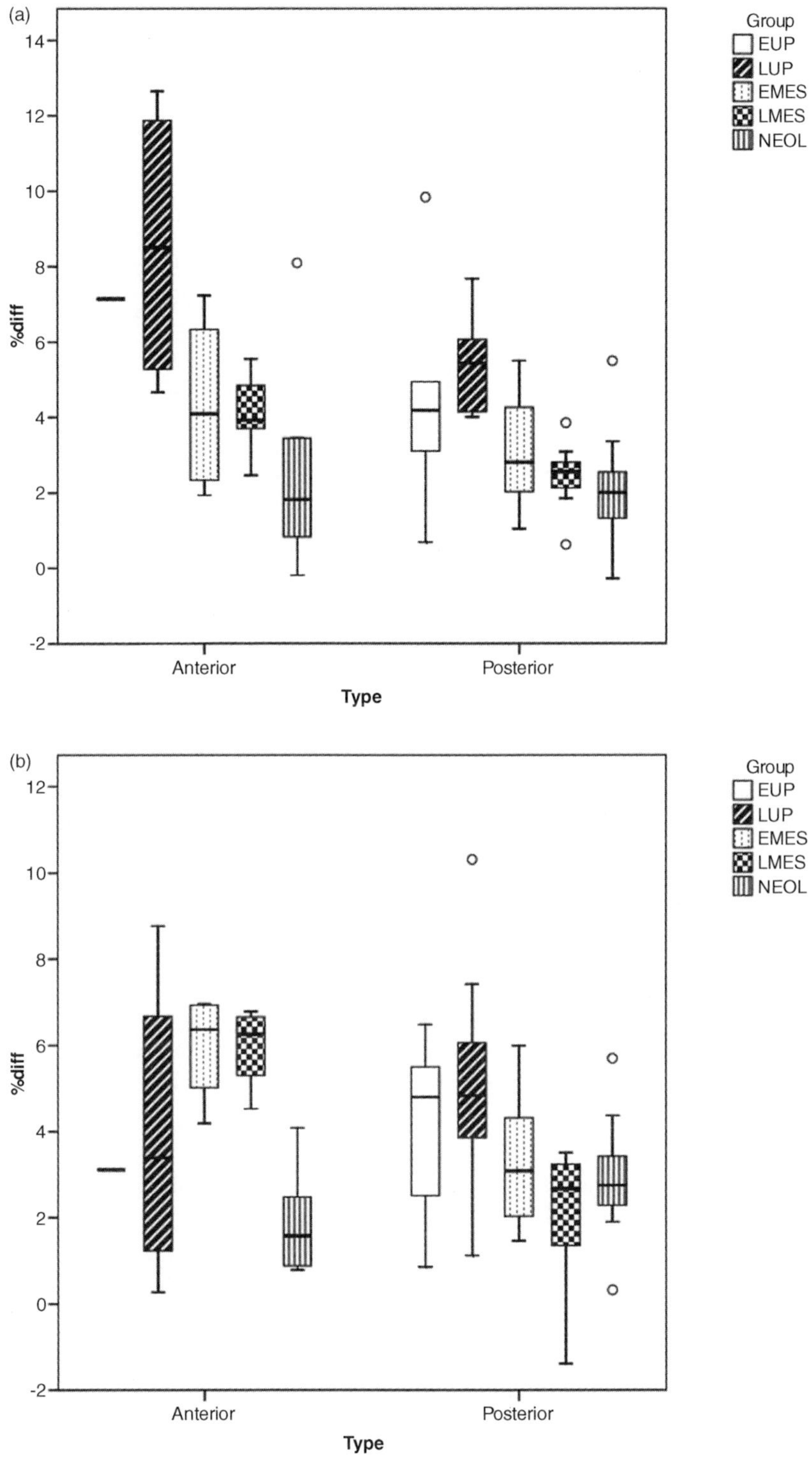

Figure 19.4 Box plots of percent dimorphism for anterior and posterior dentition of (a) the lower arcade, (b) the upper arcade (open circles: extreme values, stars: outliers)

Table 19.6 Linear regressions by tooth dimensions – Levant Epipalaeolithic to Neolithic

Upper	N	r^2	Slope (mm/1000y)	Sig.	Lower	N	r^2	Slope (mm/1000y)	Sig.
I1MD	77	0.043	0.001	ns	I1MD	71	0.002	0.023	ns
I1BL	78	0.061	0.007	*	I1BL	72	0.254	**0.177**	**
I2MD	69	0.067	0.098	*	I2MD	89	0.004	–0.027	ns
I2BL	70	0.001	0.039	ns	I2BL	91	0.136	**0.164**	**
CMD	105	0.000	0.004	ns	CMD	119	0.044	0.081	*
CBL	106	0.036	0.091	ns	CBL	120	0.099	**0.158**	**
P3MD	92	0.026	0.086	ns	P3MD	128	0.024	0.026	ns
P3BL	91	0.181	**0.228**	**	P3BL	129	0.249	**0.234**	**
P4MD	90	0.026	–0.067	ns	P4MD	125	0.002	–0.021	ns
P4BL	90	0.152	**0.191**	**	P4BL	122	0.082	**0.125**	**
M1MD	107	0.000	0.001	ns	M1MD	148	0.001	0.054	ns
M1BL	107	0.122	**0.185**	**	M1BL	151	0.144	**0.171**	**
M2MD	89	0.002	–0.029	ns	M2MD	150	0.015	0.061	ns
M2BL	89	0.065	**0.175**	*	M2BL	151	0.075	**0.151**	**
M3MD	64	0.013	–0.085	ns	M3MD	107	0.007	0.005	ns
M3BL	63	0.015	**0.026**	**	M3BL	104	0.068	**0.145**	**

[a] Values that are significant ($p \leq 0.05$) are in bold.

dimensions for central Europe showed all slopes to be significant at $p \leq 0.05$. With all slopes positive, the reduction in all crown dimensions has an average rate of 0.046 mm/ka (millimetre per 1000 years) for the upper dentition; 0.036 mm/ka for the lower. Different trends operated in Early Holocene Levantine populations with rates of dental reduction as much as 12 times higher (Tables 19.6 and 19.7). In the Levantine populations, most of the significant changes are

Table 19.7 Slope ratios of Levant/Central European by dimension. Ratios are only calculated for significant slopes

Upper	Dif.	Lower	Dif.
I1MD	ns	I1MD	ns
I1BL	2.1	I1BL	12.6
I2MD	2.5	I2MD	ns
I2BL	ns	I2BL	5.5
CMD	ns	CMD	2.0
CBL	ns	CBL	2.5
P3MD	ns	P3MD	ns
P3BL	4.7	P3BL	4.6
P4MD	ns	P4MD	ns
P4BL	5.5	P4BL	4.2
M1MD	ns	M1MD	ns
M1BL	4.9	M1BL	9.0
M2MD	ns	M2MD	ns
M2BL	3.2	M2BL	3.8
M3MD	ns	M3MD	ns
M3BL	3.7	M3BL	2.8

in canines, premolars and molars, with 12 out of 17 significant slopes on buccolingual dimensions. Moreover, the steepest slopes are from the buccolingual dimensions of the upper premolars and molars. The noted diachronic trend results in a more rectangular crown shape (narrower buccolingual crown dimension without corresponding mesiodistal reduction). Thus, while we can assert that a significant dental reduction trend occurred in both central European and southern Levantine populations, the nature of the trend is different, in outcome rather than in rate or magnitude.

Results of the analysis of CVs indicate that mesiodistal dimensions are significantly more variable than buccolingual. However, with the analysis carried out by period, the mesiodistal dimensions showed no differences. The only noticeable difference in corrected CVs was noted when comparing anterior and posterior dentition by period, with Late Mesolithic and Early Neolithic groups showing significantly lower posterior CVs. Frayer (1978) compared CV values for both dimensions of both upper and lower dentition for Early Upper Palaeolithic, Late Upper Palaeolithic and Mesolithic groups. He reports reduction in CVs over time for both mandibular and maxillary mesiodistal and buccolingual dimensions (EUP > LUP > MES). In the mandible, it is mostly molar dimensions that underwent reduction in CVs, in the maxillae those of the canines and premolars. The present analysis does not confirm this trend, but shows an overall decrease in all dental dimensions over time, without change in magnitude of the corrected CVs (the exception is the reduction in the posterior dentition of Late Mesolithic and Early Neolithic groups). In the southern Levantine analysis, the CVs of buccolingual dimensions were significantly lower than the mesiodistal. Differences were also significant when comparing upper and lower jaws, with the latter having lower average CVs than the former. How can these differences be explained? Are different mechanisms of dental reduction acting on each population? Or are both trends the outcome of a similar reduction process?

19.4.1 Models of Dental Reduction

Addressing the nature of the reduction trend requires a brief outline of the main models proposed to explain similar trends in diverse world populations following the transition to agriculture. All have in common an emphasis on the role of (micro) evolutionary processes. However, all view environmental and/or cultural factors as the dominant factor triggering crown reduction.

19.4.1.1 Probable Mutation Effect

The probable mutation effect (PME) was developed by Brace (1963, 1964). The model proposes that when a species' niche changes, the adaptive significance of some biological structures is relaxed. Brace argues that the genetic consequence of structures or traits not subject to selective pressures is that the only change is induced from accumulation of mutations no longer eliminated by selection (Calcagno, 1989). For Brace, the majority of mutations diminish the integrity of a structure. The 'probable' outcome is reduction in size and/or complexity of any given trait. Brace proposed that introduction and use of pottery during the Early Holocene, and subsequent changes in food preparation techniques, triggered the PME following relaxation of selective forces on the masticatory apparatus and consequent decrease in tooth size (Brace and Mahler, 1971; Brace, Rosenberg and Hunt, 1987).

19.4.1.2 Compensatory Interaction (CI)

Sofaer (1973; Sofaer, Bailit and MacLean, 1971) proposed a model of dental reduction as part of a process secondary to change in jaw dimensions. Sofaer points out that during the course of hominid evolution, reduction in tooth size has followed a pattern in which late developing teeth in a given tooth class (i.e. incisors, canines, premolars, molars) are considerably more variable in size (with high levels of left/right fluctuating asymmetry). The model argues that higher variability in the dimensions of later developing teeth is a secondary process in response to reduction in overall jaw size and competition between tooth germs for available local jaw space. Consequently, a compensatory interaction between adjacent tooth germs occurs due to competition on requirements for growth.

19.4.1.3 The Somatic Budget Effect (SBE)

The SBE model was proposed by Jolly (1970) to explain changes in hominid dentition and body size following transition to a seed-eating diet. Jolly argues that selection acts to reduce any morphological structure that loses its function, as its development and maintenance requires energy. The reduction in such structures is therefore the outcome of the organism's need to reduce and conserve both the ontogenetic and metabolic energy needed. Selection then favours individuals that do not allocate the 'somatic budget' to unnecessary structures.

19.4.1.4 The Selective Compromise Effect (SCE)

Calcagno (1986, 1989) developed an evolutionary model emphasizing the relationship between dental morphology and size, and susceptibility to caries and other periodontal diseases. Hard abrasive diets produce dental wear at a faster rate than softer diets. Dental wear and complex crown morphology are correlated, with high prevalence of caries and other periodontal diseases (Meiklejohn, Wyman and Schentag, 1992). These can affect an individual's fitness. Large morphologically complex crowns provide more surface area for caries, but large crown area is at the same time essential for mastication of abrasive foodstuff. In populations undergoing a subsistence shift following the transition to agriculture, a selective compromise must occur between selection for smaller teeth with less complex cusp pattern and thin enamel, and a selection for bigger teeth with thicker enamel to counter occlusal wear. A central aspect of this model is the assumption that dental crowding and high prevalence of cariogenic disease trigger selection for smaller dentition (cf Calcagno and Gibson, 1987; y'Edynak, 1983, 1989) as evident in Near Eastern (Eshed, Gopher and Hershkovitz, 2006; Smith, 1984) and Nubian early Holocene populations (Calcagno, 1986).

19.4.1.5 Increasing Population Density Effect (IPDE)

The (IPDE) mode (Macchiarelli and Bondioli, 1986) suggests that reduction in dental crown size was mainly derived from changes in population density associated with the transition to sedentism. Sedentism and the emergence of larger Neolithic groups led to poorer diet and increase in disease prevalence compared to the previous hunter-gatherer model. New post-Pleistocene environmental conditions (poor diet and increased disease load) induced positive selection for reduced nutritional and metabolic requirements, leading to reduction in body size, of which reduction in tooth size was a by-product.

19.4.2 Criticism of Evolutionary Models

There are several points to be raised in regards to the above models. First is the issue of testability. These models make assumptions about a cause-and-effect chain of events but it is not possible, for example, to envision how to test the PME, SCE, SBE or IPDE models, since the assumed causality cannot be translated into testable hypotheses. In the case of the PME model, it does not explain:

1. why reduction is non-uniform;
2. why relaxation of selection should lead to dental reduction rather than to overall increase in variation of dental dimensions; and
3. how the accumulation of mutations in the genotype results in drastic change in complex traits, in this case dental crown size, over a relatively short time period (in evolutionary terms).

The CI model is theoretically sound, but has only been tested on lab mice. Dental asymmetry correlation studied on Australian aborigines (Towsend and Brown, 1980), Upper Palaeolithic and Mesolithic Europeans (Frayer, 1978) and modern Caucasians (Garn, Lewis and Kerewsky, 1966) provides no empirical support for a compensatory interaction between developing teeth. Pinhasi, Eshed and Shaw (2008) report a pronounced dental reduction trend that predominantly affected all buccolingual dimensions of southern Levantine Epipalaeolithic and Neolithic populations. Hence, this trend did not differentially affect different teeth according to their position in a given class. Furthermore, the overall reduction in dental dimensions was not associated with reduction in the overall size of the lower jaw but only in width of the mandibular ramus and height of the mandible at the symphysis. The SBE and IPDE models regard dental reduction as the outcome of changes in metabolic requirements but need to be tested on past populations taking into account confounding effects (e.g. population size, changes in other morphological structures, etc.). The SCE model is even more difficult to test, as it brings together epidemiological, pathological, developmental and dimensional factors. It is difficult, if not impossible, to control for the interaction (or partial correlations) between these aspects, and test for possible causality.

But the main drawback in these models is to claim detection of a universal phenomenon, while a review of published regional studies shows lack of concordance. For example, studies of changes in the jaw and dental dimensions following the transition to agriculture in the Danube Gorges (y'Edynak, 1989) and Nubia (Calcagno and Gibson, 1987) showed that while softer foods resulted in jaw reduction, it was not followed by corresponding dental reduction. Thus, malocclusion and dental crowding, especially of the anterior teeth, occurred in many Neolithic populations. Malocclusions consequently resulted in elevated levels of acute dental diseases, leading to increased mortality rates of infected individuals.

Second, these models present a rather blurred definition of how natural selection (in its various forms) acts upon a specific phenotypic complex trait, dental size. Osborne (1967) raised this concern more than 40 years ago by stating that:

> All paleo and dental anthropologists know that natural selection is not merely a scientific colloquialism for Mother Nature but is in effect a synonym for the differential reproduction of different phenotypes. But the recognition of this meaning of natural selection has not always come through in the course of speculations concerning the 'possible adaptive

advantages of one fissure or cusp pattern versus another, in different fossil forms or groups of living men . . .' (pp. 945–6).

Dental reduction models need to consider the genetic basis and evolutionary causes of quantitative variation. Recent genetic research indicates that the relationship between evolutionary influences on phenotypic traits and molecular (genotypic) variation is complex and not well understood. Mackay (2001) points out that a comprehensive understanding of the 'genetic architecture' of a quantitative trait requires knowledge of:

1. the numbers and identities of all genes in the developmental, physiological and/or biochemical pathway affecting the phenotype;
2. the mutation rates at these loci;
3. the numbers and identities of the subset of loci responsible for variation in the trait within and between populations, and between species;
4. the homozygous and heterozygous effects of new mutations and segregating alleles on the trait;
5. all two-way and higher-order epistatic interaction effects;
6. the pleiotropic effects on other quantitative traits and, most importantly, reproductive fitness;
7. the extent to which additive, dominance, epistatic and pleiotropic effects vary between the sexes, and in a range of ecologically relevant environments;
8. the molecular polymorphism(s) that functionally define QTL alleles;
9. the molecular mechanism causing the differences in trait phenotype; and
10. QTL allele frequencies.

The effects of pleiotropy and epistasis are of particular relevance, as interaction/linkage between alleles implies that natural selection will normally not directly affect complex traits in isolation (Mitchell-Olds, Willis and Goldstein, 2007; Phillips, 2008). It is also likely that reduction in crown size is not necessarily the outcome of selection for this specific trait but could be the outcome of a pleiotropic side effect (Barton and Keightley, 2002). The variation in dental dimensions is then maintained either as a consequence of balancing selection (a possibility that cannot be presently tested), or due to polymorphism in other linked alleles. Genetic research also indicates that the lower the heritability and the more loci involved, the more difficult it is to dissect genotype-phenotype relationships (Ellegren and Sheldon, 2008). Research on the heritability of dental dimensions provides evidence of differences in heritabilities for different tooth groups, and for mesiodistal and buccolingual dimensions (Dempsey and Townsend, 2001; Hillson, 1996; Kieser, 1990). It is therefore possible that dimension-specific differences in rate of dental reduction, as reported for the transition to agriculture in the Levant (Pinhasi, Eshed and Shaw, 2008), need not indicate differential selection on each trait (i.e. mesiodistal, buccolingual), but differential developmental genotype-phenotype interaction.

Third, is the issue of imprecise use of the evolutionary term 'fitness'. According to Orr (2009) 'fitness involves the ability of organisms – or, more rarely, of populations or species – to survive and reproduce in the environment in which they find themselves' (p. 531). But assessing fitness in the case of crown size dimensions implies the measurement of fitness effects of all loci affecting variation in this complex trait (Mackay, 2001). Moreover, as reported by

Frayer (1978), very slow (but statistically significant) changes in dental dimensions over time could not confer higher fitness to individuals in the population, at any given generation, with smaller teeth. In any case, even if we were able to measure the fitness differences between individuals in a given population, it would be impossible to assert whether they are caused by the observed quantitative trait (Barton and Keightley, 2002).

Carlson and Van Gerven (1977) reported diachronic alteration and gracilization of the craniofacial complex in prehistoric Nubians. They proposed that this involves relaxed selection for a robust masticatory apparatus and large teeth dimensions, consequent to the shift to softer, more cariogenic, foodstuffs. They contended that since the latter resulted in increased prevalence of caries and related pathologies, selective pressures acted to reduce overall size and morphological complexity of the masticatory apparatus. While difficult to test, this model also implies that selection for smaller dental dimensions may have been a side product of selection for changes in the cranio-facial complex (and not just the mandible).

19.5 CONCLUSIONS

In conclusion, our results do not support any single-cause model, although we did not carry out statistical tests of any of these (mainly since testability of these models is problematic). That our results differ from those previously reported by Frayer (1978) for the European set, more than likely reflects previous misattribution of specimens in Upper Palaeolithic and Mesolithic context due to poor chronological control. It is interesting to point out that both the rate and nature of the dental reduction trend observed for central Europe differs from the one reported by Pinhasi, Eshed and Shaw (2008) for early agricultural populations in the southern Levant.

The transition to agriculture in Europe was a mosaic historical process that varied between regions of Europe (Meiklejohn and Zvelebil, 1991; Pinhasi and Pluciennik, 2004; Zvelebil, 2001). Our results clearly indicate major differences in dental reduction trends when comparing the process in central Europe to that of the southern Levant. We suggest that the most parsimonious mechanism behind the dental reduction is natural selection, since drift alone could not explain a significant diachronic reduction process. However, we believe that all the models specified above tend to undermine the complexity of genetic, ontogenetic and environmental mechanisms that interact in a complex and differential manner on a given phenotypic trait (i.e. crown dimension). The additive polygenic mode of inheritance of dental crown size challenges any simple cause-and-effect model of dental reduction. Moreover, heritability coefficients of dental dimensions (Potter *et al.*, 1976; Potter, Nance and Yu, 1978) suggest that crown breadth is less influenced by environmental factors than crown length. Thus, even if a single selective pressure acted on all dimensions, differential input (e.g. different diets, population densities, etc.) would result in differential reduction trends. As we are not certain of the degree to which various environmental factors affect specific dental dimensions, and affects the growth and development of the various tooth crowns, any attempt to over-interpret observed trends as the outcome of a single factor is questionable. Future research on heritability of dental dimensions and the genes that moderate their ontogenetic process may shed more light on our understanding of such trends.

Finally, our results highlight that while the transition to farming resulted in significant changes in dental dimensions, the magnitude and nature of these changes need to be addressed on a case-to-case geographical basis, before it is possible to draw conclusions about universal evolutionary trends.

ACKNOWLEDGEMENTS

We are grateful to David Frayer for his data and support. We also wish to thank the two anonymous referees for their constructive comments.

REFERENCES

Barton, N.H. and Keightley, P.D. (2002) Understanding quantitative genetic variation. *Nat. Rev. Genet.*, **3** (1), 11–21.

Brace, C.L. (1963) Structural reduction in evolution. *Am. Nat.*, **97**, 39–49.

Brace, C.L. (1964) The probable mutation effect. *Am. Nat.*, **98**, 453–455.

Brace, C.L. (1966) Structural reduction in evolution. *Am. Nat.*, **97**, 185–195.

Brace, C.L. (1976) Krapina, 'classic' Neanderthals and the evolution of the European face. *J. Hum. Evol.*, **8**, 527–550.

Brace, C.L. (1979) Dental reduction in the Orient. *Asian Perspect.*, **19**, 203–219.

Brace, C.L. (1980) Australian tooth-size clines and the death of the stereotype. *Curr. Anthropol.*, **21**, 141–153.

Brace, C.L. and Mahler, P.E. (1971) Early Holocene changes in the human dentition. *Am. J. Phys. Anthropol.*, **34**, 191–204.

Brace, C.L., Rosenberg, K.R. and Hunt, K.D. (1987) Gradual change in human tooth size in the Late Pleistocene and Early Holocene. *Evolution*, **41**, 705–720.

Brose, D.S. and Wolpoff, M.H. (1971) Early Upper Paleolithic man and the late Middle Pleistocene teeth. *Am. J. Phys. Anthropol.*, **73**, 1156–1194.

Calcagno, J.M. (1986) Dental Reduction in Post-Pleistocene Nubia. *Am. J. Phys. Anthropol.*, **70**, 349–363.

Calcagno, J.M. (1989) *Mechanisms of Human Dental Reduction. A Case Study from Post-Pleistocene Nubia*, University of Kansas, Lawrence, Kansas.

Calcagno, J.M. and Gibson, K.R. (1987) The improbable probable mutation effect. *Am. J. Phys. Anthropol.*, **72**(2), 185.

Calcagno, J.M. and Gibson, K.R. (1988a) 40 000 years of tooth size reduction: Evidence, mechanism, and controversy. *Am. J. Phys. Anthropol.*, **75**(2), 193.

Calcagno, J.M. and Gibson, K.R. (1988b) Human dental reduction: Natural selection or the probable mutation effect. *Am. J. Phys. Anthropol.*, **77**(4), 505–517.

Carlson, D.S. and Van Gerven, D.P. (1977) Masticatory function and Early Holocene evolution in Nubia. *Am. J. Phys. Anthropol.*, **46**, 495–506.

Chamla, M.C. (1980) Étude des variations métriques des couronnes dentaires du Nord-Africains de l'Épipaléolithique à l'Époque Actuelle. *L'Anthropologie*, **84**, 254–271.

Conard, N.J. and Bolus, M. (2003) Radiocarbon dating the appearance of modern humans and the timing of cultural innovations in Europe: New results and new challenges. *J. Hum. Evol.*, **44**, 331–371.

Conard, N.J., Grootes, P.M. and Smith, F.H. (2004) Unexpectedly recent dates for human remains from Vogelherd. *Nature*, **430**(6996), 198–201.

Dahlberg, A.A. (1963) Dental evolution and culture. *Hum. Biol.*, **35**, 237–249.

Dempsey, P.J. and Townsend, G.C. (2001) Genetic and environmental contributions to variation in human tooth size. *Heredity*, **86**(6), 685–693.

Ellegren, H. and Sheldon, B.C. (2008) Genetic basis of fitness differences in natural populations. *Nature*, **452**(7184), 169–175.

Eshed, V., Gopher, A. and Hershkovitz, I. (2006) Tooth wear and dental pathology at the advent of agriculture: New evidence from the Levant. *Am. J. Phys. Anthropol.*, **130**(2), 145–159.

Formicola, V., Pettitt, P.B., Maggi, R. and Hedges, R. (2005) Tempo and mode of formation of the Late Epigravettian necropolis of Arene Candide cave (Italy): Direct radiocarbon evidence. *J. Archaeol. Sci.*, **32**(11), 1598–1602.

Frayer, D.W. (1977) Metric dental change in the European Upper Paleolithic and Mesolithic. *Am. J. Phys. Anthropol.*, **46**, 109–120.

Frayer, D.W. (1978) *Evolution of the Dentition in Upper Paleolithic and Mesolithic Europe*, University of Kansas, Kansas, Lawrence.

Frayer, D.W. (1984) Biological and cultural change in the European Late Pleistocene and Early Holocene, in *The Origins of Modern Humans: A World Survey of the Fossil Evidence* (eds F.H. Smith and F. Spencer), Alan R. Liss, New York, pp. 211–215.

Garn, S.M., Lewis, A.B. and Kerewsky, R.S. (1966) The meaning of bilateral asymmetry in the permanent dentition. *Angle Orthod.*, **36**, 55–62.

Goose, D.H. (1971) The inheritance of tooth size in British families, in *Dental Morphology and Evolution* (ed. A.A. Dahlberg), University of Chicago Press, Chicago, pp. 263–270.

Guthrie, R.D. (1963) Variability in characters undergoing rapid evolution, an analysis of Microtus molars. *Evolution*, **19**, 214–233.

Harzer, W. (1987) A hypothetical model of genetic control of tooth-crown growth in man. *Arch. Oral Biol.*, **32**, 159–162.

Hillson, S. (1996) *Dental Anthropology*, Cambridge University Press, Cambridge, UK.

Jolly, C.J. (1970) The seed-eaters: a new model of hominid differentiation based on a baboon analogy. *Man, NS*, **5**, 5–26.

Kieser, J.A. (1990) *Human Adult Odontometrics*, Cambridge University Press.

Leamy, J.L. and Bader, R.S. (1970) Stabilizing selection for odontometric traits in *Peromyscus leucopus*. *Am. Midl. Nat. J.*, **84**, 521–536.

Macchiarelli, R. and Bondioli, L. (1986) Early Holocene reductions in human dental structure: a reappraisal in terms of increasing population density. *Hum. Evol.*, **1**, 405–417.

Mackay, T.F.C. (2001) The genetic architechture of quantitative traits. *Annu. Rev. Genet.*, **35**, 309–329.

Mayr, E. (1963) *Animal Species and Evolution*, Belknap Press, London.

Meiklejohn, C. (2009) Radiocarbon dating of Mesolithic human remains in Spain. *Mesolithic Miscellany*, **20**(2), 2–20.

Meiklejohn, C. and Schentag, C.T. (1988) Dental size in two Portuguese Mesolithic series and their implications. *Am. J. Phys. Anthropol.*, **75**(2), 249.

Meiklejohn, C. and Zvelebil, M. (1991) Health status of European populations at the agricultural transition and the implications for the adoption of farming, in *Health in Past Populations*, (eds H. Bush and M. Zvelebil), British Archaeological Reports, International Series, 567 Oxford, pp. 129–145.

Meiklejohn, C., Schentag, C.T., Venema, V. and Key, P. (1984) Socioeconomic change and patterns of pathology and variation in the Mesolithic and Neolithic of Western Europe: some suggestions, in *Paleopathology at the Origins of Agriculture* (eds M.N. Cohen and G.J. Armelagos), Academic Press, Orlando, pp. 75–100.

Meiklejohn, C., Wyman, J.M. and Schentag, C.T. (1992) Caries and attrition: dependent or independent variables? *Int. J. Anthropol.*, **7**(1), 17–22.

Meiklejohn, C., Roksandic, M., Jackes, M.K. and Lubell, D. (2009) Radiocarbon dating of Mesolithic human remains in Portugal. *Mesolithic Miscellany*, **20**(1), 4–16.

Mitchell-Olds, T., Willis, J.H. and Goldstein, D.B. (2007) Which evolutionary processes influence natural genetic variation for phenotypic traits? *Nat. Rev. Genet.*, **8**(11), 845–856.

Newell, R.R., Constandse-Westermann, T.S. and Meiklejohn, C. (1979) The skeletal remains of Mesolithic Man in Western Europe: an evaluative catalogue. *J. Hum. Evol.*, **8**(1), 1–228.

Oakley, K.P., Campbell, B.G. and Molleson, T.I. (1971) *Catalogue of Fossil Hominids, Part II: Europe*, British Museum (Natural History), London.

Orr, H.A. (2009) Fitness and its role in evolutionary genetics. *Nat. Rev. Genet.*, **10**(8), 531–539.

Osborne, R.H. (1967) Some genetic problems in interpreting the evolution of the human denition. *J. Dent. Res.*, **46**, 945–948.

Pettitt, P.B. and Trinkaus, E. (2000) Direct radiocarbon dating of the Brno 2 Gravettian human remains. *Anthropologie [Brno]*, **38**(2), 149–150.

Phillips, P.C. (2008) Epistasis – the essential role of gene interactions in the structure and evolution of genetic systems. *Nat. Rev. Genet.*, **9**(11), 855–867.

Pinhasi, R. (1998) An odontometric investigation of the affinities of the Nazlet Khater specimen to prehistoric, protohistoric and modern African populations. *Dental Anthropol.*, **12**, 2–10.

Pinhasi, R., Eshed, V. and Shaw, P. (2008) Evolutionary changes in the masticatory complex following the transition to farming in the southern Levant. *Am. J. Phys. Anthropol.*, **135**(2), 136–148.

Pinhasi, R. and Pluciennik, M. (2004) A regional biological approach to the spread of farming in Europe Anatolia, the Levant, South-Eastern Europe, and the Mediterranean. *Curr. Anthropol.*, **45**(54), 59–82.

Plavcan, J.M. and Cope, D. (2001) Metric variation and species recognition in the fossil record. *Evol. Anthropol.*, **10**, 204–222.

Potter, R.H.Y. and Nance, W.E. (1976) A twin study of dental dimension. I. Discordance, asymmetry, and mirror imagery. *Am. J. Phys. Anthropol.*, **44**, 391–395.

Potter, R.H.Y., Nance, W.E., Yu, P.-L. and Davis, W.B. (1976) A twin study of dental dimension II, Independent genetic determinants. *Am. J. Phys. Anthropol.*, **44**, 397–412.

Potter, R.H.Y., Nance, W.E. and Yu, P.L. (1978) Genetic determinants of dental dimensions: a twin study. *Prog. Clin. Biol. Res.*, **24**, 235–240.

Smith, B.H. (1984) Patterns of molar wear in hunter-gatherers and agriculturalists. *Am. J. Phys. Anthropol.*, **63**, 39–56.

Smith, P. (1977) Regional variation in tooth size and pathology in fossil hominids. *Am. J. Phys. Anthropol.*, **47**, 459–466.

Sofaer, J.A. (1973) A model relating developmental interaction and differential evolutionary reduction of tooth size. *Evolution*, **27**, 427–434.

Sofaer, J.A., Bailit, H.L. and MacLean, C.J. (1971) A developmental basis for differential tooth reduction during hominid evolution. *Evolution*, **25**, 509–517.

Sokal, R.R. and Braumann, C.A. (1980) Significance tests for coefficients of variations and variability porfiles. *Syst. Zool.*, **29**, 50–56.

Tillier, A.M., Mester, Z., Bocherens, H., *et al.* (2009) Direct dating of the 'Gravettian' Balla child's skeleton from Bukk Mountains (Hungary): unexpected results. *J. Hum. Evol.*, **56**(2), 209–212.

Townsend, G.C. (1992) Genetic and environmental contributions to morphometric dental variation. *J. Hum. Ecol., Special Issue*, **2**, 61–72.

Townsend, G.C. and Brown, T. (1978) Heritability of permanent tooth size. *Am. J. Phys. Anthropol.*, **49**, 497–505.

Towsend, G.C. and Brown, T. (1980) Dental asymmetry in Australian aboriginals. *Hum. Biol.*, **52**, 661–673.

Wolpoff, M.H. (1971) *Metric Trends in Hominid Dental Evolution: Studies in Anthropology*, Case Western Reserve Press, Cleveland.

y'Edynak, G. (1983) Microevolution and biological adaptability in the transition from food-collecting to food producing in the Iron Gates of Yugoslavia. *J. Hum. Evol.*, **12**, 279–296.

y'Edynak, G. (1989) Yugoslav Mesolithic dental reduction. *Am. J. Phys. Anthropol.*, **78**, 17–36.

Zvelebil, M. (2001) The agricultural transition and the origins of Neolithic society in Europe. *Documenta Praehistorica*, **28**, 1–26.

Index

Human Bioarchaeology of the Transition to Agriculture Edited by Ron Pinhasi and Jay T. Stock